Advances in Volcanology

An Official Book Series of the International Association of Volcanology and Chemistry of the Earth's Interior

Series Editor

Karoly Nemeth, Institute of Natural Resources, Massey University, Palmerston North, New Zealand

More information about this series at https://link.springer.com/bookseries/11157

Luigi Marini · Claudia Principe ·
Matteo Lelli

The Solfatara Magmatic-Hydrothermal System

Geochemistry, Geothermometry and Geobarometry of Fumarolic Fluids

Springer

Luigi Marini
STEAM srl
Pisa, Italy

Matteo Lelli
Institute of Geosciences
and Earth Resources
CNR - Italian National
Research Council
Pisa, Italy

Claudia Principe
Institute of Geosciences
and Earth Resources
CNR - Italian National
Research Council
Pisa, Italy

ISSN 2364-3277 ISSN 2364-3285 (electronic)
Advances in Volcanology
ISBN 978-3-030-98470-0 ISBN 978-3-030-98471-7 (eBook)
https://doi.org/10.1007/978-3-030-98471-7

This Springer imprint is published by the registered company Springer Nature Switzerland AG
The registered company address is: Gewerbestrasse 11, 6330 Cham, Switzerland

Preface

There are at least three reasons that pushed me to write this book on the Solfatara of Pozzuoli, with the competent help of my wife Claudia Principe for the geological-volcanological aspects and my young friend and colleague Matteo Lelli for the geochemical aspects. These three reasons are personal, but I really want to mention them.

Reason # 1. My first scientific article, written together with Roberto Cioni and Egizio Corazza, was focused on the geochemistry of the Solfatara fumarolic gases (Cioni et al. 1984). Both Roberto and Egizio died last year, the cursed 2020, and this book is dedicated to both of them. The scientific paper of Roberto, Egizio, and myself appeared in 1984 as part of a special issue of the Bulletin Volcanologique (that has been called Bulletin of Volcanology starting from the following issue) collecting several scientific articles on the bradyseismic crisis that affected the Campi Flegrei caldera in 1982–1984. In that work, a conceptual geochemical model of the Solfatara magmatic-hydrothermal system was developed, based on the limited data available at the time. The model proved inadequate a few years later, due to the evolution of the system, as clearly and correctly pointed out by Caliro

Left: Roberto Cioni is sampling the geothermal gases at Larderello http://www.societageochimica.it/wp-content/uploads/2020/05/Roberto-Cioni.pdf *Right: Egizio Corazza is performing the gas chromatographic analysis of a gas mixture* http://www.societageochimica.it/wp-content/uploads/2020/10/Egizio-Corazza.pdf

et al. (2007), who proposed a new conceptual geochemical model of the Solfatara magmatic-hydrothermal system based on a dataset much larger than that of Cioni et al. (1984). This new model will probably need to be revised and updated as well. It is the normal course of these things, in that geology is not an exact science and volcanic systems are very complex and extend to considerable depths, but can be directly studied from the surface only.

Reason # 2. The fumarolic fluids of the Solfatara and the thermal waters of the Campi Flegrei have attracted me repeatedly. This interest first blossomed in the late 1970s and early 1980s, when I participated in the geochemical monitoring of selected thermal waters of the Campi Flegrei and the short-term production tests of the deep geothermal wells drilled by Agip-ENEL at Mofete, as a geochemist of Aquater, formerly Geotecneco, two companies of the ENI group. Afterwards, the fumarolic fluids of Solfatara were the focus of other scientific articles of mine (e.g., Cioni et al. 1989; Cioni and Marini 1990). In the meantime, Giovanni Chiodini had joined the Pisa group led by Roberto Cioni and a fruitful collaboration lasting for many years was born with him. Some scientific articles I wrote together with Giovanni, who was working at the Osservatorio Vesuviano, were focused on the Solfatara, as a whole or in part, such as that on the hydrothermal gas equilibria in the H_2O-H_2-CO_2-CO-CH_4 system (Chiodini and Marini 1998) and that on the CO_2 degassing and energy release at the Solfatara (Chiodini et al. 2001). Then our paths separated and, in the following years, I must confess that I had a limited interest in the Solfatara and the Campi Flegrei, while Giovanni carried out researches of the highest quality, many of which were dedicated to the geochemistry of the Solfatara fluids and he and his talented collaborators produced an impressive amount of data, on which, as a matter of fact, this book is based almost entirely.

It must be underscored that some of the activities performed by Giovanni and coworkers at the Solfatara as well as by other geo-scientists in other active volcanic areas of the world, are not free of risks. We cannot forget, in this regard, what happened on January 14, 1993, at Galeras volcano, Colombia, when six scientists were killed by a sudden explosion.[1] Four were Colombians: Dr. Jose Arles Zapata, a geochemist at the Pasto observatory, Dr. Fernando Cuenca, a geophysicist from Bogota who had conducted a magnetic survey of the volcano, Dr. Nestor Garcia, an industrial chemist at the National University, Dr. Carlos Trujillo, a community college teacher in Pasto who liked to use the volcano as a classroom. The other two were Dr. Geoff C. Brown, a geologist at the Open University in Britain and Dr. Igor Menyailov, a Russian scientist from the Institute of Volcanology in Petropavlovsk on the Kamchatka peninsula. And again, it is worth to recall what happened on March 12 of the same year at Guagua Pichincha, Ecuador, when two volcanologists from the Geophysical Institute of the National Polytechnic School, Víctor Hugo Pérez and Álvaro Sánchez, were

[1] https://archive.nytimes.com/www.nytimes.com/books/01/04/15/specials/volcano.html (last visit on 11/12/2021).

monitoring the fumaroles of the dome inside the crater and were killed by a phreatic explosion.[2] Geo-scientific data can have a very high price.

Reason # 3. Starting from May 2019, I was assigned a free research task by INGV, the Italian National Institute of Geophysics and Volcanology, for carrying out research activities concerning the dynamics of the hydrothermal systems related to active volcanoes. Subsequently, in April 2020, I was included in the Scientific Evaluation Panel of the Volcanoes Department of the INGV for the project "From the unrest pre-eruptive dynamics to the forecasting of the expected eruption" concerning the Etna, Stromboli, and Campi Flegrei volcanoes. And so my interest in the magmatic-hydrothermal system of the Solfatara was reawakened. The understanding of its characteristics and past evolution representing the *sine qua non* condition to predict its future evolution, which is the essence of volcanic surveillance, became a stimulating challenge for me.

This renewed interest of mine represented the seed that finally led to the writing of this book with the skilled contributions of Claudia and Matteo.

Viareggio, Italy
December 2021

Luigi Marini

References

Caliro S, Chiodini G, Moretti R, Avino R, Granieri D, Russo M, Fiebig J (2007) The origin of the fumaroles of La Solfatara (Campi Flegrei, south Italy). Geochim Cosmochim Acta 71(12):3040–3055

Chiodini G, Marini L (1998) Hydrothermal gas equilibria: The H_2O-H_2-CO_2-CO-CH_4 system. Geochim Cosmochim Acta 62:2673–2687

Chiodini G, Frondini F, Cardellini C, Granieri D, Marini L, Ventura G (2001) CO_2 degassing and energy release at Solfatara volcano, Campi Flegrei, Italy. J Geophys Res 106:16213–16222

Cioni R, Corazza E, Marini L (1984) The gas/steam ratio as indicator of heat transfer at the Solfatara fumaroles, Phlegraean Fields (Italy). Bull Volcanol 47(2):295–302

Cioni R, Corazza E, Fratta M, Guidi M, Magro G, Marini L (1989) Geochemical precursors at Solfatara Volcano, Pozzuoli (Italy). In: Latter JH (ed) Volcanic Hazards. IAVCEI Proceedings in Volcanology, vol 1. Springer, Berlin, Heidelberg, pp 384–398

Cioni R, Marini L (1990) The determination of deep temperatures by means of the CO-CO_2-H_2-H_2O geothermometer: an example using fumaroles in the Campi Flegrei, Italy. A comment. Bull Volcanol 53(1):67–69

[2]http://ceniza-ecuador.over-blog.com/2018/03/a45-tragico-accidente-en-el-volcan-guagua-pichincha-12-de-marzo-de-1993.html (last visit on 11/12/2021).

Introduction

Solfatara volcano is situated close to the center of the Campi Flegrei caldera which, in the past, has alternatively experienced phases of resurgence and subsidence, traditionally indicated with the term bradyseism.

Ludovico Sicardi (1895–1987) was an Italian scientist, who carried out pioneering studies on the Solfatara and other Italian volcanoes (Vesuvius, Stromboli, Vulcano, Panarea, Salina, and Etna), developing innovative instruments and methods to sample and analyze volcanic gases (Calabrese et al. 2020). He wrote several papers on the Solfatara of Pozzuoli, one of which, in Italian, was published on the Bulletin Volcanologique in 1956. In that paper, he noted that *among the numerous craters of the Campi Flegrei, on the Gulf of Naples, that of the Solfatara of Pozzuoli is the only one that has probably preserved an uninterrupted fumarolic activity for at least two millennia. In 1538, a few kilometers away, a new volcanic vent opened and created its edifice in 48 hours. It is known as Monte Nuovo and reaches an elevation of 140 m, but the only remnant of that eruptive episode of such a short duration are some weak fumarolic emissions. The fumaroles of the Solfatara of Pozzuoli, on the other hand, reveal a truly unusual persistence, even if the thermal level of the exhalation is not excessive, since the alternation of different phases of activity is quite common in volcanic centers* (Sicardi 1956).

Today, owing to the on-going unrest of Solfatara volcano, there is an urgent need to estimate the increasingly higher temperatures and pressures present at depth by using suitable gas-geoindicators, whose implementation and testing is the major aim of this book. In fact, the currently used gas-geothermometers and geobarometers, especially those related to the H_2O-H_2-CO_2-CO-CH_4 system (e.g., Chiodini and Marini 1998) involve carbon monoxide, which is expected to re-equilibrate upon cooling at temperatures in the order of 210–250 °C which are presumably found at shallow depths. Furthermore, these currently used geochemical tools assume ideal gas behavior and, therefore, cannot be used above 250–300 °C approximately. Among the innovative approaches adopted in this work to overcome the limitations of the currently used gas-geoindicators and differentiating this book from numerous previous geochemical studies, not only those focused on the Solfatara, we recall the following ones:

(1) Methane, being characterized by a sluggish behavior, was treated separately from fast-reacting carbon monoxide, in that mixing apples and oranges is not a good idea.
(2) The deviations from the ideal gas behavior were considered, in order to implement geoindicators that can be used at the high pressures and temperatures which are presumably encountered along the decompression path of magmatic-hydrothermal fluids.
(3) The kinetics of the relevant reactions, never considered before in the development of geothermometric and geobarometric techniques, was taken into account. To this purpose, both the numerous studies carried out by the chemical engineers for the optimization of the chemical processes in industrial plants and the geochemical investigations on the origin of methane were reviewed.

These innovations were possible because of the availability, for the Solfatara, of a dataset including several geochemical parameters and extending over a time interval of 37 years, from 1983 to 2020, with a good sampling frequency. This makes the Solfatara a case history probably unique in the world.

Clearly, if the time series had been longer and the sampling frequency had been higher it would have probably been possible to better understand short-duration processes. An example is the crisis of January 15, 1935, when the outlet temperature of fumarolic fluids reached 215 °C. Unfortunately, the closest available samples were obtained in 1923 and 1938, while no geochemical data was collected at that time. Although there are no ifs, ands, or buts in history, it is very likely that the availability of geochemical data would have made it possible to provide a more solid interpretation of what we tried to propose here in chapter "Historical Data". We reiterate that long time series, with a good sampling frequency, and including all the necessary geochemical parameters are of fundamental importance and will be increasingly so in the future thanks to their continuous growth.

Focusing on the content of this book, the background information is presented in chapters "Historical Data", "Summary of the Campi Flegrei Geology", "The Magmatic–Hydrothermal System Hosted in the Campi Flegrei Caldera with Emphasis on the Solfatara", "Chemical and Isotopic Characteristics of Solfatara-Pisciarelli Fumarolic Fluids", and "Soil CO_2 Diffuse Degassing and Thermal Energy Release as Indicators of Volcanic Unrest in the Solfatara-Pisciarelli Area", namely: (a) the temperature and chemical data acquired since the half of the nineteenth century are taken into account in chapter "Historical Data"; (b) a brief summary of the geo-volcanological framework of Campi Flegrei, including the Solfatara volcanic activity and the bradyseismic phenomenon affecting the homonymous caldera, is presented in chapter "Summary of the Campi Flegrei Geology"; (c) different aspects relevant for the elaboration of the conceptual models of the magmatic-hydrothermal system (or systems) hosted in the Campi Flegrei caldera are discussed in chapter "The Magmatic–Hydrothermal System Hosted in the Campi Flegrei Caldera with Emphasis on the Solfatara", with a special focus on the magmatic-hydrothermal system of Solfatara volcano; (d) the chemical and isotopic characteristics of Solfatara-Pisciarelli fumarolic fluids are examined in chapter "Chemical and Isotopic Characteristics of Solfatara-Pisciarelli Fumarolic Fluids"; (e) the diffuse degassing of CO_2 from soil and the related thermal energy release from the Solfatara-Pisciarelli area, two powerful indicators of the on-going volcanic unrest occurring at Solfatara volcano, are considered in chapter "Soil CO_2 Diffuse Degassing and Thermal Energy Release as Indicators of Volcanic Unrest in the Solfatara-Pisciarelli Area".

A review of the different equations of state (EOS) that can be used to model the deviations from the ideality condition of the gas mixtures of interest was considered mandatory and was done in chapter "The Expansion (Decompression) of the Solfatara Fumarolic Fluids". As a result of this review, the GERG-2008 EOS and the EOS of Gallagher et al. (1993) were used for computing the fugacity coefficients of CO_2 and H_2O, that is, the major components of Solfatara fluids, whereas the fugacity coefficients of all relevant gas species, including CO_2 and H_2O, were calculated by means of the well-known Peng-Robinson EOS. Calculations were performed for several possible expansion (decompression) paths of Solfatara fluids to provide a complete picture of the deviations from the ideal gas behavior at the temperatures and pressures possibly encountered by the Solfatara fluids along their ascent towards the surface.

Chapter "The Geoindicators Involving H_2O, CO_2, CO, CH_4 and H_2" is devoted to a thorough re-examination of the five possible reactions occurring in the H_2O-H_2-CO_2-CO-CH_4 homogeneous gaseous system in order to choose the two most representative ones for the geothermometry and geobarometry of magmatic-hydrothermal fluids, obviously keeping CH_4 and CO separate the one from the other.

Chapter "The Chemical Kinetics of the Reactions Involving H_2O, CO_2, CO, CH_4 and H_2" is focused on the kinetics of relevant reactions between H_2O, H_2, CO_2, CO, and CH_4. A reaction path model was developed to simulate the heating of Solfatara fluids and applied to a congruous number of selected samples.

New geothermometric-geobarometric techniques applicable at temperatures up to ca. 1000 °C and pressures up to ca. 3 kbar, both for the H_2O-H_2-CO_2-CO-CH_4 homogeneous gaseous system and the heterogeneous gas-solid systems involving H_2S, were implemented and applied to all Solfatara fluid samples, as described in chapters "Calibration and Application of the RWG and SS4 Indicators for the Possible Expansion Paths of Solfatara Fluids" and "The Redox Potential and Sulfur Gas Species", respectively. To reach this goal, the deviations from the ideal gas behavior were taken into account in a simplified and generalizable form. Some final considerations are drawn in chapter "Conclusive Remarks" including the two possible scenarios of the future evolution of the Solfatara magmatic-hydrothermal system.

All in all, we think that the geochemical tools proposed in this book might provide useful results for the volcanic surveillance of the Solfatara volcano and Campi Flegrei as a whole, which is a densely inhabited area with ca. 500,000 inhabitants at present. Nevertheless, the assessment of the current state of risk is beyond the aims of this book.

Finally, we are sure that the case history of the magmatic-hydrothermal system of the Solfatara is a useful term of comparison for the geo-scientists engaged in the study and surveillance of other similar volcanic systems, elsewhere in the world, for the reasons presented in chapter "A Comparison Between the Solfatara and Other Magmatic-Hydrothermal Systems".

References

Calabrese S, Li Vigni L, Brugnone F, Capasso G, D'Alessandro W, Parello F, Ferla P (2020) The precious treasure of Mariano Valenza: the history of Ludovico Sicardi and the birth of geochemical volcano monitoring. Italian Journal of Geosciences 139 (3):413–435

Chiodini G, Marini L (1998) Hydrothermal gas equilibria: The H_2O-H_2-CO_2-CO-CH_4 system. Geochim Cosmochim Acta 62:2673–2687

Gallagher JS, Crovetto R, Sengers JL (1993) The thermodynamic behavior of the CO_2-H_2O system from 400 to 1000 K, up to 100 MPa and 30% mole fraction of CO_2. J Phys Chem Ref Data 22(2):431–513

Sicardi L (1956) La Solfatara di Pozzuoli. Bull Volcanol 18(1):151–158

Contents

Historical Data

Abstract

The outlet temperature and the chemical composition of Bocca Grande fumarole were measured since the half of the nineteenth century, albeit from time to time rather than in a systematic way. The temperature peak of 215°C recorded on 15 January 1935 is significantly greater than the possible uncertainty of ca. 20°C on the outlet temperature determination. Based on the thermodynamic properties of pure water, the temperature of 215°C at 1 atm pressure constrains the enthalpy of fumarolic gases at ca. 2905 kJ/kg. This value is ca. 100 kJ/kg higher than the maximum enthalpy of saturated steam, which is 2803 kJ/kg at 236°C, 31.2 bar, suggesting that, at that time, a superheated steam phase was present in the plumbing system of the Bocca Grande fumarole, possibly at all depths, due to dry up. It was already clear to the scientists of the nineteenth century that the main component of Bocca Grande fumarolic fluids is by far H_2O, accompanied by CO_2, H_2S, N_2, H_2, CH_4, NH_3, He and Ar, in order of decreasing importance. Sulfur dioxide was reported in some investigations, but De Luca (1871) already recognized that SO_2 was produced by the O_2-driven oxidation of H_2S, upon mixing of fumarolic gases with atmospheric air.■■■

As noted by Sicardi (1956) "the bibliographic roots of the Solfatara are above all literary and only Strabo, as a geographer, fits in providing descriptions of scientific interest. The Solfatara phenomenology, however, always lacked that grandeur for which the other volcanoes of the Italian peninsula are so tied to ancient mythology and myths. The place situated on the very pleasant gulf of Baia and Pozzuoli, so loved by the Romans, was well known to them and the contrast between the luxuriant richness of the nearby countryside and its acrid squalor must have appeared very vivid.

The desolation of the Solfatara is described by Petronius Arbiter with such lively lyricism that today nothing can be removed from the imagination of that time, because the same words adapt to our impressions, to the current reality very similar to that of twenty centuries ago.[1] Today, we find the same sulfur field and the stinking vapors remembered by Strabo, in the same squalid and mortal realm where Petronius saw only Chaos without any sign of spring. This concomitance is certainly not sufficient to affirm that the Solfatara has always been like this. However, there are not enough

[1] Petronio in his Satyricon wrote: "*Est locus exciso penitus demersus hiatus Parthenopen inter magnaeque Dicarchidos arva, Cocyti perfusus aqua; nam spiritus, extra qui furit effusus, funesto spargitur aestu.*" According to https://www.loebclassics.com/view/petronius-satyricon/1913/pb_LCL015.303.xml, the English translation is: "Between Parthenope [Naples] and the fields of the great town of Dicarchis [Pozzuoli] there lies a spot plunged deep in a cloven chasm, wet with the water of Cocytus [one of the six rivers of the underworld in Greek mythology]: for the air that rushes furiously outward is laden with that baleful spray".

L. Marini et al., *The Solfatara Magmatic-Hydrothermal System*, Advances in Volcanology,
https://doi.org/10.1007/978-3-030-98471-7_1

reasons to think differently since in its structure it does not reveal a tormented life.

An exhaustive examination of the historical sources from the first century BC to the nineteenth century AD was carried out by Sicardi (1944). These historical sources suggest the probable occurrence of fumarolic activity inside the Solfatara crater since Roman times, as noted by Breislak (1811) among the others.

The first measurements of the temperature and chemical composition of fumarolic fluids dates back to the nineteenth century, with the works of Charles Joseph Sainte-Claire Deville in 1856–1862, who called "Bocca della Solfatara" the fumarole that today is known as "Bocca Grande", Claude-Henri Gorceix in 1867, and Sebastiano De Luca in 1869 and 1871. Further details are given in the following two sections.

1 The Temperature of Solfatara Fumaroles from the Historical Sources

A good starting point of this section is the scientific paper of Mercalli (1907) who reported on the temperatures measured in the past as follows: "*At the 'Bocca della Solfatara', Ch. S. C. Deville found 88° C in 1856; Gorceix determined 155° C in 1867; Seb. De Luca measured just over 114° C in 1871; H. Johnston-Lavis obtained 156° C in 1889; finally Nasini observed 130 and 132° C in 1897*".

These disparate results could lead us to believe that there are greater variations in activity than actually occurring. In fact, it is necessary to recall that, between 1856 and 1867, the vapors of the 'Bocca della Solfatara' came out of a narrow cave, partly artificial, into which one could only enter on all fours and with difficulty for a few meters. It is therefore evident that the disparate temperatures obtained depend mostly on the different placement of the thermometer more or less close to the true point of origin of the fumarole.

In 1899, between my trips of July 16 and November 2, on a day that I could not specify, a portion of the vault of that small cave collapsed. For some months, the 'Bocca della Solfatara' remained almost blocked; then at least a part of the fallen material was excavated, and so now the fumarole can be approached closer and the true temperature can be determined with less uncertainty.

After that time, I measured the temperature of this fumarole 12 times with a maximum thermometer of the Müller house in Bonn, checked by Prof. O. Scarpa in the Technical Physics Cabinet of the Royal Polytechnic High School in Naples. However, I would like to point out that even now it is not easy to always put oneself in identical conditions for determining the temperature; because the vapors do not come to light from a stable opening, but come out, hissing, from a heap of sand and loose lapillus, which are in constant motion, especially when digging a small cavity a few decimeters deep, to apply the thermometer. The two photographs [in Fig. 1] show the state of the 'Bocca della Solfatara' in 1901–1903, following the landslide and indicate the precise point [A inside the black star in Fig. 1b], where I took the temperature of the fumarole.

The temperatures I observed at the 'Bocca grande'[2] are the following:

Date of measurement	Temperature (°C)
October 28, 1900	153
February 27, 1901	126
August 11, 1901	152.5
February 23, 1902	153
March 27, 1902	152
May 21, 1903	154
December 24, 1903	151
January 28, 1905	157.5
April 11, 1906	154.5
March 1, 1907	154.5
March 10, 1907	154
March 25, 1907	153.5

I must warn that the minimum of 126 °C probably depended on having applied the

[2] To be noted that here the fumarole is indicated with a different name from the one used above.

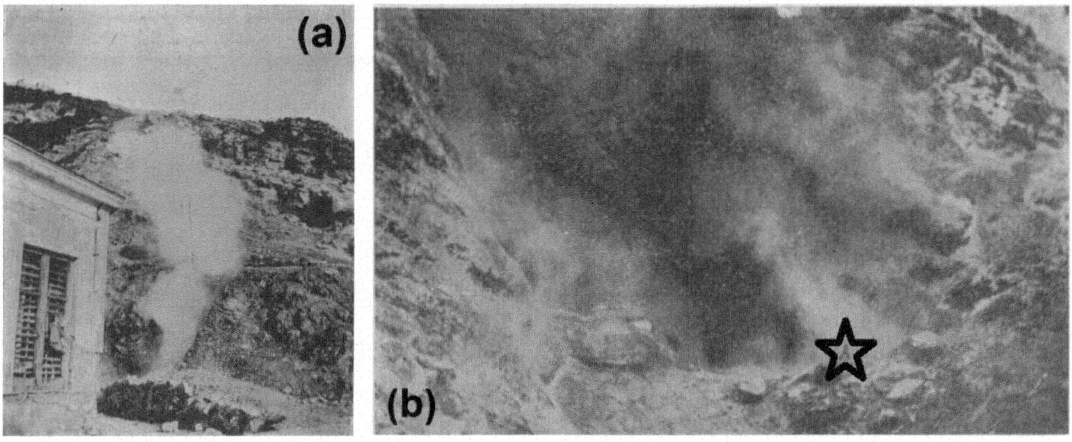

Fig. 1. **a** The 'Bocca della Solfatara', on February 27, 1901 (picture by Mercalli) and **b** on June 2, 1903 (picture by Aguilar). The A inside the black star indicates the position of the main vent after the landslide, where temperatures were measured by Mercalli (1907). Both pictures are from Mercalli (1907)

thermometric bulb too superficially; because on that day the fumarole made a loud noise and threw the grains at a height of about one meter, that is, it did not appear to have less activity than usual. Discarding this observation as not good, we see that the other 11 temperatures are quite concordant, and, in general, much higher than those given by previous observers; among which perhaps only Dr. Johnston-Lavis placed the thermometer at the true point of exit of the vapors".

Mercalli (1907) reports also that the temperatures of the other Solfatara fumaroles were up to 102.5 °C, whereas the temperature of the Pisciarelli fumaroles was 97.5 °C. To be kept in mind that the measured temperatures depend not only on the point but also on the depth reached by the probe, as reported for Bocca Grande by Dall'Aglio et al. (1972) who determined (i) a temperature of 141 °C at 20 cm depth on 23 July 1970, when M. Sato measured 162 °C at 2 m depth and (ii) a temperature of 143 °C at 20 cm depth on 7 October 1970, when 151 °C were determined at 2 m depth.

Sicardi (1941) compiled 57 outlet temperatures of Bocca Grande which were measured from 1856 to 1939 by several authors (Sainte-Claire Deville 1856; De Luca 1869; Gorceix 1872; Johnston-Lavis 1890; Nasini et al. 1904; Mercalli 1907, 1910; Signore 1924, 1929, 1935; Majo 1931; Parascandola 1936; Ginori Conti 1938) and himself, whereas Sicardi (1970) reported 12 data for the period 1949–1969 again by other authors (e.g., Parascandola 1955; Santi 1956) and himself. All the data obtained since October 1900 for the outlet temperature of Bocca Grande (apart from the value of 126 °C which was considered unreliable by the author, see above) as well as those reported by Dall'Aglio et al. (1972) are shown in the chronogram of Fig. 2.

The chronogram shows: (i) the presence of two peaks, one on 31 July 1927 with a temperature of 174.5 °C, the other on 15 January 1935 with a temperature of 215 °C and (ii) a possible downward shift of 10–20 °C in the baseline after the two peaks, especially after the second one, compared to its position before the peaks. Most of these observations must be taken with caution, considering a possible uncertainty of ca. 20 °C in the measured temperatures, depending on the point and depth at which the probe is placed, but the temperature peak of 15 January 1935 is significantly greater than the possible uncertainty on the measured temperatures and, therefore, calls for a detailed examination.

The sources quoted by Sicardi (1941) for the anomalously high temperatures of the period are the communications given by prof. G. B. Rizzo, Director of the Institute of Earth Physics of the

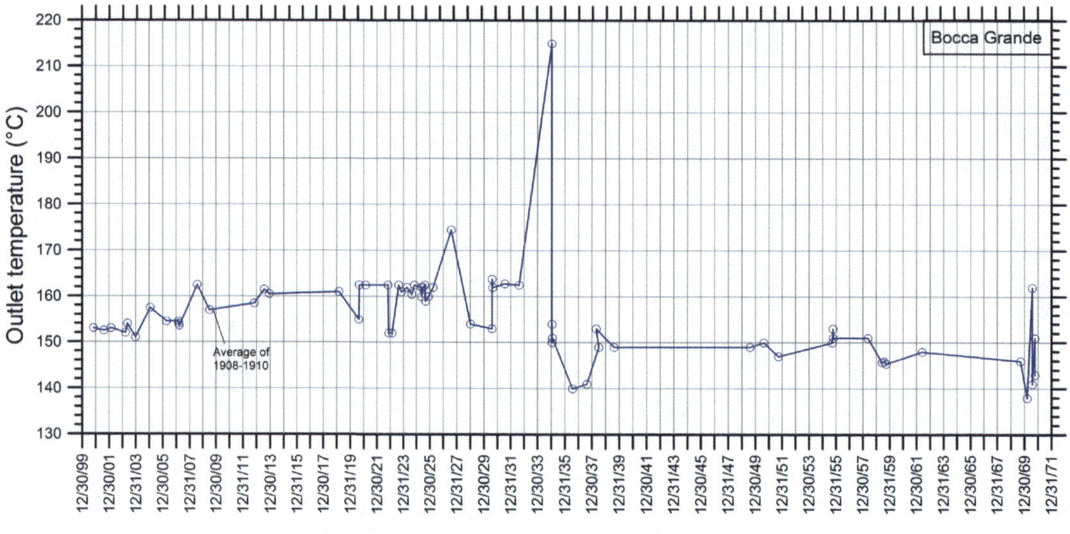

Fig. 2. Chronogram of the outlet temperature of Bocca Grande in the period 1900–1969 based on the data compiled by Sicardi (1941, 1970) and those reported by Dall'Aglio et al. (1972)

Royal University of Naples, to the local newspapers. The same sources are also reported by Signore (1935), who wrote what follows: "*Finally in January 1935 the activity of the Solfatara receives a new increase, which raises the temperature of the Bocca grande, according to the press release of the Institute of Earth Physics, first to 195 °C and then to over 215 °C and produces the opening of the new vent*" [in the southern part of the crater, close to the mud volcano known as Fangaia, see Fig. 3.11]. Further considerations are postponed to Sect. 3.

2 The Chemical Composition of Solfatara Fumarolic Fluids Based on the Historical Sources

Sicardi (1941) compiled also the chemical analyses of Solfatara fumarolic fluids, including those of Bocca Grande, which are listed in Table 1. The results of these historical analyses are given on a dry-gas basis although the prevalence of water vapor over the other gas constituents was already clear to the scientists of

Table 1 Chemical analyses (on a dry-gas basis) of the fumarolic fluids of Bocca Grande from the historical sources

Author	Year	T	CO_2	H_2S	SO_2	O_2	N_2	CH_4	H_2
		°C	vol%	vol%	vol%	vol%	vol%	vol%	vol%
Gorceix	1867	>120	84.10	6.00		1.70	8.40		
De Luca	1871	>115	99.10	0.90		0.00	0.00		
Nasini	1897	132	98.43	0.60		0.15	0.82		
Salvatore	1923	147	99.19	0.47			0.262	0.0086	0.0682
Sicardi	1928	154	98.90	0.50			0.60		
Sicardi	1937	143	99.12	0.48			0.52		
Ginori Conti	1938	148	98.66	0.94			0.265	0.0960	0.04515

the nineteenth century. Not shown in Table 1 is the chemical analysis of Sainte-Claire Deville (1856) who found that SO_2 was the only S-bearing gas species, with a concentration of 24.5° vol%. To be noted that the fluid he analyzed was heavily mixed with atmospheric air, with 14.5° vol% O_2 and 61.0◦vol% N_2. Six years later, Sainte-Claire Deville (1862) described the gases emitted from Bocca Grande as an SO_2-free mixture of CO_2 and H_2S (Sicardi 1941) in accordance with all the other historical data reported in Table 1.

A few years later, De Luca (1871), in addition to the fluid sampled from the Bocca Grande vent whose composition is given in Table 1, collected also six gas mixtures at a distance of 50 cm from the same vent and found that they were mainly made up of N_2 + O_2, with 83.1–87.0 vol%, accompanied by 12.5–15.9 vol% CO_2, and 0.4–0.6 vol% SO_2, whereas H_2S resulted to be undetectable.

Furthermore, De Luca (1871) recognized that SO_2 was produced by the O_2-driven oxidation of H_2S according to the reaction:

$$H_2S + 1.5O_2 = SO_2 + H_2O \qquad (1)$$

In spite of these early findings, Cheminée et al. (1968) found an SO_2 concentration of 1.9 vol% on a dry gas basis for a gas sample obtained in September 1965 from an unspecified Solfatara vent with outlet temperature of 140 ± 10 °C. The gas sample was mainly made up of CO_2, 95.3 vol%, accompanied by 0.9 vol% H_2S, 1.7 vol% N_2 (+rare gases), and 0.2 vol% O_2. It was collected into an evacuated gas bottle containing 50 g of anhydrous molten $CaCl_2$ used to trap water, but addition of air evidently took place, leading to partial conversion of H_2S to SO_2.

The water concentration of Bocca Grande fluids was measured by Salvatore (1923b) and Ginori Conti (1938). Salvatore (1923b) performed five measurements within 10 days. Assuming that CO_2 represents on average 99.19° vol% of dry gases, based on the analytical results of Salvatore (1923a), the average H_2O and CO_2 concentrations turn out to be 909,160 μmol/mol and 90,100 μmol/mol, respectively, with an uncertainty of 2050 μmol/mol on both values. Ginori Conti (1938) measured a gas/steam ratio of 80.4°L/kg. Hypothesizing that the gas volume is referred at 0 °C and 1 atm, this gas/steam ratio corresponds to a H_2O concentration of 939,300 μmol/mol and an incondensable gas concentration of 60,700 μmol/mol. Hence, both gas mixtures are much richer in water and much poorer in incondensable gases than those collected at Bocca Grande in the 1983–2020 period, having H_2O concentration varying from 689,200 to 867,400 μmol/mol, around a mean of 793,800 μmol/mol and a median of 804,600 μmol/mol, with a standard deviation of 42,800 μmol/mol (Chiodini et al. 2021 and references therein).

Salvatore (1923a) and Ginori Conti (1938) measured also CH_4 and H_2 (see Table 1). Salvatore (1923a) reports also a He + Ar concentration of 0.0021°vol% on a dry gas basis, whereas Ginori Conti (1938) gives also the concentrations of NH_3, He and Ar on a total fluid basis, which are as follows: NH_3 = 170 mol/mol, He = 0.5070 μmol/mol, Ar = 0.7733 mol/mol. To be noted that the He concentration measured by Ginori Conti (1938) is significantly lower than the values of Bocca Grande, from 1989 to 2020, ranging from 0.92 to 2.93 μmol/mol (Chiodini et al. 2021 and references therein). These low He and CO_2 concentrations in 1938 are reasonable based on the correlation between these two gas constituents (see Sect. 1 in Chap. "Chemical and Isotopic Characteristics of Solfatara-Pisciarelli Fumarolic Fluids").

3 The Temperature Anomaly of 15 January 1935

Let us accept that the outlet temperature of Bocca Grande achieved the value of 215 °C on 15 January 1935 (see Sect. 1) and that fumarolic fluids were water-rich, considering that a H_2O concentration of 909,160 μmol/mol was determined in 1923, that is, ca. 7 years before (see Sect. 2). If so, the enthalpy of this gas mixture at discharge conditions was close to 2905 kJ/kg, based on the thermodynamic properties of pure water (Kretzschmar and Wagner 2019). This value is ca. 100 kJ/kg higher than the maximum enthalpy of saturated steam, which is 2803 kJ/kg at 236 °C, 31.2 bar, suggesting that a superheated steam phase was present at all depths in the plumbing system of the Bocca Grande fumarole, possibly due to dry up. Since there might be different causes leading to dry up, it is preferable to refrain from speculating further on this point.

The depressurization of the system might have triggered the subsequent inflow of external cold waters (meteoric and/or marine) into the fumarolic conduits causing (a) a decrease in the outlet temperature, in agreement with the values measured on 19 January, 22 January, and 1 February 1935, namely, 154, 150, and 151 °C, respectively (Signore 1935; Sicardi 1941), and (b) an increase in H_2O concentration, in accordance with the value of 939,000 μmol/mol that was determined in 1938 by Prince Ginori Conti (see Sect. 2). It is unclear if the anomalously high temperature of 15 January 1935 is a long-term effect of the Irpinia earthquake of 23 July 1930, with an equivalent magnitude of 6.7 (Guidoboni et al. 2018, 2019). For sure, the short-term effects of that earthquake comprised explosions throwing pieces of mud at heights of 10–15 m above the rim of the crater (Signore 1935).[3]

References

Breislak S (1811) Introduzione alla geologia. Stamperia Reale, Milano, p 490

Cheminée J, Letolle R, Olive P (1968) Premiéres données isotopiques sur des fumerolles de volcans italiens. Bulletin Volcanologique 32(3):467–475

Chiodini G, Caliro S, Avino R, Bini G, Giudicepietro F, De Cesare W, Ricciolino P, Aiuppa A, Cardellini C, Petrillo Z, Selva J, Siniscalchi A, Tripaldi S (2021) Hydrothermal pressure-temperature control on CO_2 emissions and seismicity at Campi Flegrei (Italy). J Volcanol Geoth Res 414:107245

[3] Quite surprisingly, the year of the Signore paper, 1929, is prior to that of the reported event, 1930.

Dall'Aglio M, Martini M, Tonani F (1972) Rilevamento geochimico delle emanazioni vulcaniche dei Campi Flegrei. Quaderni de 'La Ricerca Scientifica'. CNR—Consiglio Nazionale delle Ricerche, Roma, vol 83, pp 152–181

De Luca S (1869) Osservazioni sulla temperatura interna della grande fumarola della Solfatara di Pozzuoli. Rendiconto dell'Accademia delle Scienze Fisiche e Matematiche. Società Reale di Napoli, Anno VIII, pp 34–38

De Luca S (1871) La composizione dei gas che si svolgono dalle fumarole della Solfatara di Pozzuoli. Rendiconto dell'Accademia delle Scienze Fisiche e Matematiche. Società Reale di Napoli, Anno X, pp 181–188; 211–221

Ginori Conti P (1938) L'attività endogena quale fonte di energia. Reale Accademia Nazionale dei Lincei. Classe di scienze fisiche, matematiche e naturali. Seduta del 3 aprile 1938, Roma, p 39

Gorceix C-H (1872) Sur les gas des solfatares des Champs Phlégréens. Annales Chimie et Physique (ser 4) 25:559–566

Guidoboni E, Ferrari G, Mariotti D, Comastri A, Tarabusi G, Sgattoni G, Valensise G (2018) CFTI5Med, Catalogo dei Forti Terremoti in Italia (461 a.C.-1997) e nell'area Mediterranea (760 a.C.-1500). Istituto Nazionale di Geofisica e Vulcanologia (INGV). https://doi.org/10.6092/ingv.it-cfti5

Guidoboni E, Ferrari G, Tarabusi G, Sgattoni G, Comastri A, Mariotti D, Ciuccarelli C, Bianchi MG, Valensise G (2019) CFTI5Med, the new release of the catalogue of strong earthquakes in Italy and in the Mediterranean area. Sci Data 6:80. https://doi.org/10.1038/s41597-019-0091-9

Johnston-Lavis HJ (1890) Osservazioni geologiche lungo il tracciato del grande emissario-fognone di Napoli dalla Pietra sino a Pozzuoli. Bollettino del Regio Comitato Geologico d'Italia

Kretzschmar HJ, Wagner W (2019). Tables of the Properties of Water and Steam. In: International Steam Tables. Springer Vieweg, Berlin, Heidelberg. https://doi.org/10.1007/978-3-662-53219-5_5

Majo E (1931) La Solfatara di Pozzuoli. Annali dell'Osservatorio Vesuviano 3:61–72

Mercalli G (1907) Sullo stato attuale della Solfatara di Pozzuoli. Atti dell'Accademia Pontaniana 37:1–16 (memoria n. 6)

Mercalli G (1910) Osservazioni sulla temperatura del vapore emanante all Solfatara di Pozzuoli. Atti Società Italiana per il Progresso delle Scienze. 4a Riunione

Nasini R, Anderlini F, Salvadori R (1904) Ricerche sulle emanazioni terrestri italiane. II—Gas del Vesuvio e dei Campi Flegrei, delle Acque Albule di Tivoli, del Bullicame di Viterbo, di Pergine, di Salsomaggiore. Memorie della Reale Accademia Nazionale dei Lincei. Classe di scienze fisiche (serie 5) 5:25–82

Parascandola A (1936) Il rione delle Mofete nei Campi Flegrei. Bollettino della Società dei Naturalisti in Napoli 48:81–94

Parascandola A (1955) Notizie sul soffione di recente apparso alla Solfatara di Pozzuoli. Ric Sci 25:3113

Sainte-Claire Deville Ch (1856) Sur les phénomènes éruptifs de du Vesuve et de l'Italie meridionale. Comptes Rendus de l'Académie des sciences 43:745–751

Sainte-Claire Deville Ch (1862) Sur les emanations volcaniques des Champs Phlégréens. Comptes Rendus de l'Académie des sciences 54:528–536

Salvatore E (1923a) Sui gas della Solfatara di Pozzuoli. Zeitschrift für Vulkanologie 7:149–154

Salvatore E (1923b) Sulla determinazione del vapor d'acqua nelle esalazioni fumaroliche della Solfatara di Pozzuoli. Zeitschrift für Vulkanologie 7:215–217

Santi B (1956) Sulla nuova manifestazione (Soffione) nel cratere della Solfatara. Boll Soc Geol It 75:282–283

Sicardi L (1944) L'attività della Solfatara di Pozzuoli attraverso la documentazione storica avanti l'ultimo ottantennio. Atti della Società Italiana di Scienze Naturali e del Museo Civico di Storia Naturale in Milano 83(2):97–114

Sicardi L (1970) Recenti misure termiche alla Solfatara di Pozzuoli e qualche ancora necessaria osservazione. Bollettino Della Società Dei Naturalisti in Napoli 79:137–144

Sicardi L. (1941) Sulle manifestazioni dell'attività fumarolica della Solfatara di Pozzuoli nell'ultimo ottantennio (1856–1939). Rivista di Scienze Naturali (Natura) 32:53–74

Sicardi L (1956) La Solfatara di Pozzuoli. Bulletin Volcanologique 18(1):151–158

Signore F (1924) Relazione su di una escursione fatta il 10 maggio 1923 nella plaga puteolana. Bollettino della Società dei Naturalisti in Napoli 36:2–8

Signore F (1929) Sur la variation d'activite du volcan de Boue ("Fangaia") de la Solfatare de Pouzzoles (Naples) par suite du grand tremblement de terre de l'Irpinia, 23 Juillet 1930. Bulletin Volcanologique 3 (2):49–52

Signore F (1935) Attività vulcanica e bradisismo nei Campi Flegrei. Annali dell'Osservatorio Vesuviano 3:173–192

Summary of the Campi Flegrei Geology

Abstract

This summary of the Campi Flegrei geology comprises: (i) the geo-volcanological and structural characteristics of Campi Flegrei and the Solfatara monogenetic edifice, (ii) the 1198 Solfatara (phantom?) eruption, which was reported for the first time about four centuries after its possible occurrence and has been a matter of debate in the volcanological literature, (iii) the bradyseism affecting the whole Campi Flegrei caldera, including alternating periods of resurgence and subsidence. Although the Campi Flegrei bradyseism is known to have occurred at least since Roman times, only two unrest episodes are well documented. One preceded the 1538 Monte Nuovo eruption, whereas the other begun in 1950 and is still ongoing, with ground deformations accompanied by seismicity and variations in the chemistry and flowrate of the fumarolic fluids discharged at Solfatara and Pisciarelli. Nevertheless, there is a lack of consensus on the processes governing the on-going unrest episode, in that the involvement of a magmatic input or the pressurization of the magma chamber was proposed by some authors, the pressurization of the overlying hydrothermal system was suggested by other authors, whereas a combination of both processes was invoked in other investigations.■■■

The Solfatara is a monogenetic volcanic edifice situated inside the Campi Flegrei volcanic field, where persistent volcanic activity has taken place since at least ~47,000 years, based on the oldest volcanic products cropping out in the area (Di Girolamo et al. 1984; Rosi and Sbrana 1987). Most erupted volcanic products are trachytic to trachyphonolitic tuffs and tuffites interstratified with sediments (silts, sands and marls), whereas lava flows and domes are subordinate (Rosi and Sbrana 1987; Fig. 1). The only historic eruption, which occurred in 1538 CE, involved a highly differentiated peralkaline phonolitic–trachytic magma and produced the ash and scoria cone of Monte Nuovo (Rosi et al. 1983).

The prominent structural element of Campi Flegrei is a subcircular depression with a diameter of 8–12 km whose inland (northern) portion is marked by a series of topographic highs between Monte di Procida to the west and the hills of Posillipo and Camaldoli to the east (Rosi et al. 1983) and whose offshore (southern) part encompasses the Gulf of Pozzuoli (Sbrana et al. 2021). According to several authors (e.g., Rittman 1950; Rosi et al. 1983; Rosi and Sbrana 1987; Lirer et al. 1987; Barberi et al. 1991; Scandone et al. 1991; Orsi et al. 1996), this depression is a caldera generated by one or more volcano-tectonic collapses. In particular, Barberi et al. (1991) identified at least three nested

L. Marini et al., *The Solfatara Magmatic-Hydrothermal System*, Advances in Volcanology,
https://doi.org/10.1007/978-3-030-98471-7_2

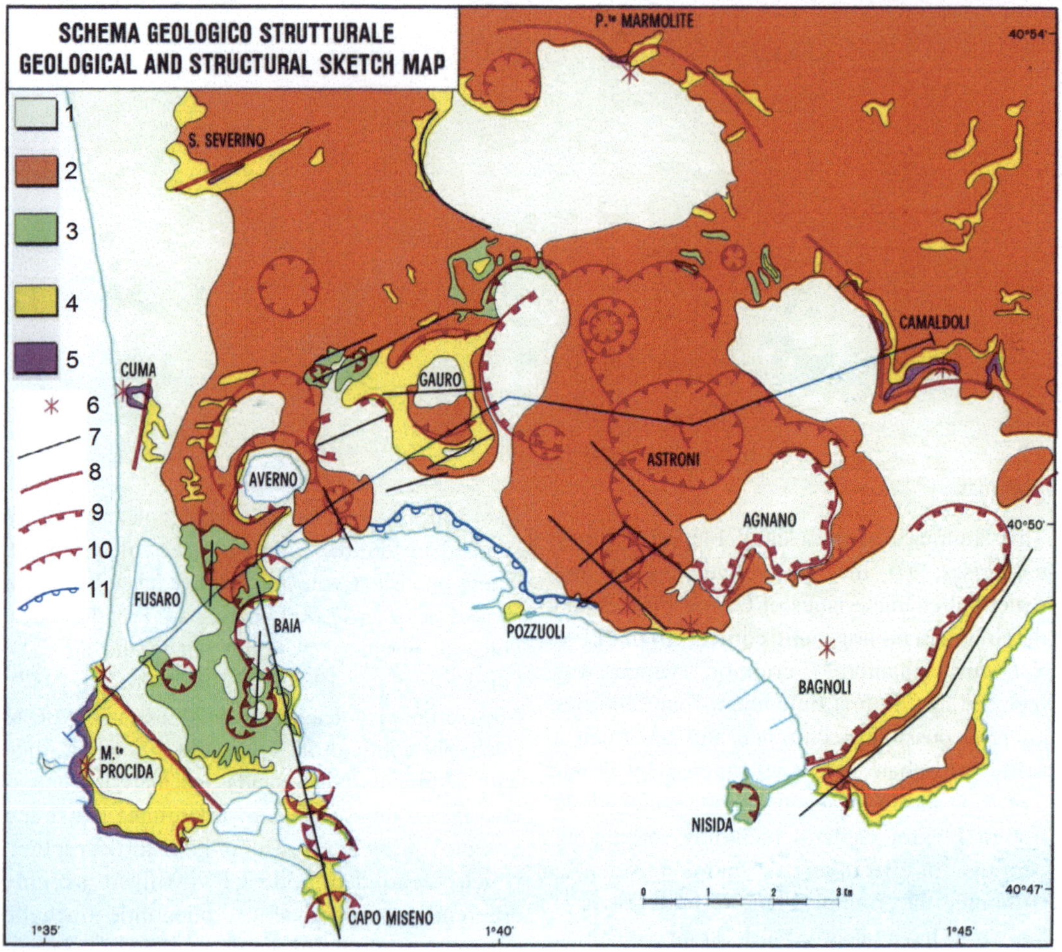

Fig. 1 Geological and structural sketch map of Campi Flegrei. *Legend* 1 = Quaternary undifferentiated deposits, areas of strong anthropization, soil cover; 2 = Products of the post-caldera subaerial recent activity; 3 = Products of the post-caldera mainly subaerial ancient activity; 4 = Products of the post-caldera mainly submarine activity (Neapolitan Yellow Tuff deposits); 5 = Products of the subaerial pre-caldera activity; 6 = Volcanic vents; 7 = Faults and fractures; 8 = Campi Flegrei caldera rim; 9 = Post-caldera volcano-tectonic collapses; 10 = Craters; 11 = Starza marine terrace (from Rosi and Sbrana 1987)

collapses, namely an outer one, ~13 km wide, an intermediate one, ~11–12 km across, and an inner one, ~4–5 km wide, by merging and elaborating surface geological–volcanological evidence (Rosi et al. 1983), subsurface information acquired by the deep geothermal wells drilled by Agip-ENEL in the Mofete, San Vito, and Licola areas (see Sect. 3 in Chap. "Summary of the Campi Flegrei Geology"), and inland and offshore gravimetric and aeromagnetic data (Fig. 2).

According to Barberi et al. (1991): (i) The outer collapse is related to the ~35,000 year-old Campanian Ignimbrite eruption, a catastrophic eruption that emitted ~80 km^3 of dense-rock equivalent (DRE) trachytic magma, producing pyroclastic flow deposits that covered an area of 7000 km^2. In this work, reference to the Campi Flegrei caldera is to the Campanian Ignimbrite collapse. (ii) The intermediate collapse is associated to the Neapolitan Yellow Tuff eruption that occurred ~12,000 a BP and emitted

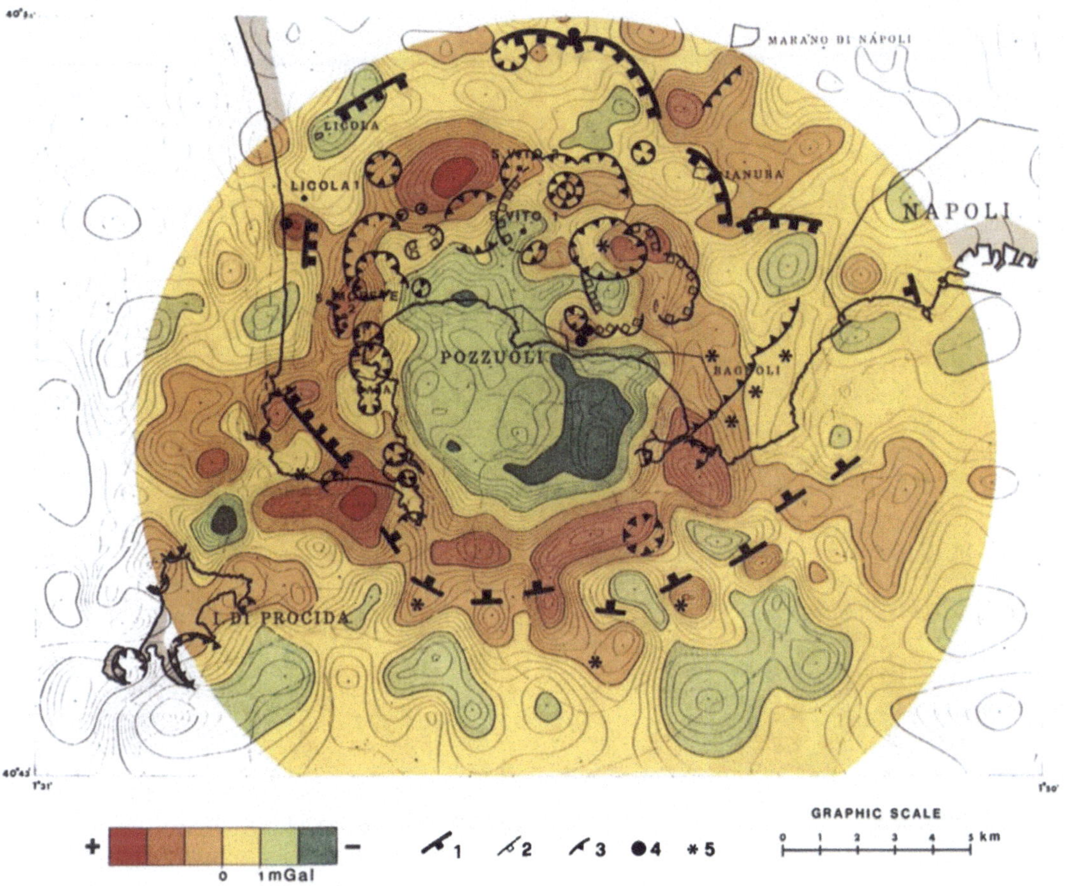

Fig. 2 Structural map of the Campi Flegrei superimposed onto the Bouguer anomaly high-pass filter map. *Legend of volcanic structures* 1 = caldera rim; 2 = minor, post Neapolitan Yellow Tuff, volcano-tectonic collapsed areas; 3 = crater rim; 4 = lava dome; 5 = eruptive center. Reprinted from Barberi et al. (1991), Copyright © 1991, with permission from Elsevier

~51 km^3 of DRE trachytic magma. This caldera is called Neapolitan Yellow Tuff caldera in this work. (iii) The inner collapse was formed in the last phase of activity, ~4,000 a BP, in the Agnano–Astroni–Solfatara sector.

Based on this evidence, Barberi et al. (1991) recognized that the volcanic and structural evolution of Campi Flegrei is characterized by the recurrence of the same eruption mechanism, including the formation of a ring fracture, magma ascent, the eruption from several centers positioned along the ring fracture, and the collapse of the central sector into the emptied magma chamber. The decrease with time of both the diameter of the collapsed area and the total sinking suggests the concurrent decrease in the volume of the magma stored in the chamber still present in the center of the caldera (Barberi et al. 1991). Accepting this explanation, it is permissible to hypothesize the current presence of a small magma batch below the Solfatara crater, that is, positioned inland, near the center of the caldera. The numerous scientific studies carried out more recently have improved the geological

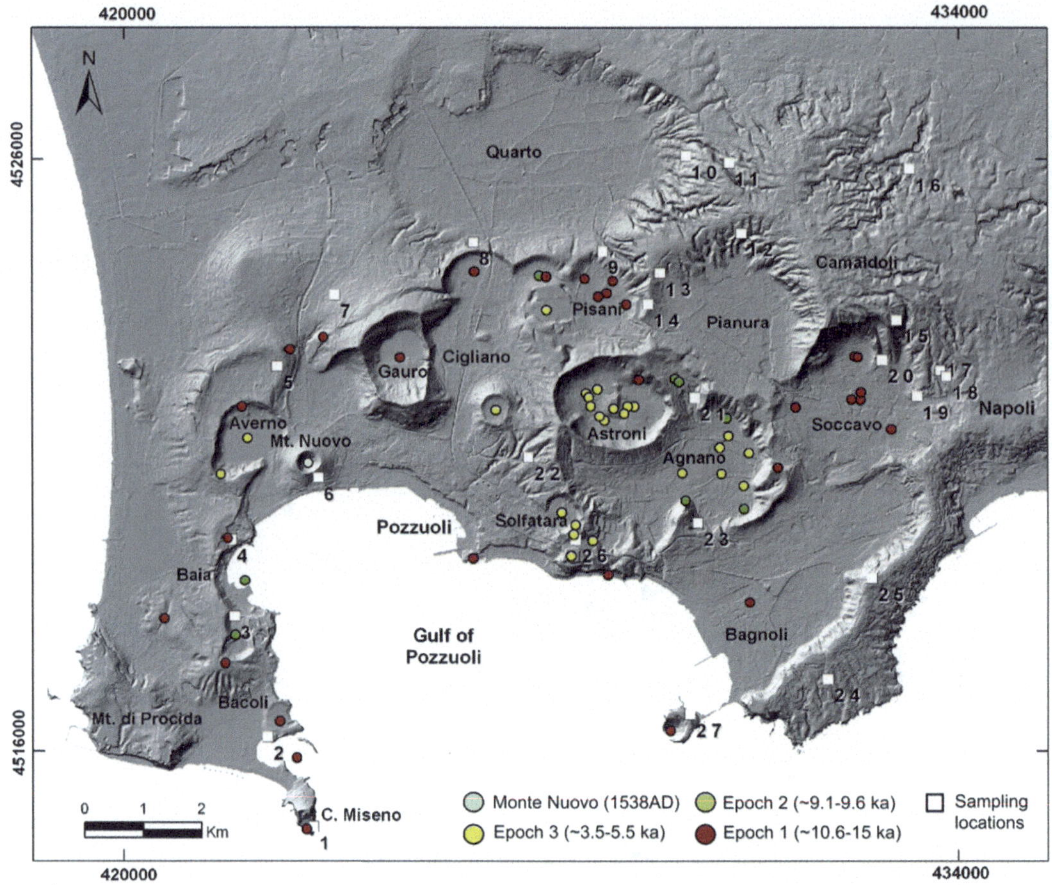

Fig. 3 Location map of the vents of the post-15 ka eruptions. Reprinted from Smith et al. (2011), Copyright © 2011, with permission from Elsevier

knowledge of Campi Flegrei providing further details although some controversial aspects still remain. For example:

(i) The age of the Campanian Ignimbrite eruption, either 34.29 ± 0.09 (1σ) ka BP, based on radiocarbon dating of charred wood or 39.85 ± 0.14 ka BP (95% confidence level) based on $^{40}Ar/^{39}Ar$ dating of sanidine phenocrysts (Giaccio et al. 2017), and the DRE volume of the erupted magma, ranging from 54 km^3 (Scarpati et al. 2014) to 180–280 km^3 (Costa et al. 2012).

(ii) The DRE volume of the Neapolitan Yellow Tuff magma, varying from 10 km^3 (Scandone et al. 1991) to at least 50 km^3 (Scarpati et al. 1993). There are controversial aspects concerning this subject, in that, according to some authors, the Neapolitan Yellow Tuff was produced during a single eruption (Deino et al. 2004) occurred 14.9 ± 0.4 (2σ) ka BP, whereas according to other authors (e.g., Rosi and Sbrana 1987) these volcanic products are grouped in a single geological unit but were emitted from several eruptive centers at different times.

(iii) A database of the compositions of the volcanic ashes (tephra) emitted by the post-15 ka eruptions in the Campi Flegrei, that is, during the post-Neapolitan Yellow Tuff activity was implemented by Smith et al. (2011) through >1900 new chemical analyses. These authors presented also an

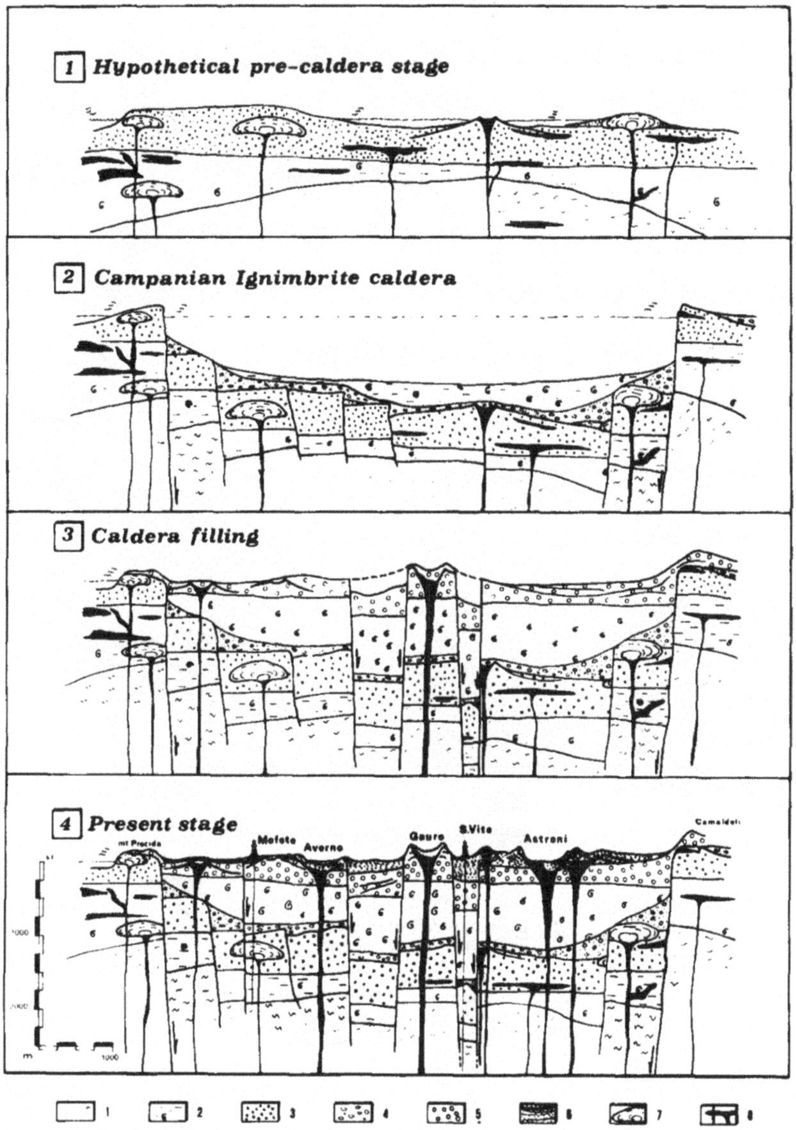

Fig. 4 Scheme of the main evolution stages of the Campi Flegrei. *Legend* 1 = Thermometamorphic rocks; 2 = Tuffites; 3 = Subaerial tuffs; 4 = Products of the Campanian Ignimbrite eruption; 5 = Neapolitan Yellow tuffs; 6 = Tephra and tuffs of the recent, mainly subaerial activity; 7 = Lava domes; 8 = Lavas, subvolcanic rocks and volcanic conduits. Reprinted from Rosi et al. (1983), Copyright © 1983, with permission from Elsevier

updated chronology based on published radiocarbon dates and $^{40}Ar/^{39}Ar$ ages and grouped the already known vents in four epochs (location in Fig. 3).

In spite of recent improvements and open questions, the works of synthesis elaborated by the volcanologists of the Pisa School (Rosi et al. 1983; Rosi and Sbrana 1987; Barberi et al. 1991) still remain the groundbreaking researches on the Campi Flegrei geology. The scheme of the main evolution stages of the Campi Flegrei (from Rosi et al. 1983) is shown in Fig. 4.

1 The Solfatara Volcano

The Solfatara volcanic edifice has a sub-circular crater, with a diameter varying between 610 and 710 m, an area of $\sim$0.35 km^2, and a perimeter of $\sim$2.15 km^2 (Isaia et al. 2015). The crater is bounded by a ring of pyroclastic products with a maximum thickness of $\sim$30 m. The diameter of the Solfatara cone varies between 1.1 and 1.5 km (Isaia et al. 2015). The well-stratified pyroclastic deposits of Solfatara volcano have primary slope from subhorizontal to gently dipping, up to 25°.

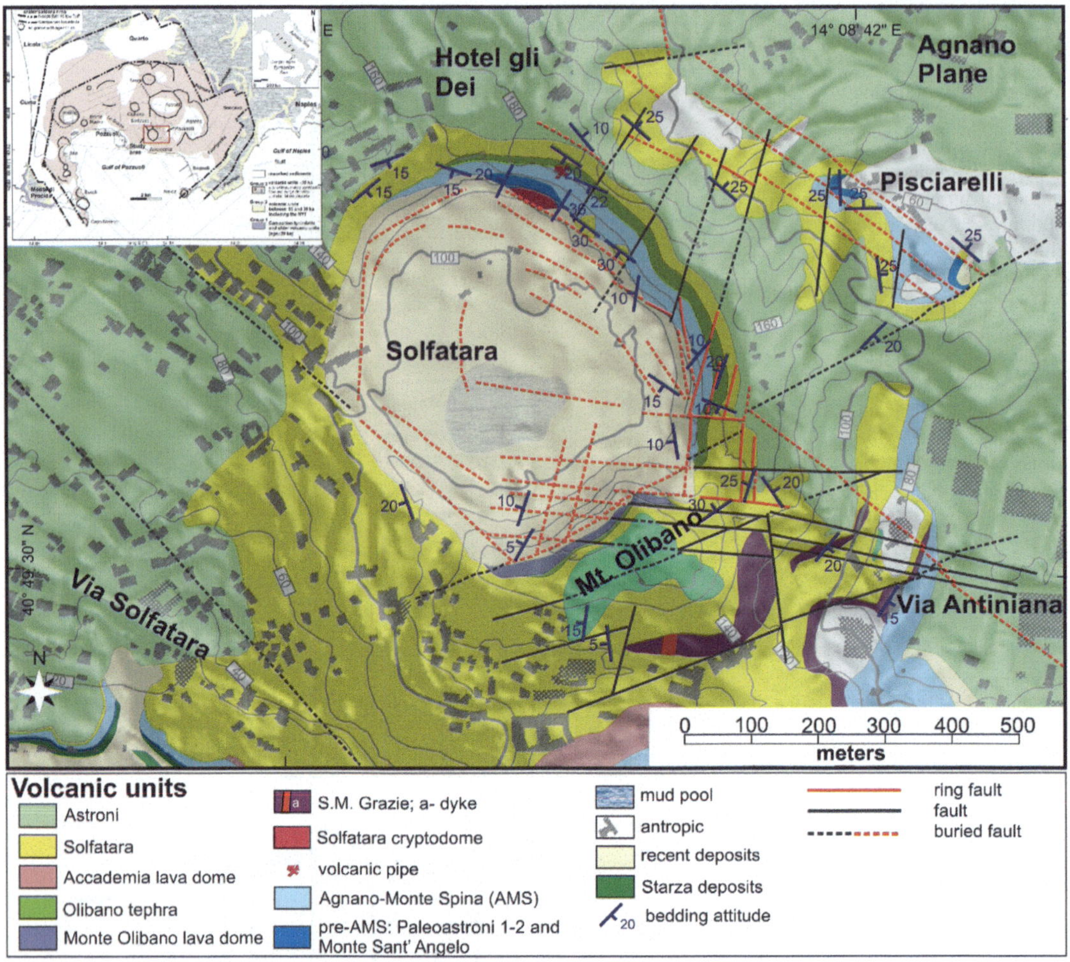

Fig. 5 Geological map of the Solfatara area. From Marotta et al. (2019), which is licensed under a Creative Commons Attribution 4.0 CC BY International License

They are interposed between the lavas of the Accademia lava dome and the pyroclastic deposits of the Astroni volcano (Fig. 5) and have a limited distribution area (Rosi and Sbrana 1987; Cipriani et al. 2008; Isaia et al. 2015).

According to Rosi and Sbrana (1987), the Solfatara tephra comprise a basal phreatomagmatic coarse breccia overlain by few meters of stratified pyroclastic surges. The breccia is very poorly sorted and constituted by a fine-grained matrix containing large blocks of green tuff, lavas, and dispersed scoriaceous bombs. The whole breccia deposit is affected by hydrothermal alteration and is weakly hardened. The pyroclastic surges are made up of layers of pisolitic ashes and coarse ashes with base surge structure enclosing some decametric beds of well sorted air-fall pumice lapilli. These surges mantle the eastern-northeastern crater wall and have overflowed the crater rim to the East and West.

The juvenile fraction of the Solfatara pyroclastic products has trachyte composition. Based on the volatile content of melt inclusions and using a suitable solubility model, Cipriani et al. (2008) proposed that magma crystallized at pressures between 200 and 685 bar, corresponding to depths between 1 and 3 km, before the eruption.

The main tectonic elements affecting the Solfatara area and controlling its crater morphology are relatable to the so-called Apenninic and anti-Apenninic trends, striking WNW and ENE, respectively (Rosi and Sbrana 1987). According to Isaia et al. (2015), morphological, stratigraphic, structural, and geoelectrical data indicate that the Solfatara tephra ring is connected with a maar-diatreme structure of phreatomagmatic origin extending to depths of 2–3 km; the upper part is made up of disaggregated rocks and collapse breccias forming a stair-step-like structure which is cut by steep concentric ring faults.

As already mentioned above, the Solfatara monogenetic edifice formed during the recent phase of post-caldera subaerial activity which begun ~4,000 years ago and is covered by the ~3,700 years old deposits of Astroni volcano (Rosi and Sbrana 1987). This post-caldera subaerial phase of volcanic activity was heralded by a tectonic event that occurred 5,400 years ago and caused the uplift of the northern sector of the Gulf of Pozzuoli, as testified by the Starza marine terrace (Rosi et al. 1983). This uplift might have been controlled by the injection of new magma at shallow depth, although this is only a conjecture.

At the end of 1500—beginning of 1600, the two chroniclers Giulio Cesare Capaccio and Scipione Mazzella reported that strong earthquakes were felt in the Neapolitan area and an eruption took place at Solfatara in 1198, that is, about four centuries earlier than the publication of these news (Scandone et al. 2010). The absence of bibliographic sources contemporary or closer to the event and, perhaps even more so, the lack of both corresponding morphological alterations in the Solfatara crater and products emitted during the eruption suggest that the two chroniclers mentioned above are scarcely reliable or completely unreliable as underscored by several authors. Some authors have questioned the authenticity of these historical sources, but have not excluded the occurrence of the 1198 eruption (e.g., Scacchi 1849; Mercalli 1883; Rittmann 1929), other authors have admitted that it was, at most, a minor phreatic (hydrothermal) explosion (Scandone et al. 2010), while still other authors have totally ruled out it (e.g., Sicardi 1944, 1956; Guidoboni 2010). Based on a diplomatic letter written in October 1498 and reporting that several earthquakes were felt in Naples and in the Campi Flegrei at that time, Guidoboni (2010) suggested a possible transcription error from MCDXCVIII (which is 1498 in Roman numbers) to MCXCVIII (which is 1198 in Roman numbers), that is a missing D, corresponding to 500 years. On the other hand, it is possible that the fumarolic activity of the Solfatara intensified following the earthquake of magnitude 7.1 that occurred in 1456 in the Neapolitan area and caused about 30,000 casualties. This event may have prompted the chroniclers of the time to create the phantom eruption of 1198, when another important seismic activity took place in the area.

As anticipated in Chap. "Historical Data", historical sources suggest that fumarolic activity inside the Solfatara crater is probably occurring since Roman times. This long-lasting and intense degassing process has determined the continuous production of relatively large amounts of steam condensates at shallow depths practically consisting of sulfuric acid solutions. These acidic solutions attacks the rocks causing the formations of authigenic minerals, mainly elemental sulfur and alunite, as well as gypsum, alunogen, pickeringite, potassium alum, hematite, and pyrite. These minerals typical of the so-called advanced argillic alteration (Pirajno 2009) were encountered in the fumarolic encrustations and in the muds of the hot pools at Solfatara and Pisciarelli (Piochi et al. 2015). Different clay minerals and amorphous silica are also formed by this pervasive water–rock interaction process, converting most of the rocks into mud. As a result, the internal crater wall was repeatedly interested by landslides which produced several debris flow deposits (Isaia et al. 2021) and exposed the pre-Solfatara volcanic substratum in the crater wall (Fig. 6). The occurrence of all these processes is very common in areas affected by fumarolic activity, but the debris flows related to landslides should not be confused with the debris flow deposits emplaced by hydrothermal

Fig. 6 Aerial view of the Solfatara Crater. Reprinted from Scandone et al. (2010), Copyright © 2010, with permission from Elsevier

(phreatic) events (Barberi et al. 1992; Marini et al. 1993) or magmatic eruptions such as that of Ruiz volcano, Colombia (Takahashi 2007).

To the best of our knowledge, no debris flow deposits related to hydrothermal eruptions were recognized at Solfatara volcano so far. As a matter of fact, one of the most energetic historical activity at Solfatara was triggered by the M 6.4 Irpinia earthquake of 23 July 1930 and was an event of mud fountains that reached maximum heights of 10–15 m above the crater rim (Signore 1929). Nevertheless, we do not intend to exclude the possible occurrence, in the future, of a hydrothermal eruption triggered by an earthquake, which appears a realistic scenario considering the current over-pressurization state at depth of the Solfatara magmatic-hydrothermal system (see Chap. "Conclusive Remarks").

2 The Bradyseism in the Campi Flegrei

Vertical ground movements known as bradyseism, comprising cycles of uplift (inflation) and subsidence (deflation), have affected the Campi Flegrei caldera since at least Roman times (Fig. 7; Parascandola 1947). Actually, the bradyseismic uplift (inflation) should be termed caldera resurgence, whereas the bradyseismic subsidence (deflation) should be called caldera subsidence, to be consistent with the modern volcanological terminology and the processes observed at other volcanoes worldwide.

Antonio Niccolini was the first to recognize the phenomenon and to describe it (Niccolini 1829, 1839, 1846), thanks to the excavation,

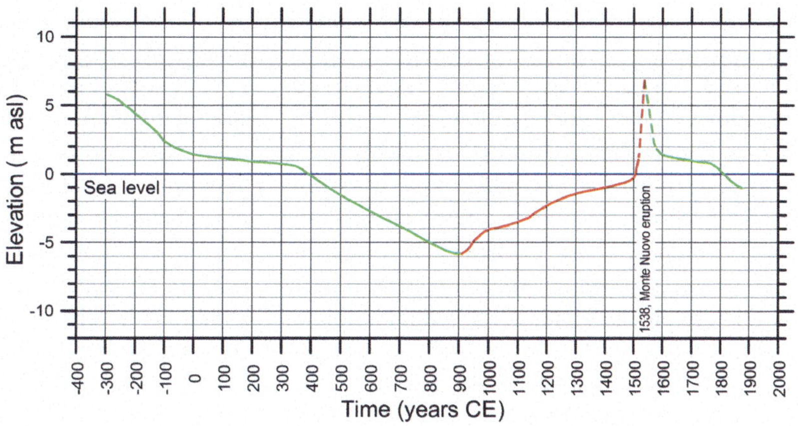

Fig. 7 Elevation changes of the Serapeum floor according to Parascandola (1947, redrawn). The green and red lines indicate caldera subsidence and resurgence, respectively

begun in 1750, of the ruins of the ancient roman market known as Serapeum of Pozzuoli (Fig. 8). Niccolini was commissioned in June 1824 to drain these ruins from the stagnant waters, but he realized that this operation was impossible because sea level was gradually rising, flooding the edifice. Moreover, Niccolini recognized the occurrence of an earlier marine submersion as testified by the holes made by the marine bivalves *Lithodomus lithophagus* (Linnaeus 1758) on the marble columns of the Roman edifice at heights of 6.30 ± 0.03 m above the second floor of the market and 8.40 ± 0.15 m above the first floor (Del Gaudio et al. 2010). The

Fig. 8 Engraving showing the Serapeum of Pozzuoli in 1845 (from Niccolini 1846)

bradyseism phenomenon was later made famous and attributed to ground movements instead of sea level variations by Charles Lyell in his "Principles of Geology" (Lyell 1850).

Two other unrest episodes are well documented. The first one preceded by several years the 1538 Monte Nuovo eruption, affected the whole Campi Flegrei caldera that raised by ~17 m in its center at Pozzuoli, where earthquakes sometimes were strong enough to collapse buildings (Di Vito et al. 1987; Dvorak and Gasparini 1991; Guidoboni and Ciuccarelli 2011; Giacomelli and Scandone 2012). In 1470–1472, the increased gas emission from Solfatara and Pisciarelli killed the vegetation in the nearby areas (Guidoboni and Ciuccarelli 2011). The frequency of the earthquakes felt by Pozzuoli inhabitants increased during the two years preceding the eruption. In the 36 h before the eruption, the vent site was affected by a sudden uplift of ~6 m (Parascandola 1947; Dvorak and Gasparini 1991; Guidoboni and Ciuccarelli 2011). Then, the Monte Nuovo eruption began on the 29th of September, lasted a week, and consisted in a small phreatomagmatic event that built the homonymous cone (Parascandola 1947; Guidoboni and Ciuccarelli 2011).

The second unrest episode begun in 1950 and is still ongoing. The deformation has been accompanied by seismicity and variations in the chemical composition and flowrate of the fumarolic fluids emitted at Solfatara and Pisciarelli. The magnitude of the recent vertical ground movements is largely based on geodetic observations, that began in 1905 (Fig. 9; Del Gaudio et al. 2010), and on GPS monitoring since 2000 (De Martino et al. 2014).

Based on the geodetic data, it seems likely that a continuous subsidence took place after the Monte Nuovo eruption, possibly until 1950, when the phenomenon inverted with a maximum uplift of 0.73 m in 1952. After a period of ~17 years in which a small deflation took place, a new inflation episode with moderate seismicity ($M < 2.5$) occurred between 1969 and 1972 resulting in a maximum uplift of 1.77 m. It was followed by a period of minor deflation lasting ~10 years and by another period of inflation from 1982 to 1984 with maximum uplift of 1.79 m at the Pozzuoli pier. This time an intense seismicity, with more than 15,000 earthquakes of magnitude up to 4.2 and depth of most hypocenters shallower than 3 km was recorded in the area encompassing the Solfatara crater, the town of Pozzuoli, and the Pozzuoli Bay (Corrado et al. 1977; Barberi et al. 1984; Aster and Meyer 1988), where the uplift gradient was maximum (Berrino et al. 1984). According to Aster and Meyer (1988), 10% of the 228 hypocenters are below the Gulf of Pozzuoli at depths between 3.5 and 5.0 km, along a steeply dipping plane, whereas 90% of the hypocenters are onshore, are found at depths of 1.8–4.0 km (apart from few exceptions), and are centered close to the Solfatara crater, about 1 km NE of the Pozzuoli pier.

After the important uplift phase of 1982–1984, a process of slow subsidence began in 1985, with minor uplift episodes of short duration in 1989 and 2000, while a temporary arrest of the subsidence took place in 1994. Since 2004, the general trend inverted and a continuous uplift has been recorded, in spite of the variable uplift rate. The vertical ground movement has been accompanied by modest seismicity, represented by swarms of small-magnitude earthquakes. As shown by Fig. 10, the uplift phase is still in progress, with an increase in elevation of ~0.8 m in June 2021 with respect to 2004 at the GPS RITE station, based on the monthly Surveillance Bulletin of June 2021 (INGV 2021).

Moreover, the overall horizontal displacements recorded by the GPS network of 25 stations during the period January 2016–June 2021 indicate that the source of deformation is located in the Gulf of Pozzuoli a few hundred meters south of the RITE station and that the whole Campi Flegrei caldera seems to be affected by this phenomenon, including the northernmost QUAR station, whereas nil to negligible displacements are recorded at the stations positioned outside the Campi Flegrei caldera, that is, LICO to the west and FRUL, MAFE and NAMM to the east (Fig. 11). Similar inferences on the stability of the ground deformation source and on the

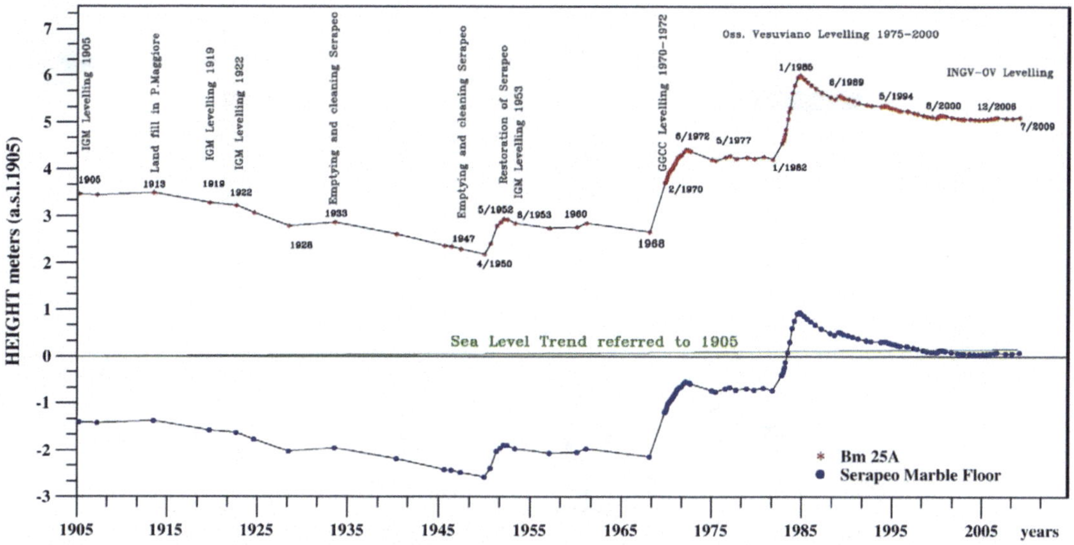

Fig. 9 Elevation changes of the Serapeum floor and benchmark BM 25A (Datum Point) from 1905 to 2009 referred to the sea level in 1905. Reprinted from Del Gaudio et al. (2010), Copyright © 2010, with permission from Elsevier

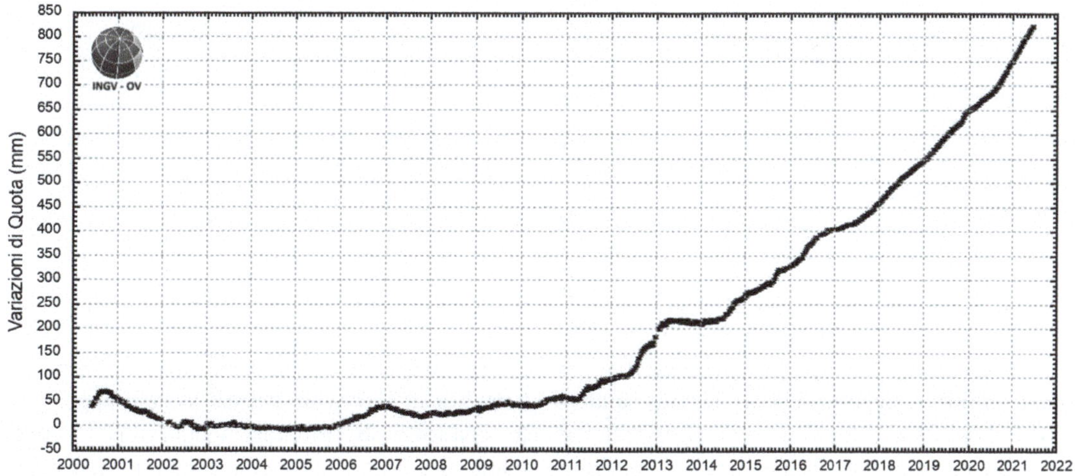

Fig. 10 Time series of the weekly changes in elevation at the GPS RITE station (Pozzuoli—Rione Terra) from 2000 to June 2021. From INGV (2021), which is licensed under a Creative Commons Attribution 4.0 CC BY International License

effects of the bradyseism over the whole Campi Flegrei caldera were previously drawn by De Martino et al. (2014) for the period 2000–2013.

In recent years, seismicity was recorded more or less in the same area and in the same depth range of the 1982–1984 uplift phase (e.g., Castaldo et al. 2019 and references therein; see Sect. 2.3 in Chap. "The Magmatic–hydrothermal System Hosted in the Campi Flegrei Caldera with Emphasis on the Solfatara"), in spite of the significantly lower magnitude of the earthquakes. For instance, in June 2021, the most energetic event, recorded on June 19, 2021 at 18:27 UTC, had $M_{dmax} = 1.7 \pm 0.3$ (INGV 2021).

Summing up, the only documented uplift episode caused by the emplacement of new

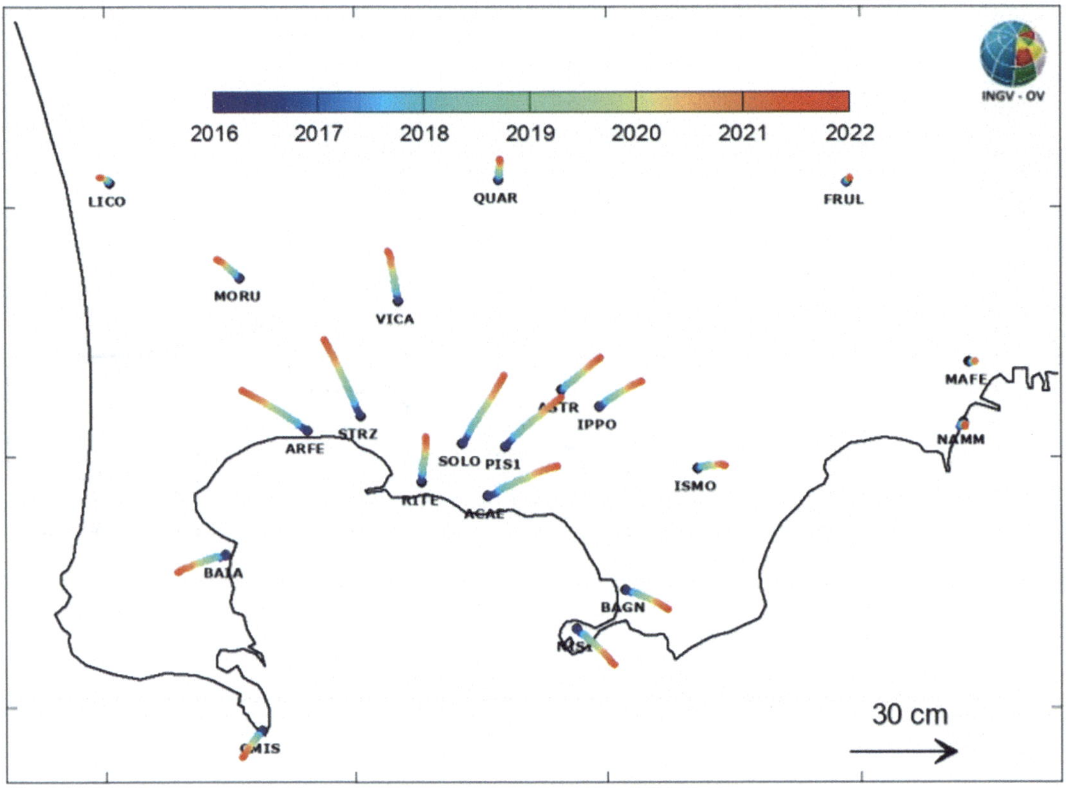

Fig. 11 Map of the horizontal GPS displacements recorded in the Campi Flegrei area from January 2016 to June 2021. From INGV (2021), which is licensed under a Creative Commons Attribution 4.0 CC BY International License

magma at shallow depth and preceding a volcanic eruption is that of Monte Nuovo in 1538 CE. In contrast, the magmatic control on the Starza uplift is hypothetical as already underscored above. Therefore, it is not surprisingly the lack of consensus on the processes controlling the 1982–1984 caldera resurgence and the ongoing uplift since 2004, in that some authors proposed the involvement of a magmatic input or the pressurization of the magma chamber, other authors suggested the pressurization of the overlying hydrothermal system, whereas a combination of both factors was invoked in other studies. In spite of this still open debate in the scientific community, the last hypothesis is considered to be the most convincing one by Todesco and coworkers, based on coupled thermo-hydro-mechanical simulations of the whole hydrothermal-magmatic system (Chiodini et al. 2003; Todesco et al. 2004; Todesco 2008, 2009). According to these models: (i) the variable behavior of the hydrothermal system appears to be controlled by the varying intensity of magmatic degassing, although peaks in CO_2/H_2O ratio in fumarolic fluids discharged at the surface are observed several months after the magmatic degassing pulses and (ii) periods of high injection rate of magmatic fluids into the overlying hydrothermal system can drive significant amounts of rock deformation, consisting of a quick caldera resurgence phase followed by a much slower caldera subsidence phase. An updated discussion of the different models which have been proposed to explain the Campi Flegrei bradyseism is given by Cannatelli et al. (2020), Smale (2020), and Lima et al. (2021).

References

Aster RC, Meyer RP (1988) Three-dimensional velocity structure and hypocenter distribution in the Campi Flegrei caldera, Italy. Tectonophysics 149(3–4):195–218

Barberi F, Corrado G, Innocenti F, Luongo G (1984) Phlegraean fields 1982–1984: brief chronicle of a volcano emergency in a densely populated area. Bulletin Volcanologique 47(2):175–185

Barberi F, Bertagnini A, Landi P, Principe C (1992) A review on phreatic eruptions and their precursors. Journal of Volcanology and Geothermal Research 52 (4):231–246

Barberi F, Cassano E, La Torre P, Sbrana A (1991) Structural evolution of Campi Flegrei caldera in light of volcanological and geophysical data. J Volcanol Geotherm Res 48(1–2):33–49

Berrino G, Corrado G, Luongo G, Toro B (1984) Ground deformation and gravity changes accompanying the 1982 Pozzuoli uplift. Bull Volcanol 47(2):187–200

Cannatelli C, Spera FJ, Bodnar RJ, Lima A, De Vivo B (2020) Ground movement (bradyseism) in the Campi Flegrei volcanic area: a review. In: De Vivo B, Belkin HE, Rolandi G (eds) Vesuvius, Campi Flegrei, and Campanian volcanism, ch 15. Elsevier, Amsterdam, pp 407–433

Castaldo R, D'Auria L, Pepe S, Solaro G, De Novellis V, Tizzani P (2019) The impact of crustal rheology on natural seismicity: Campi Flegrei caldera case study. Geosci Front 10(2):453–466

Chiodini G, Todesco M, Caliro S, Del Gaudio C, Macedonio G, Russo M (2003) Magma degassing as a trigger of bradyseismic events: the case of Phlegrean fields (Italy). Geophys Res Lett 30(8):1434. https://doi.org/10.1029/2002GL016790

Cipriani F, Marianelli P, Sbrana A (2008) Studio di una sequenza piroclastica del vulcano della Solfatara (Campi Flegrei). Considerazioni vulcanologiche e sul sistema di alimentazione. Atti della Società Toscana di Scienze Naturali, Memorie, Serie A 113:39–48

Corrado G, Guerra I, Bascio AL, Luongo G, Rampoldi R (1977) Inflation and microearthquake activity of Phlegraean fields, Italy. Bulletin Volcanologique 40 (3):169–188

Costa A, Folch A, Macedonio G, Giaccio B, Isaia R, Smith VC (2012) Quantifying volcanic ash dispersal and impact of the Campanian Ignimbrite super-eruption. Geophys Res Lett 39(10):L10310. https://doi.org/10.1029/2012GL051605

De Martino P, Tammaro U, Obrizzo F (2014) GPS time series at Campi Flegrei caldera (2000–2013). Ann Geophys 57(2):0213. https://doi.org/10.4401/ag-6431

Deino AL, Orsi G, de Vita S, Piochi M (2004) The age of the Neapolitan Yellow Tuff caldera-forming eruption (Campi Flegrei caldera-Italy) assessed by $^{40}Ar/^{39}Ar$ dating method. J Volcanol Geotherm Res 133(1–4):157–170

Del Gaudio C, Aquino I, Ricciardi GP, Ricco C, Scandone R (2010) Unrest episodes at Campi Flegrei: a reconstruction of vertical ground movements during 1905–2009. J Volcanol Geotherm Res 195(1):48–56

Di Girolamo P, Ghiara MR, Lirer L, Munno R, Rolandi G, Stanzione D (1984) Vulcanologia e petrologia dei Campi Flegrei. Bollettino della Società Geologica Italiana 103(2):349–413

Di Vito M, Lirer L, Mastrolorenzo G, Rolandi G (1987) The 1538 Monte Nuovo eruption (Campi Flegrei, Italy). Bull Volcanol 49(4):608–615

Dvorak JJ, Gasparini P (1991) History of earthquakes and vertical ground movement in Campi Flegrei caldera, southern Italy: comparison of precursory events to the AD 1538 eruption of Monte Nuovo and of activity since 1968. J Volcanol Geotherm Res 48:77–92

Giaccio B, Hajdas I, Isaia R, Deino A, Nomade S (2017) High-precision ^{14}C and $^{40}Ar/^{39}Ar$ dating of the Campanian Ignimbrite (Y-5) reconciles the time-scales of climatic-cultural processes at 40 ka. Sci Rep 7(1):1–10

Giacomelli L, Scandone R (2012) History of the exploitation of thermo-mineral resources in Campi Flegrei and Ischia, Italy. J Volcanol Geotherm Res 209:19–32

Guidoboni E (2010) History and volcanology: dialogue overdue? The case of a false eruption in medieval Italy. EOS Trans Am Geophys Union 91(26):231–231

Guidoboni E, Ciuccarelli C (2011) The Campi Flegrei caldera: historical revision and new data on seismic crises, bradyseisms, the Monte Nuovo eruption and ensuing earthquakes (twelfth century 1582 AD). Bull Volcanol 73(6):655–677

INGV (2021) Bollettino di Sorveglianza. Campi Flegrei. Giugno 2021, p 31. https://www.ov.ingv.it/ov/bollettini-mensili-campania/Bollettino_Mensile_Campi_Flegrei_2021_06.pdf. Accessed 4 Aug 2021

Isaia R, Vitale S, Di Giuseppe MG, Iannuzzi E, D'Assisi TF, Troiano A (2015) Stratigraphy, structure, and volcano-tectonic evolution of Solfatara maar-diatreme (Campi Flegrei, Italy). Geol Soc Am Bull 127(9–10):1485–1504

Isaia R, Di Giuseppe MG, Natale J, Tramparulo FDA, Troiano A, Vitale S (2021) Volcano-tectonic setting of the Pisciarelli fumarole field, Campi Flegrei caldera, Southern Italy: insights into fluid circulation patterns and hazard scenarios. Tectonics 40(5): e2020TC006227

Lima A, Bodnar RJ, De Vivo B, Spera FJ, Belkin HE (2021) Interpretation of recent unrest events (bradyseism) at Campi Flegrei, Napoli (Italy): comparison of models based on cyclical hydrothermal events versus shallow magmatic intrusive events. Geofluids 16. Article ID 2000255. https://doi.org/10.1155/2021/2000255

Linnaeus C (1758) Systema Naturae per Regna Tria Naturae, secundum classes, ordines, genera, species, cum characteribus, differentiis, synonymis, locis. Holmiae, Imprensis Laurentii Salvii

Lirer L, Luongo G, Scandone R (1987) On the volcanological evolution of Campi Flegrei. EOS Trans Am Geophys Union 68(16):226–234

Lyell C (1850) Principles of geology. J. Murray, London, pp 489–498

Marini L, Principe C, Chiodini G, Cioni R, Fytikas M, Marinelli G (1993) Hydrothermal eruptions of Nisyros (Dodecanese, Greece). Past events and present hazard. J Volcanol Geotherm Res 56:71–94

Marotta E, Peluso R, Avino R, Belviso P, Caliro S, Carandente A, Chiodini G, Macedonio G, Avvisati G, Marfè B (2019) Thermal energy release measurement with thermal camera: the case of la Solfatara Volcano (Italy). Remote Sens 11(2):167

Mercalli G (1883) Vulcani e fenomeni vulcanici in Italia. Arnaldo Forni Editore, Bologna, p 374

Niccolini A (1829) Rapporto sulle acque che invadono il pavimento dell'antico edifizio detto il Tempio di Giove Serapide. Stamperia Reale, Napoli, p 46

Niccolini A (1839) Tavola metrica-cronologica delle varie altezze tracciate dalla superficie del mare fra la costa di Amalfi ed il promontorio di Gaeta nel corso di diciannove secoli. Tipografia Flautina, Napoli, p 52

Niccolini A (1846) Descrizione della Gran Terma Puteolana volgarmente detta Tempio di Serapide preceduta da taluni cenni storici per servire alla dilucidazione de' fenomeni geologici, e de' problemi architettonici di quel celebre monumento e considerazioni su i laghi Maremmani. Stamperia Reale, Napoli, p 257

Orsi G, De Vita S, Di Vito M (1996) The restless, resurgent Campi Flegrei nested caldera (Italy): constraints on its evolution and configuration. J Volcanol Geotherm Res 74(3–4):179–214

Parascandola A (1947) I fenomeni bradisismici del Serapeo di Pozzuoli. Stabilmento tipografico G. Genovese, Napoli, p 156

Piochi M, Mormone A, Balassone G, Strauss H, Troise C, De Natale G (2015) Native sulfur, sulfates and sulfides from the active Campi Flegrei volcano (southern Italy): genetic environments and degassing dynamics revealed by mineralogy and isotope geochemistry. J Volcanol Geotherm Res 304:180–193

Pirajno F (2009) Hydrothermal processes and mineral systems. Springer, Berlin, Heidelberg, p 1250

Rittman A (1950) Sintesi geologica dei Campi Flegrei. Bollettino della Società Geologica Italiana 69(2):117–128

Rittmann A (1929) Die Solfatara. Naturwissenschaften 17 (34):659–663

Rosi M, Sbrana A, Principe C (1983) The Phlegraean fields: structural evolution, volcanic history and eruptive mechanisms. J Volcanol Geotherm Res 17(1–4):273–288

Rosi M, Sbrana A (1987) Stratigraphy. In: Rosi M, Sbrana A (eds) Phlegrean fields. Quaderni de La Ricerca Scientifica 9(114):10–39

Sbrana A, Marianelli P, Pasquini G (2021) The Phlegrean fields volcanological evolution. J Maps 17(2):545–558. https://doi.org/10.1080/17445647.2021.1982033

Scacchi A (1849) Memorie Geologiche sulla Campania e relazione dell'incendio accaduto nel Vesuvio nel mese di Febbrajo del 1850. Estratte dal Rendiconto della R. Accademia delle Scienze di Napoli, p 48

Scandone R, Bellucci F, Lirer L, Rolandi G (1991) The structure of the Campanian Plain and the activity of the Neapolitan volcanoes (Italy). J Volcanol Geotherm Res 48(1–2):1–31

Scandone R, D'Amato J, Giacomelli L (2010) The relevance of the 1198 eruption of Solfatara in the Phlegraean Fields (Campi Flegrei) as revealed by medieval manuscripts and historical sources. J Volcanol Geotherm Res 189(1–2):202–206

Scarpati C, Sparice D, Perrotta A (2014) A crystal concentration method for calculating ignimbrite volume from distal ash-fall deposits and a reappraisal of the magnitude of the Campanian Ignimbrite. J Volcanol Geotherm Res 280:67–75

Scarpati C, Cole P, Perrotta A (1993) The Neapolitan Yellow Tuff. A large volume multiphase eruption from Campi Flegrei, southern Italy. Bull Volcanol 55 (5):343–356

Sicardi L (1944) L'attività della Solfatara di Pozzuoli attraverso la documentazione storica avanti l'ultimo ottantennio. Atti della Società Italiana di Scienze Naturali e del Museo Civico di Storia Naturale in Milano 83(2):97–114

Sicardi L (1956) La Solfatara di Pozzuoli. Bulletin Volcanologique 18(1):151–158

Signore F (1929) Sur la variation d'activite du volcan de Boue ("Fangaia") de la Solfatare de Pouzzoles (Naples) par suite du grand tremblement de terre de l'Irpinia, 23 Juillet 1930. Bulletin Volcanologique 3 (2):49–52

Smale L (2020) A re-interpretation of long-term deformation at Campi Flegrei caldera, Italy and perceptions of the causes of caldera unrest. PhD thesis, Department of Earth Sciences, University College London, p 340

Smith VC, Isaia R, Pearce NJG (2011) Tephrostratigraphy and glass compositions of post-15 kyr Campi Flegrei eruptions: implications for eruption history and chronostratigraphic markers. Quat Sci Rev 30(25–26):3638–3660

Takahashi T (2007) Debris flow: mechanics, prediction and countermeasures. Taylor and Francis. London 448 pp

Todesco M (2009) Signals from the Campi Flegrei hydrothermal system: role of a "magmatic" source of fluids. J Geophys Res Solid Earth 114(B5):B05201. https://doi.org/10.1029/2008JB006134

Todesco M, Rutqvist J, Chiodini G, Pruess K, Oldenburg CM (2004) Modeling of recent volcanic episodes at Phlegrean fields (Italy): geochemical variations and ground deformation. Geothermics 33(4):531–547

Todesco M (2008) Hydrothermal fluid circulation and its effect on caldera unrest. In: Gottsmann J, Marti J (eds) Caldera volcanism: analysis, modelling and response. Dev Volcanol 10:393–416

The Magmatic–Hydrothermal System Hosted in the Campi Flegrei Caldera with Emphasis on the Solfatara

Abstract

The available information relevant to reconstruct the conceptual models of both the magmatic–hydrothermal system hosted in the Campi Flegrei caldera and the Solfatara magmatic-hydrothermal system was summarized and reviewed. This information includes (but is not limited to): the main characteristics of the deep geothermal wells drilled in the area, the hydrothermal alteration mineralogy and the geochemistry of the fluids encountered by the AGIP-ENEL geothermal wells, the tomographic modeling of active and passive seismic data, the chemical and isotopic characteristics of onshore thermal springs and shallow wells, onshore fumaroles and gas vents, offshore fumaroles and gas vents, and sub-lacustrine hydrothermal discharges. Taking into account the general conceptual model of volcano-hosted magmatic-hydrothermal systems (Fournier, 1999) and adopting the methodology commonly used by geothermal scientists (Cumming 2009, 2016), the conceptual model of the magmatic-hydrothermal system hosted in the Campi Flegrei caldera was reconstructed. The local stratigraphy comprises: (a) water-saturated volcanic products above 0.6 km depth; (b) volcanic and marine deposits extensively affected by propylitic, phyllitic, and argillic hydrothermal alteration, between 0.6 and 2.7 km depths; (c) a layer of thermometamorphic rocks at depths between 2.7 and 4 km approximately; (d) Mesozoic carbonate rocks (such as those cropping out around the Campanian Plain) and the crystalline bedrock from ca. 4 to ca. 7.5 km depths (e) a magma reservoir whose top is situated at ca. 7.5 km depth. Levels (c) and (d) host over-pressurized gases and hypersaline brines that are expelled from the crystallizing magma and CO_2-rich gases produced in-situ by different reactions. Actually, CO_2-rich fluids were discharged by well San Vito 1. Thus, levels (c) and (d) are the "engine" governing the bradyseism as proposed by several authors. Levels (b) and (c) are separated by a relatively impermeable zone produced by self-sealing, mainly quartz deposition, at temperature close to 400°C (Fournier 1999). Actually, this quartz-bearing zone was encountered in well San Vito 1, at depths between ca. 2.5 and 2.8 km and temperatures of ca. 360 to 385°C. This quartz-rich zone represents the boundary between the underlying deep-magmatic portion of the magmatic-hydrothermal system, where fluid pressure is controlled by the lithostatic regime, and the overlying shallow-hydrothermal portion of the magmatic-hydrothermal system, where fluid pressure is governed by the hydrostatic regime (Fournier 1999). In the shallow-hydrothermal portion, the circulation of hydrothermal fluids of marine origin (at least for the most) is restricted to the brittle rocks of the propylitic zone, where

L. Marini et al., *The Solfatara Magmatic-Hydrothermal System*, Advances in Volcanology,
https://doi.org/10.1007/978-3-030-98471-7_3

permeability is fracture-controlled, whereas the overlying impermeable, plastic rocks of the phyllitic and argillic zones act as the clay cap of the geothermal system. Similar to the basinal-type geothermal systems, the extension of the Campi Flegrei shallow geothermal reservoir is marked, to a first approximation, by the neutral chloride thermal waters emerging at the surface, or intercepted by shallow wells, close to the coastline. However, the onshore thermal manifestations (both thermal waters and gas emissions) cluster in the two distinct areas of Mofete and Solfatara-Agnano. Hydrothermal circulation might be restricted, almost completely, to these two areas, which would host two different, comparatively small hydrothermal systems. Alternatively (and this is our preferred interpretation), the two areas of Mofete and Solfatara-Agnano mark two separate upflows of a unique, relatively large hydrothermal system, extending between them, southwards (as indicated the offshore gas vents) and northwards (as proven by the deep boreholes of San Vito). Finally, the two conceptual models of the Solfatara magmatic-hydrothermal system proposed by Cioni et al. (1984) and Caliro et al. (2007) are recalled. Both conceptual models are realistic, in that they describe the system of interest during the 1982-1984 bradyseismic crisis and after its cessation, respectively.■■■

The purpose of this chapter is twofold. First, the present state of knowledge of the magmatic–hydrothermal system hosted in the Campi Flegrei caldera, with emphasis on the Solfatara, is summarized and briefly discussed, thus building on what was presented in the previous chapter. Second, the conceptual models of the Solfatara magmatic–hydrothermal system proposed in relevant studies are examined and discussed. Unfortunately, it was not possible to examine all the pertinent scientific works due to their considerable number and their rapid growth with time. Therefore, the most relevant works were necessarily selected, taking into account both the general conceptual model of volcano-hosted magmatic–hydrothermal systems (Fournier 1999) and the methodology commonly adopted by geothermal scientists to reconstruct the conceptual models of these systems (Cumming 2009, 2016).

According to the general conceptual model of Fournier (1999), volcano-hosted magmatic–hydrothermal systems comprise (Fig. 1a): (i) a lower part, where hypersaline brines and gases exsolved from the underlying crystallizing magma accumulate at lithostatic pressure within a volume of plastic rocks, and (ii) an upper part, where meteoric-derived hydrothermal fluids circulate through brittle rocks at hydrostatic pressure. As discussed in Sect. 1.3, marine waters may circulate into the upper part of the system at hydrostatic pressure, instead of meteoric waters or in addition to them, in islands and coastal areas, as is the case of Campi Flegrei. The two parts are separated by a thin, relatively impermeable zone produced through self-sealing (mainly quartz deposition[1]), where temperature is usually close to 400 °C.

From time to time, this self-sealed zone is breached by an upward surge in magma or other mechanisms causing the quick expulsion of hypersaline brines and gases from the usually plastic region into the brittle domain, at lower pressure and temperature (Fig. 1b). The ensuing increase in fluid pressure and temperature within the brittle domain leads to faulting and fracturing, which increase permeability and the rate of discharge of magmatic–hydrothermal fluids.

The general conceptual model of volcano-hosted magmatic–hydrothermal systems proposed by Fournier (1999) was modified and applied to the Campi Flegrei by Lima et al. (2021) and references therein. Nevertheless, there is no reason and no evidence supporting the presence of the low-permeability layer B postulated by Lima et al. (2021).

The methodology of Cumming (2009, 2016) is focused on the upper portion of volcano-hosted

[1] The reduction in permeability close to an intrusive heat source caused by quartz precipitation was discussed by several authors (e.g., Fournier 1985, 1999; Wells and Ghiorso 1991; Lowell et al. 1993; Moore et al. 1994; White and Mroczek 1998; Saishu et al. 2014; Scott and Driesner 2018).

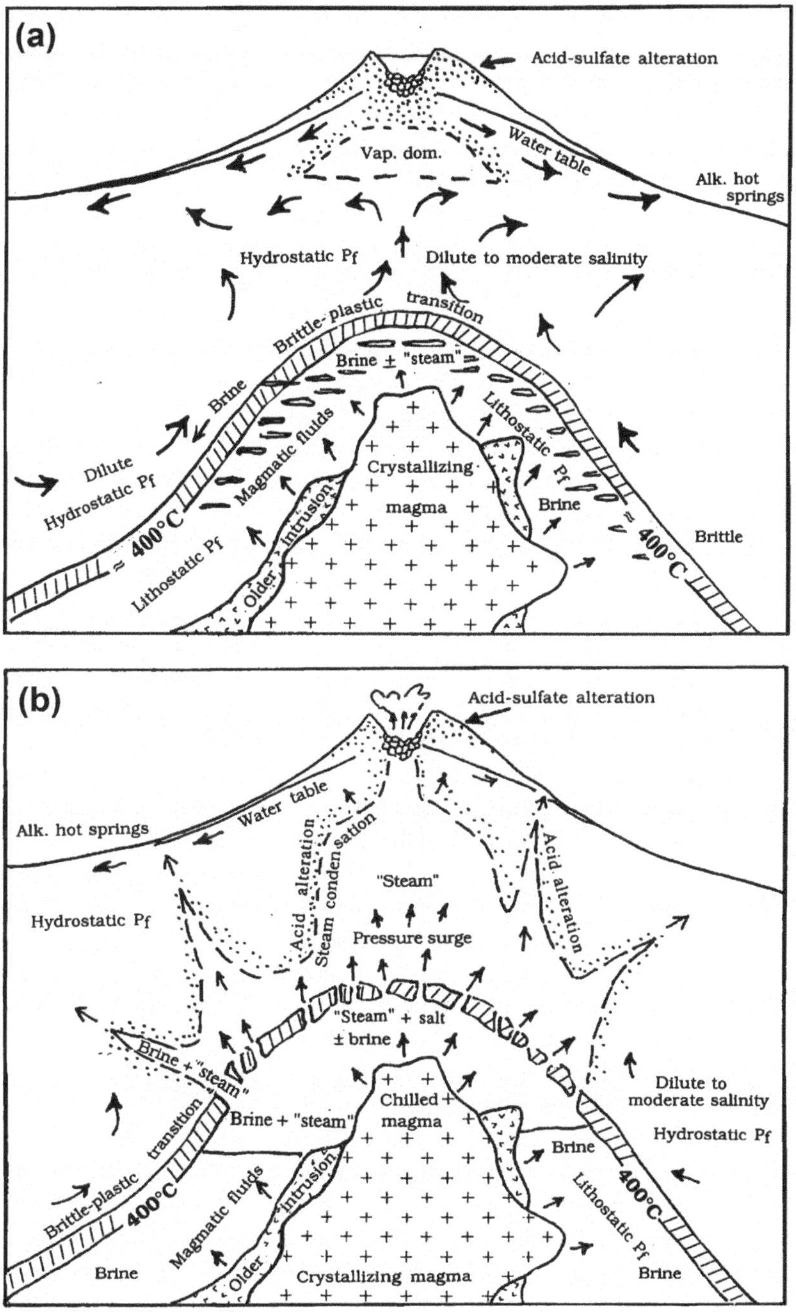

Fig. 1 Schematic conceptual model of a volcano-hosted magmatic–hydrothermal system, where the top of the crystallizing magma is at depths ranging from 1 to 3 km, and its change in state. **a** The brittle to plastic transition (BPT) occurs at ca. 370–400 °C and dilute, prevailingly meteoric waters circulate at hydrostatic pressure in brittle rocks above the BPT, while dominantly magmatic gases and highly saline brines accumulate in plastic rocks at lithostatic pressure, below the BPT. **b** Episodic and temporary breaching of the usually self-sealed zone allows magmatic fluids to escape into the overlying hydrothermal system. Reprinted from Fournier (1999), Copyright © 1999, with Fair Use Permission from the Society of Economic Geologists

magmatic–hydrothermal systems, which is that of interest for their industrial exploitation and, therefore, is investigated by means of surface exploration techniques, first, and deep drilling, afterwards. Thus, the works of Cumming (2009, 2016) provide the guidelines for the reconstruction of the conceptual models through the processing of the available data, usually including: (i) maps of surface geology also showing tectonic and volcano-tectonic structures, (ii) images of the low resistivity clay cap derived from magnetotelluric (MT) surveys, and (iii) the geochemistry of hot springs and fumaroles.

Following the general conceptual model of volcano-hosted magmatic–hydrothermal systems of Fournier (1999), the upper "shallow-hydrothermal" portion of the Campi Flegrei magmatic–hydrothermal system, where fluid pore pressure probably follow the hydrostatic regime, and the lower "deep-magmatic" portion of the system, in which fluid pore pressures is probably close to the lithostatic regime, are considered in two distinct sections. To be noted that thermometamorphic reactions occur in the "deep-magmatic" sector of the Campi Flegrei magmatic–hydrothermal system, generating important amounts of different gas species, as discussed below.

1 The Shallow-Hydrothermal Portion of the Magmatic–Hydrothermal System Hosted in the Campi Flegrei Caldera

Referring the dataset usually available for reconstructing the conceptual models of volcano-hosted geothermal systems, the bad news is the lack of the MT data over most of the Campi Flegrei caldera, although this gap was partly filled for the Solfatara crater and nearby areas (see Sect. 1.2), while the good news is the detailed knowledge of both the local stratigraphic sequence and the hydrothermal alteration mineralogy thanks to the data acquired through the deep geothermal wells drilled by AGIP-ENEL at depths of 1600–3000 m below ground level in the areas of Mofete and San Vito.

As shown by the N–S geological cross-section across the Campi Flegrei and Pozzuoli Bay (Fig. 2; from Lima et al. 2009), the local stratigraphic sequence comprises coarse-grained to fine-grained clastic sediments, and different volcanic and volcano-clastic rocks overlying the Mesozoic–Cenozoic carbonate basement cropping out in the Apennine thrust belt all around the Campanian plain.

The knowledge of the hydrothermal alteration mineralogy allows one to distinguish the clay cap from the hydrothermal reservoir and to gain indications on the P–T–X conditions that were present when the hydrothermal minerals were deposited (see Sect. 1.2). To follow a logical order, the main features of the deep geothermal wells drilled in the Campi Flegrei caldera are summarized in the next section.

1.1 Main Characteristics of the Deep Geothermal Wells Drilled in the Campi Flegrei Caldera

Several boreholes were drilled from 1939 to 1943 and from 1953 to 1955 by SAFEN at Mofete, Monte Nuovo, Agnano–Solfatara, and Licola (Penta 1949) and from 1978 to 1985 by AGIP-ENEL at Mofete, San Vito, and Licola (Carella and Guglielminetti 1983; Guglielminetti 1986; Chelini and Sbrana 1987). The location of the geothermal wells drilled at depth >500 m by SAFEN and of the deep geothermal wells drilled by AGIP-ENEL is shown in the Google Earth map of Fig. 3.

Among the wells drilled by SAFEN, the following four reached depths >500 m below ground level:

(a) Well CF23 (also called Agnano 1), which was drilled on the eastern slopes of the Solfatara relief and reached a total depth of 1841 m, where a maximum temperature of 325 °C was measured (Minucci 1964). Two aquifer zones were encountered by the well at 110 and 1445 m depth. Under optimum conditions, the well section from 1560 to 1712 m depth (i.e., the deep aquifer zone) discharged a total flow rate decreasing from 4481 to 4249 kg/h, with a concurrent decrease in the mass steam fraction from

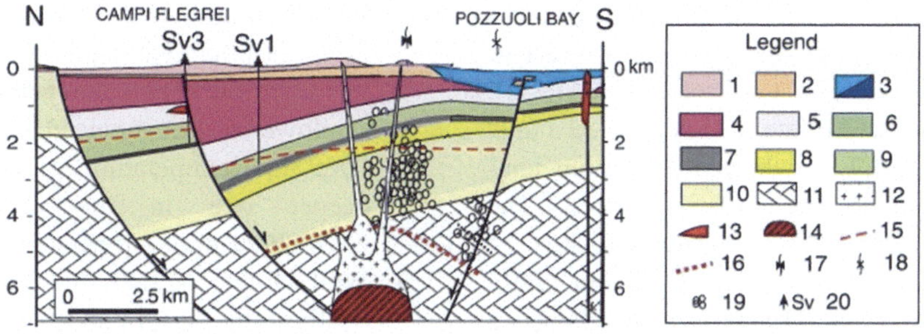

Fig. 2 N–S geological cross-section across the Campi Flegrei and Pozzuoli Bay. *Legend:* 1. Holocene volcanics; 2. Neapolitan Yellow Tuff; 3. Main sediments post 39 ka; 4. Campanian Ignimbrite and previous tuffs; 5. Middle Pleistocene sandstones, siltstones and volcanics; 6. Middle Pleistocene marine sediments (sandstones and siltstones); 7. Fine-grained Middle Pleistocene marine sediments (claystones and siltstones); 8. Middle Pleistocene deep water debris flows; 9. Lower Pleistocene marine sediments; 10. Continental deposits and conglomerates; 11. Meso-Cenozoic carbonate substrate; 12. Crystallized magma; 13. Volcanic bodies; 14. Magma body; 15. Thermometamorphic boundary; 16. Impermeable zone surrounding the crystallizing magma body; 17. Pozzuoli Anticline; 18. Pozzuoli Bay Syncline; 19. 1983–84 earthquake hypocenters; 20. Deep geothermal wells. Reprinted from Lima et al. (2009), modified, Copyright © 2009, with permission from Elsevier

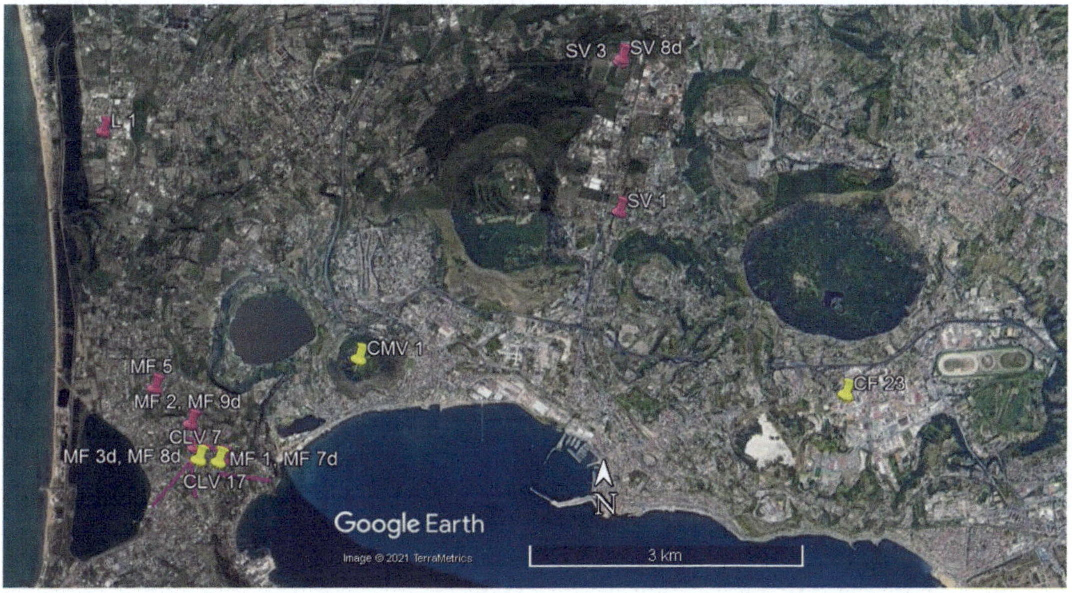

Fig. 3 Location map of the geothermal wells drilled at depth >500 m by SAFEN (yellow pins) and the deep geothermal wells drilled by AGIP-ENEL (magenta pins)

0.539 to 0.396. Assuming that these values refer to atmospheric conditions, the total discharge enthalpy decreased from 1636 to 1312 kJ/kg, indicating enthalpy-temperatures of 345 to 294 °C if a single liquid is considered to be present in the reservoir. These enthalpy temperatures are 20 °C higher and 31 °C lower than the maximum measured temperature of 325 °C, respectively.

(b) Well CF 21 at Mofete, which was drilled to a total depth of 1218 m and found a reservoir temperature of 233 °C (Chelini et al. 1982).
(c) Well CLV 7 at Mofete, which reached a total depth of 584.9 m, where a maximum temperature of 225 °C was measured. The borehole discharged a considerable flow of water and steam for 15 months (Penta 1949).
(d) Well CLV 17 at Mofete, which attained a total depth of 521.7 m, where a maximum temperature of 225 °C was determined. The borehole discharged a flow of water and steam (Penta 1949).
(e) Well CMV 1 at Monte Nuovo, which reached a total depth of 676.9 m, but measured temperatures oscillated from 62 to 77 °C (possibly 80 °C) from 15 m downwards (Penta 1949).
(f) Well CF 22 at Licola, which was drilled to a total depth of 1600 m and encountered a reservoir temperature of 170 °C (Chelini et al. 1982).

The main characteristics of the deep geothermal wells drilled by AGIP-ENEL at Mofete are as follows (Carella and Guglielminetti 1983):

(i) Well Mofete 1 (MF-1) was drilled vertically to a total depth of 1606 m and encountered two productive zones, the shallower one in the depth interval 550–896 m at temperature of 247 °C and the deeper one in the depth interval 1223–1469 m at temperature of 295 °C.
(ii) Well Mofete 2 (MF-2) was drilled vertically to a total depth of 1989 m and met a productive zone from 1272 to 1989 m depth at temperature of 337–342 °C.
(iii) Well Mofete 3d (MF-3d) was drilled vertically until 1749 m and then it was deviated south-westwards until the total depth of 1909 m. It met two productive zones, the shallower one in the depth interval 430–665 m at temperature of 230 °C and the deeper one in the depth interval 1302–1900 m at temperature of 275 °C.
(iv) Well Mofete 5 (MF-5) was drilled vertically until the total depth of 2700 m. Also this well found two productive zones, the shallower one in the depth interval 1627–1960 m at temperature of 286 °C and the deeper one in the depth interval 2310–2699 m at temperature of 362 °C.
(v) Well Mofete 7d (MF-7d) was drilled vertically until 1476 m and then it was deviated towards ESE until the total depth of 1648 m. It met a productive zone from 1042 to 1646 m depth at temperature of 301 °C.
(vi) Well Mofete 8d (MF-8d) was drilled vertically until 800 m and then it was deviated southwards until the total depth of 907 m. It encountered a productive zone from 660 to 907 m depth at temperature of 234 °C.
(vii) Well Mofete 9d (MF-9d) was drilled vertically until 1582 m and then it was deviated towards ESE until the total depth of 1745 m. This well has two productive zones, the shallower one in the depth interval 694–1004 m at temperature of 210 °C and the deeper one in the depth interval 1342–1745 m at temperature of 308 °C.

All in all, there is undoubtedly an up-doming of the isotherms in the Mofete area, with culmination between wells MF-1 and MF-2, as shown by the cross-section of Fig. 4, whereas the isotherms deepen considerably in well MF-5, which is positioned outside the Campi Flegrei caldera (Chelini and Sbrana 1987). This means that there is an upflow zone of the geothermal system in the Mofete area, consistent with the presence of weak steaming grounds and hydrothermally altered rocks at the surface. Besides, at least three different geothermal aquifers were encountered by the wells drilled in the Mofete area, a shallow one (500–900 m) at temperatures of 210–250 °C, an intermediate one (1800–2000 m) at temperatures of 295–310 °C, and a

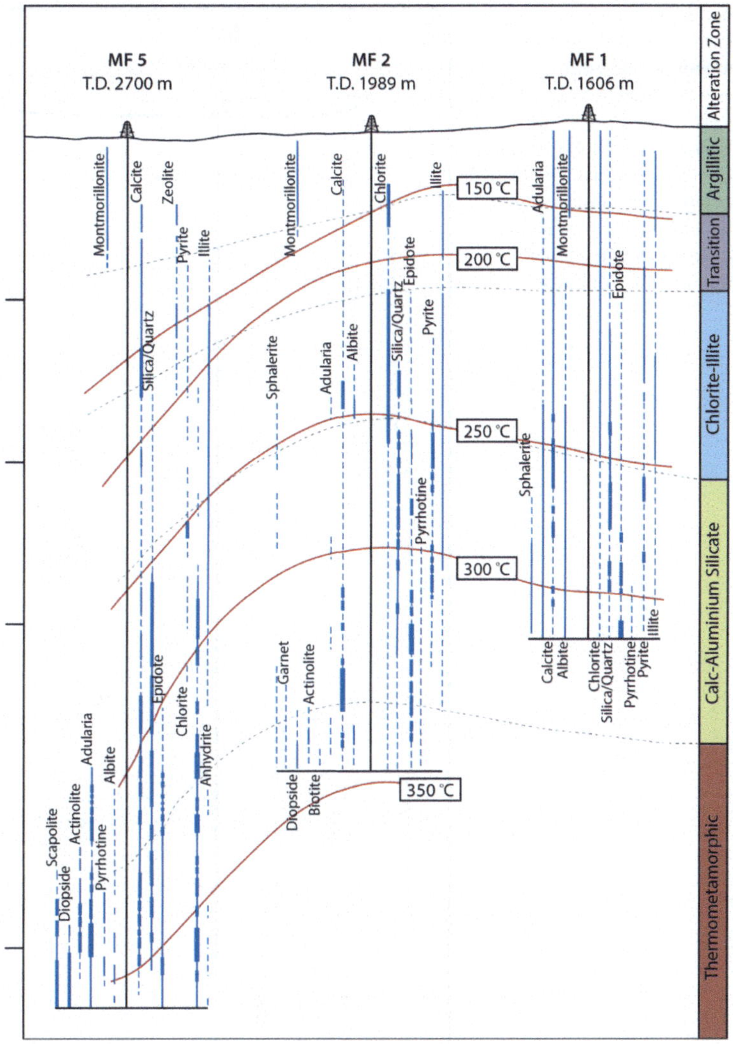

Fig. 4 Distribution of hydrothermal minerals, boundaries between hydrothermal zones (dotted gray lines) and isotherms (solid red lines) in the Mofete area. Depth in meters. The vertical scale is the same as the horizontal scale (from Chelini and Sbrana 1987, redrawn by Smale 2020)

deep one (2500–2700 m) at temperatures of 340–360 °C (Carella and Guglielminetti 1983). The lateral extension of these aquifers is poorly known, but is probably limited, also considering that well MF-2 is connected with an aquifer which was not encountered by all the other wells (Guglielminetti 1986). According to Guglielminetti (1986), the high-temperature Mofete geothermal reservoir is of limited volume, is characterized by a general lack of permeability, and has a power capacity of 10–15 MW.

The three boreholes drilled by AGIP-ENEL in the San Vito area, that is, San Vito 1 (SV-1), San Vito 8d (SV-8d), and San Vito 3 (SV-3), are known less well than those of the Mofete area. The well San Vito 1 was drilled to a total depth of 3046 m, where the temperature was estimated to be higher than 419 °C based on the partial melting of a piece of zinc left at well bottom for a sufficient time (Bruni et al. 1985). Based on the cross-section of Fig. 5 (Chelini and Sbrana 1987), showing the outcomes of wells SV-1 and SV-3, the isotherms are relatively flat in the San Vito area, at least above ~1.8 km depth, whereas at greater depths the isotherms deepen moving from well SV-3 (that reached a total depth of 2360 m) to well SV-1. Furthermore, isotherms are considerably lowered in the San

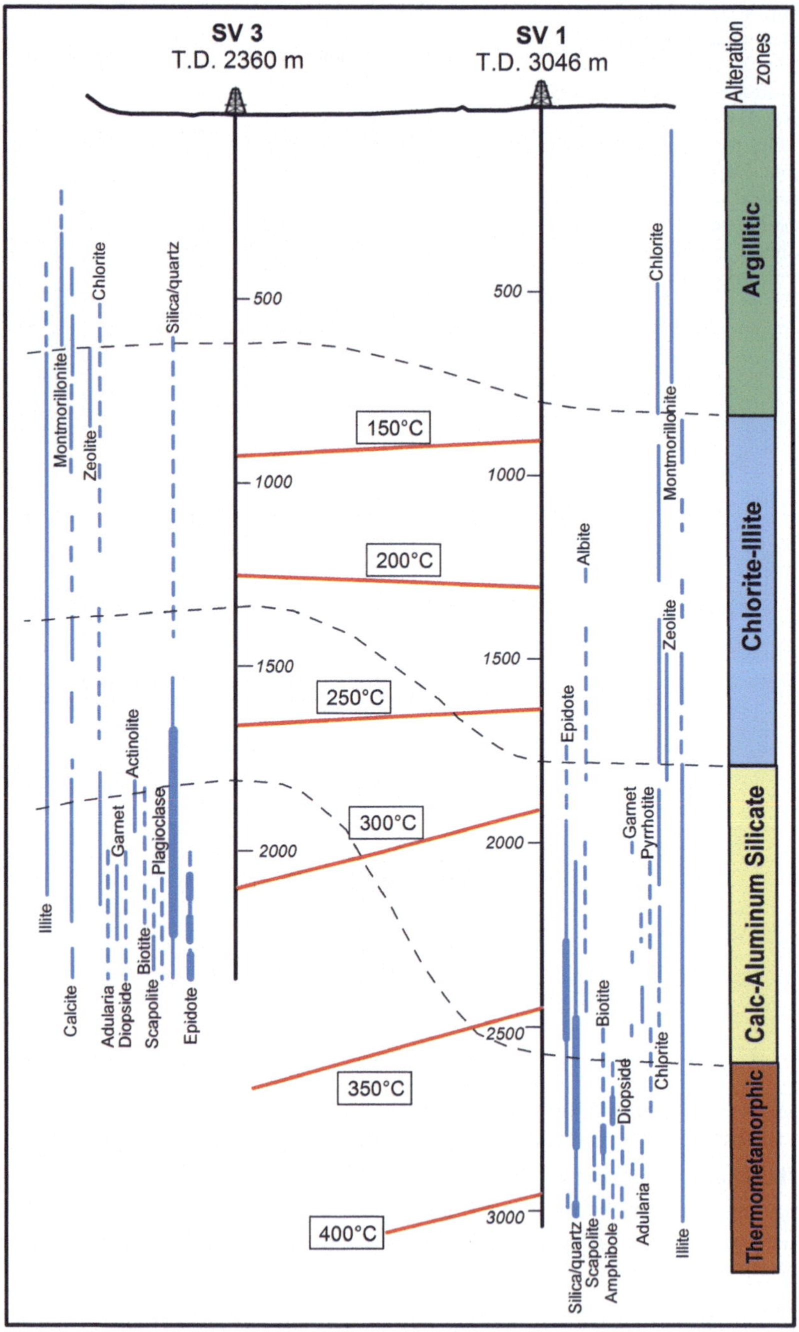

Fig. 5 Distribution of hydrothermal minerals, boundaries between hydrothermal zones (gray dashed lines) and isotherms (red solid lines) in the San Vito area. Depth in meters. The vertical scale is the same as the horizontal scale (from Chelini and Sbrana 1987, redrawn)

Vito area, compared to the Mofete area, confirming that there is an upflow zone at Mofete (see above). In spite of this lowering of the isotherms at San Vito, the deepest part of well SV-1 shows a very high geothermal gradient, from 335 °C (Chelini and Sbrana 1987, page 100) or 355 °C (Fig. 5) at 2500 m to over 420 °C at 3040 m (Chelini and Sbrana 1987). The extrapolation of this geothermal gradient of 120 or 157 °C/km at greater depths indicates the occurrence of 957 or 1122 °C at the top of the melt zone positioned at 7.5 km depth (see

Sect. 2). The circulation of hydrothermal fluids in the San Vito area is poorly known because of the low quality of the data acquired through short-term production tests. Apart from the phreatic aquifer, no other aquifer with an appreciable lateral extension is probably present above 1.8 km depth, corresponding to the bottom of the clay cap (Chelini and Sbrana 1987).

The well Licola 1, located outside the Campi Flegrei caldera, encountered temperatures substantially lower than those found inside the caldera, up to a maximum of 236 °C at the total depth of 2.5 km (Chelini 1984). Furthermore, the static temperature profile of this well shows a constant geothermal gradient close to 85 °C/km which is evidently controlled by conductive heat transfer (Wohletz et al. 1999). Carlino et al. (2012) reviewed the geothermal exploration activities carried out in the Campanian volcanic areas of Campi Flegrei, Vesuvio, and Ischia. The outcomes of a 501 m deep borehole drilled in December 2012 in the Bagnoli plain, in the eastern side of the Campi Flegrei caldera, are reported by Carlino et al. (2018) and references therein.

1.2 Hydrothermal Alteration Mineralogy

The rocks encountered by deep geothermal boreholes drilled in the Campi Flegrei area are dominated by alkali-trachytic tuffs and tuffites interbedded with subordinate marine sediments (siltites, sandstones, and marls), lava flows, and domes (Chelini and Sbrana 1987). These original lithotypes were extensively affected by the interaction with water and gases at progressively higher temperatures with increasing depth, leading to the destruction of primary, rock-forming minerals and the production of secondary, hydrothermal minerals. Based on the study of the hydrothermal alteration mineralogy found in the deep geothermal wells Mofete 1, Mofete 2, Mofete 5, San Vito 1 and San Vito 3 (Figs. 4 and 5), the following hydrothermal alteration zones (from top to bottom) were recognized by Chelini and Sbrana (1987):

(i) The argillic zone, which is characterized by abundant montmorillonite (smectite) accompanied by minor illite, chlorite, and low-temperature zeolites. The base of the argillic zone generally coincides with the 150–160 °C isotherm.

(ii) The illite–chlorite zone (or phyllitic zone), which is marked by a considerable increase in chlorite and illite as well as by the first appearance of mixed-layer clay minerals. The base of the illite–chlorite zone coincides with the 250 °C isotherm at Mofete and with temperatures of 220–270 °C in the San Vito area.

(iii) The Ca–Al–silicate zone (or propylitic zone). Epidote is the index mineral of this zone, although it begins to appear in the deepest parts of the illite–chlorite zone. Epidote is accompanied by abundant adularia, albite, silica minerals (chiefly quartz), sulfide minerals (pyrite, sphalerite, and pyrrhotite), calcite, and high-temperature zeolites. Chlorite and illite are also present, but in lower amounts compared to the illite–chlorite zone, whereas mixed-layer clay minerals disappear close to the base of the Ca–Al–silicate zone, which coincides with the 325 °C isotherm in the Mofete area and with temperatures of 270–360 °C in the San Vito area. Since the neoformed mineral phases represent more than 60 volume% of the rock, the propylitic alteration causes a strong lithification of the original solid materials and, consequently, a marked increase in both density and P-wave velocity compared to those of the illite–chlorite zone.

(iv) The thermometamorphic zone, which is marked by (a) a nearly complete textural re-arrangement, especially in the sedimentary lithotypes, (b) the appearance of amphibole, biotite, diopside, scapolite, and garnet and (c) the persistence of sulfide minerals (decreasing with increasing depth), quartz, and epidote at Mofete but not at San Vito. To be noted that the disappearance of epidote at San Vito is probably due to the high CO_2 partial pressure of the fluids hosted in this zone of the hydrothermal reservoir (see below). To be noted also that scapolite is absent in wells Mofete 1 and 2, is scarce in the deepest parts of San Vito wells and is abundant in well Mofete 5, which produced a brine with a TDS close to 200 g/kg (under reservoir conditions) from the depth interval 2310–2700 m, where scapolite occurs. The simplified general formula of scapolite is $\left[Na_xCa_{4-x}Al_ySi_{12-y}O_{24}(Cl, CO_3)\right]$ assuming that its composition varies between that of the endmembers marialite $[Na_4Al_3Si_9O_{24}Cl]$ and meionite $[Ca_4Al_6Si_6O_{24}CO_3]$ (Hassan and Buseck 1988) and neglecting the contribution of other chemical components possibly present (Shaw 1960; Evans et al. 1969). Based on these assumptions, it is possible that CO_3-rich scapolite is present at San Vito (as already supposed by Chelini and Sbrana 1987), whereas the occurrence of Cl-rich scapolite in two cores sampled in well Mofete 5 at 1714 and 2695 m depth was confirmed by SEM analyses (Piochi et al. 2015).

Taken together, the argillic zone and the underlying phyllitic zone have an extremely low permeability due to the plastic behavior of the phyllosilicate-rich rocks affected by these types of hydrothermal alteration. Therefore, these phyllosilicate-rich lithotypes constitute the impermeable clay cap of the hydrothermal system hosted in the Campi Flegrei caldera. The clay cap extends to depths of 890–1370 m below ground level at Mofete and 1350–1780 m below ground level at San Vito.

In contrast, the rocks affected by propylitic and thermometamorphic alteration have brittle behavior, permitting the development of fractures, acting as high-permeability pathways for fluid circulation. However, it is likely that the two zones are not hydraulically connected or not always hydraulically connected, being separated by a thin, relatively impermeable zone generated through self-sealing (e.g., quartz deposition), at temperature generally close to 400 °C, consistent with the conceptual model of Fournier (1999). Indeed, abundant hydrothermal quartz was found in well San Vito 1, at depths between ca. 2500 and 2800 m and temperatures of ca. 360–385 °C, at the transition between the Ca–Al-silicate (propylitic) zone (above) and the thermometamorphic zone (below), as shown by Fig. 5. Therefore, it seems likely that two distinct reservoirs occur in the deepest portion of the San Vito 1 well, namely: (a) above, a hydrothermal reservoir characterized by propylitic alteration, where fluid pressure is governed by the hydrostatic regime, and (b) below, a magmatic reservoir marked by thermometamorphic alteration, where fluid pressure is controlled by the lithostatic regime. Although the thermometamorphic zone was found also in the deepest portion of well Mofete 5, measured temperatures turned out to be significantly lower than at San Vito 1, suggesting that this zone has been affected by cooling and the authigenic minerals are reminiscent of a past situation.

The depth of the clay cap is poorly known away from the deep wells because the geoelectrical methods adopted in the geothermal exploration carried out by both SAFEN in 1939–1955 and AGIP-ENEL in the '70s had a limited penetration, partly due to the obvious logistic limitations in a densely inhabited area (Cassano and La Torre 1987). More recently, some resistivity models were implemented for the Solfatara crater area (Bruno et al. 2007; Byrdina et al. 2014; Isaia et al. 2015; Gresse et al. 2017), along a 1 km-long profile joining Solfatara and Pisciarelli (Troiano et al. 2014), and along a ~5.6 km-long profile connecting the Solfatara—Pisciarelli area and the Agnano crater (Siniscalchi et al. 2019). The latter one was generated through 2-D inversion of audiomagnetotelluric data and showed (Fig. 6):

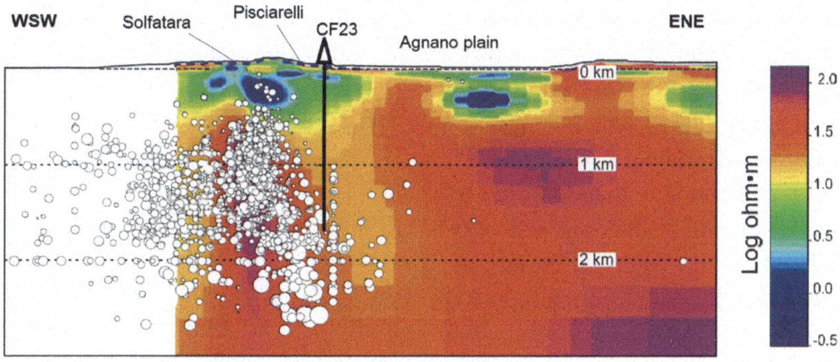

Fig. 6 Resistivity model of the Solfatara–Pisciarelli–Agnano area, also showing the local seismicity (at distance <0.6 km from the section), represented by white circles of radius proportional to the magnitude of the seismic events. Reprinted from Chiodini et al. (2021), modified, Copyright © 2021, with permission from Elsevier

(1) very low-resistivity values (<10 Ω m) above 500 m depth in the Solfatara–Pisciarelli area and in the Agnano plain;
(2) a resistive plume (>100 Ω m) extending under the Solfatara crater to 3 km depth, and
(3) intermediate resistivity values to the east of this resistive plume along the well Agnano 1/CF23, drilled to a total depth of 1841 m below ground level.

The very low resistivity values (<10 Ω m) encountered at shallow depths in the Solfatara–Pisciarelli area and along the Agnano 1/CF23 well are related to a type of hydrothermal alteration, known as advanced argillic alteration (corresponding to the kaolinite zone of the acid alteration suite sensu Reyes 1990), which is typically found around fumaroles and in high-sulphidation epithermal systems (Pirajno 2009) and is characterized by the presence of sulfate minerals such as alunite, alunogen, K-alum, and gypsum (or anhydrite above 42.8 °C at 1 bar, according to Shen et al. 2019) accompanied by elemental sulfur, kaolinite and pyrite. As already recalled in Sect. 1 in chapter "Summary of the Campi Flegrei Geology" all these minerals were reported to be present at Solfatara and Pisciarelli in the fumarolic encrustations and in the muds of the hot pools (Piochi et al. 2015). This type of hydrothermal alteration was not encountered by the AGIP-ENEL wells because they were drilled far from fumarolic areas, whereas the advanced argillic alteration was probably found in well Agnano 1/CF23. Indeed, according to Minucci (1964), this well was "*located in an area which not only displayed important smoke vents and hot springs, but was also indicated by the electrical exploration method, because of the anomalous low resistivity by comparison with the neighboring areas*" and "*fumarolized layers*" with "*veinlet of anhydrite*" were locally encountered by this well from 130 to 988 m depth.

1.3 The Geochemistry of the Geothermal Liquids Discharged by the AGIP-ENEL Boreholes

Short-term production tests were performed on the wells drilled by AGIP-ENEL discharging the fluids through an atmospheric separator. The discharge enthalpy and total flow were obtained by means of the Russel James method (James 1970; Grant et al. 1982), measuring the critical lip-pressure at lip of a pipe discharging the bi-phase mixture and the liquid flow-rate after separation. Separated brine samples for chemical analysis were collected at the weir box. Results are reported by Chelini et al. (1982), Carella and Guglielminetti (1983), and Guglielminetti (1986).

Long-term production tests were run on wells Mofete 1 and Mofete 2 using surface facilities including two pressure separators and equipment for scaling and corrosion tests. Scale inhibitors were successful for mitigating $CaCO_3$ deposition, but were unsuccessful for controlling amorphous silica precipitation which begins at ~160 °C (Guglielminetti 1986).

Since most flow and enthalpy data reported by Carella and Guglielminetti (1983) are not consistent with each other, the reservoir liquid enthalpy was computed in this work at the reported reservoir (aquifer) temperature assuming the presence of a single liquid phase under reservoir conditions for all the Mofete wells and well SV-3.

The bottom-hole temperature of well SV-1, >420 °C, is highly uncertain considering how it was estimated (see above) and there are serious doubts that it is representative of the reservoir (productive) zone. Although there is a difference of 20 °C between the Na–K–Ca temperatures of the two samples collected from well SV-1, 372 and 352 °C, they are probably more representative of the deep SV-1 aquifer temperature, considering that the Na–K–Ca geothermometer provides temperatures deviating by few degrees from the reported (physically measured) reservoir temperature for wells MF-5, MF-8d, and SV-3. Hence, the reservoir liquid enthalpy of well SV-1 was computed at these two Na–K–Ca temperatures assuming the presence of a single liquid phase under reservoir conditions.

Solute concentrations are reported in mg/L by Chelini et al. (1982) and Carella and Guglielminetti (1983), whereas the same values are given as ppm (mg/kg) by Guglielminetti (1986), who evidently assumed a density of 1 kg/dm^3. This, however, is an unrealistically low value for these high-salinity aqueous solutions. Therefore, the density of NaCl aqueous solution at 25 °C, 1 bar (Lvov and Wood 1990) was used here to convert the concentrations from mg/L to mg/kg. This correction is very important for the three samples of well MF-5, due to their very high salinity. Concentrations under reservoir conditions were then calculated for all solutes. Results are listed in Table 1 together with aquifer temperature, reservoir liquid enthalpy, the pH value of the brine separated at atmospheric conditions, pH_{asb}, and the pH of mineral-solution equilibrium, pH_{mse}, which was computed using the following relation (Chiodini et al. 1991):

$$pH_{mse} = 1.757 - 0.822 \log \Sigma_{eq} + 1846/T - 0.0171 \log f_{CO_2}, \quad (1)$$

in which Σ_{eq} is total ionic salinity in eq/kg, T is temperature in K, and f_{CO_2} is CO_2 fugacity in bar. To be noted that the pH of mineral-solution equilibrium is controlled by total ionic salinity, temperature, and CO_2 fugacity, in order of decreasing importance. In agreement with the findings of Ellis (1970), Eq. (1) indicates that pH_{mse} decreases with increasing Σ_{eq}, temperature, and CO_2 fugacity and vice versa.

Balducci and Chelini (1992) computed the reservoir chemistry for the wells Mofete 1 shallow (acronym MF-1S), Mofete 1 deep (MF-1P), and Mofete 2 (MF-2), by means of the speciation–saturation code ENTHALP (Truesdell and Singers 1974) using the data acquired during the long-term production tests, thus taking into account the CO_2 lost through steam separation, and assuming the presence of a single liquid phase under reservoir conditions. The computed pH values resulted to be 5.31 for MF-1S, 5.32 for MF-1P, and 5.27 for MF-2. These pH values differ by 0.2, 0.4, and 0.3 pH units, respectively, from the corresponding pH_{mse} values, confirming that these three reservoir liquids are close to equilibrium with pertinent hydrothermal minerals as previously shown by Balducci and Chelini (1992).

Speciation–saturation calculations cannot be performed for the other wells having been tested for short periods only. In spite of these limitations, the data reported in Table 1 allows one to draw some considerations, as discussed below. First of all, some brief comments on the fluids encountered by wells SV-1 and MF-5 are needed due to their particular characteristics.

Well San Vito 1—The two samples of well SV-1 have very high Mg concentrations, 354 and 232 mg/kg, which are unusual for high-temperature geothermal liquids, apart from those of low pH, which are often characterized by high SO_4 concentrations. Unfortunately, SO_4 concentration was not determined for these two samples,

Table 1 Chemistry and physical characteristics (under reservoir conditions) of the geothermal liquids encountered by the deep wells drilled by AGIP-ENEL at Mofete and San Vito (from Chelini et al. 1982 and Carella and Guglielminetti 1983); pH_{asb} is the pH value of the brine separated at atmospheric conditions, whereas pH_{mse} is the pH of mineral-solution equilibrium, computed using Eq. (1)

Well	pH_{asb}	pH_{mse}	T res	H	B	SiO_2	Li	Na	K	Mg	Ca
			°C	kJ/kg	ppm	ppm	ppm	ppm	ppm	ppm	ppm
MF-1S	7.56	5.5	219	939	134	426	26.8	10,752	1322	6.16	595
MF-1P	6.5	4.9	295	1317	106	400	26.7	12,096	2250	n.r	1232
MF-2	6	5.0	342	1607	137	435	13.0	4920	1145	n.r	466
MF-3d	7.5	5.2	276	1216	67.0	269	21.5	8720	709	n.r	451
MF-5	4.5	4.1	360	1762	(70.3)	63.9	146	25,910	13,199	n.r	16,414
MF-5	n.r	4.0	360	1762	512	81.5	185	33,003	16,815	12.0	20,913
MF-5	n.r	3.9	360	1762	599	83.9	217	40,817	20,090	13.3	24,901
MF-7d	7.2	5.0	301	1351	82.4	366	32.0	8444	1493	n.r	452
MF-8d	7.7	5.4	234	1009	64.9	327	26.7	10,514	1172	n.r	542
MF-9d	6.9	4.8	308	1391	157	315	30.5	11,612	2404	n.r	1919
SV-1	3.2	4.7	372	1935	n.r	117	14.9	3713	2528	354	1040
SV-1	4.38	5.0	352	1688	n.r	106	11.2	2698	1729	232	851
SV-3	6.8	4.9	315	1432	n.r	360	16.3	8162	2666	20.8	941

Well	Sr	Mn	Fe	Cu	Zn	Pb	As	Total alkalinity	SO_4	Cl	TDS
	ppm	ppm	ppm	ppm	ppm	ppm	ppm	mg HCO_3/kg	ppm	ppm	ppm
MF-1S	36.8	7.66	0.75	n.r	n.r	n.r	9.91	87.1	54.1	19,000	32,458
MF-1P	39.4	16.2	1.74	n.r	n.r	n.r	9.86	44.6	4.06	21,919	38,145
MF-2	13.9	24.1	0.46	n.r	n.r	n.r	10.2	39.4	5.57	9825	17,035
MF-3d	27.2	2.53	1.90	n.r	n.r	n.r	10.1	69.6	98.6	14,792	25,239
MF-5	399	1676	2875	n.r	n.r	n.r	n.r	<d.l	<d.l	95,490	156,244
MF-5	507	1155	1980	1.25	201	170	0.03	<d.l	<d.l	117,507	193,044
MF-5	597	1162	2031	1.25	188	183	0.03	< d.l	<d.l	139,315	230,199
MF-7d	14.9	5.72	12.01	n.r	n.r	n.r	14.9	112	40.0	15,246	26,314
MF-8d	29.5	5.77	0.72	n.r	n.r	n.r	10.8	70.6	59.1	18,140	30,963
MF-9d	34.9	30.0	1.09	n.r	n.r	n.r	17.4	39.8	7.6	23,931	40,500
SV-1	n.r	n.r	n.r	n.r	n.r	n.r	n.r	<d.l	n.r	11,932	19,699
SV-1	n.r	n.r	n.r	n.r	n.r	n.r	n.r	<d.l	n.r	8601	14,227
SV-3	n.r	n.r	n.r	n.r	n.r	n.r	n.r	192	n.r	16,698	29,056

but the two brines of well SV-1 separated at atmospheric conditions have low pH values, 3.20 and 4.38. The lower the pH, the higher is Mg concentration. It is likely that these low pH values are controlled by very high CO_2 partial pressures, which can be roughly estimated based on the well-head temperature and pressure values recorded during the short-term production test, 222 °C and 70 (kg cm^{-2})-gauge, corresponding to 69.6 bar-absolute. Since water saturation pressure at 222 °C is 24.1 bar, a gas pressure of 69.6–24.1 = 45.5 bar can be inferred. It is likely that gas pressure is chiefly explained by CO_2 partial pressure, that is, $P_{gas} \approx P_{CO_2}$. Besides, based on the Dalton's law, the separated gas phase is characterized by $X_{H_2O} = 24.1/69.6 =$

0.35 and $X_{gas} \approx X_{CO_2}$ = 45.5/69.6 = 0.65. These high P_{CO_2} and X_{CO_2} values are probably due to both the occurrence of decarbonation reactions at depth >2.8 km and the presence of an effective, 300 m-thick impervious cap, marked by the occurrence of abundant authigenic quartz, in the area of well SV-1 above the thermometamorphic rocks (see Sect. 1.1).

Well Mofete 5—Owing to heavy salt deposition, it was necessary to stop the short-term production test of the deep reservoir encountered by this well at depths of 2310–2699 m. Not surprisingly, the three brine samples of well MF-5 have very high salinity, varying from 156 to 230 g/kg, and high concentrations of most solutes, namely B, Li, Na, K, Ca, Sr, Mn, Fe, Zn, Pb, and Cl, while alkalinity, and the concentrations of SO_4, As, and SiO_2 are low. A pH value of 4.5 was measured in one of these three brines separated at atmospheric conditions and alkalinity was found below detection limit in all three samples. The computed pH values of mineral-solution equilibrium are in the range 4.0 ± 0.1 and are significantly lower than those of the other Mofete and San Vito wells. The low pH values of these brines and the high concentrations of most solutes, including the first transition metals, are explained by prolonged water–rock interaction at high temperature, 360 °C. These characteristics are similar to those of other high-temperature hypersaline brines, such as those of Salton Sea (e.g., Williams and McKibben 1989), Reykjanes (e.g., Hardardóttir et al. 2009) and Asal (e.g., Sanjuan et al. 1990; D'Amore et al. 1998). The low SO_4 concentrations of the MF-5 aqueous solutions are probably controlled by the high Ca concentrations at saturation with anhydrite, which occurs in the deepest part of the borehole as hydrothermal mineral. The high concentrations of Fe, Zn, and Pb are probably governed by saturation with sulfide minerals, such as pyrite and/or pyrrhotite, sphalerite, and galena, respectively. If so, sulfide concentration is expected to be relatively low and the low As concentration is unlikely to be due to saturation with an As-bearing sulfide mineral, such as orpiment and/or realgar. Instead, the low As concentration is probably ascribable to As loss through prolonged boiling, although it is unclear why a concurrent B loss did not occurred.

Finally, the low SiO_2 concentrations of these brines (63.9–83.9 mg/kg) are likely due to the precipitation of amorphous silica. In fact: (i) quartz solubility in pure water at the critical point is higher, although poorly constrained, being either 278 mg/kg, according to Von Damm et al. (1991), or 260 mg/kg, according to Fournier and Potter (1982), or 240 mg/kg, according to Shibue (1996), or 215 mg/kg, according to Manning (1994), and (ii) quartz solubility is expected to increase with increasing salinity (Cioni and Marini 2020 and references therein).

Enthalpy–chloride relationships among all Mofete and San Vito reservoir liquids (Fig. 7) suggest that the geothermal aquifers are recharged by seawater, which is conductively heated acquiring enthalpies in the range 720–1600 kJ/kg, corresponding to temperatures of 170–340 °C. Seawater heating explains the enthalpy and Cl concentration of the brines encountered by well MF-1 at shallow levels and by well MF-8d. These conductively-heated brines may experience either steam gain, as is the case of the reservoir liquids of wells MF-2, MF-3d, MF-7d, SV-1, and SV-3, or alternatively steam loss, as is the case of the brines met by wells MF-1 at deep levels, MF-9d, and especially MF-5. In particular, the hypersaline brines of well MF-5 experience a cumulative enthalpy gain of 2231–2277 kJ/kg, through conductive heat transfer, and an enthalpy loss of 574–620 kJ/kg, through steam separation, with a net enthalpy gain of 1657 kJ/kg. Summing the enthalpy of seawater at 25 °C, 104.8 kJ/kg, to this value, the enthalpy of saturated water at 360 °C, 1762 kJ/kg is obviously obtained. Thus, the seawater recharge model explains convincingly the enthalpy–chloride relationships among all Mofete and San Vito reservoir liquids.

The importance of seawater boiling was already recognized by Carella and Guglielminetti (1983), who proposed that: (i) the hypersaline

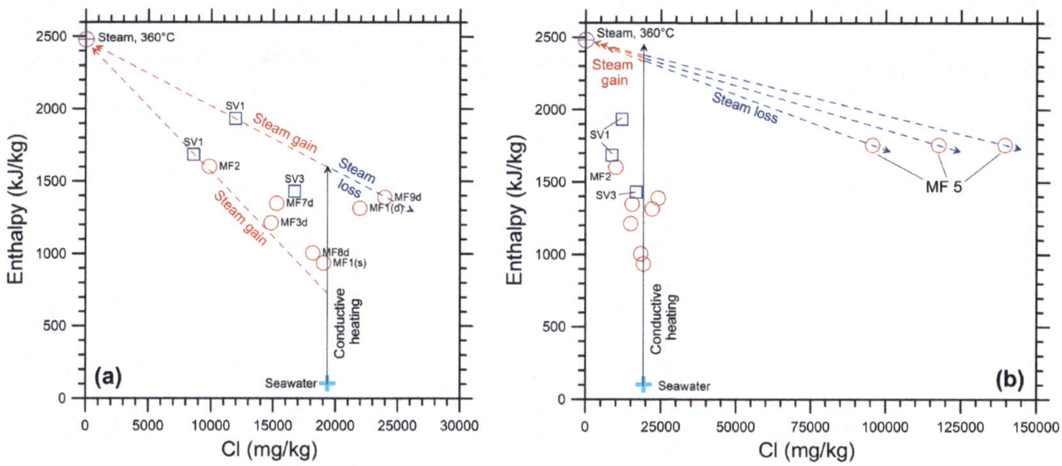

Fig. 7 Diagrams of chloride concentration under reservoir conditions versus reservoir liquid enthalpy for the geothermal liquids encountered by the deep wells drilled by AGIP-ENEL at Mofete and San Vito, **a** excluding the three brines of well MF-5 and **b** including the three brines of well MF-5

liquids of the deep reservoir are generated by boiling of seawater in a zone of limited recharge and (ii) the vapor phase thus produced migrates towards the surface, interacting with and heating the fluids circulating in the intermediate and shallow reservoirs.

The widespread occurrence of steam loss/gain in the Campi Flegrei magmatic–hydrothermal system is confirmed by the characteristics of the primary fluid inclusions hosted in authigenic K-feldspar, quartz, calcite, and epidote recovered from the geothermal wells Mofete 1, Mofete 2, Mofete 5, San Vito 1, and San Vito 3 (De Vivo et al 1989). In fact, three types of primary fluid inclusions were found in these hydrothermal minerals, namely: (A) two-phase, liquid + vapor (L + V) inclusions, rich in liquids of varying salinity; (B) two-phase (L + V), vapor-rich inclusions of low salinity (when it was possible to measure it); and (C) three-phase (L + V + halite) inclusions, rich in hypersaline liquids. The coexistence of type A and B inclusions in many samples suggests the occurrence of boiling, at some stage, possibly leading to mineral deposition with the consequent inclusion trapping.

Oxygen and hydrogen isotopes provide evidence on the origin of the Campi Flegrei geothermal fluids, despite the scarcity of available data. Based on δ^2H and δ^{18}O values of the H_2O molecule, Giggenbach (1992) convincingly demonstrated that most hydrothermal systems situated along convergent plate boundaries host mixtures rich in meteoric H_2O with subordinate amounts, usually less than 40%, of arc-type magmatic H_2O, which is characterized by δ^2H of -20 ± 10‰. However, marine waters may contribute to recharge the geothermal systems situated in coastal areas and islands, such as the hydrothermal system of Nisyros Island in Greece (Marini and Fiebig 2005). Consistent with this possibility, the total fluid discharged from the shallow reservoir met by the Mofete 1 geothermal well, with a δ^2H value of −3‰ and a δ^{18}O value of +0.8‰ (Ghiara et al. 1988), appears to be a mixture mainly made up by local seawater with lower amounts of meteoric water and magmatic water, as highlighted by the binary plot of δ^2H versus δ^{18}O values of Fig. 8, in which the total fluid of Mofete 1 is found close to the marine components, that is, SMOW and local seawater.

In this work, the fractions of the three end-members of interest were calculated by means of simple isotope mass balances on deuterium and oxygen-18, imposing that the sum of the three fractions is equal to 1 and considering the isotopic values of (1) local seawater, with δ^2H of +8.2‰ and δ^{18}O of + 1.1‰ (Bolognesi et al.

Fig. 8 Diagram of the δ^2H versus $\delta^{18}O$ values of the H_2O molecule for the total fluid discharged from the shallow reservoir met by the Mofete 1 geothermal well, also showing the marine, magmatic, and meteoric endmembers involved in its origin

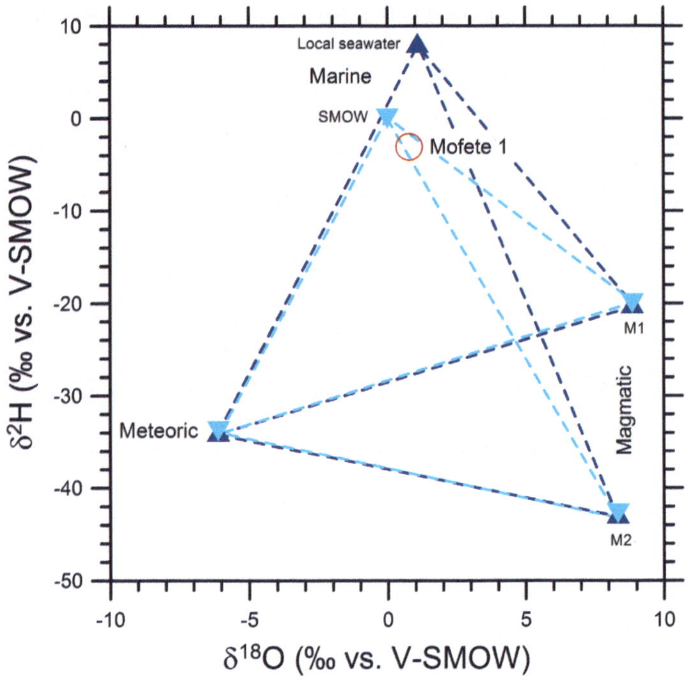

1986) or, alternatively, of the Vienna standard mean ocean water (V-SMOW, see below), (2) local meteoric water ($\delta^2H = -33.8$‰ and $\delta^{18}O = -6.1$‰; Valentino et al. 1999), and (3) magmatic water, assuming alternatively that the latter is represented either by the magmatic endmember M1 ($\delta^2H = -20$‰ and $\delta^{18}O = +8.87$‰) or by the magmatic endmember M2 ($\delta^2H = -42.75$‰ and $\delta^{18}O = +8.34$‰), which are discussed in Sect. 10.2 in chapter "Chemical and Isotopic Characteristics of Solfatara-Pisciarelli Fumarolic Fluids".

If the magmatic endmember M1 is taken into account in the calculations, the fraction of local seawater is 0.6911, the fraction of local meteoric water is 0.1804 and the fraction of magmatic water is 0.1285. Alternatively, if the magmatic endmember M2 is involved in the calculations, the fraction of local seawater is 0.7550, the fraction of local meteoric water is 0.1436 and the fraction of magmatic water is 0.1014.

If the V-SMOW ($\delta^2H = 0$‰ and $\delta^{18}O = 0$‰) is considered in the isotope mass balances instead of local seawater, the fraction of the marine component increases to 0.8674–0.8925, the fraction of local meteoric water decreases to 0.0000–0.252, whereas the fraction of the magmatic water does not change significantly being 0.1075.

In any case, the total fluid discharged from the shallow reservoir met by the Mofete 1 geothermal well results to be a mixture mainly constituted by local seawater with lower quantities of meteoric water and magmatic water, as recognized earlier by Ghiara et al. (1988) and in satisfactory agreement with the seawater recharge model outlined above based on enthalpy–chloride relationships. To the best of our knowledge, the δ^2H and $\delta^{18}O$ values were not measured for the fluids encountered by the other geothermal wells drilled at Mofete, San Vito, and Licola by AGIP-ENEL.

Caprarelli et al. (1997) measured sixteen $\delta^{18}O$ values and four δ^2H values (as well as two $\delta^{13}C$ values) on the cores recovered from the geothermal wells MF-1, MF-5, SV-1 and SV-3 at depths of 1150–2865 m, obtaining $\delta^{18}O$ values varying from +7.5 to +10‰ versus V-SMOW for MF-1, from +12.5 to +16.5‰ for MF-5, from +4.3 to +6.5‰ for SV-1, and +6.7 to +10.8‰ for SV-3, as well as δ^2H values of −46‰ versus V-SMOW for MF-1, −72 and −79‰ for MF-5, and −75‰ for SV-1. Then,

Caprarelli et al. (1997) computed the $\delta^{18}O$ and δ^2H values of the liquid water in equilibrium with these rocks considering the O isotope fractionation factors of the systems calcite-H_2O (Friedman and O'Neil 1977) and aluminosilicate-H_2O (O'Neil and Taylor 1967) and the H isotope fractionation factor of the system epidote-H_2O (Graham et al. 1980) at the measured temperatures, thus obtaining $\delta^{18}O$ values varying from +1.2 to +3.6‰ for MF-1, from +5.7 to +8.3‰ for MF-5, from +1.0 to +1.5‰ for SV-1, and +2.7‰ for SV-3, as well as δ^2H values of +1‰ for MF-1, −52 and −46‰ for MF-5, and −41‰ for SV-1. Based on these results, Caprarelli et al. (1997) concluded that the Campi Flegrei geothermal system hosts two different fluid types. The shallower fluid has a marine origin, whereas the deeper fluid is a mixture of magmatic and meteoric water, possibly modified by interaction with evaporite rocks. The two geothermal aquifers are hydrologically independent due to the higher density of the deeper, hypersaline liquid (Caprarelli et al. 1997).

Assuming isotopic equilibrium between the mineral of interest (indicated by subscript min) and liquid water and knowing the temperature and the isotope values of the mineral, δ_{min}, it is straightforward to calculate the isotope values of liquid water, $\delta_{H_2O(L)}$, by means of relations of the following type:

$$1000 \ln \alpha_{min-H_2O_{(L)}} \approx \delta_{min} - \delta_{H_2O_{(L)}} = f(T), \tag{2}$$

as done by Caprarelli et al. (1997). Relations of the type of Eq. (2) were derived in the '50s and, since then, they were used in a countless number of studies due to their simplicity and elegance (Horita 2005). However, the concept of isotope thermometry has experienced a revolutionary change, from Eq. (2) to:

$$1000 \ln \alpha_{min-H_2O_{(L)}} \approx \delta_{min} - \delta_{H_2O_{(L)}} = f(T, P, X_{fluid}), \tag{3}$$

as underscored by Horita (2005): "*In fact, in some geological settings, the effect of pressure and dissolved salts/minerals seems greater than that of temperature. Moreover, the three variables (T, P and X_{fluid}) depend on each other. The solubility of a mineral varies as a function not only of temperature and pressure, but also of the composition of dissolved salts and other minerals.*"

Furthermore, also the effect of dissolved gases, mainly CO_2 and H_2S, might be important, as well as the occurrence of relevant processes, such as steam loss or gain. In the Campi Flegrei magmatic–hydrothermal system, steam loss is expected to be considerable for the fluids of well MF-5 and could explain the high $\delta^{18}O_{H_2O_{(L)}}$ values computed by Caprarelli et al. (1997) for this well (see above). Nevertheless, being affected by considerable uncertainties, the $\delta^{18}O_{H_2O_{(L)}}$ and $\delta^2H_{H_2O_{(L)}}$ values computed by Caprarelli et al. (1997) will not be used here in the reconstruction of the conceptual model of the Campi Flegrei magmatic–hydrothermal system.

The triangular diagram of Cl–B–Li (Fig. 9; from Giggenbach 1991) provides ambiguous evidence on the origin of the Campi Flegrei reservoir liquids, also due the lack of B data for the San Vito wells SV-1 and SV-3, for which only the Li/Cl ratio is therefore shown in the plot. In addition to the rock compositions (from Turekian and Wedepohl 1961) and average seawater, the compositions of the high-temperature volcanic gases from several volcanoes positioned along convergent plate boundaries, including Vulcano, Satsuma Iwojima, Kudryavy, Colima, White Island and Galeras (data from Bernard et al. 2011 and references therein) are also displayed in the ternary plot. Volcanic gas condensates have Cl/B weight ratio of 650 ± 100 apart from Vulcano fluids, which are characterized by a higher B content and lower Cl/B ratios, 120 ± 30, probably because Vulcano magmas are contaminated by crustal materials (Bernard et al. 2011). The relative concentrations of Cl, B and Li of all Mofete reservoir liquids are similar to each other, apart from Mofete 2 which is weakly enriched in B, probably due to steam gain, in agreement with the indications given by enthalpy–chloride relations (see above). Moreover, the reservoir liquids of wells SV-1 and SV-

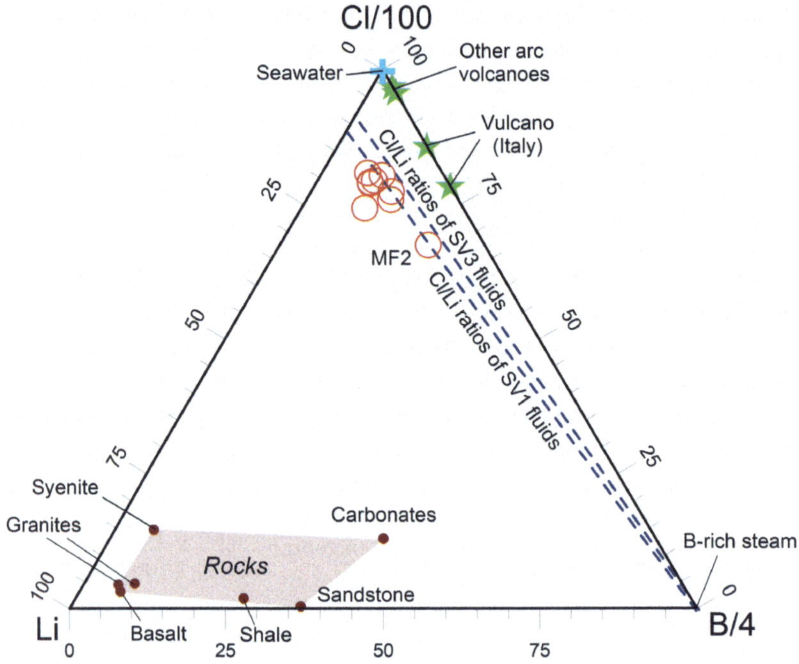

Fig. 9 Cl–B–Li triangular diagram (from Giggenbach 1991), showing the reservoir liquids of Mofete, average rocks (from Turekian and Wedepohl 1961), and high-temperature volcanic gases (from Bernard et al. 2011). Also shown are the Cl/Li ratios of the reservoir liquids of San Vito. Reprinted from Apollaro et al. (2015), modified, Copyright © 2015, with permission from Elsevier

3 have Cl/Li ratios similar to those of Mofete fluids. Unfortunately, it is unclear if the enrichment in Cl of Mofete reservoir liquids derives from the marine component or the magmatic endmember based on Cl–B–Li relationships.

Summing up, in spite of some ambiguities, the evidence provided by the available physical, chemical, and isotopic data suggests that the geothermal aquifers encountered by the deep wells drilled at Mofete and San Vito by AGIP-ENEL are recharged by marine waters, *in toto* or at least for the most. In spite of some doubts, this seems to be a reasonable conclusion also because a considerable part of the Campi Flegrei caldera is submerged by the Gulf of Pozzuoli waters.

2 The Deep-Magmatic Portion of the Magmatic–Hydrothermal System Hosted in the Campi Flegrei Caldera

Indications on the deep-magmatic portion of the magmatic–hydrothermal system currently hosted in the Campi Flegrei caldera are mainly provided by the active seismic reflection data acquired by the SERAPIS survey (Zollo et al. 2003, 2008; Judenherc and Zollo 2004) and by the tomographic modeling of both the active seismic data of the SERAPIS survey and the passive seismic data obtained during the 1982–1984 bradyseismic crisis (Vanorio et al. 2005; Chiarabba and Moretti 2006; Battaglia et al 2008). In particular, according to Zollo et al. (2008), the inferred schematic stratigraphy comprises, from top to bottom (Fig. 10):

(a) Water-saturated volcanic deposits above 0.6 km depths.
(b) Pre- and post-caldera volcanic and sedimentary-marine deposits from 0.6 to 2.7 km depths.
(c) Thermometamorphic rocks hosting brines and/or gases from 2.7 to ~4 km depths.
(d) Mesozoic carbonate rocks and the crystalline bedrock from ~4 to ~7.5 km depths.
(e) A melt zone from ~7.5 to ~8.5 km depths.

Levels (a) and (b) were crossed by the deep geothermal wells drilled inside the Campi Flegrei caldera, the upper part of unit (c) was met by

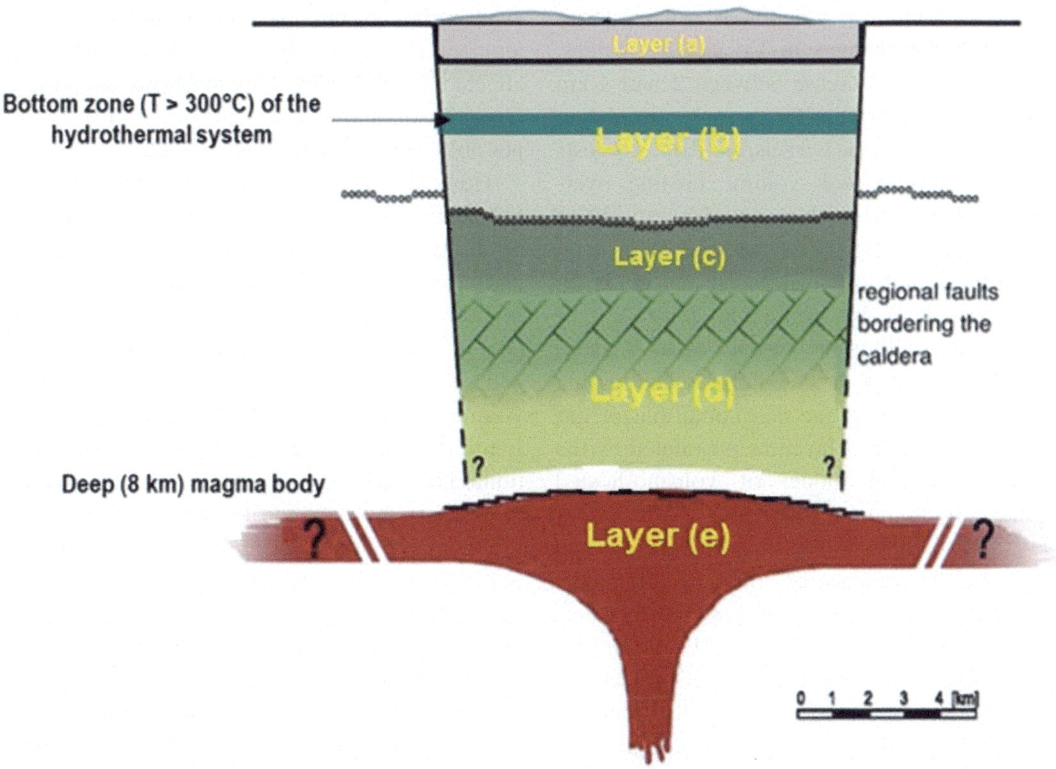

Fig. 10 Geological schematic model of the Campi Flegrei caldera (from Zollo et al. 2008 and Moretti et al. 2020). Reprinted from Moretti et al. 2020, modified, Copyright © 2020, with permission from Elsevier

these wells, whereas levels (d) and (e) are situated at depths not reached by drilling. Therefore, the following discussion is focused on levels (c), (d), and (e).

2.1 Layer (C)—Thermometamorphic Rocks

The active seismic reflection data reveal a sharp and considerable increase in the velocities of both P and S waves (Vp and Vs, respectively) at a depth of 2,700 m, indicating the transition from (i) volcanic and marine deposits (tuffites, marls and siltites) affected by different types of hydrothermal alteration, including the so-called Ca–Al–silicate zone in the deepest parts, to (ii) underlying similar primary lithotypes, whose original rock texture was completely reorganized by thermometamorphic processes.

The occurrence of thermometamorphic reactions on rocks with a carbonate matrix (like these tuffites, marls and siltites) determines a considerable increase in porosity due to the decrease in the volume of mineral products and the release of CO_2 (Chelini and Sbrana 1987). Furthermore, other gas species (e.g., CH_4, H_2S, N_2, NH_3, and other compounds of C, S, and N) may be produced through different reactions (e.g., breakdown of organic matter and decomposition of S-bearing minerals) presumably occurring in the thermometamorphic layer, whereas magmatic gases and hypersaline brines are expected to enter the thermometamorphic layer from below.

Consequently, these thermometamorphic rocks represent a gas-and-brine-bearing layer, probably extending to a depth of ~4 km, consistent with the changes in Vp and Vs values. The sharp Vp/Vs decrease between 2 and 4 km depth, confirmed by the tomographic modeling of both active and passive seismic data, suggests the presence of a rock volume hosting over-pressurized gas (Chiarabba and Moretti 2006; Battaglia et al 2008). These findings agree with the high P_{CO_2} and X_{CO_2} values estimated for the fluids discharged by well San Vito 1 (see Sect. 1.2) and call for the presence of an effective cap-rock at the top of the thermometamorphic layer, which is a *sine qua non* condition for gas accumulation into this layer, as postulated by the general conceptual model of volcano-hosted magmatic–hydrothermal systems of Fournier (1999).[2] Actually, a 300 m-thick impervious cap marked by the presence of abundant authigenic quartz, is present in well SV-1 above the thermometamorphic rocks (see Sect. 1.2). Furthermore, owing to the presence of over-pressurized gas, this layer of thermometamorphic rocks is considered to be the source of ground uplift and related shallow seismicity (see Sect. 2 in chapter "Summary of the Campi Flegrei Geology") according to several authors (e.g. Vanorio et al. 2005; Chiarabba and Moretti 2006; Zollo et al. 2008; Battaglia et al 2008).

2.2 Layer (D): Mesozoic Carbonate Rocks and the Crystalline Bedrock

Carbonate rocks whose top is situated at ~4 km depth, according to Chiarabba and Moretti (2006) and Zollo et al. (2008), or 4–5 km depth, according to Battaglia et al (2008), are present below the thermometamorphic layer and above the crystalline bedrock. Similar to the rocks of the overlying thermometamorphic layer, and even more so due to the presumably higher temperatures, also these carbonate rocks are affected by decarbonation reactions and, therefore, constitute an additional source of CO_2 and, possibly, other gas species as well.

However, the presence of carbonate rocks below the Campi Flegrei caldera was questioned by D'Antonio (2011) based on volcanological and petrological evidence and mass-balance modeling of the fractional crystallization process of a parent shoshonite magma, first to latite and second to trachyte, trying to reproduce the volume of magma involved in the Campanian Ignimbrite and Neapolitan Yellow Tuff eruptions. The final aim of this modeling exercise was to estimate the volume of crystallized solids left in the crust by these eruptions that resulted to be ~2130 km^3. According to D'Antonio (2011), this volume of cumulate material, likely monzonitic to syenitic from base upwards, is more than enough to completely fill the basement volume in the 4–8 km depth interval underneath the Campi Flegrei caldera, which was estimated to be ~1250 km^3. Thus, D'Antonio (2011) proposed that the basement of the Campi Flegrei caldera below 4 km is mainly made up of crystalline igneous rocks. According to D'Antonio (2011), these rocks would meet the physical properties requirements for the interpretation of seismic data, explain some geochemical and isotopic features of the volcanic products erupted during the past 15 ka, and justify the carbon isotopic composition of Solfatara fumaroles. In our opinion, the last two points are rather weak pieces of evidence. In particular, the carbon isotopic composition of CO_2 in Solfatara fluids suggests that the local carbon source is chiefly controlled by thermometamorphic decarbonation of marine carbonates and magma degassing (see Sect. 9 in chapter "Chemical and Isotopic Characteristics of Solfatara-Pisciarelli Fumarolic Fluids"). Furthermore, some sentences of D'Antonio (2011) like "*the lack of carbonate lithics in any Campi Flegrei caldera volcanic rocks does not support the hypothesis of a limestone basement*" and "*not one limestone fragment has ever been found in any CF* [Campi Flegrei] *volcanic products*" are questionable, not to say false, in

[2] A similar situation occurs, *mutatis mutandis*, in the Torre Alfina geothermal system, where a flysch cap-rock controls the occurrence of CO_2-rich gases at a pressure of ca. 40 bar at the top of a carbonate reservoir hosting a liquid phase (Buonasorte et al. 1988).

that marble lithics are present, although they are not abundant, in the lithic-rich breccia and spatter agglutinate (unit C) of the Campanian Ignimbrite deposits (Rosi et al. 1996).

2.3 Layer (E): The Melt Zone

According to Zollo et al. (2008), the deeper seismic discontinuity at depth of ~7,500 m is the top of a low-velocity layer, whose Vp and Vs values are consistent with those expected for a magma body hosted in a densely fractured volume of rocks. The melt fraction in the layer below the 7,500 m reflector was estimated to range from 65–75% up to possibly 100%. Based on the extrapolation of the geothermal gradient present in the deepest part of the SV-1 well (see Sect. 1.1), the temperature at the top of this melt zone is estimated to be 957–1122 °C. These temperatures compare with those occurring during the crystallization of a shoshonite magma, initially at 1120 °C, to a trachyte magma at 950 °C, as indicated by petrological evidence (e.g., Cannatelli et al. 2007; Arienzo et al. 2010). The minimum horizontal extension of this melt layer and its approximate thickness were estimated to be 30 km^2 and 1 km, respectively. Assuming a cylindrical shape, the minimum diameter of the base is little more than 6 km. Seismic tomography and reflection amplitude analysis excludes the occurrence of magma batches greater than ~1 km in diameter at shallow depths, whereas smaller magma bodies are not detectable.

The presence of a magma reservoir below 7.5 km and the possible presence of small magma batches at shallower levels is consistent with the evidence provided by the melt inclusions in the volcanic products of the post-Neapolitan Yellow Tuff activity (calibrated ages from Smith et al. 2011), including Minopoli 2 (~10.5–11.9 ka BP; Mangiacapra et al. 2008; Vetere et al. 2011), Fondo Riccio (~10.5–11.9 ka BP; Mangiacapra et al. 2008), Astroni (4.1–4.3 ka BP; Cecchetti et al. 2003), Agnano–Monte Spina (4.5–4.6 ka BP; Roach 2005; Arienzo et al. 2010), Averno 2 (~4.2 ka BP; Fourmentraux et al. (2012), Nisida (3.2–4.2 ka BP; Arienzo et al. 2016), Solfatara (4.2–4.4 ka BP; Cipriani et al. 2008), and Monte Nuovo (1538 CE; Piochi et al. 2005), in spite of the considerable uncertainties in the calculation of the pressure and depth of melt entrapment.

The melt layer with top at 7.5 km depth can be considered the primary magma reservoir, acting as source of heat and fluids to the overlying magmatic–hydrothermal system, although CO_2 and other gas species are also supplied by reactions of decarbonation, breakdown of organic matter, and other type, occurring in the thermometamorphic layer (see Sect. 2.1) and in the underlying carbonate rocks (see Sect. 2.2). This flux of energy and mass from the primary magma reservoir to the hydrothermal system is controlled by the permeability distribution in the volume of rocks between 7.5 and 4.0 km, which is poorly constrained.

An attempt to reduce these uncertainties was performed by Castaldo et al. (2019) who first modeled the thermal field assuming a purely conductive regime by using a finite element method. In this way, the 800 °C isotherm, assumed to be representative of a rhyolitic melt, was found to be situated at depths of 5 and 20 km beneath the Pozzuoli Bay and outside the Campi Flegrei caldera, respectively. Next, Castaldo et al. (2019) used this thermal field to image the 3D rheology in the domain of interest. It turns out that the brittle/ductile transition is very deep (15–20 km) outside the caldera, whereas it is located at ~4 km depth within the caldera.

Assuming that the top of the magma reservoir is actually positioned at 7.5 km (as indicated by seismic data, see above) instead of 5 km (as computed by Castaldo et al. 2019), it follows that the brittle/ductile transition is expected to be found at deeper levels, let us say 6.0–6.5 km, just to give a couple of figures.

Keeping in mind these figures, let us now consider the seismicity recorded in the Campi Flegrei from 1982 to 2016, as reported by

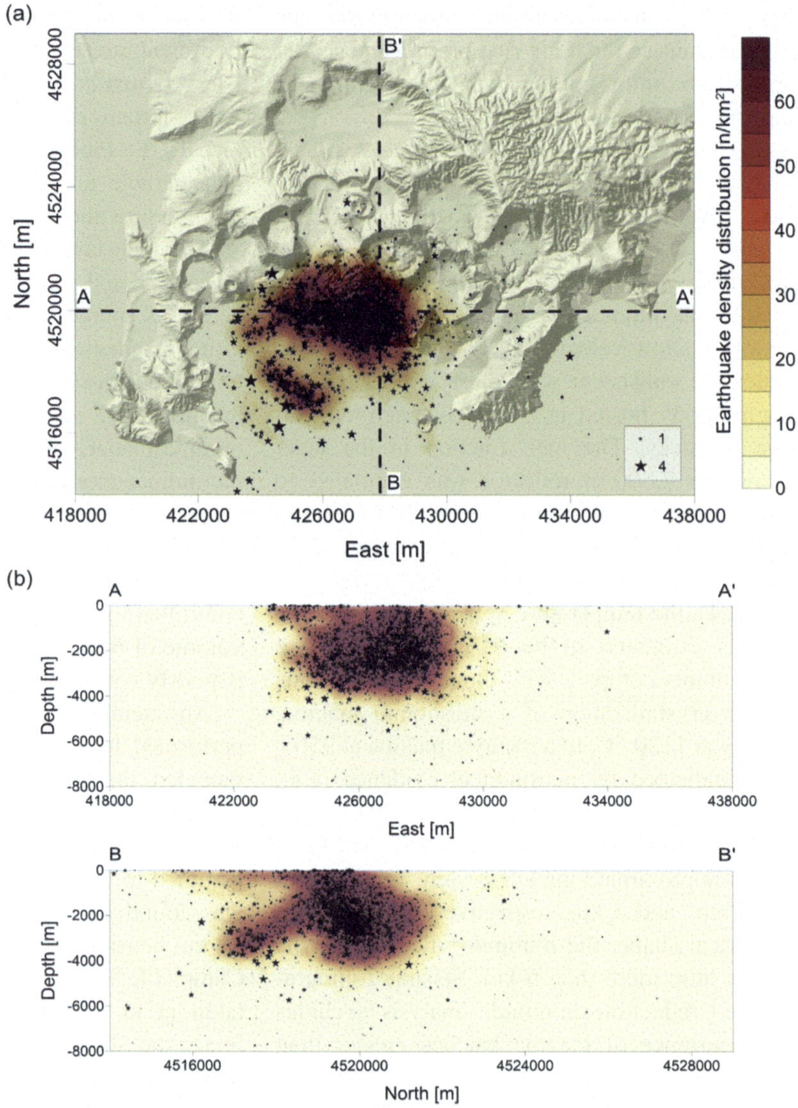

Fig. 11 Seismicity of $M \geq 1$ in the Campi Flegrei caldera during the period 1982–2016. **a** Distribution map of the epicenters density superimposed on the SRTM DEM of the area. Single events are indicated by black stars of size proportional to the magnitude. **b** Vertical cross sections A–A′ and B–B′ showing the distribution of seismicity with depth. From Castaldo et al. (2019), which was performed under a Creative Commons license

Castaldo et al. (2019). The earthquakes of magnitude ≥ 1 distribute in two clusters, one extending from the Solfatara crater to the town of Pozzuoli and the other within the Bay of Pozzuoli, with a sharp decrease in the number of events at ~4 km depth for both clusters (Fig. 11). Since this depth corresponds to the brittle/ductile transition depth according to Castaldo et al. (2019), the obvious conclusion would be that the lack of seismicity below ~4 km depth is due to the ductile behavior of the crust. However, accepting that the brittle/ductile transition is much deeper (~6.0–6.5 km) this conclusion must be rejected. The alternative explanation is that the seismicity is controlled by reasons others than the brittle/ductile transition, probably the upflow of fluids which is expected to occur in the thermometamorphic zone hosting over-pressured gas (e.g. Vanorio et al. 2005; Chiarabba and Moretti 2006; Zollo et al. 2008; Battaglia et al 2008) as well as at shallower depths.

3 The Surface Discharges of the Campi Flegrei Magmatic–Hydrothermal System

The surface discharges of the Campi Flegrei magmatic–hydrothermal system comprise: (i) onshore thermal springs and shallow wells of different composition, (ii) onshore fumaroles and gas vents, (iii) offshore fumaroles and gas vents and (iv) sub-lacustrine hydrothermal discharges. Most of these thermal manifestations are controlled by the presence of active faults and fractures acting as high-permeability conduits for the upflow of deep fluids and locally breaching the clay cap, if any.

3.1 The Onshore Thermal Springs and Shallow Wells

The chemical and isotopic characteristics of onshore thermal waters have been the subject of several studies (e.g., Dall'Aglio et al. 1972; Baldi et al. 1975; Cortecci et al. 1978; Corniello and Nicotera 1982; Antrodicchia et al. 1985; Ghiara and Stanzione 1988; Ghiara et al. 1988; Martini et al. 1991; Celico et al. 1992; Valentino et al. 1999; Valentino and Stanzione 2003, 2004; Aiuppa et al. 2006).

Most onshore thermal springs and shallow wells discharge neutral sodium-chloride to sodium-chloride (bicarbonate) waters, apart from:

(a) The acid–sulfate waters of Pisciarelli (samples 7A and 7B; temperature 84–95 °C; pH 1.5–2.7; Valentino and Stanzione 2003; green pin in Fig. 12), which are steam condensates, whose chemistry is chiefly controlled by oxidation of H_2S to H_2SO_4 driven by atmospheric O_2 (Giggenbach 1988). Similar waters are present in the Fangaia mud pool, which is situated on an E–W fault close to the center of the Solfatara crater (location in Fig. 13), and in other bubbling pools of the Solfatara crater (Valentino and Stanzione 2004; Aiuppa et al. 2006). In addition to the enrichment in SO_4, these acidic waters are also characterized by high concentrations of NH_3, B, and As, which are transported in the vapor phase and are trapped in condensed steam due to their high solubility, as well as by high $\delta^{18}O$ values due to steady-state evaporation from pools continuously heated by the injection of fumarolic steam (Giggenbach 1978; Giggenbach and Stewart 1982).

(b) The bicarbonate water of Agnano (sample 2; temperature 31–46 °C; pH 6.5–6.8; Valentino and Stanzione 2003; yellow pin in Fig. 12), whose chemical characteristics are mainly controlled by rock titration governed by conversion of CO_2 to bicarbonate. In addition to the enrichment in HCO_3, these waters have high concentrations of NH_3, B and As and low Cl/B ratios due to absorption of hydrothermal gases (Valentino and Stanzione 2003, 2004; Aiuppa et al. 2006).

The neutral sodium-chloride to sodium-chloride (bicarbonate) thermal waters are produced through different mixing processes, which were quantified by Antrodicchia et al. (1985) by means of a mixing model involving the Na–K, Ca–K, and quartz geothermometers. It turned out that: (i) Some sodium-chloride to sodium-chloride (bicarbonate) thermal waters are ternary mixtures of geothermal water, seawater, and groundwater (magenta pins in Fig. 12). (ii) Others are binary mixtures of geothermal water and groundwater (purple pins in Fig. 12). (iii) Still others are mixtures involving geothermal water but they cannot be properly modeled (light blue pins in Fig. 12). The computed aquifer temperatures resulted to be in the range 210–280 °C for all the ternary and binary mixtures, apart from a single case with aquifer temperature of 187 °C, suggesting that the geothermal waters involved in these mixtures come from the first geothermal reservoir identified by deep drilling in the Mofete area (see Sect. 1.1).

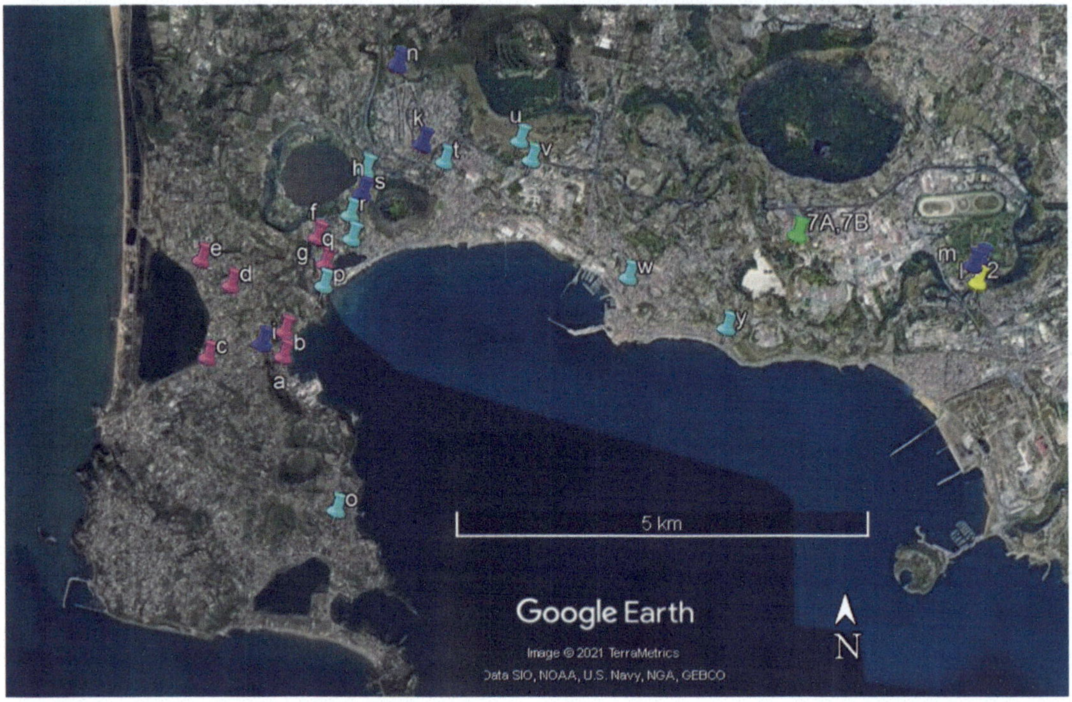

Fig. 12 Location map of the onshore thermal springs and shallow wells showing the acid-sulfate waters of Pisciarelli (samples 7A and 7B, green pin; Valentino and Stanzione 2003), the bicarbonate water of Agnano (sample 2, yellow pin; Valentino and Stanzione 2003), and different neutral sodium-chloride to sodium-chloride (bicarbonate) thermal waters (Antrodicchia et al. 1985) including: ternary mixtures of geothermal water, seawater, and groundwater (magenta pins), binary mixtures of geothermal water and groundwater (purple pins), indefinable mixtures involving geothermal water (light blue pins)

3.2 The Onshore Fumaroles and Gas Vents

The main area of onshore fumarolic activity and the most important place of energy release in the whole Campi Flegrei caldera is the Solfatara–Pisciarelli Diffuse Degassing Structure (SPDDS), comprising the Solfatara crater (whose geological characteristics are summarized in Sect. 1 in chapter "Summary of the Campi Flegrei Geology") and the adjacent thermal manifestations of Pisciarelli (Chiodini et al. 2001). The SPDDS is a high-diffuse-degassing area, most of which is devoid of vegetation. The total amount of CO_2 released daily from the SPDDS has been repeatedly determined since 1998, when measurements began by means of the accumulation chamber method and related data processing (Chiodini et al. 1998). The minimum value, 745 ± 61 t/d, was recorded in July 2003 (Cardellini et al. 2017 and references therein), while a substantial increase occurred afterwards, from ~1000 t/d in 2008–2010 up to 3000–4000 t/d in 2019–2020 (Chiodini et al. 2021).

Considering the H_2O/CO_2 mass ratios of Bocca Grande effluents (1.99 in July 2003, 1.37 in 2008–2010, and 1.05 in 2019–2020; Chiodini et al. 2021), these CO_2 flux values can be converted: (i) in the corresponding flux of condensed H_2O, which results to be 1480 t/d in July 2003, 1370 t/d in 2008–2010, and 3150–4200 t/d in 2019–2020 and (ii) in the energy released through steam condensation, which turns out to be 39 MWt in July 2003, 36 MWt in 2008–2010, and 82–110 MWt in 2019–2020, based on the latent heat of 2257 J/g at 100 °C, 1 bar (see

Fig. 13 Google Earth image of the Solfatara–Pisciarelli area. The red circles indicate the main fumarolic vents of Bocca Grande (BG), Bocca Nuova (BN), and Fangaia (F) at Solfatara and Soffione at Pisciarelli (P)

Chiodini et al. 2001 for details). These figures indicate that the energy dissipated through steam condensation in 2019–2020 is two or even three times higher than the values of 2008–2010.

Focused degassing takes place from some fumarolic vents of outlet temperature varying from 95 to 165 °C, that are positioned along the south-eastern inner crater wall of the Solfatara and at Pisciarelli (Fig. 13).

At Solfatara, the main fumarolic vents are Bocca Grande (BG in Fig. 13; outlet temperature of 150–165 °C) and Bocca Nuova (BN in Fig. 13; outlet temperature of 140–150 °C), emitting 107–158 t/d of CO_2 and 52_57 t/d of CO_2, respectively, according to Aiuppa et al. (2013). Both vents are positioned at the intersection of the NNW-SSE and ENE-WSW faults bordering the Solfatara crater to the E and to the S, respectively, and are directly connected through a ~10 m-thick conduit to a gas-dominated reservoir located at ~60 m depth, with a volume of ~25,000 m^3 (Byrdina et al. 2014; Gresse et al. 2017). The large quantity of condensed steam produced around the two main fumaroles flows towards the Fangaia mud pool through a subterranean conduit probably controlled by a buried NW-SE-trending fault.

The discharged gases are mainly constituted by H_2O and CO_2, accompanied by smaller amounts of H_2S, N_2, H_2, CH_4, He, Ar and CO, in decreasing order of importance (Chiodini et al. 2021 and references therein). The acidic–magmatic gas species, namely SO_2, HCl and HF, are virtually absent possibly because of removal by scrubbing (Symonds et al. 2001; Marini and Gambardella 2005) upon minor steam condensation at depth during the close-to-isenthalpic expansion of the fluids of interest (Caliro et al.

2007; Moretti et al. 2013; Aiuppa et al. 2013; Chiodini et al. 2012, 2021). Nevertheless, Vaselli et al. (2011) reported small concentrations of HCl, 130–204 μmol/mol, and HF, 16–22 μmol/mol, both on a dry gas basis, for six dry gas samples collected in 2004 from Bocca Grande and Bocca Nuova. The SO_2 reported in some analyses (e.g., Cheminée et al. 1968) is produced by the O_2-driven oxidation of H_2S, according to reaction (1.1), upon mixing of fumarolic gases with atmospheric air, as already proven by the experiments and analyses of De Luca (1871; see Sect. 2 in chapter Historical Data).

At Pisciarelli, there are large, intensely-bubbling mud pools with water temperature close to 90 °C and high-flow-rate fumaroles discharging a gas mixture rich in H_2O and CO_2 with subordinate H_2S, as reported by Tamburello et al. (2019). The major fumarolic vent of the Pisciarelli area is called Soffione (P in Fig. 13; outlet temperature 114 °C). It appeared in 2009 (Chiodini et al. 2015) and, since then, its degassing activity increased substantially attaining a CO_2 output >600 t d^{-1} in 2019 (Tamburello et al. 2019). Considering the H_2O/CO_2 mass ratio obtained by sampling and analysis of fumarolic effluents (~ 1.6 in 2017–2018), the H_2O output results >960 t/d. Based on the enthalpy of superheated steam at 114 °C, 1.013 bar, which is 2704 kJ/kg, this water output is equivalent to >29 MWt of energy released from a single fumarole (Tamburello et al. 2019).

The chemical and isotopic characteristics of the fumarolic manifestations of Solfatara–Pisciarelli are thoroughly discussed in chapter "Chemical and Isotopic Characteristics of Solfatara-Pisciarelli Fumarolic Fluids". In contrast to the findings relating to well Mofete 1 (see above), the Solfatara fumaroles discharge mixtures mainly constituted by magmatic H_2O (and magmatic-thermometamorphic CO_2), with subordinate quantities of meteoric H_2O, based on isotopic evidence (see Sect. 10.4 in chapter "Chemical and Isotopic Characteristics of Solfatara-Pisciarelli Fumarolic Fluids"). This is probably a local situation, because of the important upflow of deep magmatic fluids that occupy the shallow hydrothermal system under the Solfatara crater, probably working as a heat pipe with recirculation of condensed steam at shallow depths (see Sect. 10.3 in chapter "Chemical and Isotopic Characteristics of Solfatara-Pisciarelli Fumarolic Fluids").

Among the other onshore fumaroles and gas vents, whose location is shown in Fig. 14, the hottest ones are those situated close to the Monte Nuovo cone, with an outlet temperature of 80–100 °C, and at the Cavone dell'Inferno (Cave of Hell) in the Mofete area, which is a ~ 10 m-deep, ~ 60 m-long fracture cutting rocks affected by argillic alteration (Rosi and Sbrana 1987). The gas vents of Agnano Spa and that of the Serapeum of Pozzuoli have outlet temperatures of 24–52 °C and 34 °C, respectively, but the propane–propene equilibrium temperature of the Agnano Spa gases is 244–251 °C (Vaselli et al. 2011). In the Agnano area, there is also an anthropogenic cave, known as Grotta del Cane, which is Italian for Dog Cave, where the temperature increases up >60 °C towards the bottom which, however, cannot be reached being saturated by a CO_2-rich gas mixture (Halliday and Cigna 2006; Ferrante 2018).

3.3 The Offshore Fumaroles and Gas Vents

The offshore fumaroles and gas vents, whose location is also displayed in Fig. 14, are mainly found in shallow waters, close to the coast, rarely at depths of 50–100 m, never at >100 m depth, indicating that gas pressure at the submarine vents is always <10 bar (Carmisciano et al. 2013). Most gas vents are found in the western sector of the Pozzuoli Gulf, between the Mofete hill and the Pozzuoli pier and close to Bacoli–Capo Miseno. The $\delta^{13}C$ values and $^3He/^4He$ ratios of sampled submarine vents compare with those of onshore fumaroles, suggesting a common origin for the magmatic (or deep) end-member present in hydrothermal fluids throughout the Campi Flegrei caldera (Tedesco et al. 1990; Vaselli et al. 2011). The eleven gas samples collected by Vaselli et al. (2011) from

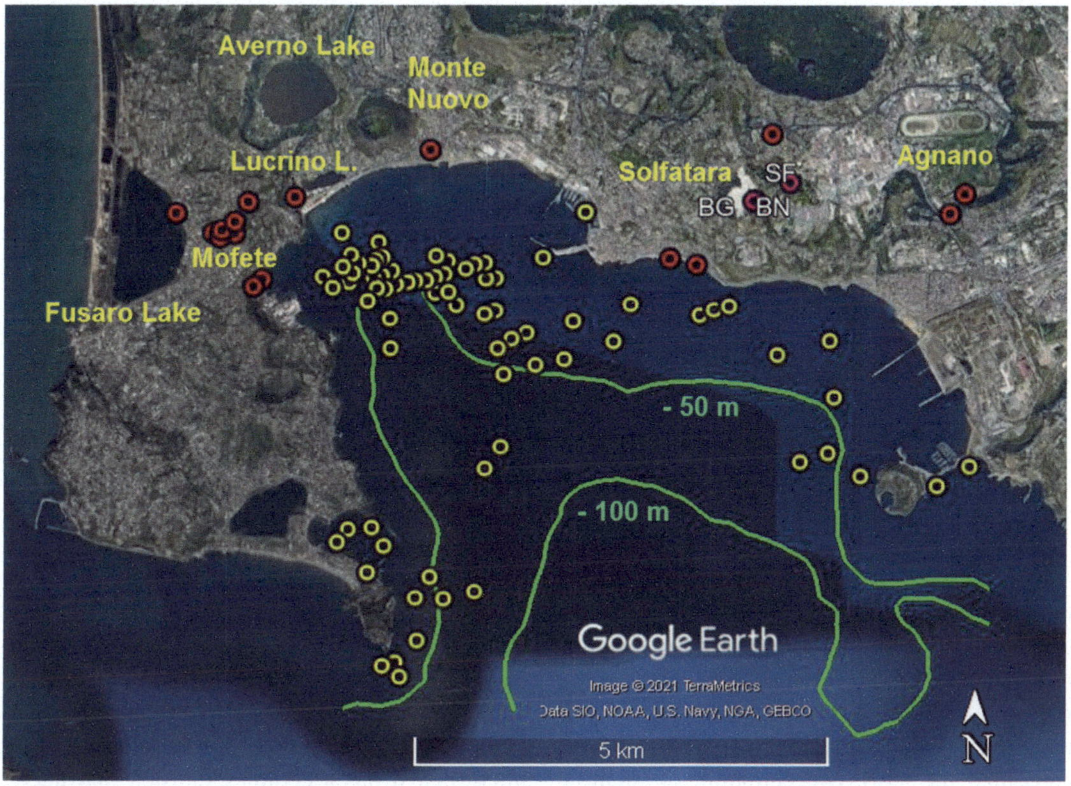

Fig. 14 Location map of onshore fumaroles and gas vents (from Rosi and Sbrana 1987 and Vaselli et al. 2011) and offshore fumaroles and gas vents (from Carmisciano et al. 2013), indicated by red circles and yellow circles, respectively

seven different offshore sites are mainly constituted by CO_2 with lower concentrations of N_2 and CH_4, and highly variable H_2 and H_2S. The propane–propene equilibrium temperature varies from 258 to 326 °C. This temperature values are within the range measured in the deep geothermal wells drilled at Mofete, San Vito, and Agnano, suggesting that circulation of magmatic–hydrothermal fluids occurs also in the submerged portion of the Campi Flegrei caldera.

Among these submarine gas discharges, those of highest outlet temperature, 93 °C, are located at Secca delle Fumose, a small, largely anthropogenic, submarine relief situated at 800 m from the coast in the northwestern sector of Pozzuoli Bay. The relief is controlled by high-angle faults at the eastern border of a N–S-trending graben that probably formed during the Monte Nuovo eruption (Di Napoli et al. 2016). Based on the CO_2 anomaly in seawater, the daily CO_2 and heat outputs from the Secca delle Fumose were estimated to be ~50 t/d and ~80 MWt, respectively (Di Napoli et al. 2016). This heat flux compares with that of the Solfatara–Pisciarelli Diffuse Degassing Structure (see previous section). Based on the Na–K geothermometer of Giggenbach (1988), the discharged brines come from a hydrothermal reservoir at 179 °C (Di Napoli et al. 2016). This temperature should be considered as a upper limit based on the characteristics of the different Na–K geothermometers (Cioni and Marini 2020).

3.4 The Sub-Lacustrine Hydrothermal Discharges in Lake Averno

Hydrothermal Na-Cl waters discharge at depth in Lake Averno as indicated by chemical and isotopic data (Caliro et al. 2008; Cabassi et al. 2013; Tassi et al. 2018). Although lacustrine waters act as a sink for deep CO_2-rich gases, the concentrations of dissolved CO_2, CH_4, and H_2S in Lake Averno waters are mainly controlled by bacterial processes. In fact, the concentration of SO_4 decreases with depth, whereas the concentrations of H_2S and CH_4 increase with depth probably due to the occurrence of bacterially-mediated sulfate reduction and methanogenesis, both involving organic matter in the sediments and in the water column.

Usually, the Lake Averno waters are stratified, with an oxic epilimnion, from the surface to 6 m depth, overlying an anoxic hypolimnion extending to lake bottom at ~33 m. However, cooling of the epilimnetic waters below 7 °C in winter may trigger the overturn of lacustrine waters, increasing the H_2S concentration of shallow waters, which causes the death of numerous fishes as in February 2005. During the fish kill event, surface lacustrine waters became white, probably due to the precipitation of elemental sulfur produced by O_2-driven oxidation of dissolved H_2S according to the reaction:

$$H_2S + 1/2O_{2(g)} \rightarrow S_{(s)} + H_2O. \quad (4)$$

Other similar overturns took place in January 2002 and February 2003. Sub-lacustrine discharges of hydrothermal fluids causing fish killing events occurred also at Lake Fusaro and Lake Lucrino, as reported by Signore (1935).

4 The Conceptual Model of the Magmatic–Hydrothermal System Hosted in the Campi Flegrei Caldera

A magma reservoir whose top is found at ~7.5 km depth acts as the source of heat and magmatic fluids for the magmatic–hydrothermal system hosted in the Campi Flegrei caldera. Although different magma batches ascended and stationed at shallower levels, in the past, the presence of magma batches of size >1 km^3 at depths shallower than that of the principal reservoir can be ruled out, at present.

At depths between 2.7 and 4 km approximately, there is a layer of thermometamorphic rocks at temperatures and pressures higher than the critical point of water (374 °C, 222 bar) hosting gases and hypersaline brines that are expelled from the crystallizing magma as well as gaseous CO_2 produced by decarbonation reactions and other gas species (e.g., CH_4, H_2S, N_2, NH_3, and other compounds of C, S, and N) produced through different reactions (e.g., breakdown of the organic matter and decomposition of S-bearing minerals). These reactions occur within the thermometamorphic layer and even more so below it, where carbonate rocks are presumably present at higher temperatures. Although this thermometamorphic layer was directly reached by the San Vito 1 well only, it is likely that it has a certain extension and lateral continuity, at least over the whole central sector of the caldera, owing to the extension and lateral continuity of the underlying melt zone. Furthermore, owing to the presence of over-pressurized gas, it is likely that these thermometamorphic rocks and the underlying carbonate rocks are the "engine" controlling the bradyseismic phenomenon, that is, the source of ground uplift and related shallow seismicity as proposed by several authors (e.g. Vanorio et al. 2005; Chiarabba and Moretti 2006; Zollo et al. 2008; Battaglia et al 2008).

According to Fournier (1999), a thin, relatively impermeable zone generated through self-sealing (e.g., quartz deposition), at temperature generally close to 400 °C, is present at the top of the thermo-metamorphic layer (see above). Actually, abundant hydrothermal quartz was observed in well San Vito 1, at depths between ca. 2500 and 2800 m and temperatures of ca. 360–385 °C, at the transition between the Ca–Al–silicate zone (above, the bottom of which is marked by the disappearance of epidote) and the thermometamorphic zone (below, the top of which is marked by the appearance of scapolite, biotite, hornblende, diopside, garnet, and adularia), as shown by Fig. 5. This quartz-rich zone could actually mark the impermeable transition zone considered to be the boundary between the underlying deep-magmatic portion of the magmatic–hydrothermal system, where fluid pressure is controlled by the lithostatic regime, and the overlying shallow-hydrothermal portion of the magmatic–hydrothermal system, where fluid pressure is governed by the hydrostatic regime (Fournier 1999).

At depths lower than ~2.7 km, the original volcanic and marine deposits (tuffites, marls and siltites) have been pervasively affected by different types of hydrothermal alteration: propylitic, phyllitic, and argillic, from bottom to top. The circulation of hydrothermal fluids of marine origin (*in toto* or at least for the most) is restricted to the brittle rocks of the propylitic zone, where permeability is fracture-controlled, whereas the overlying impervious and plastic rocks of the phyllitic and argillic zones constitute the clay cap of the geothermal system. In the fumarolic areas of Solfatara–Pisciarelli and at Agnano (where this type of activity probably occurred in the past), a different type of hydrothermal alteration, known as advanced argillic alteration, affects the rocks present at the surface and at shallow depths.

The simple distribution of the Na-Cl thermal waters provide some reliable clues on the extension of the hydrothermal system hosted in the Campi Flegrei caldera. It must be recalled that, in volcanic regions of high topographic relief, the Na-Cl geothermal water may flow laterally for up to 20 km from the center of the geothermal field. Consequently, neutral chloride springs may be situated at considerable distances from the geothermal reservoir, whose position may be identified by fumarole activity and/or the occurrence of steam-heated sulfate–bicarbonate thermal waters higher on the volcano flanks (Henley and Ellis 1983). In contrast, in the basinal-type geothermal systems, such as those located in the Taupo Volcanic Zone of New Zealand, the boiling Na-Cl springs are positioned directly above the geothermal reservoir, while fumaroles and steaming grounds are situated in nearby areas of higher grounds at short distances from the boiling Na-Cl springs (Bogie et al. 2005).

In the Campi Flegrei caldera, there are limited differences in elevation and the maximum elevations (usually less than 200 m asl) are related to small or relatively small volcanic cones. The geomorphological framework of the area is, therefore, similar to that of the basinal-type geothermal systems, where the extension of the geothermal reservoir is marked, to a first approximation, by the boiling Na-Cl springs discharging at the surface. Furthermore, at Campi Flegrei, the depth of the water table decreases progressively approaching the coast and, consequently, the neutral chloride thermal waters coming from the geothermal reservoir are expected to emerge at the surface or to be intercepted by shallow wells close to the coastline, as it is actually observed (Fig. 12).

Nevertheless, the onshore thermal manifestations (both thermal waters and gas emissions) cluster in the two distinct areas of Mofete and Solfatara–Agnano, both characterized by high heat flow, ~160 MW m^{-2} and ~120 MW m^{-2}, respectively (Corrado et al. 1998; Wohletz et al. 1999). It is possible, therefore, that hydrothermal circulation is restricted, almost completely, to these two areas, which would host two distinct, relatively small hydrothermal systems. Alternatively (and this is our preferred option), it is possible that the two areas of Mofete and Solfatara–Agnano correspond to two separate upflows of a unique, relatively large hydrothermal system, which would extend:

(i) Between them, as suggested by the presence of some (albeit not numerous) thermal manifestations situated between the two main sites of hydrothermal activity.
(ii) To the south, as indicated by the distribution of offshore gas exhalations, representing at least outflows of the hydrothermal system.
(iii) To the north, as demonstrated by the outcomes of the deep boreholes of San Vito, especially well SV-1. In fact, it must be recalled that the short-term production test

of well SV-1 was interrupted due to the very high well head temperature and pressure and the unusual chemical characteristics of the discharged fluids (see Sect. 1.2). Well SV-1 was probably not exploitable in the '70–'80s using the technologies and materials available at that time, but it would probably be usable today, also considering the ongoing drilling activities in several supercritical geothermal systems.

5 The Conceptual Geochemical Models of the Solfatara Magmatic–Hydrothermal System

The first geochemical model of the Solfatara magmatic–hydrothermal system was probably elaborated by Cioni et al. (1984) and was largely based on the condition of full equilibrium among H_2O, CO_2, CH_4, H_2 and CO in a pure saturated vapor phase. A mean equilibrium temperature of 234 °C was obtained considering the log K of the reverse water–gas shift reaction:

$$CO_2 + H_2 = CO + H_2O. \quad (5)$$

Based on this mean equilibrium temperature, a mean water pressure of 28 bar was computed from the log K of the reaction:

$$4CO + 2H_2O = CH_4 + 3CO_2. \quad (6)$$

Cioni et al. (1984) underscored that (i) these pressure and temperature values are very close to the conditions of maximum enthalpy of saturated steam, namely 236 °C and 31 bar and (ii) the isenthalpic expansion (decompression) of steam from these T and P would produce dry steam with temperature close to 160 °C, at atmospheric pressure. This temperature value agrees with the outlet temperature of the Bocca Grande fumarole. Based on these findings, the local geological setting, and the results of deep geothermal boreholes drilled in the area, Cioni et al. (1984) proposed a geochemical model (Fig. 15) comprising: (i) a shallow magma body located at 3–5 km depth, (ii) a main boiling aquifer positioned at approximately 1450 m depth, whose temperature was estimated to be about 280 °C in 1983, based on the isotopic and chemical evidence that was available at the time, and close to 270 °C before 1981, considering the change in the CO_2/H_2O ratio in this time interval, and (iii) a relatively shallow steam zone where temperature is close to 236 °C and pressure is ca. 31 bar.

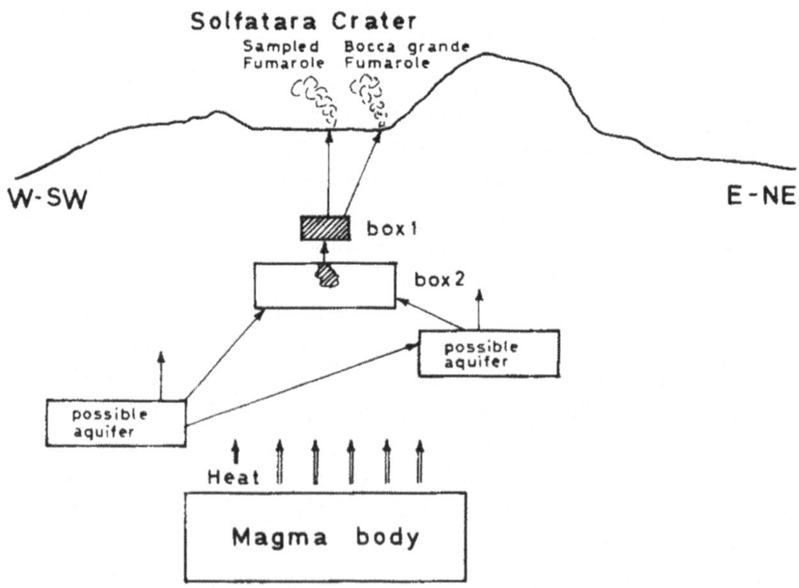

Fig. 15 Simplified sketch showing the conceptual geochemical model of the Solfatara magmatic–hydrothermal proposed by Cioni et al. (1984). The shaded areas correspond to intensively fractured zones where the vapor phase is separated from liquid water. Reprinted from Cioni et al. (1984), Copyright © 1984, with permission from Springer Nature

Cioni et al. (1984) underscored that the gas/steam ratio of the Solfatara fumarolic fluids is the most effective parameter to detect changes in heat flow at depth and that this parameter is expected to decrease in response to an increase of the heat flow from the underlying magma body. Nevertheless, they also recognized that monitoring of gas chemistry must be continued, to improve the general knowledge of the hydrothermal–volcanic system and to detect possible important changes.

The geochemical model of Cioni et al. (1984) was modified by Caliro et al. (2007) who processed the large database of chemical and isotopic data acquired during 24 years of geochemical monitoring of the Solfatara fumaroles and recognized the presence of two distinct temperatures and pressures of gas equilibration. The fast reacting gas species, i.e., H_2 and CO, equilibrates at low temperatures, 200–240 °C, which are present at shallow depths (Caliro et al. 2007). The chemical and isotopic geoindicators involving CH_4 and the H_2–Ar geothermometer provide instead equilibrium temperatures close to 360 °C, which were considered by Caliro et al. (2007) to be representative of a deep mixing zone where magmatic gases mix with the hydrothermal liquid causing its flashing and the formation a single-phase gas plume (Fig. 16). The computed pressure in this deep mixing zone, 200–250 bar, suggests that it is positioned at a depth of 2.0–2.5 km assuming

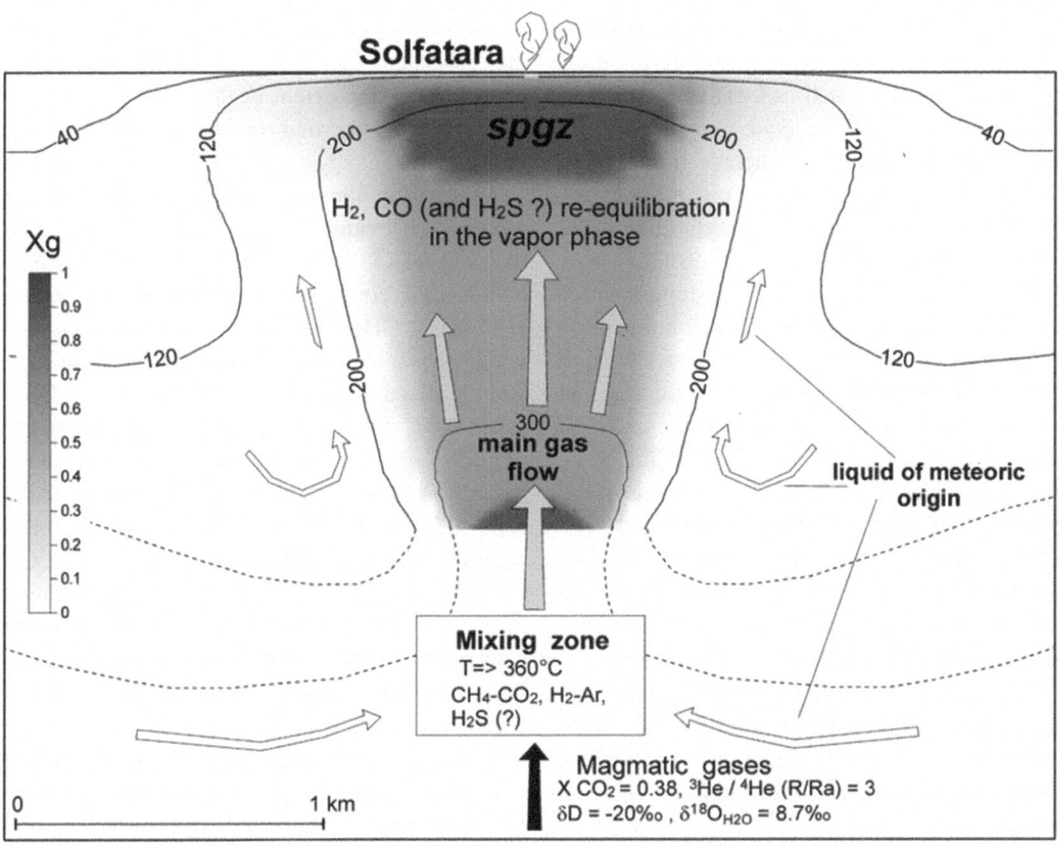

Fig. 16 Geochemical conceptual model of the Solfatara, as modified by Caliro et al. (2007), showing the deep mixing zone and the re-equilibration zones of different gas species (from Caliro et al. 2007). Gas/liquid mass fractions and isotherms are also reported, as returned by the physico-numerical simulations of Chiodini et al. (2003) and Todesco et al. (2003). Reprinted from Caliro et al. (2007), Copyright © 2007, with permission from Elsevier

a near-hydrostatic regime. Since this depth compares with those of the hypocenters of the fluid-triggered seismic swarms (e.g., Saccorotti et al. 2001), Caliro et al. (2007) proposed that the seismicity recorded at Solfatara might be controlled by the processes occurring in the deep mixing zone.

According to Caliro et al. (2007), the magmatic endmember involved in this mixing process has the isotopic fingerprint of a typical 'andesitic' or 'arc-type' water type, with δ^2H close to −20‰, and $\delta^{18}O$ close to +10‰, and is rich in CO_2, having a CO_2 mole fraction close to 0.4, while the hydrothermal endmember has a meteoric origin and a CO_2 fugacity fixed by fluid-rock reaction at high temperatures. Furthermore, Caliro et al. (2007) suggested that this deep mixing zone divides the Solfatara system in two portions. Reducing hydrothermal conditions dominate above it, favoring the formation of reduced gas species, such as CH_4 and H_2S, and the disappearance of the acid magmatic gases (i.e., SO_2, HCl, and HF), whereas higher temperatures and oxidizing magmatic conditions are expected to be present below it, stabilizing the acid magmatic gases.

Also shown in Fig. 16 are the temperatures and the gas/liquid mass fractions obtained by modeling the periodic injection of hot CO_2-rich fluids (at 350 °C) at the base of the hydrothermal portion of the system for 2000 years. These simulations were performed by Chiodini et al. (2003) and Todesco et al. (2003) by means of the geothermal simulator TOUGH2 (Pruess 1991). These modeling exercises predict that the ascending fluid present in the central column above the injection point is either a single-phase gas (at depths of 1500–1400 m and 300–100 m) or a bi-phase (gas + liquid) mixture. The simulations calculate a temperature of 190–230 °C for the shallow single-phase gas zone, in excellent agreement with the equilibrium temperatures of 200–240 °C of H_2 and CO (see above), which re-equilibrate during the ascent of the single-phase gas plume.

Caliro et al. (2007) also underscored that the CO_2/H_2O mole fraction ratio, acquired during 24 years of geochemical monitoring of fumarolic fluids, experienced three clear peaks in 1985, 1990 and 1995, which occurred a few months after three inflation episodes, as already noted by Chiodini et al. (2003). These CO_2/H_2O peaks were attributed by Chiodini and coworkers to increases in the fraction of the magmatic endmember entering the Solfatara system, with a maximum magmatic fraction of 0.45 in 1985, possibly because of the occurrence of increased magma degassing during the inflation episodes. If so, the bradyseismic crises would be triggered by periodic injections of CO_2-rich magmatic fluids at the bottom of the hydrothermal system. This interpretation of Chiodini and coworkers is completely different from that previously proposed by Cioni et al. (1984), who suggested that the gas/steam ratio or the CO_2/H_2O ratio is controlled by the extent of the boiling process of the deep hydrothermal aquifer, as already noted above.

Comparing the two conceptual geochemical models, it is evident that in addition to the role of the CO_2/H_2O ratio, there is also a contrast between the condition of full equilibrium among H_2O, CO_2, CH_4, H_2 and CO in a pure saturated vapor phase, as observed by Cioni et al. (1984), and the disequilibrium between CH_4 and CO, as noted by Caliro et al. (2007), leaving aside H_2, requiring further considerations that are postponed to chapters "The Geoindicators Involving H_2O, CO_2, CO, CH_4 and H_2", "The Chemical Kinetics of the Reactions Involving H_2O, CO_2, CO, CH_4 and H_2", "Calibration and Application of the RWG and SS4 Indicators for the Possible Expansion Paths of Solfatara Fluids" and "The Redox Potential and Sulfur Gas Species" of this book.

For the moment we would like to emphasize that this contrast is real and indicates that the Solfatara magmatic–hydrothermal system underwent profound changes, probably because of the bradyseismic crisis of 1982–1984, which was characterized by a considerable inflation and intense seismicity. In other terms, the magmatic system acted only as a source of heat to the overlying hydrothermal system before the first '80s, as postulated by the conceptual model of Cioni et al. (1984), whereas in the following years, the magmatic system contributed not only heat but also magmatic fluids to the hydrothermal system above it, and increasingly so with time, according to the conceptual model of Caliro et al. (2007). Therefore, neither of the two models is wrong. Although they are different, they are both correct (apart from a few missing details, see below), since these differences are real being due to a change of state in the Solfatara magmatic–hydrothermal system. Furthermore, this change of state in the Solfatara system is not surprising, being consistent with the general conceptual model of Fournier (1999). Thanks to the efforts of Cioni and collaborators, first, and of Chiodini and coworkers afterwards, the Solfatara magmatic–hydrothermal system, therefore, represents a well-studied case history, which fits perfectly into the general conceptual model of volcano-hosted magmatic–hydrothermal systems of Fournier (1999).

After 2007, the conceptual geochemical model of Caliro et al. (2007) was adopted for the interpretation of the data gradually collected over time, concerning not only the geochemical and isotopic compositions of fumarolic fluids, but also soil CO_2 flux measurements, ground deformation, swarm events, background seismicity, and fumarolic tremor. Thus, the conceptual geochemical model of Caliro et al. (2007) has been enriched with new details that have further consolidated it (e.g., Caliro et al. 2014; Chiodini 2009; Chiodini et al. 2010, 2011, 2012, 2015, 2016, 2017a, b, 2021). In particular:

(a) Chiodini et al. (2015) recognized that ground deformation is controlled by two processes, namely (i) short-time pulses that are caused by injection of magmatic fluids into the deepest portion of the hydrothermal system and are marked by the increase in fumarolic CO_2/CH_4 and He/CH_4 ratios and (ii) a long-term process of rock heating.

(b) Caliro et al. (2014) used an improved version of the magma degassing model of Nuccio and Paonita (2001) and explained the continuous decreases in both the N_2/He and N_2/CO_2 ratios as well as the concurrent increase in the He/CO_2 ratio occurring since 1985, in terms of magmatic degassing.

Although so much and so good has been done by Chiodini and coworkers, there are some missing details in the model of Caliro et al. (2007). One is the lack of the thermometamorphic/carbonate layer hosting over-pressurized gases, partly coming from the underlying degassing melt zone and partly generated in-situ by decarbonation and other reactions. Another missing detail concerns the CH_4 geothermometers, which have been disregarded by Chiodini and coworkers after the work of Caliro et al. (2007) and will be reconsidered here in chapters "The Geoindicators Involving H_2O, CO_2, CO, CH_4 and H_2", "The Chemical Kinetics of the Reactions Involving H_2O, CO_2, CO, CH_4 and H_2" and "The Chemical Kinetics of the Reactions Involving H_2O, CO_2, CO, CH_4 and H_2 " of this book.

References

Aiuppa A, Avino R, Brusca L, Caliro S, Chiodini G, D'Alessandro W, Favara R, Federico C, Ginevra W, Inguaggiato S, Longo M, Pecoraino G, Valenza M (2006) Mineral control of arsenic content in thermal waters from volcano-hosted hydrothermal systems: insights from island of Ischia and Phlegrean Fields (Campanian Volcanic Province, Italy). Chem Geol 229(4):313–330

Aiuppa A, Tamburello G, Di Napoli R, Cardellini C, Chiodini G, Giudice G, Grassa F, Pedone M (2013) First observations of the fumarolic gas output from a restless caldera: implications for the current period of unrest (2005–2013) at Campi Flegrei. Geochem Geophys Geosyst 14(10):4153–4169

Antrodicchia E, Cioni R, Chiodini G, Gagliardi R, Marini L (1985) Geochemical temperatures of the thermal waters of Phlegraean Fields (Naples, Italy). Geotherm Resour Counc Trans 9(I):287–292

Apollaro C, Vespasiano G, De Rosa R, Marini L (2015) Use of mean residence time and flowrate of thermal waters to evaluate the volume of reservoir water contributing to the natural discharge and the related geothermal reservoir volume. Application to Northern Thailand hot springs. Geothermics 58:62–74

Arienzo I, Moretti R, Civetta L, Orsi G, Papale P (2010) The feeding system of Agnano-Monte Spina eruption (Campi Flegrei, Italy): dragging the past into present activity and future scenarios. Chem Geol 270(1–4):135–147

Arienzo I, Mazzeo FC, Moretti R, Cavallo A, D'Antonio M (2016) Open-system magma evolution and fluid transfer at Campi Flegrei caldera (Southern Italy) during the past 5 ka as revealed by geochemical and isotopic data: the example of the Nisida eruption. Chem Geol 427:109–124

Baldi P, Ferrara GC, Panichi C (1975) Geothermal research in western Campania (southern Italy): chemical and isotopic studies of thermal fluids in the Campi Flegrei. In: Proceedings of the second United Nations symposium on the development and use of geothermal resources, vol 1. San Francisco, 20–29 May 1975, pp 687–697

Balducci S, Chelini W (1992) Hydrothermal equilibria in the active Mofete geothermal system, Phlegraean Fields, Naples, Italy. Acta Vulcanol (Marinelli) 2:17–34

Battaglia J, Zollo A, Virieux J, Dello Iacono D (2008) Merging active and passive data sets in traveltime tomography: the case study of Campi Flegrei caldera southern Italy. Geophys Prospect 56(4):555–573

Bernard R, Taran Y, Pennisi M, Tello E, Ramirez A (2011) Chloride and boron behavior in fluids of Los Humeros geothermal field (Mexico). A model based on the existence of deep acid brine. Appl Geochem 26:2064–2073

Bogie I, Lawless JV, Rychagov S, Belousov V (2005) Magmatic-related hydrothermal systems: classification of the types of geothermal systems and their ore mineralization. In: Proceedings of geoconference in Russia, Kuril

Bolognesi L, Noto P, Nuti S (1986) Studio chimico ed isotopico della solfatara di Pozzuoli: ipotesi sull'origine e sulle temperature profonde dei fluidi. Rendiconti della Società Italiana di Mineralogia e Petrologia 41 (2):281–295

Bruni P, Chelini W, Sbrana A, Verdiani G (1985) Deep exploration of the S. Vito area (Pozzuoli-NA) Well S. Vito 1. In: Strub AS, Ungemach P (eds) European geothermal update, proceedings of the third international seminar on the results of EC geothermal energy research, Munich, 29 November–1 December, 1983, pp 390–406

Bruno PG, Ricciardi GP, Petrillo Z, Di Fiore V, Troiano A, Chiodini G (2007) Geophysical and hydrogeological experiments from a shallow hydrothermal system at Solfatara Volcano, Campi Flegrei, Italy: response to caldera unrest. J Geophys Res 112:B06201

Buonasorte G, Cataldi R, Ceccarelli A, Costantini A, D'Offizi S, Lazzarotto A, Ridolfi A, Baldi P, Barelli A, Bertini G, Bertrami R, Calamai A, Cameli G, Corsi R, D'Acquino C, Fiordelisi A, Gezzo A, Lovari F (1988) Ricerca ed esplorazione nell'area geotermica di Torre Alfina (Lazio–Umbria). Boll Soc Geol Ital 107:265–337

Byrdina S, Vandemeulebrouck J, Cardellini C, Legaz A, Camerlynck C, Chiodini G, Lebourg T, Gresse M, Bascou P, Motos G, Carrier A, Caliro S (2014) Relations between electrical resistivity, carbon dioxide flux, and self-potential in the shallow hydrothermal system of Solfatara (Phlegrean Fields, Italy). J Volcanol Geotherm Res 283:172–182

Cabassi J, Tassi F, Vaselli O, Fiebig J, Nocentini M, Capecchiacci F, Rouwet D, Bicocchi G (2013) Biogeochemical processes involving dissolved CO_2 and CH_4 at Albano, Averno, and Monticchio meromictic volcanic lakes (Central-Southern Italy). Bull Volcanol 75(1):1–19

Caliro S, Chiodini G, Moretti R, Avino R, Granieri D, Russo M, Fiebig J (2007) The origin of the fumaroles of La Solfatara (Campi Flegrei, south Italy). Geochim Cosmochim Acta 71(12):3040–3055

Caliro S, Chiodini G, Izzo G, Minopoli C, Signorini A, Avino R, Granieri D (2008) Geochemical and biochemical evidence of lake overturn and fish kill at Lake Averno, Italy. J Volcanol Geotherm Res 178 (2):305–316

Caliro S, Chiodini G, Paonita A (2014) Geochemical evidences of magma dynamics at Campi Flegrei (Italy). Geochim Cosmochim Acta 132:1–15

Cannatelli C, Lima A, Bodnar RJ, De Vivo B, Webster JD, Fedele L (2007) Geochemistry of melt inclusions from the Fondo Riccio and Minopoli 1 eruptions at Campi Flegrei (Italy). Chem Geol 237(3–4):418–432

Caprarelli G, Tsutsumi M, Turi B (1997) Chemical and isotopic signatures of the basement rocks from the Campi Flegrei geothermal field (Naples, southern Italy): inferences about the origin and evolution of its hydrothermal fluids. J Volcanol Geotherm Res 76(1–2):63–82

Cardellini C, Chiodini G, Frondini F, Avino R, Bagnato E, Caliro S, Lelli M, Rosiello A (2017) Monitoring diffuse volcanic degassing during volcanic unrests: the case of Campi Flegrei (Italy). Sci Rep 7(1):6757. https://doi.org/10.1038/s41598-017-06941-2

Carella R, Guglielminetti M (1983) Multiple reservoirs in the Mofete field, Naples, Italy. In: Proceedings ninth workshop geothermal reservoir engineering. Stanford University, Stanford, California, December 1983, SGP-TU-74, pp 53–64

Carlino S, Somma R, Troise C, De Natale G (2012) The geothermal exploration of Campanian volcanoes: historical review and future development. Renew Sustain Energy Rev 16(1):1004–1030

Carlino S, Piochi M, Tramelli A, Mormone A, Montanaro C, Scheu B, Klaus M (2018) Field-scale permeability and temperature of volcanic crust from borehole data: Campi Flegrei, southern Italy. J Volcanol Geotherm Res 357:276–286

Carmisciano C, Grassi M, Cocchi L, Masetti G, Filippone M, Ricci E, Pratellesi M, Ivaldi R, Iannaccone G,

Berrino G, De Natale G, Chiodini G, Caliro S, Dialti L, Guideri M, Muccini F, Stefanelli P, Orsi G, Iafolla V, Italiano F, Buongiorno F, Stramondo S, Bignami C, Polcari M, Silvestri M (2013) RICAMAR2013: rilievi per la caratterizzazione dell'ambiente marino nel Golfo di Pozzuoli. Rapporto sull'attività 13–31 maggio 2013. Rapporti Tecnici INGV 262:54

Cassano E, La Torre P (1987) Geophysics. In: Rosi M, Sbrana A (eds) Phlegrean Fields, Progetto Finalizzato 'Geodinamica', Monografie Finali, vol 9. Quaderni de "La Ricerca Scientifica", AGIP-ENEL Joint-Venture for Geothermal Research and Progetto Finalizzato 'Geodinamica'—CNR, Consiglio Nazionale delle Ricerche, Roma, pp 103–131

Castaldo R, D'Auria L, Pepe S, Solaro G, De Novellis V, Tizzani P (2019) The impact of crustal rheology on natural seismicity: Campi Flegrei caldera case study. Geosci Front 10(2):453–466

Cecchetti A, Marianelli P, Sbrana A (2003) L'eruzione di Astroni (caldera dei Campi Flegrei): dati preliminari dallo studio di inclusioni silicatiche. Atti della Società Toscana di Scienze Naturali (Memorie, Serie A) 108:59–63

Celico P, Dall'Aglio M, Ghiara MR, Stanzione D, Brondi M, Prosperi M (1992) Geochemical monitoring of the thermal fluids in the Phlegraean Fields from 1970 to 1990. Boll Soc Geol Ital 111(3–4):409–422

Chelini W (1984) Alcuni aspetti geologico-petrografici sul sistema geotermico flegreo. Rendiconti della Società Italiana di Mineralogia e Petrologia 39:387–391

Chelini W, Sbrana A (1987) Subsurface geology. In: Rosi M, Sbrana A (eds) Phlegrean Fields, Progetto Finalizzato 'Geodinamica', Monografie Finali, vol 9. Quaderni de "La Ricerca Scientifica", AGIP-ENEL Joint-Venture for Geothermal Research and Progetto Finalizzato 'Geodinamica'—CNR, Consiglio Nazionale delle Ricerche, Roma, pp 94–103

Chelini W, Cioppi D, Sbrana A, La Torre P (1982) Schema geologico-geochimico dell'area flegrea. AGIP, Settore Esplorazione e Ricerche Geotermiche, San Donato Milanese, Italy. Unpublished report

Cheminée J, Letolle R, Olive P (1968) Premiéres données isotopiques sur des fumerolles de volcans italiens. Bull Volcanol 32(3):467–475

Chiarabba C, Moretti M (2006) An insight into the unrest phenomena at the Campi Flegrei caldera from Vp and Vp/Vs tomography. Terra Nova 18(6):373–379

Chiodini G (2009) CO_2/CH_4 ratio in fumaroles a powerful tool to detect magma degassing episodes atn quiescent volcanoes. Geophys Res Lett 36(2):L02302. https://doi.org/10.1029/2008GL036347

Chiodini G, Cioni R, Guidi M, Marini L (1991) Chemical geothermometry and geobarometry in hydrothermal aqueous solutions: a theoretical investigation based on a mineral-solution equilibrium model. Geochim Cosmochim Acta 55:2709–2727

Chiodini G, Cioni R, Guidi M, Raco B, Marini L (1998) Soil CO_2 flux measurements in volcanic and geothermal areas. Appl Geochem 13:543–552

Chiodini G, Frondini F, Cardellini C, Granieri D, Marini L, Ventura G (2001) CO_2 degassing and energy release at Solfatara volcano, Campi Flegrei, Italy. J Geophys Res 106:16213–16222

Chiodini G, Todesco M, Caliro S, Del Gaudio C, Macedonio G, Russo M (2003) Magma degassing as a trigger of bradyseismic events: the case of Phlegrean Fields (Italy). Geophys Res Lett 30(8):1434. https://doi.org/10.1029/2002GL016790

Chiodini G, Caliro S, Cardellini C, Granieri D, Avino R, Baldini A, Donnini M, Minopoli C (2010) Long-term variations of the Campi Flegrei, Italy, volcanic system as revealed by the monitoring of hydrothermal activity. J Geophys Res Solid Earth 115(B3):B03205. https://doi.org/10.1029/2008JB006258

Chiodini G, Caliro S, De Martino P, Avino R, Gherardi F (2012) Early signals of new volcanic unrest at Campi Flegrei caldera? Insights from geochemical data and physical simulations. Geology 40(10):943–946

Chiodini G, Vandemeulebrouck J, Caliro S, D'Auria L, De Martino P, Mangiacapra A, Petrillo Z (2015) Evidence of thermal-driven processes triggering the 2005–2014 unrest at Campi Flegrei caldera. Earth Planet Sci Lett 414:58–67

Chiodini G, Paonita A, Aiuppa A, Costa A, Caliro S, De Martino P, Acocella V, Vandemeulebrouck J (2016) Magmas near the critical degassing pressure drive volcanic unrest towards a critical state. Nat Commun 7 (1):1–9

Chiodini G, Giudicepietro F, Vandemeulebrouck J, Aiuppa A, Caliro S, De Cesare W, Tamburello G, Avino R, Orazi M, D'Auria L (2017b) Fumarolic tremor and geochemical signals during a volcanic unrest. Geology 45(12):1131–1134

Chiodini G, Selva J, Del Pezzo E, Marsan D, De Siena L, D'Auria L, Bianco F, Caliro S, De Martino P, Ricciolino P, Petrillo Z (2017a) Clues on the origin of post-2000 earthquakes at Campi Flegrei caldera (Italy). Sci Rep 7(1):1–10

Chiodini G, Avino R, Caliro S, Minopoli C (2011) Temperature and pressure gas geoindicators at the Solfatara fumaroles (Campi Flegrei). Ann Geophys 54 (2). https://doi.org/10.4401/ag-5002

Chiodini G, Caliro S, Avino R, Bini G, Giudicepietro F, De Cesare W, Ricciolino P, Aiuppa A, Cardellini C, Petrillo Z, Selva J, Siniscalchi A, Tripaldi S (2021) Hydrothermal pressure-temperature control on CO_2 emissions and seismicity at Campi Flegrei (Italy). J Volcanol Geotherm Res 414:107245

Cioni R, Corazza E, Marini L (1984) The gas/steam ratio as indicator of heat transfer at the Solfatara fumaroles, Phlegraean Fields (Italy). Bull Volcanol 47(2):295–302

Cioni R, Marini L (2020) A thermodynamic approach to water geothermometry. In: Springer geochemistry. Springer Nature, Cham, Switzerland, p 415

Cipriani F, Marianelli P, Sbrana A (2008) Studio di una sequenza piroclastica del vulcano della Solfatara (Campi Flegrei). Considerazioni vulcanologiche e sul sistema di alimentazione. Atti della Società Toscana di Scienze Naturali (Memorie, Serie A) 113:39–48

Corniello A, Nicotera P (1982) Geologia, idrogeologia e idrochimica della zona sud-occidentale dei Campi Flegrei. Memorie e note dell'Istituto di Geologia Applicata di Napoli, vol 16, p 45

Corrado G, De Lorenzo S, Mongelli F, Tramacere A, Zito G (1998) Surface heat flow density at the Phlegrean Fields caldera (Southern Italy). Geothermics 27(4):469–484

Cortecci G, Noto P, Panichi C (1978) Environmental isotopic study of the Campi Flegrei (Naples, Italy) geothermal field. J Hydrol 36(1–2):143–159

Cumming W (2016) Resource conceptual models of volcano-hosted geothermal reservoirs for exploration well targeting and resource capacity assessment: Construction, pitfalls and challenges. Geotherm Resour Counc Trans 40:623–637

Cumming W (2009) Geothermal resource conceptual models using surface exploration data. In: Proceedings of the 34th workshop on geothermal reservoir engineering. Stanford University, Stanford, California, February 9–11, 2009 SGP-TR-187, p 6

D'Amore F, Giusti D, Abdallah A (1998) Geochemistry of the high-salinity geothermal field of Asal, Republic of Djibouti, Africa. Geothermics 27:197–210

D'Antonio M (2011) Lithology of the basement underlying the Campi Flegrei caldera: volcanological and petrological constraints. J Volcanol Geotherm Res 200 (1–2):91–98

Dall'Aglio M, Martini M, Tonani F (1972) Rilevamento geochimico delle emanazioni vulcaniche dei Campi Flegrei. In: Quaderni de 'La Ricerca Scientifica', vol 83. CNR—Consiglio Nazionale delle Ricerche, Roma, pp 152–181

De Luca S (1871) La composizione dei gas che si svolgono dalle fumarole della Solfatara di Pozzuoli. Rendiconto dell'Accademia delle Scienze Fisiche e Matematiche (Società Reale Di Napoli, Anno X) 181–188:211–221

De Vivo B, Belkin HE, Barbieri M, Chelini W, Lattanzi P, Lima A, Tolomeo L (1989) The Campi Flegrei (Italy) geothermal system: a fluid inclusion study of the Mofete and San Vito fields. J Volcanol Geotherm Res 36(4):303–326

Di Napoli R, Aiuppa A, Sulli A, Caliro S, Chiodini G, Acocella V, Ciraolo G, Di Vito MA, Interbartolo F, Nasello C, Valenza M (2016) Hydrothermal fluid venting in the offshore sector of Campi Flegrei caldera: a geochemical, geophysical, and volcanological study. Geochem Geophys Geosyst 17(10):4153–4178

Ellis AJ (1970) Quantitative interpretation of chemical characteristics of hydrothermal systems. Geothermics 2:516–528

Evans BW, Shaw DM, Haughton DR (1969) Scapolite stoichiometry. Contrib Miner Petrol 24(4):293–305

Ferrante A (2018) La Grotta del Cane, Agnano, Napoli: storie e prospettive di valorizzazione. Thesis, Università Federico II, Napoli, p 52

Fourmentraux C, Métrich N, Bertagnini A, Rosi M (2012) Crystal fractionation, magma step ascent, and syn-eruptive mingling: the Averno 2 eruption (Phlegraean Fields, Italy). Contrib Mineral Petrol 163:1121–1137

Fournier RO (1985) The behavior of silica in hydrothermal solutions. Rev Econ Geol 2:45–72

Fournier RO (1999) Hydrothermal processes related to movement of fluid from plastic into brittle rock in the magmatic-epithermal environment. Econ Geol 94 (8):1193–1211

Fournier RO, Potter RWII (1982) An equation correlating the solubility of quartz in water from 25 to 900 C at pressures up to 10,000 bars. Geochim Cosmochim Acta 46:1969–1973

Friedman I, O'Neil JR (1977) Compilation of stable isotope fractionation factors of geochemical interest. In: Fleischer M (ed) Data of geochemistry. Geological survey professional paper 440-KK, 6th edn, p 12

Ghiara MR, Stanzione D (1988) Studio geochimico sul sistema idrotermale dei Campi Flegrei, Campania, Italia. Rendiconto dell'Accademia delle Scienze Fisiche e Matematiche (Napoli, Ser IV) 55:61–83

Ghiara MR, Stanzione D, Bellucci L, Panichi C (1988) Preliminary observations on the chemical variations of the solutes monitored in the thermal waters of the Phlegraean Fields in the period 1983–1985. Rendiconti della Società Italiana di Mineralogia e Petrologia 43:1139–1150

Giggenbach WF (1978) The isotopic composition of waters from the El Tatio geothermal field, Northern Chile. Geochim Cosmochim Acta 42(7):979–988

Giggenbach WF, Stewart MK (1982) Processes controlling the isotopic composition of steam and water discharges from steam vents and steam-heated pools in geothermal areas. Geothermics 11(2):71–80

Giggenbach WF (1988) Geothermal solute equilibria. Derivation of Na–K–Mg–Ca geoindicators. Geochim Cosmochim Acta 52(12):2749–2765

Giggenbach WF (1991) Chemical techniques in geothermal exploration. In: D'Amore F (ed) Application of geochemistry in geothermal reservoir development. UNITAR, Geneva, pp 119–144

Giggenbach WF (1992) Isotopic shifts in waters from geothermal and volcanic systems along convergent plate boundaries and their origin. Earth Planet Sci Lett 113(4):495–510

Graham CM, Sheppard SM, Heaton TH (1980) Experimental hydrogen isotope studies—I. Systematics of hydrogen isotope fractionation in the systems epidote-H_2O, zoisite-H_2O and AlO(OH)–H_2O. Geochimica et Cosmochimica Acta 44(2):353–364

Grant MA, Donaldson IG, Bixley PF (1982) Geothermal reservoir engineering. Academic Press, New York, p 369

Gresse M, Vandemeulebrouck J, Byrdina S, Chiodini G, Revil A, Johnson TC, Ricci T, Vilardo G, Mangiacapra A, Lebourg T, Grangeon J, Bascou P, Metral L (2017) Three-dimensional electrical resistivity tomography of the Solfatara crater (Italy): implication for the multiphase flow structure of the shallow hydrothermal system. J Geophys Res Solid Earth 122(11):8749–8768

Guglielminetti M (1986) Mofete geothermal field. Geothermics 15(5–6):781–790

Halliday WR, Cigna AA (2006) The Grotta Del Cane (Dog Cave), Naples, Italy. Cave Karst Sci 33(3):131–136

Hardardóttir V, Brown KL, Fridriksson T, Hedenquist JW, Hannington MD, Thorhallsson S (2009) Metals in deep liquid of the Reykjanes geothermal system, southwest Iceland: implications for the composition of seafloor black smoker fluids. Geology 37:1103–1106

Hassan I, Buseck PR (1988) HRTEM characterization of scapolite solid solutions. Am Miner 73(1–2):119–134

Henley RW, Ellis AJ (1983) Geothermal systems ancient and modern: a geochemical review. Earth Sci Rev 19 (1):1–50

Horita J (2005) Stable isotope thermometry: there is more to it than temperature. Geochem J 39(6):481–496

Isaia R, Vitale S, Di Giuseppe MG, Iannuzzi E, D'Assisi TF, Troiano A (2015) Stratigraphy, structure, and volcano-tectonic evolution of Solfatara maar-diatreme (Campi Flegrei, Italy). Geol Soc Am Bull 127(9–10):1485–1504

James R (1970) Factors controlling borehole performance. Geothermics 2:1502–1515

Judenherc S, Zollo A (2004) The Bay of Naples (southern Italy): constraints on the volcanic structures inferred from a dense seismic survey. J Geophys Res 109: B10312. https://doi.org/10.1029/2003JB002876

Lima A, De Vivo B, Spera FJ, Bodnar RJ, Milia A, Nunziata C, Belkin HE, Cannatelli C (2009) Thermodynamic model for uplift and deflation episodes (bradyseism) associated with magmatic–hydrothermal activity at the Campi Flegrei (Italy). Earth-Science Reviews, 97(1-4), 44-58

Lima A, Bodnar RJ, De Vivo B, Spera FJ, Belkin HE (2021) Interpretation of recent unrest events (bradyseism) at Campi Flegrei, Napoli (Italy): comparison of models based on cyclical hydrothermal events versus shallow magmatic intrusive events. Geofluids 16. Article ID 2000255. https://doi.org/10.1155/2021/2000255

Lowell RP, Van Cappellen P, Germanovich LN (1993) Silica precipitation in fractures and the evolution of permeability in hydrothermal upflow zones. Science 260(5105):192–194

Lvov SN, Wood RH (1990) Equation of state of aqueous NaCl solutions over a wide range of temperatures, pressures and concentrations. Fluid Phase Equilibria 60(3):273–287

Mangiacapra A, Moretti R, Rutherford M, Civetta L, Orsi G, Papale P (2008) The deep magmatic system of the Campi Flegrei caldera (Italy). Geophys Res Lett 35 (21). https://doi.org/10.1029/2008GL035550

Manning CE (1994) The solubility of quartz in H_2O in the lower crust and upper mantle. Geochim Cosmochim Acta 58:4831–4839

Marini L, Fiebig J (2005) Fluid geochemistry of the magmatic-hydrothermal system of Nisyros (Greece). In: Hunziker JC, Marini L (eds) The geology, geochemistry and evolution of Nisyros Volcano (Greece). Implications for the volcanic hazards, vol 44. Memoires de Géologie, Lausanne, pp 121–163

Marini L, Gambardella B (2005) Geochemical modeling of magmatic gas scrubbing. Ann Geophys 48:739–753

Martini M, Giannini L, Buccianti A, Prati F, Cellini LP, Iozzelli P, Capaccioni B (1991) Ten years of geochemical investigation at Phlegrean Fields, Italy. J Volcanol Geotherm Res 48:161–171

Minucci L (1964) Rotary drilling for geothermal energy. In: Proceedings of the U.N. conference on new sources of energy. United Nations, New York, II.A.2 Harnessing geothermal energy—Eletricity production, G/66, pp 234–244

Moore DE, Lockner DA, Byerlee JD (1994) Reduction of permeability in granite at elevated temperatures. Science 265(5178):1558–1561

Moretti R, Arienzo I, Civetta L, Orsi G, Papale P (2013) Multiple magma degassing sources at an explosive volcano. Earth Planet Sci Lett 367:95–104

Moretti R, De Natale G, Troise C (2020) Hydrothermal versus magmatic: geochemical views and clues into the unrest dilemma at Campi Flegrei. In: De Vivo B, Belkin HE, Rolandi G (eds) Vesuvius, Campi Flegrei, and Campanian volcanism. Elsevier, Amsterdam, pp 371–406

Nuccio PM, Paonita A (2001) Magmatic degassing of multicomponent vapors and assessment of magma depth: application to Vulcano Island (Italy). Earth Planet Sci Lett 193(3–4):467–481

O'Neil JR, Taylor Jr HP (1967) The oxygen isotope and cation exchange chemistry of feldspars. Am Mineral 52(9–10):1414–1437

Penta F (1949) Temperature nel sottosuolo della regione Flegrea. Ann Geophys 2(3):328–346

Piochi M, Mormone A, Balassone G, Strauss H, Troise C, De Natale G (2015) Native sulfur, sulfates and sulfides from the active Campi Flegrei volcano (southern Italy): genetic environments and degassing dynamics revealed by mineralogy and isotope geochemistry. J Volcanol Geotherm Res 304:180–193

Piochi M, Bruno PP, De Astis G (2005) Relative roles of rifting tectonics and magma ascent processes: inferences from geophysical, structural, volcanological, and geochemical data for the Neapolitan volcanic region (southern Italy). Geochem Geophys Geosyst 6 (7). https://doi.org/10.1029/2004GC000885

Pirajno F (2009) Hydrothermal processes and mineral systems. Springer, Berlin, Heidelberg, p 1250

Pruess K (1991) TOUGH2—a general purpose numerical simulator for multiphase fluid and heat flow. Report LBL-29400/UC-251, Lawrence Berkeley Laboratory, University of California, Earth Sciences Division

Reyes AG (1990) Petrology of Philippine geothermal systems and the application of alteration mineralogy to their assessment. J Volcanol Geotherm Res 43:279–309

Roach AL (2005) The evolution of silicic magmatism in the post-caldera volcanism of the Phlegrean Fields, Italy. PhD thesis, Department of Geological Sciences at Brown University, Providence, Rhode Island, p 151

Rosi M, Vezzoli L, Aleotti P, De Censi M (1996) Interaction between caldera collapse and eruptive dynamics during the Campanian Ignimbrite eruption, Phlegraean Fields, Italy. Bull Volcanol 57(7):541–554

Rosi M, Sbrana A (1987) Stratigraphy. In: Rosi M, Sbrana A (eds) Phlegrean Fields. Quaderni de La Ricerca Scientifica 9(114):10–39

Saccorotti G, Bianco F, Castellano M, Del Pezzo E (2001) The July–August 2000 seismic swarms at Campi Flegrei volcanic complex, Italy. Geophys Res Lett 28 (13):2525–2528

Saishu H, Okamoto A, Tsuchiya N (2014) The significance of silica precipitation on the formation of the permeable–impermeable boundary within Earth's crust. Terra Nova 26(4):253–259

Sanjuan B, Michard G, Michard A (1990) Origine des substances dissoutes dans les eaux des sources thermales et des forages de la région Asal-Ghoubbet (République de Djibouti). J Volcanol Geotherm Res 43:333–352

Scott SW, Driesner T (2018) Permeability changes resulting from quartz precipitation and dissolution around upper crustal intrusions. Geofluids, Article ID 6957306

Shaw DM (1960) The geochemistry of scapolite part I. Previous work and general mineralogy. J Petrol 1 (1):218–260

Shen L, Sippola H, Li X, Lindberg D, Taskinen P (2019) Thermodynamic modeling of calcium sulfate hydrates in the $CaSO_4$–H_2O system from 273.15 to 473.15 K with extension to 548.15 K. J Chem Eng Data 64 (6):2697–2709

Shibue Y (1996) Empirical expressions of quartz solubility in H_2O, $H_2O + CO_2$, and H_2O + NaCl fluids. Geochem J 30:339–354

Signore F (1935) Attività vulcanica e bradisismo nei Campi Flegrei. Annali dell'Osservatorio Vesuviano 3:173–192

Siniscalchi A, Tripaldi S, Romano G, Chiodini G, Improta L, Petrillo Z, D'Auria L, Caliro S, Avino R (2019) Reservoir structure and hydraulic properties of the Campi Flegrei geothermal system inferred by audiomagnetotelluric, geochemical, and seismicity study. J Geophys Res Solid Earth 124(6):5336–5356

Smale L (2020) A re-interpretation of long-term deformation at Campi Flegrei caldera, Italy and perceptions of the causes of caldera unrest. PhD thesis, Department of Earth Sciences, University College London, p 340

Smith VC, Isaia R, Pearce NJG (2011) Tephrostratigraphy and glass compositions of post-15 kyr Campi Flegrei eruptions: implications for eruption history and chronostratigraphic markers. Quat Sci Rev 30(25–26):3638–3660

Symonds RB, Gerlach TM, Reed MH (2001) Magmatic gas scrubbing: implications for volcano monitoring. J Volcanol Geotherm Res 108(1–4):303–341

Tamburello G, Caliro S, Chiodini G, De Martino P, Avino R, Minopoli C, Carandente A, Rouwet D, Aiuppa A, Costa A, Bitetto M, Giudice G, Francofonte V, Ricci T, Sciarra A, Bagnato E, Capecchiacci F (2019) Escalating CO_2 degassing at the Pisciarelli fumarolic system, and implications for the ongoing Campi Flegrei unrest. J Volcanol Geotherm Res 384:151–157

Tassi F, Fazi S, Rossetti S, Pratesi P, Ceccotti M, Cabassi J, Capecchiacci F, Venturi S, Vaselli O (2018) The biogeochemical vertical structure renders a meromictic volcanic lake a trap for geogenic CO_2 (Lake Averno, Italy). PLoS ONE 13(3):e0193914

Tedesco D, Allard P, Sano Y, Wakita H, Pece R (1990) Helium-3 in subaerial and submarine fumaroles of Campi Flegrei caldera, Italy. Geochim Cosmochim Acta 54(4):1105–1116

Todesco M, Chiodini G, Macedonio G (2003) Monitoring and modelling hydrothermal fluid emission at La Solfatara (Phlegrean Fields, Italy). An interdisciplinary approach to the study of diffuse degassing. J Volcanol Geotherm Res 125(1–2):57–79

Troiano A, Di Giuseppe MG, Patella D, Troise C, De Natale G (2014) Electromagnetic outline of the Solfatara-Pisciarelli hydrothermal system, Campi Flegrei (southern Italy). J Volcanol Geotherm Res 277:9–21

Truesdell AH, Singers W (1974) The calculation of aquifer chemistry in hot-water geothermal systems. J Res US Geol Surv 2:271–278

Turekian KK, Wedepohl KH (1961) Distribution of the elements in some major units of the earth's crust. Geol Soc Am Bull 72:175–192

Valentino GM, Stanzione D (2003) Source processes of the thermal waters from the Phlegraean Fields (Naples, Italy) by means of the study of selected minor and trace elements distribution. Chem Geol 194(4):245–274

Valentino GM, Stanzione D (2004) Geochemical monitoring of the thermal waters of the Phlegraean Fields. J Volcanol Geotherm Res 133(1–4):261–289

Valentino GM, Cortecci G, Franco E, Stanzione D (1999) Chemical and isotopic compositions of minerals and waters from the Campi Flegrei volcanic system, Naples, Italy. J Volcanol Geotherm Res 91(2–4):329–344

Vanorio T, Virieux J, Capuano P, Russo G (2005) Three-dimensional seismic tomography from P wave and S wave microearthquake travel times and rock physics characterization of the Campi Flegrei Caldera. J Geophys Res Solid Earth 110(B3)

Vaselli O, Tassi F, Tedesco D, Poreda JR, Caprai A (2011) Submarine and inland gas discharges from the Campi Flegrei (southern Italy) and the Pozzuoli Bay: geochemical clues for a common hydrothermal-magmatic source. In: "Research in Shallow Marine and Fresh Water Systems". Procedia Earth Planet Sci 4:57–73

Vetere F, Botcharnikov RE, Holtz F, Behrens H, De Rosa R (2011) Solubility of H_2O and CO_2 in shoshonitic melts at 1250 °C and pressures from 50 to 400 MPa: implications for Campi Flegrei magmatic systems. J Volcanol Geotherm Res 202(3–4):251–261

Von Damm KL, Bischoff JL, Rosenbauer RJ (1991) Quartz solubility in hydrothermal seawater: an experimental study and equation describing quartz solubility for up to 0.5 M NaCl solutions. Am J Sci 291:977–1007

Wells JT, Ghiorso MS (1991) Coupled fluid flow and reaction in mid-ocean ridge hydrothermal systems: the behavior of silica. Geochim Cosmochim Acta 55 (9):2467–2481

White SP, Mroczek EK (1998) Permeability changes during the evolution of a geothermal field due to the dissolution and precipitation of quartz. Transp Porous Media 33(1):81–101

Williams AE, McKibben MA (1989) A brine interface in the Salton Sea geothermal system, California: fluid geochemical and isotopic characteristics. Geochim Cosmochim Acta 53:1905–1920

Wohletz K, Civetta L, Orsi G (1999) Thermal evolution of the Phlegraean magmatic system. J Volcanol Geotherm Res 91(2–4):381–414

Zollo A, Maercklin N, Vassallo M, Dello Iacono D, Virieux J, Gasparini P (2008) Seismic reflections reveal a massive melt layer feeding Campi Flegrei caldera. Geophys Res Lett 35(12):L12306

Zollo A, Judenherc S, Auger E, D'Auria L, Virieux J, Capuano P, Chiarabba C, de Franco R, Makris J, Michelini A, Musacchio G (2003) Evidence for the buried rim of Campi Flegrei caldera from 3-d active seismic imaging. Geophys Res Lett 30(19). https://doi.org/10.1029/2003GL018173

The Expansion (Decompression) of the Solfatara Fumarolic Fluids

Abstract

The Bocca Grande and Bocca Nuova fumaroles have mean outlet temperatures of 161.4°C and 145.4°C, respectively, indicating that they discharge superheated (dry) gas mixtures. Water (688,900 to 873,100 μmol/mol) and CO_2 (125,000 to 308,700 μmol/mol) constitute together 99.3 to 100 mol % of these fluids, which can be considered binary CO_2-H_2O gas mixtures to a first approximation. Therefore, first of all, the main characteristics of the binary CO_2-H_2O system were summarized in this chapter. Although a general review of the equations of state (EOS) is beyond the scope of this book, an excursus on this topic was deemed necessary to contextualize the three distinct approaches adopted to evaluate the deviations from the ideal gas behavior of H_2O and CO_2 as well as the effects of steam condensation during the isenthalpic expansion of Bocca Grande and Bocca Nuova fluids, namely, the EOS of Gallagher et al. (1993), the GERG-2008 EOS together with the virial EOS, and the Peng-Robinson EOS. It turned out that deviations from ideality cannot be neglected at temperatures higher than 195-210°C and pressures greater than 15-30 bar approximately. The Peng-Robinson EOS was also used to compute the fugacity coefficients not only of H_2O and CO_2 but also of H_2, CH_4, and CO in the gas mixtures with CO_2 mole fraction of 0.05, 0.10, 0.20, 0.30, and 0.40. Calculations were performed separately for the ternary mixtures H_2O-CO_2-H_2, H_2O-CO_2-CH_4, and H_2O-CO_2-CO, adopting the mean concentrations of H_2, CH_4, and CO of Solfatara fumarolic fluids and computing the H_2O mole fraction by difference to unity. Fugacity coefficients were calculated (i) at the P,T conditions of the isenthalpic expansion path, (ii) at the P,T conditions fixed by coexistence of the vapor phase with a brine containing either 21 wt% or 33.5 wt% NaCl, and (iii) for the linear P-T decompression path, that is, assuming that total fluid pressure and temperature increase linearly from the critical point of water to the values of the magmatic endmember M1 of Caliro et al. (2014), 2879 bar and 1120°C. In all these decompression paths, the fugacity coefficients of non-polar gases increase with increasing P,T deviating gradually from unity, whereas the fugacity coefficient of H_2O decreases with increasing P,T, departing progressively from one. The ensuing practical implication is that the analytical mole fraction ratios of carbon gases may be utilized in geothermometric-geobarometric functions without incurring excessive errors, whereas use of the analytical H_2/H_2O mole fraction ratio leads to significant errors in computed equilibrium temperatures and pressures.

L. Marini et al., *The Solfatara Magmatic-Hydrothermal System*, Advances in Volcanology,
https://doi.org/10.1007/978-3-030-98471-7_4

1 The Outlet Temperature and Related Implications

The chronogram of Fig. 1, which is based on the data of Chiodini and coworkers (Chiodini et al. 2021 and references therein) shows that:

- From June 1983 to October 2020, the outlet temperature of the Bocca Grande fumarole varied between 150.0 and 165.4 °C around a mean of 161.4 °C, a median of 162.0 °C, with a standard deviation of 2.5 °C.
- From March 1995 to October 2020, the outlet temperature of the Bocca Nuova fumarole oscillated between 135.2 and 154.3 °C around a mean of 145.4 °C, almost coinciding with the median, 145.3 °C, with a standard deviation of 2.3 °C.

The outlet temperature of the sampled Pisciarelli fumarole (Chiodini 2015) ranged between 90.5 and 105.6 °C from March 1999 to September 2010, around a mean of 97.1 °C, a median of 96.0 °C, with a standard deviation of 3.3 °C. The major fumarolic vent of Pisciarelli, called Soffione, appeared in 2009 and has outlet temperature of 114 °C (Tamburello et al. 2019; see Sect. 3.2 in chapter "The Magmatic–hydrothermal System Hosted in the Campi Flegrei Caldera with Emphasis on the Solfatara").

In spite of the variable frequency of data acquisition and the availability of a single datum from August 1989 to November 1993, the chronogram of Fig. 1 shows that the Bocca Grande temperature experienced limited variations before 1998, whereas significant changes were recorded afterwards, possibly because of the different measurement method.

Since the Bocca Grande and Bocca Nuova fumaroles discharge superheated vapors, the outlet temperature of these two fumarolic vents can be used to evaluate the temperature and pressure of the saturated vapor zones, moving backwards (from surface to depth) along the fluid flowpath and assuming isenthalpic decompression. As already recalled in Sect. 5 in chapter "The Magmatic–Hydrothermal System Hosted in the Campi Flegrei Caldera with Emphasis on the Solfatara", this exercise was done by Cioni et al. (1984) who used the Mollier's diagram of pure water (Fig. 2) to infer the presence of a saturated vapor zone below the Bocca Grande fumarole at temperature of ~236 °C and pressure of ~31 bar, corresponding to the conditions of maximum enthalpy of saturated steam. Temperature and pressure of the saturated vapor zone below the Bocca Nuova fumarole are expected to be

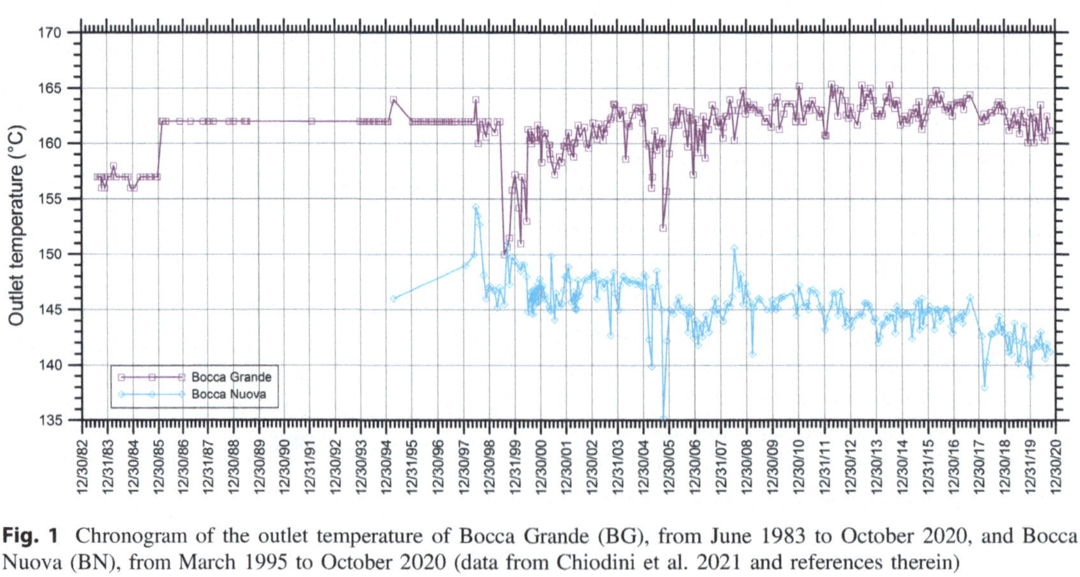

Fig. 1 Chronogram of the outlet temperature of Bocca Grande (BG), from June 1983 to October 2020, and Bocca Nuova (BN), from March 1995 to October 2020 (data from Chiodini et al. 2021 and references therein)

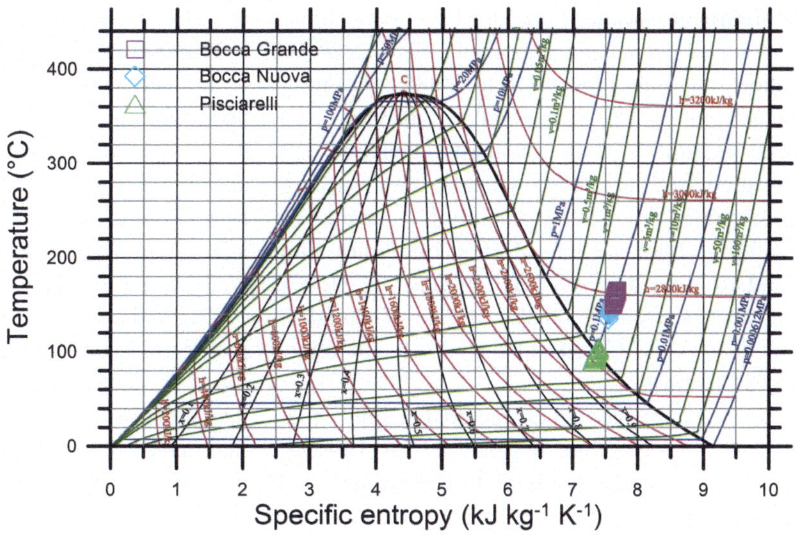

Fig. 2 Temperature-entropy Mollier's diagram of pure water showing the Solfatara fumaroles Bocca Grande, BG, and Bocca Nuova, BN (data from Chiodini et al. 2021 and references therein) and Pisciarelli, Pi (data from Chiodini 2015). Reprinted from Kim et al. (2020), Copyright © 2020, with permission from Elsevier

somewhat lower than those of the Bocca Grande, although these different P,T conditions might be due to loss of heat during steam expansion from the saturated vapor zone to the surface outlet. Heat losses are probably larger at Pisciarelli, as suggested by the position of the Pisciarelli fumarolic fluids close to the saturated vapor curve.

Two points related to the thermodynamic properties of Solfatara (mainly Bocca Grande) fumarolic fluids deserve further consideration.

First, the point of maximum enthalpy of steam represents a special condition, which is met only by the steam of enthalpy equal to 2803 kJ/kg (Fig. 2). In fact, moving along this isenthalpic path, steam attains saturation at 236 °C, 31.2 bars, but becomes superheated both at higher and lower temperatures and pressures. In contrast, the steam of enthalpy <2803 kJ/kg, initially superheated at 1 bar, becomes saturated at temperature <236 °C upon isenthalpic compression, but further isenthalpic compression determines partial condensation until the saturation curve is met again at temperature >236 °C. For example, the steam with enthalpy of 2700 kJ/kg becomes saturated at 162–163 °C, upon isenthalpic compression, and encounters again the saturation curve at 294 °C, upon further isenthalpic compression. This means that it is possible to recognize both a lower saturation temperature and an upper saturation temperature for any steam with enthalpy <2803 kJ/kg.

Second, the pure steam hypothesis is not completely satisfactory for the superheated fumarolic fluids discharged at Solfatara, because their H_2O concentration is much lower than 1,000,000 μmol/mol, varying from 689,200 to 867,400 μmol/mol at Bocca Grande, around a mean of 793,800 μmol/mol and a median of 804,600 μmol/mol, with a standard deviation of 42,800 μmol/mol, and oscillating from 688,900 to 873,100 μmol/mol at Bocca Nuova, with a mean of 792,400 μmol/mol, a median of 785,500 μmol/mol, and a standard deviation of 45,800 μmol/mol, based on the analytical data of Chiodini and coworkers (Chiodini et al. 2021 and references therein).

This departure from pure steam is due to the considerable CO_2 concentration, from 130,300 to 306,000 μmol/mol at Bocca Grande, with a mean of 203,700 μmol/mol, a median of 193,000 μmol/mol, and a standard deviation of 42,400 μmol/mol, and from 125,000 to 308,700 μmol/mol at Bocca Nuova, with a mean of 205,400 μmol/mol, a median of 212,500 μmol/mol, and a standard deviation of 45,500 μmol/mol, based on the analytical data of Chiodini and coworkers (Chiodini et al. 2021 and references therein).

Hydrogen sulfide is the third gas constituent in order of decreasing importance, with concentration ranging from 960 to 3988 μmol/mol, mean of 1520 μmol/mol, median of 1420 μmol/mol, and standard deviation of 459 μmol/mol at Bocca Grande, and varying from 602 to 5014 μmol/mol, mean of 1150 μmol/mol, median of 1060 μmol/mol, and standard deviation of 401 μmol/mol at Bocca Nuova, based on the analytical data of Chiodini and coworkers (Chiodini et al. 2021 and references therein).

These compositional characteristics imply that H_2O and CO_2 together constitute 99.3 to 100 mol % of the fumarolic fluids of Bocca Grande and Bocca Nuova. Therefore, it is reasonable to assume that these fluids are binary mixtures of H_2O and CO_2, neglecting the other constituents which are present in lower concentrations, in order to evaluate the lower and upper saturation temperatures possibly met by these fluids during the isenthalpic decompression from high pressure, temperature conditions to the surface discharge points.

This approach was already adopted by Moretti et al. (2018) for a binary gas mixture made up of 80 mol% H_2O and 20 mol% CO_2. Considering the thermodynamic properties calculated by Gallagher et al. (1993) for the H_2O-CO_2 system (see below for further details), Moretti et al. (2018) recognized that condensation cannot occur during the isenthalpic expansion of the 80:20 H_2O-CO_2 gas mixture with specific enthalpies higher than 48.3kJ/mol and claimed that the post-1984 subsidence results from single-gas isenthalpic decompression, from the base of the hydrothermal system upward. However, the specific enthalpy of the 80:20 H_2O-CO_2 gas mixture is 47.18 kJ/mol, at 161.7 °C and atmospheric pressure (i.e., at the mean/median T, P conditions of Bocca Grande), and 46.47 kJ/mol, at 145.35 °C and atmospheric pressure (i.e., at the mean/median T, P conditions of Bocca Nuova), based on the thermodynamic properties of the H_2O-CO_2 system computed by Gallagher et al. (1993). Therefore, some adjustments are needed to reconcile the single-gas isenthalpic expansion (as suggested by Moretti et al. 2018) with the outlet temperature evidence.

In geothermal production wells, constant operation for a long time leads to achievement of thermal equilibrium with negligible heat losses and the consequent occurrence of isenthalpic flow (Palacio 1990). However, heat losses might affect fluid properties at low flow rates and if the well has not reached steady state.

Giggenbach (1987) investigated the magmatic-hydrothermal system of White Island (New Zealand), and suggested that the rapid decompression of volcanic gases from the magma body to the fumaroles, with outlet temperatures of 100–800 °C, is likely to approach a free, isenthalpic expansion, and that conductive cooling of ascending gases is likely to be negligible for systems having reached thermal equilibrium or steady-state conditions.

Starting from the Giggenbach's hypotheses, Stevenson (1993) modeled the steady-state flow of an ideal, steam-like volcanic gas in: (1) a rough, cylindrical, constant area conduit, with conductive heat loss through the walls; and (2) a similar conduit, but filled with spheres, thus representing the flow in a porous pipe. Model (1) applies to isolated vents only, because groups of vents will have a tendency to insulate each other, whereas model (2) is valid for fumarolic fields. The boundary conditions assumed in the two models are magmatic temperature at the base of the conduit, atmospheric pressure at its surface end, and an ambient geotherm at a given distance from the conduit. Two ambient geotherms were taken into account by Stevenson (1993): (a) a linear gradient from atmospheric to magmatic temperatures and (b) a hydrothermal geotherm, involving water saturation temperatures and a hydrostatic pressure gradient. It was found that outlet temperature decreases with increasing conductive losses, increasing depth of the magmatic body, decreasing conduit size, decreasing mass flowrate, and decreasing distance to ambient temperatures. The porous pipe model was applied to the fumarole fields of Poás and Vulcano Island to evaluate the maximum depth of the magmatic source. Computed pressures at the base of the conduit, that is, at the top of the magma body, resulted to be close to the atmospheric value, suggesting that most gas decompression occurs as the gas leaves the magma. Stevenson (1993) reiterated that gas decompression is likely to approximate the isenthalpic condition, with the gas making nil to negligible work, but this statement remains a hypothesis instead of a proven conclusion.

Considering the high flux of hydrothermal fluids ascending below the Solfatara crater (e.g., Chiodini et al. 2001; Cardellini et al. 2017) and the persistence of this process at least since Roman times (see Chapter "Historical Data"), the attainment of steady state appears .to be very likely in the Solfatara fumarolic system. Nevertheless, the possible occurrence of cooling processes at shallow depths, such as the addition of cold external waters to the rising hydrothermal fluids, especially during the rainy seasons, as well as the heat transfer from hot deep rocks to the ascending hydrothermal fluids cannot be completely excluded. If so, the decompression of hydrothermal fluids would depart from the isenthalpic condition. In spite of the possible occurrence of these interfering processes, the isenthalpic decompression represents a useful reference model for the application of geothermometric-geobarometric techniques.

Summing up, recognizing to Moretti et al. (2018) the merit of having extended the approach of Cioni et al. (1984) to the H_2O-CO_2 system, it is worthwhile to broaden the discussion of this topic, first of all recalling the most important characteristics of the H_2O-CO_2 binary system in the relevant ranges of temperature, pressure and composition.

2 The H_2O-CO_2 Binary System

2.1 Experimentally Measured P-T-X and P-T-X-V Data of CO_2-H_2O Mixtures

Phase relations in the binary H_2O-CO_2 system have been established based on the results obtained by a relatively limited number of experimental studies, in which the gas and liquid phases coexisting at different temperatures and pressures were analyzed for their composition. Khitarov and Malinin (1958) reported preliminary results in the temperature interval 200–330 °C and pressures up to 588 bar. The detailed data of Malinin (1959) showed, for the first time, that the critical temperatures of the binary H_2O-CO_2 system are lower than the critical temperature of pure H_2O. Tödheide and Franck (1963) measured the composition of coexisting equilibrium vapors and liquids at temperatures of 50, 100, 150, 200, 250, 260, 265, 267, 268, 270, 275, 300, and 350 °C and pressures generally ranging from 200 to 3500 bar and found that the critical curve of the binary H_2O-CO_2 system extends from the critical point of pure water to the minimum critical temperature of 266 °C, where the critical pressure is 2450 bar and the critical CO_2 mole fraction is 0.415. Then, the critical curve bends back to higher temperatures, as the pressure continues to rise, as shown in Fig. 3. Therefore, at temperatures in the range 266–374 °C and pressures ≤ 2450 bar, the isothermal bi-phase envelops[1] are closed and sub-circular in shape, whereas at lower temperatures and higher pressures, the isothermal bi-phase envelops are open or undefined upwards.

Takenouchi and Kennedy (1964) studied the binary H_2O-CO_2 system in the temperature interval 110 to 350 °C providing detailed results up to pressures of 1600 bar and preliminary outcomes up to 2500 bars. The data of Takenouchi and Kennedy (1964) qualitatively confirm that the critical curve of the binary H_2O-CO_2 system moves from the critical point of pure water towards lower temperatures with only a minor pressure dependence, passes through the minimum temperature of 265 °C at 2150 bar and extends to high pressures with only a minor temperature increment, as already recognized by Tödheide and Franck (1963).

[1] The critical point divides each isothermal bi-phase envelop into the bubble-point curve, comprising the points where the first drop of the liquid phase begins to vaporize, and the dew-point curve, encompassing the points where the first drop of the gaseous mixture begins to condense.

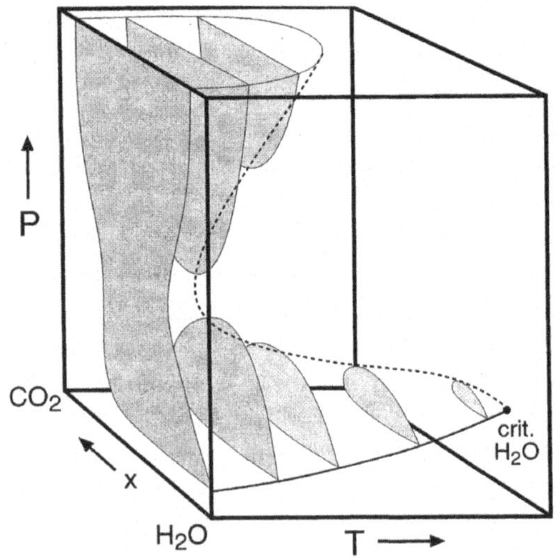

Fig. 3 Qualitative P-T-X diagram of the binary H_2O-CO_2 system (from Tödheide and Franck 1963) showing some isothermal bi-phase envelops (shaded) within the high-temperature portion of the immiscibility region and the upper critical curve (dashed line) ending at the critical point of H_2O. Reprinted from Bakker and Diamond (2000), Copyright © 2000, with permission from Elsevier

Nevertheless, some of the data of Takenouchi and Kennedy (1964) differ considerably from those of Tödheide and Franck (1963) as shown by the comparison of the isothermal bi-phase envelops of both studies (Fig. 4). The discrepancy between these two datasets was later resolved by Mather and Franck (1992) through the determination of the dew points for three mixtures (53.0, 62.5, and 78.9 mol % CO_2) at temperatures of 225 to 275 °C and pressures from 1140 to 3110 bar, thus substantiating the data of Tödheide and Franck (1963).

Furthermore, the results of Mather and Franck (1992) are supported by the phase transition temperatures measured by Sterner and Bodnar (1991) using the synthetic fluid inclusion technique. Sterner and Bodnar (1991) synthesized fluid inclusions of known compositions at selected temperatures and pressures in the single-phase region. The bulk molar volumes of the mixtures were obtained from determinations of the temperature at which CO_2 liquid and CO_2 vapor homogenize in the presence of liquid H_2O, together with the calculated molar volumes and compositions of the coexisting phases at this temperature. Pressures on the solvus were obtained indirectly based on both the measured homogenization temperatures and the isochores calculated using a modified Redlich-Kwong equation of state (see Sect. 2.2). Thus, the data of Sterner and Bodnar (1991) describes the P–T-X-V of the solvus over the whole compositional range, whereas Tödheide and Franck (1963), Takenouchi and Kennedy (1964), and Mather and Franck (1992) provided P–T-X descriptions of the solvus, as their techniques were not designed to measure molar volumes. However, the conclusions of Mather and Franck (1992) were questioned by Blencoe et al. (2001) based on the experimental data acquired using a vibrating U-tube technique[2] to constrain the high-pressure liquid–vapor equilibrium boundary curve at high temperatures. In fact, the experimental results obtained by Blencoe et al. (2001) are in excellent agreement with those of Takenouchi and Kennedy (1964) and Seitz and Blencoe (1997), but are highly discordant with those of Tödheide and Franck (1963) and Sterner and Bodnar (1991).

Volumetric data for H_2O-CO_2 mixtures were directly determined among the others by Franck and Tödheide (1959), Greenwood (1969), Gehrig (1980), Zakirov (1984), Patel et al. (1987) and Patel and Eubank (1988), Abdulagatov et al. (1995), Fenghour et al. (1996), and Seitz and Blencoe (1999).

Franck and Tödheide (1959) determined the compressibility factor, which is defined as:

$$Z_{mixt} = \frac{PV_{mixt}}{RT}, \tag{1}$$

for four different H_2O-CO_2 mixtures, of H_2O mole fraction 0.2, 0.4, 0.6, and 0.8, at

[2] In this experimental technique, first, an isobaric-isothermal, physically isolated and chemically homogeneous CO_2-H_2O sample of known composition is produced. Fluid pressure is then lowered slowly at constant temperature, recording the pressure readings and the corresponding values of the period of vibration of the U-tube, τ, at 1 or 2 bar intervals. When the single-phase fluid separates into two phases (liquid + vapor), a distinct inflection is observed in the P vs. τ curve.

Fig. 4 Isothermal pressure-composition diagrams showing the experimental data of Tödheide and Franck (1963) and Takenouchi and Kennedy (1964) as well as the results of thermodynamic modeling (see legend) for the binary H_2O-CO_2 system at **a** 300 °C and **b** 270 °C. Reprinted from Bakker and Diamond (2000), Copyright © 2000, with permission from Elsevier

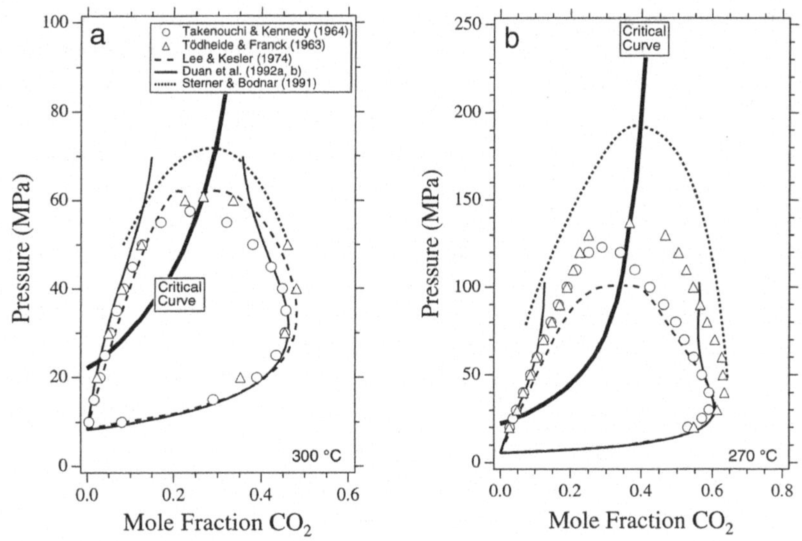

temperatures from 400 to 750 °C, and pressure from 300 to 2000 bar. Data were obtained by injecting H_2O and CO_2 into an autoclave of known volume and measuring pressure as a function of temperature. The errors on the obtained compressibility factors are considered by the authors to be less than 2.5%.

The approach of Franck and Tödheide (1959) was adopted by Greenwood (1969) to determine the compressibility factor of gas mixtures of H_2O and CO_2 in the pressure range 50 to 500 bar and the temperature range 450 to 800 °C. Values of Z_{mixt} are reported for mol fraction of CO_2 varying from 0 to 1 at 0.1 steps. The author underscored that in these experiments it is difficult to know the exact volume of the autoclave and the exact amount of gas it contains during the run.

Gehrig (1980) obtained volumetric data for CO_2-H_2O mixtures using a volume-calibrated piston-cylinder device. After injection of the fluid mixture into the apparatus, pressure was measured as a function of temperature at constant cell volume. The transition from the initial bi-phase to a single-phase condition causes a discontinuity of the measured isochoric temperature-pressure curves. The smoothed and interpolated molar volume data are reported for nine CO_2-H_2O mixtures at pressures from 150 to 600 bar and temperatures from 400 to 500 °C. These volumetric determinations have an uncertainty of ca. ±0.4% according to the author.

Molar volume and excess molar volume data for CO_2-H_2O mixtures were measured by Zakirov (1984) at 300 and 400 °C and pressures from 50 to 1000 bar. Data were obtained along each isotherm for known amounts of CO_2 and H_2O using a piezometer of known volume and determining the pressure. The overall relative error of measurement was estimates to be ±0.2% by the author.

Vapor-phase density was determined by Patel et al. (1987) and Patel and Eubank (1988) for H_2O-CO_2 mixtures with H_2O mole fractions of 0.02, 0.05, 0.10, 0.25, and 0.50, by using a Burnet-isochoric apparatus, at temperatures in the range 50–225 °C, at 25 °C interval, and pressures in the range 0.27–103.4 bar or the dew point at each temperature. Experimental dew points were published by Patel et al. (1987).

A constant-volume piezometer immersed in a precision thermostat was utilized by Abdulagatov et al. (1995) to measure the molar volume of pure H_2O and three H_2O-CO_2 mixtures, with H_2O mole fraction 0.40, 0.60, and 0.80, at 400 °C and pressure from 100 to 400 bar. The reported uncertainties are ±0.01 °C on temperature, ±0.15% on pressure and density, and ±0.002 mol fraction units on composition.

Fenghour et al. (1996) measured the density of eleven H_2O-CO_2 mixtures with H_2O mole fractions ranging from 0.0612 to 0.7913 in the temperature range 142–426 °C, at pressures from 59 to 346 bar, using a fully-automated isochoric instrument, into which known quantities of H_2O and CO_2 were introduced and pressure measured as a function of temperature. The dew points of these H_2O-CO_2 mixtures were also determined. The uncertainty in density was estimated by the authors to be ±0.08% for the CO_2-rich mixtures and ±0.14% for the H_2O-rich mixtures.

Seitz and Blencoe (1999) utilized a high-pressure, high-temperature vibrating-tube densimeter to determine the densities of pure CO_2 and CO_2-H_2O mixtures at 400 °C and pressures from 100 to 1000 bar. The accuracy was estimated to be ±0.2 bar for pressure, ±0.05 °C for temperature, and from ±0.0005 to 0.0010 g/cm^3 for density, at low and high values, respectively. Excess molar volumes resulted to be positive at all pressures and to increase with pressure below 300 bar but to decrease with pressure above 300 bar. Excess molar volumes close to 300 bar represent 37% of the total volume.

2.2 The Dimensionless Thermodynamic Properties of CO_2-H_2O Mixtures

Among the research works mentioned in the previous section, only Patel et al. (1987) and Patel and Eubank (1988) used the experimental vapor-phase densities of CO_2-H_2O mixtures (see previous section for experimental details) to obtain the dimensionless thermodynamic properties, that is, internal energy, enthalpy, entropy, Helmholtz free energy, and Gibbs free energy, as well as the compressibility factor, and the interaction second and third virial coefficients of the truncated virial EOS, which is discussed in Sect. 3.4. Each thermodynamic property was calculated using the corresponding residual property, which is defined as the difference between the property value for the real fluid, at the temperature and density of the fluid, and the property value for the hypothetical perfect gas state, at the reference temperature, T_{ref}, and reference pressure, P_{ref}, following Hall et al. (1983). Thus, for a generic property, Ξ, the corresponding residual property is defined as:

$$\Xi - \Xi^*_{ref} \equiv \Xi(T, \rho) - \Xi^*(T_{ref}, P_{ref}/RT_{ref}), \quad (2)$$

where the asterisk indicates the perfect gas state. The path chosen for integration is:

$$\Xi(T, \rho) \rightarrow \Xi^*(T, 0) \rightarrow \Xi^*(T_{ref}, 0) \rightarrow \Xi^*(T_{ref}, P_{ref}/RT_{ref}).$$

The residual internal energy and the residual entropy were first determined by the relations:

$$\frac{U - U^*_{ref}}{RT} = \frac{1}{T}\int_0^{\rho}\left(\frac{\partial Z}{\partial 1/T}\right)_{\rho}\frac{d\rho}{\rho} + \frac{1}{T}\int_{T_{ref}}^{T}\frac{C^*_v}{R}dT \quad (3)$$

and

$$\frac{S - S^*_{ref}}{RT} = \int_0^{\rho}\left[1 - Z + \frac{1}{T}\left(\frac{\partial Z}{\partial 1/T}\right)_{\rho}\right]\frac{d\rho}{\rho} + \int_{T_{ref}}^{T}\frac{C^*_v}{R}\frac{dT}{T} - \ln\frac{\rho RT_{ref}}{P_{ref}}. \quad (4)$$

Afterwards, the other residual properties were computed by using:

$$\frac{H - H^*_{ref}}{RT} = \frac{U - U^*_{ref}}{RT} + Z - \frac{T_{ref}}{T}, \quad (5)$$

$$\frac{A - A^*_{ref}}{RT} = \frac{U - U^*_{ref}}{RT} - \frac{S - S^*_{ref}}{R} - \frac{S^*_{ref}}{R}\left(1 - \frac{T_{ref}}{T}\right) \quad (6)$$

$$\frac{G - G^*_{ref}}{RT} = \frac{H - H^*_{ref}}{RT} - \frac{S - S^*_{ref}}{R} - \frac{S^*_{ref}}{R}\left(1 - \frac{T_{ref}}{T}\right). \quad (7)$$

The only integrals needed to compute these residual properties are $\frac{1}{T}\int_0^{\rho}\left(\frac{\partial Z}{\partial 1/T}\right)_{\rho}\frac{d\rho}{\rho}$, $\int_0^{\rho}(Z-1)\frac{d\rho}{\rho}$, $\frac{1}{T}\int_{T_{ref}}^{T}\frac{C_v^*}{R}dT$, and $\int_{T_{ref}}^{T}\frac{C_v^*}{R}\frac{dT}{T}$. The first two call for simple applications of the EOS, while the latter two involve only perfect-gas isochoric specific heats. The isochoric perfect-gas specific heats were taken from Angus et al. (1976) for pure CO_2 and from Haar et al. (1984) for pure H_2O. All residual properties are based upon a reference state for which both H^*_{ref} and S^*_{ref} are set to zero for P_{ref} = 1 kPa and T_{ref} = 273.16 K. Therefore, $G^*_{ref} = 0$, $U^*_{ref} = -RT_{ref}$, and $A^*_{ref} = -RT_{ref}$. The estimated accuracies are 0.05% for densities and compressibility factors, 0.15% for internal energies and enthalpies, 0.17% for entropies, and 0.20% for Helmholtz and Gibbs free energies. Patel and Eubank (1988) reported in separate tables the experimental densities and the derived energy properties for the five CO_2-H_2O mixtures and prepared a pressure-enthalpy diagram for each mixture, showing entropy, density and temperature.

2.3 The Excess Molar Enthalpy of CO_2-H_2O Mixtures

Wormald and coworkers and Bottini and Saville (1985) performed experimental measurements by flow calorimetry to obtain the excess molar enthalpy of binary CO_2-H_2O mixtures. The approach adopted in these experimental measurements comprises an isothermal Joule–Thomson expansion to a pressure lower than water saturation pressure, followed by an enthalpy-of-mixing experiment at this pressure to relate the molar enthalpy of the mixture to that of the pure components (Bottini and Saville 1985). The excess molar enthalpy of CO_2-H_2O mixtures was then computed based on the measured molar enthalpy changes and the isothermal Joule_Thomson coefficient (also known as isothermal throttling coefficient) of H_2O and CO_2 (Bottini and Saville 1985).

The following relation was obtained by Bottini and Saville (1985) for computing the excess molar enthalpy of H_2O-CO_2 mixtures (H^E_{mix} in J/mol, P in MPa, T in K):

$$H^E_{mix} = P\cdot\begin{pmatrix}47576.48 - 227.8140\cdot T + 0.3652747\\ \cdot T^2 - 0.0001949688\cdot T^3\end{pmatrix}\cdot X_{H_2O}\cdot(1-X_{H_2O}) + P\cdot(-802.6036 + 1.346434\cdot T)\cdot X_{H_2O}\cdot(1-X_{H_2O})\cdot(1-2X_{H_2O}) + P^2\cdot(424.1407 - 0.6771757\cdot T)\cdot X_{H_2O}\cdot(1-X_{H_2O}). \quad (8)$$

Equation (8) is based on 90 experimental data acquired using a flow mixing calorimeter at temperatures from 247 to 347 °C, pressures from 1.36 to 4.54 MPa, and H_2O mol fractions from 0.229 to 0.796 and, therefore, applies to these conditions.

Wormald and coworkers (Wormald et al. 1983 1986 1997; Smith and Wormald 1984; Wormald and Lloyd 1994; Lancaster and Wormald 1990) performed several measurements, again using flow mixing calorimeters, to obtain the excess molar enthalpy of CO_2-H_2O mixtures. Among these studies, those of Smith and Wormald (1984) and Wormald et al. (1986) are of interest for the Solfatara fumaroles. Smith and Wormald (1984) report 40 data acquired at atmospheric pressure, temperatures of 90.25, 102.05, 110.05, and 119.45 °C and X_{H_2O} varying from 0.350 to 0.683. Wormald et al. (1986) provide 152 data measured at temperatures from 175 to 425 °C and pressures from 0.4 to 13.4 MPa, whereas X_{H_2O} was varied from 0.3 to 0.7 in 12 cases but was fixed at 0.5 in the other 140 cases.

In this work, we processed the experimental results of Smith and Wormald (1984) and Wormald et al. (1986) to obtain two simple relations applicable to the Solfatara fumarolic fluids, one at atmospheric pressure, the other at water saturation pressure. First, simple log–log plots of H^E_{mix} vs. pressure were prepared and used to obtain the H^E_{mix} values at atmospheric pressure and different temperatures, in the interval 125–200 °C, and at water saturation pressure and distinct temperatures, in the range 175–300 °C, through linear fitting of the experimental data at constant temperature and X_{H_2O} equal to 0.5. Second, assuming that the excess molar enthalpy of H_2O-CO_2 mixtures is a symmetrical function of composition, the H^E_{mix} values thus obtained were fitted against the equation:

$$H^E_{mix} = \alpha \cdot X_{H_2O} \cdot (1 - X_{H_2O}) \quad (9)$$

to compute the coefficient α at different temperatures for both atmospheric pressure and water saturation pressure. Third, the decimal logarithms of coefficient α were regressed against temperature and the following two relations were finally obtained (T in K, H^E_{mix} in J/mol):

$$H^E_{mix} = 10\wedge(1.126434E-4 \cdot T^2 - 0.1047675 \cdot T + 25.95471) \cdot X_{H_2O} \cdot (1 - X_{H_2O}) \quad (10)$$

$$H^E_{mix} = 10\wedge(6.317364E-3 \cdot T + 0.1340958) \cdot X_{H_2O} \cdot (1 - X_{H_2O}). \quad (11)$$

Equation (10) is valid at atmospheric pressure in the temperature interval 125–200 °C, while Eq. (11) holds true at water saturation pressure in the temperature range 175–325 °C.

3 An Excursus on the Equations of State (EOS)

A general review of EOS is beyond the scope of this work and can be found elsewhere, such as in the first chapter of the textbook of Walas (1985), which explains several aspects of different EOS, and in the first volume of the monumental work of Partington (1949), which is largely devoted to gas properties and is very rich of historical documentations and quotations. Both sources do not cover the most recent advancements, which can be found in other works, such as the book of Span (2000) and the book chapter on multiparameter equations of state for pure fluids and mixtures written by Lemmon and Span (2010). All these sources were extensively used to prepare this section.

The state of one mole of a given substance is known when fixed values of any two of the three variables pressure P, temperature T, and molar volume V = M/ρ (where ρ is the density and M is the molecular weight of the substance) are assigned. Actually, only two variables have to be specified because the three variables are connected by a relation, which is called equation of state (EOS) or characteristic equation, f (P, V, T) = 0. Thermodynamic fundamental equations likewise are relations between certain groups of three properties—for instance, f (U, S, V) = 0—from which all other thermodynamic properties can be found, including the PVT relation. Conversely, the PVT information, in combination with knowledge of the ideal gas heat capacity as a function of temperature, also constitutes a fundamental equation.

By itself, an appropriate EOS can be utilized to estimate many important properties of pure substances and mixtures, including: (i) densities of liquid and vapor phases; (ii) vapor pressure; (iii) critical properties of mixtures; (iv) vapor–liquid equilibrium relations; (v) deviation of enthalpy from ideality; (vi) deviation of entropy from ideality.

3.1 The First Steps of an Incomplete History that Began Over 350 Years Ago

The EOS history began with the experiments performed by Robert Boyle with air (Boyle 1662), from which he deduced that the volume of a given mass of gas, at constant temperature, is inversely proportional to the pressure, that is, the so-called Boyle's law:

$$PV = \text{constant}. \quad (12)$$

It seems that the effect of temperature on the gas volume was first quantified by Jacques

Alexandre César Charles in 1787, although he did not published the results of his experiments. Actually, several scientists investigated this topic at that time. Alessandro Giuseppe Antonio Anastasio Volta was among them and published a paper in 1793 on an Italian scientific journal of limited circulation (Volta 1793). Joseph Louis Gay-Lussac seems to have been the first to refer the expansions to a standard initial volume and temperature, e.g. 0° C (Gay-Lussac 1802). He wrote that "*Charles had noticed for 15 years the same property in these gases, but never having published his results, it was by the greatest chance that I learned them*". Today, the proportionality between ΔV and ΔT at constant P, that is:

$$V = V_0(1 + kT), \tag{13}$$

is known as law of Charles or Charles and Gay-Lussac or Charles, Volta, and Gay-Lussac, depending on personal preference or nationality.

One of the first persons, if not the first, who combined the Boyle's and Gay-Lussac's laws in a single equation was Benoit-Pierre-Émile Clapeyron. However, Clapeyron attributed the volume-pressure law, Eq. (12), to the French scientist Edmé Mariotte rather than to Robert Boyle. In his famous memoir on the Carnot cycles (Clapeyron 1834), he wrote the combined equation as:

$$Pv = R(267 + t) \tag{14}$$

where t is the temperature in °C and v is the volume of the unit mass.

In 1850, Rudolf Clausius, using the experimental data of the French chemist Henri Victor Regnault, reevaluated the constant inside the parentheses and rewrote Eq. (14) as:

$$Pv = R(273 + t). \tag{15}$$

Clausius (1850), following Clapeyron, explicitly proposed that the combined equation be called the Mariotte–Gay-Lussac law. Both Clapeyron and Clausius utilized the volume per unit mass of gas rather than the volume per mole of gas in their relations. Therefore, R was different from one gas to another and was inversely proportional to the density of the gas, to a first approximation, as noted by Clausius.

Later, the gas-specific constant of Clapeyron and Clausius was converted into a universal gas constant by a Clausius' student, the German chemist August F. Horstmann, who in 1873 rewrote Eq. (15) as:

$$PV = RT \tag{16}$$

where V is the molar volume (Horstmann 1873). Equation (16) is the ideal gas law as we know and use it today, but the process that led to its discovery lasted more than two centuries, during which it was recognized that the ideal gas law is very often an approximation of the behavior of real gases. We recall that the deviations from the ideal behavior of real substances can be conveniently expressed by means of the residual volume

$$\Delta V = V - \frac{RT}{P}, \tag{17}$$

whose difference from zero measures the deviation from the ideal behavior, or the compressibility factor, given by Eq. (1), whose difference from unity is a measure of non-ideality.

3.2 The Van Der Waals's Equation

Deviations from the ideal gas law were attributed to the finite volumes occupied by the molecules themselves and to forces of repulsion and attraction between the molecules. Both these factors were quantitatively taken into consideration by the Dutch scientist Johannes Diderik van der Waals, who deduced the equation:

$$\left(P + \frac{a}{V^2}\right) \cdot (V - b) = RT \tag{18}$$

in his dissertation (van der Waals 1873).[3] In the van der Waals's equation, *a* and *b* are constants related to the molecular attractions and the finite size of the molecules, respectively. The parameter *b* is the effective volume of the molecules present in one mole of gas and the difference V − *b* represents the unoccupied or free molar volume, which is available for the movements of gas molecules. The ratio a/V^2 is a correction for molecular attractions, which are proportional to $1/V^2$, with the *a* coefficient representing a sort of proportionality constant. The most important features of van der Waals's equation are that:

(i) it applies to gaseous, liquid, and supercritical states, with the same constants *a* and *b*, and
(ii) it provides a semi-quantitative description of vapor–liquid phase equilibria.

However, it is only an approximate equation, and, as recognized by van der Waals himself, it deviates more and more from the experimental data with increasing pressure. In spite of its limitations, it had and still has a considerable practical interest and is likely to remain so, in that it also provides a qualitative survey of the main properties of gases and liquids where experimental data are lacking.

For gas mixtures, the parameters *a* and *b* of the van der Waals's equation are related to those of pure components by the combining rules:

$$a = \left(\sum_i X_i \cdot \sqrt{a_i}\right)^2 = \sum_i \sum_j X_i \cdot X_j \cdot \sqrt{a_i \cdot a_j} = \sum_i \sum_j X_i \cdot X_j \cdot a_{ij} \quad (19)$$

$$b = \sum_i X_i \cdot b_i. \quad (20)$$

Returning to consider a single component, the van der Waals Eq. (18) can be rearranged as follows:

$$V^3 - \left(b + \frac{RT}{P}\right)V^2 + \left(\frac{a}{P}\right)V - \frac{a \cdot b}{P} = 0. \quad (21)$$

Equation (21) is a cubic equation in V, has the three roots V_1, V_2, and V_3, and may be written as:

$$(V - V_1) \cdot (V - V_2) \cdot (V - V_3) = 0 \quad (22)$$

At the critical point, Eq. (21) becomes:

$$V^3 - \left(b + \frac{RT_c}{P_c}\right)V^2 + \left(\frac{a}{P_c}\right)V - \frac{a \cdot b}{P_c} = 0. \quad (23)$$

Since the three roots are equal at the critical point, $V_1 = V_2 = V_3 = V_c$, Eq. (22) becomes:

$$(V - V_c)^3 = V^3 - 3 \cdot V_c \cdot V^2 + 3 \cdot V_c^2 \cdot V - V_c^3 = 0 \quad (24)$$

By equating the coefficients of the same powers of V in Eqs. (23) and (24) and rearranging, it turns out that:

$$a = 3 \cdot P_c \cdot V_c^2 = \frac{27 \cdot R^2 T_c^2}{64 \cdot P_c} \quad (25)$$

$$b = \frac{V_c}{3} = \frac{RT_c}{8 \cdot P_c}. \quad (26)$$

The same results are obtained by differentiating Eq. (21) twice with respect to V, at constant T, and recalling that at the critical point, both $(\partial P/\partial V)_T$ and $(\partial^2 P/\partial V^2)_T$ are equal to zero (see Partington 1949 for further details), as required by the so-called condition of criticality. Equations (25) and (26) relate the parameters *a* and *b* of the van der Waals's equation and the universal gas constant R to the critical properties of a given substance.

[3] The main focus of van der Waals was to provide a description of critical phenomena (as shown by the Dissertation title) rather than to find an equation representing the imperfect behavior of real gases. Nevertheless, in 1910, the Nobel Prize in Physics was awarded to him "*for his work on the equation of state for gases and liquids*", see https://www.nobelprize.org/prizes/physics/1910/summary/ (last visit on March 27th, 2022).

Let us now consider the three relations:

$$a = 3 \cdot P_c \cdot V_c^2, \qquad b = \frac{V_c}{3}, \qquad \text{and} \qquad R = \frac{8 \cdot P_c V_c}{3 \cdot T_c}$$

and substitute them in the van der Waals's equation, namely in Eq. (18). The following relation is obtained through some rearrangements:

$$\left[\frac{P}{P_c} + 3\left(\frac{V_c}{V}\right)^2\right] \cdot \left(3\frac{V}{V_c} - 1\right) = 8\frac{T}{T_c}. \tag{27}$$

The ratios between pressure, volume, and temperature and the corresponding critical values, are the so-called reduced pressure $P_r = P/P_c$, reduced volume $V_r = V/V_c$, and reduced temperature $T_r = T/T_c$. Thus, Eq. (27) can be rewritten as:

$$\left[P_r + \frac{3}{V_r^2}\right] \cdot (3V_r - 1) = 8T_r, \tag{28}$$

as already noted by van der Waals himself, Eq. (28) is called the reduced van der Waals's equation, and is a general equation for all gases and liquids. In other words, it applies to any substance, if the proper values of P_r, V_r, and T_r are used, although deviations from experimental data become more and more important with increasing pressure. The reduced van der Waals's equation is the mathematical form of the so-called principle of corresponding states: *substances at equal reduced pressures and temperatures have equal reduced volumes.*

Related to the principle of corresponding states is the rule of dimensional similitude. According to it, any relation between physical variables can be rearranged as a relation involving a limited number of dimensionless ratios. Thus an EOS, $f_1(P, V, T, P_c, V_c, T_c) = 0$, is equivalent to another EOS, $f_2(P_r, V_r, T_r) = 0$. Also descending from the principle of corresponding states is the reference substance method (see Sect. 3.6), in which the properties of the substance of interest are related to those of a reference substance whose properties are well known.

Elaborating on the principle of corresponding states, Pitzer (1955) defined the acentric factor, ω, representing a measure of the non-sphericity of molecules. To this purpose, he proposed the following relation:

$$\omega = -\log P_r - 1.000, \tag{29}$$

in which P_r is the reduced vapor pressure at the reduced temperature $T_r = 0.7$. The acentric factor is zero for simple fluids (e.g., Ar, Kr, Xe, CH_4) and is very close to zero for spherical molecules. The acentric factor is higher than zero for non-spherical molecules, but only the most highly non-spherical have ω approaching unity. According to Pitzer, *any group of substances with equal values of the acentric factor should conform among themselves to the principle of corresponding states.*

Summing up, the van der Waals's equation has been a milestone in the EOS history and it is still a good example of a simple model incorporating some corrections to the ideal gas law for attractions and repulsions between molecules. Although references to it still appear in recent textbooks and papers, it is now out-of-date and, in fact, other EOS of not much greater complexity are far superior to it. This is the case of the cubic EOS, representing the subject of the next section.

3.3 The Cubic EOS

Most of the so-called "cubic equations of state" are empirical and arbitrary, with adjustable parameters obtained by fitting some experimental data such as vapor pressures, densities, or enthalpies. However, some statistical mechanical theory has been applied toward improving cubic EOS (e.g., Carnahan and Starling 1972; Donohue and Prausnitz 1978). Although knowledge of the behavior of liquids is less quantitative than that of gases, many EOS, such as the Peng-Robinson EOS also provide good representation of liquid densities. The parameters of cubic EOS can be related to critical properties but include changes for temperature and another property such as the critical compressibility or the acentric factor.

The van der Waals's equation was the starting point for the development of the cubic EOS, such as those proposed by Redlich and Kwong (1949), Soave (1972), and Peng and Robinson (1976a), which are explicit in pressure and third-degree in volume. Of the three roots, the largest one is that of the vapor phase, the smallest one corresponds to the liquid phase, and the intermediate one has no physical meaning. All three roots are the same at the critical point. As already noted above for the van der Waals's equation, an equation of at least the third degree in volume is needed to satisfy the condition of criticality, that is, $(\partial P/\partial V)_T = (\partial^2 P/\partial V^2)_T = 0$ at the critical point, and to permit representation of liquid and vapor phases. Similar to the van der Waals's equation, the cubic EOS satisfy the condition of criticality, are simple and effective, and are still used. In turn, the condition of criticality allows one to evaluate the two basic parameters a and b of these cubic EOS, as shown in Sect. 3.2 for the van der Waals's equation. In fact, several cubic equations, as the Redlich-Kwong EOS, the Soave EOS, and the Peng-Robinson EOS, have only two basic parameters, although other EOS have three or more parameters.

The Redlich-Kwong EOS—The Redlich-Kwong EOS is:

$$P = \frac{RT}{V - b} - \frac{a}{T^{0.5} \cdot V \cdot (V + b)}. \quad (30)$$

Redlich (1975) noted that "*there is no really good theoretical justification for the two changes in the denominator of the a-term but the agreement with observed data is considerably improved*" compared to the van der Waals's equation.

Again, by application of the criticality condition, the parameters a and b are linked to the critical properties of the gas by the relations:

$$a = \frac{\Omega_a R^2 T_c^{2.5}}{P_c} \quad (31)$$

$$b = \frac{\Omega_b R T_c}{P_c} \quad (32)$$

in which Ω_a and Ω_b are two coefficients used to improve the fit of data to the EOS. These two coefficients have been correlated to the reduced temperature and the acentric factor, which is defined by Eq. (29) in Sect. 3.2.

The third-degree polynomial in volume corresponding to the Redlich-Kwong EOS is:

$$V^3 - \frac{RT}{P} \cdot V^2 + \frac{1}{P}\left(\frac{a}{T^{0.5}} - bRT - Pb^2\right) \cdot V - \frac{a \cdot b}{PT^{0.5}} = 0. \quad (33)$$

The reduced form of the Redlich-Kwong EOS is:

$$P_r = \frac{3T_r}{V_r - 3\Omega_b} - \frac{9\Omega_a}{T_r^{0.5} V_r (V_r + 3\Omega_b)}. \quad (34)$$

The Redlich-Kwong EOS and its modified forms were discussed by Horvath (1974) and Tsonopoulos and Heidman (1985), but other studies focused on the so-called modified Redlich-Kwong EOS have appeared in the subsequent years. The original combining rules for the Redlich-Kwong parameters of mixtures were the same as those for the van der Waals's equation [i.e., Eqs. (19) and (20)], but different approaches were suggested afterwards (see Walas 1985 for details). Since the Redlich-Kwong EOS is not at all adequate for the liquid phase, it cannot be used by itself for modeling vapor–liquid equilibria.

Some EOS for CO_2-H_2O mixtures, like those of de Santis et al. (1974), Holloway (1977), later corrected by Flowers (1979), Shmulovich et al. (1980), Kerrick and Jacobs (1981), Bowers and Helgeson (1983), Flowers and Helgeson (1983),

Holland and Powell (1991), King et al. (1992), and Spycher et al. (2003) were obtained modifying the Redlich-Kwong EOS.

In particular, Holland and Powell (1991), proposed a compensated Redlich-Kwong equation of state, indicated by the acronym CORK EOS, which can be used to compute the volumes and fugacities of H_2O and CO_2 at pressures ranging from 1 bar to 50 kbar and temperatures varying from 100 to 1400 °C for H_2O and 100 to 1600°C for CO_2. The virial-type extension of this EOS has been conceived to: (a) balance the tendency of the modified Redlich-Kwong EOS to overestimate volumes at high pressures, and (b) describe the volume of coexisting vapor and liquid phases at saturation conditions, which is a known limitation of the Redlich-Kwong EOS, as already mentioned above.

Holland and Powell (1991) addressed the first of these two issues by addition of two virial-type coefficients to the volume computed by means of the Redlich-Kwong EOS, V_{RK}, according to the relation:

$$V = V_{RK} + (c_0 + c_1 \cdot T) \cdot \sqrt{(P - P^\circ)} + (d_0 + d_1 \cdot T) \cdot (P - P^\circ), \tag{35}$$

in which P° is the pressure at which the RK equation starts to overestimate the molar volume considerably. The value of P° can be determined from the available experimental data.

The second issue was addressed by Holland and Powell (1991) using a new coefficient, called a_{gas}, to describe the properties of the vapor phase below the critical temperature of water, instead of the a coefficient of the Redlich-Kwong EOS, whereas the a coefficient is used to describe the properties of the liquid phase below the critical temperature of water as well as at supercritieal conditions. In practice, the two coefficients a and a_{gas} converge at 695 K (422 °C) which is the critical temperature (where the liquid and gas phase are indistinguishable) according to the CORK EOS. This T_c value is 48 °C higher than the critical temperature of pure water, 647.3 K (374.15 °C).

Furthermore, the equation developed by Holland and Powell (1991) for CO_2 may be also utilized to calculate volumes and fugacities of CH_4, H_2, CO, N_2, O_2, and possibly other gases as well, adopting the principle of corresponding states. Calculations of volumes, fugacities and activities of H_2O, CO_2, CH_4, H_2, and CO can be performed using the CHO program written by T. J.B. Holland, which is available at the website https://filedn.com/lU1GlyFhv3UuXg5E9dbnWFF/TJBHpages/cho.html (last visit on March 27th, 2022).

On the one hand, the semi-empirical Redlich-Kwong EOS and its modified versions have the advantage that they can be extrapolated above the P–T range covered by the experimental data. On the other hand, there are some inconveniences in the use of these EOS, as noted by Spycher and Reed (1988). In fact, since the Redlich-Kwong equation is a cubic equation in volume, it is challenging to use it when only P and T are known. Furthermore, the computation of the derivatives of cubic EOS with respect to P, T, and X is not a simple exercise. When these EOS are used with gas mixtures, another issue arises due to the adoption of mixing rules which are controversial being empirical in nature.

The Soave EOS—The Soave EOS was obtained by Soave (1972) replacing the temperature-dependent term $a/T^{0.5}$ of the Redlich-Kwong EOS with a function involving the temperature and the acentric factor, ω:

$$P = \frac{RT}{V - b} - \frac{a(T, \omega)}{V \cdot (V + b)}. \tag{36}$$

The parameter $a(T, \omega)$ was mainly designed to make the equation fit the vapor pressure data of hydrocarbons and is defined by the relation:

$$a(T, \omega) = a \cdot \alpha = 0.42748 \frac{R^2 T_c^2}{P_c} \alpha. \tag{37}$$

with:

$$\alpha^{0.5} = 1 + \left(1 - T_r^{0.5}\right) \cdot \left(0.480 + 1.574\,\omega - 0.176\,\omega^2\right). \quad (38)$$

Graboski and Daubert (1978a, b, 1979) proposed to modify Eq. (38) as follows:

$$\alpha^{0.5} = 1 + \left(1 - T_r^{0.5}\right) \cdot \left(0.48508 + 1.55171\,\omega - 0.15613\,\omega^2\right) \quad (39)$$

and to use the following relation for hydrogen:

$$\alpha^{0.5} = 1.096 \cdot \exp(-0.15114 \cdot T_r). \quad (40)$$

For water, alcohols, and other strongly polar substances, Soave (1979) suggested to use the relation:

$$\alpha = 1 + (1 - T_r) \cdot (m + n \cdot T_r). \quad (41)$$

In Eq. (41), m and n are two adjustable parameters which must be derived from experimental vapor pressure data for each pure substance, following the indications of Soave (1979, 1980). The parameter b is given by the relation:

$$b = \frac{0.08664 \cdot RT_c}{P_c}. \quad (42)$$

The Soave EOS can be rewritten as a third-degree polynomial in volume:

$$V^3 - \frac{RT}{P} \cdot V^2 + \frac{1}{P}\left(a \cdot \alpha - bRT - Pb^2\right) \cdot V - \frac{a \cdot \alpha \cdot b}{P} = 0. \quad (43)$$

The partly reduced form of the Soave EOS is:

$$P_r = \frac{3\,T_r}{V_r - 0.2599} - \frac{3.8473 \cdot \alpha}{V_r(V_r + 0.2599)}. \quad (44)$$

See Walas (1985) for details on the mixing rules needed to apply the Soave EOS to mixtures.

The Peng-Robinson EOS—The Peng-Robinson EOS was implemented by Peng and Robinson (1976a) to achieve the following aims:

(i) To involve parameters expressible in terms of P_c, T_c, and the acentric factor, ω.
(ii) To obtain an improved performance near the critical point, especially for the critical compressibility, Z_c, and the density of the liquid phase.
(iii) To use mixing rules with only one binary interaction parameter, independent of temperature, pressure, and composition.
(iv) To develop an equation for calculating all fluid properties in natural gas processes.

The obtained Peng-Robinson EOS is:

$$P = \frac{RT}{V - b} - \frac{a(T)}{V \cdot (V + b) + b \cdot (V - b)}. \quad (45)$$

At the critical point:

$$a(T_c) = 0.45724 \frac{R^2 T_c^2}{P_c} \quad (46)$$

$$b(T_c) = \frac{0.07780 \cdot RT_c}{P_c} \quad (47)$$

$$Z_c = P_c V_c / RT_c = 0.307. \quad (48)$$

Similar to the Soave approach, the parameter a (T) at $T \neq T_c$ is modified as follows:

$$a(T) = a(T_c) \cdot \alpha(T_r, \omega) \quad (49)$$

with:

$$\alpha^{0.5} = 1 + \left(1 - T_r^{0.5}\right) \cdot \left(0.37464 + 1.5422\omega - 0.26992\omega^2\right). \quad (50)$$

Several applications of the Peng-Robinson EOS were described by Robinson et al. (1977). Application to multiphase equilibria was made by Peng and Robinson (1976b) and Robinson and Peng (1980). The Peng-Robinson EOS generally predicts liquid densities more precisely than the Soave equation, which is otherwise quite comparable in performance, including fugacity coefficients.

The Peng-Robinson EOS together with the mixing rules of Wong and Sandler (1992), as modified by Eubank et al. (1995), was used by Shyu et al. (1997) to describe phase equilibria in the CO_2-H_2O binary system from 25 to 350 °C and up to 1000 bar. A modified Peng-Robinson equation together with the mixing rule of Panagiotopoulos and Reid (1986) was also used by Bamberger et al. (2000) to reproduce their experimental solubility data at 50, 60, and 80 °C and pressures up to 141 bars.

3.4 The Virial EOS

The virial EOS expresses the molar compressibility factor, Z, of a gas as either an infinite power series in the reciprocal molar volume, 1/V:

$$Z = 1 + \frac{B}{V} + \frac{C}{V^2} + \frac{D}{V^3} + \ldots \qquad (51)$$

or an infinite power series in the density, ρ:

$$Z = 1 + B \cdot \rho + C \cdot \rho^2 + D \cdot \rho^3 + \ldots \qquad (52)$$

or an infinite power series in the pressure, P:

$$Z = 1 + B^* \cdot P + C^* \cdot P^2 + D^* \cdot P^3 + \ldots, \qquad (53)$$

with $B^* = B/R \cdot T$, $C^* = \left(C - B^2\right)/(R \cdot T)^2$, $D^* = \left(D - 3 \cdot B \cdot C + 2B^3\right)/(R \cdot T)^3$.

Since the representation of Z by infinite series is impossible in practice, virial equations truncated to two or three terms are usually adopted, although the range of applicability of the equation increases with the number of considered coefficients. For a given truncation, the power series in density is usually more accurate than the one in pressure. By convention, B (or B^*) is known as the second virial coefficient, C (or C^*) is called the third virial coefficient, D (or D^*) is termed the fourth, and so on, implying that the first virial coefficient is unity. In theory, for a given substance, all virial coefficients are independent of pressure and density and depend on temperature only. An exhaustive compilation of the experimental data on the second and third virial coefficients is given by Dymond et al. (2002a; b). Several theoretical methods for the evaluation of virial coefficients have been proposed (e.g., Walas 1985 and references therein). Lee et al. (1977) presented both generalized diagrams suggesting optimal truncation of the virial equation and generalized diagrams indicating probable errors associated with the most common truncations.

The virial EOS is the only one that has a rigorous theoretical basis, although this fact was understood about 40 years after it was first proposed by Thiesen (1885) on a merely empirical basis. The virial EOS was developed exhaustively by Heike Kamerlingh Onnes[4] (Kamerlingh Onnes 1901). Afterwards, a statistical-mechanical derivation of the virial EOS was provided by Ursell (1927), through the analysis of the forces between molecules, and further theoretical developments were given by Mayer and coworkers (Mayer 1937, 1939; Mayer and Ackermann 1937; Mayer and Harrison 1938; Harrison and Mayer 1938; Mayer and Mayer 1940).[5] Assuming a suitable mathematical form for the intermolecular potentials, it was thus shown that the virial coefficients are not empirical parameters, but are linked to non-ideal intermolecular forces, with the second virial coefficient B accounting for two body interactions, the third virial coefficient C accounting for three body interactions, and so on. Since the perfect gas has low enough pressure or density, all virial coefficients can be ignored and Z is unity.

In recent studies, thanks to the increased computing capacity of modern computers, virial

---

[4] Heike Kamerlingh Onnes (21 September 1853–21 February 1926) was a famous Dutch physicist who received the 1913 Nobel Prize in Physics for "*his investigations on the properties of matter at low temperatures which led, inter alia, to the production of liquid helium*".

[5] As noted by Benedict et al. (1940), Lennard–Jones and coworkers (e.g., Lennard–Jones 1937; Lennard–Jones and Devonshire 1937) adopted a less rigorous approach, leading to a single relation describing the thermodynamic properties of both liquid and gaseous phases.

coefficients were obtained from quantitative molecular models, such as ab initio-based molecular models and Gaussian charge polarizable models. Examples of applications to the CO_2-H_2O system are recalled in Sect. 4.2.

A valuable implication of the theoretical basis of the virial EOS is the exact relation between the virial coefficients of a mixture and those of the pure components. In fact, for a gas mixture of m components, the composition dependence of virial coefficients is known exactly from statistical mechanics, as expressed by the following relations (Prausnitz et al. 1999):

$$B_{mixt} = \sum_{i=1}^{m}\sum_{j=1}^{m} X_i \cdot X_j \cdot B_{ij} \tag{54}$$

$$C_{mixt} = \sum_{i=1}^{m}\sum_{j=1}^{m}\sum_{k=1}^{m} X_i \cdot X_j \cdot X_k \cdot C_{ijk}, \tag{55}$$

in which X_i, X_j and X_k are the mole fractions of the i-th, j-th and k-th components of the mixture, while B_{ij} and C_{ijk} are the second and third virial coefficients, respectively. A limitation of the virial EOS, even with a large number of terms, is its inability to reproduce the liquid state and the coexistence of liquid and vapor phases.

Several well-known equations of state may be recast into virial form. For instance, the virial form of the van der Waals's equation is:

$$Z = 1 + \left(b - \frac{a}{RT}\right) \cdot \rho + b^2 \cdot \rho^2 + b^3 \cdot \rho^3 + \ldots \tag{56}$$

Virial EOS in pressure truncated to the third term were developed by Spycher and Reed (1988) for pure H_2O, CO_2, H_2, and CH_4, as well as for H_2O-CO_2, H_2O-CH_4, and CO_2-CH_4 binary mixtures, with the B* and C* virial coefficients expressed as function of temperature. These B* and C* virial coefficients were used to calculate the fugacity coefficients of pure H_2O, CO_2, H_2, and CH_4, and of H_2O-CO_2-CH_4 mixed gases and to estimate the enthalpies of these gases. The pressure–temperature range of application is from 25 to 600 °C and up to 3000 bar for H_2 and from 16 to 350 °C and up to 500 bar for CH_4, whereas two pressure–temperature ranges were considered: below 350 °C, up to 500 bar, and from 450 to 1000 °C, up to 1000 bar. The virial EOS of Spycher and Reed (1988) does not reproduce accurately the PVT data close to the critical region. According to Spycher and Reed (1988), ideal gas mixing is a reasonable approximation for modeling boiling in hydrothermal and epithermal systems. However, deviations from ideal gas behavior occur at pressures much greater than the saturation pressure of pure water.

3.5 The Benedict-Webb-Rubin EOS and Its Modified Versions

The so-called Benedict-Webb-Rubin equation of state, often indicated by the acronym BWR EOS, is somewhat related to the virial equation being given by the relation (Benedict et al. 1940):

$$P = RT\rho + \left(B_0RT - A_0 - \frac{C_0}{T^2}\right) \cdot \rho^2 + (bRT - a) \cdot \rho^3 + a\alpha \cdot \rho^6 + \frac{c \cdot \rho^3}{T^2} \cdot \left(1 + \gamma \cdot \rho^2\right) \cdot \exp\left(-\gamma \cdot \rho^2\right). \tag{57}$$

where ρ indicates the molal density. The BWR EOS was conceived by Benedict et al. (1940) as a modification Beattie-Bridgeman EOS (Beattie and Bridgeman 1927):

$$P = RT\rho + \left(B_0RT - A_0 - \frac{Rc}{T^2}\right) \cdot \rho^2 + \left(-bB_0RT + aA_0 - \frac{cRB_0}{T^2}\right) \cdot \rho^3 + \frac{bcRB_0}{T^2} \cdot \rho^4 \tag{58}$$

to enable it to represent more accurately the properties of fluids at high densities. However, the BWR EOS was the first equation of state that involved an exponential term in density and can be thus considered the predecessor of almost all modern EOS. Values for the eight empirical parameters A_0, B_0, C_0, a, b, c, α, and γ of the BWR EOS were reported by Benedict et al.

(1951) for 12 hydrocarbons, methane through n-heptane. Several details and references are given in Walas (1985). A thorough treatment of all aspects of the reduced form of the BWR EOS, including the methods for determining the empirical parameters from experimental data and evaluation of thermodynamic properties, is given by Opfell et al. (1956).

Among the modifications of the BWR EOS, the most widely accepted is that of Starling (1973):

$$P = RT\rho + \left(B_0RT - A_0 - \frac{C_0}{T^2} + \frac{D_0}{T^3} - \frac{E_0}{T^4}\right) \cdot \rho^2 + \left(bRT - a - \frac{d}{T}\right) \cdot \rho^3 + \alpha \cdot \left(a + \frac{d}{T}\right) \cdot \rho^6 + \frac{c \cdot \rho^3}{T^2} \cdot (1 + \gamma \cdot \rho^2) \cdot \exp(-\gamma \cdot \rho^2), \tag{59}$$

which is often indicated by the acronym BWRS EOS. Three parameters were added (D_0, E_0, and d) to enhance the temperature dependence of the coefficients of ρ^2, ρ^3, and ρ^6. The book by Professor Kenneth E. Starling of the University of Oklahoma (Starling 1973) gives the values of the eleven parameters of the BWRS EOS for fifteen substances and the binary interaction parameters for mixtures. As recalled by Lemmon and Span (2010), other notable modifications of the BWR EOS were proposed by Bender (1973) and Jacobsen and Stewart (1973).

The BWR EOS equation and its modified versions yield a sufficiently accurate representation of the thermodynamic surface for typical technical applications in the gas phase and at supercritical states with low to medium densities, but computed energy at liquid or liquid-like supercritical states may be in error by more than 10% (Lemmon and Span 2010).

Benedict et al. (1940) presented also the following relation for the residual molar Helmholtz free energy, $\tilde{A}$:

$$\tilde{A} = \left(B_0RT - A_0 - \frac{C_0}{T^2}\right) \cdot \rho + \frac{bRT - a}{2} \cdot \rho^2 + \frac{a\alpha \cdot \rho^5}{5} + \frac{c \cdot \rho^2}{T^2}\left[\frac{1 - \exp(-\gamma\rho^2)}{\gamma\rho^2} - \frac{\exp(-\gamma\rho^2)}{2}\right] \tag{60}$$

Although the formulation in pressure of the BWR EOS, Eq. (57), had considerable success and was used in many works, the Helmholtz free energy formulation of the BWR EOS, Eq. (60), was seldom utilized, probably because the ideal gas contribution was missing (Lemmon and Span 2010).

3.6 The Lee-Kesler EOS and Its Modified Forms

A successful class of EOS is based on the deviations of the PVT properties of the substance of interest from those of suitably selected reference substances. This approach was adopted by Pitzer et al. (1955) who expressed the compressibility factor of a fluid whose acentric factor is ω (see Sect. 3.2 for the definition of the acentric factor) as follows:

$$Z = Z^{(0)} + \omega Z^{(1)}, \tag{61}$$

in which term $Z^{(0)}$ is the compressibility of a so-called simple fluid, such as argon, and $Z^{(1)}$ represents the deviation of the compressibility factor of the real fluid from $Z^{(0)}$. Pitzer et al. (1955) initially expanded the compressibility factor function as a power series in the acentric factor, but then showed that it is advisable to retain the linear term only.

Lee and Kesler (1975) expressed the compressibility factor of any fluid in terms of both the compressibility factor of a simple fluid, $Z^{(0)}$, and the compressibility factor of a reference fluid, $Z^{(r)}$, by the relation:

$$Z = Z^{(0)} + \left(\frac{\omega}{\omega^{(r)}}\right)\left(Z^{(r)} - Z^{(0)}\right), \tag{62}$$

in which the deviation term $Z^{(1)}$ of Eq. (61) is obviously equivalent to $\left(Z^{(r)} - Z^{(0)}\right)/\omega^{(r)}$. For $Z^{(0)}$, Lee and Kesler (1975) used mainly data of methane, argon, and krypton. As a reference fluid, they used n-octane, the heaviest hydrocarbon for which accurate P–V-T and enthalpy data were available over a wide range of conditions. Its acentric factor is $\omega^{(r)} = 0.3978$.

The compressibility factors of both the simple fluid, $Z^{(0)}$, and the reference fluid $Z^{(r)}$, were represented by the following reduced form of a modified BWR EOS:

$$Z = \frac{P_r V_r}{T_r} = 1 + \frac{B}{V_r} + \frac{C}{V_r^2} + \frac{D}{V_r^5} + \frac{c_4}{T_r^3 V_r^2}\left(\beta + \frac{\gamma}{V_r^2}\right)\exp\left(-\frac{\gamma}{V_r^2}\right), \tag{63}$$

where

$$B = b_1 - \frac{b_2}{T_r} - \frac{b_3}{T_r^2} - \frac{b_4}{T_r^3} \tag{64}$$

$$C = c_1 - \frac{c_2}{T_r} + \frac{c_3}{T_r^3} \tag{65}$$

$$D = d_1 + \frac{d_2}{T_r}. \tag{66}$$

However, the constants of the simple fluid are different from those of the reference fluid. In detail, the values of the twelve constants for the simple fluid are: $b_1 = 0.1181193$, $b_2 = 0.265728$, $b_3 = 0.154790$, $b_4 = 0.030323$, $c_1 = 0.0236744$, $c_2 = 0.0186984$, $c_3 = 0.0$, $c_4 = 0.042724$, $d_1 \cdot 10^4 = 0.155488$, $d_2 \cdot 10^4 = 0.623689$, $\beta = 0.65392$, and $\gamma = 0.060167$, while the values of the twelve constants for the reference fluid are: $b_1 = 0.2026579$, $b_2 = 0.331511$, $b_3 = 0.027655$, $b_4 = 0.203488$, $c_1 = 0.0313385$, $c_2 = 0.0503618$, $c_3 = 0.016901$, $c_4 = 0.041577$, $d_1 \cdot 10^4 = 0.48736$, $d_2 \cdot 10^4 = 0.0740336$, $\beta = 1.226$, and $\gamma = 0.03754$.

To calculate Z for the fluid of interest at given T and P, it is first necessary to compute the values of T_r=T/T_c and P_r=P/ P_c utilizing the critical properties of the fluid. Equation (63) is then used to compute V_r using the simple fluid constants given above. To be noted that V_r is not the reduced volume, but the so-called pseudo-reduced volume, which is defined as:

$$V_r = \frac{VP_c}{RT_c}. \tag{67}$$

Introducing V_r in the first equality of Eq. (63), $Z^{(0)}$ is computed for the simple fluid. The next step is the same of the first, except that the reference fluid constants are used with the same T_r and P_r values of the fluid of interest. In this way, $Z^{(r)}$ is obtained. Finally, the compressibility factor Z for the fluid of interest is calculated, inserting the $Z^{(0)}$ from the first step and the $Z^{(r)}$ from the second into Eq. (62).

Lee and Kesler (1975) gave also relations for computing the fugacity coefficient, the enthalpy departure, the entropy departure, the isochoric heat capacity departure and the isobaric heat capacity departure for pure substances as well as a set of mixing rules for hydrocarbon mixtures.

Joffe (1976) extended the Lee-Kesler EOS for calculating the fugacity coefficients of components and of vapor–liquid equilibria in multicomponent systems. To this purpose, the Lee-Kesler mixing rules were modified by the introduction of binary interaction constants, which are evaluated from experimental pseudocritical constants. The critical compressibility factor was introduced as a correlating parameter. The Lee-Kesler EOS was extensively applied to the study of vapor–liquid equilibria of mixtures by Plocker et al. (1978), who modified the combining rules and gave interaction factors for about 140 binary systems. Lin and Daubert (1979) used the Lee-Kesler EOS to calculate partial molal volumes which generally resulted to be within 15 percent of the experimentally measured data.

Duan et al. (1992a) proposed a modified form of the Lee-Kesler EOS for the endmembers of the CH_4-CO_2-H_2O system:

$$Z = \frac{PV}{RT} = \frac{P_r V_r}{T_r} = 1 + \frac{B}{V_r} + \frac{C}{V_r^2} + \frac{D}{V_r^4} + \frac{E}{V_r^5} + \frac{F}{V_r^2} \left(\beta + \frac{\gamma}{V_r^2}\right) \exp\left(-\frac{\gamma}{V_r^2}\right), \quad (68)$$

where

$$B = a_1 + \frac{a_2}{T_r^2} + \frac{a_3}{T_r^3} \quad (69)$$

$$C = a_4 + \frac{a_5}{T_r^2} + \frac{a_6}{T_r^3} \quad (70)$$

$$D = a_7 + \frac{a_8}{T_r^2} + \frac{a_9}{T_r^3} \quad (71)$$

$$E = a_{10} + \frac{a_{11}}{T_r^2} + \frac{a_{12}}{T_r^3} \quad (72)$$

$$F = \frac{\alpha}{T_r^3} \quad (73)$$

In Eqs. (68)–(73), T_r, P_r, and V_r have the same meaning as seen above for the Lee-Kesler EOS. Thus, the EOS of Duan et al. (1992a) has fifteen adjustable parameters, namely, α, β, γ, and a_1 to a_{12}. Although these adjustable parameters were evaluated from the PVT data in the temperature range from 0 to 1000°C for CO_2 and H_2O and from 0 to 450°C for CH_4, and for pressures from 0 to 3500 bar, the EOS reproduces the experimental data of the pure systems from 0 to 1000 °C and from 0 to ~8000 bar within experimental uncertainty.

Duan et al. (1992b) extended the EOS of Duan et al. (1992a) to mixtures in the CH_4-CO_2-H_2O system, introducing an empirical mixing rule, which is based on the virial EOS expansion. The developed EOS predicts both phase equilibrium and volumetric properties of binary mixtures with accuracy close to that of experimental data from 50 to 1000 °C and from 0 to 1000 bars, but it can be used up to 3000 bar with less accuracy.

3.7 The EOS of Keenan, Keyes, Hill, and Moore

Keenan et al. (1969) described the thermodynamic properties of liquid water and steam using the following equation of state:

$$A = A_0^{pg}(T) + RT \cdot [\ln \rho + \rho Q(\rho, \psi)], \quad (74)$$

which expresses the Helmholtz free energy, A, in terms of the density, ρ, and the absolute temperature, T, with $\psi = 1000/T$. The function $A_0(T)$ represents the ideal gas behavior of the fluid and is given by:

$$A_0^{pg}(T) = \sum_{i=1}^{6} \frac{C_i}{\psi^{i-1}} + C_7 \cdot \ln T + \frac{C_8 \cdot \ln T}{\psi} \quad (75)$$

The empirical function $Q(\rho, \psi)$ of Eq. (74) designates the residual contribution to the Helmholtz free energy and is given by:

$$Q(\rho, \psi) = (\psi - \psi_c) \sum_{j=1}^{7} (\psi - \psi_{aj})^{j-2} \left[\sum_{i=1}^{8} a_{ij} (\rho - \rho_{aj})^{i-1} + e^{-E \cdot \rho} \sum_{i=9}^{10} a_{ij} \cdot \rho^{i-9}\right] \quad (76)$$

with $E = 4.8$, $\psi_{aj} = \psi_c$ for $j = 1$, but $\psi_{aj} = 2.5$ for $j > 1$, and $\rho_{aj} = 0.634$ for $j = 1$, but $\rho_{aj} = 1.0$ for $j > 1$. The numerical coefficients a_{ij} are fifty.

Since the Helmholtz energy formulation of the BWR EOS, Eq. (60), was seldom utilized (as already noted in Sect. 3.6), the EOS of Keenan et al. (1969) can be considered the first equation of state that was formulated exclusively in terms of the Helmholtz free energy.

3.8 The EOS of Haar, Gallagher, and Kell

Haar and coworkers implemented an EOS for ammonia (Haar and Gallagher 1978) and an EOS for water (Haar et al. 1984), in which the Helmholtz free energy, A, was described as a function of the density, ρ, and the absolute temperature, T. Both equations were accepted as standards and are still often utilized in industrial works, although they have been superseded by the EOS generated by Tillner-Roth et al. (1993) for ammonia and Wagner and Pruß (2002) for water.

Haar and Gallagher (1978) followed the approach of Keenan et al. (1969) and developed a Helmholtz free energy function consisting in the sum of two terms, one describing the properties of the ideal gas as a function of temperature only, the other corresponding to the residual contribution depending on both temperature and density.

The EOS of Haar et al. (1984) is more complicated than those of Keenan et al. (1969) and Haar and Gallagher (1978) as it comprises three terms, namely a base function, which accurately describes the liquid and supercritical states, a residual function providing a proper fitting in the mixed states, and the ideal gas part which is a function of temperature only:

$$A(\rho, T) = A_{Base}(\rho, T) + A_{Residual}(\rho, T) + A_{Id.Gas}(T). \quad (77)$$

The base function was derived from the Ursell-Mayer virial theory. To obtain a theoretical equation that is accurate for high density values, the virial series was transformed into a perturbation expansion in density, for which the reference term is the EOS of a fluid whose molecules are considered to be hard ellipsoids to a first approximation. The equation of state thus obtained is:

$$A_{Base}(\rho, T) = RT\left[-\ln(1-y) - \frac{\beta-1}{1-y} + \frac{\alpha+\beta+1}{2\cdot(1-y)^2} + 4y\left(\frac{\overline{B}}{b} - 3.5\right) - \frac{\alpha-\beta+3}{2} + \ln\frac{\rho RT}{P_0}\right]. \quad (78)$$

where $y = [b(T)\cdot\rho]/4$ is a dimensionless density, $\alpha = 11$, $\beta = 133/3$ and $\gamma = 7/2$ are geometric constants, and $P_0 = 1.01325$ bar = 1 atm. Equation (78) involves only two molecular parameters, b (representing the second virial coefficient) and $\overline{B}$ (representing the excluded volume), which are given by the following temperature functions:

$$b = b_l \cdot \ln\frac{T}{T_0} + \sum_{j=0,1,3,5} b_j \cdot \left(\frac{T_0}{T}\right)^j \quad (79)$$

$$\overline{B} = \sum_{j=0,2,4} B_j \cdot \left(\frac{T_0}{T}\right)^j \quad (80)$$

with $T_0 = 647.073$ K, while b_j and B_j are eight numerical coefficients.

The residual part comprises 40 numerical coefficients and has the form:

$$A_{Residual}(\rho, T) = \sum_{i=1}^{36} \frac{g_i}{k(i)}\left(\frac{T_0}{T}\right)^{l(i)} \cdot [1-\exp(-\rho)]^{k(i)} + \sum_{i=37}^{40} g_i \cdot \delta_i^{l(i)} \exp\left(-\alpha_i \cdot \delta_i^{k(i)} - \beta_i \cdot \tau_i^2\right) \quad (81)$$

where g_i are coefficients obtained by fitting the available data, α_i, β_i, k(i) and l(i) are integers, while δ_i and τ_i are reduced densities and temperatures, namely $\delta_i = (\rho - \rho_i)/\rho_i$ and $\tau_i = (T - T_i)/T_i$, with $\rho_i = 0.319$ g cm^{-3} for i = 37, 38, and 39, $\rho_i = 1.55$ g cm^{-3} for i = 40, $T_i = 640$ K for i = 37 and 38, $T_i = 641.6$ K for i = 39, and $T_i = 270$ K for i = 40. The first 36 terms are needed to fit the experimental data in the low-temperature mixed states, while the other terms correct the function in the proximity of the critical point.

The ideal gas part is given by the relation:

$$A_{Id.Gas}(T) = -RT\left[1 + \left(\frac{C_1}{T_R} + C_2\right)\ln T_R + \sum_{i=3}^{18} C_i \cdot T_R^{i-6}\right] \tag{82}$$

in which T_R = T/100 K and C_i are 18 numerical coefficients. Equation (82) approximates very closely the contributions of the vibrational and rotational states reported by Woolley (1980) as part of a detailed analysis of the rotational-vibrational structure of the water molecule. The EOS of Haar et al. (1984) was adopted to implement the NBS/NRC Steam Tables of the National Bureau of Standards of USA and the National Research Council of Canada.

3.9 The Schmidt-Wagner EOS

Schmidt and Wagner (1985) and Jacobsen et al. (1986) developed two wide-range equations of state that can be considered as the origin of most, so-called reference equations of state implemented afterwards. Complex optimization algorithms were used for the first time to determine the functional form of these equations of state.

In the empirical Schmidt-Wagner EOS, the dimensionless Helmholtz energy, $\alpha(\delta, \tau)$, is described as a function of the reduced density, $\delta = V_c/V = \rho/\rho_c$, and the inverse reduced temperature, $\tau = T_c/T$ and comprises a perfect-gas part, $\alpha^{pg}(\delta, \tau)$, and a residual part, $\alpha^r(\delta, \tau)$, as follows:

$$\frac{A_m(\rho, T)}{RT} = \alpha(\delta, \tau) = \alpha^{pg}(\delta, \tau) + \alpha^r(\delta, \tau). \tag{83}$$

The Schmidt-Wagner EOS was implemented for describing the whole pure-fluid region within the experimental uncertainty of the available data and oxygen was chosen as the test fluid. Based on a bank of 336 terms, the evolutionary optimization algorithm by Ewers and Wagner (1982) was used to establish the functional form of the residual part of the Schmidt-Wagner EOS, thus obtaining the following relation:

$$\alpha^r(\delta, \tau) = \sum_{i=1}^{13} n_i \cdot \delta^{r_i} \cdot \tau^{s_i} + \exp(-\delta^2) \cdot \sum_{i=14}^{24} n_i \cdot \delta^{r_i} \cdot \tau^{s_i} + \exp(-\delta^4) \cdot \sum_{i=25}^{32} n_i \cdot \delta^{r_i} \cdot \tau^{s_i} \tag{84}$$

where n_i, r_i, and s_i are numerical coefficients.

The ideal part of the dimensionless Helmholtz free energy, $\alpha^{pg}(\delta, \tau)$, was obtained from an expression of the ideal gas heat capacity of molecular oxygen developed by Wagner et al. (1982):

$$\alpha^{pg}(\delta, \tau) + k_6 \cdot \ln\left(1 + \frac{2}{3}e^{-k_8\tau}\right) + k_9 + \frac{h^{pg}(\tau_0)}{RT} - \frac{s^{pg}(\delta_0, \tau_0)}{R} + \ln\left(\frac{\delta}{\delta_0}\right) \tag{85}$$

where k_i are numerical coefficients, $\tau_0 = T_c/T_0$ and $\delta_0 = P_0/(RT_0\rho_c)$, T_0 = 298.15 K and P_0 = 1.01325 bar. Both $h^{pg}(\tau_0)$ and $s^{pg}(\delta_0, \tau_0)$ can be chosen arbitrarily and were taken equal to zero.

Thus, the Schmidt-Wagner EOS contains not only pure polynomial and BWR-terms but also new exponential functions which are especially convenient for the representation of fluid properties in the critical region. As a result, this EOS is capable of representing the thermodynamic surface of oxygen for temperatures from 54 to 300 K, pressures from 0 to 818 bar and molar densities from 0 to 41 mol dm^{-3} within the experimental uncertainty of the available data. Apart from very few exceptions, this is also true for the entire vapor–liquid coexistence curve and the critical region. Furthermore, extrapolation of the Schmidt-Wagner EOS beyond the range of available data yields physically meaningful results.

3.10 The Reference EOS of Wagner and Coworkers

Functional forms more complex than those of the Schmidt-Wagner EOS are necessary to describe accurately the fluid properties in the extended critical region, and a higher number of terms was needed when the EOS was required to improve the representation of measured values for some substances. New Gaussian bell-shaped terms were added in the work carried out at the Ruhr-University Bochum under the leadership of Prof. Wolfgang Wagner in the 1990's and 2000's. In this way, the functional form of the $\alpha^r(\delta, \tau)$ part of the EOS was modified as follows:

$$\alpha^r(\delta, \tau) = \sum_{k=1}^{k_1} n_k \cdot \delta^{d_k} \cdot \tau^{t_k} + \sum_{k=k_1+1}^{k_2} n_k \cdot \delta^{d_k} \cdot \tau^{t_k} \cdot \exp(-\delta^{l_k}) + \sum_{k=k_2+1}^{k_3} n_k \cdot \delta^{d_k} \cdot \tau^{t_k} \cdot \exp\begin{bmatrix} -\eta_k \cdot (\delta - \varepsilon_k)^2 \\ -\beta_k \cdot (\tau - \gamma_k)^2 \end{bmatrix}. \quad (86)$$

The third term on the right-hand side of Eq. (86), where η and β identify adjustable parameters, makes this equation different from earlier functions. This new functional form of the $\alpha^r(\delta, \tau)$ term was used in the reference EOS for methane (Setzmann and Wagner 1991), carbon dioxide (Span and Wagner 1996), argon (Tegeler et al. 1999), ethene (Smukala et al. 2000), nitrogen (Span et al. 2000), water (Wagner and Pruß 2002), ethane, butane, and methylpropane (Bücker and Wagner 2006a, b), and propane (Lemmon et al. 2009).

3.11 The Technical EOS of Span and Lemmon

In high-level technical applications are also used the so-called technical EOS of some fluids for which state-of-the-art measurements are not available and very low uncertainties are not mandatory. Technical EOS have well-established functional forms and less numerical coefficients than reference EOS which often involve 20 to 50 adjustable parameters to describe the fluid densities with an uncertainty between 0.01 and 0.1%. Furthermore, EOS with fewer parameters leads to considerably reduced computational time and decreases both the correlation between terms and the risk of over-fitting. Technical equations can be extrapolated above and below the temperature range of available data.

Span and Wagner (2003a, b, c) implemented two 12-term EOS of fixed functional forms, one for non-polar or slightly polar fluids and the other for polar substances. Testing these two EOS with precise data sets for fluids already described with highly accurate EOS, Span and Wagner were able to evaluate the uncertainty of the two obtained EOS. The EOS for polar substances resulted unsuitable for some associating fluids, such as water (probably due to hydrogen bonding), whereas it turned out to be suitable for other associating fluid, such as ammonia.

Technical EOS were developed for several organic compounds, as well as for a number of inorganic substances, including carbon monoxide, carbonyl sulfide, hydrogen sulfide, sulfur dioxide, neon, krypton, and xenon (Lemmon and Span 2006), fluorine (de Reuck 1990), and helium (McCarty and Arp 1990). For several fluids, these technical EOS were the first multiparameter equations used to describe their thermodynamic properties. For other fluids, these technical EOS replaced the modified BWR EOS and those based on the principle of corresponding states, locally characterized by physically impossible behavior. Further details on technical EOS are given by Lemmon and Span (2010).

3.12 Equations of State for Mixtures

For the simple EOS, such as that of van der Waals (Sect. 3.2), the cubic equations of Redlich-Kwong, Soave, and Peng-Robinson (Sect. 3.3), the virial EOS (Sect. 3.4), and the complex EOS, such as the Benedict-Webb-Rubin EOS (Sect. 3.5) and the Lee-Kesler EOS (Sect. 3.6), mixing rules were discussed above, at least in some cases.

In contrast, no mention was done, so far, on how to deal with mixtures for the equations of state formulated in terms of the Helmholtz free energy, like the EOS of Keenan, Keyes, Hill, and Moore (Sect. 3.7), the EOS of Haar, Gallagher, and Kell (Sect. 3.8), the Schmidt-Wagner EOS (Sect. 3.9), the reference equations of state of Wagner and coworkers (Sect. 3.10) and the technical equations of state of Span and Lemmon (Sect. 3.11). To fill this gap, the two methods describing the present state of the art in mixture modeling are presented in the following two sections.

3.13 The Generalization and Extension of the Principle of Corresponding States

About one century after the publication of the van der Waals's Dissertation, the principle of corresponding states (see Sect. 3.2) was extended by Leach et al. (1968), Leland and Chappelear (1968), and Rowlinson and Watson (1969), to map the Helmholtz free energy, A, and the compressibility factor, Z, of a substance of interest (indicated by the subscript β) onto those of a reference substance (denoted by the subscript α). This exercise is performed using the two scale factors f° and h°, respectively, which are obtained from the critical molar volumes and temperatures by means of the simple equations:

$$f^\circ = \frac{T_{c,\beta}}{T_{c,\alpha}}. \tag{87}$$

$$h^\circ = \frac{V_{c,\beta}}{V_{c,\alpha}}. \tag{88}$$

Since the principle of corresponding states is not obeyed exactly by any two distinct substances, Leach et al. (1968), Leland and Chappelear (1968), and Rowlinson and Watson (1969) generalized the principle introducing the generalized shape factors $\xi(V_\alpha, T_\alpha)$ and $\theta(V_\alpha, T_\alpha)$, which are defined by the relations:

$$f(V_\alpha, T_\alpha) = \frac{T_{c,\beta}}{T_{c,\alpha}} \cdot \theta(V_\alpha, T_\alpha) \tag{89}$$

$$h(V_\alpha, T_\alpha) = \frac{V_{c,\beta}}{V_{c,\alpha}} \cdot \xi(V_\alpha, T_\alpha). \tag{90}$$

The generalized shape factors are functions of the volume and temperature of the reference fluid. If the deviations from the principle of corresponding states are limited, the temperature and volume dependence of these functions is expected to be weak and simple to describe. For instance, the following two relations were adopted by Gallagher et al. (1993):

$$\theta(V_\alpha, T_\alpha) = 1 + \theta_V \cdot (V_r - 1) + \theta_T \cdot (T_r - 1) + \theta_{VT} \cdot (V_r - 1) \cdot (T_r - 1) \tag{91}$$

$$\xi(V_\alpha, T_\alpha) = \xi_0 + \xi_V \cdot (V_r - 1) + \xi_T \cdot (T_r - 1) + \xi_{VT} \cdot (V_r - 1) \cdot (T_r - 1) \tag{92}$$

in which the adjustable coefficients θ_V, θ_T, θ_{VT}, ξ_V, ξ_T, and ξ_{VT} are presumably small and ξ_0 is close to unity.

The principle of corresponding states was extended to mixtures by van der Waals himself who postulated that the Helmholtz free energy of a binary mixture of constant composition X (indicated by the subscript X) can be mapped onto that of the pure components. The scale factors f_X^o and h_X^o are related to the so-called pseudo-critical parameters of the mixture of composition X and to the critical parameters of the reference substance by the equations:

$$f_X^o = \frac{T_{c,X}}{T_{c,\alpha}} \tag{93}$$

$$h_X^o = \frac{V_{c,X}}{V_{c,\alpha}}. \tag{94}$$

which are similar to Eqs. (87) and (88). The pseudo-critical parameters of the mixture of composition X can be obtained from those of the pure reference fluid α and the pure substance of interest β, by means of the van der Waals-type mixing rules:

$$V_{c,X} = (1 - X)^2 \cdot V_{c,\alpha} + 2 \cdot X \cdot (1 - X) \cdot V_{c,\alpha\beta} + X^2 \cdot V_{c,\beta} \tag{95}$$

$$T_{c,X} = (1 - X)^2 \cdot T_{c,\alpha} + 2 \cdot X \cdot (1 - X) \cdot T_{c,\alpha\beta} + X^2 \cdot T_{c,\beta}. \tag{96}$$

The two constants $V_{c,\alpha\beta}$ and $T_{c,\alpha\beta}$ are obtained from the critical parameters of the two pure substances by means of the Lorentz-Berthelot-type combining rules:

$$V_{c,\alpha\beta} = \frac{k \cdot \left[V_{c,\alpha}^{1/3} + V_{c,\beta}^{1/3}\right]^3}{2^3} \tag{97}$$

$$T_{c,\alpha\beta} = j \cdot \left[T_{c,\alpha} \cdot T_{c,\beta}\right]^{1/2} \tag{98}$$

in which j and k are two adjustable empirical constants or interaction parameters, allowing for some flexibility. Equations (95) and (96) define a reference curve interpolating smoothly between the critical point of the reference fluid α and the critical point of the pure substance of interest β.

The principle of corresponding states for mixtures is generalized by defining the shape factors $\xi_X(V_\alpha, T_\alpha)$ and $\theta_X(V_\alpha, T_\alpha)$, which are assumed to be weakly-varying functions of volume and temperatures, for a given composition. In analogy with Eqs. (89) and (90), they are defined by the relations:

$$f_X(V_\alpha, T_\alpha) = \frac{T_{c,X}}{T_{c,\alpha}} \cdot \theta_X(V_\alpha, T_\alpha) \tag{99}$$

$$h_X(V_\alpha, T_\alpha) = \frac{V_{c,X}}{V_{c,\alpha}} \cdot \xi_X(V_\alpha, T_\alpha), \tag{100}$$

whereas the functions ξ_X and θ_X are given by the empirical equations:

$$\theta_X = (1 - X)^2 + 2 \cdot X \cdot (1 - X) \cdot \theta^{1/2} + X^2 \cdot \theta \tag{101}$$

$$\xi_X = (1 - X) + X \cdot \xi \tag{102}$$

with $\theta(V_\alpha, T_\alpha)$ and $\xi(V_\alpha, T_\alpha)$ from Eqs. (91) and (92).

All in all, the binary-mixture model based on the generalized principle of corresponding states comprises: the Helmholtz free energy of the reference substance α, the critical parameters of both the reference substance α and the fluid of interest β, as well as eight adjustable parameters, namely θ_V, θ_T, θ_{VT}, ξ_V, ξ_T, and ξ_{VT} in the shape factors, and j and k in the combining rules.

This model was applied to the binary CO_2-H_2O system by Gallagher et al. (1993) who adopted the NBS/NRC Steam Tables (which are based on the EOS of Haar et al. 1984; see Sect. 3.8) as reference state for pure H_2O and used the corresponding-states principle to obtain the Helmholtz free energy of H_2O-CO_2 mixtures. Computed phase boundaries, molar volumes, enthalpies, and fugacity coefficients of CO_2 and H_2O were tabulated along 35 isobars, from 0.5 to 1000 bar, for CO_2 mole fractions of 0.05, 0.10, 0.20, and 0.30, at 19 temperatures in the interval 400 to 1000 K, that is, from 127 to 727 °C (see Table 3 in Appendix A of Gallagher et al. 1993). The FORTRAN code used to compute these properties was also given by Gallagher et al. (1993).

3.14 Mixture Properties from Helmholtz Energy EOS

More or less at the same time of the Gallagher et al. (1993) work, Tillner-Roth (1993), Lemmon (1996), Lemmon and Tillner-Roth (1999) and Lemmon and Jacobsen (1999) implemented a generalized mixture model that is based on the EOS for one of the pure fluids in the mixture and an excess function to account for the interaction between the different components of the mixture. The generalized mixture model of Lemmon and Tillner-Roth is largely empirical in nature and is an extension of the excess property model which accounts for the difference between an ideal solution and the real property value at constant pressure and temperature. It follows, to some degree, the multi-fluid model for mixtures of

Leland and Chappelear (1968). The model may be used to calculate the thermodynamic properties of mixtures at various compositions including dew and bubble point properties and critical points.

In the generalized mixture model of Lemmon and Tillner-Roth, the Helmholtz free energy for mixtures of fluids, A_m, is given by the following relation:

$$A_m = A_m^{id} + A_m^E. \quad (103)$$

where A_m^{id} is the ideal mixing contribution and A_m^E is the excess contribution (Lemmon and Span 2010).

The ideal mixing contribution is defined by the relation:

$$A_m^{id} = \sum_{i=1}^{C} X_i \cdot \left[A_{m,i}^{pg}(\rho_n, T) + A_{m,i}^{r}(\delta, \tau) + RT \ln X_i\right] \quad (104)$$

in which, C is the number of the components forming the mixture, $A_{m,i}^{pg}$ is the perfect gas Helmholtz free energy of the i-th component, and $A_{m,i}^{r}$ indicates is the residual Helmholtz free energy of the i-th component at the reduced density δ and reduced temperature τ, which are defined below.

The excess Helmholtz free energy of mixing is:

$$\frac{A_m^E}{RT} = \alpha^E = \sum_{i=1}^{C-1} \sum_{j=i+1}^{C} X_i \cdot X_j \cdot F_{ij} \times \left\{ \sum_{k=1}^{K_{pol}} N_k \cdot \delta^{d_k} \cdot \tau^{t_k} + \sum_{k=K_{pol}+1}^{K_{pol}+K_{exp}} N_k \cdot \delta^{d_k} \cdot \tau^{t_k} \cdot \exp\left[-\eta_k \cdot (\delta - \varepsilon_k)^2 - \beta_k \cdot (\delta - \gamma_k)\right] \right\} \quad (105)$$

in which the adjustable coefficients and exponents are obtained from nonlinear regression of the experimental data available for the mixture. The generalized parameter F_{ij} is utilized to relate the excess properties of one binary mixture to those of another. Thanks to this generalized parameter, the same set of mixture coefficients can be utilized for several binary mixtures in the model. However, the generalized parameter is not employed for some binary systems (CH_4-N_2, CH_4-CO_2, CH_4-C_2H_6, CH_4-C_3H_8, CH_4-H_2, N_2-CO_2, and N_2-C_2H_6) whose coefficients and exponents of binary specific excess functions are derived from experimental data.

All the single-phase thermodynamic properties can then be computed from the Helmholtz free energy. The reduced density δ and reduced temperature τ of the mixture are given by the relations:

$$\delta = \rho/\rho_r(X) \quad (106)$$

$$\tau = T_r(X)/T \quad (107)$$

in which ρ and T are the density and temperature of the mixture, while $\rho_r(X)$ and $T_r(X)$ are the reducing values obtained by:

$$\frac{1}{\rho_r(X)} = \sum_{i=1}^{C} \sum_{j=1}^{C} X_i \cdot X_j \cdot \beta_{v,ij} \cdot \gamma_{v,ij} \cdot \frac{X_i + X_j}{\beta_{v,ij}^2 \cdot X_i + X_j} \cdot \frac{1}{8} \cdot \left(\frac{1}{\rho_{c,i}^{1/3}} + \frac{1}{\rho_{c,j}^{1/3}}\right)^3 \quad (108)$$

$$T_r(X) = \sum_{i=1}^{C} \sum_{j=1}^{C} X_i \cdot X_j \cdot \beta_{T,ij} \cdot \gamma_{T,ij} \cdot \frac{X_i + X_j}{\beta_{T,ij}^2 \cdot X_i + X_j} \cdot \sqrt{T_{c,i} \cdot T_{c,j}} \quad (109)$$

The parameters $\beta_{v,ij}$, $\gamma_{v,ij}$, $\beta_{T,ij}$, and $\gamma_{T,ij}$ are not the same as the critical parameters of the mixture and are computed concurrently in the nonlinear fit of experimental data with the other parameters of the mixture model. They are utilized to describe the shapes of the reducing temperature lines and reducing density lines, respectively.

Paulus and Penoncello (2006) adopted the approach of Lemmon and Tillner-Roth to represent the thermodynamic properties of CO_2-H_2O single-phase vapors in the temperature interval 50–801 °C, at pressures up to 1000 bar, and covering the whole composition range. The experimental data used by Paulus and Penoncello (2006) are pressure, density, temperature, composition, second virial coefficients, and excess enthalpy. The model reproduces the density of the mixture to within ±0.1%. Due to the lack of single-phase liquid data for the mixture, the Peng-Robinson equation of state was utilized to predict the vapor–liquid equilibrium properties of the mixture at temperatures from 50 to 351 °C and pressures up to 1000 bar.

The approach of Lemmon and Tillner-Roth was adopted to develop the GERG-2004 (Kunz et al. 2007) and the GERG-2008 (Kunz and Wagner 2012) wide-range EOS, which were conceived for computing the thermodynamic properties of natural gases and other binary and multi-component mixtures. GERG-2008 is an extended version of GERG-2004, and includes 21 components, namely water, carbon dioxide, hydrogen sulfide, nitrogen, oxygen, argon, helium, hydrogen, carbon monoxide, methane, ethane, propane, n-butane, i-butane, n-pentane, i-pentane, n-hexane, n-heptane, n-octane, n-nonane, and n-decane. GERG-2004 is based on a multi-fluid approximation. The mixture model utilizes accurate EOS in the form of fundamental equations for each component as well as functions developed for the binary mixtures of the components to describe the residual mixture behavior.

GERG-2008 includes the EOS of Klimeck (2000) for CO_2, which is valid from −57 to 627 ° C and up to 3000 bar and the EOS of Kunz et al. (2007) for H_2O, with range of validity from 0 to 1000 °C and up to 1000 bar. According to the GERG Technical Monograph 15 (Kunz et al. 2007), for the binary CO_2-H_2O mixtures, the GERG database includes the P-V-T-X data points of Fenghour et al. (1996) and Patel and Eubank (1988) spanning the temperature range 50–426 °C, the pressure interval 0.27–346 bar and the H_2O mol fraction 0.02 to 0.79, as well as the vapor–liquid equilibrium data of Tödheide and Franck (1963) and Takenouchi and Kennedy (1964), covering the temperature range 110–350 °C, the pressure interval 100–3500 bar and the H_2O mol fraction 0.57–1.00 in the liquid phase and 0.04–0.95 in the vapor phase. It is unclear how the discrepancies between the two datasets of Tödheide and Franck (1963) and Takenouchi and Kennedy (1964), which were underscored in Sect. 2.1, are treated in the GERG database.

Over the whole composition range, the GERG-2008 wide-range EOS is valid in the gas phase, in the liquid phase, in the supercritical region, and covers vapor–liquid equilibrium states for mixtures of these components.

The normal interval of validity of GERG-2008 encompasses temperatures from −183 to 177 °C and pressures up to 350 bar, in which the most accurate experimental data of the thermal and caloric properties are reproduced within their accuracy. For instance, the uncertainties in the gas phase density are lower than 0.1%. The extended range validity covers temperatures from −213 to 427 °C and pressures up to 700 bar, with uncertainties in the gas phase density which are estimated to be less than 0.2–0.5% approximately. Uncertainties increase outside the extended range of validity, but the EOS describes the data of some binary mixtures within ±1% at pressures up to 3000 bar. Further details on the range of validity and uncertainty estimates are given in Sect. 7.13 of the GERG Technical Monograph (Kunz et al. 2007) and in the paper of Kunz and Wagner (2012).

The GERG-2008 wide-range EOS was adopted as ISO Standard, namely as ISO 20765–2:2015. The applicability and performance of GERG-2008 was tested by several studies (e.g., Dauber and Span 2012a, b; Mazzoccoli et al. 2012; Hernández-Gómez et al. 2014; Yuan et al. 2015; Elshahomi et al. 2015; Gernert and Span (2016); Nguyen and Elmegaard 2016; Pérez-Sanz et al. 2016; Rowland et al. 2016; Varzandeh et al. 2017; Guerrero-Zárate et al. 2017; Lentner et al. 2017; Hassanpouryouzband et al. 2020) with generally positive outcomes, although further improvements were suggested in some cases.

3.15 Conclusive Considerations and Way Forward

Based on what was discussed in previous sections, the Solfatara fluids were assumed to be simple binary mixtures of H_2O and CO_2, to a first approximation, and their isenthalpic expansion was modeled adopting the GERG-2008 EOS, although previous studies (mentioned at the end of the previous section) were focused on mixtures rich in hydrocarbons or hydrogen or carbon dioxide, while the GERG-2008 EOS was never applied to hydrothermal and volcanic gas mixtures, so far and to the best of our knowledge. In this way, the temperature, molar volume, and entropy of the fluids during their isenthalpic decompression were obtained for gas mixtures with CO_2 mole fractions of 0.05, 0.10, 0.20, 0.30, and 0.40 (see Sect. 4). The fugacity coefficients of H_2O and CO_2 were then calculated using the virial EOS (see Sect. 4.2).

4 The Isenthalpic Decompression of Solfatara Fluids: Results of the GERG-2008 Thermodynamic Module of the CONVAL Software Package

The temperature, molar volume, and entropy of CO_2-H_2O gas mixtures similar to Bocca Grande and Bocca Nuova fluids, along their isenthalpic decompression paths, were computed by means of the GERG-2008 thermodynamic module of the CONVAL® 11 software package, see https://conval.de/en/features(last visit on March 27th, 2022). Five distinct CO_2-H_2O gas mixtures were considered, with CO_2 mole fractions of 0.05, 0.10, 0.20, 0.30, and 0.40 thus covering the whole composition range of the Bocca Grande and Bocca Nuova fluids in the period 1982–2020. The obtained temperature, molar volume, and entropy values are listed as a function of pressure in Tables 1 and 2, respectively.

Table 1 Pressures, temperatures, entropies, and molar volumes along the isenthalpic expansion path of Bocca Grande fluids of different composition. Values in bold-italics are at saturation conditions

X_{CO2} = 0.05; H = 4568.392 kJ/mol				X_{CO2} = 0.10; H = 4624.713 kJ/mol				X_{CO2} = 0.20; H = 4731.757 kJ/mol			
P	T	S	V	P	T	S	V	P	T	S	V
bar	°C	kJ/(mol·K)	m^3/kmol	bar	°C	kJ/(mol·K)	m^3/kmol	bar	°C	kJ/(mol·K)	m^3/kmol
1	161.700	13.179	35.857	1	161.700	15.670	35.908	1	161.700	17.395	35.955
2	165.067	8.777	17.948	2	164.635	9.963	17.956	2	163.921	11.678	17.972
4	171.378	3.134	8.980	4	170.193	4.309	8.979	4	168.200	6.004	8.980
6	177.219	−0.123	5.990	6	175.399	1.042	5.986	6	172.287	2.719	5.983
8	182.674	−2.406	4.494	8	180.307	−1.250	4.489	8	176.204	0.411	4.484
10	187.806	−4.156	3.595	10	184.959	−3.008	3.590	10	179.970	−1.362	3.584
12	192.662	−5.571	2.996	12	189.391	−4.430	2.991	12	183.600	−2.798	2.984
14	197.279	−6.755	2.568	14	193.628	-5.620	2.563	14	187.108	−4.002	2.556
16	201.687	−7.771	2.247	16	197.693	−6.642	2.241	16	190.504	−5.035	2.234
18	205.909	−8.658	1.997	18	201.604	-7.536	1.991	***17***	***192.822***	***−5.502***	***2.104***
20	209.966	−9.445	1.796	***19***	***204.036***	***−7.943***	***1.888***	***25***	***210.937***	***−8.418***	***1.440***
22	***214.289***	***−10.151***	***1.634***	***30***	***227.120***	***−11.306***	***1.201***	***50***	***247.202***	***−13.397***	***0.719***
30	***230.556***	***−12.404***	***1.200***	***40***	***242.873***	***−13.345***	***0.899***	***75***	***270.743***	***−16.124***	***0.474***
40	***246.693***	***−14.428***	***0.899***	***50***	***255.788***	***−14.879***	***0.717***	***90***	***281.876***	***−17.297***	***0.391***
50	***259.973***	***−15.761***	***0.718***	***65***	***271.786***	***−16.624***	***0.547***	***113***	***296.203***	***−18.713***	***0.307***
60	***271.310***	***−17.157***	***0.593***	***79***	***284.265***	***−17.875***	***0.445***	114	296.787	−18.766	0.304

(continued)

Table 1 (continued)

X_{CO2} = 0.05; H = 4568.392 kJ/mol				X_{CO2} = 0.10; H = 4624.713 kJ/mol				X_{CO2} = 0.20; H = 4731.757 kJ/mol			
P	T	S	V	P	T	S	V	P	T	S	V
bar	°C	kJ/(mol·K)	m^3/kmol	bar	°C	kJ/(mol·K)	m^3/kmol	bar	°C	kJ/(mol·K)	m^3/kmol
65	277.132	−17.677	0.546	80	285.239	−17.954	0.439	115	297.549	−18.819	0.301
70	282.915	−18.152	0.507	85	290.248	−18.334	0.413	120	301.296	−19.077	0.289
80	293.860	−18.994	0.442	90	295.098	−18.689	0.390	130	308.501	−19.556	0.266
90	304.081	−19.722	0.392	100	304.361	−19.333	0.350	140	315.345	−19.995	0.247
100	313.677	−20.360	0.352	125	325.324	−20.659	0.279	150	321.856	−20.398	0.231
125	335.380	−21.672	0.281	150	343.709	−21.704	0.232	175	336.836	−21.284	0.198
150	354.425	−22.705	0.233	175	360.012	−22.562	0.199	200	350.211	−22.037	0.174
175	371.335	−23.553	0.200	200	374.594	−23.289	0.174	250	373.118	−23.268	0.141
200	386.483	−24.270	0.175	250	399.646	−24.472	0.141	300	392.056	−24.257	0.120
250	412.566	−25.435	0.141	300	420.480	−25.417	0.119	400	421.478	−25.809	0.094
300	434.320	−26.362	0.119	400	453.325	−26.886	0.092	500	442.955	−27.029	0.079
400	468.821	-27.796	0.092	500	477.959	−28.025	0.077	600	459.044	−28.056	0.070
500	495.037	−28.901	0.076	600	496.884	−28.971	0.067	700	471.386	−28.957	0.064
600	515.464	−29.813	0.066	700	511.741	−29.793	0.061	800	481.009	−29.771	0.059
700	531.686	−30.600	0.059	800	523.630	−30.529	0.056	900	488.591	−30.521	0.055
800	544.815	−31.300	0.054	900	533.255	−31.203	0.052	1000	494.615	−31.221	0.052
900	555.593	−31.938	0.051	1000	541.137	−31.828	0.049	1250	504.894	−32.816	0.047
1000	564.529	-32.527	0.048	1250	555.402	−33.237	0.044	1500	510.602	−34.257	0.043
1250	581.056	−33.850	0.042	1500	564.487	−34.498	0.040	1750	513.355	−35.592	0.041
1500	592.035	−35.025	0.039	1750	570.214	−35.658	0.038	2000	514.081	−36.851	0.039
1750	599.431	−36.102	0.036	2000	573.612	−36.744	0.036				
2000	604.332	−37.106	0.034	2250	575.325	−37.775	0.034				
2250	607.418	−38.056	0.033	2500	575.775	−38.762	0.033				
2500	609.139	-38.963	0.031								
2750	609.804	−39.835	0.030								

X_{CO2} = 0.30; H = 4832.096 kJ/mol				X_{CO2} = 0.40; H = 4927.200 kJ/mol			
P	T	S	V	P	T	S	V
bar	°C	kJ/(mol·K)	m^3/kmol	bar	°C	kJ/(mol·K)	m^3/kmol
1	161.700	18.568	35.993	1	161.700	19.329	36.023
2	163.392	12.842	17.987	2	163.010	13.596	18.000
4	166.692	7.152	8.984	4	165.587	7.893	8.989
6	169.890	3.852	5.983	6	168.108	4.581	5.985
8	172.994	1.530	4.482	8	170.577	2.247	4.484
10	176.011	−0.256	3.582	10	172.996	0.449	3.583
12	178.948	−1.705	2.982	12	175.369	−1.010	2.982
14	181.812	−2.921	2.553	14	177.697	−2.236	2.553

(continued)

Table 1 (continued)

X_{CO2} = 0.30; H = 4832.096 kJ/mol				X_{CO2} = 0.40; H = 4927.200 kJ/mol			
P	T	S	V	P	T	S	V
bar	°C	kJ/(mol·K)	m^3/kmol	bar	°C	kJ/(mol·K)	m^3/kmol
16	184.606	−3.966	2.231	16	179.984	−3.292	2.231
17	***186.476***	***−4.438***	***2.101***	18	182.230	−4.217	1.981
25	***203.808***	***−7.392***	***1.439***	***19***	***184.191***	***−4.640***	***1.879***
50	***238.197***	***−12.456***	***0.721***	***25***	***196.054***	***−6.761***	***1.436***
75	***260.246***	***−15.247***	***0.477***	***50***	***228.539***	***−11.911***	***0.722***
100	***276.658***	***−17.138***	***0.354***	***75***	***249.143***	***−14.767***	***0.479***
125	***289.726***	***−18.550***	***0.280***	***100***	***264.351***	***−16.714***	***0.357***
148	***299.738***	***−19.586***	***0.234***	***125***	***276.375***	***−18.177***	***0.284***
149	300.148	−19.627	0.232	***150***	***286.265***	***−19.341***	***0.235***
150	300.725	−19.667	0.231	***180***	***296.123***	***−20.478***	***0.195***
160	306.351	−20.055	0.217	190	300.153	−20.811	0.185
170	311.718	−20.417	0.204	200	304.356	−21.125	0.176
180	316.843	−20.755	0.193	250	322.882	−22.481	0.144
190	321.743	−21.073	0.184	300	337.996	−23.583	0.123
200	326.432	−21.373	0.175	400	360.767	−25.343	0.098
250	347.141	−22.663	0.142	500	376.628	−26.757	0.084
300	364.150	−23.706	0.121	600	387.958	−27.968	0.075
400	390.155	−25.357	0.096	700	396.178	−29.047	0.069
500	408.673	−26.670	0.082	800	402.176	−30.033	0.064
600	422.225	−27.787	0.073	900	406.541	−30.949	0.060
700	432.354	−28.775	0.066	1000	409.669	−31.813	0.057
800	440.016	−29.674	0.061	1250	413.683	−33.801	0.052
900	445.852	−30.506	0.058	1500	414.079	−35.620	0.048
1000	450.307	−31.287	0.055				
1250	457.193	−33.077	0.050				
1500	460.018	−34.706	0.046				
1750	460.242	−36.226	0.043				

Table 2 Pressures, temperatures, entropies, and molar volumes along the isenthalpic decompression path of Bocca Nuova fluids of different composition. Values in bold-italics are at saturation conditions

X_{CO2} = 0.05; H = 3977.393 kJ/mol				X_{CO2} = 0.10; H = 4030.623 kJ/mol				X_{CO2} = 0.20; H = 4130.127 kJ/mol			
P	T	S	V	P	T	S	V	P	T	S	V
bar	°C	kJ/(mol·K)	m^3/kmol	bar	°C	kJ/(mol·K)	m^3/kmol	bar	°C	kJ/(mol·K)	m^3/kmol
1	145.350	13.093	34.481	1	145.350	14.277	34.514	1	145.350	15.985	34.570
2	149.251	7.404	17.253	2	148.740	8.581	17.262	2	147.897	10.275	17.281
4	156.462	1.782	8.636	4	155.087	2.945	8.635	4	152.765	4.616	8.636
6	163.038	−1.457	5.762	6	160.953	−0.305	5.758	6	157.370	1.345	5.754
8	169.110	−3.722	4.324	8	166.428	−2.580	4.319	8	161.749	−0.950	4.312
8.5	***170.714***	***−4.195***	***4.071***	***8.5***	***168.350***	***−3.056***	***4.069***	***8.5***	***163.393***	***−1.430***	***4.062***
15	***195.636***	***−8.540***	***2.341***	***15***	***192.852***	***−7.423***	***2.341***	***15***	***187.036***	***−5.844***	***2.339***
25	***220.800***	***−12.285***	***1.414***	***25***	***217.516***	***−11.192***	***1.415***	***25***	***210.691***	***−9.661***	***1.415***
50	***259.877***	***−17.057***	***0.702***	***50***	***255.622***	***−16.003***	***0.703***	***50***	***246.859***	***−14.554***	***0.706***
75	***285.752***	***−19.646***	***0.460***	***75***	***280.680***	***−18.618***	***0.461***	***100***	***287.942***	***−19.031***	***0.343***
93	***300.457***	***−20.946***	***0.365***	***109***	***305.672***	***−20.872***	***0.308***	***120***	***299.482***	***−20.127***	***0.282***
100	307.178	−21.373	0.339	125	319.030	−21.657	0.268	***143***	***310.791***	***−21.144***	***0.233***
125	329.440	−22.648	0.270	150	337.808	−22.672	0.223	145	311.842	−21.228	0.23
150	348.878	−23.651	0.224	175	354.380	−23.505	0.191	150	315.159	−21.415	0.222
175	366.061	−24.474	0.192	200	369.141	−24.210	0.168	175	330.413	−22.276	0.190
200	381.392	−25.169	0.169	250	394.375	−25.359	0.136	200	343.973	−23.007	0.167
250	407.662	−26.300	0.136	300	415.256	−26.277	0.115	250	367.075	−24.204	0.136
300	429.465	−27.200	0.115	400	447.993	−27.707	0.089	300	386.068	−25.166	0.116
400	463.871	−28.595	0.089	500	472.368	−28.819	0.075	400	415.364	−26.679	0.091
500	489.850	−29.673	0.074	600	490.986	−29.746	0.066	500	436.568	−27.873	0.077
600	509.978	−30.565	0.065	700	505.553	−30.553	0.059	600	452.367	−28.881	0.068
700	525.904	−31.337	0.058	800	517.174	−31.278	0.055	700	464.438	−29.769	0.062
800	538.760	−32.025	0.053	900	526.563	−31.942	0.051	800	473.811	−30.573	0.058
900	549.287	−32.653	0.050	1000	534.225	−32.559	0.048	900	481.173	−31.315	0.054
1000	557.990	−33.235	0.047	1250	548.056	−33.956	0.043	1000	487.009	−32.009	0.051
1250	574.044	−34.544	0.042	1500	556.819	−35.209	0.040	1250	496.918	−33.594	0.046
1500	584.676	−35.711	0.038	1750	562.294	−36.364	0.037	1500	502.350	−35.029	0.043
1750	591.799	−36.782	0.036	2000	565.492	−37.448	0.035	1750	504.893	−36.361	0.040
2000	596.481	−37.783	0.034	2250	567.045	−38.477	0.034	2000	505.455	−37.620	0.038
2250	599.388	−38.730	0.032	2500	567.363	−39.464	0.033				
2500	600.962	−39.636	0.031								
2750	601.504	−40.508	0.030								

X_{CO2} = 0.30; H = 4221.361 kJ/mol				X_{CO2} = 0.40; H = 4306.185 kJ/mol			
P	T	S	V	P	T	S	V
bar	°C	kJ/(mol·K)	m^3/kmol	bar	°C	kJ/(mol·K)	m^3/kmol
1	145.350	17.137	34.614	1	145.350	17.873	34.648

(continued)

Table 2 (continued)

X_{CO2} = 0.30; H = 4221.361 kJ/mol				X_{CO2} = 0.40; H = 4306.185 kJ/mol			
P	T	S	V	P	T	S	V
bar	°C	kJ/(mol·K)	m^3/kmol	bar	°C	kJ/(mol·K)	m^3/kmol
2	147.274	11.417	17.298	2	146.827	12.145	17.313
4	151.006	5.738	8.639	4	149.721	6.451	8.645
6	154.599	2.449	5.753	6	152.540	3.148	5.756
8	158.066	0.138	4.310	8	155.289	0.823	4.311
9	***160.221***	***−0.802***	***3.832***	9	156.639	−0.123	3.830
15	***180.790***	***−4.806***	***2.334***	***10***	***158.159***	***−0.967***	***3.446***
25	***203.411***	***−8.673***	***1.414***	***15***	***173.938***	***−4.169***	***2.325***
50	***237.656***	***−13.651***	***0.708***	***30***	***203.622***	***−9.447***	***1.178***
100	***275.881***	***−18.249***	***0.347***	***50***	***227.762***	***−13.150***	***0.709***
150	***299.551***	***−20.733***	***0.226***	***100***	***263.273***	***−17.870***	***0.351***
183	***311.235***	***−21.891***	***0.183***	***150***	***284.941***	***−20.452***	***0.231***
185	312.138	−21.953	0.181	***224***	***306.106***	***−22.881***	***0.153***
190	314.601	−22.106	0.176	250	315.225	−23.529	0.139
200	319.362	−22.398	0.168	300	330.432	−24.606	0.119
250	340.294	−23.653	0.137	400	353.124	−26.329	0.095
300	357.373	−24.669	0.117	500	368.782	−27.719	0.082
400	383.262	−26.282	0.093	600	379.893	−28.914	0.073
500	401.533	−27.571	0.080	700	387.906	−29.981	0.067
600	414.826	−28.670	0.071	800	393.722	−30.958	0.063
700	424.711	−29.646	0.065	900	397.932	−31.869	0.059
800	432.156	−30.535	0.060	1000	400.927	−32.728	0.056
900	437.807	−31.360	0.057	1250	404.679	−34.711	0.051
1000	442.103	−32.136	0.054	1500	404.886	−36.530	0.048
1250	448.675	−33.919	0.049				
1500	451.269	−35.544	0.045				
1750	451.321	−37.063	0.043				

First, the enthalpy of the considered gas mixture was obtained at atmospheric pressure and averaged mean and median outlet temperature, namely 161.7 °C for Bocca Grande and 145.35 °C for Bocca Nuova (see Sect. 1). Second, the enthalpy was kept constant and temperature, molar volume, and entropy were computed at increasing pressure values, up to (i) 2750 bar for the gas mixture with X_{CO_2} of 0.05, (ii) 2500 bar for the gas mixture with X_{CO_2} of 0.10, (iii) 2000 bar for the gas mixture with X_{CO_2} of 0.20, (iv) 1750 bar for the gas mixture with X_{CO_2} of 0.30, and (v) 1500 bar for the gas mixture with X_{CO_2} of 0.40.

These different upper pressure values were adopted because temperature does not change appreciably with varying pressure close to these values, as shown in the two correlation diagrams of Fig. 5.

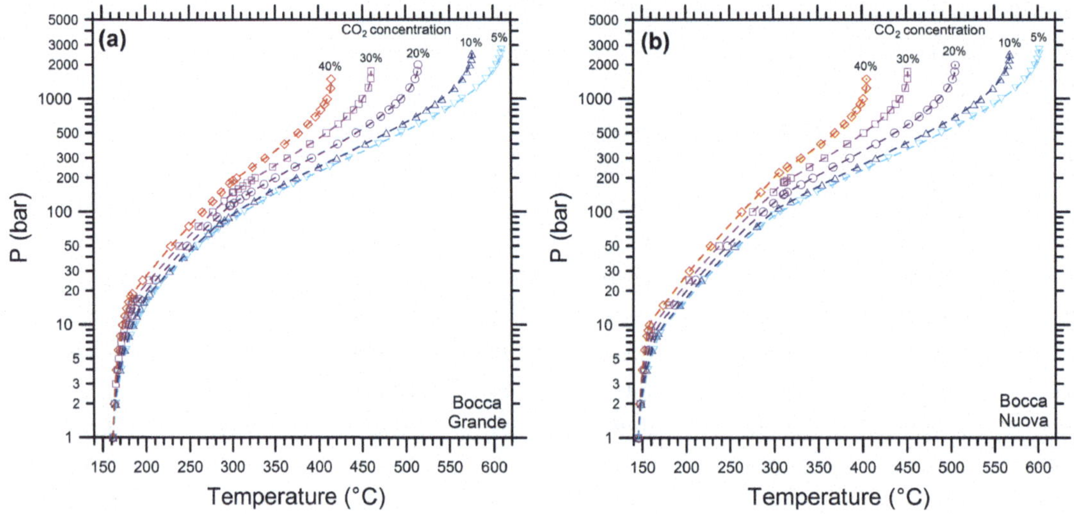

Fig. 5 Plots of pressure versus temperature, along the isenthalpic decompression paths of **a** Bocca Grande and **b** Bocca Nuova fluids of different composition

4.1 The PTV Values Computed by the GERG-2008 Thermodynamic Module of the CONVAL Software Package for the Isenthalpic Expansion of Bocca Grande and Bocca Nuova Fluids and Related Implications

The two plots of Fig. 5 are similar in shape and consist of three distinct sections. A single-phase gas mixture is present both at low temperatures and pressures and high temperatures and pressures, whereas a bi-phase (liquid + vapor) system occurs at intermediate temperatures and pressures.

In detail, based on the isenthalpic hypothesis, a single-phase gas mixture is present in the plumbing system of Bocca Grande: (i) below 22 bar–214.3 °C and above 60 bar–271.3 °C for the gas mixture with X_{CO_2} of 0.05 and enthalpy of 4568.4 kJ/mole; (ii) below 19 bar–204.0 °C and above 79 bar–284.3 °C for the gas mixture with X_{CO_2} of 0.10 and enthalpy of 4624.7 kJ/mole; (iii) below 17 bar–192.8 °C and above 113 bar–296.2 °C for the gas mixture with X_{CO_2} of 0.20 and enthalpy of 4731.8 kJ/mole; (iv) below 17 bar–186.5 °C and above 148 bar–299.7 °C for the gas mixture with X_{CO_2} of 0.30 and enthalpy of 4832.1 kJ/mole; (v) below 19 bar–184.2 °C and above 180 bar–296.1 °C for the gas mixture with X_{CO_2} of 0.40 and enthalpy of 4927.2 kJ/mole.

Based on the isenthalpic assumption, a single-phase gas mixture is present in the plumbing system of Bocca Nuova: (i) below 8.5 bar–170.7 °C and above 93 bar–300.5 °C for the gas mixture with X_{CO_2} of 0.05 and enthalpy of 3977.4 kJ/mole; (ii) below 8.5 bar–168.4 °C and above 109 bar–305.7 °C for the gas mixture with X_{CO_2} of 0.10 and enthalpy of 4030.6 kJ/mole; (iii) below 8.5 bar–163.4 °C and above 143 bar–310.8 °C for the gas mixture with X_{CO_2} of 0.20 and enthalpy of 4130.1 kJ/mole; (iv) below 9 bar–160.2 °C and above 183 bar–311.2 °C for the gas mixture with X_{CO_2} of 0.30 and enthalpy of 4221.4 kJ/mole; (v) below 10 bar–158.2 °C and above 224 bar–306.1 °C for the gas mixture with X_{CO_2} of 0.40 and enthalpy of 4306.2 kJ/mole.

For Bocca Grande, the lower saturation temperature, T_{LST}, (in °C) depends on the CO_2 mole fraction of the gas mixture according to the following equation:

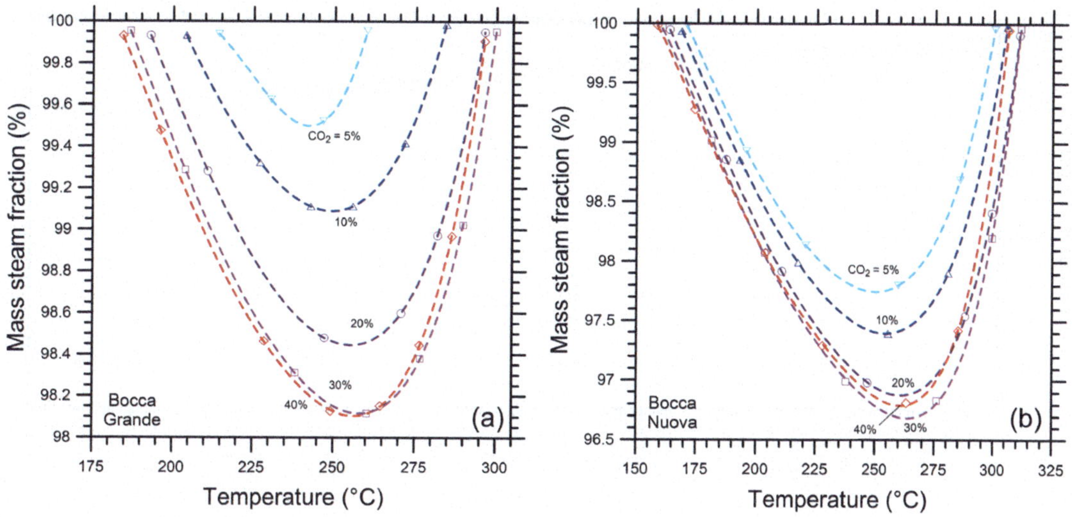

Fig. 6 Correlation diagrams of the mass steam fraction (in percent units) versus temperature for **a** Bocca Grande and **b** Bocca Nuova fumaroles

$$T_{LST} = 3911.69 \cdot X_{CO_2}^4 - 4046.36 \cdot X_{CO_2}^3 + 1693.35 \cdot X_{CO_2}^2 - 395.58 \cdot X_{CO_2} + 230.32, \quad (110)$$

whereas the following relation links the upper saturation temperature, T_{UST}, (in °C) and the CO_2 mole fraction of the gas mixture:

$$T_{UST} = -5246.30 \cdot X_{CO_2}^4 + 5455.21 \cdot X_{CO_2}^3 - 2381.72 \cdot X_{CO_2}^2 + 530.73 \cdot X_{CO_2} + 250.08. \quad (111)$$

For Bocca Nuova, the lower saturation temperature (in °C) is a function of the CO_2 mole fraction of the gas mixture as expressed by the following relation:

$$T_{LST} = -1517.52 \cdot X_{CO_2}^4 + 1404.87 \cdot X_{CO_2}^3 - 374.26 \cdot X_{CO_2}^2 - 12.87 \cdot X_{CO_2} + 172.13, \quad (112)$$

whereas the following equation relates the upper saturation temperature (in °C) and the CO_2 mole fraction of the gas mixture:

$$T_{UST} = -1803.90 \cdot X_{CO_2}^4 + 1654.01 \cdot X_{CO_2}^3 - 775.14 \cdot X_{CO_2}^2 + 195.01 \cdot X_{CO_2} + 292.45. \quad (113)$$

At these upper and lower saturation temperatures, the mass percentage of steam, y%, is higher than 99.9, whereas the y% value decreases moderately at temperatures and pressures intermediate between the upper and lower saturation temperatures, as shown in the two correlation diagrams of Fig. 6.

The lowest y% value is attained at the X_{CO_2} value given by the following relation for Bocca Grande:

$$y(\%) = -5.28919E+01 \cdot X_{CO_2}^4 + 5.16898E+01 \cdot X_{CO_2}^3 - 1.83221E+00 \cdot X_{CO_2}^2 - 8.72045E+00 \cdot X_{CO_2} + 9.99330E+01 \quad (114)$$

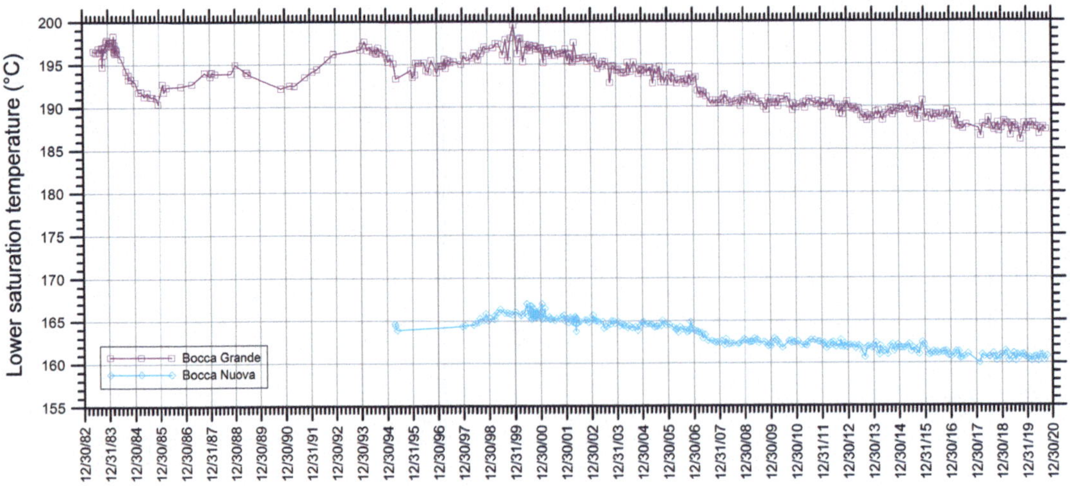

Fig. 7 Chronograms of the lower saturation temperatures of Bocca Grande and Bocca Nuova fumaroles

and by the following equation for Bocca Nuova:

$$y(\%) = -5.35839E+00 \cdot X_{CO_2}^4 + 5.98580E+00 \cdot X_{CO_2}^3 + 1.24116E+01 \cdot X_{CO_2}^2 - 9.00317E+00 \cdot X_{CO_2} + 9.81676E+01. \quad (115)$$

The complement to 100 of the lowest y% value given by Eqs. (114) and (115) represents the maximum mass fraction of condensed steam (in % units), which might be separated during the isenthalpic decompression of Solfatara fluids. Therefore, the lowest y% values given by Eqs. (114) and (115) can be used to correct the analytical data for the maximum effects of steam condensation, under the hypothesis of isenthalpic expansion.

The lower saturation temperatures computed for Bocca Grande and Bocca Nuova fumaroles using Eqs. (110) and (112), respectively, are reported in the chronogram of Fig. 7. The lower saturation temperature of the Bocca Grande fumarole, in the whole considered period, i.e., from June 1983 to October 2020, ranged between 186.1 and 199.7 °C, around a mean of 192.8 °C and a median of 193.3 °C, with a standard deviation of 3.3 °C. The lower saturation temperature of the Bocca Nuova fumarole, from March 1995 to October 2020, varied between 160.0 and 167.0 °C, around a mean of 163.4 °C and a median of 162.9 °C, with a standard deviation of 1.8 °C.

The upper saturation temperatures computed for Bocca Grande and Bocca Nuova fumaroles using Eqs. (111) and (113), respectively, are reported in the chronogram of Fig. 8. The upper saturation temperature of Bocca Grande, from June 1983 to October 2020, oscillated between 289.4 and 299.8 °C around a mean of 295.8 °C, which is equal to the median, with a standard deviation of 2.6 °C. The upper saturation temperature of Bocca Nuova, from March 1995 to October 2020, fluctuated between 307.5 and 311.6 °C, around a mean of 310.5 °C and a median of 311.1 °C, with a standard deviation of 1.1 °C.

The chronograms of the lower and upper saturation temperatures of both fumaroles are the mirror images of each other and, being polynomial functions of the CO_2 concentration, are obviously strictly related to the chronograms of CO_2 and H_2O concentrations, which are shown in Figs. 9 and 10.

Also the latter two chronograms are the mirror images of each other due to the inverse relation linking the concentrations of CO_2 and H_2O and the preponderance of these two constituents, representing 99.3 to 100 mol % of the gas mixtures of interest (as already noted in Sect. 1). Thus, the chronogram of the lower saturation

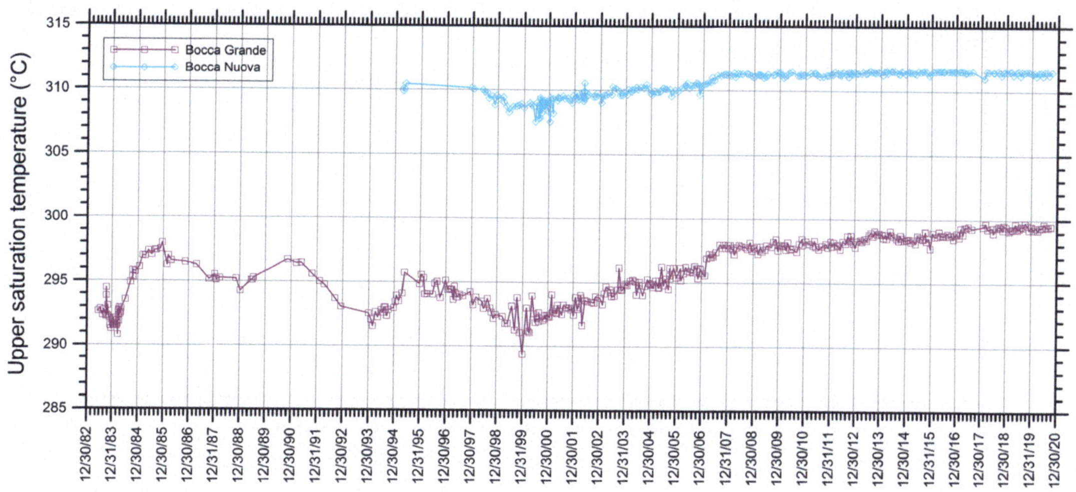

Fig. 8 Chronograms of the upper saturation temperatures of Bocca Grande and Bocca Nuova fumaroles

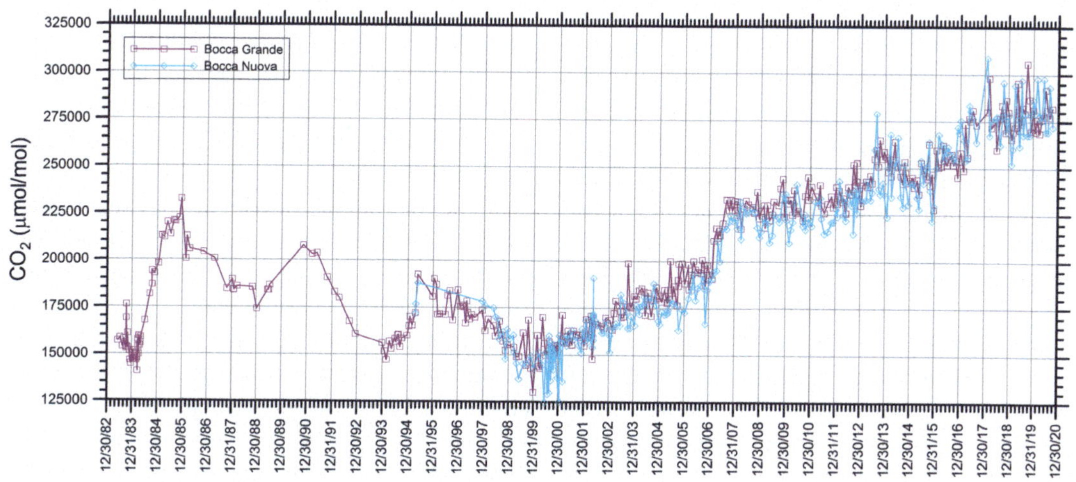

Fig. 9 Chronograms of the CO_2 concentration of Bocca Grande and Bocca Nuova fumaroles

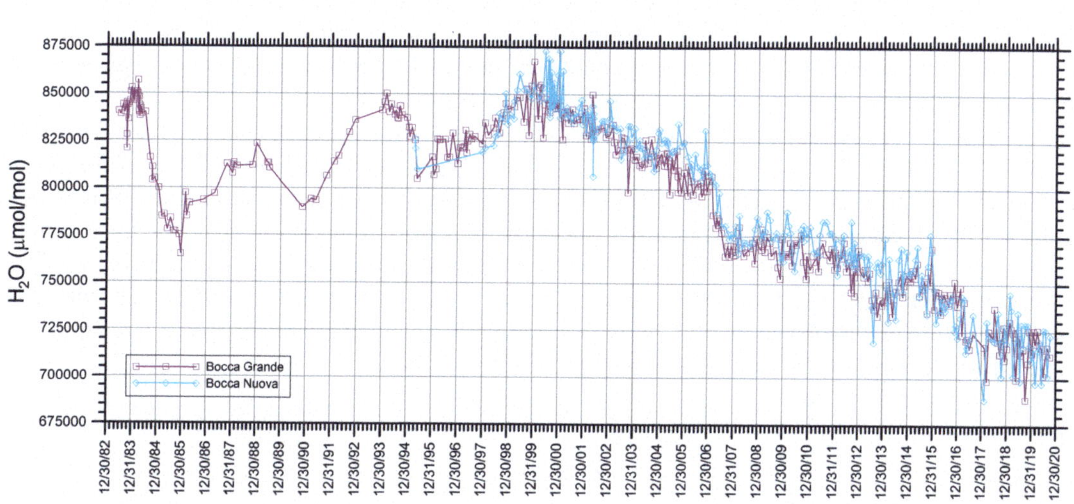

Fig. 10 Chronograms of the H_2O concentration of Bocca Grande and Bocca Nuova fumaroles

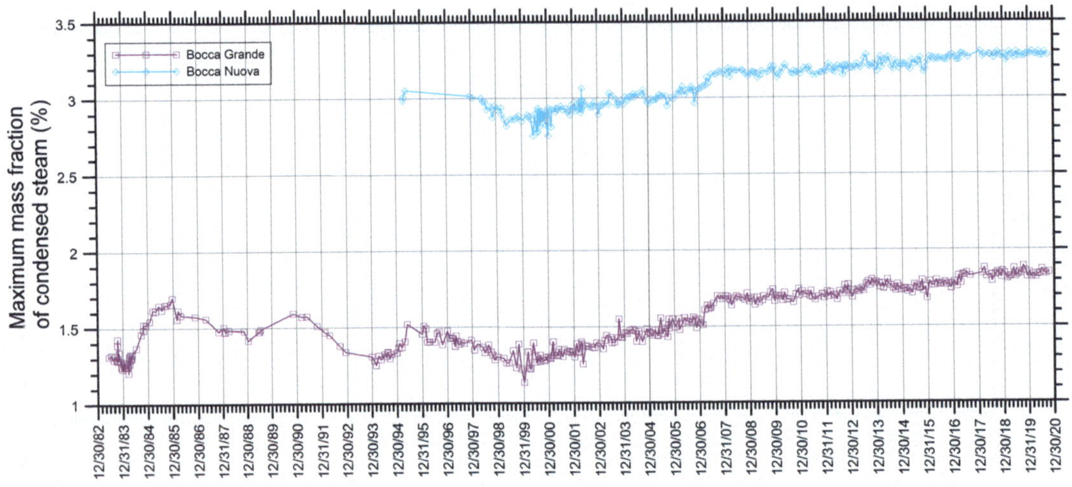

Fig. 11 Chronograms of the maximum mass fraction of condensed steam for Bocca Grande and Bocca Nuova fluids

temperature is a direct transposition of the H_2O chronogram, whereas the chronogram of the upper saturation temperature is a direct transposition of the CO_2 chronogram.

Owing to the continuous decrease of the lower saturation temperature of both fumaroles since January 2000 and the uninterrupted increase of the upper saturation temperatures of both fumaroles in the same period of time (Figs. 7 and 8), the bi-phase (liquid + vapor) region possibly encountered by the fumarolic fluids along their isenthalpic decompression has progressively expanded, leading to a gradual decrease of the minimum mass steam fraction and to a corresponding increase in the maximum mass fraction of condensed steam. The latter variable for Bocca Grande and Bocca Nuova fluids, obtained taking the complement to 100 of the minimum mass steam fraction given by Eqs. (113) and (114), respectively, is reported in the chronograms of Fig. 11.

The maximum mass fraction of condensed steam for the Bocca Grande fluids, in the whole considered period, i.e., from June 1983 to October 2020, ranged between 1.137 and 1.892%, around a mean of 1.546% and a median of 1.522%, with a standard deviation of 0.195% units. The maximum mass fraction of condensed steam for the Bocca Nuova fluids, from March 1995 to October 2020, varied between 2.755 and 3.302%, around a mean of 3.087% and a median of 3.140%, with a standard deviation of 0.147% units. Not surprisingly, also the chronograms of the maximum mass fraction of condensed steam for Bocca Grande and Bocca Nuova fluids are straightforward transpositions of the CO_2 chronogram.

Turning back to Fig. 5, it can be noted that the isenthalpic expansion paths of the five considered gas mixtures diverge progressively as pressure and temperature increase, not only in the upper and lower dry regions but also in the intermediate saturation field. The isenthalpic decompression paths diverge less from each other, in the diagrams in which the P_{H_2O} (instead of the total P) is contrasted with temperature, as shown in Fig. 12.

Hence, it is convenient to derive functions describing the dependence of P_{H_2O} (instead of total P) on the temperature, at least for the saturation region. In the bi-phase (saturation) region, the partial pressure of H_2O depends on the CO_2 mole fraction and temperature as described by the following equations for Bocca Grande (P_{H_2O} in bar, T in K):

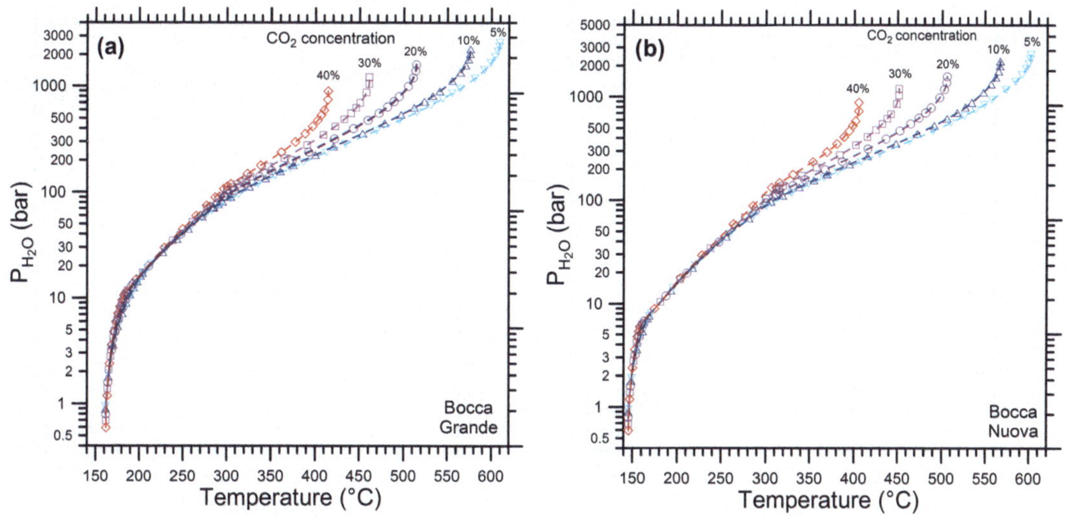

Fig. 12 Plots of water partial pressure vs. temperature, along the isenthalpic expansion paths of **a** Bocca Grande and **b** Bocca Nuova fluids of different composition

$$\log P_{H_2O} = \left(\begin{array}{c} 3766.28 \cdot X_{CO_2}^4 - 2888.86 \cdot X_{CO_2}^3 \\ +54.9517 \cdot X_{CO_2}^2 - 433.599 \cdot X_{CO_2} - 2005.70 \end{array} \right) / T$$
$$-7.14286 \cdot X_{CO_2}^4 + 5.47619 \cdot X_{CO_2}^3 + 0.115000 \cdot X_{CO_2}^2 + 0.922310 \cdot X_{CO_2} + 5.43296. \quad (116)$$

Utilizing Eq. (116), the partial pressure of H_2O at saturation is readily calculated as a function of both the CO_2 mole fraction and temperature, for Bocca Grande fluids with an uncertainty on $P_{H_2O} \leq 3.0\%$(absolute value).

Similarly, in the bi-phase (saturation) region, the partial pressure of H_2O depends on the CO_2 mole fraction and temperature as described by the following equations for Bocca Nuova (P_{H_2O} in bar, T in K):

$$\log P_{H_2O} = \left(\begin{array}{c} 2228.40 \cdot X_{CO_2}^4 - 788.267 \cdot X_{CO_2}^3 \\ -755.365 \cdot X_{CO_2}^2 - 320.711 \cdot X_{CO_2} - 2018.92 \end{array} \right) / T$$
$$-4.86667 \cdot X_{CO_2}^4 + 1.88333 \cdot X_{CO_2}^3 + 1.62667 \cdot X_{CO_2}^2 + 0.738167 \cdot X_{CO_2} + 5.45672. \quad (117)$$

Using Eq. (117), the partial pressure of H_2O at saturation is readily calculated as a function of both the CO_2 mole fraction and temperature, for Bocca Nuova fluids with an uncertainty on $P_{H_2O} \leq 5.5\%$(absolute value).

In the single-phase gas region present at low and high temperatures and pressures along the isenthalpic decompression paths, the relations linking $\log P_{H_2O}$ and the inverse of the absolute temperature diverge strongly from linearity (Fig. 12).

The temperatures, pressures, and molar volumes of the considered Bocca Grande and Bocca Nuova fluids of different compositions, along their isenthalpic expansion paths (Tables 1 and 2), were used to compute the corresponding values of the molar compressibility factor of these gas mixtures, Z_{mixt}, using Eq. (1), with R = 0.0831446261815324 $m^3 \cdot bar \cdot K^{-1} \cdot kmol^{-1}$. The computed Z_{mixt} values are plotted as a function of the molar volume reciprocal in Fig. 13, showing a series of U-shaped curves, one for each mixture composition.

Each curve (i) begins at Z_{mixt} close to 1 and 1/V close to 0, (ii) passes through a minimum Z_{mixt} value that varies from 0.595 to 0.727 for Bocca Grande and from 0.580 to 0.710 for Bocca Nuova, and (iii) attains a maximum Z_{mixt} value ranging from 1.13 to 1.27 for both Bocca Grande and Bocca Nuova at its right-end, that is, at the maximum 1/V value. Both the minimum and maximum values of Z_{mixt} increase with the CO_2

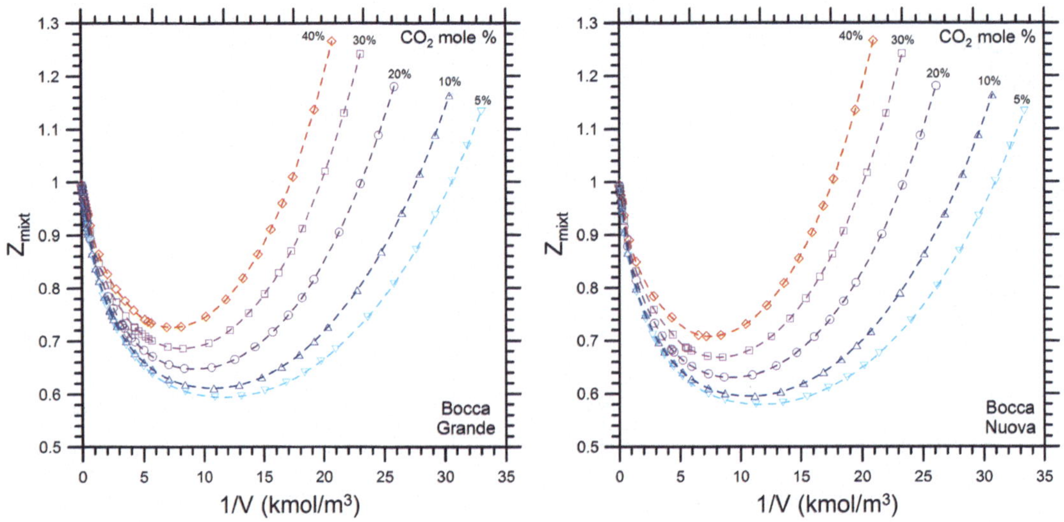

Fig. 13 Plots of the molar compressibility factor of the gas mixture, Z_{mixt}, versus the reciprocal of the molar volume, along the isenthalpic decompression paths of (left) Bocca Grande and (right) Bocca Nuova fluids of different composition

concentration of the gas mixture. The ensuing implication is that the considered gas mixtures have close to ideal behavior only at low 1/V values, that is, at low pressures and temperatures. Assuming that deviation from ideality begins to be significant for Z_{mixt} lower than 0.9, it can be noted that this condition is met along the descending branch of the Z_{mixt} - 1/V curves at (i) 1/V higher than 0.517–0.910 kmol/m^3, corresponding to P higher than 18.6–32.8 bar and T higher than 207–208 °C, for Bocca Grande, and (ii) 1/V higher than 0.433–0.760 kmol/m^3, corresponding to P higher than 15.2–26.8 bar, and T higher than 196–199 °C, for Bocca Nuova. These thresholds in 1/V, P, and T increase with the CO_2 concentration of the gas mixture. Because of these considerable deviations from the ideal gas behavior, it is of fundamental importance to know the fugacity coefficients of CO_2, H_2O and other gas constituents to be used in geothermometric-geobarometric calculations.

4.2 The Fugacity Coefficients of CO_2 and H_2O Computed by Using the GERG-2008 Thermodynamic Module of the CONVAL Software Package and the Virial EOS for the Isenthalpic Decompression of Bocca Grande and Bocca Nuova Fluids

According to Kunz and Wagner (2012), the fugacity coefficients of CO_2 and H_2O are among the properties computed by GERG-2008 but, quite surprisingly, they are not among the results of the GERG-2008 thermodynamic module of the CONVAL software package. Therefore, the fugacity coefficients of CO_2 and H_2O in the binary CO_2-H_2O mixtures of interest were calculated from: (i) the P-V-T-X properties given by the GERG-2008 thermodynamic module of

CONVAL and (ii) the virial coefficients of CO_2, H_2O and CO_2-H_2O mixtures,[6] appearing in the virial EOS (see Sect. 3.4 for details).

The fugacity coefficients of CO_2 and H_2O in binary CO_2-H_2O mixtures are readily computed using the following relations (Prausnitz et al. 1999):

$$\ln \varphi_1 = \frac{2}{V_{mixt}} \cdot (X_1 \cdot B_{11} + X_2 \cdot B_{12}) + \frac{3}{2} \cdot \frac{1}{V^2_{mixt}} \times (X_1^2 \cdot C_{111} + 2 \cdot X_1 \cdot X_2 \cdot C_{112} + X_2^2 \cdot C_{122}) - \ln Z_{mixt} \tag{118}$$

$$\ln \varphi_2 = \frac{2}{V_{mixt}} \cdot (X_2 \cdot B_{22} + X_1 \cdot B_{12}) + \frac{3}{2} \cdot \frac{1}{V^2_{mixt}} \cdot \times (X_2^2 \cdot C_{222} + 2 \cdot X_1 \cdot X_2 \cdot C_{122} + X_1^2 \cdot C_{112}) - \ln Z_{mixt} \tag{119}$$

in which *indices 1 and 2 identify CO_2 and H_2O*, respectively. Equations (118) and (119) refer to the virial EOS for binary mixtures truncated after the third term, which is written as follows (Prausnitz et al. 1999):

$$Z_{mixt} = \frac{P \cdot V_{mixt}}{R \cdot T} = 1 + \frac{B_{mixt}}{V} + \frac{C_{mixt}}{V^2}. \tag{120}$$

The virial EOS is truncated after the third term because only the second virial coefficient B_{12} and the third virial coefficients C_{112} and C_{122} were measured and predicted theoretically for the binary CO_2-H_2O system (see below).

The virial coefficients B_{mixt} and C_{mixt} appearing in Eq. (120) are given by the relations:

$$B_{mixt} = X_1^2 \cdot B_{11} + 2 \cdot X_1 \cdot X_2 \cdot B_{12} + X_2^2 \cdot B_{22} \tag{121}$$

$$C_{mixt} = X_1^3 \cdot C_{111} + 3 \cdot X_1^2 \cdot X_2 \cdot C_{112} + 3 \cdot X_1 \cdot X_2^2 \cdot C_{122} + X_2^3 \cdot C_{222}. \tag{122}$$

The scientific literature on virial coefficients up to the end of 1998 was covered by the compilations of Dymond et al. (2002a; b), who screened the available data to identify the "best" (i.e., most accurate) values. According to Dymond et al. (2002a), the best values of the second virial coefficient of pure CO_2, B_{11}, are from Michels and Michels (1935), Vukalovich and Masalov (1966), Waxman et al. (1973), Holste et al. (1987), Jaeschke (1987), Duschek et al. (1990), and Główka (1990). The following equation was obtained by Dymond et al. (2002a) through least square fit of these selected experimental values (B_{11} in $cm^3 \cdot mol^{-1}$, T in K):

$$B_{11} = 57.400 - 3.8829 \cdot 10^4/T + 4.2899 \cdot 10^5/T^2 - 1.4661 \cdot 10^9/T^3. \tag{123}$$

Equation (123) can be used from −53 to 497 °C. Further references of interest are Leduc and Sacerdote (1897), Strutt (1905), Cawood and Patterson (1933), Cawood and Patterson (1936), Schäfer (1937), Bottomley et al. (1950), Lambert et al. (1950), MacCormack and Schneider (1950), Pfefferle et al. (1955), Cottrell et al. (1956), Cottrell and Hamilton (1956), Cook (1957), Mullins et al. (1963), Butcher and Dadson (1964), Dadson et al. (1967), Ku and Dodge (1967), Sass et al. (1967), Lichtenthaler and Schäfer (1969), Bose and Cole (1970), Holste et al. (1980), Katayama et al. (1980), Schmiedel et al. (1980), Bender et al. (1981), Ohgaki et al. (1982), Schramm and Müller (1982), Ohgaki et al. (1984), Elias et al. (1986), Mallu et al. (1987), Patel et al. (1988), Lopatinskiii et al. (1991), Schramm et al. (1991), and Weber (1992).

As reported by Dymond et al. (2002a), the third virial coefficient of pure CO_2, C_{111}, was experimentally measured by Michels and Michels (1935), MacCormack and Schneider (1950), Pfefferle et al. (1955), Butcher and Dadson (1964), Vukalovich and Masalov (1966), Ku and Dodge (1967), Holste et al. (1980), Katayama et al. (1980), Ohgaki et al. (1982),

[6] Again, GERG-2008 should calculate also the second and third virial coefficients of the binary CO_2-H_2O mixtures of interest but, quite unexpectedly, they are not among the outcomes of CONVAL.

Holste et al. (1987), Patel et al. (1988), Duschek et al. (1990), Główka (1990), and Weber (1992).

According to Dymond et al. (2002a), the best values of the second virial coefficient of pure H_2O, B_{22}, are from Eubank et al. (1988), Kell et al. (1989), and Hendl et al. (1997). The following relation was found by Dymond et al. (2002a) by least square fit of these selected experimental values (B_{22} in $cm^3 \cdot mol^{-1}$, T in K):

$$B_{22} = 158.83 - 3.0107 \cdot 10^5/T + 1.8189 \cdot 10^8/T^2 - 5.6932 \cdot 10^{10}/T^3. \quad (124)$$

Equation (124) applies from 77 to 497 °C. Further references of interest are Collins and Keyes (1938), McCullough et al. (1952), Kell et al. (1965), Vukalovich et al. (1967), Kell et al. (1968), Treiner et al. (1973), Lane and Leidecker (1974), Bohlander and Gebbie (1975), Lie and Clementi (1976), Yamada et al. (1991), and Abdulagatov et al. (1993a, b, 1996).

As reported by Dymond et al. (2002a), the third virial coefficient of pure H_2O, C_{222}, was experimentally determined by McCullough et al. (1952), Kell et al. (1965), Vukalovich et al. (1967), Kell et al. (1968), Eubank et al. (1988), Kell et al. (1989), and Abdulagatov et al. (1993a, b, 1996).

According to Dymond et al. (2002b), the best values of the second virial coefficient of CO_2-H_2O mixtures, B_{12}, are from King and Coan (1971), Skripka (1979), Malesinska (1980),[7] Wormald and Colling (1980), Smith and Wormald (1984), Patel et al. (1987), and Wormald and Lancaster (1988). The following equation, which can be utilized from 27 to 227 °C, was derived by Dymond et al. (2002b) through least square fit of these selected experimental values (B_{12} in $cm^3 \cdot mol^{-1}$, T in K):

$$B_{12} = -107.44 + 1.1123 \cdot 10^5/T - 4.0394 \cdot 10^7/T^2. \quad (125)$$

[7] See also Zawisza and Malesinska (1981).

As reported by Dymond et al. (2002b), the third virial coefficients of CO_2-H_2O mixtures, C_{112} and C_{122} were experimentally measured by Patel et al. (1987) in the temperature intervals 100 to 225 °C and 75 to 225 °C, respectively.

Values of virial coefficients can be obtained not only by appropriate analysis of suitable experimental data but also, and progressively more and more, by computations based on accurate molecular models. Wheatley and Harvey (2011) implemented a high-quality CO_2-H_2O pair potential and utilized it to compute the cross second virial coefficient B_{12} over an extensive interval of temperatures, both below and above the range of available experimental data (see also Wheatley and Harvey 2016). Their values were validated by the dew-point data experimentally determined by Meyer and Harvey (2015) in the temperature interval 10–80 °C.

Schultz et al. (2015)[8] computed the second and third virial coefficients for the binary system CO_2-H_2O, via cluster integrals using quantitative molecular models obtained from the pertinent scientific literature, namely: (a) the rigid-monomer "CCpol2" model of Góra et al. (2014) for H_2O-H_2O pair interactions, (b) the 2-body rigid-monomer potential of Hellmann (2014) for CO_2-CO_2 pair interactions, and (c) the high-quality CO_2-H_2O pair potential of Wheatley and Harvey (2011) for the related cross interaction (see above). Schultz et al. (2015) used both (i) fits to very accurate ab initio calculations of the potential energy surfaces, and (ii) semiempirical Gaussian Charge Polarizable Models (GCPM), taking into account both 2-body and 3-body potentials. The computed virial coefficients of pure CO_2 and H_2O as well as the cross second virial coefficient B_{12} are in good agreement with the experimental data. In contrast, the two cross third virial coefficients C_{112} and C_{122} computed by Schultz et al. (2015) do not agree well with

[8] The same notation adopted so far is also used to identify the virial coefficients of Schultz et al. (2015), which are based on a different notation. So, B_{20} and B_{02} of Schultz et al. (2015) are B_{11} and B_{22}, respectively; B_{11} of Schultz et al. (2015) is B_{12}; B_{30} and B_{03} of Schultz et al. (2015) are C_{111} and C_{222}, respectively; B_{21} and B_{12} of Schultz et al. (2015) are C_{112} and C_{122}, respectively.

the scarce experimental data available (see above). According to Schultz et al. (2015), their C_{112} and C_{122} values could be more accurate than experimental data because of the mutual consistency between values from ab initio and GCPM models. Since the conclusions of Schultz et al. (2015) are reasonable and acceptable, their results were used to obtain the following polynomial equations (T in °C):

$$\begin{aligned}B_{11}(\mathrm{cm^3\cdot mol^{-1}}) = &-3.336153\mathrm{E}-15\cdot T^6 \\ &+9.849270\mathrm{E}-12\cdot T^5 \\ &-1.217753\mathrm{E}-08\cdot T^4 \\ &+8.300463\mathrm{E}-06\cdot T^3 \\ &-3.490897\mathrm{E}-03\cdot T^2 \\ &+9.758693\mathrm{E}-01\cdot T \\ &-1.425066\mathrm{E}+02\end{aligned} \tag{126}$$

$$\begin{aligned}B_{22}(\mathrm{cm^3\cdot mol^{-1}}) = &-7.374906\mathrm{E}-14\cdot T^6 \\ &+2.101704\mathrm{E}-10\cdot T^5 \\ &-2.463764\mathrm{E}-07\cdot T^4 \\ &+1.535516\mathrm{E}-04\cdot T^3 \\ &-5.477024\mathrm{E}-02\cdot T^2 \\ &+1.110430\mathrm{E}+01\cdot T \\ &-1.113021\mathrm{E}+03\end{aligned} \tag{127}$$

$$\begin{aligned}B_{12}(\mathrm{cm^3\cdot mol^{-1}}) = &-6.036180\mathrm{E}-15\cdot T^6 \\ &+1.784973\mathrm{E}-11\cdot T^5 \\ &-2.195114\mathrm{E}-08\cdot T^4 \\ &+1.465719\mathrm{E}-05\cdot T^3 \\ &-5.853551\mathrm{E}-03\cdot T^2 \\ &+1.465024\mathrm{E}+00\cdot T \\ &-1.998803\mathrm{E}+02\end{aligned} \tag{128}$$

$$\begin{aligned}C_{111}(\mathrm{cm^6\cdot mol^{-2}}) = &\,2.250104\mathrm{E}-13\cdot T^6 \\ &-6.251394\mathrm{E}-10\cdot T^5 \\ &+7.239038\mathrm{E}-07\cdot T^4 \\ &-4.585660\mathrm{E}-04\cdot T^3 \\ &+1.749942\mathrm{E}-01\cdot T^2 \\ &-4.073669\mathrm{E}+01\cdot T \\ &+6.392494\mathrm{E}+03\end{aligned} \tag{129}$$

$$\begin{aligned}C_{222}(\mathrm{cm^6\cdot mol^{-2}}) = &-3.378517\mathrm{E}-20\cdot T^{10} \\ &+1.471705\mathrm{E}-16\cdot T^9 \\ &-2.842419\mathrm{E}-13\cdot T^8 \\ &+3.205655\mathrm{E}-10\cdot T^7 \\ &-2.338661\mathrm{E}-07\cdot T^6 \\ &+1.154048\mathrm{E}-04\cdot T^5 \\ &-3.906145\mathrm{E}-02\cdot T^4 \\ &+8.974644\cdot T^3 \\ &-1344.502\cdot T^2+119309.2 \\ &\cdot T-4807498\end{aligned} \tag{130}$$

$$\begin{aligned}C_{112}(\mathrm{cm^6\cdot mol^{-2}}) = &-6.809784\mathrm{E}-14\cdot T^6 \\ &+1.178379\mathrm{E}-10\cdot T^5 \\ &-2.577929\mathrm{E}-08\cdot T^4 \\ &-7.412705\mathrm{E}-05\cdot T^3 \\ &+6.897072\mathrm{E}-02\cdot T^2 \\ &-2.527856\mathrm{E}+01\cdot T \\ &+4.560480\mathrm{E}+03\end{aligned} \tag{131}$$

$$\begin{aligned}C_{122}(\mathrm{cm^6\cdot mol^{-2}}) = &-2.965648\mathrm{E}-01 \\ &\cdot T^2+1.109957\mathrm{E}+02\cdot T \\ &-8.847992\mathrm{E}+03\end{aligned} \tag{132}$$

Table 3 Fugacity coefficients of H_2O and CO_2 along the isenthalpic expansion path of Bocca Grande fluids of different composition. Values in bold-italics are at saturation conditions

X_{CO2} = 0.05; H = 4568.392 kJ/mol				X_{CO2} = 0.10; H = 4624.713 kJ/mol				X_{CO2} = 0.20; H = 4731.757 kJ/mol			
P	T	φ_{CO_2}	φ_{H_2O}	P	T	φ_{CO_2}	φ_{H_2O}	P	T	φ_{CO_2}	φ_{H_2O}
bar	°C			bar	°C			bar	°C		
1	161.700	1.0046	0.9950	1	161.700	1.0032	0.9941	1	161.700	1.0020	0.9938
2	165.067	1.0077	0.9890	2	164.635	1.0063	0.9884	2	163.921	1.0040	0.9878
4	171.378	1.0147	0.9790	4	170.193	1.0122	0.9779	4	168.200	1.0078	0.9765
6	177.219	1.0212	0.9698	6	175.399	1.0178	0.9682	6	172.287	1.0116	0.9659
8	182.674	1.0273	0.9615	8	180.307	1.0230	0.9592	8	176.204	1.0151	0.9560
10	187.806	1.0331	0.9537	10	184.959	1.0281	0.9509	10	179.970	1.0186	0.9467
12	192.662	1.0386	0.9466	12	189.391	1.0329	0.9432	12	183.600	1.0220	0.9379
14	197.279	1.0438	0.9399	14	193.628	1.0375	0.9359	14	187.108	1.0254	0.9297
16	201.687	1.0489	0.9336	16	197.693	1.0420	0.9291	16	190.504	1.0287	0.9219
18	205.909	1.0537	0.9277	18	201.604	1.0464	0.9227	***17***	***192.822***	***1.0309***	***0.9194***
20	209.966	1.0585	0.9222	***19***	***204.036***	***1.0489***	***0.9207***	***25***	***210.937***	***1.0483***	***0.9045***
22	***214.289***	***1.0633***	***0.9177***	***30***	***227.120***	***1.0740***	***0.9005***	***50***	***247.202***	***1.0870***	***0.8575***
30	***230.556***	***1.0819***	***0.9031***	***40***	***242.873***	***1.0927***	***0.8821***	***75***	***270.743***	***1.1166***	***0.8126***
40	***246.693***	***1.1019***	***0.8849***	***50***	***255.788***	***1.1094***	***0.8642***	***90***	***281.876***	***1.1327***	***0.7870***
50	***259.973***	***1.1165***	***0.8655***	***65***	***271.786***	***1.1322***	***0.8381***	***113***	***296.203***	***1.1562***	***0.7491***
60	***271.310***	***1.1364***	***0.8501***	***79***	***284.265***	***1.1522***	***0.8147***	114	296.787	1.1572	0.7476
65	277.132	1.1448	0.8435	80	285.239	1.1537	0.8135	115	297.549	1.1583	0.7466
70	282.915	1.1532	0.8378	85	290.248	1.1612	0.8083	120	301.296	1.1639	0.7419
80	293.860	1.1696	0.8273	90	295.098	1.1686	0.8033	130	308.501	1.1748	0.7329
90	304.081	1.1856	0.8179	100	304.361	1.1831	0.7940	140	315.345	1.1853	0.7244
100	313.677	1.2011	0.8094	125	325.324	1.2179	0.7737	150	321.856	1.1956	0.7164
125	335.380	1.2384	0.7911	150	343.709	1.2505	0.7567	175	336.836	1.2200	0.6980
150	354.425	1.2737	0.7759	175	360.012	1.2813	0.7420	200	350.211	1.2429	0.6816
175	371.335	1.3072	0.7632	200	374.594	1.3105	0.7292	250	373.118	1.2849	0.6538
200	386.483	1.3392	0.7523	250	399.646	1.3648	0.7077	300	392.056	1.3228	0.6310
250	412.566	1.3992	0.7337	300	420.480	1.4152	0.6895	400	421.478	1.3872	0.5936
300	434.320	1.4552	0.7170	400	453.325	1.5067	0.6563	500	442.955	1.4359	0.5602
400	468.821	1.5596	0.6844	500	477.959	1.5857	0.6245	600	459.044	1.4709	0.5287
500	495.037	1.6546	0.6544	600	496.884	1.6511	0.5955	700	471.386	1.4952	0.4995
600	515.464	1.7383	0.6289	700	511.741	1.7048	0.5698	800	481.009	1.5114	0.4726
700	531.686	1.8110	0.6065	800	523.630	1.7490	0.5469	900	488.591	1.5213	0.4480
800	544.815	1.8742	0.5845	900	533.255	1.7853	0.5255	1000	494.615	1.5265	0.4255
900	555.593	1.9294	0.5614	1000	541.137	1.8148	0.5051	1250	504.894	1.5259	0.3768
1000	564.529	1.9776	0.5369	1250	555.402	1.8678	0.4563	1500	510.602	1.5133	0.3364
1250	581.056	2.0735	0.4754	1500	564.487	1.9008	0.4119	1750	513.355	1.4934	0.3017
1500	592.035	2.1451	0.4228	1750	570.214	1.9204	0.3729	2000	514.081	1.4689	0.2716
1750	599.431	2.1999	0.3824	2000	573.612	1.9304	0.3394				
2000	604.332	2.2419	0.3513	2250	575.325	1.9331	0.3107				
2250	607.418	2.2738	0.3258	2500	575.775	1.9302	0.2862				

(continued)

Table 3 (continued)

X_{CO2} = 0.05; H = 4568.392 kJ/mol				X_{CO2} = 0.10; H = 4624.713 kJ/mol				X_{CO2} = 0.20; H = 4731.757 kJ/mol			
P	T	φ_{CO_2}	φ_{H_2O}	P	T	φ_{CO_2}	φ_{H_2O}	P	T	φ_{CO_2}	φ_{H_2O}
bar	°C			bar	°C			bar	°C		
2500	609.139	2.2977	0.3033								
2750	609.804	2.3148	0.2822								

X_{CO2} = 0.30; H = 4832.096 kJ/mol				X_{CO2} = 0.40; H = 4927.200 kJ/mol			
P	T	φ_{CO_2}	φ_{H_2O}	P	T	φ_{CO_2}	φ_{H_2O}
bar	°C			bar	°C		
1	161.700	1.0011	0.9938	1	161.700	1.0004	0.9940
2	163.392	1.0022	0.9877	2	163.010	1.0007	0.9881
4	166.692	1.0043	0.9761	4	165.587	1.0015	0.9766
6	169.890	1.0064	0.9651	6	168.108	1.0023	0.9657
8	172.994	1.0085	0.9547	8	170.577	1.0031	0.9552
10	176.011	1.0106	0.9448	10	172.996	1.0040	0.9452
12	178.948	1.0126	0.9354	12	175.369	1.0049	0.9355
14	181.812	1.0147	0.9264	14	177.697	1.0058	0.9263
16	184.606	1.0167	0.9179	16	179.984	1.0067	0.9174
17	***186.476***	***1.0182***	***0.9147***	18	182.230	1.0076	0.9089
25	***203.808***	***1.0331***	***0.8994***	***19***	***184.191***	***1.0090***	***0.9064***
50	***238.197***	***1.0640***	***0.8508***	***25***	***196.054***	***1.0177***	***0.8945***
75	***260.246***	***1.0858***	***0.8043***	***50***	***228.539***	***1.0408***	***0.8444***
100	***276.658***	***1.1039***	***0.7602***	***75***	***249.143***	***1.0552***	***0.7964***
125	***289.726***	***1.1203***	***0.7185***	***100***	***264.351***	***1.0659***	***0.7512***
148	***299.738***	***1.1345***	***0.6821***	***125***	***276.375***	***1.0748***	***0.7086***
149	300.148	1.1351	0.6806	***150***	***286.265***	***1.0826***	***0.6686***
150	300.725	1.1358	0.6797	***180***	***296.123***	***1.0910***	***0.6240***
160	306.351	1.1430	0.6707	190	300.153	1.0948	0.6139
170	311.718	1.1501	0.6622	200	304.356	1.0988	0.6055
180	316.843	1.1568	0.6540	250	322.882	1.1165	0.5681
190	321.743	1.1634	0.6462	300	337.996	1.1309	0.5367
200	326.432	1.1698	0.6387	400	360.767	1.1494	0.4853
250	347.141	1.1992	0.6055	500	376.628	1.1558	0.4437
300	364.150	1.2247	0.5779	600	387.958	1.1537	0.4083
400	390.155	1.2638	0.5331	700	396.178	1.1458	0.3773
500	408.673	1.2881	0.4960	800	402.176	1.1340	0.3498
600	422.225	1.3009	0.4633	900	406.541	1.1197	0.3251
700	432.354	1.3056	0.4337	1000	409.669	1.1038	0.3027
800	440.016	1.3043	0.4067	1250	413.683	1.0600	0.2547
900	445.852	1.2987	0.3820	1500	414.079	1.0139	0.2155
1000	450.307	1.2903	0.3594				
1250	457.193	1.2614	0.3103				
1500	460.018	1.2263	0.2698				
1750	460.242	1.1881	0.2357				

Table 4 Fugacity coefficients of H_2O and CO_2 along the isenthalpic expansion path of Bocca Nuova fluids of different composition. Values in bold-italics are at saturation conditions

X_{CO2} = 0.05; H = 3977.393 kJ/mol				X_{CO2} = 0.10; H = 4030.623 kJ/mol				X_{CO2} = 0.20; H = 4130.127 kJ/mol			
P	T	φ_{CO_2}	φ_{H_2O}	P	T	φ_{CO_2}	φ_{H_2O}	P	T	φ_{CO_2}	φ_{H_2O}
bar	°C			bar	°C			bar	°C		
1	145.350	1.0049	0.9933	1	145.350	1.0040	0.9930	1	145.350	1.0025	0.9926
2	149.251	1.0094	0.9870	2	148.740	1.0077	0.9863	2	147.897	1.0048	0.9855
4	156.462	1.0178	0.9754	4	155.087	1.0148	0.9739	4	152.765	1.0094	0.9721
6	163.038	1.0255	0.9649	6	160.953	1.0213	0.9627	6	157.370	1.0138	0.9597
8	169.110	1.0325	0.9555	8	166.428	1.0274	0.9525	8	161.749	1.0180	0.9482
8.5	***170.714***	***1.0343***	***0.9535***	***8.5***	***168.350***	***1.0293***	***0.9510***	***8.5***	***163.393***	***1.0195***	***0.9464***
15	***195.636***	***1.0601***	***0.9430***	***15***	***192.852***	***1.0540***	***0.9405***	***15***	***187.036***	***1.0416***	***0.9354***
25	***220.800***	***1.0883***	***0.9244***	***25***	***217.516***	***1.0807***	***0.9217***	***25***	***210.691***	***1.0650***	***0.9161***
50	***259.877***	***1.1391***	***0.8775***	***50***	***255.622***	***1.1281***	***0.8741***	***50***	***246.859***	***1.1047***	***0.8668***
75	***285.752***	***1.1809***	***0.8337***	***75***	***280.680***	***1.1664***	***0.8292***	***100***	***287.942***	***1.1615***	***0.7751***
93	***300.457***	***1.2096***	***0.8040***	***109***	***305.672***	***1.2149***	***0.7724***	***120***	***299.482***	***1.1819***	***0.7412***
100	307.178	1.2211	0.7977	125	319.030	1.2380	0.7596	***143***	***310.791***	***1.2050***	***0.7037***
125	329.440	1.2604	0.7792	150	337.808	1.2721	0.7422	145	311.842	1.2072	0.7011
150	348.878	1.2973	0.7636	175	354.380	1.3040	0.7270	150	315.159	1.2124	0.6972
175	366.061	1.3321	0.7504	200	369.141	1.3340	0.7137	175	330.413	1.2375	0.6785
200	381.392	1.3651	0.7390	250	394.375	1.3896	0.6914	200	343.973	1.2608	0.6617
250	407.662	1.4267	0.7198	300	415.256	1.4409	0.6729	250	367.075	1.3034	0.6332
300	429.465	1.4841	0.7029	400	447.993	1.5336	0.6398	300	386.068	1.3415	0.6099
400	463.871	1.5906	0.6703	500	472.368	1.6122	0.6080	400	415.364	1.4051	0.5722
500	489.850	1.6865	0.6396	600	490.986	1.6761	0.5784	500	436.568	1.4517	0.5391
600	509.978	1.7698	0.6132	700	505.553	1.7277	0.5521	600	452.367	1.4839	0.5081
700	525.904	1.8412	0.5905	800	517.174	1.7694	0.5289	700	464.438	1.5053	0.4793
800	538.760	1.9026	0.5691	900	526.563	1.8029	0.5076	800	473.811	1.5185	0.4527
900	549.287	1.9556	0.5474	1000	534.225	1.8297	0.4876	900	481.173	1.5255	0.4283
1000	557.990	2.0012	0.5248	1250	548.056	1.8762	0.4413	1000	487.009	1.5281	0.4060
1250	574.044	2.0905	0.4670	1500	556.819	1.9034	0.3996	1250	496.918	1.5219	0.3579
1500	584.676	2.1560	0.4149	1750	562.294	1.9179	0.3628	1500	502.350	1.5045	0.3180
1750	591.799	2.2051	0.3721	2000	565.492	1.9232	0.3308	1750	504.893	1.4807	0.2841
2000	596.481	2.2417	0.3377	2250	567.045	1.9219	0.3033	2000	505.455	1.4530	0.2547
2250	599.388	2.2688	0.3093	2500	567.363	1.9156	0.2795				
2500	600.962	2.2882	0.2850								
2750	601.504	2.3014	0.2636								

X_{CO2} = 0.30; H = 4221.361 kJ/mol				X_{CO2} = 0.40; H = 4306.185 kJ/mol			
P	T	φ_{CO_2}	φ_{H_2O}	P	T	φ_{CO_2}	φ_{H_2O}
bar	°C			bar	°C		
1	145.350	1.0013	0.9925	1	145.350	1.0004	0.9928

(continued)

Table 4 (continued)

X_{CO2} = 0.30; H = 4221.361 kJ/mol				X_{CO2} = 0.40; H = 4306.185 kJ/mol			
P	T	φ_{CO_2}	φ_{H_2O}	P	T	φ_{CO_2}	φ_{H_2O}
bar	°C			bar	°C		
2	147.274	1.0026	0.9853	2	146.827	1.0009	0.9857
4	151.006	1.0051	0.9715	4	149.721	1.0018	0.9721
6	154.599	1.0076	0.9586	6	152.540	1.0027	0.9592
8	158.066	1.0100	0.9464	8	155.289	1.0037	0.9469
9	***160.221***	***1.0117***	***0.9413***	9	156.639	1.0042	0.9410
15	***180.790***	***1.0293***	***0.9306***	***10***	***158.159***	***1.0049***	***0.9355***
25	***203.411***	***1.0491***	***0.9106***	***15***	***173.938***	***1.0169***	***0.9261***
50	***237.656***	***1.0805***	***0.8595***	***30***	***203.622***	***1.0391***	***0.8947***
100	***275.881***	***1.1203***	***0.7640***	***50***	***227.762***	***1.0562***	***0.8524***
150	***299.551***	***1.1515***	***0.6789***	***100***	***263.273***	***1.0804***	***0.7540***
183	***311.235***	***1.1708***	***0.6279***	***150***	***284.941***	***1.0957***	***0.6674***
185	312.138	1.1721	0.6260	***224***	***306.106***	***1.1138***	***0.5593***
190	314.601	1.1754	0.6221	250	315.225	1.1222	0.5403
200	319.362	1.1818	0.6146	300	330.432	1.1357	0.5092
250	340.294	1.2112	0.5811	400	353.124	1.1514	0.4582
300	357.373	1.2362	0.5532	500	368.782	1.1546	0.4169
400	383.262	1.2732	0.5081	600	379.893	1.1495	0.3822
500	401.533	1.2945	0.4712	700	387.906	1.1388	0.3520
600	414.826	1.3042	0.4390	800	393.722	1.1245	0.3252
700	424.711	1.3058	0.4102	900	397.932	1.1080	0.3013
800	432.156	1.3016	0.3839	1000	400.927	1.0901	0.2797
900	437.807	1.2935	0.3599	1250	404.679	1.0425	0.2336
1000	442.103	1.2828	0.3379	1500	404.886	0.9937	0.1960
1250	448.675	1.2492	0.2905				
1500	451.269	1.2104	0.2514				
1750	451.321	1.1694	0.2185				

$$C_{122}(\mathrm{cm}^6 \cdot \mathrm{mol}^{-2}) = -6.627770\mathrm{E}-06 \cdot T^3 + 1.385232\mathrm{E}-02 \cdot T^2 - 9.917372 \cdot T + 2938.278. \quad (133)$$

Equations (126)–(131) are valid from 127 to 727 °C, while the cross third virial coefficient C_{122} is computed using either Eq. (132) from 127 to 200 °C or Eq. (133) from 200 to 727 °C.

Since virial coefficients are function of temperature only (as already noted in Sect. 3.4), Eqs. (126)–(133) were used to compute the virial coefficients for the Bocca Grande and Bocca Nuova fluids of different composition at the temperatures encountered along their isenthalpic decompression paths. Virial coefficients were then inserted into Eqs. (118) and (119) together with the molar volumes computed by the GERG-2008 thermodynamic module of CONVAL (Tables 1 and 2) and the compressibility factor values given by Eq. (1), thus obtaining the fugacity coefficients of CO_2 and H_2O for the binary CO_2-H_2O mixtures of interest. The

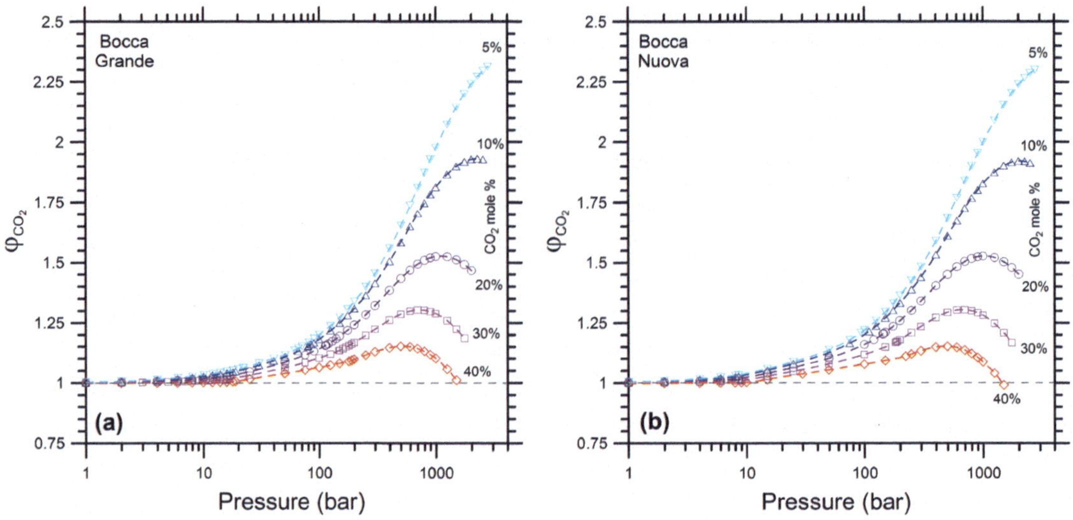

Fig. 14 Plots of the fugacity coefficient of carbon dioxide vs. pressure (in log-scale), along the isenthalpic expansion paths of **a** Bocca Grande and **b** Bocca Nuova fluids of different composition

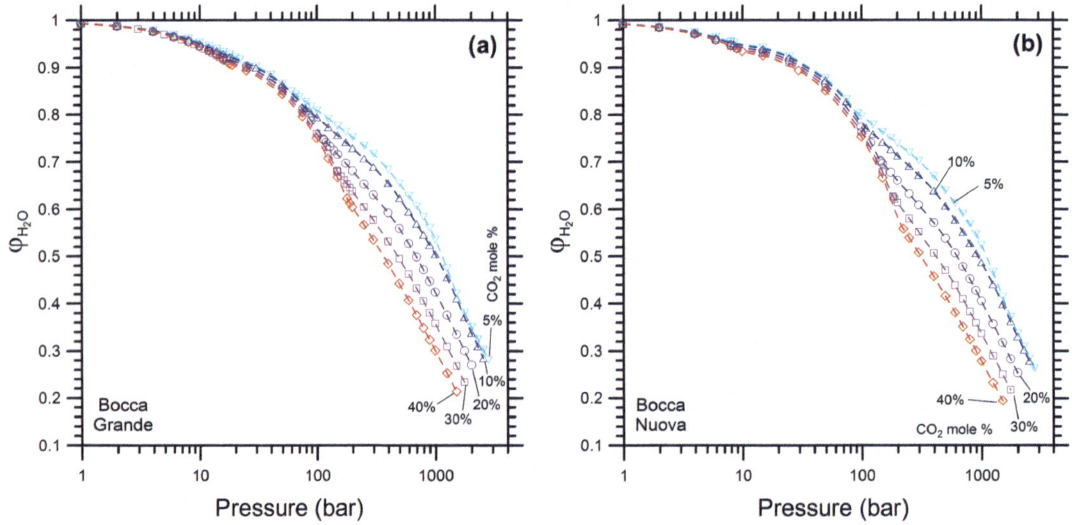

Fig. 15 Plots of the fugacity coefficient of water vs. pressure (in logarithmic scale), along the isenthalpic expansion paths of **a** Bocca Grande and **b** Bocca Nuova fluids of different composition

computed φ_{CO_2} and φ_{H_2O} values are listed in Tables 3 and 4 and are reported as a function of pressure in the diagrams of Figs. 14 and 15, respectively, showing that the fugacity coefficients of CO_2 and H_2O deviate significantly from the unitary value for pressures higher than a few tens of bars.

Figure 14 also shows that the CO_2 fugacity coefficient increases with increasing pressure, moving further and further away from unity and reaching maximum values of: (i) 2.31 for Bocca Grande and 2.30 for Bocca Nuova, at 2750 bar, for the gas mixture with X_{CO_2} of 0.05; (ii) 1.93 at 2250 bar for Bocca Grande and 1.92 at 2000 bar for Bocca Nuova, for the gas mixture with X_{CO_2} of 0.10; (iii) 1.53 at 1000 bar, for both Bocca Grande and Bocca Nuova, for the gas mixture with X_{CO_2} of 0.20; (iv) 1.31 at 700 bar, for both Bocca Grande and Bocca Nuova, for the gas mixture with X_{CO_2} of 0.30; (v) 1.16 for Bocca Grande and 1.15 for Bocca Nuova, at 500 bar, for the gas mixture with X_{CO_2} of 0.40. The maximum φ_{CO_2} value is attained at the maximum considered pressure by the gas mixture with X_{CO_2} of 0.05 and nearly so by the gas mixture with X_{CO_2} of 0.10, while the CO_2 fugacity coefficient of the other three gas mixtures increases with pressure, attains the maximum value, and then it decreases with a further increase in pressure. Thus, the final values of 1.47 for Bocca Grande and 1.45 for Bocca Nuova are reached at 2000 bar by the gas mixture with X_{CO_2} of 0.20, the final values of 1.19 for Bocca Grande and 1.17 for Bocca Nuova are achieved at 1750 bar by the gas mixture with X_{CO_2} of 0.30, and the final values of 1.01 for Bocca Grande and 0.99 for Bocca Nuova are attained at 1500 bar by the gas mixture with X_{CO_2} of 0.40. The CO_2 fugacity coefficient is always higher than unity for all considered gas mixtures, apart from the value of 0.99 attained at 1500 bar by the Bocca Nuova gas mixture with X_{CO_2} of 0.40.

In contrast, as shown in Fig. 15, the fugacity coefficient of H_2O decreases continuously with increasing pressure deviating more and more from unity and attaining the lowest value of: (i) 0.28 for Bocca Grande and 0.26 for Bocca Nuova, at 2750 bar, for the gas mixture with X_{CO_2} of 0.05; (ii) 0.29 for Bocca Grande and 0.28 for Bocca Nuova, at 2500 bar, for the gas mixture with X_{CO_2} of 0.10; (iii) 0.27 for Bocca Grande and 0.26 for Bocca Nuova, at 2000 bar, for the gas mixture with X_{CO_2} of 0.20; (iv) 0.24 for Bocca Grande and 0.22 for Bocca Nuova, at the 1750 bar, for the gas mixture with X_{CO_2} of 0.30; (v) 0.22 for Bocca Grande and 0.20 for Bocca Nuova, at 1500 bar, for the gas mixture with X_{CO_2} of 0.40.

5 The Fugacity Coefficients of H_2O, CO_2, H_2, CH_4, and CO Computed by Using the Peng-Robinson EOS for the Isenthalpic Decompression of Bocca Grande Fluids

The fugacity coefficients of all gas components in the H_2O-CO_2-H_2-CH_4-CO system, at the pressure and temperature values of the isenthalpic decompression of Bocca Grande fluids (see Sect. 4), were computed by means of the Peng-Robinson EOS. These calculations were performed by means of the Microsoft Excel computer code Prfug.xlsx, which is supplied as appendix of the textbook written by Elliott and Lira (2012). The use of this computer program is clearly explained by the YouTube video accessible at the following web address: https://www.youtube.com/watch?v=jLIIkVnEFpY (last accessed on March 27th, 2022). This Excel spreadsheet can be used to compute the compressibility factor and volume of gas mixtures as well as the component fugacities and fugacity coefficients, for up to three components only. Therefore, calculations were carried out separately for the ternary mixtures H_2O-CO_2-H_2, H_2O-CO_2-CH_4, and H_2O-CO_2-CO, for X_{CO_2} equal to 0.05, 0.10, 0.20, 0.30, and 0.40, fixed concentrations of H_2, CH_4, and CO, namely, $X_{H_2} = 4.69 \cdot 10^{-4}$, $X_{CH_4} = 1.66 \cdot 10^{-5}$, and $X_{CO} = 9.88 \cdot 10^{-7}$ (which are the mean values of the Solfatara fluids sampled by Chiodini and coworkers; Chiodini et al. 2021), and X_{H_2O} computed by difference to unity. This simplified approach is reasonable because the mole fractions of H_2, CH_4, and CO of the Bocca Grande (and Bocca Nuova) fumarolic fluids are much smaller than those of CO_2 and H_2O, whose cumulated concentration varies from 99.3 to 100 mol %. Thus, a molecule of H_2 has a high probability to interact with H_2O and CO_2 but a nil to negligible probability to interact with CH_4 and CO. Consequently, considering the ternary system H_2O-CO_2-H_2, neglecting CH_4 and CO, is

Table 5 Compressibility factor, volume, and fugacity of the gas mixture with CO_2 mole fraction of 0.05 and fugacity coefficients of H_2O, CO_2, H_2, CH_4, and CO in the same gas mixture computed by means of the Peng-Robinson EOS at the P,T conditions presumably met by Bocca Grande fluids along their isenthalpic expansion path. Values in bold-italics are at saturation conditions

P	T	Z_{mixt}	V_{mixt}	f_{mixt}	φ_{H_2O}	φ_{CO_2}	φ_{CH_4}	φ_{H_2}	φ_{CO}
bar	°C		cm^3/mol	bar					
1	161.7	0.9947	35,961.01	0.99	0.9944	1.0007	1.0039	1.0053	1.0041
4	171.4	0.9798	9052.46	3.92	0.9789	1.0031	1.0153	1.0207	1.0161
10	187.8	0.9534	3653.82	9.55	0.9525	1.0087	1.0373	1.0497	1.0394
16	201.7	0.9303	2295.47	14.95	0.9303	1.0153	1.0586	1.0772	1.0622
20	210.0	0.9163	1840.20	18.44	0.9173	1.0200	1.0726	1.0949	1.0772
22	***214.3***	***0.9098***	***1676.01***	***20.17***	***0.9113***	***1.0225***	***1.0793***	***1.1034***	***1.0845***
30	***230.6***	***0.8865***	***1237.46***	***26.91***	***0.8904***	***1.0328***	***1.1057***	***1.1360***	***1.1130***
40	***246.7***	***0.8595***	***928.74***	***35.02***	***0.8672***	***1.0469***	***1.1395***	***1.1769***	***1.1497***
50	***260.0***	***0.8344***	***739.70***	***42.81***	***0.8464***	***1.0625***	***1.1747***	***1.2187***	***1.1879***
60	***271.3***	***0.8106***	***611.54***	***50.31***	***0.8274***	***1.0795***	***1.2117***	***1.2621***	***1.2281***
65	277.1	0.8001	563.12	54.01	0.8192	1.0882	1.2297	1.2829	1.2479
80	293.9	0.7720	454.90	64.90	0.7977	1.1152	1.2836	1.3438	1.3070
100	313.7	0.7401	361.10	78.96	0.7740	1.1534	1.3569	1.4243	1.3877
150	354.4	0.6823	237.33	112.62	0.7306	1.2558	1.5442	1.6193	1.5956
200	386.5	0.6484	177.79	144.99	0.7012	1.3598	1.7274	1.7965	1.8009
300	434.3	0.6279	123.10	208.02	0.6647	1.5433	2.0401	2.0648	2.1551
500	495.0	0.6646	84.89	332.99	0.6318	1.8052	2.4578	2.3414	2.6334
700	531.7	0.7226	69.08	459.21	0.6182	2.0194	2.7844	2.5095	3.0035
900	555.6	0.7828	59.93	587.90	0.6122	2.2303	3.1057	2.6599	3.3614
1000	564.5	0.8129	56.62	653.54	0.6110	2.3385	3.2715	2.7347	3.5439
1500	592.0	0.9616	46.11	999.79	0.6164	2.9293	4.1929	3.1284	4.5386
2000	604.3	1.1081	40.42	1386.42	0.6351	3.6355	5.3224	3.5732	5.7271
2500	609.1	1.2537	36.78	1824.73	0.6631	4.4994	6.7398	4.0886	7.1874
2750	609.8	1.3264	35.41	2066.72	0.6799	5.0050	7.5850	4.3780	8.0459

Table 6 Compressibility factor, volume, and fugacity of the gas mixture with CO_2 mole fraction of 0.10 and fugacity coefficients of H_2O, CO_2, H_2, CH_4, and CO in the same gas mixture computed by means of the Peng-Robinson EOS at the P,T conditions presumably met by Bocca Grande fluids along their isenthalpic expansion path. Values in bold-italics are at saturation conditions

P	T	Z_{mixt}	V_{mixt}	f_{mixt}	φ_{H_2O}	φ_{CO_2}	φ_{CH_4}	φ_{H_2}	φ_{CO}
bar	°C		cm^3/mol	bar					
1	161.7	0.9950	35,972.30	1.00	0.9944	1.0005	1.0036	1.0050	1.0038
4	170.2	0.9808	9038.20	3.92	0.9788	1.0022	1.0141	1.0195	1.0151
10	185.0	0.9555	3639.32	9.57	0.9519	1.0065	1.0346	1.0472	1.0371
16	197.7	0.9332	2283.13	14.99	0.9289	1.0118	1.0547	1.0735	1.0587
18	201.6	0.9262	2031.11	16.75	0.9220	1.0137	1.0613	1.0821	1.0658

(continued)

Table 6 (continued)

P	T	Z_{mixt}	V_{mixt}	f_{mixt}	φ_{H_2O}	φ_{CO_2}	φ_{CH_4}	φ_{H_2}	φ_{CO}
bar	°C		cm^3/mol	bar					
19	***204.0***	***0.9231***	***1927.57***	***17.63***	***0.9188***	***1.0147***	***1.0644***	***1.0861***	***1.0692***
30	***227.1***	***0.8922***	***1236.95***	***27.05***	***0.8886***	***1.0264***	***1.0978***	***1.1281***	***1.1057***
50	***255.8***	***0.8433***	***741.68***	***43.12***	***0.8435***	***1.0517***	***1.1610***	***1.2050***	***1.1750***
79	***284.3***	***0.7816***	***458.50***	***64.54***	***0.7906***	***1.0974***	***1.2630***	***1.3246***	***1.2872***
80	285.2	0.7798	452.54	65.26	0.7891	1.0990	1.2664	1.3285	1.2911
100	304.4	0.7496	359.91	79.46	0.7638	1.1328	1.3343	1.4040	1.3665
150	343.7	0.6958	237.90	113.49	0.7172	1.2225	1.5050	1.5835	1.5575
200	374.6	0.6653	179.14	146.28	0.6853	1.3127	1.6691	1.7433	1.7429
300	420.5	0.6481	124.59	210.20	0.6450	1.4729	1.9499	1.9860	2.0633
500	478.0	0.6849	85.54	336.76	0.6070	1.7143	2.3480	2.2561	2.5207
700	511.7	0.7427	69.24	464.14	0.5890	1.9207	2.6770	2.4352	2.8941
900	533.3	0.8038	59.88	593.90	0.5793	2.1254	3.0040	2.5966	3.2588
1000	541.1	0.8346	56.50	660.10	0.5764	2.2305	3.1732	2.6765	3.4453
1500	564.5	0.9888	45.91	1010.23	0.5745	2.8069	4.1165	3.0943	4.4647
2000	573.6	1.1426	40.22	1403.00	0.5865	3.5023	5.2877	3.5677	5.6975
2250	575.3	1.2196	38.24	1619.23	0.5961	3.9089	5.9874	3.8332	6.4222
2500	575.8	1.2966	36.61	1850.71	0.6076	4.3628	6.7804	4.1213	7.2348

Table 7 Compressibility factor, volume, and fugacity of the gas mixture with CO_2 mole fraction of 0.20 and fugacity coefficients of H_2O, CO_2, H_2, CH_4, and CO in the same gas mixture computed by means of the Peng-Robinson EOS at the P,T conditions presumably met by Bocca Grande fluids along their isenthalpic expansion path. Values in bold-italics are at saturation conditions

P	T	Z_{mixt}	V_{mixt}	f_{mixt}	φ_{H_2O}	φ_{CO_2}	φ_{CH_4}	φ_{H_2}	φ_{CO}
bar	°C		cm^3/mol	bar					
1	161.7	0.9956	35,993.52	1.00	0.9945	1.0001	1.0030	1.0044	1.0033
4	168.2	0.9829	9016.59	3.93	0.9788	1.0005	1.0119	1.0173	1.0131
10	180.0	0.9598	3615.69	9.61	0.9511	1.0023	1.0295	1.0422	1.0324
16	190.5	0.9390	2262.30	15.07	0.9268	1.0051	1.0469	1.0662	1.0517
17	***192.8***	***0.9360***	***2133.07***	***15.97***	***0.9234***	***1.0057***	***1.0496***	***1.0698***	***1.0547***
25	***210.9***	***0.9150***	***1473.01***	***23.02***	***0.8994***	***1.0109***	***1.0702***	***1.0970***	***1.0778***
50	***247.2***	***0.8597***	***743.83***	***43.72***	***0.8387***	***1.0320***	***1.1358***	***1.1800***	***1.1516***
75	***270.7***	***0.8132***	***490.32***	***62.84***	***0.7902***	***1.0591***	***1.2067***	***1.2657***	***1.2315***
113	***296.2***	***0.7527***	***315.32***	***89.63***	***0.7290***	***1.1108***	***1.3293***	***1.4082***	***1.3699***
114	296.8	0.7513	312.28	90.30	0.7276	1.1123	1.3327	1.4121	1.3738
120	301.3	0.7451	296.56	94.49	0.7209	1.1206	1.3502	1.4312	1.3937
150	321.9	0.7199	237.40	115.11	0.6917	1.1622	1.4363	1.5226	1.4922
200	350.2	0.6944	179.94	148.64	0.6550	1.2311	1.5725	1.6580	1.6490
300	392.1	0.6829	125.89	214.12	0.6078	1.3556	1.8072	1.8648	1.9213

(continued)

Table 7 (continued)

P	T	Z_{mixt}	V_{mixt}	f_{mixt}	φ_{H_2O}	φ_{CO_2}	φ_{CH_4}	φ_{H_2}	φ_{CO}
bar	°C		cm^3/mol	bar					
500	443.0	0.7219	85.96	343.53	0.5596	1.5590	2.1695	2.1213	2.3420
700	471.4	0.7816	69.11	473.44	0.5336	1.7433	2.4912	2.3100	2.7099
900	488.6	0.8461	59.54	605.95	0.5173	1.9292	2.8174	2.4835	3.0760
1000	494.6	0.8791	56.12	673.72	0.5115	2.0253	2.9874	2.5696	3.2646
1250	504.9	0.9627	49.82	849.05	0.5017	2.2793	3.4418	2.7892	3.7623
1500	510.6	1.0472	45.49	1034.68	0.4969	2.5584	3.9494	3.0213	4.3097
1750	513.4	1.1323	42.31	1232.56	0.4956	2.8686	4.5243	3.2712	4.9211
2000	514.1	1.2179	39.86	1444.52	0.4970	3.2158	5.1803	3.5432	5.6103

Table 8 Compressibility factor, volume, and fugacity of the gas mixture with CO_2 mole fraction of 0.30 and fugacity coefficients of H_2O, CO_2, H_2, CH_4, and CO in the same gas mixture computed by means of the Peng-Robinson EOS at the P,T conditions presumably met by Bocca Grande fluids along their isenthalpic expansion path. Values in bold-italics are at saturation conditions

P	T	Z_{mixt}	V_{mixt}	f_{mixt}	φ_{H_2O}	φ_{CO_2}	φ_{CH_4}	φ_{H_2}	φ_{CO}
bar	°C		cm^3/mol	bar					
1	161.7	0.9961	36,012.95	1.00	0.9946	0.9997	1.0024	1.0038	1.0028
4	166.7	0.9849	9003.77	3.94	0.9791	0.9990	1.0098	1.0151	1.0112
10	176.0	0.9640	3599.88	9.65	0.9510	0.9985	1.0245	1.0372	1.0280
15	183.2	0.9480	2398.02	14.25	0.9300	0.9988	1.0367	1.0552	1.0421
25	***203.8***	***0.9233***	***1464.54***	***23.19***	***0.8977***	***1.0017***	***1.0588***	***1.0861***	***1.0677***
50	***238.2***	***0.8744***	***743.45***	***44.27***	***0.8352***	***1.0144***	***1.1135***	***1.1581***	***1.1312***
75	***260.2***	***0.8340***	***493.16***	***63.91***	***0.7849***	***1.0321***	***1.1712***	***1.2307***	***1.1984***
100	***276.7***	***0.7993***	***365.37***	***82.42***	***0.7417***	***1.0537***	***1.2335***	***1.3061***	***1.2708***
150	300.7	0.7430	236.33	116.82	0.6698	1.1067	1.3725	1.4667	1.4326
200	326.4	0.7216	179.86	151.06	0.6287	1.1583	1.4869	1.5838	1.5675
300	364.2	0.7150	126.29	218.10	0.5753	1.2544	1.6860	1.7639	1.8036
500	408.7	0.7580	85.93	350.62	0.5179	1.4215	2.0121	2.0044	2.1886
600	422.2	0.7886	75.98	417.01	0.4996	1.5005	2.1637	2.1013	2.3651
700	432.4	0.8215	68.84	483.97	0.4851	1.5795	2.3151	2.1924	2.5396
800	440.0	0.8559	63.43	551.75	0.4732	1.6596	2.4692	2.2810	2.7155
900	445.9	0.8911	59.19	620.64	0.4634	1.7416	2.6278	2.3688	2.8950
1000	450.3	0.9270	55.76	690.86	0.4553	1.8262	2.7923	2.4569	3.0795
1250	457.2	1.0184	49.47	873.54	0.4403	2.0518	3.2365	2.6836	3.5712
1500	460.0	1.1116	45.17	1068.46	0.4306	2.3025	3.7406	2.9263	4.1202
1750	460.2	1.2061	42.02	1277.94	0.4247	2.5843	4.3201	3.1911	4.7424

Table 9 Compressibility factor, volume, and fugacity of the gas mixture with CO_2 mole fraction of 0.40 and fugacity coefficients of H_2O, CO_2, H_2, CH_4, and CO in the same gas mixture computed by means of the Peng-Robinson EOS at the P,T conditions presumably met by Bocca Grande fluids along their isenthalpic expansion path. Values in bold-italics are at saturation conditions

P	T	Z_{mixt}	V_{mixt}	f_{mixt}	φ_{H_2O}	φ_{CO_2}	φ_{CH_4}	φ_{H_2}	φ_{CO}
bar	°C		cm^3/mol	bar					
1	161.7	0.9966	36,030.58	1.00	0.9948	0.9994	1.0020	1.0033	1.0024
4	165.6	0.9867	8997.86	3.95	0.9797	0.9977	1.0078	1.0131	1.0095
10	173.0	0.9681	3590.93	9.69	0.9517	0.9951	1.0197	1.0325	1.0238
14	177.7	0.9565	2560.94	13.41	0.9346	0.9939	1.0277	1.0452	1.0335
16	180.0	0.9509	2239.06	15.24	0.9264	0.9935	1.0318	1.0515	1.0384
19	***184.2***	***0.9432***	***1887.53***	***17.97***	***0.9151***	***0.9930***	***1.0377***	***1.0606***	***1.0455***
25	***196.1***	***0.9307***	***1452.26***	***23.35***	***0.8966***	***0.9933***	***1.0484***	***1.0763***	***1.0586***
50	***228.5***	***0.8872***	***740.15***	***44.77***	***0.8327***	***0.9985***	***1.0935***	***1.1391***	***1.1133***
75	***249.1***	***0.8521***	***493.35***	***64.88***	***0.7809***	***1.0082***	***1.1405***	***1.2011***	***1.1703***
100	***264.4***	***0.8225***	***367.56***	***83.95***	***0.7366***	***1.0212***	***1.1903***	***1.2640***	***1.2306***
150	***286.3***	***0.7760***	***240.61***	***119.71***	***0.6628***	***1.0542***	***1.2982***	***1.3936***	***1.3610***
180	***296.1***	***0.7551***	***198.55***	***139.95***	***0.6255***	***1.0772***	***1.3675***	***1.4730***	***1.4447***
200	304.4	0.7491	179.85	153.80	0.6085	1.0921	1.4058	1.5130	1.4914
300	338.0	0.7472	126.55	222.61	0.5496	1.1636	1.5738	1.6702	1.6964
400	360.8	0.7667	101.02	290.84	0.5117	1.2310	1.7218	1.7899	1.8761
500	376.6	0.7952	85.92	359.09	0.4843	1.2962	1.8614	1.8912	2.0439
600	388.0	0.8283	75.88	427.79	0.4631	1.3611	1.9987	1.9835	2.2071
700	396.2	0.8640	68.69	497.32	0.4462	1.4267	2.1374	2.0714	2.3700
900	406.5	0.9397	59.00	640.14	0.4205	1.5634	2.4272	2.2436	2.7052
1000	409.7	0.9788	55.57	713.97	0.4107	1.6355	2.5811	2.3305	2.8808
1250	413.7	1.0791	49.30	907.44	0.3918	1.8298	3.0013	2.5567	3.3537
1500	414.1	1.1820	45.02	1115.99	0.3785	2.0484	3.4852	2.8024	3.8890

expected to have negligible effects on the compressibility factor and volume of the gas mixture as well as on the fugacity coefficients of H_2O, CO_2, and H_2, for a given P–T-X condition (see Sect. 7.3 for further details). Similar considerations apply to the ternary systems H_2O-CO_2-CH_4 and H_2O-CO_2-CO.

Results are listed in Tables 5, 6, 7, 8 and 9, while the fugacity coefficients of H_2O, CO_2, H_2, CH_4, and CO are presented in the diagrams of Fig. 16. Inspection of these diagrams and tables shows that the fugacity coefficients of H_2, CH_4, and CO are relatively similar to each other and are somewhat higher than the fugacity coefficient of CO_2 in general. The fugacity coefficients of all these nonpolar gas species increase with increasing pressure (and temperature), in contrast with the fugacity coefficient of the polar H_2O molecule, which decreases with increasing pressure (and temperature). At the same total pressure, the fugacity coefficients of H_2, CH_4, CO, and CO_2 decrease with increasing CO_2 mole fraction and vice versa. For instance, at 1000 bar total fluid pressure: (i) the φ_{H_2} is 2.7347 for $X_{CO_2} = 0.05$, 2.6765 for $X_{CO_2} = 0.10$, 2.5696 for $X_{CO_2} = 0.20$, 2.4569 for $X_{CO_2} = 0.30$, and 2.3305 for $X_{CO_2} = 0.40$; (ii) the φ_{CH_4} is 3.2715 for $X_{CO_2} = 0.05$, 3.1732 for $X_{CO_2} = 0.10$, 2.9874 for $X_{CO_2} = 0.20$, 2.7923 for $X_{CO_2} = 0.30$, and 2.5811 for $X_{CO_2} = 0.40$;

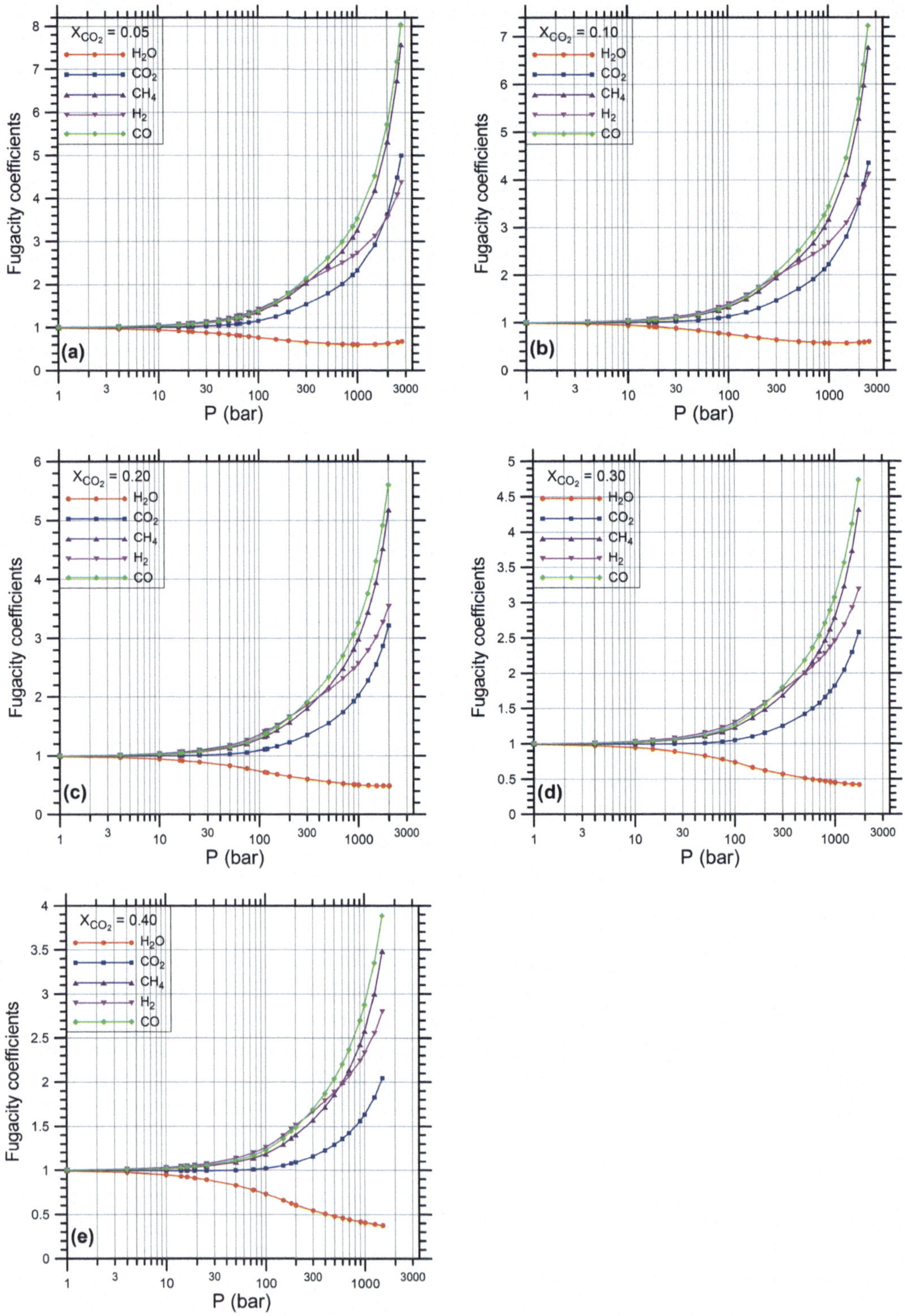

Fig. 16 Fugacity coefficients of H_2O, CO_2, H_2, CH_4, and CO in the gas mixtures with CO_2 mole fraction of (a) 0.05, (b) 0.10, (c) 0.20, (d) 0.30, and (e) 0.40, computed by means of the Peng-Robinson EOS at the P,T conditions met by Bocca Grande fluids along their isenthalpic expansion path

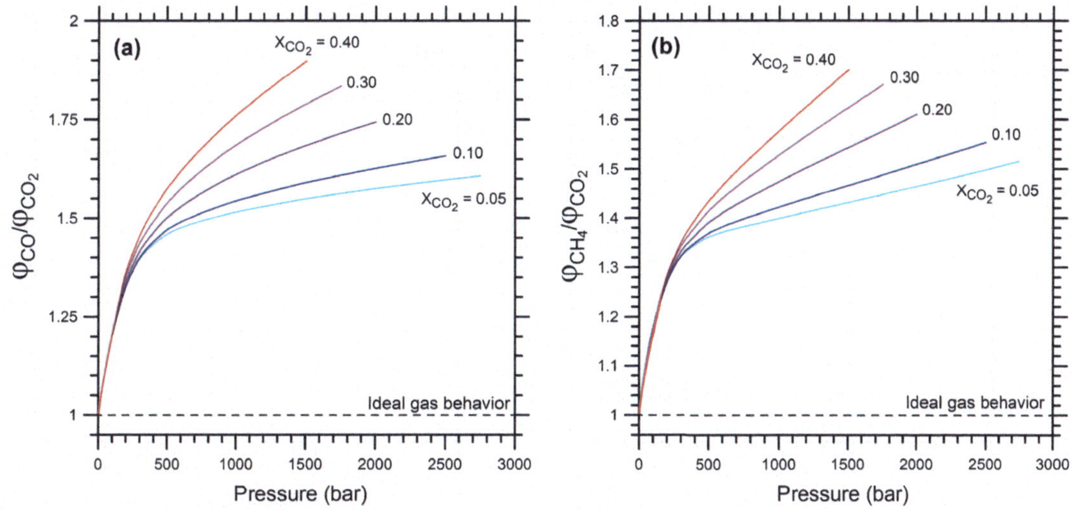

Fig. 17 Ratio of the fugacity coefficients of **a** CO and CO_2 and **b** CH_4 and CO_2 computed by means of the Peng-Robinson EOS at the P,T conditions presumably met by Bocca Grande fluids with different CO_2 mole fractions along their isenthalpic expansion path

(iii) the φ_{CO} is 3.5439 for $X_{CO_2} = 0.05$, 3.4453 for $X_{CO_2} = 0.10$, 3.2646 for $X_{CO_2} = 0.20$, 3.0795 for $X_{CO_2} = 0.30$, and 2.8808 for $X_{CO_2} = 0.40$; (iv) the φ_{CO_2} is 2.3385 for $X_{CO_2} = 0.05$, 2.2305 for $X_{CO_2} = 0.10$, 2.0253 for $X_{CO_2} = 0.20$, 1.8262 for $X_{CO_2} = 0.30$, and 1.6355 for $X_{CO_2} = 0.40$.

Further indications are given by the ratios of fugacity coefficients for the Bocca Grande fluids with different CO_2 mole fractions. The ratio of the fugacity coefficients of CO and CO_2 increases with increasing total fluid pressure and CO_2 mole fraction, departing more and more from unity (Fig. 17a). For instance, at the total fluid pressure of 1000 bar, the $\varphi_{CO}/\varphi_{CO_2}$ ratio is 1.52 for $X_{CO_2} = 0.05$, 1.54 for $X_{CO_2} = 0.10$, 1.61 for $X_{CO_2} = 0.20$, 1.69 for $X_{CO_2} = 0.30$, and 1.76 for $X_{CO_2} = 0.40$. The $\varphi_{CO}/\varphi_{CO_2}$ ratio attains the value of 1.90, representing the maximum deviation from unity, at 1500 bar and 414.1 °C for the gas mixture with CO_2 mole fraction of 0.40.

Also the ratio of the fugacity coefficients of CH_4 and CO_2 increases with increasing total fluid pressure and CO_2 mole fraction, deviating progressively from unity (Fig. 17b). For example, at the total fluid pressure of 1000 bar, the $\varphi_{CH_4}/\varphi_{CO_2}$ ratio is 1.40 for $X_{CO_2} = 0.05$, 1.42 for $X_{CO_2} = 0.10$, 1.48 for $X_{CO_2} = 0.20$, 1.53 for $X_{CO_2} = 0.30$, and 1.58 for $X_{CO_2} = 0.40$. The maximum deviation from unity is reached at 1500 bar and 414.1 °C, for the gas mixture having CO_2 mole fraction of 0.40, with a $\varphi_{CH_4}/\varphi_{CO_2}$ ratio of 1.70.

In general, the ratio of the fugacity coefficients of CO and CH_4 exhibits small deviation from unity (Fig. 18a). In detail, the $\varphi_{CO}/\varphi_{CH_4}$ ratio increases with total fluid pressure, up to maximum values of 1.083 at 1000 bar for $X_{CO_2} = 0.05$, 1.086 at 1000 bar for $X_{CO_2} = 0.10$, 1.093 at 1250 bar for $X_{CO_2} = 0.20$, 1.103 at 1250 bar for $X_{CO_2} = 0.30$, and 1.117 at 1250 bar for $X_{CO_2} = 0.40$. Upon further increments in total fluid pressure, the $\varphi_{CO}/\varphi_{CH_4}$ ratio decreases slightly, attaining the values of 1.061 at 2750 bar for $X_{CO_2} = 0.05$, 1.067 at 2500 bar for

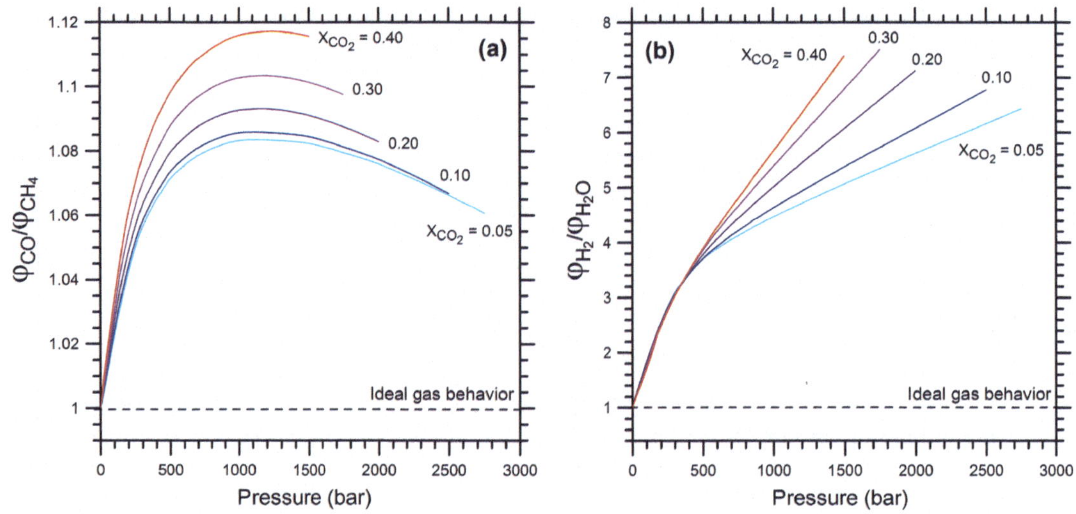

Fig. 18 Ratio of the fugacity coefficients of **a** CO and CH_4 and **b** H_2 and H_2O computed by means of the Peng-Robinson EOS at the P,T conditions presumably met by Bocca Grande fluids with different CO_2 mole fractions along their isenthalpic expansion path

$X_{CO_2} = 0.10$, 1.083 at 2000 bar for $X_{CO_2} = 0.20$, 1.098 at 1750 bar for $X_{CO_2} = 0.30$, and 1.116 at 1500 bar for $X_{CO_2} = 0.40$.

In contrast, the ratio of the fugacity coefficients of H_2 and H_2O departs significantly from unity, with deviations becoming more and more conspicuous with increasing total fluid pressure and CO_2 mole fraction (Fig. 24). The $\varphi_{H_2}/\varphi_{H_2O}$ ratio is close to unity up to pressures of a few hundred bars, but it grows quickly and continuously with increasing pressure achieving the maximum values of 6.44 at 2750 bar for $X_{CO_2} = 0.05$, 6.78 at 2500 bar for $X_{CO_2} = 0.10$, 7.13 at 2000 bar for $X_{CO_2} = 0.20$, 7.51 at 1750 bar for $X_{CO_2} = 0.30$, and 7.40 at 1500 bar for $X_{CO_2} = 0.40$.

All in all, neglecting the $\varphi_{CO}/\varphi_{CH_4}$ ratio produces nil to negligible effects on the geothermometers and geobarometers calibrated for the isenthalpic expansion of Solfatara fluids, disregarding the $\varphi_{CO}/\varphi_{CO_2}$ and $\varphi_{CH_4}/\varphi_{CO_2}$ ratios may cause moderate errors, whereas ignoring the $\varphi_{H_2}/\varphi_{H_2O}$ ratio may lead to significant errors.

6 Comparison of the Fugacity Coefficients of CO_2 and H_2O from the Peng-Robinson EOS, GERG-2008/CONVAL, and Gallagher et al. (1993) for the Isenthalpic Expansion of Bocca Grande Fluids

It is instructive to compare the fugacity coefficients of CO_2 and H_2O computed using the Peng-Robinson EOS, at the pressure, temperature conditions encountered by the Bocca Grande fluids with CO_2 mole fractions of 0.05, 0.10, 0.20, 0.30, and 0.40 along their isenthalpic expansion path, with both: (i) those obtained elaborating the results of the GERG-2008 thermodynamic module of the CONVAL software package, and (ii) those given by Gallagher et al. (1993). The three series of φ_{CO_2} and φ_{H_2O} values are plotted as a function of pressure in Figs. 19 and 20, respectively, in which the logarithmic scale is adopted for pressure, to accommodate values distributing over three orders of magnitude. The φ_{CO_2} and φ_{H_2O} were tabulated by Gallagher et al. (1993) up to a maximum pressure

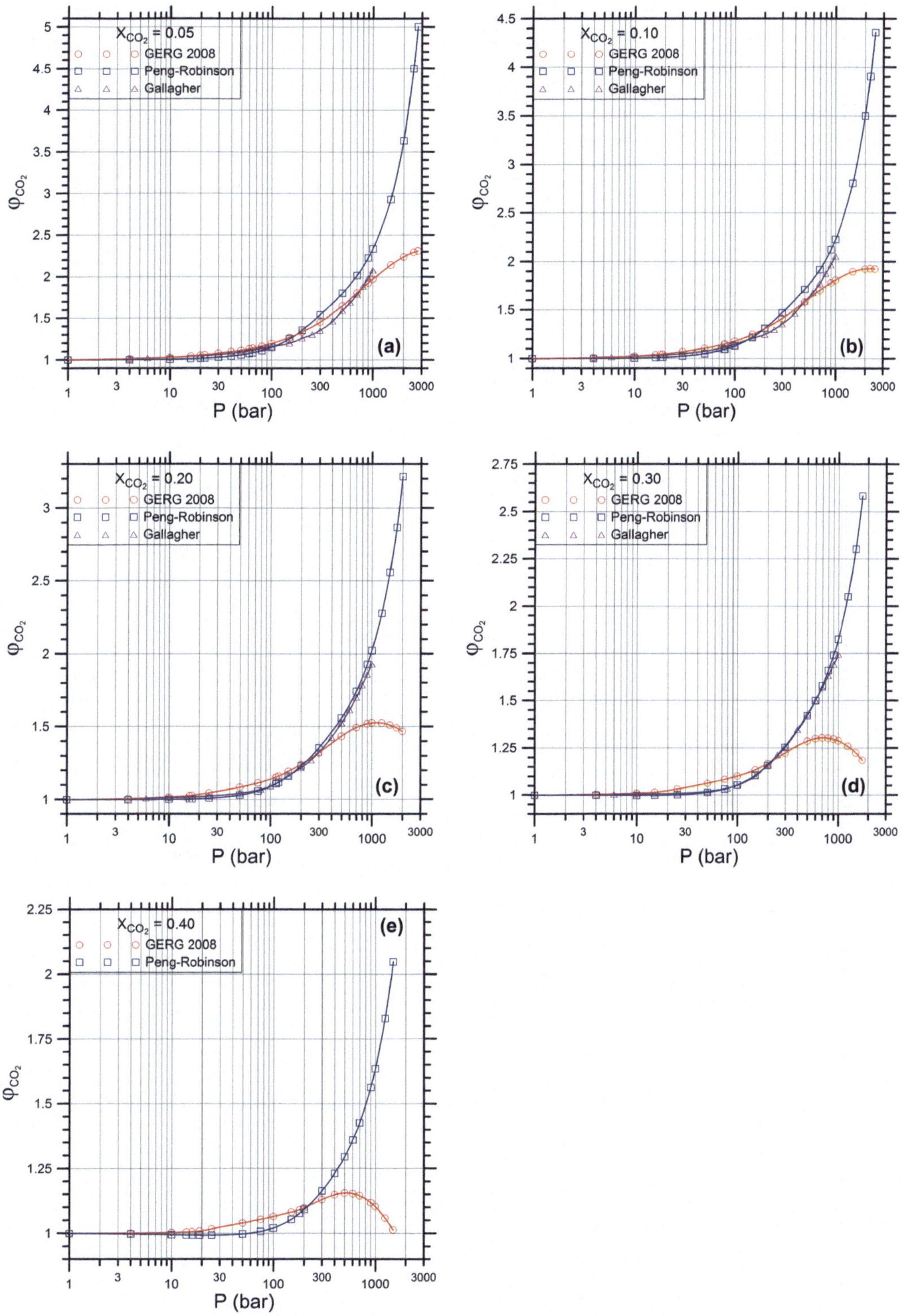

Fig. 19 Comparison of the fugacity coefficients of CO_2 from the Peng-Robinson EOS, GERG-2008/CONVAL, and Gallagher et al. (1993) at the pressure, temperature conditions encountered by the Bocca Grande fluids with CO_2 mole fractions of **a** 0.05, **b** 0.10, **c** 0.20, **d** 0.30 and **e** 0.40 along their isenthalpic decompression path

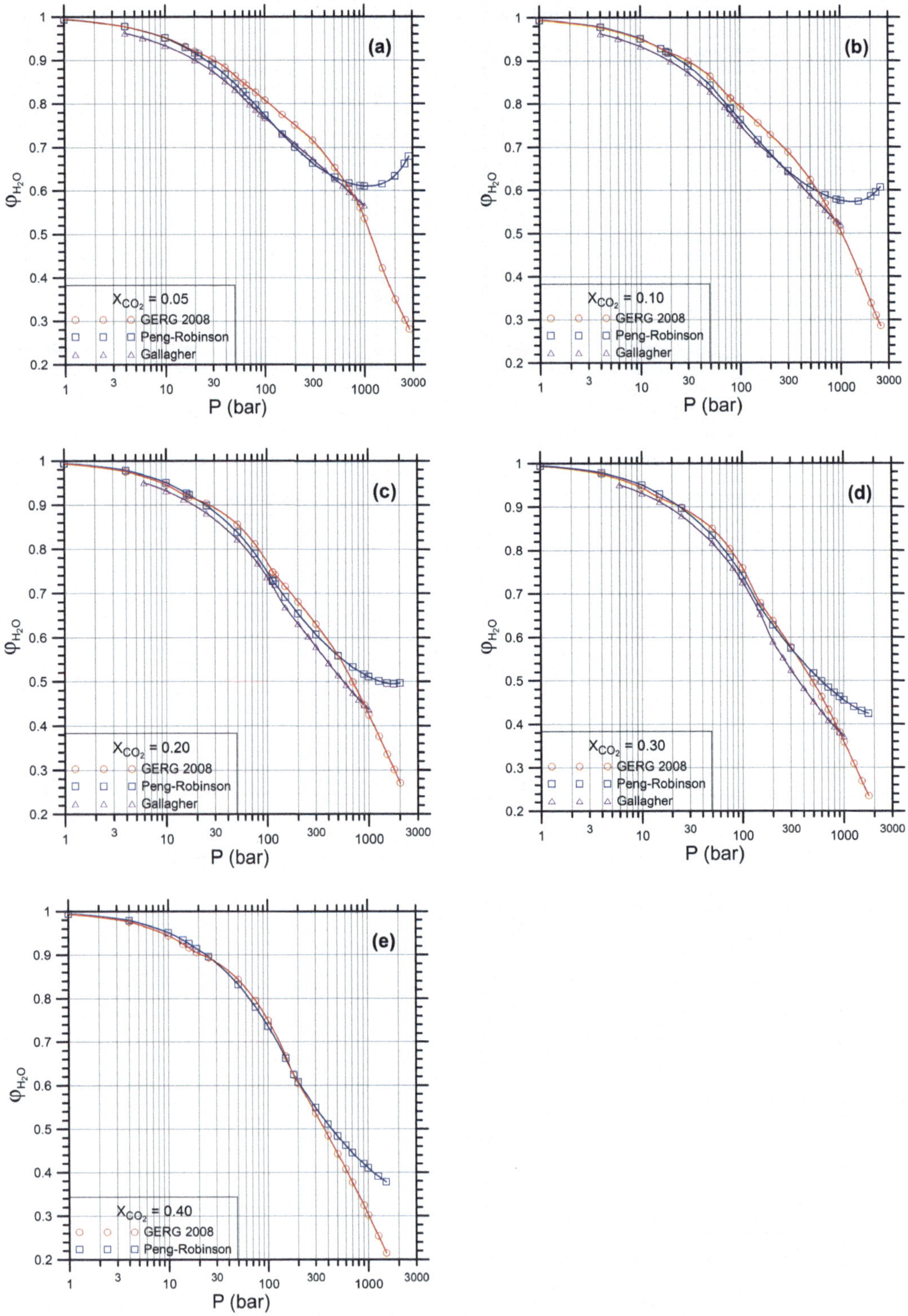

Fig. 20 Comparison of the fugacity coefficients of H_2O from the Peng-Robinson EOS, GERG-2008/CONVAL, and Gallagher et al. (1993) at the pressure, temperature conditions encountered by the Bocca Grande fluids with CO_2 mole fractions of **a** 0.05, **b** 0.10, **c** 0.20, **d** 0.30 and **e** 0.40 along their isenthalpic decompression path

of 1000 bar, whereas the fugacity coefficients calculated by means of the Peng-Robinson EOS and GERG-2008/CONVAL achieve higher pressures as mentioned in Sects. 5 and 4.2, respectively.

Some differences in the three series of fugacity coefficients are expected because they are derived from three different EOS, although the GERG-2008 EOS, and the EOS of Gallagher et al. (1993) are both explicit in the Helmholtz free energy. Furthermore, Gallagher et al. (1993) derived the φ_{CO_2} and φ_{H_2O} values directly from the Helmholtz free energy of the considered gas mixture, through numerical differentiation, whereas the molar volumes and compressibility factors from GERG-2008/CONVAL were used together with the virial coefficients of the CO_2-H_2O gas mixtures to obtain the fugacity coefficients of CO_2 and H_2O (see Sect. 4.2).

The diagrams of Fig. 19 show that the CO_2 fugacity coefficients from the Peng-Robinson EOS and those of Gallagher et al. (1993) increase continuously with pressure, behaving as an increasing monotonic function, and are in good agreement with each other. In contrast, the CO_2 fugacity coefficients of GERG-2008/CONVAL increase with pressure, passes through a maximum, and then decreases with further increasing pressure (as already mentioned in Sect. 4.2), deviating more and more from the other two data series. The discrepancies between the φ_{CO_2} values of GERG-2008/CONVAL and those of the other two datasets are limited below pressures close to 1000 bar for the gas mixtures with X_{CO_2} of 0.05, 800 bar for X_{CO_2} of 0.10, 600 bar for X_{CO_2} of 0.20, and 400 bar for X_{CO_2} of 0.30, whereas the φ_{CO_2} values of GERG-2008/CONVAL deviate significantly from those of the Peng-Robinson EOS above 400 bar approximately for the gas mixtures with X_{CO_2} of 0.40.

The plots of Fig. 20 indicate that the H_2O fugacity coefficients according to both Gallagher et al. (1993) and GERG-2008/CONVAL decrease continuously with increasing pressure, behaving as a decreasing monotonic function.

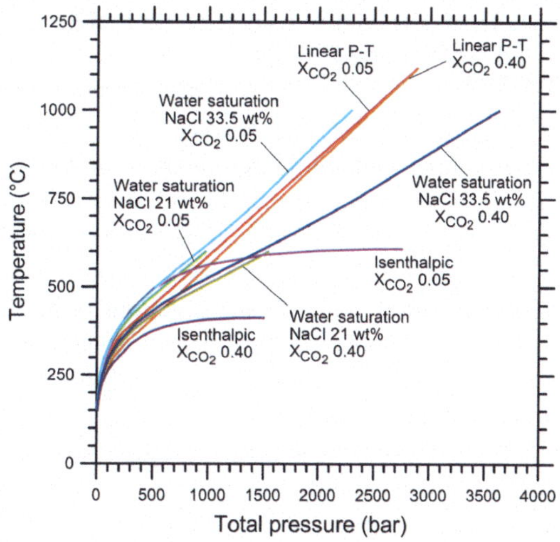

Fig. 21 Temperature and total fluid pressure along possible decompression paths of Solfatara fluids

The two datasets show moderate deviations. In contrast, the φ_{H_2O} from the Peng-Robinson EOS for the gas mixtures with X_{CO_2} of 0.05, 0.10, and 0.20 decreases with pressure, passes through a minimum, and then increases moderately with further increasing pressure, deviating somewhat from the other two φ_{H_2O} series at pressures higher than 1000–2000 bar. The φ_{H_2O} from the Peng-Robinson EOS decreases continually with increasing pressure for the gas mixtures with X_{CO_2} of 0.30 and 0.40, but it departs from the other φ_{H_2O} values at pressures higher than several hundred bar.

Summing up, at pressures greater than some tens of bar, corresponding to temperatures higher than 230–250 °C, the φ_{CO_2} values obtained from the three EOS are significantly higher than unity, while the three series of φ_{H_2O} values are considerably lower than unity. Therefore, above these P,T thresholds, it is advisable, not to say mandatory, to take into account deviations from ideality, of both CO_2 and H_2O in geothermometric-geobarometric calculations. However, the fugacity coefficients of both CO_2 and H_2O might be affected by some uncertainties at pressures greater than 400–1000 bar depending on the CO_2 mole fraction.

7 Other Possible Expansion Paths of Solfatara Fluids

Previous sections of this chapter were largely focused on the isenthalpic expansion of Solfatara fluids for several reasons. As already recalled in Sect. 1, the isenthalpic flow is closely approached by geothermal fluids in production wells and presumably in natural manifestations of high flowrate and, more specifically, the isenthalpic decompression was already suggested for the Solfatara fluids by previous studies (e.g., Cioni et al 1984; Moretti et al. 2018, 2020).

However, as already underscored in Sect. 4, the isenthalpic paths can be modeled up to maximum pressures and temperatures of: (i) 2750 bar and 609.8 °C for the gas mixture with X_{CO_2} of 0.05, (ii) 2500 bar and 575.8 °C for the gas mixture with X_{CO_2} of 0.10, (iii) 2000 bar and 514.1 °C for the gas mixture with X_{CO_2} of 0.20, (iv) 1750 bar and 460.2 °C for the gas mixture with X_{CO_2} of 0.30, and (v) 1500 bar and 414.1 °C for the gas mixture with X_{CO_2} of 0.40. This is because temperature does not change appreciably close to these upper pressure values, as shown for the two limiting isenthalpic paths reported in the diagram of Fig. 21, which refer to CO_2 mole fractions of 0.05 and 0.40. Actually, the range of applicability of the isenthalpic expansion model is further limited by the non-monotonic behavior of the sum of log-ratios of reaction SS4 (including mole fraction ratios, fugacity coefficient ratios and other terms; see Sect. 2 in chapter "Calibration and Application of the RWG and SS4 Indicators for the Possible Expansion Paths of Solfatara Fluids"). Due to these limitations of the isenthalpic modeling, two alternative approaches, already considered by Stevenson (1993; see Sect. 1), were taken into account in this work, as discussed in the next two sections.

7.1 The Saturation Decompression Paths

One of the alternative approaches mentioned at the end of the previous section assumes the occurrence of saturation conditions, that is, equilibrium coexistence of a liquid phase and steam. If the liquid phase is pure H_2O, the maximum attainable temperature and pressure are its critical values, that is, 373.946 °C and 220.640 bar (Wagner and Pruß 2002), which are probably too low compared to the temperatures and pressures presumably present at depth below the Solfatara. Since the critical temperatures of the binary H_2O- CO_2 system are lower than the critical temperature of pure H_2O (see Sect. 2.1), referring to the H_2O- CO_2 system makes things worse. In contrast, since the critical temperatures found along the critical curve of the binary H_2O-NaCl system are higher than the critical temperature of pure H_2O (e.g., Driesner and Heinrich 2007 and references therein), a NaCl brine and the related vapor phase coexist at temperatures and pressures higher than those of pure water. Based on these considerations, Giggenbach (1987) adopted the following simple equation:

$$\log f_{H_2O} = 4.90 - 1820/T \qquad (134)$$

to describe the coexistence of a NaCl brine with a vapor phase up to a temperature of 600 °C and a f_{H_2O} of 654 bar. The possible shifts of the brine-vapor co-occurrence condition to lower temperatures or higher H_2O fugacities due to the presence of CO_2 were neglected by Giggenbach (1987) because they were considered to be negligible for a vapor phase containing up to 20 mol % CO_2 at 6 weight % NaCl. Although the CO_2 concentration attains values higher than 20 mol % in the Solfatara vapors, the binary H_2O-NaCl system is considered here, instead of the ternary H_2O-NaCl-CO_2 system, to describe brine-vapor coexistence both: (1) to avoid complicating the calculations too much and (2) because of the incomplete knowledge of the P-V-T-X properties of the H_2O-NaCl-CO_2 system, which is largely based on the experimental work by Gehrig (1980) and the P-V-T-X data obtained through the synthetic fluid inclusion technique (e.g., Kotelnikov and Kotelnikova 1990; Frantz et al. 1992; Johnson 1992; Shmulovich and Graham 1999; Schmidt and Bodnar 2000).

In addition to the determinations relevant for the binary H_2O-CO_2 system (see Sect. 2.1), Gehrig (1980) measured the molar volume and defined the immiscibility boundaries for the (H_2O +6 wt.% NaCl)-CO_2 pseudobinary system at pressures from 0 to 3000 bar and temperatures from 200 to 560 °C using a high-pressure variable volume autoclave. The data of Gehrig (1980) represent the main experimental foundations of the modified Redlich–Kwong EOS of Bowers and Helgeson (1983) as well as of the EOS of Duan et al. (1995), both for H_2O-CO_2-NaCl fluids. Bowers and Helgeson (1983) computed fugacity coefficients of H_2O and CO_2 at temperatures of 400, 450, 500, 550, and 600 °C and pressures of 500, 1000, and 2000 bar and used these fugacity coefficients together with solubility data to establish the compositions of the coexisting immiscible phases. However, the EOS of Bowers and Helgeson (1983) was questioned by Duan et al. (1995). According to Duan et al. (1995), their EOS predicts P–V-T-X data, immiscibility/phase equilibria, solubilities, and activities with an accuracy similar to that of the experimental data at temperatures from 300 to ~1000 °C and pressures from 0 to 6000 bar for NaCl concentrations to ~30 wt% (relative to NaCl + H_2O) and to ~50 wt% with lower accuracy. However, the EOS of Duan et al. (1995) was considered poorly reliable by other authors (e.g., Schmidt and Bodnar 2000).

Given this situation, the binary H_2O-NaCl system was considered in this work, instead of the ternary H_2O-NaCl-CO_2 system, to describe the equilibrium coexistence of the vapor phase with a brine containing either 21 wt% NaCl (data from Tanger and Pitzer 1989) or 33.5 wt% NaCl (data from Driesner and Heinrich 2007). Reported pressures were considered to be representative of H_2O partial pressures. Then, the total fluid pressure of the system including CO_2 was obtained by means of the simple relation:

$$P_{tot} = P_{H_2O} \cdot \left(1 + \frac{X_{CO_2}}{1 - X_{CO_2}}\right) \quad (135)$$

for CO_2 mole fractions in the vapor phase of 0.05, 0.10, 0.20, 0.30, and 0.40. The two limiting water-saturation decompression paths, for CO_2 mole fractions in the vapor phase of 0.05 and 0.40, are reported in the diagram of Fig. 21, for the two systems comprising a brine with 21 and 33.5 wt% NaCl relative to water. Although the temperatures and pressures of vapor–liquid coexistence for these two systems were taken from two distinct sources, the two expansion paths at the same CO_2 mole fraction in the vapor phase are almost superimposed and differ only for their extension, which is greater for the system involving a brine with 33.5 wt% NaCl than for that including a brine with 21 wt% NaCl. To be noted that the slope of the saturation expansion paths decreases progressively with increasing pressure up to some hundreds bar, but it becomes almost constant at higher pressures. Thus, for the system including a brine with 21 wt % NaCl, at 600 °C, the increase from 0.05 to 0.40 in the CO_2 mole fraction of the vapor phase causes an increment from 688 to 1090 bar in the maximum total fluid pressure. In contrast, for the system involving a brine with 33.5 wt% NaCl, at 1000 °C, the increment from 0.05 to 0.40 in the CO_2 mole fraction of the vapor phase determines an increase from 2291 to 3627 bar in the maximum total fluid pressure. These two maximum pressures bracket the value of 2879 bar presumably present in the magmatic sector of the Solfatara magmatic-hydrothermal system, whereas the maximum temperature is 120 °C lower than that of the magmatic endmember M1, 1120 °C (Caliro et al. 2014; see Sect. 10.2 in chapter "Chemical and Isotopic Characteristics of Solfatara-Pisciarelli Fumarolic Fluids" for further details).

7.2 The Linear P–T Decompression Path

The other alternative decompression path considered in this work assumes that both total fluid pressure and temperature increase linearly from the critical point of water to the values of the magmatic endmember M1 (Caliro et al. 2014; see above). Below the critical point of water, P_{H_2O} and temperature are assumed to be constrained by the saturation condition for pure water, while total fluid pressure is computed using Eq. (135).

Under this hypothesis, which is called "linear P–T expansion path", total fluid pressure (P_{tot} in bar) is a function of both temperature (T in °C) and X_{CO_2}, above the critical point of pure water, according to the relation:

$$P_{tot} = \begin{pmatrix} -1.07619 \cdot X^4_{CO_2} + 0.09285714 \cdot X^3_{CO_2} - 0.3716667 \\ \cdot X^2_{CO_2} - 0.2898571 \cdot X_{CO_2} + 3.563117 \end{pmatrix} \cdot T + \begin{pmatrix} 1152.829 \cdot X^4_{CO_2} - 57.5119 \cdot X^3_{CO_2} + 403.51 \\ \cdot X^2_{CO_2} + 325.7244 \cdot X_{CO_2} - 1111.676 \end{pmatrix}. \quad (136)$$

This assumption, although extremely crude, allows one to overcome the deadlock of the isenthalpic models and to "approach the magma" which is presumably present at depth underneath the Solfatara volcano. To be noted also that the linear P–T expansion path for X_{CO_2} of 0.05 in the vapor phase, is intermediate between the two limiting water-saturation decompression paths for the system involving a brine with 33.5 wt% NaCl, having CO_2 mole fractions of 0.05 and 0.40 in the vapor phase, whereas the linear P–T expansion path for X_{CO_2} of 0.40 in the vapor phase is in between these two limiting water-saturation decompression paths above a total pressure of ca. 750 bar. Thus, also the linear P–T expansion path may be controlled by saturation conditions in a system comprising a high-salinity brine.

7.3 Fugacity Coefficients at the Pressures and Temperatures of Coexistence of a Brine Containing 21 Wt% NaCl and the Related Vapor Phase

The fugacity coefficients of H_2O, CO_2, H_2, CH_4, and CO were computed by means of the Peng-Robinson EOS, at temperatures varying from 150 to 600 °C, at steps of 25 °C, and at the corresponding total pressures given by Eq. (135), for a vapor phase with CO_2 mole fraction of 0.05, 0.10, 0.20, 0.30, and 0.40 coexisting with a brine containing 21 wt% NaCl. Calculations were done by means of the Microsoft Excel computer code Prfug.xlsx (Elliott and Lira 2012) adopting the same approach described in Sect. 5.

The obtained fugacity coefficients are listed in Tables 10, 11, 12, 13 and 14 and are reported in the diagrams of Fig. 22, whose inspection suggests what follows:

(i) The fugacity coefficient of H_2O is close to one at low pressures and temperatures, ranging from 0.9725, at 4.54 bar, 150 °C, and CO_2 mole fraction of 0.05, to 0.9597, at 7.20 bar, 150 °C, and CO_2 mole fraction of 0.40. It decreases with increasing pressure and temperature, reaching minimum values varying from 0.5969, at 497 bar, 475 °C, and CO_2 mole fraction of 0.05, to 0.5691, at 645 bar, 450 °C, and CO_2 mole fraction of 0.40. It rises again due to a further increase in pressure and temperature, up to values fluctuating from 0.6686, at 1026 bar, 600 °C, and CO_2 mole fraction of 0.10, to 0.7115, at 1539 bar, 600 °C, and CO_2 mole fraction of 0.40.

(ii) The fugacity coefficient of CO_2 is close to unity at low pressures and temperatures,

Table 10 Compressibility factor, volume, and fugacity of the gas mixture with CO_2 mole fraction of 0.05 and fugacity coefficients of H_2O, CO_2, H_2, CH_4, and CO in the same gas mixture computed by means of the Peng-Robinson EOS at the P,T conditions fixed by coexistence with a brine containing 21 wt% NaCl

P	T	Z_{mixt}	V_{mixt}	f_{mixt}	φ_{H_2O}	φ_{CO_2}	φ_{CH_4}	φ_{H_2}	φ_{CO}
bar	°C		cm^3/mol	bar					
4.5	150	0.9735	7535.77	4.43	0.9725	1.0035	1.0197	1.0270	1.0206
8.3	175	0.9581	4288.68	7.99	0.9572	1.0071	1.0328	1.0441	1.0345
14.3	200	0.9374	2578.43	13.45	0.9371	1.0133	1.0520	1.0686	1.0552
23.3	225	0.9107	1620.58	21.35	0.9120	1.0235	1.0797	1.1032	1.0852
36.2	250	0.8773	1055.48	32.15	0.8821	1.0397	1.1193	1.1513	1.1282
53.9	275	0.8368	707.00	46.26	0.8478	1.0653	1.1759	1.2181	1.1900
77.7	300	0.7894	484.23	63.91	0.8100	1.1047	1.2571	1.3111	1.2789
109	325	0.7349	336.10	85.42	0.7693	1.1659	1.3772	1.4441	1.4108
148	350	0.6768	237.27	110.59	0.7280	1.2587	1.5532	1.6315	1.6048
196	375	0.6204	170.30	139.84	0.6874	1.3960	1.8103	1.8917	1.8894
255	400	0.5795	126.99	173.74	0.6509	1.5728	2.1412	2.2033	2.2580
326	425	0.5677	101.01	213.65	0.6218	1.7480	2.4658	2.4750	2.6233
407	450	0.5855	86.43	260.40	0.6040	1.8741	2.6832	2.6171	2.8721
497	475	0.6207	77.64	315.04	0.5969	1.9570	2.8012	2.6553	3.0106
592	500	0.6645	72.14	377.23	0.5996	2.0087	2.8450	2.6270	3.0653
689	525	0.7112	68.54	446.42	0.6101	2.0416	2.8439	2.5658	3.0687
783	550	0.7576	66.19	521.34	0.6269	2.0611	2.8134	2.4886	3.0378
878	575	0.8024	64.47	602.55	0.6476	2.0762	2.7744	2.4116	2.9962
972	600	0.8451	63.13	690.27	0.6710	2.0901	2.7345	2.3399	2.9525

Table 11 Compressibility factor, volume, and fugacity of the gas mixture with CO_2 mole fraction of 0.10 and fugacity coefficients of H_2O, CO_2, H_2, CH_4, and CO in the same gas mixture computed by means of the Peng-Robinson EOS at the P,T conditions fixed by coexistence with a brine containing 21 wt% NaCl

P	T	Z_{mixt}	V_{mixt}	f_{mixt}	φ_{H_2O}	φ_{CO_2}	φ_{CH_4}	φ_{H_2}	φ_{CO}
bar	°C		cm^3/mol	bar					
4.80	150	0.9736	7140.33	4.67	0.9711	1.0025	1.0191	1.0268	1.0202
8.79	175	0.9585	4064.39	8.43	0.9550	1.0055	1.0318	1.0436	1.0339
15.1	200	0.9381	2444.44	14.21	0.9338	1.0109	1.0505	1.0679	1.0542
24.6	225	0.9119	1537.35	22.56	0.9075	1.0200	1.0774	1.1019	1.0837
38.2	250	0.8794	1002.43	33.99	0.8761	1.0349	1.1158	1.1489	1.1259
56.9	275	0.8407	672.87	48.96	0.8403	1.0582	1.1702	1.2136	1.1860
82.0	300	0.7961	462.63	67.76	0.8009	1.0942	1.2473	1.3022	1.2714
115	325	0.7465	323.42	90.77	0.7586	1.1490	1.3583	1.4252	1.3947
156	350	0.6965	231.31	117.94	0.7162	1.2286	1.5131	1.5894	1.5673
207	375	0.6524	169.65	149.86	0.6753	1.3380	1.7216	1.7983	1.8010
270	400	0.6249	129.75	187.34	0.6394	1.4688	1.9673	2.0256	2.0783
344	425	0.6214	104.75	231.89	0.6116	1.5992	2.2055	2.2201	2.3496

(continued)

Table 11 (continued)

P	T	Z_{mixt}	V_{mixt}	f_{mixt}	φ_{H_2O}	φ_{CO_2}	φ_{CH_4}	φ_{H_2}	φ_{CO}
bar	°C		cm^3/mol	bar					
430	450	0.6411	89.66	284.27	0.5950	1.7053	2.3843	2.3363	2.5555
525	475	0.6759	80.10	345.55	0.5890	1.7878	2.5027	2.3846	2.6935
625	500	0.7188	73.93	415.24	0.5928	1.8495	2.5679	2.3826	2.7707
727	525	0.7647	69.81	492.71	0.6045	1.8967	2.5961	2.3516	2.8050
827	550	0.8103	67.06	576.42	0.6223	1.9318	2.5968	2.3036	2.8071
926	575	0.8544	65.04	667.07	0.6441	1.9618	2.5865	2.2522	2.7959
1026	600	0.8966	63.46	764.94	0.6686	1.9894	2.5721	2.2023	2.7791

Table 12 Compressibility factor, volume, and fugacity of the gas mixture with CO_2 mole fraction of 0.20 and fugacity coefficients of H_2O, CO_2, H_2, CH_4, and CO in the same gas mixture computed by means of the Peng-Robinson EOS at the P,T conditions fixed by coexistence with a brine containing 21 wt% NaCl

P	T	Z_{mixt}	V_{mixt}	f_{mixt}	φ_{H_2O}	φ_{CO_2}	φ_{CH_4}	φ_{H_2}	φ_{CO}
bar	°C		cm^3/mol	bar					
5.40	150	0.9738	6347.99	5.26	0.9680	1.0003	1.0179	1.0264	1.0196
9.88	175	0.9589	3614.45	9.49	0.9501	1.0021	1.0299	1.0429	1.0329
17.0	200	0.9391	2175.16	16.00	0.9267	1.0059	1.0476	1.0667	1.0526
27.6	225	0.9139	1369.62	25.42	0.8977	1.0129	1.0731	1.0997	1.0813
42.9	250	0.8834	895.05	38.36	0.8633	1.0248	1.1092	1.1449	1.1221
64.1	275	0.8479	603.27	55.36	0.8243	1.0442	1.1597	1.2059	1.1794
92.3	300	0.8089	417.86	76.86	0.7818	1.0739	1.2295	1.2868	1.2588
129	325	0.7686	296.01	103.46	0.7371	1.1180	1.3254	1.3930	1.3681
176	350	0.7328	216.33	135.34	0.6935	1.1783	1.4491	1.5224	1.5098
233	375	0.7074	163.51	173.57	0.6534	1.2544	1.5991	1.6686	1.6825
303	400	0.6987	128.94	219.41	0.6201	1.3401	1.7609	1.8111	1.8700
387	425	0.7087	106.19	274.85	0.5955	1.4282	1.9176	1.9306	2.0522
484	450	0.7357	91.45	340.74	0.5821	1.5103	2.0492	2.0107	2.2058
591	475	0.7742	81.55	418.31	0.5790	1.5861	2.1546	2.0565	2.3283
703	500	0.8192	74.89	506.78	0.5854	1.6540	2.2320	2.0734	2.4173
818	525	0.8663	70.30	605.17	0.5997	1.7146	2.2861	2.0706	2.4784
930	550	0.9126	67.14	711.35	0.6200	1.7673	2.3197	2.0541	2.5148
1042	575	0.9572	64.77	826.30	0.6444	1.8158	2.3431	2.0327	2.5386
1154	600	0.9999	62.90	950.40	0.6716	1.8618	2.3607	2.0101	2.5550

Table 13 Compressibility factor, volume, and fugacity of the gas mixture with CO_2 mole fraction of 0.30 and fugacity coefficients of H_2O, CO_2, H_2, CH_4, and CO in the same gas mixture computed by means of the Peng-Robinson EOS at the P,T conditions fixed by coexistence with a brine containing 21 wt% NaCl

P	T	Z_{mixt}	V_{mixt}	f_{mixt}	φ_{H_2O}	φ_{CO_2}	φ_{CH_4}	φ_{H_2}	φ_{CO}
bar	°C		cm^3/mol	bar					
6.17	150	0.9737	5553.75	6.01	0.9642	0.9979	1.0168	1.0263	1.0190
11.3	175	0.9589	3162.71	10.85	0.9444	0.9983	1.0281	1.0427	1.0321
19.4	200	0.9395	1904.17	18.29	0.9184	1.0004	1.0448	1.0661	1.0516
31.6	225	0.9153	1200.23	29.07	0.8864	1.0051	1.0690	1.0986	1.0798
49.1	250	0.8866	786.03	43.93	0.8488	1.0143	1.1032	1.1424	1.1198
73.2	275	0.8545	531.96	63.53	0.8064	1.0299	1.1505	1.2004	1.1752
105	300	0.8213	371.19	88.50	0.7612	1.0543	1.2143	1.2748	1.2501
148	325	0.7901	266.26	119.72	0.7147	1.0900	1.2983	1.3673	1.3490
201	350	0.7671	198.15	157.72	0.6712	1.1372	1.4006	1.4720	1.4699
266	375	0.7570	153.11	204.13	0.6332	1.1949	1.5171	1.5806	1.6080
347	400	0.7632	123.24	260.83	0.6034	1.2598	1.6387	1.6803	1.7522
443	425	0.7855	102.98	330.60	0.5830	1.3299	1.7586	1.7638	1.8943
553	450	0.8214	89.35	414.61	0.5736	1.4019	1.8676	1.8244	2.0222
675	475	0.8667	79.89	514.54	0.5744	1.4754	1.9652	1.8657	2.1352
804	500	0.9168	73.34	629.23	0.5846	1.5478	2.0480	1.8897	2.2287
935	525	0.9679	68.72	757.32	0.6024	1.6177	2.1164	1.9008	2.3038
1063	550	1.0172	65.48	895.79	0.6262	1.6828	2.1697	1.9016	2.3602
1191	575	1.0643	63.02	1045.97	0.6541	1.7453	2.2147	1.8981	2.4059
1319	600	1.1092	61.06	1208.41	0.6850	1.8061	2.2546	1.8928	2.4448

Table 14 Compressibility factor, volume, and fugacity of the gas mixture with CO_2 mole fraction of 0.40 and fugacity coefficients of H_2O, CO_2, H_2, CH_4, and CO in the same gas mixture computed by means of the Peng-Robinson EOS at the P,T conditions fixed by coexistence with a brine containing 21 wt% NaCl

P	T	Z_{mixt}	V_{mixt}	f_{mixt}	φ_{H_2O}	φ_{CO_2}	φ_{CH_4}	φ_{H_2}	φ_{CO}
bar	°C		cm^3/mol	bar					
7.20	150	0.9731	4757.61	7.01	0.9597	0.9950	1.0156	1.0265	1.0188
13.2	175	0.9583	2709.21	12.65	0.9373	0.9940	1.0263	1.0430	1.0318
22.6	200	0.9392	1631.51	21.33	0.9084	0.9941	1.0423	1.0665	1.0513
36.8	225	0.9158	1029.26	33.92	0.8729	0.9966	1.0655	1.0988	1.0796
57.2	250	0.8889	675.47	51.30	0.8315	1.0029	1.0981	1.1419	1.1195
85.4	275	0.8604	459.10	74.35	0.7857	1.0150	1.1428	1.1977	1.1741
123	300	0.8334	322.86	103.93	0.7380	1.0349	1.2018	1.2667	1.2462
172	325	0.8119	234.53	141.37	0.6906	1.0642	1.2769	1.3480	1.3379
234	350	0.8018	177.52	187.67	0.6484	1.1029	1.3643	1.4341	1.4447
311	375	0.8062	139.77	245.28	0.6136	1.1504	1.4606	1.5179	1.5622
404	400	0.8265	114.40	317.05	0.5884	1.2058	1.5609	1.5926	1.6838
517	425	0.8615	96.82	407.00	0.5732	1.2696	1.6637	1.6566	1.8068

(continued)

Table 14 (continued)

P	T	Z_{mixt}	V_{mixt}	f_{mixt}	φ_{H_2O}	φ_{CO_2}	φ_{CH_4}	φ_{H_2}	φ_{CO}
bar	°C		cm^3/mol	bar					
645	450	0.9082	84.67	517.03	0.5691	1.3401	1.7642	1.7072	1.9246
787	475	0.9626	76.05	649.66	0.5751	1.4173	1.8623	1.7474	2.0368
938	500	1.0203	69.96	803.47	0.5905	1.4980	1.9537	1.7774	2.1384
1090	525	1.0779	65.60	976.58	0.6134	1.5796	2.0367	1.7990	2.2277
1240	550	1.1324	62.48	1164.60	0.6423	1.6588	2.1085	1.8128	2.3023
1390	575	1.1842	60.09	1369.46	0.6754	1.7368	2.1740	1.8231	2.3682
1539	600	1.2331	58.18	1591.96	0.7115	1.8140	2.2351	1.8317	2.4280

varying from 1.0035, at 4.54 bar, 150 °C, and CO_2 mole fraction of 0.05, to 0.9950, at 7.20 bar, 150 °C, and CO_2 mole fraction of 0.40. It increases significantly and continuously with increasing pressure and temperature, reaching maximum values ranging from 2.0901, at 972 bar, 600 °C, and CO_2 mole fraction of 0.05, to 1.8061, at 1319 bar, 600 °C, and CO_2 mole fraction of 0.30. A similar maximum φ_{CO_2} value, 1.8140, is attained at 1539 bar, 600 °C, and CO_2 mole fraction of 0.40.

(iii) The fugacity coefficients of CH_4 and CO are similar to each other, being close to unity at low pressures and temperatures but becoming much greater than unity at high pressures and temperatures. They increase continuously with pressure and temperature at CO_2 mole fractions of 0.20, 0.30, and 0.40, whereas they pass through a maximum at intermediate P,T values and decrease weakly with a further increase in pressure and temperature at CO_2 mole fractions of 0.05 and 0.10. The minimum values are 1.0156 for CH_4 and 1.0188 for CO, at 7.20 bar, 150 °C, and CO_2 mole fraction of 0.40, whereas the maximum values are 2.8450 at 592 bar, 500 °C, and CO_2 mole fraction of 0.05 for CH_4 and 3.0687 at 654 bar, 525 °C, and CO_2 mole fraction of 0.05 for CO.

(iv) Also the fugacity coefficient of H_2, similar to CH_4 and CO, is close to unity at low pressures and temperatures and increases with increasing pressure and temperature, but less markedly than the fugacity coefficients of CH_4 and CO. Furthermore, the fugacity coefficient of H_2 passes through a maximum at intermediate P,T values and decrease slightly with a further increase in pressure and temperature, at CO_2 mole fractions of 0.05, 0.10, 0.20, and 0.30, whereas it rises continuously with increasing pressure and temperature at CO_2 mole fraction of 0.40. The minimum value is 1.0263 at 6.17 bar, 150 °C, and CO_2 mole fraction of 0.30, while the maximum value is 2.6553 at 972 bar, 600 °C, and CO_2 mole fraction of 0.05.

Consequently, the ratios of the fugacity coefficients of CO, CH_4, and CO_2 do not deviate much from unity. In detail, the $\varphi_{CO}/\varphi_{CO_2}$ ratio oscillates from 1.02 to 1.54, the $\varphi_{CH_4}/\varphi_{CO_2}$ ratio varies from 1.02 to 1.43, while the $\varphi_{CO}/\varphi_{CH_4}$ ratio ranges from 1.00 to 1.09. Therefore, the uncorrected mole fraction ratios of carbon gases may be utilized in geothermometric-geobarometric functions calibrated for pressures and temperatures controlled by equilibrium coexistence of a brine containing 21 wt% NaCl and the related vapor phase without incurring excessive errors. This inference was already drawn by Giggenbach (1987) for a separate gas phase containing water vapor as the major component, but with $X_{H_2O} > 0.8$, whereas X_{H_2O} values as low as 0.689 were recorded in Solfatara vapors (Chiodini et al. 2021 and references therein).

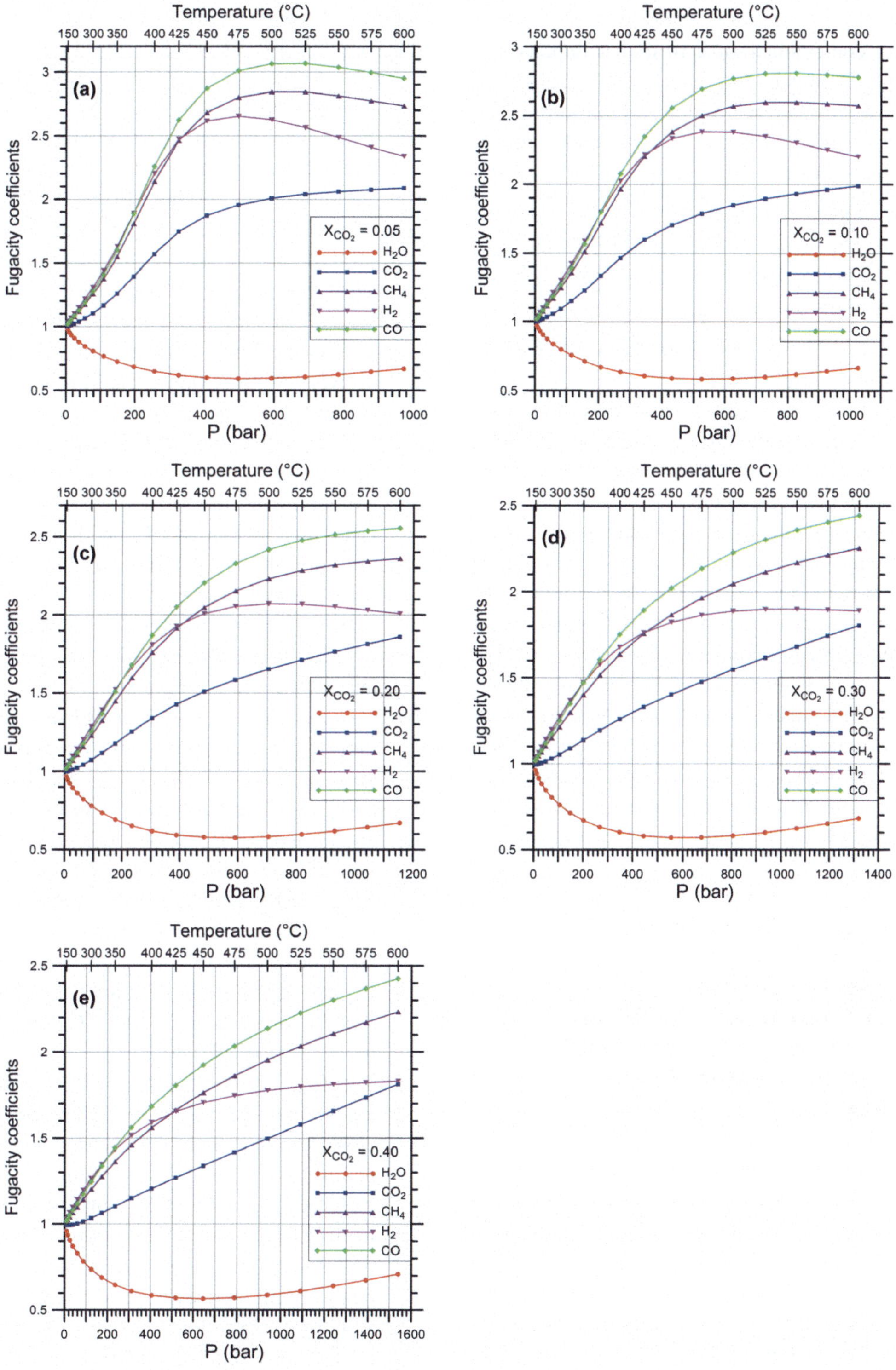

Fig. 22 Fugacity coefficients of H_2O, CO_2, H_2, CH_4, and CO in the gas mixtures with CO_2 mole fraction of **a** 0.05, **b** 0.10, **c** 0.20, **d** 0.30, and **e** 0.40, at the P,T conditions fixed by coexistence with a brine containing 21 wt% NaCl. Calculations performed by means of the Peng-Robinson EOS

The $\varphi_{H_2}/\varphi_{H_2O}$ ratio is different from other fugacity coefficient ratios, in that it departs significantly from one with increasing P,T, although it is close to unity at low P,T. Maximum deviations from unity occur at 475 °C and different total fluid pressure depending on X_{CO_2}, with: (i) $\varphi_{H_2}/\varphi_{H_2O}$ of 4.45 at 497 bar and X_{CO_2} of 0.05; (ii) $\varphi_{H_2}/\varphi_{H_2O}$ of 4.05 at 525 bar and X_{CO_2} of 0.10; (iii) $\varphi_{H_2}/\varphi_{H_2O}$ of 3.55 at 591 bar and X_{CO_2} of 0.20; (iv) $\varphi_{H_2}/\varphi_{H_2O}$ of 3.25 at 675 bar and X_{CO_2} of 0.30; (v) $\varphi_{H_2}/\varphi_{H_2O}$ of 3.04 at 787 bar and X_{CO_2} of 0.40.

Therefore, use of the uncorrected X_{H_2}/X_{H_2O} ratio in geothermometric-geobarometric functions may lead to significant errors in computed temperatures and pressures.

The uncertainties on the computed fugacity coefficients were evaluated for $X_{CO_2} = 0.20$, at the minimum and maximum temperature and pressure of interest, considering alternatively the minimum and maximum concentrations of H_2, CH_4 and CO measured for Solfatara fluids. Results are compared with those obtained for the average concentrations of these three gas components in Table 15, whose inspection suggests that the differences in the computed parameters are negligible for the aims of this work.

7.4 Fugacity Coefficients at the Pressures and Temperatures of Coexistence of a Brine Containing 33.5 Wt% NaCl and the Related Vapor Phase

Again, the Peng-Robinson EOS was utilized to calculate the fugacity coefficients of H_2O, CO_2, H_2, CH_4, and CO at temperatures ranging from 150 to 1000 °C, at steps of 50 °C, and at the corresponding total pressures computed by means of Eq. (135) for CO_2 mole fractions in the vapor phase of 0.05, 0.10, 0.20, 0.30, and 0.40. The Microsoft Excel computer code Prfug.xlsx (Elliott and Lira 2012) was used following the same approach delineated in Sect. 5. The computed fugacity coefficients are listed in Tables 16, 17, 18, 19 and 20 and are shown in the graphs of Fig. 23, whose inspection leads to the following considerations:

(i) The fugacity coefficient of H_2O is relatively close to unity at low pressures and temperatures, varying from 0.9795, at 3.39 bar, 150 °C, and CO_2 mole fraction of 0.05, to 0.9698, at 5.37 bar, 150 °C, and CO_2 mole fraction of 0.40. It decreases considerably with increasing pressure and temperature, attaining minimum values varying from 0.6269, at 528 bar, 500 °C, and CO_2 mole fraction of 0.05, to 0.5950, at 572 bar, 450 °C, and CO_2 mole fraction of 0.40. At higher pressures and temperatures, the fugacity coefficient of H_2O increases markedly with increasing pressure and temperature, reaching maximum values ranging from 1.1068, at 2291 bar, 1000 °C, and CO_2 mole fraction of 0.05, to 1.3676, at 3627 bar, 1000 °C, and CO_2 mole fraction of 0.40.

(ii) The fugacity coefficient of CO_2 is close to unity at low pressures and temperatures, varying from 0.9949, at 5.37 bar, 150 °C, and CO_2 mole fraction of 0.40, to 1.0025, at 3.39 bar, 150 °C, and CO_2 mole fraction of 0.05. It increases continuously and significantly with increasing pressure and temperature, reaching maximum values ranging from 2.1837, at 2291 bar, 1000 °C, and CO_2 mole fraction of 0.05, to 2.6575, at 3627 bar, 1000 °C, and CO_2 mole fraction of 0.40.

(iii) The fugacity coefficients of CH_4 and CO are similar to each other. They are both close to unity at low pressures and temperatures, with values ranging from 1.0115 to 1.0145 (CH_4) and oscillating from 1.0138 and 1.0152 (CO). In CO_2-rich gas mixtures, they increase continuously with pressure and temperature attaining maximum values of 2.5037 (CH_4) and 2.5925 (CO) at 1000 °C, 3109 bar, for CO_2 mole fraction of 0.30 and 2.7481(CH_4) and 2.8162 (CO) at 1000 °C, 3627 bar, for CO_2 mole fraction

Table 15 Compressibility factor, volume, and fugacity of the gas mixture with CO_2 mole fraction of 0.20 and fugacity coefficients of H_2O, CO_2, H_2, CH_4, and CO in the same gas mixture computed by means of the Peng-Robinson EOS at the minimum and maximum temperature and pressure of interest and minimum, mean, and maximum concentrations of H_2, CH_4 and CO in Solfatara fluids

X_{H_2O}	X_{H_2}	T	P	Z_{mixt}	V_{mixt}	f_{mixt}	φ_{H_2O}	φ_{CO_2}	φ_{H_2}
		°C	bar		cm^3/mol	bar			
0.79985	1.46E-04	150	5.40	0.9738	6347.99	5.26	0.9680	1.0003	1.0264
0.79953	4.69E-04	150	5.40	0.9738	6348.12	5.26	0.9680	1.0003	1.0264
0.79860	1.40E-03	150	5.40	0.9739	6348.48	5.26	0.9680	1.0003	1.0263
0.79985	1.46E-04	600	1154	0.9998	62.90	950.38	0.6716	1.8619	2.0110
0.79953	4.69E-04	600	1154	1.0000	62.92	950.72	0.6717	1.8611	2.0101
0.79860	1.40E-03	600	1154	1.0006	62.95	951.69	0.6719	1.8588	2.0074
X_{H_2O}	X_{CH_4}	T	P	Z_{mixt}	V_{mixt}	f_{mixt}	φ_{H_2O}	φ_{CO_2}	φ_{CH_4}
		°C	bar		cm^3/mol	bar			
0.79999	5.29E-06	150	5.40	0.9738	6347.93	5.26	0.9680	1.0003	1.0179
0.79998	1.66E-05	150	5.40	0.9738	6347.94	5.26	0.9680	1.0003	1.0179
0.79996	4.14E-05	150	5.40	0.9738	6347.95	5.26	0.9680	1.0003	1.0179
0.79999	5.29E-06	600	1154	0.9998	62.90	950.24	0.6716	1.8622	2.3608
0.79998	1.66E-05	600	1154	0.9998	62.90	950.25	0.6716	1.8622	2.3607
0.79996	4.14E-05	600	1154	0.9998	62.90	950.28	0.6716	1.8621	2.3606
X_{H_2O}	X_{CO}	T	P	Z_{mixt}	V_{mixt}	f_{mixt}	φ_{H_2O}	φ_{CO_2}	φ_{CO}
		°C	bar		cm^3/mol	bar			
0.80000	1.04E-07	150	5.40	0.9738	6347.93	5.26	0.9680	1.0003	1.0196
0.80000	9.88E-07	150	5.40	0.9738	6347.93	5.26	0.9680	1.0003	1.0196
0.80000	3.78E-06	150	5.40	0.9738	6347.93	5.26	0.9680	1.0003	1.0196
0.80000	1.04E-07	600	1154	0.9998	62.90	950.23	0.6716	1.8622	2.5550
0.80000	9.88E-07	600	1154	0.9998	62.90	950.23	0.6716	1.8622	2.5550
0.80000	3.78E-06	600	1154	0.9998	62.90	950.23	0.6716	1.8622	2.5550

Table 16 Compressibility factor, volume, and fugacity of the gas mixture with CO_2 mole fraction of 0.05 and fugacity coefficients of H_2O, CO_2, H_2, CH_4, and CO in the same gas mixture computed by means of the Peng-Robinson EOS at the P,T conditions fixed by coexistence with a brine containing 33.5 wt% NaCl

P	T	Z_{mixt}	V_{mixt}	f_{mixt}	φ_{H_2O}	φ_{CO_2}	φ_{CH_4}	φ_{H_2}	φ_{CO}
bar	°C		cm^3/mol	bar					
3.39	150	0.9803	10,167.59	3.33	0.9795	1.0025	1.0145	1.0199	1.0152
11.4	200	0.9506	3278.89	10.87	0.9497	1.0100	1.0403	1.0533	1.0428
30.4	250	0.8984	1284.83	27.58	0.9006	1.0311	1.0957	1.1218	1.1030
67.6	300	0.8211	578.74	57.14	0.8343	1.0825	1.2069	1.2516	1.2247
130	350	0.7261	289.78	100.96	0.7603	1.1944	1.4217	1.4858	1.4617
226	400	0.6351	156.93	161.52	0.6877	1.4170	1.8284	1.8903	1.9143
361	450	0.6153	102.49	242.21	0.6392	1.6807	2.2905	2.2747	2.4365
528	500	0.6697	81.51	349.58	0.6269	1.8452	2.5253	2.3854	2.7096

(continued)

Table 16 (continued)

P	T	Z_{mixt}	V_{mixt}	f_{mixt}	φ_{H_2O}	φ_{CO_2}	φ_{CH_4}	φ_{H_2}	φ_{CO}
bar	°C		cm^3/mol	bar					
717	550	0.7512	71.71	487.59	0.6435	1.9415	2.5969	2.3432	2.7972
914	600	0.8357	66.40	655.50	0.6794	2.0076	2.5975	2.2559	2.8009
1108	650	0.9144	63.33	848.88	0.7272	2.0552	2.5656	2.1613	2.7646
1294	700	0.9840	61.54	1061.70	0.7812	2.0884	2.5182	2.0719	2.7085
1469	750	1.0444	60.49	1288.58	0.8377	2.1104	2.4650	1.9917	2.6444
1634	800	1.0966	59.86	1526.42	0.8943	2.1251	2.4127	1.9220	2.5804
1795	850	1.1423	59.43	1774.99	0.9496	2.1365	2.3662	1.8634	2.5221
1954	900	1.1831	59.05	2036.95	1.0033	2.1481	2.3286	1.8156	2.4731
2118	950	1.2205	58.60	2317.42	1.0555	2.1630	2.3019	1.7781	2.4356
2291	1000	1.2559	58.02	2623.61	1.1068	2.1837	2.2875	1.7505	2.4109

Table 17 Compressibility factor, volume, and fugacity of the gas mixture with CO_2 mole fraction of 0.10 and fugacity coefficients of H_2O, CO_2, H_2, CH_4, and CO in the same gas mixture computed by means of the Peng-Robinson EOS at the P,T conditions fixed by coexistence with a brine containing 33.5 wt% NaCl

P	T	Z_{mixt}	V_{mixt}	f_{mixt}	φ_{H_2O}	φ_{CO_2}	φ_{CH_4}	φ_{H_2}	φ_{CO}
bar	°C		cm^3/mol	bar					
3.58	150	0.9804	9633.58	3.51	0.9784	1.0018	1.0141	1.0197	1.0149
12.0	200	0.9511	3107.94	11.47	0.9472	1.0082	1.0391	1.0527	1.0421
32.1	250	0.9001	1219.49	29.15	0.8956	1.0272	1.0930	1.1200	1.1012
71.4	300	0.8262	551.68	60.53	0.8262	1.0744	1.1999	1.2455	1.2196
137	350	0.7400	279.79	107.44	0.7495	1.1749	1.3986	1.4621	1.4413
239	400	0.6677	156.30	173.36	0.6761	1.3589	1.7399	1.7979	1.8256
381	450	0.6624	104.52	262.78	0.6294	1.5685	2.1033	2.0919	2.2409
558	500	0.7194	82.94	382.52	0.6193	1.7216	2.3184	2.1954	2.4910
757	550	0.8008	72.43	536.66	0.6382	1.8309	2.4160	2.1837	2.6053
964	600	0.8852	66.63	724.17	0.6765	1.9161	2.4521	2.1298	2.6459
1170	650	0.9636	63.23	939.88	0.7267	1.9834	2.4541	2.0640	2.6449
1366	700	1.0328	61.19	1176.84	0.7832	2.0349	2.4362	1.9978	2.6194
1550	750	1.0926	59.96	1428.93	0.8423	2.0733	2.4075	1.9359	2.5806
1725	800	1.1443	59.18	1692.66	0.9014	2.1021	2.3754	1.8808	2.5372
1894	850	1.1894	58.63	1967.93	0.9592	2.1256	2.3452	1.8338	2.4953
2063	900	1.2297	58.14	2257.89	1.0155	2.1476	2.3210	1.7952	2.4597
2236	950	1.2668	57.62	2568.48	1.0703	2.1717	2.3056	1.7653	2.4332
2418	1000	1.3021	56.99	2908.03	1.1243	2.2008	2.3010	1.7440	2.4179

Table 18 Compressibility factor, volume, and fugacity of the gas mixture with CO_2 mole fraction of 0.20 and fugacity coefficients of H_2O, CO_2, H_2, CH_4, and CO in the same gas mixture computed by means of the Peng-Robinson EOS at the P,T conditions fixed by coexistence with a brine containing 33.5 wt% NaCl

P	T	Z_{mixt}	V_{mixt}	f_{mixt}	φ_{H_2O}	φ_{CO_2}	φ_{CH_4}	φ_{H_2}	φ_{CO}
bar	°C		cm^3/mol	bar					
4.03	150	0.9805	8564.16	3.95	0.9761	1.0002	1.0132	1.0194	1.0144
13.5	200	0.9518	2764.76	12.92	0.9415	1.0042	1.0369	1.0519	1.0409
36.1	250	0.9031	1087.61	32.87	0.8847	1.0191	1.0879	1.1171	1.0984
80.3	300	0.8360	496.19	68.55	0.8092	1.0584	1.1871	1.2351	1.2113
154	350	0.7667	257.68	122.80	0.7282	1.1401	1.3592	1.4220	1.4084
269	400	0.7249	150.85	201.44	0.6556	1.2750	1.6166	1.6677	1.7056
429	450	0.7439	104.35	311.64	0.6141	1.4274	1.8740	1.8616	2.0056
627	500	0.8108	83.09	461.76	0.6096	1.5647	2.0588	1.9476	2.2205
851	550	0.8968	72.10	656.48	0.6339	1.6879	2.1809	1.9654	2.3586
1085	600	0.9841	65.84	894.16	0.6778	1.7990	2.2610	1.9522	2.4441
1316	650	1.0644	62.08	1167.68	0.7338	1.8960	2.3106	1.9255	2.4917
1536	700	1.1346	59.76	1467.58	0.7963	1.9772	2.3373	1.8931	2.5114
1744	750	1.1948	58.28	1785.67	0.8614	2.0432	2.3479	1.8596	2.5121
1941	800	1.2464	57.30	2117.43	0.9264	2.0970	2.3493	1.8278	2.5019
2131	850	1.2912	56.57	2462.98	0.9902	2.1427	2.3474	1.7998	2.4879
2321	900	1.3313	55.96	2826.78	1.0525	2.1846	2.3473	1.7771	2.4755
2515	950	1.3685	55.33	3217.05	1.1134	2.2268	2.3531	1.7607	2.4691
2721	1000	1.4042	54.63	3645.11	1.1739	2.2730	2.3677	1.7512	2.4716

Table 19 Compressibility factor, volume, and fugacity of the gas mixture with CO_2 mole fraction of 0.30 and fugacity coefficients of H_2O, CO_2, H_2, CH_4, and CO in the same gas mixture computed by means of the Peng-Robinson EOS at the P,T conditions fixed by coexistence with a brine containing 33.5 wt% NaCl

P	T	Z_{mixt}	V_{mixt}	f_{mixt}	φ_{H_2O}	φ_{CO_2}	φ_{CH_4}	φ_{H_2}	φ_{CO}
bar	°C		cm^3/mol	bar					
4.60	150	0.9804	7492.84	4.51	0.9733	0.9983	1.0123	1.0194	1.0140
15.5	200	0.9521	2419.91	14.76	0.9348	0.9999	1.0348	1.0515	1.0401
41.3	250	0.9055	954.16	37.63	0.8722	1.0105	1.0832	1.1154	1.0968
91.7	300	0.8453	439.02	78.82	0.7905	1.0426	1.1759	1.2272	1.2058
176	350	0.7929	233.16	142.54	0.7065	1.1098	1.3277	1.3903	1.3850
307	400	0.7777	141.60	237.80	0.6372	1.2162	1.5345	1.5789	1.6304
490	450	0.8190	100.51	375.55	0.6028	1.3427	1.7401	1.7201	1.8737
717	500	0.8994	80.65	567.27	0.6057	1.4759	1.9113	1.7952	2.0716
973	550	0.9945	69.96	819.31	0.6374	1.6124	2.0507	1.8295	2.2259
1240	600	1.0884	63.72	1129.46	0.6889	1.7464	2.1634	1.8424	2.3435

(continued)

Table 19 (continued)

P	T	Z_{mixt}	V_{mixt}	f_{mixt}	φ_{H_2O}	φ_{CO_2}	φ_{CH_4}	φ_{H_2}	φ_{CO}
bar	°C		cm^3/mol	bar					
1504	650	1.1736	59.90	1487.82	0.7528	1.8705	2.2506	1.8431	2.4283
1756	700	1.2473	57.48	1881.14	0.8234	1.9797	2.3146	1.8362	2.4846
1993	750	1.3098	55.90	2297.91	0.8965	2.0728	2.3595	1.8249	2.5186
2218	800	1.3629	54.82	2731.91	0.9696	2.1517	2.3913	1.8122	2.5374
2436	850	1.4088	54.01	3183.46	1.0413	2.2204	2.4162	1.8006	2.5484
2652	900	1.4499	53.32	3659.08	1.1115	2.2836	2.4400	1.7920	2.5580
2874	950	1.4882	52.65	4170.51	1.1807	2.3461	2.4678	1.7882	2.5713
3109	1000	1.5254	51.93	4733.92	1.2499	2.4123	2.5037	1.7902	2.5925

Table 20 Compressibility factor, volume, and fugacity of the gas mixture with CO_2 mole fraction of 0.40 and fugacity coefficients of H_2O, CO_2, H_2, CH_4, and CO in the same gas mixture computed by means of the Peng-Robinson EOS at the P,T conditions fixed by coexistence with a brine containing 33.5 wt% NaCl

P	T	Z_{mixt}	V_{mixt}	f_{mixt}	φ_{H_2O}	φ_{CO_2}	φ_{CH_4}	φ_{H_2}	φ_{CO}
bar	°C		cm^3/mol	bar					
5.37	150	0.9800	6419.63	5.26	0.9698	0.9962	1.0115	1.0196	1.0138
18.1	200	0.9518	2073.42	17.22	0.9267	0.9949	1.0328	1.0518	1.0399
48.2	250	0.9070	819.23	43.92	0.8573	1.0011	1.0792	1.1153	1.0967
107	300	0.8543	380.33	92.42	0.7692	1.0266	1.1669	1.2223	1.2040
206	350	0.8198	206.65	168.93	0.6837	1.0830	1.3034	1.3661	1.3711
359	400	0.8309	129.67	287.04	0.6200	1.1738	1.4792	1.5159	1.5853
572	450	0.8954	94.20	463.73	0.5950	1.2918	1.6605	1.6272	1.8016
836	500	0.9931	76.33	716.24	0.6078	1.4320	1.8331	1.6993	1.9989
1135	550	1.1016	66.42	1054.71	0.6498	1.5886	1.9956	1.7484	2.1750
1447	600	1.2063	60.53	1476.93	0.7122	1.7511	2.1433	1.7838	2.3266
1754	650	1.2998	56.86	1969.03	0.7877	1.9081	2.2705	1.8086	2.4500
2048	700	1.3798	54.50	2511.63	0.8700	2.0512	2.3744	1.8247	2.5443
2325	750	1.4469	52.93	3087.60	0.9548	2.1771	2.4567	1.8342	2.6131
2588	800	1.5035	51.84	3687.63	1.0395	2.2867	2.5222	1.8398	2.6628
2842	850	1.5522	51.01	4312.27	1.1228	2.3841	2.5776	1.8441	2.7011
3094	900	1.5958	50.30	4971.41	1.2046	2.4745	2.6297	1.8499	2.7353
3354	950	1.6366	49.63	5682.86	1.2857	2.5637	2.6847	1.8594	2.7718
3627	1000	1.6767	48.93	6471.26	1.3676	2.6575	2.7481	1.8743	2.8162

of 0.40. In CO_2-poor gas mixtures, the fugacity coefficients of CH_4 and CO increase with pressure and temperature up to intermediate P,T values and decrease moderately with further increments in pressure and temperature. The maximum fugacity coefficient of CH_4 in CO_2-poor gas mixtures varies from 2.5975, at 914 bar, 600 °C, for X_{CO_2} of 0.05, to 2.4541, at 1170 bar, 650 °C, for X_{CO_2} of 0.10, whereas the maximum fugacity coefficient of CO ranges from 2.8009, at 914 bar, 600 °C, for X_{CO_2} of 0.05, to 2.6459, at 964 bar, 600 °C, for X_{CO_2} of

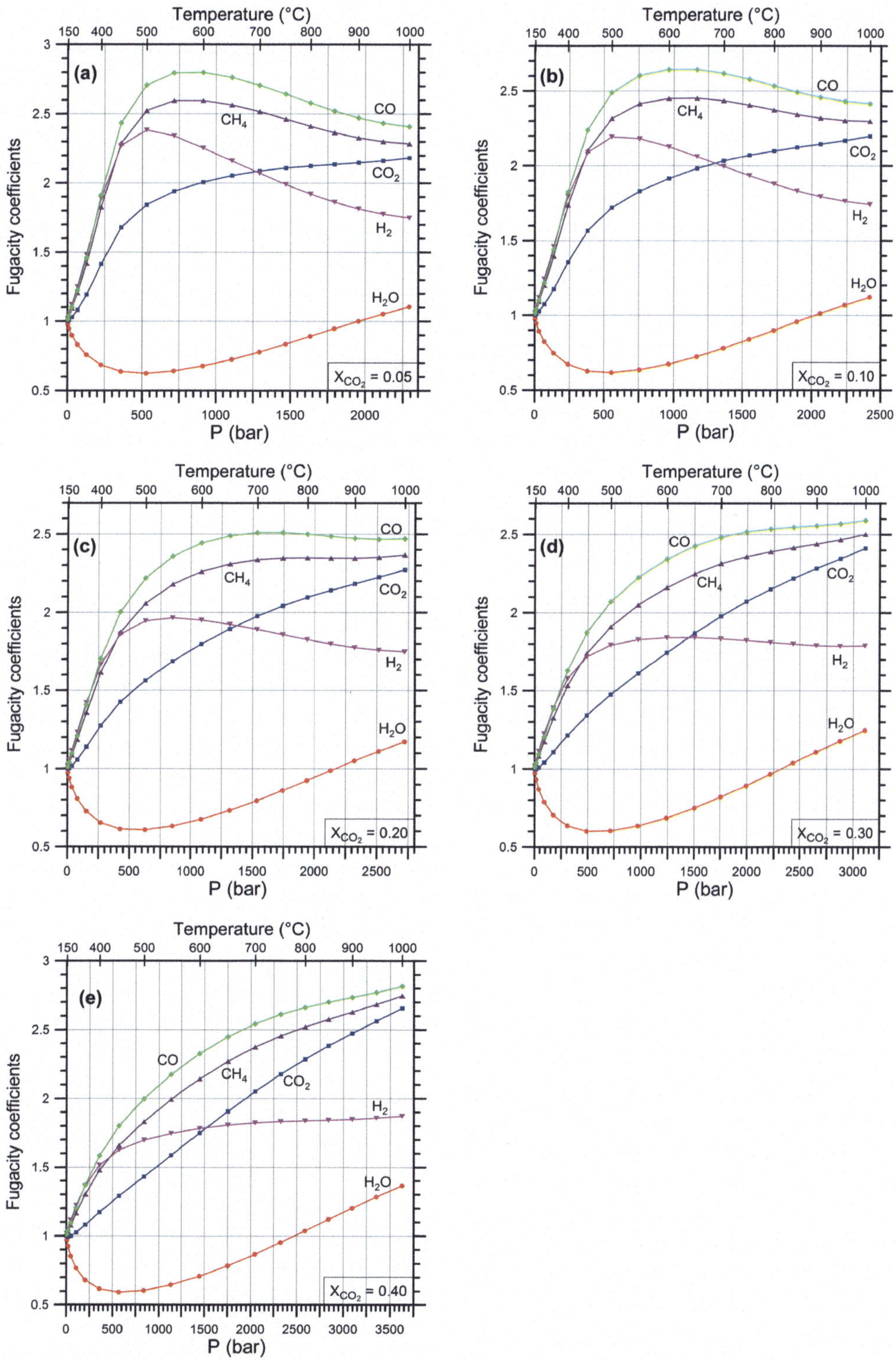

Fig. 23 Fugacity coefficients of H_2O, CO_2, H_2, CH_4, and CO in the gas mixtures with CO_2 mole fraction of **a** 0.05, **b** 0.10, **c** 0.20, **d** 0.30, and **e** 0.40, at the P,T conditions fixed by coexistence with a brine containing 33.5 wt% NaCl. Calculations performed by means of the Peng-Robinson EOS

0.10. At X_{CO_2} of 0.20, the fugacity coefficient of CH_4 attains its maximum value of 2.3677 at 2721 bar, 1000 °C (which is the final P,T condition considered here), whereas the fugacity coefficient of CO reaches its maximum value of 2.5121 at the intermediate P,T of 1744 bar, 750 °C and declines weakly with a further increase in pressure and temperature.

(iv) The fugacity coefficient of H_2 is similar to those of CH_4 and CO. In detail: for X_{CO_2} of 0.05, 0.10, 0.20 and 0.30, it is close to unity at low pressures and temperatures, it increases with pressure and temperature up to intermediate P,T values, and it decrease to some extent with further increments in pressure and temperature. For X_{CO_2} of 0.40, the fugacity coefficient of H_2 rises continuously with pressure and temperature. Thus, the minimum value varies from 1.0194 at 4.60 bar, 150 °C, and CO_2 mole fraction of 0.30, to 1.0199 at 3.39 bar, 150 °C, and CO_2 mole fraction of 0.05, while the maximum value fluctuates from 1.8431 at 1504 bar, 650 °C, and CO_2 mole fraction of 0.30, to 2.3854 at 528 bar, 500 °C, and CO_2 mole fraction of 0.05.

As a consequence, the ratios of the fugacity coefficients of CO, CH_4, and CO_2 depart moderately from unity. In detail, the $\varphi_{CO}/\varphi_{CO_2}$ ratio varies from 1.01 to 1.47, the $\varphi_{CH_4}/\varphi_{CO_2}$ ratio ranges from 1.01 to 1.37, and the $\varphi_{CO}/\varphi_{CH_4}$ ratio oscillates from 1.00 to 1.09. Hence, the uncorrected mole fraction ratios of carbon gases may be used in geothermometric-geobarometric tools calibrated for pressures and temperatures controlled by equilibrium coexistence of a brine containing 33.5 wt% NaCl and the related vapor phase without excessive errors, as already noted by Giggenbach (1987; see Sect. 7.3).

In contrast, the $\varphi_{H_2}/\varphi_{H_2O}$ ratio has a different behavior. In fact, it is close to unity (1.04–1.05) at low pressures and low temperatures, but it increases rapidly with increasing pressure and temperature, attaining maximum values varying from 2.80, at 836 bar, 500 °C, and CO_2 mole fraction of 0.40, to 3.80, at 528 bar, 500 °C, and CO_2 mole fraction of 0.05. Then, at higher pressures, the $\varphi_{H_2}/\varphi_{H_2O}$ ratio decreases with increasing pressure, reaching again low values, ranging from 1.37, at 3627 bar, 1000 °C, and CO_2 mole fraction of 0.40, to 1.58, at 2291 bar, 1000 °C, and CO_2 mole fraction of 0.05. Therefore, the use of the uncorrected X_{H_2}/X_{H_2O} ratio in geothermometric-geobarometric functions may cause important errors in computed temperatures and pressures.

7.5 Fugacity Coefficients Along the Linear P–T Decompression Path

Also in this case, the Peng-Robinson EOS was used to compute the fugacity coefficients of H_2O, CO_2, CO, CH_4, and H_2. The Microsoft Excel spreadsheet Prfug.xlsx (Elliott and Lira 2012) was utilized as described in Sect. 5. Calculations were performed for CO_2 mole fractions in the vapor phase of 0.05, 0.10, 0.20, 0.30, and 0.40, temperatures varying from 150 to 1120 °C, and total fluid pressures ranging from minimum values of 5.01–7.94 bar, depending on X_{CO_2}, and a maximum value of 2879 bar, irrespective of X_{CO_2} (see Sect. 7.2 for further details). The obtained fugacity coefficients are listed in Tables 21, 22, 23, 24 and 25 and are displayed in the plots of Fig. 24, whose inspection allows one to draw the following observations:

(i) The fugacity coefficient of H_2O is close to unity at low pressures and temperatures, varying from 0.9555, at 7.94 bar, 150 °C, and X_{CO_2} of 0.40, to 0.9697, at 5.01 bar, 150 °C, and X_{CO_2} of 0.05. First, it decreases with increasing pressure and temperature, attaining minimum values ranging from 0.5435, at 502 bar, 450 °C, and X_{CO_2} of 0.05, to 0.5588, at 405 bar, 400 °C, and X_{CO_2} of 0.30. Second, it increases with further increments in pressure and temperature, reaching maximum values fluctuating from 1.2466, for X_{CO_2}

Table 21 Compressibility factor, volume, and fugacity of the gas mixture with CO_2 mole fraction of 0.05 and fugacity coefficients of H_2O, CO_2, H_2, CH_4, and CO in the same gas mixture computed by means of the Peng-Robinson EOS at the P,T conditions met by Solfatara fluids along the linear P–T decompression path

P	T	Z_{mixt}	V_{mixt}	f_{mixt}	φ_{H_2O}	φ_{CO_2}	φ_{CH_4}	φ_{H_2}	φ_{CO}
bar	°C		cm^3/mol	bar					
5.01	150	0.9707	6813.09	4.87	0.9697	1.0040	1.0219	1.0300	1.0229
16.4	200	0.9278	2230.00	15.26	0.9280	1.0158	1.0607	1.0801	1.0644
41.9	250	0.8554	888.94	36.50	0.8637	1.0496	1.1453	1.1837	1.1560
90.4	300	0.7463	393.39	71.81	0.7794	1.1397	1.3346	1.4023	1.3624
174	350	0.5925	176.43	122.72	0.6800	1.4063	1.8544	1.9585	1.9320
232	373.946	0.5111	118.40	153.48	0.6285	1.7075	2.4532	2.5500	2.5914
325	400	0.4606	79.40	196.77	0.5659	2.2132	3.5239	3.4930	3.7771
502	450	0.5577	66.79	293.30	0.5435	2.2861	3.5732	3.3307	3.8563
679	500	0.6684	63.24	413.40	0.5681	2.2274	3.2856	2.9376	3.5535
857	550	0.7683	61.37	558.22	0.6110	2.1924	3.0539	2.6423	3.3040
1034	600	0.8564	60.11	727.87	0.6630	2.1796	2.8837	2.4284	3.1172
1212	650	0.9341	59.17	921.88	0.7197	2.1810	2.7581	2.2703	2.9764
1389	700	1.0027	58.41	1139.37	0.7788	2.1908	2.6638	2.1504	2.8679
1566	750	1.0635	57.76	1379.15	0.8388	2.2055	2.5918	2.0575	2.7825
1744	800	1.1177	57.19	1639.87	0.8987	2.2228	2.5361	1.9842	2.7140
1921	850	1.1661	56.68	1920.03	0.9577	2.2413	2.4926	1.9255	2.6580
2099	900	1.2096	56.22	2218.10	1.0154	2.2600	2.4584	1.8780	2.6117
2276	950	1.2487	55.80	2532.51	1.0715	2.2785	2.4313	1.8392	2.5728
2453	1000	1.2841	55.40	2861.73	1.1256	2.2963	2.4100	1.8073	2.5398
2631	1050	1.3161	55.04	3204.30	1.1775	2.3132	2.3932	1.7810	2.4691
2808	1100	1.3452	54.69	3558.80	1.2273	2.3292	2.3802	1.7593	2.4034
2879	1120	1.3561	54.56	3703.64	1.2466	2.3353	2.3758	1.7517	2.3803

of 0.05, to 1.2933, for X_{CO_2} of 0.40, at 2879 bar, 1120 °C.

(ii) Also the fugacity coefficient of nonpolar gases is close to unity at low temperatures and pressures, with values of 0.9945 to 1.0040 for CO_2, 1.0173 to 1.0219 for CH_4, 1.0291 to 1.0300 for H_2, and 1.0208 to 1.0229 for CO, at 150 °C and pressures of 5.01 to 7.94, depending on X_{CO_2}. Different from water, the fugacity coefficient of nonpolar gases increases with increasing pressure and temperature, although in distinct ways.

(iii) The fugacity coefficient of H_2 increases quickly at low pressures and temperatures, attains a peak value varying from 3.4930, at 325 bar, 400 °C, and X_{CO_2} of 0.05, to 1.6845, at 624 bar, 450 °C, and X_{CO_2} of 0.40, and then it decreases significantly in CO_2-poor gas mixtures and to a minor extent in CO_2-rich gas mixtures.

(iv) Also the evolution of the fugacity coefficient of CO_2 is different in CO_2-poor and CO_2-rich gas mixtures. At X_{CO_2} of 0.05, the fugacity coefficient of CO_2 increases sharply with increasing temperature and pressure, attains a peak value of 2.2861, at 502 bar, 450 °C, and does not change significantly upon further increments in pressure and temperature. In contrast, at X_{CO_2} of 0.40, the fugacity coefficient of CO_2 rises gradually with increasing

Table 22 Compressibility factor, volume, and fugacity of the gas mixture with CO_2 mole fraction of 0.10 and fugacity coefficients of H_2O, CO_2, H_2, CH_4, and CO in the same gas mixture computed by means of the Peng-Robinson EOS at the P,T conditions met by Solfatara fluids along the linear P–T decompression path

P	T	Z_{mixt}	V_{mixt}	f_{mixt}	φ_{H_2O}	φ_{CO_2}	φ_{CH_4}	φ_{H_2}	φ_{CO}
bar	°C		cm^3/mol	bar					
5.29	150	0.9708	6455.70	5.14	0.9682	1.0028	1.0212	1.0297	1.0225
17.3	200	0.9286	2114.42	16.12	0.9243	1.0130	1.0590	1.0791	1.0633
44.2	250	0.8582	844.90	38.61	0.8569	1.0436	1.1408	1.1804	1.1529
95.4	300	0.7557	377.36	76.22	0.7689	1.1251	1.3193	1.3875	1.3497
184	350	0.6270	176.87	131.42	0.6673	1.3397	1.7496	1.8473	1.8271
245	373.946	0.5727	125.69	165.74	0.6172	1.5316	2.1341	2.2241	2.2563
337	400	0.5457	90.59	213.50	0.5637	1.8005	2.6875	2.7094	2.8782
514	450	0.6238	73.01	318.05	0.5450	1.9468	2.9022	2.7575	3.1340
690	500	0.7237	67.41	446.02	0.5708	1.9719	2.8113	2.5556	3.0426
867	550	0.8168	64.50	598.89	0.6143	1.9889	2.6999	2.3697	2.9220
1043	600	0.9000	62.63	776.69	0.6666	2.0105	2.6067	2.2228	2.8175
1220	650	0.9738	61.27	978.82	0.7235	2.0363	2.5338	2.1089	2.7327
1396	700	1.0392	60.22	1204.27	0.7827	2.0643	2.4773	2.0198	2.6645
1573	750	1.0973	59.35	1451.75	0.8427	2.0932	2.4338	1.9495	2.6092
1749	800	1.1491	58.61	1719.79	0.9025	2.1219	2.4002	1.8932	2.5639
1926	850	1.1955	57.97	2006.86	0.9614	2.1498	2.3742	1.8479	2.5264
2102	900	1.2372	57.40	2311.34	1.0190	2.1765	2.3542	1.8110	2.4949
2279	950	1.2748	56.89	2631.67	1.0748	2.2017	2.3390	1.7809	2.4684
2455	1000	1.3087	56.42	2966.29	1.1287	2.2253	2.3276	1.7563	2.4458
2632	1050	1.3395	55.99	3313.73	1.1805	2.2474	2.3192	1.7361	2.3871
2808	1100	1.3674	55.59	3672.59	1.2300	2.2679	2.3134	1.7195	2.3320
2879	1120	1.3779	55.43	3819.03	1.2492	2.2757	2.3117	1.7138	2.3127

Table 23 Compressibility factor, volume, and fugacity of the gas mixture with CO_2 mole fraction of 0.20 and fugacity coefficients of H_2O, CO_2, H_2, CH_4, and CO in the same gas mixture computed by means of the Peng-Robinson EOS at the P,T conditions met by Solfatara fluids along the linear P–T decompression path

P	T	Z_{mixt}	V_{mixt}	f_{mixt}	φ_{H_2O}	φ_{CO_2}	φ_{CH_4}	φ_{H_2}	φ_{CO}
bar	°C		cm^3/mol	bar					
5.95	150	0.9710	5739.46	5.78	0.9647	1.0004	1.0199	1.0293	1.0217
19.4	200	0.9298	1881.94	18.15	0.9163	1.0072	1.0556	1.0777	1.0614
49.7	250	0.8633	755.45	43.59	0.8422	1.0315	1.1324	1.1750	1.1479
107	300	0.7735	343.37	86.66	0.7475	1.0976	1.2921	1.3623	1.3287
207	350	0.6835	171.38	151.94	0.6442	1.2479	1.6122	1.7012	1.6936
276	373.946	0.6584	128.44	194.26	0.5977	1.3572	1.8370	1.9145	1.9518
367	400	0.6617	100.98	249.64	0.5609	1.4759	2.0704	2.1057	2.2219
541	450	0.7332	81.46	370.05	0.5529	1.5979	2.2380	2.1641	2.4209

(continued)

Table 23 (continued)

P	T	Z_{mixt}	V_{mixt}	f_{mixt}	φ_{H_2O}	φ_{CO_2}	φ_{CH_4}	φ_{H_2}	φ_{CO}
bar	°C		cm^3/mol	bar					
716	500	0.8210	73.74	514.16	0.5818	1.6688	2.2607	2.0937	2.4499
890	550	0.9043	69.53	683.40	0.6268	1.7268	2.2490	2.0094	2.4353
1065	600	0.9796	66.80	877.59	0.6799	1.7805	2.2316	1.9348	2.4113
1239	650	1.0468	64.84	1095.83	0.7371	1.8315	2.2163	1.8732	2.3875
1413	700	1.1067	63.34	1336.87	0.7964	1.8798	2.2046	1.8233	2.3664
1588	750	1.1600	62.14	1599.20	0.8563	1.9255	2.1966	1.7830	2.3484
1762	800	1.2077	61.14	1881.22	0.9159	1.9683	2.1918	1.7504	2.3331
1937	850	1.2504	60.28	2181.25	0.9744	2.0082	2.1895	1.7241	2.3203
2111	900	1.2889	59.54	2497.62	1.0315	2.0453	2.1895	1.7030	2.3095
2286	950	1.3235	58.88	2828.72	1.0868	2.0796	2.1911	1.6859	2.3003
2460	1000	1.3549	58.29	3173.00	1.1401	2.1113	2.1942	1.6723	2.2925
2635	1050	1.3833	57.76	3528.97	1.1912	2.1404	2.1984	1.6616	2.2522
2809	1100	1.4092	57.27	3895.28	1.2400	2.1673	2.2036	1.6532	2.2136
2879	1120	1.4189	57.08	4044.40	1.2589	2.1774	2.2059	1.6504	2.2002

Table 24 Compressibility factor, volume, and fugacity of the gas mixture with CO_2 mole fraction of 0.30 and fugacity coefficients of H_2O, CO_2, H_2, CH_4, and CO in the same gas mixture computed by means of the Peng-Robinson EOS at the P,T conditions met by Solfatara fluids along the linear P–T decompression path

P	T	Z_{mixt}	V_{mixt}	f_{mixt}	φ_{H_2O}	φ_{CO_2}	φ_{CH_4}	φ_{H_2}	φ_{CO}
bar	°C		cm^3/mol	bar					
6.80	150	0.9709	5021.31	6.61	0.9606	0.9977	1.0186	1.0291	1.0211
22.2	200	0.9304	1647.73	20.75	0.9068	1.0008	1.0524	1.0770	1.0602
56.8	250	0.8675	664.28	49.96	0.8257	1.0188	1.1248	1.1714	1.1446
123	300	0.7906	307.08	100.00	0.7248	1.0719	1.2692	1.3423	1.3136
236	350	0.7325	160.72	178.26	0.6225	1.1835	1.5227	1.6058	1.6119
315	373.946	0.7274	124.15	230.77	0.5802	1.2590	1.6808	1.7481	1.7987
405	400	0.7490	103.57	293.27	0.5588	1.3277	1.8031	1.8343	1.9444
577	450	0.8236	85.88	429.27	0.5637	1.4244	1.9164	1.8631	2.0790
748	500	0.9053	77.76	589.74	0.5976	1.5020	1.9608	1.8300	2.1286
920	550	0.9820	73.03	775.48	0.6450	1.5721	1.9825	1.7867	2.1484
1092	600	1.0513	69.89	985.91	0.6993	1.6376	1.9967	1.7466	2.1572
1264	650	1.1132	67.60	1219.85	0.7573	1.6993	2.0087	1.7128	2.1616
1436	700	1.1684	65.84	1475.76	0.8168	1.7570	2.0201	1.6851	2.1642
1608	750	1.2176	64.43	1751.98	0.8766	1.8108	2.0317	1.6630	2.1663
1779	800	1.2616	63.26	2046.74	0.9359	1.8609	2.0435	1.6454	2.1681
1951	850	1.3011	62.27	2358.29	0.9939	1.9073	2.0557	1.6316	2.1698
2123	900	1.3366	61.41	2684.91	1.0504	1.9501	2.0681	1.6210	2.1715
2295	950	1.3686	60.65	3024.99	1.1049	1.9896	2.0806	1.6131	2.1732

(continued)

Table 24 (continued)

P	T	Z_{mixt}	V_{mixt}	f_{mixt}	φ_{H_2O}	φ_{CO_2}	φ_{CH_4}	φ_{H_2}	φ_{CO}
bar	°C		cm^3/mol	bar					
2467	1000	1.3976	59.97	3376.97	1.1573	2.0259	2.0932	1.6073	2.1750
2638	1050	1.4238	59.37	3739.42	1.2075	2.0593	2.1059	1.6033	2.1476
2810	1100	1.4477	58.81	4111.01	1.2554	2.0900	2.1185	1.6009	2.1208
2879	1120	1.4567	58.61	4261.92	1.2739	2.1015	2.1236	1.6004	2.1117

Table 25 Compressibility factor, volume, and fugacity of the gas mixture with CO_2 mole fraction of 0.40 and fugacity coefficients of H_2O, CO_2, H_2, CH_4, and CO in the same gas mixture computed by means of the Peng-Robinson EOS at the P,T conditions met by Solfatara fluids along the linear P–T decompression path

P	T	Z_{mixt}	V_{mixt}	f_{mixt}	φ_{H_2O}	φ_{CO_2}	φ_{CH_4}	φ_{H_2}	φ_{CO}
bar	°C		cm^3/mol	bar					
7.94	150	0.9703	4301.25	7.71	0.9555	0.9945	1.0173	1.0294	1.0208
25.9	200	0.9301	1411.82	24.19	0.8954	0.9936	1.0494	1.0775	1.0598
66.3	250	0.8708	571.50	58.36	0.8062	1.0053	1.1184	1.1703	1.1439
143	300	0.8075	268.85	117.69	0.6997	1.0473	1.2507	1.3278	1.3050
275	350	0.7799	146.67	213.51	0.6009	1.1348	1.4608	1.5388	1.5612
368	373.946	0.7923	115.91	280.00	0.5639	1.1942	1.5843	1.6415	1.7114
455	400	0.8259	101.49	349.80	0.5572	1.2426	1.6545	1.6772	1.7963
624	450	0.9041	87.15	501.68	0.5761	1.3261	1.7342	1.6845	1.8891
792	500	0.9821	79.71	678.78	0.6166	1.4038	1.7812	1.6644	1.9385
960	550	1.0538	75.10	881.05	0.6675	1.4776	1.8154	1.6402	1.9697
1129	600	1.1180	71.91	1107.42	0.7240	1.5477	1.8438	1.6182	1.9922
1297	650	1.1753	69.55	1356.35	0.7830	1.6137	1.8694	1.6002	2.0099
1465	700	1.2262	67.71	1626.07	0.8430	1.6756	1.8933	1.5860	2.0248
1634	750	1.2716	66.21	1914.73	0.9029	1.7333	1.9159	1.5752	2.0377
1802	800	1.3121	64.97	2220.48	0.9618	1.7868	1.9376	1.5674	2.0491
1970	850	1.3485	63.91	2541.50	1.0193	1.8363	1.9585	1.5620	2.0593
2138	900	1.3812	62.99	2876.07	1.0749	1.8820	1.9786	1.5587	2.0685
2307	950	1.4106	62.19	3222.59	1.1285	1.9240	1.9981	1.5570	2.0768
2475	1000	1.4373	61.47	3579.55	1.1798	1.9627	2.0169	1.5568	2.0844
2643	1050	1.4615	60.82	3945.57	1.2288	1.9982	2.0350	1.5578	2.0661
2812	1100	1.4835	60.23	4319.40	1.2754	2.0308	2.0526	1.5598	2.0478
2879	1120	1.4917	60.01	4470.86	1.2933	2.0430	2.0595	1.5609	2.0418

temperature and pressure, reaching the maximum value, 2.0430, at 2879 bar, 1120 °C. At X_{CO_2} of 0.10, 0.20, and 0.30, the behavior of φ_{CO_2} is intermediate between those at X_{CO_2} of 0.05 and 0.40.

(v) The fugacity coefficients of CH_4 and CO behave similar to H_2 in CO_2-poor gas mixtures and similar to CO_2 in CO_2-rich gas mixtures. For instance, at X_{CO_2} of 0.05, they attain peak values of 3.5732

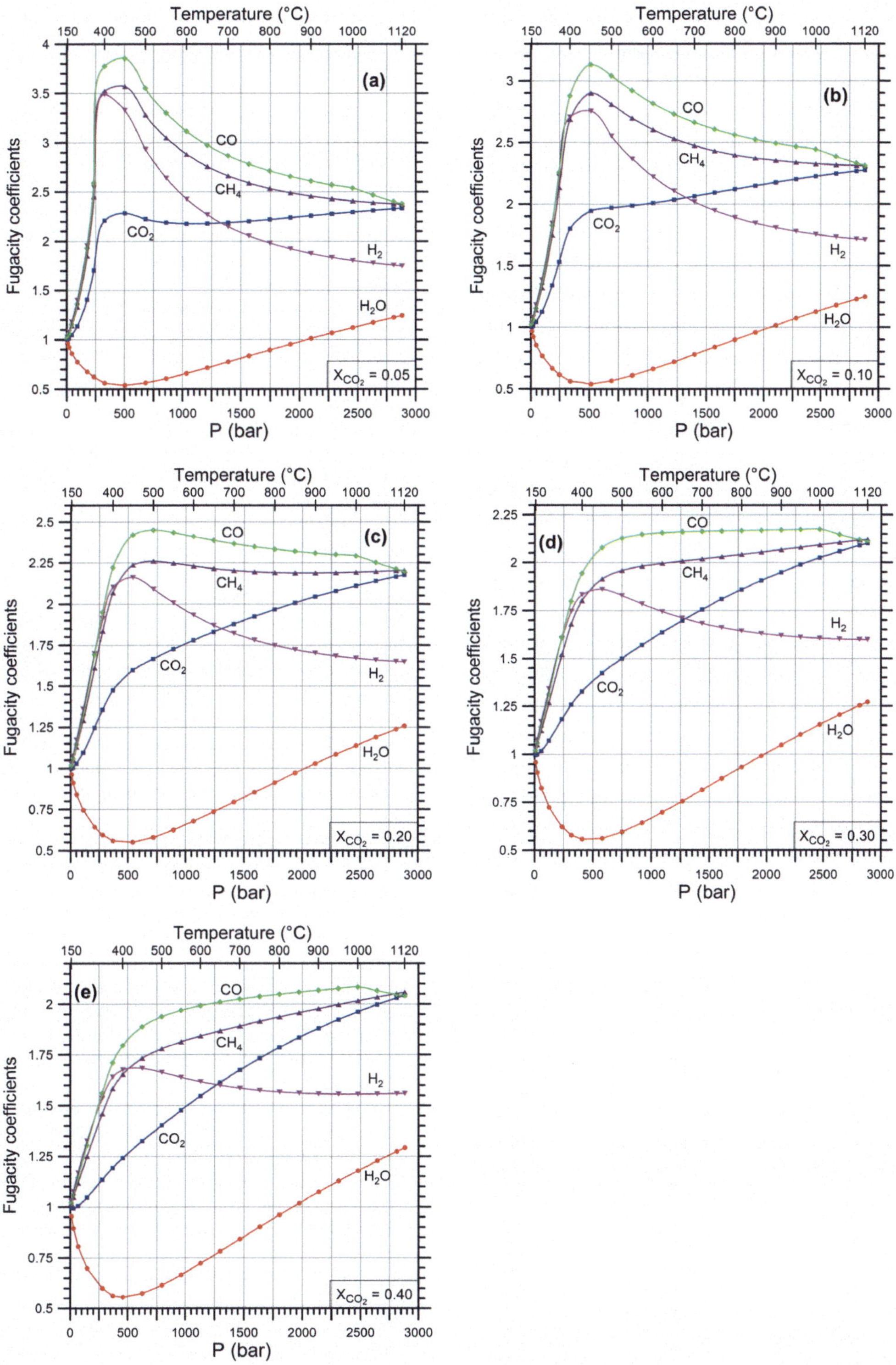

Fig. 24 Fugacity coefficients of H_2O, CO_2, H_2, CH_4, and CO in the gas mixtures with CO_2 mole fraction of **a** 0.05, **b** 0.10, **c** 0.20, **d** 0.30, and **e** 0.40, at the P,T conditions met by Solfatara fluids along the linear P–T decompression path. Calculations performed by means of the Peng-Robinson EOS

(CH_4) and 3.8563 (CO), at the same, intermediate P,T, 502 bar, 450 °C. In contrast, at X_{CO_2} of 0.40, they reach the maximum values of 2.0595 (CH_4) and 2.0844 (CO), at different and high P,T, 2879 bar, 1120 °C, and 2475 bar, 1000 °C, respectively.

Thus: (a) The $\varphi_{CO}/\varphi_{CH_4}$ ratio oscillates from 0.99 to 1.09 and, therefore, it remains close to unity at all P-T-X_{CO_2} conditions. (b) The $\varphi_{CO}/\varphi_{CO_2}$ ratio fluctuates from 1.00 to 1.71 and the $\varphi_{CH_4}/\varphi_{CO_2}$ ratio varies from 1.01 to 1.59, and, consequently, they depart moderately from unity. (c) The $\varphi_{H_2}/\varphi_{H_2O}$ ratio ranges from 1.06 to 6.17 at X_{CO_2} of 0.05, from 1.06 to 5.06 at X_{CO_2} of 0.10, from 1.07 to 3.91 at X_{CO_2} of 0.20, from 1.07 to 3.31 at X_{CO_2} of 0.30, and from 1.08 to 3.01 at X_{CO_2} of 0.40. Therefore, it deviates considerably from unity.

Summing up, the uncorrected mole fraction ratios of carbon gases may be used in geothermometric-geobarometric functions calibrated for the pressures and temperatures of the linear P–T decompression path with limited errors in calculated temperatures and pressures, whereas use of the uncorrected X_{H_2}/X_{H_2O} ratio in geothermometric-geobarometric functions may cause substantial errors in computed temperatures and pressures.

References

Abdulagatov IM, Bazaev AR, Ramazanova AE (1993a) Volumetric properties and virial coefficients of (water +methane) The J Chem Thermodyn 25(2):249–259

Abdulagatov IM, Bazaev AR, Ramazanova AE (1993b) Volume properties and virial coefficients of water-methane binary mixtures. Zhurnal Fizicheskoi Khimii 67(1):13–17

Abdulagatov IM, Bazaev AR, Ramazanova AE, Gasanov RK (1995) Experimental study of the PVTX-properties of the binary system carbon dioxide-water in the supercritical range. In: White HJ, Sengers JV, Neumann DB, Bellows JC (eds) Physical chemistry of aqueous systems: meeting the needs of industry. Begell, New York, pp 558–564

Abdulagatov IM, Bazaev AR, Gasanov RK, Ramazanova AE (1996) Measurements of the (p, ρ, T) properties and virial coefficients of pure water, methane, n-hexane, n-octane, benzene, and of their aqueous mixtures in the supercritical region. J Chem Thermodyn 28(9):1037–1057

Angus S, Armstrong B, de Reuck K M (1976) International Thermodynamic Tables of the Fluid State, Vol. 3, Carbon Dioxide. Pergamon Press, Oxford, U.K. xviii + 385

Bakker RJ, Diamond LW (2000) Determination of the composition and molar volume of H_2O-CO_2 fluid inclusions by microthermometry. Geochim Cosmochim Acta 64(10):1753–1764

Bamberger A, Sieder G, Maurer G (2000) High-pressure (vapor+ liquid) equilibrium in binary mixtures of (carbon dioxide+ water or acetic acid) at temperatures from 313 to 353 K. The J Supercritical Fluids 17 (2):97–110

Beattie JA, Bridgeman OC (1927) A new equation of state for fluids. I. Application to gaseous ethyl ether and carbon dioxide1. J Am Chem Soc 49(7):1665–1667

Bender E (1973) An equation of state for predicting vapour-liquid equilibria of the system N_2-Ar-O_2. Cryogenics 13(1):11–18

Bender R, Bier K, Manrer G (1981) Messung der spezifischen Wärme und des Joule-Thomson-Koeffizienten von Kohlendioxid bei Temperaturen von–40° C bis 200 °C und Drücken bis 15 bar. Ber Bunsenges Phys Chem 85(8):778–784

Benedict M, Webb GB, Rubin LC (1940) An empirical equation for thermodynamic properties of light hydrocarbons and their mixtures. I. Methane, ethane, propane and n-butane. The J Chem Phys 8(4):334–345

Benedict M, Webb GB, Rubin LC (1951) An empirical equation for thermodynamic properties of light hydrocarbons and their mixtures. III. Constants for Twelve Hydrocarbons. Chem Eng Progr 47(8):419–422

Blencoe JG, Naney MT, Anovitz LM (2001) The CO_2-H_2O system: III. A new experimental method for determining liquid-vapor equilibria at high subcritical temperatures. Am Mineralogist 86(9):1100–1111

Bohlander RA, Gebbie HA (1975) Molecular complexity of water vapour and the speed of sound. Nature 253 (5492):523–525

Bose TK, Cole RH (1970) Dielectric and pressure virial coefficients of imperfect gases. II. CO_2-argon mixtures. The J Chem Phys 52(1):140–147

Bottini SB, Saville G (1985) Excess enthalpies for (water + nitrogen)(g) and (water+ carbon dioxide)(g) at 520 to 620 K and up to 4.5 MPa. The J Chem Thermodyn 17(1):83–97

Bottomley GA, Massie DS, Whytlaw-Gray R (1950) A comparison of the compressibilities of some gases with that of nitrogen at pressures below one atmosphere. Proc Roy Soc Lond. Series A. Math Phys Sci 200(1061):201–218

Bowers TS, Helgeson HC (1983) Calculation of the thermodynamic and geochemical consequences of nonideal mixing in the system H_2O-CO_2-NaCl on phase relations in geologic systems: Equation of state for H_2O-CO_2-NaCl fluids at high pressures and temperatures. Geochim Cosmochim Acta 47 (7):1247–1275

Boyle R (1662) New experiments physico-mechanical, touching the spring of the air, and its effects, (made, for the most part, in a new pneumatical engine) written by the way of letter to the right honorable Charles Lord Vicount of Dungarvan, eldest son to the Earle of Corke, by the honorable Robert Boyle. Oxford: printed by H. Hall, printer to the University, for Tho: Robinson. Downloaded from Gutenberg.beic

Bücker D, Wagner W (2006a) A reference equation of state for the thermodynamic properties of ethane for temperatures from the melting line to 675 K and pressures up to 900 MPa. J Phys Chem Ref Data 35 (1):205–266

Bücker D, Wagner W (2006b) Reference equations of state for the thermodynamic properties of fluid phase n-butane and isobutane. J Phys Chem Ref Data 35 (2):929–1019

Butcher EG, Dadson RS (1964) The virial coefficients of the carbon dioxide-ethylene system I. Pure gases. Proc Roy Soc Lond. Series A. Math Phys Sci 277 (1371):448–467

Caliro S, Chiodini G, Paonita A (2014) Geochemical evidences of magma dynamics at Campi Flegrei (Italy). Geochim Cosmochim Acta 132:1–15

Cardellini C, Chiodini G, Frondini F, Avino R, Bagnato E, Caliro S, Lelli M, Rosiello A (2017) Monitoring diffuse volcanic degassing during volcanic unrests: the case of Campi Flegrei (Italy). Sci Rep 7(1):6757. https://doi.org/10.1038/s41598-017-06941-2

Carnahan NF, Starling KE (1972) Intermolecular repulsions and the equation of state for fluids. AIChE J 18 (6):1184–1189

Cawood W, Patterson HS (1933) The compressibilites of certain gases at low pressures and various temperatures. J Chem Soc (Resumed) 619–624

Cawood W, Patterson HS (1936) The determination of atomic weights by means of the microbalance, and the values obtained for carbon, nitrogen, and fluorine. Philos Trans Roy Soc Lond. Series A, Math Phys Sci 236(760):77–102

Chiodini G, Frondini F, Cardellini C, Granieri D, Marini L, Ventura G (2001) CO_2 degassing and energy release at Solfatara volcano, Campi Flegrei, Italy. J Geophys Res 106:16213–16222

Chiodini G, Caliro S, Avino R, Bini G, Giudicepietro F, De Cesare W, Ricciolino P, Aiuppa A, Cardellini C, Petrillo Z, Selva J, Siniscalchi A, Tripaldi S (2021) Hydrothermal pressure-temperature control on CO_2 emissions and seismicity at Campi Flegrei (Italy). J Volcanol Geothermal Res 414:107245

Cioni R, Corazza E, Marini L (1984) The gas/steam ratio as indicator of heat transfer at the Solfatara fumaroles, Phlegraean Fields (Italy). Bull Volcanol 47(2):295–302

Clapeyron É (1834) Mémoire sur la puissance motrice de la chaleur. Journal de l'École Polytechnique 14:153–190. https://www3.nd.edu/~powers/ame.20231/clapeyron1834.pdf

Clausius R (1850) Ueber die bewegende Kraft der Wärme und die Gesetze, welche sich daraus für die Wärmelehre selbst ableiten lassen. Ann Phys 79:368–397

Collins SC, Keyes FG (1938) The heat capacity and pressure variation of the enthalpy for steam from 38° to 125° C. Part V. Steam research program. In: Proceedings of the American academy of arts and sciences 72(8):283–299

Cook D (1957) The second virial coefficient of carbon dioxide at low temperatures. Can J Chem 35(3):268–275

Cottrell TL, Hamilton RA (1956) The second virial coefficient of gases and mixtures. Part 1. Carbon dioxide+ helium mixtures. Trans Faraday Soc 52:156–160

Cottrell TL, Hamilton RA, Taubinger RP (1956) The second virial coefficients of gases and mixtures. Part 2. Mixtures of carbon dioxide with nitrogen, oxygen, carbon monoxide, argon and hydrogen. Trans Faraday Soc 52:1310–1312

Dadson RS, Evans EJ, King JH (1967) The second virial coefficient of carbon dioxide. Proc Phys Soc (1958–1967) 92(4):1115

Dauber F, Span R (2012a) Modelling liquefied-natural-gas processes using highly accurate property models. Appl Energy 97:822–827

Dauber F, Span R (2012b) Achieving higher accuracies for process simulations by implementing the new reference equation for natural gases. Comput Chem Eng 37:15–21

De Reuck KM (1990) International thermodynamic tables of the fluid state-11 fluorine. Pergamon Press, Oxford, International Union of Pure and Applied Chemistry

De Santis R, Breedveld GJF, Prausnitz JM (1974) Thermodynamic properties of aqueous gas mixtures at advanced pressures. Ind Eng Chem Process Des Dev 13(4):374–377

Donohue MD, Prausnitz JM (1978) Perturbed hard chain theory for fluid mixtures: thermodynamic properties for mixtures in natural gas and petroleum technology. AIChE J 24(5):849–860

Driesner T, Heinrich CA (2007) The system H_2O–NaCl. Part I: Correlation formulae for phase relations in temperature–pressure–composition space from 0 to 1000°C, 0 to 5000 bar, and 0 to 1 X_{NaCl}. Geochimica et Cosmochimica Acta 71(20):4880–4901

Duan Z, Møller N, Weare JH (1992a) An equation of state for the CH_4-CO_2-H_2O system: I. Pure systems from 0 to 1000 C and 0 to 8000 bar. Geochimica et Cosmochimica Acta 56(7):2605–2617

Duan Z, Møller N, Weare JH (1992b) An equation of state for the CH_4-CO_2-H_2O system: II. Mixtures from 50 to 1000 C and 0 to 1000 bar. Geochimica et Cosmochimica Acta 56(7):2619–2631

Duan Z, Møller N, Weare JH (1995) Equation of state for the NaCl-H_2O-CO_2 system: prediction of phase equilibria and volumetric properties. Geochim Cosmochim Acta 59(14):2869–2882

Duschek W, Kleinrahm R, Wagner W (1990) Measurement and correlation of the (pressure, density, temperature) relation of carbon dioxide. I. The homogeneous gas and liquid regions in the temperature range from 217 K to 340 K at pressures up to 9 MPa. The J Chem Thermodyn 22(9):827–840

Dymond JH, Marsh KN, Wilhoit RC, Wong KC (2002a) Virial coefficients of pure gases and mixtures. Subvolume A—virial coefficients of pure gases. In: Frenkel M, Marsh KN (eds) Landolt-Börnstein. Numerical data and functional relationships in science and technology. New Series. Editor in Chief W. Martienssen. Group IV: Physical Chemistry. Volume 21. Springer, Berlin, p 309

Dymond JH, Marsh KN, Wilhoit RC, Wong KC (2002b) Virial coefficients of pure gases and mixtures. Subvolume B—virial coefficients of mixtures. In: Frenkel M, Marsh KN (eds) Landolt-Börnstein. Numerical data and functional relationships in science and technology. New Series. Editor in Chief W. Martienssen. Group IV: Physical Chemistry. Volume 21. Springer, Berlin, p 365

Elias E, Hoang N, Sommer J, Schramm B (1986) Die zweiten Virialkoeffizienten von Helium-Gasmischungen im Bereich unterhalb Zimmertemperatur. Ber Bunsenges Phys Chem 90(4):342–351

Elliott JR, Lira CT (2012) Introductory chemical engineering thermodynamics, 2nd edn. Prentice Hall, Upper Saddle River, NJ. ISBN 978-0-13-606854-9, p 876

Elshahomi A, Lu C, Michal G, Liu X, Godbole A, Venton P (2015) Decompression wave speed in CO_2 mixtures: CFD modelling with the GERG-2008 equation of state. Appl Energy 140:20–32

Eubank PT, Joffrion LL, Patel MR, Warowny W (1988) Experimental densities and virial coefficients for steam from 348 to 498 K with correction for adsorption effects. J Chem Thermodyn 20(9):1009–1034

Eubank PT, Shyu GS, Hanif NS (1995) New procedures for application of the Wong-Sandler mixing rules to the prediction of vapor-liquid equilibria. Ind Eng Chem Res 34(1):314–323

Ewers J, Wagner W (1982) A method for optimizing the structure of equations of state and its application to an equation of state for oxygen. VDI Forschungsheft, 609, 27–34

Fenghour A, Wakeham WA, Watson JTR (1996) Densities of (water+ carbon dioxide) in the temperature range 415 K to 700 K and pressures up to 35 MPa. J Chem Thermodyn 28(4):433–446

Flowers GC (1979) Correction of Holloway's (1977) adaptation of the modified Redlich-Kwong equation of state for calculation of the fugacities of molecular species in supercritical fluids of geologic interest. Contrib Miner Petrol 69(3):315–318

Flowers G C, Helgeson H C (1983) Equilibrium and mass transfer during progressive metamorphism of siliceous dolomites. American Journal of Science, 283(3), 230–286.

Franck EU, Tödheide K (1959) Thermische Eigenschaften überkritischer Mischungen von Kohlendioxyd und Wasser bis zu 750 °C und 2000 Atm. Zeitschrift Für Physikalische Chemie Neue Folge 22:232–245

Frantz JD, Popp RK, Hoering TC (1992) The compositional limits of fluid immiscibility in the system H_2O-$NaCl$-CO_2 as determined with the use of synthetic fluid inclusions in conjunction with mass spectrometry. Chem Geol 98:237–255

Gallagher JS, Crovetto R, Sengers JL (1993) The thermodynamic behavior of the CO_2-H_2O system from 400 to 1000 K, up to 100 MPa and 30% mole fraction of CO_2. J Phys Chem Ref Data 22(2):431–513

Gay-Lussac JL (1802) Recherches sur la dilatation des gaz et des vapeurs, lues à l'Institut national, le 11 pluviôse an 10. Annales de Chimie. Fuchs, Paris, Tome 43, pp 137–175. Downloaded from Google Books

Gehrig M (1980) Phasengleichgewichte und pVT-daten ternärer Mischungen aus Wasser, Kohlendioxid und Natriumchlorid bis 3 kbar und 550 °C. PhD dissertation, Univ. Karlsruhe, Hochschul Verlag, Freiburg

Gernert J, Span R (2016) EOS-CG: A Helmholtz energy mixture model for humid gases and CCS mixtures. J Chem Thermodyn 93:274–293

Giggenbach WF (1987) Redox processes governing the chemistry of fumarolic gas discharges from White Island, New Zealand. Appl Geochem 2:143–161

Główka S (1990) Determination of volumetric properties of ammonia between 298 and 473 K and carbon dioxide between 304 and 423 K using the Burnett method. Pol J Chem 64:699–709

Góra U, Cencek W, Podeszwa R, van der Avoird A, Szalewicz K (2014) Predictions for water clusters from a first-principles two-and three-body force field. The J Chem Phys 140(19):194101

Graboski MS, Daubert TE (1978a) A modified Soave equation of state for phase equilibrium calculations. 1. Hydrocarbon systems. Indus Eng Chem Process Des Dev 17(4):443–448

Graboski MS, Daubert TE (1978b) A modified Soave equation of state for phase equilibrium calculations. 2. Systems containing CO_2, H_2S, N_2, and CO. Indus Eng Chem Process Des Dev 17(4):448–454

Graboski MS, Daubert TE (1979) A modified Soave equation of state for phase equilibrium calculations. 3. Systems containing hydrogen. Indus Eng Chem Process Des Dev 18(2):300–306

Greenwood HJ (1969) The compressibility of gaseous mixtures of carbon dioxide and water between 0 and 500 bars pressure and 450° and 800° centigrade. Am J Sci 267:191–208

Guerrero-Zárate D, Estrada-Baltazar A, Iglesias-Silva G A (2017) Calculation of critical points for natural gas mixtures with the GERG-2008 equation of state. Fluid Phase Equilibria, 437, 69–82.

Haar L, Gallagher JS (1978) Thermodynamic properties of ammonia. J Phys Chem Ref Data 7:635–792

Haar L, Gallagher JS, Kell GS (1984) NBS/NRC steam tables: thermodynamic and transport properties and computer programs for vapour and liquid states of water in SI units. Hemisphere Publishing Corporation, Washington, p 320

Hall KR, Eubank PT Holste JC (1983) Residual functions and fugacity. Chemical Engineering Education, 17(3), 124-128.

Harrison SF, Mayer JE (1938) Statistical mechanics of condensing systems IV. The J Chem Phys 6(2):101–104

Hassanpouryouzband A, Joonaki E, Edlmann K, Heinemann N, Yang J (2020) Thermodynamic and transport properties of hydrogen containing streams. Sci Data 7 (1):1–14

Hellmann R (2014) Ab initio potential energy surface for the carbon dioxide molecule pair and thermophysical properties of dilute carbon dioxide gas. Chem Phys Lett 613:133–138

Hendl H, Bich E, Vogel E (1997) A new evaluation of (p, ρ, T) measurements on steam with corrections for the effects of physical and chemical adsorption. J Chem Thermodyn 29(7):765–784

Hernández-Gómez R, Tuma D, Villamañán MA, Mondéjar ME, Chamorro CR (2014) Accurate thermodynamic characterization of a synthetic coal mine methane mixture. J Chem Thermodyn 68:253–259

Holloway JR (1977) Fugacity and activity of molecular species in supercritical fluids. In: Fraser DG (ed) Thermodynamics in geology. NATO advanced study institutes series (Series C—Mathematical and Physical Sciences), vol 30. Springer, Dordrecht, pp 161–181

Holste JC, Watson MQ, Bellomy MT, Eubank PT, Hall KR (1980) Determination of interaction second virial coefficients; He-CO_2 system. AIChE J 26 (6):954–964

Holste JC, Hall KR, Eubank PT, Esper G, Watson MQ, Warowny W, Bailey DM, Young JG, Bellomy MT (1987) Experimental (p, V_m, T) for pure CO_2 between 220 and 450 K. J Chem Thermodyn 19(12):1233–1250

Holland T, Powell R (1991) A Compensated-Redlich-Kwong (CORK) equation for volumes and fugacities of CO_2 and H_2O in the range 1 bar to 50 kbar and 100–1600°C. Contributions to Mineralogy and Petrology, 109(2), 265–273.

Horstmann A (1873) Theorie der dissociation. Justus Liebigs Ann Chem 170(1–2):192–210. https://doi.org/10.1002/jlac.18731700118

Horvath AL (1974) Redlich-Kwong equation of state: review for chemical engineering calculations. Chem Eng Sci 29(5):1334–1340

Jacobsen RT, Stewart RB (1973) Thermodynamic properties of nitrogen including liquid and vapor phases from 63 K to 2000 K with pressures to 10,000 bar. J Phys Chem Ref Data 2(4):757–922

Jacobsen RT, Stewart RB, Jahangiri M, Penoncello SG (1986) A new fundamental equation for thermodynamic property correlations. In: Fast RW (ed) Advances in cryogenic engineering. Springer, Boston, MA, pp 1161–1168

Jaeschke M (1987) Determination of the interaction second virial coefficients for the carbon dioxide-ethane system from refractive index measurements. Int J Thermophys 8(1):81–95

Joffe J (1976) Vapor-liquid equilibria by the pseudocritical method. Industrial & Engineering Chemistry Fundamentals, 15(4), 298-303.

Johnson EL (1992) An assessment of the accuracy of isochore location techniques for H_2O-CO_2-NaCl fluids at granulite facies pressure-temperature conditions. Geochim Cosmochim Acta 56:295–302

Kamerlingh Onnes H (1901) Expression of the equation of state of gases and liquids by means of series. In: Gavroglu K, Goudaroulis Y (eds) Through measurement to knowledge. The selected papers of Heike Kamerlingh Onnes 1853–1926. Kluwer, Dordrecht (1991), pp 146–163

Katayama T, Ohgaki K, Ohmori H (1980) Measurement of interaction second virial coefficients with high accuracy. J Chem Eng Jpn 13(4):257–262

Keenan JH, Keyes FG, Hill PG, Moore JG (1969) Steam tables. Thermodynamic properties of water including vapour, liquid, and solid phases. Wiley, New York, p 156

Kell GS, McLaurin GE, Whalley E (1965) Advances in thermophysical properties at extreme pressures and temperatures. American Society of Mechanical Engineers, New York, p 104

Kell GS, McLaurin GE, Whalley E (1968) PVT properties of water. II. Virial coefficients in the range 150–450 ° C without independent measurement of vapor volumes. The J Chem Phys 48(8):3805–3813

Kell GS, McLaurin GE, Whalley E (1989) PVT properties of water. VII. Vapour densities of light and heavy water from 150 to 500 °C. Proc Roy Soc Lond. A. Math Phys Sci 425(1868):49–71

Kerrick DM, Jacobs GK (1981) A modified Redlich-Kwong equation for H_2O, CO_2, and H_2O-CO_2 mixtures at elevated pressures and temperatures. Am J Sci 281(6):735–767

Khitarov NI, Malinin SD (1958) Phase equilibria in the system H_2O-CO_2. Geokhimiya (USSR) 7:846–848

Kim C, Kim H, Mun K (2020) Use of the international association for the properties of water and steam (IAPWS) formulations, IAPWS-95 & IAPWS-IF97: making of Mollier diagram and Ts diagram of water and steam. Thermal Sci Eng Progress 20:100691

King AD Jr, Coan CR (1971) Solubility of water in compressed carbon dioxide, nitrous oxide, and ethane. Evidence for hydration of carbon dioxide and nitrous oxide in the gas phase. J Am Chem Soc 93(8):1857–1862

King MB, Mubarak A, Kim JD, Bott TR (1992) The mutual solubilities of water with supercritical and liquid carbon dioxides. The J Supercritical Fluids 5 (4):296–302

Klimeck R (2000) Entwicklung einer Fundamentalgleichung für Erdgase für das Gas- und Flüssigkeitsgebiet sowie das Phasengleichgewicht. Ph.D. Dissertation, Fakultät für Maschinenbau, Ruhr-Universität Bochum, Germany

Kotelnikov AR, Kotelnikova ZA (1990) An experimental study of the phase state of the system H_2O-CO_2-NaCl using synthetic fluid inclusions in quartz. Geokhimiya 4:526–537 (in Russian)

Ku PS, Dodge BF (1967) Compressibility of the binary systems: helium-nitrogen and carbon dioxide-ethylene. J Chem Eng Data 12(2):158–164

Kunz O, Wagner W (2012) The GERG-2008 wide-range equation of state for natural gases and other mixtures: an expansion of GERG-2004. J Chem Eng Data 57 (11):3032–3091

Kunz O, Klimeck R, Wagner W, Jaeschke M (2007) The GERG-2004 wide-range equation of state for natural gases and other mixtures. GERG Technical Monograph 15, p 535

Lambert JD, Staines EN, Woods SD (1950) Thermal conductivities of organic vapours. Proc Roy Soc Lond. Series A. Math Phys Sci 200(1061):262–271

Lancaster NM, Wormald CJ (1990) Excess molar enthalpies of nine binary steam mixtures: new and corrected values. J Chem Eng Data 35(1):11–16

Lane JG, Leidecker HW (1974) Theoretical calculations of the second virial coefficient of water. J Chem Phys 60(2):705–707

Leach JW, Chappelear PS, Leland TW (1968) Use of molecular shape factors in vapor-liquid equilibrium calculations with the corresponding states principle. AIChE J 14(4):568–576

Leduc A, Sacerdote P (1897) Constantes Critiques de quelques Gaz. C.R. Hebd Seances Acad Sci 125:397–398

Lee BI, Kesler MG (1975) A generalized thermodynamic correlation based on three-parameter corresponding states. AIChE J 21(3):510–527

Lee SM, Eubank PT, Hall KR (1977) Truncation errors associated with the virial equation. Fluid Phase Equilib 1(3):219–224

Leland TW, Chappelear PS (1968) The corresponding states principle. A review of current theory and practice. Indus Eng Chem 60(7):15–43

Lemmon EW (1996) A generalized model for the prediction of the thermodynamic properties of mixtures including vapor–liquid equilibrium. PhD Dissertation, University of Idaho, Moscow

Lemmon EW, Jacobsen RT (1999) A generalized model for the thermodynamic properties of mixtures. Int J Thermophys 20(3):825–835

Lemmon EW, Span R (2006) Short fundamental equations of state for 20 industrial fluids. J Chem Eng Data 51(3):785–850

Lemmon EW, Span R (2010) Multi-parameter equations of state for pure fluids and mixtures. In: Goodwin ARH, Sengers JV, Peters CJ (eds) Applied thermodynamics of fluids, Chapter 12. International Union of Pure and Applied Chemistry. Royal Society of Chemistry, Cambridge, UK, pp 394–432

Lemmon EW, Tillner-Roth R (1999) A Helmholtz energy equation of state for calculating the thermodynamic properties of fluid mixtures. Fluid Phase Equilibria 165(1):1–21

Lemmon EW, McLinden MO, Wagner W (2009) Thermodynamic properties of propane. III. A reference equation of state for temperatures from the melting line to 650 K and pressures up to 1000 MPa. J Chem Eng Data 54(12):3141–3180

Lennard-Jones JE (1937) The equation of state of gases and critical phenomena. Physica 4(10):941–956

Lennard-Jones JE, Devonshire AF (1937) Critical phenomena in gases. I. Proc Roy Soc Lond. Series A-Math Phys Sci 163(912):53–70

Lentner R, Richter M, Kleinrahm R, Span R (2017) Density measurements of liquefied natural gas (LNG) over the temperature range from (105 to 135) K at pressures up to 8.9 MPa. J Chem Thermodyn 112:68–76

Lichtenthaler RN, Schäfer K (1969) Zwischenmolekulare Kräfte kugelförmiger und nicht kugelförmiger Molekelnermittelt aus zweiten Virialkoeffizienten. Ber Bunsenges Phys Chem 73(1):42–48

Lie GC, Clementi E (1976) Calculation of the second virial coefficients for water using a recent "ab initio" potential. J Chem Phys 64(12):5308–5309

Lin CT, Daubert TE (1979) Prediction of partial molar volume from the Lee-Kesler equation of state. AIChE J 25(2):365–367

Lopatinskiii ES, Rozhnov MS, Zhdanov VI, Parnovskii SL, Kudrya YN (1991) Second virial coefficient of gas mixtures of hydrocarbons, carbon dioxide and hydrogen with nitrogen and argon. Zh Fiz Khim 65:2060–2065

MacCormack KE, Schneider WG (1950) Compressibility of gases at high temperatures. IV. Carbon dioxide in the temperature range 0–600 °C, and pressures up to 50 atmospheres. The J Chem Phys 18(9):1269–1272

Malesinska B (1980) Determination of the second cross virial coefficient from gas-liquid solubility data. Pol J Chem 54(7–8):1527–1537

Malinin SD (1959) System water-carbon dioxide at high temperatures and pressures. Geokhimiya (USSR) 3:292–306

Mallu BV, Natarajan G, Viswanath DS (1987) Compression factors and second virial coefficients of CO_2, CO, and {xCO+ (1–x)CO2}. J Chem Thermodyn 19 (5):549–554

Mather AE, Franck EU (1992) Phase equilibria in the system carbon dioxide-water at elevated pressures. J Phys Chem 96(1):6–8

Mayer JE (1937) The statistical mechanics of condensing systems. I. the Journal of Chemical Physics 5(1):67–73

Mayer JE (1939) Statistical mechanics of condensing systems. V. Two-component systems. J Phys Chem 43(1):71–95

Mayer JE, Ackermann PG (1937) The statistical mechanics of condensing systems.II. The J Chem Phys 5 (1):74–83

Mayer JE, Harrison SF (1938) Statistical mechanics of condensing systems III. The J Chem Phys 6(2):87–100

Mayer JE, Mayer MG (1940) Statistical mechanics. Wiley, New York, p 495

Mazzoccoli M, Bosio B, Arato E (2012) Analysis and comparison of equations-of-state with p-ρ-T experimental data for CO_2 and CO_2-mixture pipeline transport. Energy Procedia 23:274–283

McCarty RD, Arp VD (1990) A new wide range equation of state for helium. In: Fast RW (eds) Advances in cryogenic engineering. Advances in cryogenic engineering, vol 35. Springer, Boston, MA

McCullough JP, Pennington RE, Waddington G (1952) A calorimetric determination of the vapor heat capacity and gas imperfection of water. J Am Chem Soc 74 (17):4439–4442

Meyer CW, Harvey AH (2015) Dew-point measurements for water in compressed carbon dioxide. AIChE J 61 (9):2913–2925

Michels A, Michels C (1935) Isotherms of CO_2 between 0 and 150° and Pressures from 16 to 250 Atm (Amagat densities 18–206). Proc Roy Soc Lond. Series A-Math Phys Sci 153(878):201–214

Moretti R, Troise C, Sarno F, De Natale G (2018) Caldera unrest driven by CO_2-induced drying of the deep hydrothermal system. Sci Rep 8(1):1–11

Moretti R, De Natale G, Troise C (2020) Hydrothermal versus magmatic: geochemical views and clues into the unrest dilemma at Campi Flegrei. In: De Vivo B, Belkin HE, Rolandi G (eds) Vesuvius, Campi Flegrei, and Campanian Volcanism. Elsevier, Amsterdam, pp 371–406

Mullins JC, Kirk BS, Ziegler WT (1963) Calculation of the vapor pressure and heats of vaporization and sublimation of liquids and solids, especially below one atmosphere, v. carbon monoxide and carbon dioxide. Tech. Rept. No. 2, Project A-663, Engineering Experiment Station, Georgia Institute of Technology, August 15, 1963 (Contract No. CST-7404, National Bureau of Standards, Boulder, Colorado)

Nguyen TV, Elmegaard B (2016) Assessment of thermodynamic models for the design, analysis and optimization of gas liquefaction systems. Appl Energy 183:43–60

Ohgaki K, Nakamura Y, Ariyasu H, Katayama T (1982) Interaction second virial coefficients for six binary systems containing carbon dioxide, methane, ethylene and propylene at 125 °C. J Chem Eng Jpn 15(2):85–90

Ohgaki K, Sakai N, Kano Y, Katayama T (1984) Experimental study of second virial coefficients for carbon dioxide, methane, ethylene and propylene at 423.15 K and 473.15 K. J Chem Eng Jpn 17(5):545–547

Opfell JB, Sage BH, Pitzer KS (1956) Application of Benedict equation to theorem of corresponding states. Ind Eng Chem 48(11):2069–2076

Palacio A (1990) Effect of heat transfer on the performance of geothermal wells. Geothermics 19(4):311–328

Panagiotopoulos AZ, Reid RC (1986) New mixing rule for cubic equations of state for highly polar, asymmetric systems. In: Chao KC, Robinson RL (eds) Equations of state: theories and applications. ACS Symposium Series, American Chemical Society, Washington, DC, vol 300, pp 571–582

Partington JR (1949) An advanced treatise on physical chemistry, vol One. The properties of gases. Longmans, Green and Co., London, Fundamental Principles, p 943

Patel MR, Eubank PT (1988) Experimental densities and derived thermodynamic properties for carbon dioxide-water mixtures. J Chem Eng Data 33(2):185–193

Patel MR, Holste JC, Hall KR, Eubank PT (1987) Thermophysical properties of gaseous carbon dioxide-water mixtures. Fluid Phase Equilib 36:279–299

Patel MR, Lane JL, Eubank PT (1988) A simple procedure for estimating virial coefficients from Burnett PVT data. AIChE J 34(7):1229–1232

Paulus ME, Penoncello SG (2006) Correlation for the carbon dioxide and water mixture based on the Lemmon-Jacobsen mixture model and the Peng-Robinson equation of state. Int J Thermophys 27 (5):1373–1386

Peng DY, Robinson DB (1976a) A new two-constant equation of state. Ind Eng Chem Fundam 15(1):59–64

Peng DY, Robinson DB (1976b) Two and three phase equilibrium calculations for systems containing water. The Can J Chem Eng 54(5):595–599

Pérez-Sanz FJ, Martín MC, Chamorro CR, Fernández-Vicente T, Segovia JJ (2016) Heat capacities and acoustic virial coefficients for a synthetic coal mine methane mixture by speed of sound measurements at T=(273.16 and 250.00) K. J Chem Thermodyn 97:137–141

Pfefferle WC Jr., Goff JA, Miller JG (1955) Compressibility of gases. I. The Burnett method. An improved method of treatment of the data. Extension of the method to gas mixtures. The J Chem Phys 23(3):509–513

Pitzer KS (1955) The volumetric and thermodynamic properties of fluids. I. Theoretical basis and virial coefficients. J Am Chem Soc 77(13):3427–3433

Pitzer KS, Lippmann DZ, Curl RF Jr., Huggins CM, Petersen DE (1955) The volumetric and thermodynamic properties of fluids. II. Compressibility factor, vapor pressure and entropy of vaporization. J Am Chem Soc 77(13):3433–3440

Plocker U, Knapp H, Prausnitz J (1978) Calculation of high-pressure vapor-liquid equilibria from a corresponding-states correlation with emphasis on asymmetric mixtures. Ind Eng Chem Process Des Dev 17(3):324–332

Prausnitz JM, Lichtenthaler RN, De Azevedo EG (1999) Molecular thermodynamics of fluid-phase equilibria. Pearson Education. Prentice Hall, Upper Saddle River, New Jersey, p 860

Redlich O (1975) On the three-parameter representation of the equation of state. Ind Eng Chem Fundam 14 (3):257–260

Redlich O., Kwong, J.N. (1949) On the thermodynamics of solutions. V. An equation of state. Fugacities of gaseous solutions. Chem Rev 44(1):233–244

Robinson DB, Peng DY (1980) The use of equations of state in multi-phase equilibrium calculations. In: Phase equilibria and fluid properties in the chemical industry, 2nd international conference, Berlin, vol 335, p 353

Robinson DB, Peng D-Y, Ng HJ (1977) Applications of the Peng-Robinson equation of state. In: Storvick TS, Sandler SI (eds) Phase equilibria and fluid properties in the chemical industry. ACS Symposium Series, vol 60, pp 200–220

Rowland D, Hughes TJ, May EF (2016) Extending the GERG-2008 equation of state: improved departure function and interaction parameters for (methane+ butane). J Chem Thermodyn 97:206–213

Rowlinson JS, Watson ID (1969) The prediction of the thermodynamic properties of fluids and fluid mixtures. I. The principle of corresponding states and its extensions. Chem Eng Sci 24(10):1565–1574

Sass A, Dodge BF, Bretton RH (1967) Compressibility of gas mixtures. Carbon dioxide-ethylene system. J Chem Eng Data 12(2):168–176

Schäfer K (1937) Der zweite Virialkoeffizient der verschiedenen Modifikationen des leichten und schweren Wasserstoffs. Z Phys Chem 36(1):85–104

Schmidt C, Bodnar RJ (2000) Synthetic fluid inclusions: XVI. PVTX properties in the system H_2O-NaCl-CO_2 at elevated temperatures, pressures, and salinities. Geochimica et Cosmochimica Acta 64 (22):3853–3869

Schmidt R, Wagner W (1985) A new form of the equation of state for pure substances and its application to oxygen. Fluid Phase Equilib 19:175–200

Schmiedel H, Gehrmann R, Schramm B (1980) Die zweiten Virialkoeffizienten verschiedener Gasmischungen im Temperaturbereich von 213 bis 475 K. Ber Bunsenges Phys Chem 84(8):721–724

Schramm B, Müller W (1982) Messungen des zweiten Virialkoeffizienten von Gasen und Gasmischungen bei Zimmertemperatur mit einer Expansionsapparatur. Ber Bunsenges Phys Chem 86(2):110–112

Schramm B, Elias E, Kern L, Natour G, Schmitt A, Weber C (1991) Precise measurements of second virial coefficients of simple gases and gas mixtures in the temperature range below 300 K. Ber Bunsenges Phys Chem 95(5):615–621

Schultz AJ, Kofke DA, Harvey AH (2015) Molecular-based virial coefficients of CO_2-H_2O mixtures. AIChE J 61(9):3029–3037

Seitz JC, Blencoe JG (1997) Experimental determination of the volumetric properties and solvus relations of H_2O-CO_2 mixtures at 300–400 °C and 75–1000 bars. Proceedings of the Fifth International Symposium on Hydrothermal Reactions, 109–112. Oak Ridge National Laboratory, Oak Ridge, TN

Seitz JC, Blencoe JG (1999) The CO_2-H_2O system. I. Experimental determination of volumetric properties at 400°C, 10–100 MPa. Geochimica et Cosmochimica Acta 63(10):1559–1569

Setzmann U, Wagner W (1991) A new equation of state and tables of thermodynamic properties for methane covering the range from the melting line to 625 K at pressures up to 100 MPa. J Phys Chem Ref Data 20 (6):1061–1155

Shmulovich K I, Shmonov V M, Mazur V A, Kalinichev A G (1980) PVT activity concentration relations in the H_2O-CO_2 systems (homogeneous solutions). Geochemistry International, 17(6), 123-139.

Shmulovich K, Graham C (1999) An experimental study of phase equilibria in the system H_2O-CO_2-NaCl at 800 °C and 9 kbar. Contrib Mineral Petrol 136:247–257

Shyu GS, Hanif NS, Hall KR, Eubank PT (1997) Carbon dioxide-water phase equilibria results from the Wong-Sandler combining rules. Fluid Phase Equilib 130(1–2):73–85

Skripka VG (1979) Volume behavior of nonpolar component-water gaseous mixtures at high temperatures. Zh Fiz Khim 53:1407–1409

Smith GR, Wormald CJ (1984) The excess molar enthalpies of {xH2O + (1−x)CO}(g) and {xH2O + (1−x)CO2}(g). J Chem Thermodyn 16(6):543–550

Smukala J, Span R, Wagner W (2000) New equation of state for ethylene covering the fluid region for temperatures from the melting line to 450 K at pressures up to 300 MPa. J Phys Chem Ref Data 29 (5):1053–1121

Soave G (1972) Equilibrium constants from a modified Redlich-Kwong equation of state. Chem Eng Sci 27 (6):1197–1203

Soave GS (1979) Application of a cubic equation of state to vapor-liquid equilibria of systems containing polar compounds. Inst Chem Eng Symp Series 56 (Distillation, vol. 1), S. 1.2/1–1.2/16

Soave G (1980) Rigorous and simplified procedures for determining the pure-component parameters in the Redlich-Kwong-Soave equation of state. Chem Eng Sci 35(8):1725–1730

Span R (2000) Multiparameter equations of state. An Accurate Source of Thermodynamic Property Data. Springer, Berlin, 367 pp.

Span R, Wagner W (1996) A new equation of state for carbon dioxide covering the fluid region from the triple-point temperature to 1100 K at pressures up to 800 MPa. J Phys Chem Ref Data 25(6):1509–1596

Span R, Wagner W (2003a) Equations of state for technical applications. I. Simultaneously optimized functional forms for nonpolar and polar fluids. Int J Thermophys 24(1):1–39

Span R, Wagner W (2003b) Equations of state for technical applications. II. Results for nonpolar fluids. Int J Thermophys 24(1):41–109

Span R, Wagner W (2003c) Equations of state for technical applications. III. Results for polar fluids. Int J Thermophys 24(1):111–162

Span R, Lemmon EW, Jacobsen RT, Wagner W, Yokozeki A (2000) A reference equation of state for the thermodynamic properties of nitrogen for temperatures from 63.151 to 1000 K and pressures to 2200 MPa. J Phys Chem Ref Data 29(6):1361–1433

Spycher NF, Reed MH (1988) Fugacity coefficients of H_2, CO_2, CH_4, H_2O and of H_2O-CO_2-CH_4 mixtures: a virial equation treatment for moderate pressures and temperatures applicable to calculations of hydrothermal boiling. Geochim Cosmochim Acta 52(3):739–749

Spycher N, Pruess K, Ennis-King J (2003) CO_2-H_2O mixtures in the geological sequestration of CO_2. I. Assessment and calculation of mutual solubilities from 12 to 100 °C and up to 600 bar. Geochimica et Cosmochimica Acta 67(16):3015–3031

Starling KE (1973) Fluid thermodynamic properties for light petroleum systems. Gulf Publishing Company, Houston, p 270

Sterner SM, Bodnar RJ (1991) Synthetic fluid inclusions. X. Experimental determination of P-V-T-X properties in the CO_2-H_2O system to 6 kb and 700°C. Am J Sci 291(1):1–54

Stevenson DS (1993) Physical models of fumarolic flow. J Volcanol Geoth Res 57(3–4):139–156

Strutt J W (1905) On the compressibility of gases between one atmosphere and half an atmosphere of pressure. Philosophical Transactions of the Royal Society of London, Series A, 204, 351–372.

Takenouchi S, Kennedy GC (1964) The binary system H_2O-CO_2 at high temperatures and pressures. Am J Sci 262(9):1055–1074

Tamburello G, Caliro S, Chiodini G, De Martino P, Avino R, Minopoli C, Carandente A, Rouwet D, Aiuppa A, Costa A, Bitetto M, Giudice G, Francofonte V, Ricci T, Sciarra A, Bagnato E, Capecchiacci F (2019) Escalating CO_2 degassing at the Pisciarelli fumarolic system, and implications for the ongoing Campi Flegrei unrest. J Volcanol Geoth Res 384:151–157

Tanger JC IV, Pitzer KS (1989) Thermodynamics of NaCl-H_2O: a new equation of state for the near-critical region and comparisons with other equations for adjoining regions. Geochim Cosmochim Acta 53 (5):973–987

Tegeler C, Span R, Wagner W (1999) A new equation of state for argon covering the fluid region for temperatures from the melting line to 700 K at pressures up to 1000 MPa. J Phys Chem Ref Data 28(3):779–850

Thiesen M (1885) Untersuchungen über die Zustandsgleichung. Ann Phys 260(3):467–492

Tillner-Roth R (1993) Die Thermodynamischen Eigenschaften von R-152a, R-134a und ihren Gemischen-Messungen und Fundamentalgleichungen, Forsch.-Ber. DKV No. 41, DKV-Verlag, Stuttgart

Tillner-Roth R, Harms-Watzenberg F, Baehr HD (1993) Eine neue Fundamentalgleichung fur Ammoniak. DKV-Tagungsbericht, 20:167–181

Tödheide K, Franck EU (1963) Das Zweiphasengebiet und die kritische Kurve im System Kohlendioxid-Wasser bis zu Drucken von 3500 bar. Z Phys Chem 37 (5–6):387–401

Treiner C, Bocquet JF, Chemla M (1973) Seconds coefficients du viriel des mélanges eau-tétrahydrofurane (THF). Influence sur les coefficients d'activité de l'eau et du THF à 25 °C. J De Chimie Physique 70:72–79

Tsonopoulos C, Heidman JL (1985) From Redlich-Kwong to the present. Fluid Phase Equilib 24(1–2):1–23

Ursell HD (1927) The evaluation of Gibbs' phase-integral for imperfect gases. In: Mathematical proceedings of the Cambridge philosophical society. Cambridge University Press, vol 23, no 6, pp 685–697

Van der Waals JD (1873) Over de Continuiteit van den Gas-en Vloeistoftoestand. A.W. Sijthoff, Leiden, p 134. https://www.lorentz.leidenuniv.nl/IL-publications/dissertations/sources/vanderWaals_1873.pdf

Varzandeh F, Stenby EH, Yan W (2017) Comparison of GERG-2008 and simpler EoS models in calculation of phase equilibrium and physical properties of natural gas related systems. Fluid Phase Equilib 434:21–43

Volta A (1793) Della uniforme dilatazione dell'aria per ogni grado di calore, cominciando sotto la temperatura del ghiaccio fin sopra quella dell'ebollizione dell'acqua; e di ciò, che sovente fa parer non equabile tal dilatazione, entrando ad accrescer a dismisura il volume dell'aria. Annali di Chimica e Storia Naturale di L. Brugnatelli. Tomo IV. Stamperia Bolzani, Pavia, pp 227–294. Downloaded from Google Books

Vukalovich MP, Masalov YF (1966) Equation of state for gaseous carbon dioxide. Teploenergetika 13:58–62

Vukalovich MP, Trakhtengerts MS, Spiridonov GA (1967) An equation of state for steam at temperatures above 500 °C. Teploenergetika 14:65–70

Wagner W, Ewers J, Schmidt R (1982) An equation for the ideal-gas heat capacity of molecular oxygen for temperatures from 30 K to 3000 K. Berichte der Bunsengesellschaft für Physikalische Chemie, 86(6), 538-540.

Wagner W, Pruß A (2002) The IAPWS formulation 1995 for the thermodynamic properties of ordinary water substance for general and scientific use. J Phys Chem Ref Data 31(2):387–535

Walas SM (1985) Phase equilibria in chemical engineering. Butterworth Publishers, Boston, p 671

Waxman M, Davis HA, Hastings JR (1973) A new determination of the second virial coefficient of carbon dioxide at temperatures between 0 and 150 C, and an evaluation of its reliability. In: Symposium on thermophysical properties, 6 th, Atlanta, Ga, pp 245–255

Weber LA (1992) Measurements of the virial coefficients and equation of state of the carbon dioxide+ ethane system in the supercritical region. Int J Thermophys 13(6):1011–1032

Wheatley RJ, Harvey AH (2011) Intermolecular potential energy surface and second virial coefficients for the water–CO2 dimer. The J Chem Phys 134(13):134309

Wheatley RJ, Harvey AH (2016) Erratum:"Intermolecular potential energy surface and second virial coefficients for the water–CO2 dimer"[J Chem Phys 134:134309 (2011)]. The J Chem Phys 145(18):189901

Wooley HW (1980) Thermodynamic properties for H2O in the ideal gas state. In: Water and Steam. J Straub and K Scheffler (Eds.) Pergamon Press, Oxford, 166–175.

Wong DSH, Sandler SI (1992) A theoretically correct mixing rule for cubic equations of state. AIChE J 38 (5):671–680

Wormald CJ, Colling CN (1980) Excess enthalpies of some mixtures containing steam. In: Proceeding of the 9th international conference on water steam: their properties of current industrial application. Oxford, Pergamon, pp 655–663

Wormald CJ, Lancaster NM (1988) Excess enthalpies and cross-term second virial coefficients for mixtures containing water vapour. J Chem Soc, Faraday Trans 1: Phys Chem Condensed Phases 84(9):3141–3158

Wormald CJ, Lloyd MJ (1994) Excess molar enthalpies for ($0.5H_2O + 0.5CO_2$) at temperatures from 598.2 K to 698.2 K and pressures up to 20 MPa. The J Chem Thermodyn 26(1):101–110

Wormald CJ, Colling CN, Smith G (1983) Thermodynamics of supercritical steam+ carbon dioxide mixtures. Fluid Phase Equilib 10(2–3):223–231

Wormald CJ, Lancaster NM, Sellars AJ (1986) The excess molar enthalpies of {xH2O + (1–x)CO}(g) and {xH2O + (1–x)CO2}(g) at high temperatures and pressures. J Chem Thermodyn 18(2):135–147

Wormald C, Lloyd MJ, Fenghour A (1997) The excess enthalpy of (steam+ carbon dioxide) in the supercritical region up to T = 698.2 K and p = 25.5 MPa. The J Chem Thermodyn 29(11):1253–1260

Yamada T, Kawae N, Shigechi T, Kanemaru K (1991) On the second virial coefficient of light water. Nagasaki Univ Eng Faculty Res Report 21(37):101–110 (in Japanese)

Yuan Z, Cui M, Song R, Xie Y (2015) Evaluation of prediction models for the physical parameters in natural gas liquefaction processes. J Nat Gas Sci Eng 27:876–886

Zakirov IV (1984) The P-V-T relations in the H_2O-CO_2 system at 300 and 400 °C up to 1000 bar. Geochem Int 21(6):13–20

Zawisza A, Malesinska B (1981) Solubility of carbon dioxide in liquid water and of water in gaseous carbon dioxide in the range 0.2–5 MPa and at temperatures up to 473 K. J Chem Eng Data 26(4):388–391

Chemical and Isotopic Characteristics of the Solfatara-Pisciarelli Fumarolic Fluids

Abstract

The chemical characteristics of the Solfatara-Pisciarelli fumarolic fluids sampled and analyzed by Chiodini and coworkers were investigated by means of simple binary diagrams, chronograms, and Principal Component Analysis (PCA). Based on these different approaches, it turns out that CO_2, H_2, CO, and He (and to some extent H_2S, N_2 and Ar as well) are tracers of deep gases. The concentrations of these gas species and related PC's depend on the proportions of deep gases in the considered gas mixtures as well as on the temperatures and/or redox conditions in the gas production zone(s). Methane concentration and the PC mainly receiving its contribution are probably governed by CH_4 production somewhere within the Solfatara magmatic-hydrothermal system, owing to the nil to negligible methane concentrations in high-temperature volcanic-magmatic fluids. In the triangular diagram of N_2-Ar-He, all the Solfatara-Pisciarelli fumarolic fluids are positioned between the He-rich gases from the mantle and/or the crust and the N_2-rich compositions typical of most arc-type magmatic gases, apart from a few samples which are slightly enriched in Ar due to air contamination. Further indications are given by the isotopic values of these three gases. In detail: (i) Argon derives from mixing of non-atmospheric ^{40}Ar and atmospheric air, with nil to negligible proportions of air-saturated water (ASW), and this implies that the H_2-Ar geothermometer cannot be applied to the Solfatara-Pisciarelli fluids. (ii) Referring to the endmembers Caliro et al. (2014), helium originates from mixing of magmatic component M1, with $^3He/^4He$ isotope ratio of 3.53 R/R_A, magmatic component M2, with $^3He/^4He$ isotope ratio of 2.41 R/R_A, and the hydrothermal component, with $^3He/^4He$ isotope ratio of 1.79 R/R_A. (iii) Nitrogen comes mainly from breakdown of sediments, but this process might occur in different places and depths. The $\delta^{34}S$ value of H_2S was close to the mantle value in the eighties and resulted to be somewhat higher in 1998 - 2001, possibly due to either (i) occurrence of gradual degassing from a single magma batch, without the involvement of new magma or (ii) separation of isotopically light elemental sulfur upstream of the sampling point. The δ^{13} C value of CO_2 indicates that carbon dioxideis mainly produced through thermometamorphic decarbonation of marine carbonates and subordinately by magmatic degassing, with relative proportions close to 80 and 20%, respectively, assuming the equality of CO_2 concentrations in these two endmembers. Mantle contamination seems

Supplementary Information The online version contains supplementary material available at (https://doi.org/10.1007/978-3-030-98471-7_5).

L. Marini et al., *The Solfatara Magmatic-Hydrothermal System*, Advances in Volcanology,
https://doi.org/10.1007/978-3-030-98471-7_5

unlikely due to the large proportion of marine carbonate sediments involved in the mixing process. The $\delta^{13}C$ values of CO_2 and CH_4 suggest that these two carbon gases attain isotopic equilibrium at temperatures of 365 to 476°C, in satisfactory agreement with the temperatures of CO_2-CH_4 chemical equilibrium. The hydrogen isotopes of H_2O and the oxygen isotopes of H_2O and CO_2 were interpreted (i) considering the exchange of oxygen isotopes between H_2O vapor and gaseous CO_2, (ii) modeling the effects of magmatic degassing, and (iii) taking into account the possible entrainment of secondary steam into the uprising deep fluids. It turns out that the H_2O released by magmatic degassing prevails by far over the H_2O coming from other sources.■■■

The chemical dataset produced by Chiodini and coworkers (Chiodini et al. 2021 and references therein) comprises 742 chemical analyses, of which 375 are of Bocca Grande fluids, 300 are of Bocca Nuova fluids, and 67 are of Pisciarelli fluids. These samples were collected in different time intervals, namely: (i) June 1983 to October 2020, at Bocca Grande, (ii) March 1995 to October 2020, at Bocca Nuova, and (iii) March 1999 to September 2010, at Pisciarelli. The outlet temperature of the fumarolic vent was also measured at the time of collection. All these data are reported as electronic supplementary material (file Chapter_5&6.XLS). The main statistical parameters of the measured variables for each of these three fumaroles are listed in Table 1, showing that the major components of the Solfatara-Pisciarelli fumarolic fluids are by far H_2O and CO_2, accompanied by smaller quantities of H_2S, N_2, H_2, CH_4, He, CO, and Ar (according to the general decreasing order), as already mentioned in previous chapters.

Table 1 Main statistical parameters of the chemical variables (in μmol/mol) and temperature (in °C) measured by Chiodini and coworkers (Chiodini et al. 2021 and references therein) for the Bocca Grande, Bocca Nuova, and Pisciarelli fumarolic samples collected from June 1983 to October 2020

Fum	Stat. Par	T	H_2O	CO_2	H_2S	Ar	N_2	CH_4	H_2	He	CO
BG	N	330	375	375	375	262	364	365	365	311	330
BG	Mean	161.4	793,772	203,709	1520	0.644	526	18.3	446	1.96	0.95
BG	Median	162.0	804,600	193,000	1420	0.567	506	17.3	431	2.09	0.68
BG	Min	150.0	689,243	130,300	960	0.321	293	7.2	146	0.89	0.10
BG	Max	165.4	867,400	305,975	3988	4.425	1072	41.4	1230	2.93	3.53
BG	Std.dev	2.5	42,792	42,402	459	0.357	125	5.6	196	0.53	0.66
BN	N	287	300	300	300	293	299	299	299	299	294
BN	Mean	145.4	792,389	205,415	1150	0.644	495	14.7	501	1.95	1.04
BN	Median	145.3	785,500	212,500	1060	0.572	492	14.0	456	2.10	0.83
BN	Min	135.2	688,888	125,000	602	0.308	387	5.3	237	0.92	0.17
BN	Max	154.3	873,100	308,695	5014	3.751	742	31.4	1402	2.93	3.78
BN	Std.dev	2.3	45,799	45,472	401	0.282	48	4.8	193	0.51	0.68
Pi	N	64	67	67	67	67	67	67	67	67	67
Pi	Mean	97.1	835,955	162,806	599	0.629	430	11.5	244	1.63	0.10
Pi	Median	96.0	843,000	156,000	591	0.577	427	11.7	243	1.60	0.077
Pi	Min	90.5	767,000	116,000	467	0.323	262	7.6	137	0.73	0.023
Pi	Max	105.6	883,000	231,000	909	1.160	632	21.5	336	2.24	0.79
Pi	Std.dev	3.3	22,648	22,562	66	0.204	46	3.0	30	0.28	0.10

1 Chemistry of the Solfatara-Pisciarelli Fumarolic Fluids Based on Binary Diagrams

A simple but effective way to inspect the chemical characteristics of the Solfatara-Pisciarelli fumarolic fluids is by means of binary diagrams in which CO_2 is considered as reference variable. Instead of the measured analytical data, the corresponding standardized values are shown in these binary diagrams. The standardized values, Z_i, are computed subtracting the mean, μ_i, from each concentration value, X_i, and dividing the obtained difference by the standard deviation, σ_i, as expressed by the following well-known relation:

$$Z_i = \frac{X_i - \mu_i}{\sigma_i}. \quad (1)$$

Thus, standardized variables have mean equal to zero and standard deviation equal to one and, therefore, the comparison of different standardized variables is easier than the comparison of the corresponding original variables. Furthermore, the distribution of the sample points in the binary diagrams and the shape of the chronograms are identical whether using the analytical data or the standardized values. Equation (1) was used considering the mean and standard deviation of each variable for the overall dataset of Chiodini and coworkers.

The diagram of H_2O versus CO_2 highlights a perfect inverse correlation ($R = -0.99988$) between these two variables (Fig. 1a) which is not surprising because H_2O and CO_2 are by far the major constituents of the considered gas mixtures and the sum of all mole fractions is equal to unity. Therefore, it follows that $X_{H_2O} \cong 1 - X_{CO_2}$ and $Z_{H_2O} \cong -Z_{CO_2}$. As further consequence, the correlation diagram of any considered variable versus H_2O is the mirror image of the correlation diagram of the same variable versus CO_2. To be noted also that H_2O and CO_2 of Bocca Grande and Bocca Nuova fluids span ranges slightly higher than 4 standardized units whereas H_2O and CO_2 of Pisciarelli fluids distribute over a smaller interval of ca. 2.5 standardized units, corresponding to the upper range of H_2O and the lower range of CO_2 of Bocca Grande and Bocca Nuova fluids.

As shown by the binary diagram of Fig. 1b, He is relatively well correlated with CO_2, with $R = +0.898$ for Bocca Grande, $R = +0.934$

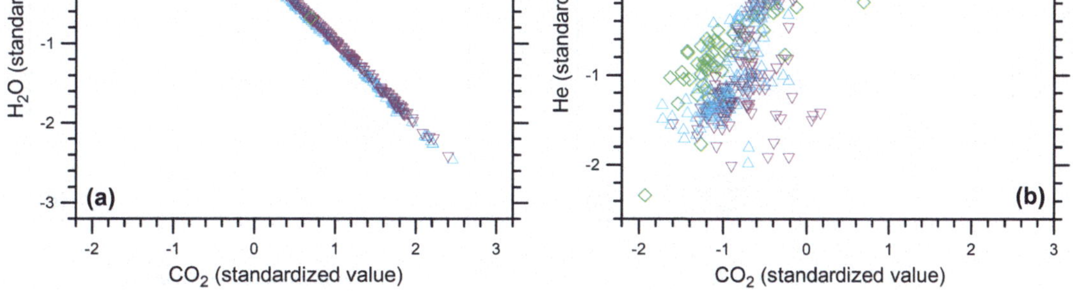

Fig. 1 Correlation diagrams of CO_2 versus **a** H_2O and **b** He for all the Solfatara-Pisciarelli fumarolic fluids analyzed by Chiodini and coworkers (Chiodini et al. 2021 and references therein). Standardized values are considered in both diagrams

for Bocca Nuova, and R = + 0.810 for Pisciarelli. According to Caliro et al. (2014), He and CO_2 are both magmatic tracers and changes in the He/CO_2 ratio with time are chiefly controlled by variations in the deep magmatic source (e.g., magma degassing), although Caliro et al. (2014) recognized also the influence of the mixing process between magmatic and meteoric-hydrothermal fluids. Since CO_2 is also produced by decarbonation reactions occurring in the thermometamorphic layer and ^{4}He is generated by radioactive decay of U and Th occurring in crustal rocks, it is reasonable to assume that CO_2 and He are tracers of "deep" gases, using this more general term instead of the more specific term "magmatic" gases. Thus, high CO_2 and He concentrations indicate the presence of higher proportions of deep gases in fumarolic fluids or the greater extent of the processes producing CO_2 and He, whereas low CO_2 and He concentrations suggest the occurrence of lower fractions of deep gases in fumarolic fluids or the lower extent of the processes generating CO_2 and He.

As shown by the two diagrams of Fig. 2, the two reactive gases hydrogen and carbon monoxide are also fairly well correlated with CO_2, in general. The H_2-CO_2 correlation coefficient is +0.795 for Bocca Grande, +0.798 for Bocca Nuova, and +0.777 for Pisciarelli, whereas the CO-CO_2 correlation coefficient is + 0.794 and +0.858 for Bocca Grande and Bocca Nuova, respectively, but is much lower for Pisciarelli, + 0.346, because of the nearly horizontal spread of the samples from this site. Disregarding Pisciarelli and allowing for some dispersion of the Bocca Grande and Bocca Nuova samples, these fumarolic gas samples appear to fit a parabolic trend in the correlation diagrams of CO_2 versus H_2 and CO_2 versus CO, rather than a linear trend as in the correlation plot of CO_2 versus He (Fig. 1b). In fact, the correlation coefficient of the second-degree polynomial fit of H_2-CO_2 data is +0.849 for Bocca Grande and +0.861 for Bocca Nuova, whereas that of the CO-CO_2 data is +0.894 for Bocca Grande and + 0.908 for Bocca Nuova. These four R values of the parabolic fit are significantly higher than the corresponding R values of the linear fit (see above) confirming the evidence given by the eyeball inspection of the two diagrams of Fig. 2.

In turn, the parabolic fit in the two graphs of Fig. 2 can be considered as the envelope of different mixing lines, between an endmember poor

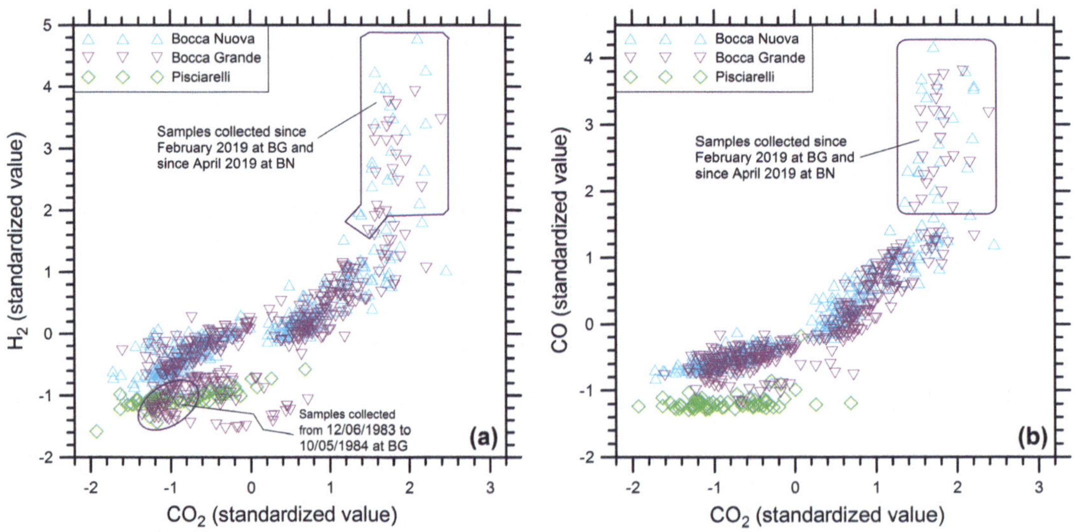

Fig. 2 Correlation diagrams of CO_2 versus **a** H_2 and **b** CO for all the Solfatara-Pisciarelli fumarolic fluids analyzed by Chiodini and coworkers (Chiodini et al. 2021 and references therein). Standardized values are considered in both diagrams

in CO_2, H_2 and CO, characterized by constant concentrations of these three gaseous species, and an endmember rich in CO_2, H_2 and CO, characterized by constant concentration of CO_2 but variable concentration of both H_2 and CO. Higher H_2 and CO concentrations indicate higher temperatures and/or more reducing conditions in the zone where the CO_2-rich endmember originates, whereas lower H_2 and CO concentrations suggest lower temperatures and/or more oxidizing conditions in the zone of production of the CO_2-rich endmember, in spite of possible re-equilibration upon cooling especially for CO.

Alternatively, building further on what was noted above for CO_2 and He, the different concentrations of CO_2, H_2 and CO could be controlled by the different extent of the processes producing CO_2 and He as well as by the temperature and redox conditions present in the gas production zone, although re-equilibration upon cooling appears to be very likely for CO. We also anticipate that the highest CO_2, He, H_2, and CO concentrations were detected in 2019–2020, as highlighted by the chronograms of these three gas species (see below).

In the correlation diagram of H_2S versus CO_2 (Fig. 3a), the samples collected at Bocca Grande from 12/06/1983 to 10/05/1984 have H_2S concentration somewhat higher than the samples collected during the following years, at least before February-April 2019. In fact, the samples collected from July 1984 to January 2019 at Bocca Grande and from March 1995 (beginning of sampling) to March 2019 at Bocca Nuova exhibit a limited increase in H_2S concentration, whereas the samples obtained since February 2019 at Bocca Grande (apart from that of 08/04/2019) and since April 2019 at Bocca Nuova have significantly higher H_2S concentration. In other terms, the diagram suggests that: (i) a major change in the H_2S concentration of the fluids discharged from Bocca Grande (Bocca Nuova was not sampled at that time) and probably in the underlying plumbing system occurred between May and July 1984, when the H_2S concentration experienced a significant decrease and (ii) another important variation was recorded by both Bocca Grande and Bocca Nuova fumaroles on February 2019 and April 2019, respectively, with very high H_2S concentrations in the following period, at least until October 2020. Since the high H_2S concentrations recorded in 2019–2020 are accompanied by high concentrations of CO_2, He, H_2, and CO (see

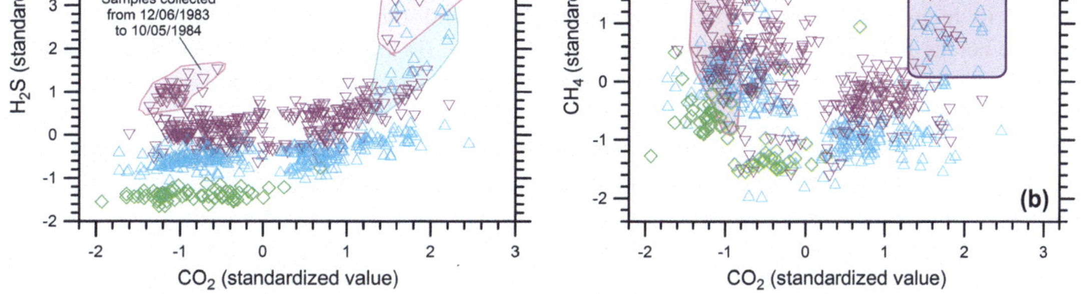

Fig. 3 Correlation diagrams of CO_2 versus **a** H_2S and **b** CH_4 for all the Solfatara-Pisciarelli fumarolic fluids analyzed by Chiodini and coworkers (Chiodini et al. 2021 and references therein). Standardized values are considered in both diagrams

above), it is reasonable to deduce that H_2S, which is stable under reducing conditions and has high reactivity, confirms the presence of higher proportions of deep gases in fumarolic fluids or higher temperatures and/or more reducing conditions in the gas production zone.

The different concentrations of H_2S could also be controlled by the different extent of the reactions occurring in the gas production zone, involving one or more H_2S partners, either gaseous, such as SO_2 and S_2, or solid, such as pyrite [FeS_2], pyrrhotite [FeS], and anhydrite [$CaSO_4$]. These reactions are examined in chapter "The Redox Potential and Sulfur Gas Species". To be noted also that, at any given time, H_2S concentration is highest at Bocca Grande, intermediate at Bocca Nuova, and lowest at Pisciarelli (Fig. 3a).

Methane is poorly correlated with CO_2, as shown by the diagram of Fig. 3b. Similar to what was observed for H_2S, at any given time, CH_4 concentration is usually greatest at Bocca Grande, intermediate at Bocca Nuova, and smallest at Pisciarelli, although the three dataset are not clearly separated as for H_2S. Furthermore, a considerable decrease in CH_4 concentration occurred at Bocca Grande from 12/06/1983 to 10/05/1984, whereas a substantial increase in CH_4 concentration took place since February 2019 at Bocca Grande and since April 2019 at Bocca Nuova. It is probable that these considerable variations experienced by methane concentration are controlled by its production somewhere within the magmatic-hydrothermal system of the Solfatara. In fact, methane is normally absent or is present in very low concentrations in high-temperature volcanic-magmatic fluids, whereas it occurs in relatively high concentrations in hydrothermal gases, as underscored by Chiodini (2009). Accordingly, Caliro et al. (2007) proposed that CH_4 is generated in the deeper and hottest zones of the Solfatara hydrothermal system by CO_2 reduction, as suggested by the agreement between the temperatures calculated by means of the CO_2-CH_4 isotopic geothermometer, 360 to 436 °C, and those computed by means of the CO_2-CH_4 chemical geothermometer (>360 °C), assuming that H_2O fugacity is fixed by equilibrium coexistence of liquid water and steam, that CO_2 fugacity is controlled by equilibrium coexistence of calcite and Ca-Al-silicates, that O_2 fugacity is fixed by the redox buffer specific for the Campanian volcanoes (Chiodini and Marini 1998; Caliro et al. 2007), and gas equilibration is achieved in a single saturated liquid phase. Although these considerations are reasonable, two issues concerning methane are worth further discussion, namely, CH_4 geothermometers and the kinetics of the reactions involving methane.

For what concerns the first issue, it is important to reconsider the chemical geothermometers involving CH_4, without placing constraints on CO_2 and O_2 fugacities, because of possible uncertainties on such constraints. This topic will be treated in chapters "The Geoindicators involving H_2O, CO_2, CO, CH_4 and H_2" and "Calibration and Application of the RWG and SS4 Indicators for the Possible Expansion Paths of Solfatara Fluids". Here, we anticipate that we agree on the virtual absence of methane in magmatic gases, as suggested by Chiodini and coworkers, and we underscore that a good place for methane generation might be the thermometamorphic layer. In fact, since the original lithotypes of this layer are tuffites, marls, and silitites with a carbonate matrix (Chelini and Sbrana 1987), the thermometamorphic reactions occurring in this rock layer are expected to produce not only CO_2, through decarbonation reactions (see Marini 2007), but also CH_4, H_2S, N_2, NH_3, and other compounds of C, S, and N, through (i) breakdown of the organic matter contained in these marine sedimentary and volcano-sedimentary lithotypes and (ii) decomposition of S-bearing minerals (see chapter "The Redox Potential and Sulfur Gas Species"). Among the CH_4-generating reactions involving the organic matter (represented by the simplified formula CH_2O), we recall its hydrogenation:

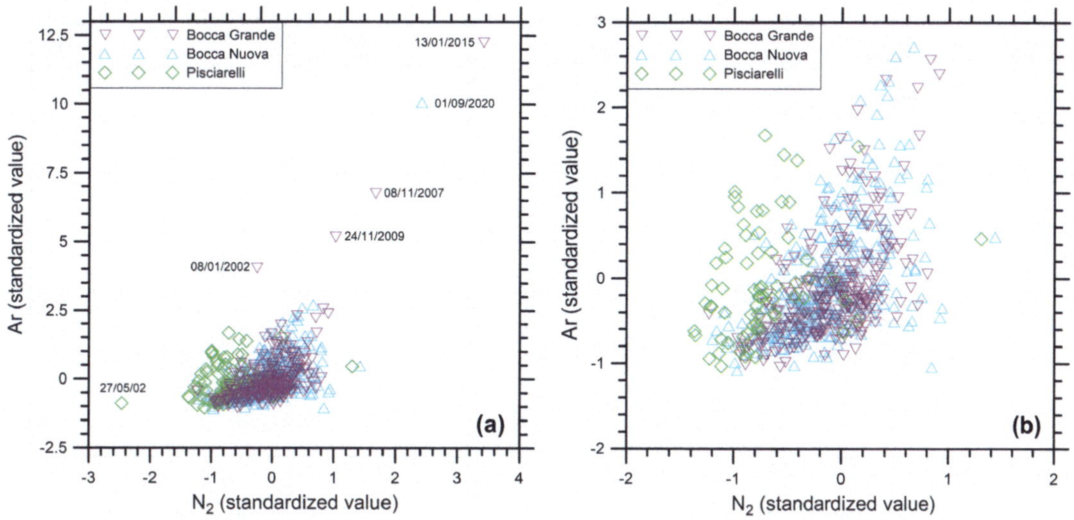

Fig. 4 Correlation diagrams of N_2 versus Ar **a** for all the Solfatara-Pisciarelli fumarolic fluids analyzed by Chiodini and coworkers (Chiodini et al. 2021 and references therein) and **b** upon correction of the six outliers. Standardized values are considered in both diagrams

$$CH_2O + 2H_2 = CH_4 + H_2O \quad (2)$$

and its disproportionation:

$$2CH_2O = CH_4 + CO_2. \quad (3)$$

Both reactions (2) and (3) occur under very reducing conditions. The second issue represents the subject of chapter "The Chemical Kinetics of the Reactions Involving H_2O, CO_2, CO, CH_4 and H_2". Here, we anticipate that the considerations of Giggenbach (1997a) on the kinetics of CO_2 methanation, which were adopted by Chiodini (2009), are based on the experimental results of Berndt et al. (1996) that were later proven to be wrong by McCollom and Seewald (2001), as discussed in Sect. 2 in chapter "The Chemical Kinetics of the Reactions Involving H_2O, CO_2, CO, CH_4 and H_2". After clarifying the thermodynamics and kinetics of the reactions involving methane, it will be possible to use the CH_4 geothermometers to estimate the pressure and temperature conditions where it originates, adopting different hypotheses and taking into account the deviations from the ideal gas behavior.

Before inspecting the correlation diagrams of CO_2 versus N_2 and CO_2 versus Ar, it is advisable to examine the Ar versus N_2 plot (Fig. 4a) looking for possible outliers, that is, samples with anomalous concentrations of these two atmospheric gases. Six outliers can be recognized in this graph, namely, the samples collected at Bocca Grande on 08/01/2002, 08/11/2007, 24/11/2009, and 13/01/2015, the sample obtained at Bocca Nuova on 01/09/2020, and the sample taken at Pisciarelli on 27/05/2002. Since the five samples of Bocca Grande and Bocca Nuova are probably affected by addition of atmospheric air, their N_2 and Ar concentrations were substituted with the values obtained by averaging the N_2 and Ar concentrations of the samples collected on the immediately preceding and following sampling dates. The same was done for the sample taken at Pisciarelli on 27/05/2002, but only its N_2 concentration was corrected, being anomalously low.

The Ar versus N_2 plot redrawn upon correction of these six outliers is shown in Fig. 4b. The spread of samples in this plot and the moderate correlation between the two variables, at least for Bocca Grande, with R = + 0.560, and Bocca Nuova, with R = + 0.537, (R = + 0.279 for

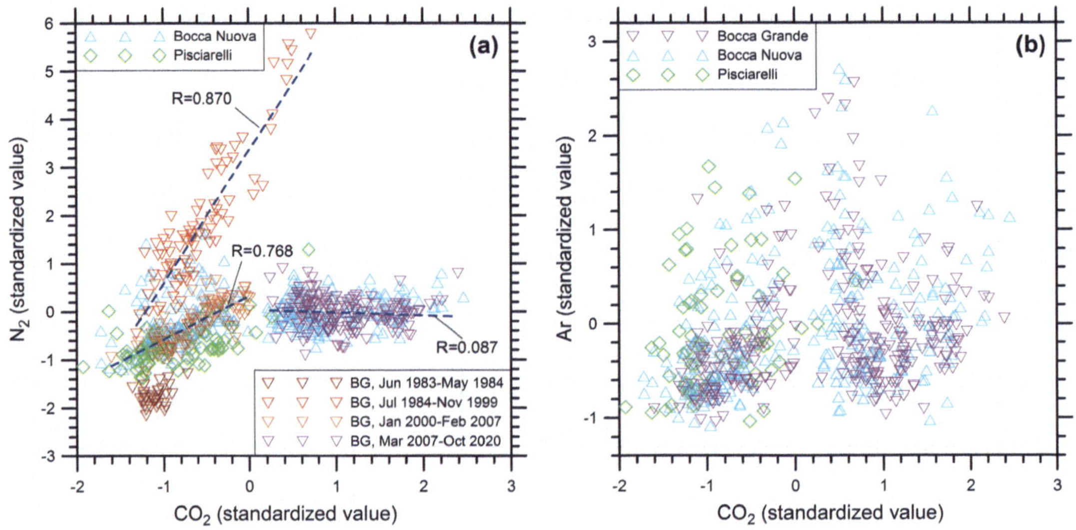

Fig. 5 Correlation diagrams of CO_2 versus **a** N_2 and **b** Ar for all the Solfatara-Pisciarelli fumarolic fluids analyzed by Chiodini and coworkers (Chiodini et al. 2021 and references therein). Standardized values are considered in both diagrams

Pisciarelli) suggest that occurrence of air contamination cannot be excluded, especially for the samples with Ar higher than ca. 0–0.5 standardized units. Nevertheless, the amounts of air possibly added to these samples are much smaller than those of the five outliers of Bocca Grande and Bocca Nuova. Furthermore, it is possible that air addition takes place in the natural system, that is, in the fumarolic conduits upstream of the sampling point.

As shown by the correlation diagram of CO_2 versus N_2 (Fig. 5a), the fumarolic fluids discharged from Bocca Grande in June 1983–May 1984 had low concentrations of both CO_2 and N_2. Then, a sudden increase in both parameters was recorded in July 1984. From July 1984 to November 1999, the concentrations of CO_2 and N_2 exhibited a relatively good correlation (R = + 0.870), controlled by the gradual decrease of both parameters. A relatively good correlation between CO_2 and N_2 (R = + 0.768) was also observed in the following years, from January 2000 to February 2007, but the trend was completely different from the previous one and both parameters increased with time. The samples collected afterwards, from March 2007 to October 2020 are characterized by increasing CO_2 at nearly constant N_2 and, due to this flat trend, the correlation between the two variables is poor (R = + 0.087). Accepting that high CO_2 concentrations indicate high fractions of deep gases in fumarolic fluids and vice versa, it is possible to recognize the involvement of three distinct deep endmembers, characterized by increasingly higher N_2 concentrations in the order: (March 2007 to October 2020) < (January 2000 to February 2007) < (July 1984 to November 1999). In contrast, the fumarolic fluids of June 1983–May 1984 come from the shallow hydrothermal system without the involvement of deep gases (Cioni et al. 1984). To be noted that the Bocca Nuova fluids, and to some extent those of Pisciarelli, have N_2 and CO_2 concentrations similar to those of the Bocca Grande fluids sampled at the same time, confirming previous considerations and adding little to them.

Caliro et al. (2014) proposed that CO_2-N_2 relations of Solfatara fumaroles are controlled by mixing between three endmembers (see Fig. 9b of Caliro et al. 2014), namely:

(i) An undegassed magmatic endmember M1, rich in both N_2 and CO_2, practically coinciding with the deep endmember of July 1984 to November 1999.

(ii) A degassed magmatic endmember, M2, somewhat depleted in CO_2 and strongly depleted in N_2 compared to M1, virtually corresponding to the deep endmember of March 2007 to October 2020.
(iii) A hydrothermal component with low concentrations of both N_2 and CO_2.

Adopting the approach of Caliro et al. (2014), the deep endmember of January 2000 to February 2007 could be a partially degassed magmatic endmember, with chemical characteristics intermediate between those of M1 and M2.

However, in addition to magma degassing, N_2 as well as NH_3 and other N-bearing gaseous species [e.g., mono-methylamine, CH_3NH_2; dimethylamine, $(CH_3)_2NH$; tri-methylamine, $(CH_3)_{3\ N})$] could also be produced by decomposition of the organic matter present in sedimentary lithotypes, such as those of the thermometamorphic layer. Unfortunately, ammonia has seldom been analyzed in the Solfatara fumarolic fluids, whereas other N-bearing gaseous species were never analyzed, to the best of our knowledge. Therefore, it is not possible to use N_2 and a suitable N-bearing reaction partner for geothermometric-geobarometric purposes.

The correlation diagram of CO_2 versus Ar (Fig. 5b) shows that Ar is poorly correlated with CO_2, with R values of +0.221 for Bocca Grande, +0.321 for Bocca Nuova, and +0.236 for Pisciarelli. In the samples with Ar higher than ca. 0—0.5 standardized units, argon is probably of atmospheric origin, at least for the most, as suggested by its relation with nitrogen (see above). Other possible sources of argon, nitrogen, and helium are indicated, to some extent, by the triangular diagram involving these three gas species (see Sect. 4) and, more reliably, by their isotopic characteristics (see Sects. 5, 6, and 7). Before considering these topics, it is worth to take into account the chronograms of analyzed gas species in Bocca Grande and Bocca Nuova fluids (representing the subject of the next section) to complement the information obtained by means of the binary diagrams discussed so far.

2 Chronograms of Gas Concentrations in Bocca Grande and Bocca Nuova Fumarolic Fluids

To facilitate the comparison between different gas species, the standardized concentrations are plotted in the chronograms of Figs. 6, 7, 8 and 9 instead of the analytical values. For the same purpose, the gases well correlated with CO_2, that is, H_2S, He, H_2, and CO (see Sect. 1), as well as H_2O, are taken into account together in the chronograms of Figs. 6 and 7, whereas the gases poorly correlated with CO_2, that is, CH_4, N_2, and Ar (see Sect. 1), are considered separately in the chronograms of Figs. 8 and 9. Nevertheless, CO_2 is also reported in these two figures for reference. All the chronograms were drawn for running averages with window width of 5 in order to smooth the effects of anomalous values, if any. Inspection of Figs. 6 and 7 show that:

(i) The chronogram of H_2O is the mirror image of that of CO_2, as expected based on the binary diagram of Fig. 1a.
(ii) The chronograms of He, H_2, and CO are, in general, similar to that of CO_2, as also expected from the correlation plots of Fig. 1b, 2.
(iii) Rather surprisingly, the chronogram of H_2S is similar to that of H_2O until 1999–2001, presents contrasting aspects in 2002–2012, and compares with that of CO_2 and related gases from 2012 onwards.

Based on Figs. 8 and 9:

(i) The chronogram of N_2 compares with that of CO_2 until 1999 and possibly until 2007, when the divergence between the two chronograms becomes clear, due to the continuous increase in CO_2 and the generally flat trend of N_2.
(ii) There is a similarity between the chronograms of N_2 and Ar, possibly due addition of atmospheric air to the fumarolic fluids, but the second one presents short-term

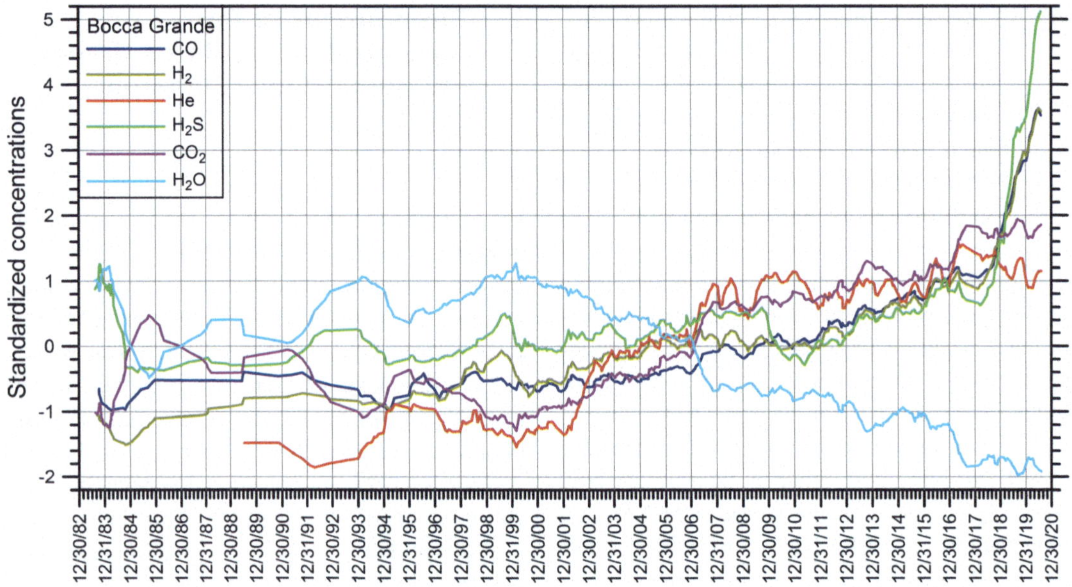

Fig. 6 Chronogram of the standardized concentrations of H_2O, CO_2, H_2S, He, H_2, and CO in the fumarolic fluids of Bocca Grande analyzed by Chiodini and coworkers (Chiodini et al. 2021 and references therein)

Fig. 7 Chronogram of the standardized concentrations of H_2O, CO_2, H_2S, He, H_2, and CO in the fumarolic fluids of Bocca Nuova analyzed by Chiodini and coworkers (Chiodini et al. 2021 and references therein)

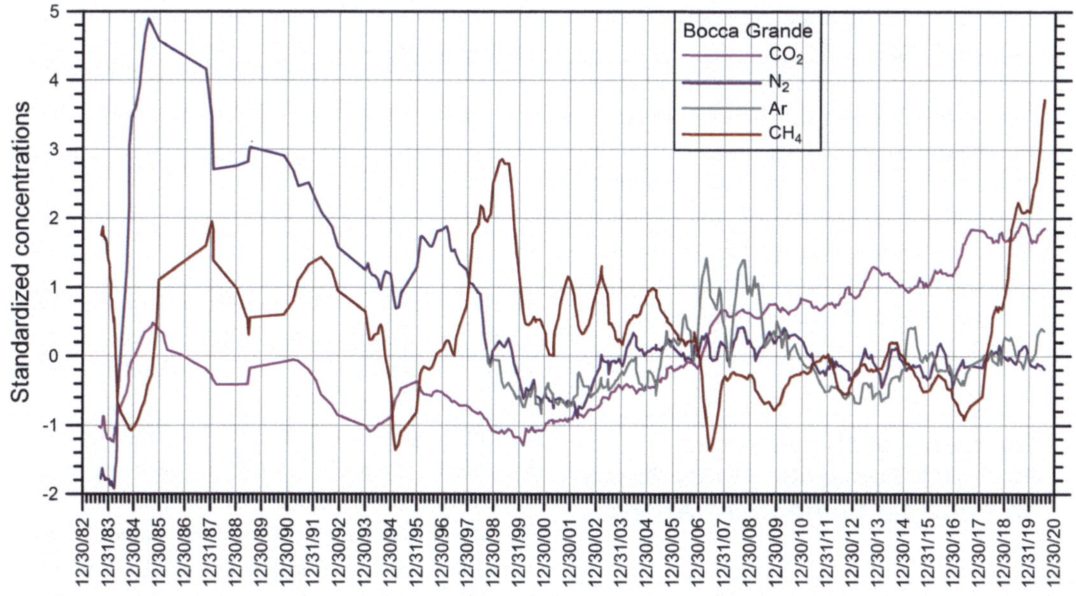

Fig. 8 Chronogram of the standardized concentrations of N_2, Ar, CH_4, and CO_2, in the fumarolic fluids of Bocca Grande analyzed by Chiodini and coworkers (Chiodini et al. 2021 and references therein)

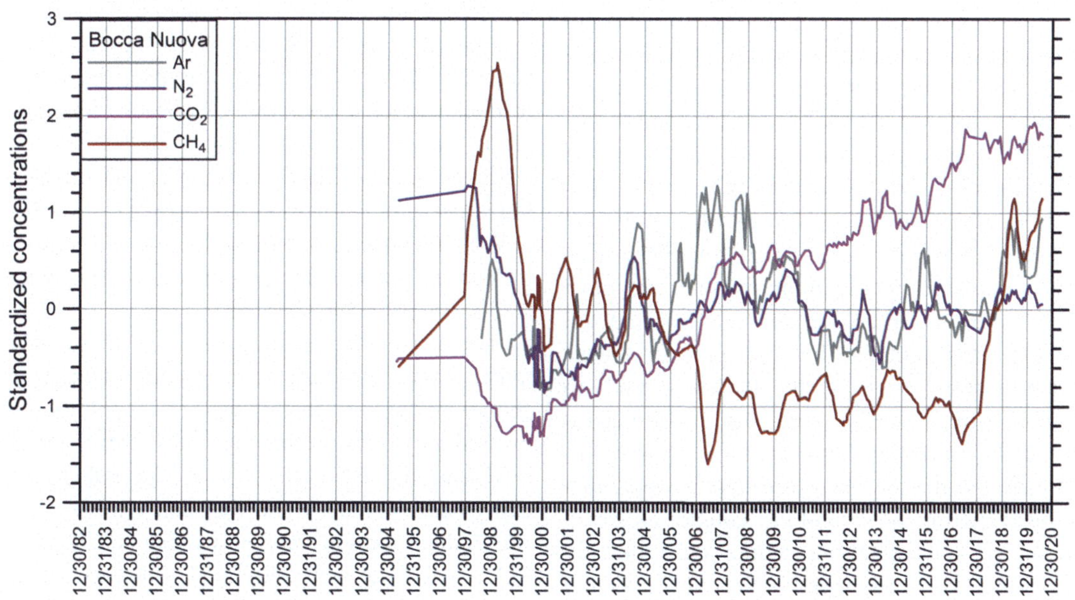

Fig. 9 Chronogram of the standardized concentrations of N_2, Ar, CH_4, and CO_2, in the fumarolic fluids of Bocca Nuova analyzed by Chiodini and coworkers (Chiodini et al. 2021 and references therein)

fluctuations which are generally absent in the first one.

(iii) The chronogram of CH_4 is unique for the marked short-term fluctuations, which are probably controlled by methane production somewhere within the magmatic-hydrothermal system of the Solfatara, due to the nil to negligible concentrations in high-temperature volcanic-magmatic fluids, as already recalled in Sect. 1.

3 Chemistry of the Solfatara-Pisciarelli Fumarolic Fluids Based on Principal Component Analysis

Principal Component Analysis (PCA) is a widely used multivariate statistical technique which transforms the original variables into a smaller number of new variables, called Principal Components (PC). The PCs have two important characteristics, being uncorrelated to each other and being linear combinations of the original variables. The first PC, or PC1, accounts for the largest proportion of the total variance, the second PC, or PC2, explains a proportion of the total variance lower than PC1 but higher than PC3, and so on.

It is advisable to involve in PCA only the samples for which the complete chemical analysis is available. In case of missing data, the most used approach is to use the mean value computed for the other samples, but this practice leaves much to desire as it bias the PCA and leads to unreliable results. Therefore, for the Solfatara-Pisciarelli fumarolic fluids, we have preferred to estimate the missing data in different ways, that is, by averaging the data of the samples collected during the immediately preceding and following samplings (as already noted in Sect. 1 for N_2 and Ar) or by taking into account the equilibrium values of CO (see Sect. 5 in chapter "The geoindicators involving H_2O, CO_2, CO, CH_4 and H_2"), for some samples collected in 1983–1989, or by computing the missing Ar and He data, for all the samples collected in 1983–1998 and in 1983–1989, respectively, by means of the following relations:

$$X_{Ar} = 1.220 - 8.639E - 7 \cdot X_{H_2O} \quad (4)$$

$$X_{He} = -2.951 + 3.609E - 5 \cdot X_{CO_2} - 5.811E - 11 \cdot X^2_{CO_2} \quad (5)$$

where X_i indicates the mol fraction in μmol/mol. These computed concentrations were used in PCA, which was performed considering a total of 742 active cases.

Before performing the PCA of the Solfatara-Pisciarelli fumarolic fluids, it must be underscored that the chemical data of interest are so-called "closed data", because the values of each variable for each sample is a part of a total sum. Since the data of this type are affected by induced correlations, it is necessary to transform them in a suitable way, for instance, first computing the ratios with respect to a reference variable and second calculating the logarithms of these ratios (Signorelli et al. 1998 and references therein). Choosing H_2O as reference variable, the transformed variables are: log (CO_2/H_2O), log (H_2/H_2O), log (CO/H_2O), log (He/H_2O), log (H_2S/H_2O), log (CH_4/H_2O), log (N_2/H_2O), and log (Ar/H_2O). However, the original variables will be mentioned in the following discussion, instead of these corresponding log-ratios with respect to H_2O, for simplicity.

The correlation matrix between these log-ratios, representing the starting point of PCA, is shown in Table 2. In this table, to highlight the most significant correlations, the $R > 0.7$ are in bold and the $0.5 < R < 0.7$ are in italics. As expected based on what was noted in Sects. 1 and 2, the tracers of deep gases, that is, CO_2, H_2, CO, and He are well correlated with each other, in spite of the somewhat lower R values of the couple CO-He. Furthermore:

(i) H_2S is well correlated with CO and to a lower extent with H_2, CO_2, and CH_4. In other words, H_2S is related not only to the tracers of deep gases but also to CH_4.

Table 2 Correlation matrix between the log-ratios of the chemical variables with respect to H_2O in Solfatara-Pisciarelli fumarolic fluids

	CO_2	He	H_2	CO	H_2S	N_2	Ar	CH_4
CO_2	1.000	**0.929**	**0.833**	**0.789**	*0.597*	*0.520*	*0.506*	−0.020
He		1.000	**0.783**	*0.676*	0.438	0.468	*0.554*	−0.127
H_2			1.000	**0.863**	*0.660*	0.348	0.465	0.220
CO				1.000	**0.754**	0.431	0.345	0.274
H_2S					1.000	0.277	0.249	*0.555*
N_2						1.000	0.333	0.124
Ar							1.000	−0.075
CH_4								1.000

(ii) For what concerns the so-called atmospheric gases, the correlation between N_2 and Ar is poor, but Ar has a relatively good correlation with CO_2 and He, while N_2 has a relatively good correlation with CO_2 only, suggesting that also N_2 and Ar come from deep sources, at least in part.

Before performing the PCA, the log-ratios are transformed in the corresponding standardized variables (which have a mean of zero and a standard deviation of one, as already mentioned in Sect. 1). Then the PCA begins with the manipulation of the correlation matrix, from which the eigenvalues and eigenvectors are extracted (e.g., Davis 2002).

Since, in the case of interest, there are eight original variables, each one of variance equal to one, the total variance of the system is eight and, in fact, the sum of the eigenvalues is eight (Table 3). The total number of PCs is also eight, being equal to the number of the considered variables, but the significant PCs are those that explain a significant part of the total variance of the system, that is mainly PC1 and PC2 and to some extent PC3 and PC4. The first four PCs altogether account for over 92% of the total variance.

Table 4 shows the correlation coefficients between the four significant PCs and the original variables in which, again, the $R > 0.7$ are in bold and the $0.5 < R < 0.7$ are in italics, in order to highlight the most significant correlations. PC1 appears to be strongly correlated with the tracers of deep gases, not only CO_2, H_2, CO, and He, but also H_2S, and to a lower extent with N_2 and Ar as well, whereas CH_4 is poorly related to PC1. Furthermore, since all correlation coefficients are positive:

(a) CO_2, H_2, CO, He, H_2S, N_2 and Ar contribute in a synergistic way to PC1 and

Table 3 Eigenvalues extracted from the correlation matrix of Table 2 and % of the total variance explained by the principal components. The cumulated values are also given

Principal components	Eigenvalues	% total variance	Cumulated eigenvalues	Cumulated % of tot. variance
PC1	4.54	56.77	4.54	56.77
PC2	1.48	18.51	6.02	75.28
PC3	0.77	9.59	6.79	84.86
PC4	0.63	7.86	7.42	92.73
PC5	0.24	3.06	7.66	95.79
PC6	0.21	2.64	7.87	98.43
PC7	0.08	1.02	7.96	99.44
PC8	0.04	0.56	8.00	100.00

Table 4 Correlation coefficients between the four significant PCs and the original variables

	PC1	PC2	PC3	PC4
CO_2	**0.935**	−0.213	0.081	0.148
He	**0.870**	−0.369	0.098	0.092
H_2	**0.916**	0.055	0.231	0.004
CO	**0.899**	0.193	0.134	0.146
H_2S	**0.746**	*0.537*	0.110	−0.057
N_2	*0.572*	−0.113	**−0.774**	0.234
Ar	*0.579*	−0.398	−0.149	*−0.692*
CH_4	0.215	**0.894**	−0.214	−0.202

(b) high PC1 values (or scores) indicate high proportions of deep gases in the considered gas mixtures or higher temperatures and/or more reducing conditions in the gas production zone whereas low scores suggest low proportions of deep gases or lower temperatures and/or more oxidizing conditions in the gas production zone.

The three variables which are not related or scarcely related to PC1 contribute to PC2, PC3, and PC4. In detail:

(i) PC2 is well related with CH_4 and to a lower degree with H_2S. The R values PC2-CH_4 and PC2-H_2S are both positive and, therefore, CH_4 and H_2S contribute in a synergistic way to PC2. Thus, high PC2 scores correspond to high concentrations of CH_4 and H_2S.
(ii) PC3 is significantly related with N_2 only, and PC4 is meaningfully related with Ar only. Low PC3 and PC4 scores correspond to high concentrations of N_2 and Ar, respectively, and vice versa, as indicated by the negative values of the pertinent correlation coefficients. These results are not surprising based on the characteristics of the correlation matrix of Table 2.

The correlation diagrams of PC1 versus log (CO_2/H_2O) and PC1 versus log (H_2S/H_2O) are shown in Fig. 10, to appreciate the correlation between PC1 and the two original variables with the highest and the lowest R values. Significant deviations from the main groups of Bocca Grande and Bocca Nuova samples are observed in Fig. 10b, mainly due to the Bocca Grande samples of June 1983-May 1984 and the samples collected in 2019–2020 from both fumaroles. To be noted also that the spread of sample points in the diagram of Fig. 10b is very similar to that in the diagram of CO_2 versus H_2S of Fig. 3a.

The good correlation between PC2 and log (CH_4/H_2O) is evident in the correlation diagram of Fig. 11a, whereas the correlation between PC2 and log (H_2S/H_2O) is largely due to the Bocca Grande and Bocca Nuova samples of 2019–2020, as shown by the correlation diagram of Fig. 11b. Without these data, the R values would decrease from 0.537 to 0.438.

In the two correlation diagrams of Fig. 12, PC3 and PC4 are contrasted with the variables that are most correlated with them, that is, log (N_2/H_2O) and log (Ar/H_2O), respectively. In spite of the good correlation between PC3 and log (N_2/H_2O), the samples collected from Bocca Grande at different times are clearly separated from each other also in this graph, as already observed in the correlation diagram of CO_2 versus N_2 (Fig. 5a).

As already mentioned above, the analytical data of Ar are available only for the samples collected since July 1998, whereas Ar concentrations were calculated using Eq. (4) for the samples collected before this date. Nevertheless, these early samples, mostly from Bocca Grande, have PC4 values generally within the range of the subsequent samples, as shown by the correlation diagram of PC4 versus log (Ar/H_2O) of

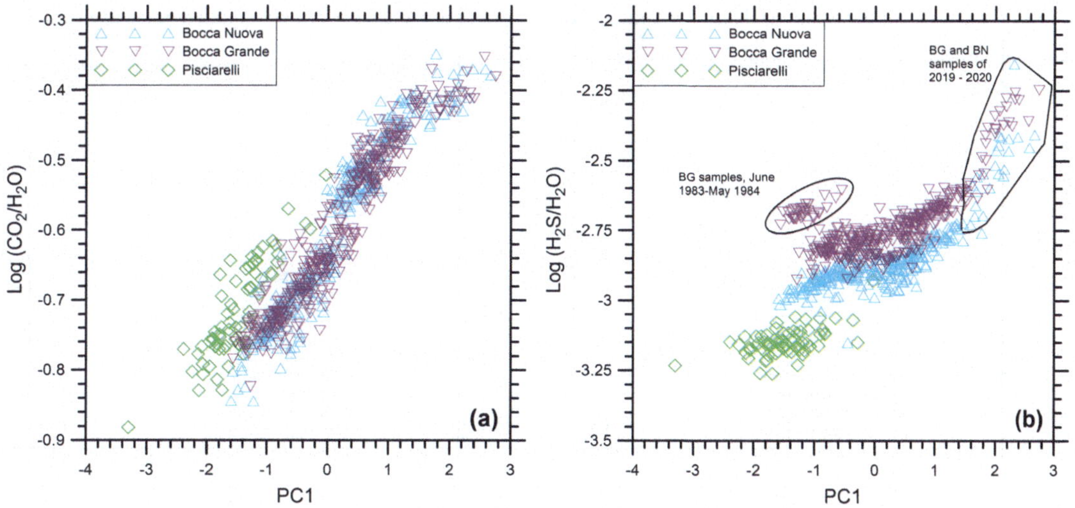

Fig. 10 Correlation diagrams of PC1 versus **a** log (CO_2/H_2O) and **b** log (H_2S/H_2O) for all the Solfatara-Pisciarelli fumarolic fluids analyzed by Chiodini and coworkers (Chiodini et al. 2021 and references therein)

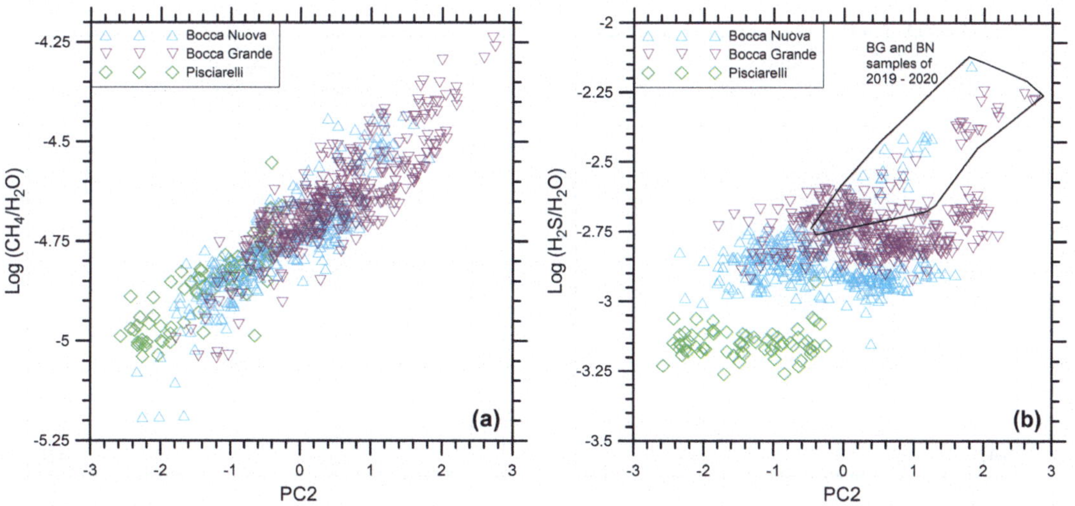

Fig. 11 Correlation diagrams of PC2 versus **a** log (CH_4/H_2O) and **b** log (H_2S/H_2O) for all the Solfatara-Pisciarelli fumarolic fluids analyzed by Chiodini and coworkers (Chiodini et al. 2021 and references therein)

Fig. 5a. Therefore, the spread of sample points in this plot does not seem to be significantly distorted by these computed Ar data. Besides, it must be noted that the samples obtained from Bocca Grande at distinct times are clearly separated from each other also in this plot.

Having ascertained the meaning of the four significant PCs, it is useful to investigate their relations, in order to highlight the most significant compositional variations experienced by the Solfatara - Pisciarelli fumarolic fluids from 1983 to 2020. To this purpose, PC1 is contrasted with PC2 in the plot of Fig. 13a and PC3 is contrasted with PC4 in the diagram of Fig. 13b.

Leaving aside the Bocca Grande and Bocca Nuova samples of 2019–2020, all other samples exhibit a good inverse correlation between PC1

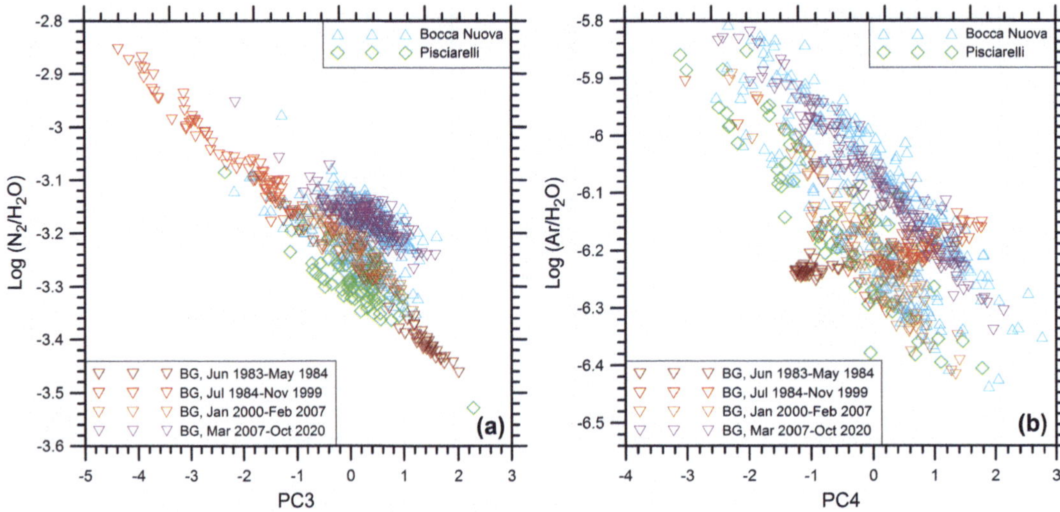

Fig. 12 Correlation diagrams of **a** PC3 versus log (N_2/H_2O) and **b** PC4 versus log (Ar/H_2O) for all the Solfatara-Pisciarelli fumarolic fluids analyzed by Chiodini and coworkers (Chiodini et al. 2021 and references therein)

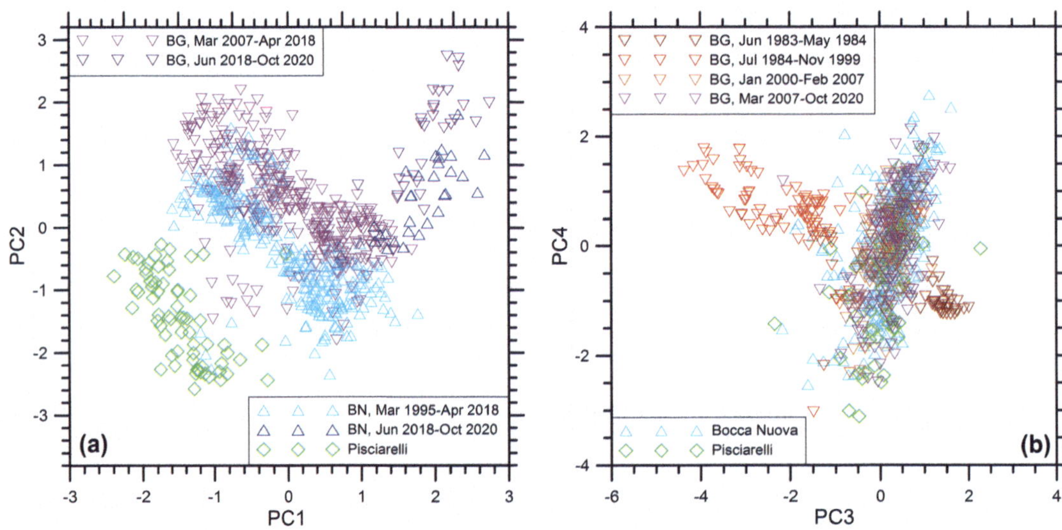

Fig. 13 Correlation diagrams of **a** PC1 versus PC2 and **b** PC3 versus PC4 for all the Solfatara-Pisciarelli fumarolic fluids analyzed by Chiodini and coworkers (Chiodini et al. 2021 and references therein)

and PC2, with the Bocca Grande samples positioned to the right-above the others, the Pisciarelli samples situated to the left-below the others, and the Bocca Nuova samples in between (Fig. 13a). In contrast, the Bocca Grande and Bocca Nuova samples of 2019–2020 are characterized by a direct correlation between PC1 and PC2 and high scores of both PCs, due to the higher concentrations of H_2S and CH_4 compared to previous samples. Therefore, the information given by the PC1 versus PC2 plot is similar to that provided by the two diagrams of H_2S versus CO_2 (Fig. 3a) and CH_4 versus CO_2 (Fig. 3a) and probably by the diagrams in which H_2S and CH_4 are contrasted with H_2, CO, and He, in spite of some doubts because these diagrams were not examined. Hence, the single plot of PC1 versus PC2 gives at once the same information that would have been obtained examining eight different diagrams.

Leaving aside the Bocca Grande samples collected from June 1983 to November 1999, all other samples exhibit a good direct correlation between PC3 and PC4, in the diagram of Fig. 13b, probably due to the presence of a common source or process controlling both N_2 and Ar, although PC4, which is explained by Ar, experiences changes larger than PC3, which is explained by N_2. Incidentally, as already mentioned above, high concentrations of N_2 and Ar correspond to low PC3 and PC4 scores, respectively, and vice versa, as indicated by the negative values of the pertinent correlation coefficients in Table 4. For what concerns the Bocca Grande samples of June 1983–November 1999: (i) those collected from June 1983 to May 1984 have high PC3 scores (i.e., low N_2) and relatively low PC4 scores (i.e., relatively high Ar) and, therefore, are situated to the right of the main sample cluster, whereas (ii) those collected from July 1984 to November 1999 have low PC3 scores (i.e., high N_2) and relatively high PC4 scores (i.e., relatively low Ar) and, consequently, are found to the left of the main sample group. Although Ar data are available since July 1998 and previous values were computed, the spread of samples points in Fig. 13b suggests that a major change in the source (or sources)—process (or processes) controlling N_2 and Ar might have occurred in 1999–2000. Unfortunately, there are few Ar analytical data prior to this date to draw well-founded conclusions on this matter.

Summing up, PCA performs a linear combination of the original variables, computing new variables, which are called principal components (PCs). The PCs are less than the original variables (in the Solfatara-Pisciarelli case, there are four PCs instead of eight original variables), but they explain a high proportion of the total variance of the system (over 92%, in the case of interest). Therefore, use of the PCs simplifies the assessment of the natural processes/sources controlling the chemical characteristics of the fluids of interest.

4 The Triangular Diagram of N_2-Ar-He

The triangular diagram of N_2-Ar-He, initially proposed by Giggenbach (1979) and later modified by Giggenbach (1992a), involves three chemical components having low solubility in water. Therefore, boiling and condensation do not change appreciably their relative concentrations. Moreover, these three gas species are unaffected by chemical reactions or nearly so. In fact, Ar and He are chemically inert noble gases, whereas gaseous N_2 is reactive but is the main nitrogen species and, consequently, N_2 concentration is not significantly altered by chemical reactions, such as the synthesis of ammonia from N_2 and H_2 or the reverse reaction, that is, NH_3 dissociation into molecular N_2 and H_2. Thanks to these characteristics of N_2, Ar, and He, this diagram can be used to individuate, at least to some extent, the possible source (or sources) of the considered chemical components. According to Giggenbach (1992a), the relative concentrations of He, Ar, and N_2 delineate the following four major sources:

(a) A meteoric source, represented by air-saturated water (ASW), with He/Ar ratio < 0.001 and N_2/Ar ratios close to 38 or somewhat higher, due to entrainment of atmospheric air with N_2/Ar ratios of 84.
(b) A magmatic "basaltic" source, rich in He and having He/Ar $\geq$ 0.1. The gases from this source are also enriched in ^{3}He. In particular, along divergent plate boundaries, ^{3}He/^{4}He ratios are typically close to 8 R/R_A, whereas in hot spot areas, ^{3}He/^{4}He ratios > 20 R/R_A occur.[1]
(c) A crustal source, also rich in He, but due to addition of radiogenic ^{4}He from the radioactive decay of crustal U and Th. Consequently, crustal gases have ^{3}He/^{4}He ratios < 0.1 R/R_A.
(d) A magmatic, "andesitic" source, typical of gas discharges from geothermal and volcanic

[1] R_A indicates the ^{3}He/^{4}He ratio of atmospheric air, which is equal to 1.39×10^{-6} (Boucher et al. 2018 and references therein).

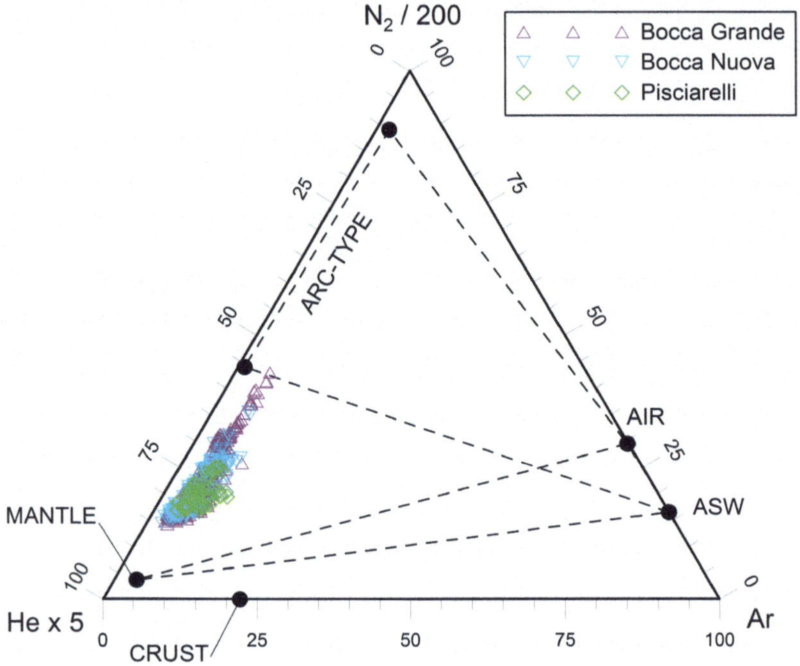

Fig. 14 Triangular diagram of N_2-Ar-He (from Giggenbach 1979, 1992a) for all the Solfatara-Pisciarelli fumarolic fluids analyzed by Chiodini and coworkers (Chiodini et al. 2021 and references therein) also showing the compositions of air, air-saturated water (ASW), crustal gases, mantle gases, and arc-type magmatic gases. Reprinted from Giggenbach and Poreda (1993), modified, Copyright © 1993, with permission from Elsevier

centers positioned along convergent plate boundaries. The high N_2/Ar ratios of these gases, often in the range 800–2000, were attributed to addition of N_2 from thermal decomposition of organic material contained in subducted sediments (Matsuo et al., 1978). However, N_2 release from the rocks situated above the subducted slab is also possible.

In some arc areas, hydrothermal-volcanic gas discharges do not show any excess of N_2, being situated along the lines connecting the He-rich compositions of mantle and/or crustal gases and the two He-poor atmospheric endmembers. This is the case of the Aegean Island arc, including Nisyros (Kavouridis et al. 1999; Brombach et al. 2003; Marini and Fiebig 2005), Milos (Giggenbach 1997b), and Kos (La Ruffa et al. 1999). These peculiar characteristics of the gases from the Aegean Island arc are ascribable to both the lack of N-rich sediments over the subducted slab and the presence of N-poor lithotypes above the mantle wedge, including the rocks hosting the hydrothermal system. Therefore, in the hydrothermal gas discharges of the Aegean Island arc, N_2 is entirely of atmospheric origin or nearly so.

All the Solfatara-Pisciarelli fumarolic fluids analyzed by Chiodini and coworkers (Chiodini et al. 2021 and references therein) are shown in the triangular diagram of N_2-Ar-He (Fig. 14). Missing values of Ar and He were computed as discussed in Sect. 3. The Solfatara-Pisciarelli fumarolic fluids form a cluster situated between the He-rich gases from the mantle and/or the crust and the N_2-rich compositions typical of most arc-type magmatic gases, with a few samples weakly enriched in Ar likely due to air contamination.

Actually, each one of the three chemical components considered in the triangular diagram of Fig. 14 can originate from one or more sources that are distinct from those of the other two chemical components. This possibility, far from remote, represents one of the main limitations of this diagram. Since isotopic data could provide useful information to better understand the possible sources of the three chemical elements of interest, the following three Sects. 5, 6, and 7, are centered on the isotopic characteristics of Ar, He, and N, respectively.

5 Argon Isotopes

The isotopic composition of atmospheric argon currently adopted by the International Union of Pure and Applied Chemistry, IUPAC (e.g., Meija et al. 2016) and the Committee on Data of the International Science Council, CODATA (Mohr et al. 2018) is based on the measurements of Lee et al. (2006), who obtained a $^{40}Ar/^{36}Ar$ ratio of 298.56 ± 0.31. This $^{40}Ar/^{36}Ar$ ratio is approximately 1% higher than the previously accepted value of 295.5 that was derived by Steiger and Jäger (1977) from the measurements of Nier (1950). Lee et al. (2006) also obtained a $^{38}Ar/^{36}Ar$ ratio of 0.1885 ± 0.0003 and percent abundances of 0.3336 ± 0.0004 for ^{36}Ar, 0.0629 ± 0.0001 for ^{38}Ar, and 99.6035 ± 0.0004 for ^{40}Ar.

The first argon isotope data for the Solfatara fumarolic fluids were probably presented by Cioni et al. (1989) in a chronogram showing that the $^{40}Ar/^{36}Ar$ ratio of the Solfatara fumarolic fluids oscillated from the atmospheric value up to 425 approximately during the 1979–1984 period.

The most numerous dataset on the argon isotope of Solfatara fumarolic fluids was obtained by Caliro et al. (2014) who measured the ^{36}Ar and ^{40}Ar concentrations in 88 fumarolic fluids samples collected in 2007–2012 at Bocca Grande and Bocca Nuova. Since the ^{36}Ar and ^{40}Ar concentrations were measured twice for most samples of Bocca Grande and four times for most samples of Bocca Nuova, the argon isotope dataset of Caliro et al. (2014) comprises 208 entries. Based on these data, the $^{40}Ar/^{36}Ar$ ratio fluctuated from 368 to 1365 around a mean of 549 and a median of 541, with a standard deviation of 149, thus extending over a range much wider than that reported by Cioni et al. (1989) for the 1979–1984 interval of time. The ^{36}Ar and ^{40}Ar concentrations measured by Caliro et al. (2014) are shown in the two binary plots of Fig. 15. For an optimal representation of the available analytical data, the linear scale was adopted on both axes of the diagram of Fig. 15a, whereas to make room for the three possible endmembers (see below), the logarithmic scale was used on both axes of the diagram of Fig. 15b.

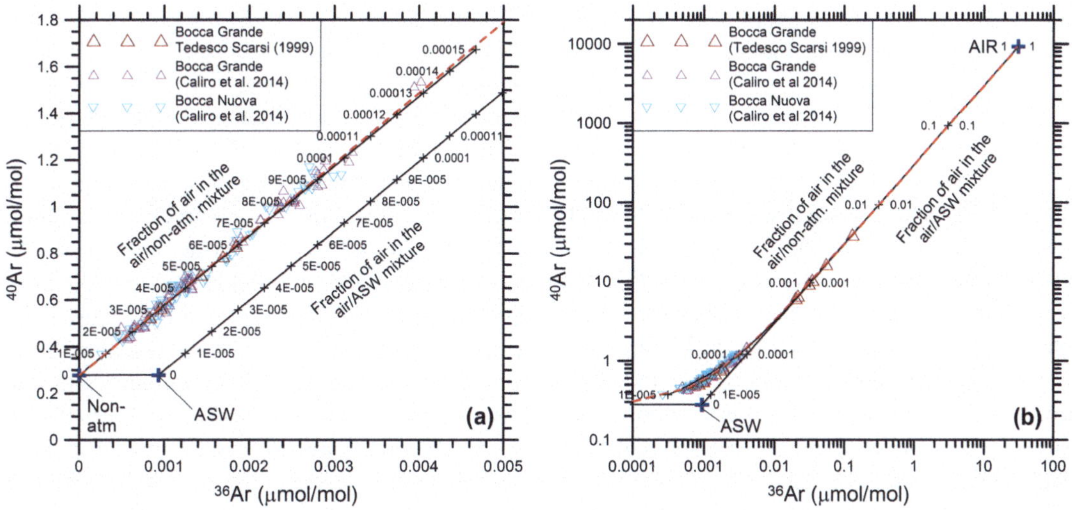

Fig. 15 Correlation plots between the ^{40}Ar and ^{36}Ar concentrations of the Solfatara fumarolic fluids (symbols, see legend) also showing the compositions of air, air-saturated water (ASW), and non-atmospheric ^{40}Ar (blue crosses) as well as the mixing lines between these three endmembers (black lines with small crosses) and the linear regression equation fitting the Ar isotope data of Caliro et al. (2014; dashed red line). The linear scale is utilized on both axes in (**a**) whereas the logarithmic scale is used on both axes in (**b**)

The argon isotope dataset of Caliro et al. (2014) fit the linear regression equation ($R^2 = 0.986$):

$$^{40}Ar = 302.87 \cdot {}^{36}Ar + 0.2784, \quad (6)$$

with the samples distributing very close to the corresponding regression line in the two binary plots of Fig. 15. As shown in the diagram of Fig. 15a, the intercept of Eq. (6) is representative of non-atmospheric ^{40}Ar, comprising (i) radiogenic argon generated in the hydrothermal system by decay of ^{40}K, mainly through electron capture and subordinately through β^+ emission, and (ii) possibly the so-called excess argon, already present in the system of interest (Kelley 2002).

The mixing line between air and non-atmospheric ^{40}Ar is also shown in the two diagrams of Fig. 15. The ^{40}Ar and ^{36}Ar concentrations of air, which are needed to derive this mixing line, are obtained from the isotopic composition of atmospheric argon (see above) and the Ar concentration of air, 9340 μmol/mol. Based on these data, the ^{40}Ar concentration of air is $9340 \cdot 0.996035 = 9302.9669$ μmol/mol, the ^{36}Ar concentration of air is $9340 \cdot 0.003336 = 31.15824$ μmol/mol, and the mixing equation between air and non-atmospheric ^{40}Ar is:

$$^{40}Ar = 298.56 \cdot {}^{36}Ar + 0.2784. \quad (7)$$

Equation (6) and (7) are very similar to each other and the corresponding lines are practically superimposed in the two graphs of Fig. 15, suggesting that the Solfatara fumarolic fluids studied by Caliro et al. (2014) are binary mixtures made up almost exclusively by non-atmospheric ^{40}Ar and to a very subordinate extent by atmospheric air, the mass fraction of which varies from $8.82 \cdot 10^{-6}$ to 1.3210^{-4} .

Also displayed in the two plots of Fig. 15 is the third possible endmember, ASW, as well as the mixing lines of ASW with air and non-atmospheric ^{40}Ar. Following Giggenbach and Goguel (1989), the Ar concentration of ASW at 15 °C, in μmol/mol, is given by:

$$X_{Ar,ASW} \cong \frac{m_{Ar,ASW}}{m_{H_2O,ASW}} \cdot 10^6 = \frac{P_{Ar,air}/K_{H,Ar}}{m_{H_2O,ASW}} \cdot 10^6 \quad (8)$$

with $P_{Ar,air} = 0.00934$ bar ,$K_{H,Ar} = 600.9$ bar $kg\ mol^{-1}$, and $m_{H_2O,ASW} = {}^{1000}/_{18.015}$. Based on Eq. (8) and the isotopic composition of atmospheric argon (see above), ASW has Ar concentration of 0.28 μmol/mol, ^{40}Ar concentration of $0.28 \cdot 0.996035 = 0.279$ μmol/mol, and ^{36}Ar concentration of $0.28 \cdot 0.003336 = 0.000934$ μmol/mol. Since the Solfatara fumarolic fluid samples studied by Caliro et al. (2014) are not positioned along the theoretical mixing lines involving ASW, it follows that the contribution of ASW to these fluids is nil to negligible.

Tedesco and Scarsi (1999) determined several isotopic and chemical parameters for 19 fumarolic fluid samples collected at Bocca Grande in 1990–1994, including the $^{40}Ar/^{36}Ar$ ratio and the Ar concentration on a dry gas basis. The approximate Ar concentration in total fluids (including H_2O) was recalculated here assuming that the $^{40}Ar/^{36}Ar$ ratios measured by Tedesco and Scarsi (1999) are consistent with Eq. (6). Meaningful results were obtained for 11 samples, obviously distributing along the air/non-atmospheric ^{40}Ar mixing line. Five of these samples have air mass fractions in the same range of the fluids studied by Caliro et al. (2014), whereas the other six samples have higher air mass fractions varying from $6.37 \cdot 10^{-4}$ to $4.23 \cdot 10^{-3}$. The computed Ar concentrations in total fluids and the Ar concentrations in dry gases reported by Tedesco and Scarsi (1999) can be used to calculate H_2O concentrations. Results are at variance with the H_2O concentrations of Chiodini et al. (2015) for the same time interval, suggesting that it is advisable not to interpret the data of Tedesco and Scarsi (1999) together with those of Chiodini and coworkers to avoid mixing apples and oranges.

Summing up, the Solfatara fumarolic fluids are binary mixtures of non-atmospheric ^{40}Ar and atmospheric air, with nil to negligible proportions of ASW. The ensuing implication is that the

H_2-Ar geothermometer proposed by Giggenbach and Goguel (1989) cannot be applied to the Solfatara fluids because this geothermometer was derived assuming that the Ar concentration of geothermal fluids is equal to that of air-saturated groundwater. The H_2-Ar temperatures estimated in previous studies (e.g., Tedesco and Scarsi 1999 and Caliro et al. 2007) may seem realistic as they compare with those obtained by means of other geothermometric techniques, but this agreement is purely coincidental and misleading.

6 Helium Isotopes

Several authors measured the He isotope ratio of the Solfatara fumarolic fluids. The first determination was probably carried out on a sample collected on 1 September 1978 from the Soffione fumarole (also called Fumarola Circolare, that collapsed in March 1984) and an uncorrected R/R_A of 2.71 was obtained, as reported by Tedesco et al. (1990). This sample is not considered any further here because its helium concentration is unknown and cannot be reconstructed. Moreover, Tedesco et al. (1990) analyzed the He isotope ratios and concentrations of other six Solfatara fumarolic fluids, two from the Soffione and four from Bocca Grande.[2] For these six samples, the He concentrations revised by Chiodini et al. (1996) are considered here, being at variance with the values given by Tedesco et al. (1990), again, to avoid mixing apples and oranges.

Afterwards, the He isotope ratios and concentrations of Bocca Grande fumarolic fluids were measured by Marty et al. (1994) for one sample collected on April 1989, by Tedesco and Scarsi (1999) for 20 samples obtained from September 1990 to October 1994, and by Chiodini et al. (1996) for 5 samples collected from December 1993 to August 1994. Again, the He concentrations of Marty et al. (1994) and Tedesco and Scarsi (1999) were recalculated to made them consistent with those of Chiodini and coworkers.

The concentrations of ^{3}He and ^{4}He, computed from the ^{3}He/^{4}He isotope ratios and He concentrations, are contrasted in the correlation diagram of Fig. 16. As a whole, the Solfatara fumarolic fluids fit the linear regression equation (N = 32; $R^2 = 0.970$):

$$^3\text{He} = 3.0364\text{E} - 006 \cdot {^4\text{He}} + 1.3828\text{E} - 006. \quad (9)$$

Assuming that the three endmembers reconstructed by Caliro et al. (2014), namely the two magmatic fluids M1 and M2 and the hydrothermal component (see Sect. 10.2) are positioned along this regression line, it turns out that the ^{3}He/^{4}He isotope ratio of the hydrothermal component, 4.87 R/R_A, is higher than those of the magmatic endmembers M1 and M2, 2.47 and 2.41 R/R_A, respectively. This is an unlikely result, as magmatic fluids are expected to have

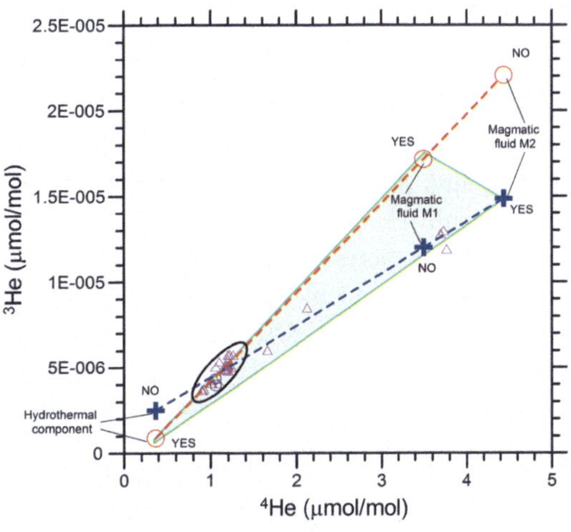

Fig. 16 Correlation plot between the ^{3}He and ^{4}He concentrations of the fumarolic fluids from Bocca Grande (small magenta triangles) and Fumarola Circolare (small blue circles) also showing the possible ^{3}He and ^{4}He concentrations of the three endmembers of Caliro et al. (2014), namely the two magmatic fluids M1 and M2 and the hydrothermal component based on two distinct mixing models indicated by blue crosses and line and red circles and line (see text). The green shaded area encloses the most plausible He isotope ratio for the three endmembers

[2] Tedesco et al. (1990) analyzed also the helium isotope ratio of the submarine fumaroles of the Pozzuoli Gulf, obtaining values similar to those of Solfatara fumarolic fluids.

helium isotope ratios greater than the hydrothermal component.

Actually, the Bocca Grande fumarolic fluids with ^{4}He lower than 1.4, that were all collected in 1990–1994, distribute along a trend steeper than the mixing line defined by Eq. (9) and fit the following linear regression equation ($N = 25; R^2 = 0.777$):

$$^3He = 5.1971E - 006 \cdot {}^4He - 1.0046E - 006. \quad (10)$$

Hypothesizing that the two magmatic fluids M1 and M2 and the hydrothermal component of Caliro et al. (2014) are situated along this regression line, it turns out that the ^{3}He/^{4}He isotope ratio of the hydrothermal component, 1.79 R/R_A, is lower than those of the magmatic endmembers M1 and M2, 3.53 and 3.58 R/R_A, respectively. However, these results do not explain the He isotope ratios and concentrations of the five Bocca Grande fumarolic fluids samples with ^{4}He higher than 1.4, that were sampled in 1983–1989 and are positioned close to the mixing line defined by Eq. (9).

All in all, it is reasonable to assume that the magmatic fluid M1 is encountered along the mixing line defined by Eq. (10), with ^{3}He/^{4}He isotope ratio of 3.53 R/R_A, the magmatic fluid M2 is found along the mixing line defined by Eq. (9), with ^{3}He/^{4}He isotope ratio of 2.41 R/R_A, and the hydrothermal component is along the mixing line defined by Eq. (10), with ^{3}He/^{4}He isotope ratio of 1.79 R/R_A. In this way it is possible to explain all the He isotope ratios and concentrations of Solfatara fumarolic fluids which are displayed in Fig. 16. The range of ^{3}He/^{4}He isotope ratios estimated for the magmatic fluids, 2.41 to 3.53 R/R_A, compare with the range of values measured in the olivine and clinopyroxene crystals of the volcanic rocks of the Phlegraean Fields, 2.5 to 3.0 R/R_A (Martelli et al. 2004) corroborating previous considerations. The different ^{3}He/^{4}He isotope ratios of the two magmatic endmembers might be caused by different processes, including (i) diffusion-controlled isotopic fractionation, (ii) addition of radiogenic ^{4}He through crustal assimilation, and (iii) contamination as a result of magma ageing (e.g., Shaw et al. 2006 and references therein).

7 Nitrogen Isotopes

Six N isotope data are available for the Solfatara fumarolic fluids from Chiodini et al. (2010). They refer to four Bocca Grande samples and two Bocca Nuova samples, all collected in 2007–2008. These δ^{15}N isotope data, ranging from +6.0 to +6.7 ‰, and the corresponding $N_2/^{36}Ar$ ratios, varying from 262,000 to 623,000, are shown in the diagram of Fig. 17, also reporting the relevant endmembers, namely: (i) The upper mantle, with $\delta^{15}N = -5 \pm 2‰$ and $N_2/^{36}Ar = 6,000,000 \pm 2,000,000$. (ii) The sediments, with $\delta^{15}N = +7 \pm 4‰$ and $N_2/^{36}Ar = 6,000,000 \pm 2,000,000$ (i.e., same $N_2/^{36}Ar$ ratio as the upper mantle). (iii) The atmospheric air and ASW, with $\delta^{15}N = 0‰$ and $N_2/^{36}Ar = 18,000$ (Sano et al. 2001).

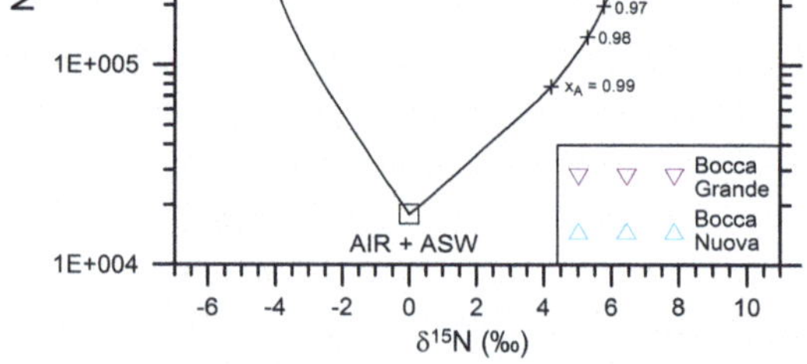

Fig. 17 Correlation plot between the δ^{15}N values and the $N_2/^{36}Ar$ ratios of the Solfatara fumarolic fluids (data from Chiodini et al. 2010). Also shown are the three endmembers and the binary mixing lines between them. Reprinted from Sano et al. (2001), modified, Copyright © 2001, with permission from Elsevier

The binary mixing lines between any two of these endmembers were calculated using the following relation (Faure 1986):

$$\delta^{15}N_M = \frac{\delta^{15}N_A \cdot C_A \cdot x_A + \delta^{15}N_B \cdot C_B \cdot (1 - x_A)}{C_M}, \quad (11)$$

in which the subscripts M, A, and B identify the mixture, the generic endmember A, and the generic endmember B, whereas x_A is the fraction of endmember A in the binary mixture. The curvature of the mixing lines depends on the nitrogen concentration ratio, C_A/C_B. If $C_A = C_B$, as in the sediments—upper mantle case, the mixing line is a straight line, whereas if $C_A \neq C_B$, as in the sediments—(air + ASW) and upper mantle—(air + ASW) cases, the mixing line is a hyperbola.

The spread of sample points in Fig. 17 suggests that the Bocca Grande and Bocca Nuova samples are binary mixtures of atmospheric N_2 (endmember A) and N_2 produced through breakdown of sediments (endmember B), although a minor contribution of upper mantle N_2 cannot be excluded. Based on this interpretation, it follows that the C_B/C_A ratio must be close to 150, to fit the six samples with the sediments—(air + ASW) mixing curve. As indicated in the diagram, the fraction of atmospheric N_2, x_A, varies from 0.90 to 0.96, but the product of x_A times the C_A/C_M ratio varies from 0.06 to 0.14, indicating that most N_2 comes from the sediments.

As noted by Chiodini et al. (2010), the $\delta^{15}N$ isotope data of the Solfatara fumarolic fluids compare with those of the volcanic fluids in subduction zones, which have $\delta^{15}N = +7 \pm 4$‰ (Sano et al. 2001), probably due to the prevailing contribution of N_2 derived from the breakdown of sediments. However, in convergent-plate zones, the sediment breakdown process can occur in different places, i.e., at the top of the subducted slab and/or in the mantle wedge and/or above it, including the rocks hosting the magmatic-hydrothermal system. Therefore, N isotopes do not allow one to clarify the doubts on the origin of N_2 raised by the analysis of the triangular diagram of N_2-Ar-He (Sect. 4).

8 Sulfur Isotopes of H_2S

According to Allard et al. (1991), the H_2S contained in the Solfatara fumarolic fluids has a mean $\delta^{34}S$ value of -0.3 ± 0.3‰ (N = 17), which is very close to the 0 ‰ value of the reference material, assumed to be representative of the mantle, in spite of the presence of not negligible heterogeneities (Marini et al. 2011 and references therein). This average $\delta^{34}S_{H_2S}$ value refers to fumarolic samples collected from 1983 to 1988, i.e., during the 1982–1984 bradyseismic crisis and some years after it. Based on these data, the H_2S of the Solfatara fumarolic fluids was assumed to proceed from an igneous source, either deep magma degassing or decomposition of sulfur-bearing minerals (e.g., sulfur, pyrite, pyrrhotite, and anhydrite) contained in the volcanic rocks of the Campi Flegrei (Allard et al. 1991).

Of the seventeen available $\delta^{34}S$ values, the fifteen data from Bocca Grande are shown in the chronogram of Fig. 18, together with the H_2S concentrations measured by Allard et al. (1991) and those determined by Chiodini and coworkers (e.g., Chiodini et al. 2021 and references therein). The other two samples were obtained in 1983, on October 29th and December 29th, from the Soffione fumarole (see Sect. 6) and have $\delta^{34}S$ values of -0.7 and -0.1 ‰ and H_2S concentrations of 824 and 772 μmol/mol, respectively. These two isotope values compare with the concomitant Bocca Grande data, whereas these two concentrations are significantly lower with the concurrent Bocca Grande data. In spite of some differences between the H_2S data of Allard et al. (1991) and those of Chiodini and coworkers, Fig. 18 shows that the H_2S concentration of Bocca Grande fumarolic fluids experienced a significant decrease during the time window in which the $\delta^{34}S$ values were acquired by Allard et al. (1991), that is, in May–September 1984, when substantial changes in other chemical parameters were also recorded (Cioni et al. 1989).

In contrast, the $\delta^{34}S_{H_2S}$ value did not suffer significant variations, probably because the S

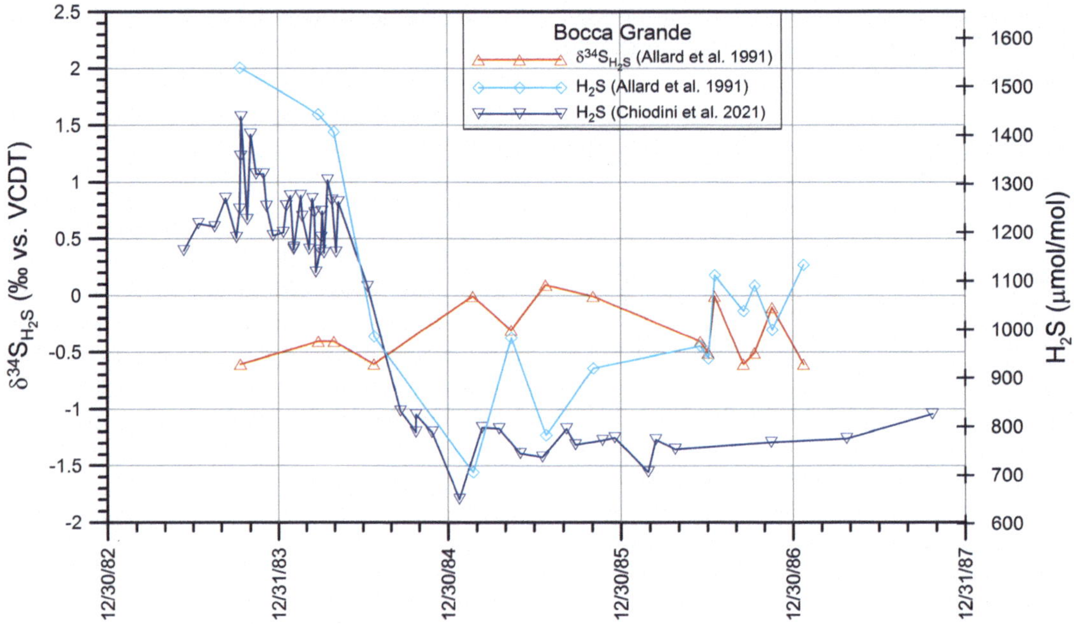

Fig. 18 Chronogram of the $\delta^{34}S_{H_2S}$ value of Bocca Grande fumarolic fluids in 1983–1987 (Allard et al. 1991). The H_2S concentrations measured by Allard et al. (1991) and Chiodini and coworkers (Chiodini et al. 2021 and references therein) are also reported

source remained the same. For instance, assuming that H_2S is produced through decomposition of pyrite occurring at shallower levels before May–September 1984 and at deeper levels afterwards, the H_2S concentration of fumarolic fluids is expected to be different because thermodynamic conditions change with depth, while the $\delta^{34}S_{H_2S}$ value is expected to be the same hypothesizing that pyrite-sulfur comes from the same magmatic source. Similar considerations can be done if H_2S were generated through a different process, although details would be different.

Afterwards, from 1998 to 2001, nineteen gas samples were collected from the Solfatara fumaroles and their average $\delta^{34}S_{H_2S}$ value resulted to be somewhat higher than that of the previous samples, namely, + 1.7 ± 0.3‰, (Barbara Gambardella, personal communication to Marini et al. 2011). This higher value could be due to (Marini et al. 2011): (i) The occurrence of gradual degassing from a single magma batch, without the involvement of new magma. (ii) Separation of isotopically light elemental sulfur upstream of the sampling point, similar to what was observed at White Island (Robinson 1987).

9 Carbon Isotopes of CO_2 and CH_4

A total of 182 measurements are available for the $\delta^{13}C_{CO_2}$ value of Solfatara fumarolic fluids (Cheminée et al. 1968; Panichi and Tongiorgi 1975; Carapezza et al. 1984; Nuti et al. 1985; Bolognesi et al. 1986; Allard et al. 1991, also reporting previous unpublished data of Allard and Létolle; Panichi and Volpi 1999; Caliro et al. 2007; Chiodini et al. 2010; Tassi et al. 2012; Fiebig et al. 2013; Caliro et al. 2014). Of these data, 86 refer to Bocca Grande, 70 to Bocca Nuova, 11 to Pisciarelli, 8 to Forum Vulcani, 6 to Soffione, and 1 to an unspecified vent. The CO_2 concentration on a total fluid basis (i.e., including H_2O) is available for 168 of these 182 samples. The diagram of the $\delta^{13}C_{CO_2}$ value versus CO_2 concentration (Fig. 19a) shows that the correlation between these two variables is poor, with

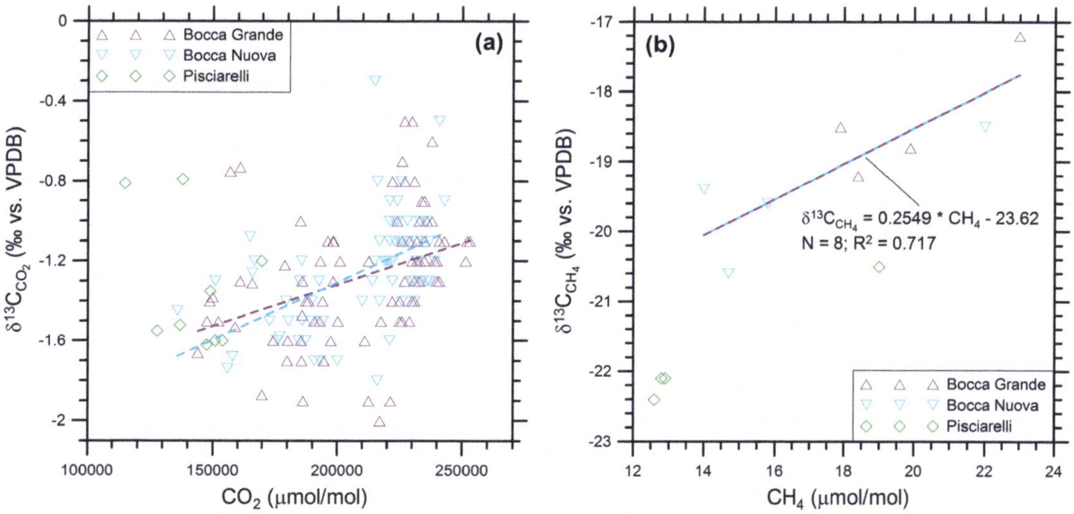

Fig. 19 Correlation diagrams of **a** the $\delta^{13}C_{CO_2}$ value versus CO_2 concentration and **b** the $\delta^{13}C_{CH_4}$ value versus CH_4 concentration of Solfatara fumarolic fluids (see text for data sources)

R = + 0.376 for Bocca Grande (N = 83) and R = 0.535 (N = 70) for Bocca Nuova. Based on this poor correlation, it seems unlikely that the $\delta^{13}C_{CO_2}$ value and CO_2 concentration are uniquely controlled by a binary mixing process.

Only 16 data are available for the $\delta^{13}C_{CH_4}$ value of Solfatara fumarolic fluids (Bolognesi et al. 1986; Caliro et al. 2007; Tassi et al. 2012; Fiebig et al. 2013), of which 6 are from Bocca Grande, 4 each are from Bocca Nuova and Pisciarelli, and 2 are from Forum Vulcani. The CH_4 concentration on a total fluid basis was determined for 14 of these 16 samples data. The diagram of the $\delta^{13}C_{CH_4}$ value versus CH_4 concentration (Fig. 19b) shows that there is a good correlation between these two variables, considering together the Bocca Grande and Bocca Nuova fluids, with R = + 0.847, while the Pisciarelli samples were kept separated being isotopically lighter for the same concentration of CH_4. Furthermore, the CH_4 of the Solfatara fumarolic fluids is significantly heavier than the organic matter source, which has typically $\delta^{13}C$ value of −27‰ (Giggenbach 1997a).

The $\delta^{13}C$ values of CO_2 and CH_4 of the Solfatara fluids are contrasted in the δ-δ plot of Fig. 20 (modified from Giggenbach 1997a), also showing three common carbon sources, namely

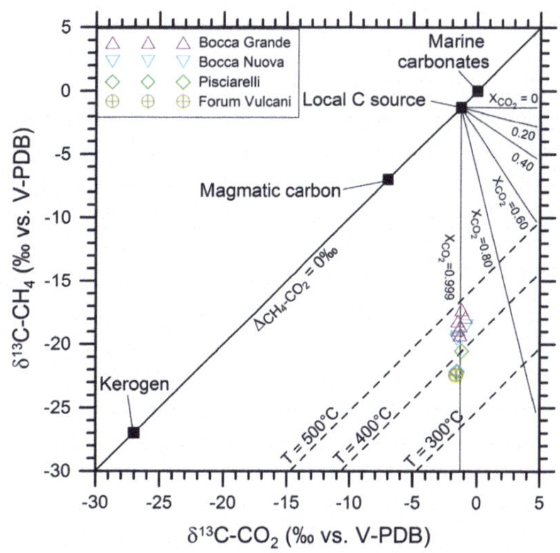

Fig. 20 Diagram of the $\delta^{13}C$ values of CO_2 versus the $\delta^{13}C$ values of CH_4 for the Solfatara fumarolic fluids (see text for data sources). The three dashed lines represent the isotope compositions expected for attainment of isotopic equilibrium at 300, 400, and 500 °C. The solid lines radiating from the point representative of the local C source, with $\delta^{13}C$ of total C of −1.3 ‰, correspond to the isotope compositions expected for the formation of CH_4 and CO_2 from this C source in a closed system, as a function of the fraction of CO_2 in the gas mixture. Also shown are three typical C sources, namely, the marine carbonates, with $\delta^{13}C$ of total C of 0‰, the magmatic (mantle) C, with $\delta^{13}C$ of total C of −7‰, and the kerogen, with $\delta^{13}C$ of total C of −27‰. Reprinted from Giggenbach (1997a), modified, Copyright © 1997, with permission from Elsevier

kerogen with $\delta^{13}C$ of −27 ‰, magmatic (mantle) carbon with $\delta^{13}C$ of −7 ‰, and marine carbonates with $\delta^{13}C$ of 0 ‰, as well as the local carbon source. The $\delta^{13}C$ value of the local carbon source is assumed to be equal to the mean $\delta^{13}C_{CO_2}$ of Solfatara fumarolic fluids, −1.3 ± 0.2 (1σ) ‰, since the fraction of CO_2 in gaseous carbon, $x_{CO_2} = X_{CO_2}/(X_{CO_2} + X_{CH_4})$, varies from 0.9998 to 0.99997.

These carbon sources are positioned along the line of slope −1, which is constrained by the condition of equality between the $\delta^{13}C$ values of CO_2 and CH_4. The other three lines of slope −1 are representative of the CO_2-CH_4 equilibrium fractionation factors at 300, 400, and 500 °C, which are calculated by means of the following relation:

$$10^3 \ln \alpha(CO_2 - CH_4) = 26.70 - 49.137\left(10^3/T\right) + 40.828\left(10^6/T^2\right) - 7.512\left(10^9/T^3\right). \quad (12)$$

Equation (12) reproduces the forty four CO_2-CH_4 equilibrium fractionation factors experimentally determined by Horita (2001) at temperatures of 200 to 600 °C with an uncertainty of ± 0.14 ‰ (1σ).

The lines radiating from the local carbon source represent the isotopic compositions of CH_4 and CO_2 as a function of the fraction of CO_2 in the gases formed from this source in a closed system. These lines were calculated by means of the mass balance:

$$\delta^{13}C_{C\_source} = x_{CO_2} \cdot \delta^{13}C_{CO_2} + (1 - x_{CO_2}) \cdot \delta^{13}C_{CH_4}. \quad (13)$$

The eighteen samples of the Solfatara fluids are positioned along the line for x_{CO_2} close to 1, radiating from the local carbon source, at isotope equilibrium temperatures of 365 to 476 °C. It will be shown in chapter "Calibration and Application of the RWG and SS4 Indicators for the Possible Expansion Paths of Solfatara Fluids" that the temperature of CO_2-CH_4 isotopic equilibrium is in satisfactory agreement with the temperature of CO_2-CH_4 chemical equilibrium (or SS4 temperature) of Bocca Grande and Bocca Nuova fluids, suggesting that these two gaseous C species achieve or approach the condition of isotopic and chemical equilibrium at these temperatures, although further isotope data are needed to substantiate this inference.

In turn, this inference supports the occurrence of the local carbon source, whose $\delta^{13}C$ value of −1.3 ‰ is intermediate between the isotope values of the marine carbonates and the magmatic source. The relative contributions of these two sources can be evaluated, to a first approximation, by means of Eq. (11), rewritten as follows:

$$\delta^{13}C_L = \frac{\delta^{13}C_C \cdot C_C \cdot x_C + \delta^{13}C_M \cdot C_M \cdot (1 - x_C)}{C_L}, \quad (14)$$

in which the subscripts L, C, and M identify the local carbon source, the marine carbonate endmember, and the magmatic endmember, whereas x_C is the fraction of the marine carbonate endmember in the local carbon source. Since the three carbon (≈ CO_2) concentrations C_L, C_C, and C_M are unknown, the only way to proceed is to assume that they are equal or not significantly different. If so, Eq. (14) reduces to:

$$\delta^{13}C_L = \delta^{13}C_C \cdot x_C + \delta^{13}C_M \cdot (1 - x_C). \quad (15)$$

For $\delta^{13}C_L = -1.3\%$, $\delta^{13}C_C = 0\%$ and $\delta^{13}C_M = -7\%$, it follows that $x_C = 0.81$ and $x_M = 0.19$. Owing to the overwhelming prevalence of the CO_2 from marine carbonates over magmatic CO_2, the contamination of the local mantle caused by incorporation of carbonate sediments seems unlikely, whereas the production of CO_2 from thermometamorphic decarbonation of marine carbonates appears to be a plausible explanation, of course, under the assumption of equality of CO_2 concentrations.

The fact that the local carbon source is chiefly controlled by the CO_2 from marine carbonates and magma degassing does not exclude the possible occurrence of organic matter degradation and the consequent supply of CH_4 and CO_2 in subordinate quantities. In fact, organic matter is present in the soils interstratified with the volcanic deposits (e.g., Bellucci 1994; Principe et al. 2018 and references therein). Moreover, it is likely that the submarine tuffites, siltites, sandstones and shales found below the volcanic products by the deep wells drilled at Mofete and San Vito (Chelini and Sbrana 1987) contain appreciable amounts of organic matter although no such data are reported.

10 Hydrogen Isotopes of H_2O and Oxygen Isotopes of H_2O and CO_2

The interpretation of the hydrogen isotopes of H_2O and the oxygen isotopes of H_2O and CO_2 is discussed in Sect. 10.4. Before tackling it, it is necessary (i) to consider the exchange of oxygen isotopes between H_2O vapor and gaseous CO_2, (ii) to model the effects of magmatic degassing, and (iii) to take into account the entrainment of secondary steam (i.e., steam produced through complete evaporation of down-flowing condensed steam) into the uprising fumarolic fluids. These topics are considered in the following three sub-sections.

10.1 The Exchange of Oxygen Isotopes Between H_2O Vapor and Gaseous CO_2

The exchange of oxygen isotopes between H_2O vapor and gaseous CO_2 is so quick in a wide temperature interval (from 100 to over 1000 °C) to cause a continuous isotopic re-equilibration during the cooling of natural gas mixtures (Chiodini et al. 2000), including the Solfatara fumarolic fluids (Caliro et al. 2007). This re-equilibration process ends upon separation of gaseous CO_2 from the condensed steam and rapid cooling from outlet temperature to sampling temperature, if suitable equipment is used for gas sampling. However, if the gaseous CO_2 is kept in contact with the condensed steam, the exchange of oxygen isotopes between gaseous CO_2 and the H_2O molecules of condensed steam proceeds very quickly even at ambient temperature.

If the analytical $\delta^{18}O$ value is available for both H_2O and CO_2, the $\delta^{18}O_{H_2O+CO_2}$ value can be computed in two different ways. First, one can use the mass balance (Chiodini et al. 2000):

$$\delta^{18}O_{H_2O+CO_2} = \delta^{18}O_{CO_2} \cdot Y_{CO_2} + \delta^{18}O_{H_2O} \cdot (1 - Y_{CO_2}), \quad (16)$$

in which Y_{CO_2} is the fraction of oxygen atoms contained in CO_2 over the total number of oxygen atoms present in the H_2O-CO_2 system, as expressed by the relation:

$$Y_{CO_2} = 2 \cdot X_{CO_2}/(2 \cdot X_{CO_2} + X_{H_2O}) \cong 2 \cdot X_{CO_2}/(X_{CO_2} + 1). \quad (17)$$

Second, one can utilize the relation (Chiodini et al. 2000):

$$\delta^{18}O_{H_2O+CO_2} = \delta^{18}O_{H_2O} + Y_{CO_2} \cdot 1000 \ln \alpha_{CO_{2(g)}-H_2O_{(g)}}, \quad (18)$$

in which $1000 \ln \alpha_{CO_{2(g)}-H_2O_{(g)}}$ represents the $CO_{2(g)}$-$H_2O_{(g)}$ fractionation factor. This is obtained through the equation (Richet et al. 1977):

$$1000 \ln \alpha_{CO_{2(g)}-H_2O_{(g)}} = -1.206 \cdot 10^9/T^3 + 10.50 \cdot 10^6/T^2 - 7.634 \cdot 10^3/T + 1.461 \quad (19)$$

at the temperature of interest, T (in K), that is, the outlet temperature. If the analytical $\delta^{18}O$ value is available only for H_2O, the $\delta^{18}O_{H_2O+CO_2}$ value can be calculated by means of the second approach only, namely, Eq. (18). This implies to assume that (1) $H_2O_{(g)}$ and $CO_{2(g)}$ attain isotopic equilibrium at the outlet temperature and (2) do not re-equilibrate during quenching from outlet temperature to sampling temperature and separation of condensed steam from gaseous CO_2. In contrast, the second assumption only is involved in the first approach, that is, Eq. (16). However, the outlet temperature is taken as reference in both approaches. In principle, Eq. (18) can be used also to compute the $\delta^{18}O$ value of water at any selected temperature different from the outlet temperature, considering the $CO_{2(g)}$-$H_2O_{(g)}$ fractionation factor at the selected temperature. Nevertheless, to avoid errors in the interpretation of the hydrogen and oxygen isotopes, it is advisable, not to say compulsory, to use the $\delta^{18}O_{H_2O+CO_2}$ value, which is constant (i.e., it does not changes with temperature) instead of the $\delta^{18}O$ value of water, which is a function of temperature due to the continuous readjustment of the $CO_{2(g)}$-$H_2O_{(g)}$ isotopic equilibrium during cooling of the gas mixture of interest (Chiodini et al. 2000).

The analytical $\delta^{18}O$ values of both H_2O and CO_2 are available for only 10 samples of the Solfatara fumarolic fluids, which were collected in 1982–1985 (Nuti et al. 1985; Bolognesi et al. 1986), whereas the 66 samples obtained afterwards, from 1985 to 2006, were analyzed for the $\delta^{18}O$ value of H_2O only (Panichi and Volpi 1999; Chiodini et al. 1996; Caliro et al. 2007). For the 10 samples analyzed for the $\delta^{18}O$ values of both H_2O and CO_2, it is instructive to calculate the $\delta^{18}O_{H_2O+CO_2}$ value by means of both Eqs. (16) and (18). The obtained results are contrasted in the correlation diagram of Fig. 21.

This binary diagram shows that Eq. (18) gives results higher than those computed by means of Eq. (16) in most cases. Nevertheless, the two series of $\delta^{18}O_{H_2O+CO_2}$ values agree within ± 0.5 ‰ units, with a mean absolute deviation of ± 0.15 ‰ units. This value is about twice the expected error of ± 0.07 ‰, which is computed for an analytical uncertainty of ± 0.05 ‰ on the separate oxygen isotope determinations of H_2O and CO_2 using the relation:

$$\sigma_{^{18}O_{H_2O+CO_2}} = \sqrt{\sigma^2_{^{18}O_{H_2O}} + \sigma^2_{^{18}O_{CO_2}}}. \quad (20)$$

This error higher than the analytical uncertainty might be due to limited deviations from the condition of perfect isotopic equilibrium between $H_2O_{(g)}$ and $CO_{2(g)}$ which is implied in the use of Eq. (18).

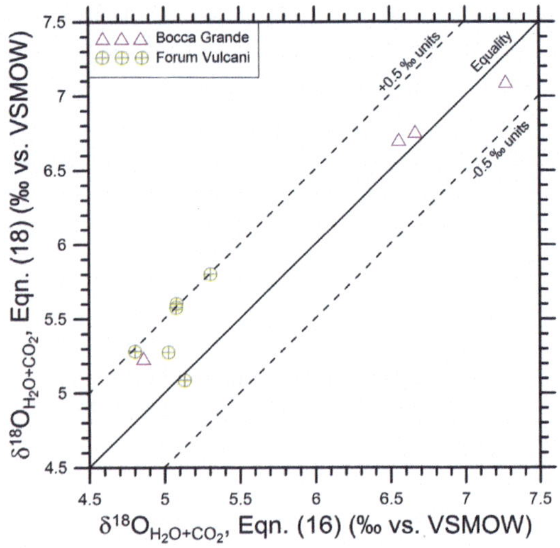

Fig. 21 Correlation diagram between the overall oxygen isotopic composition of the H_2O-CO_2 system, $\delta^{18}O_{H_2O+CO_2}$, computed through a simple mass balance, i.e., Eq. (16) and that obtained from the $\delta^{18}O$ value of H_2O and the $CO_{2(g)}$-$H_2O_{(g)}$ fractionation factor at the outlet temperature, i.e., Eq. (18). Data from Nuti et al. (1985) and Bolognesi et al. (1986)

10.2 Modeling of Magmatic Degassing

Caliro et al. (2014) used the magma degassing model of Nuccio and Paonita (2001) and considered the solubilities of noble gases from Iacono Marziano et al. (2010) and Paonita et al. (2012) to model open-system degassing of four different magma types possibly present underneath the Solfatara, namely a trachybasalt, a shoshonite, and a tephri-phonolite at 1120 °C, as well as a trachyte at 950 °C. Initial pressure was assumed to be 1, 2, and 3 kbar, in separate runs, to take into account distinct sources possibly situated at different depths. Several runs were performed to cover the whole interval of H_2O-CO_2 compositions of melt inclusions (i.e., H_2O up to 3.2 wt% and CO_2 up to 0.16 wt%) and the CO_2 mole fraction of the magmatic fluids entering the Solfatara magmatic-hydrothermal system from below, i.e., X_{CO_2} = 0.38 ± 0.06. This X_{CO_2} value was estimated by Caliro et al. (2007) on the basis of the relationship between the CO_2 concentration and the $\delta^{18}O$ value of the H_2O-CO_2

gaseous mixtures discharged by Solfatara fumaroles.

Caliro et al. (2014) noted that the N_2/CO_2 and N_2/He ratios of the gas phase separated through open-system degassing of the trachybasalt and the shoshonite, from 1 kbar to 300–450 bar (corresponding to nearly complete loss of CO_2), are consistent with fumarolic compositions. In contrast, the corresponding outcomes for the tephri-phonolite and the trachyte are at variance with fumarole chemistry. Therefore, based on these findings, Caliro et al. (2014) proposed that a degassing trachybasaltic magma is probably positioned below the Solfatara crater. Moreover, Caliro et al. (2014) underscored that fumarolic chemistry is explained by mixing between a hydrothermal component, HC, and two magmatic endmembers, M1 and M2, the chemical characteristics of which are reported in Table 5.

According to Caliro et al. (2014): (i) Endmembers M1 and M2 are representative of an undegassed magma and a magma that has experienced substantial gas losses, respectively. (ii) Nevertheless, both endmembers M1 and M2 can be also produced through different extents of open-system degassing from a single magma batch or two distinct magma batches. (iii) Mixing between these two magmatic endmembers is also possible.

The theoretical effects of magma degassing and mixing between the three endmembers reconstructed by Caliro et al. (2014) are displayed in the correlation plots of CO_2 versus N_2 and CO_2 versus He of Fig. 22, also showing the analytical concentrations of Bocca Grande fumarolic fluids. The analytical data of Bocca Nuova and Pisciarelli are not reported in these diagrams because they cover smaller time intervals compared to Bocca Grande and confirm *pro parte* the information given by Bocca Grande data without adding anything new.

As shown by the two diagrams of Fig. 22, the separated gas phase is expected to experience a considerable decrease in the concentrations of both N_2 and CO_2, as well as a limited increase in He concentration, with increasing magma degassing.

Before considering the distribution of the Bocca Grande fumarolic fluids in these diagrams, it must be recalled that the N_2 and CO_2 data of Chiodini and coworkers (e.g., Chiodini et al. 2011, 2015, 2021) are available from June 1983 to October 2020, whereas systematic He analysis began later, in June 1989. Therefore, the early evolution experienced by fumarolic fluids in the period June 1983–May 1989 can be seen in the diagram of CO_2 versus N_2 (Fig. 22a), but not in the diagram of CO_2 versus He (Fig. 22b). In the diagram of CO_2 versus N_2 (Fig. 22a), the Bocca Grande fumarolic fluids are bracketed by the two mixing lines involving the hydrothermal component HC and one of the two magmatic endmember M1 and M2. In detail, the June 1983–July 1984 samples are situated close to the M2-HC mixing line, the September 1984–December 1994 samples are positioned along the M1-HC mixing line, and the composition of the subsequent samples moves back towards the M2-HC mixing line. The spread of samples points is similar in the diagram of CO_2 versus He (Fig. 22b).

Here, we used the magma degassing model of Caliro et al. (2014) to understand the effects of magma degassing on the hydrogen isotopes of H_2O and the oxygen isotopes of H_2O and CO_2. Therefore, following Caliro et al. (2014), the initial magma was assumed to be a trachybasalt of composition equal to that the melt inclusion

Table 5 Chemical composition of the hydrothermal component and the two magmatic endmembers reconstructed by Caliro et al. (2014)

Endmember	CO_2 µmol/mol	N_2 µmol/mol	He µmol/mol
M2	360,000	570	4.44
M1	600,000	2870	3.50
HC	70,000	150	0.370

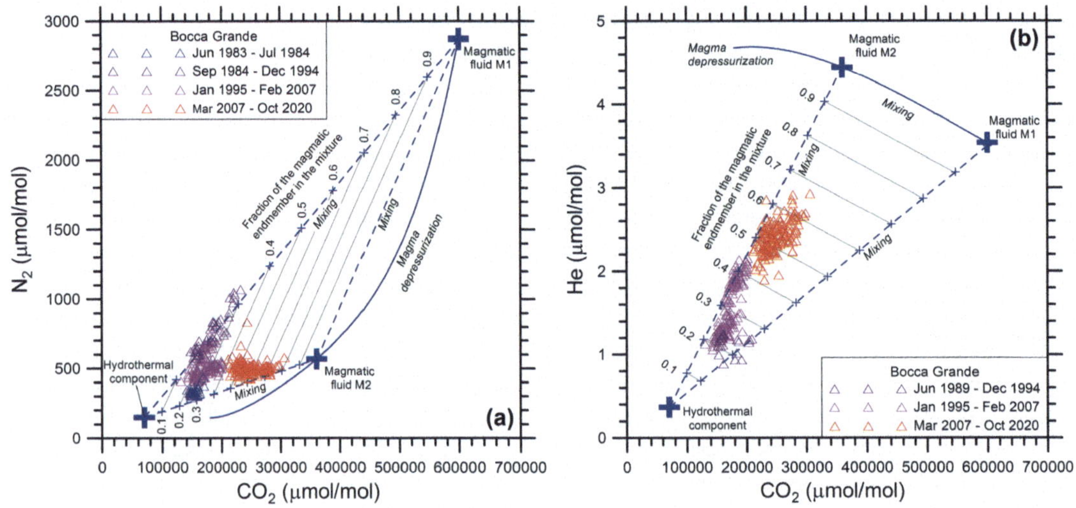

Fig. 22 Correlation diagrams of **a** CO_2 versus N_2 and **b** CO_2 versus He (both from Caliro et al. 2014, redrawn), showing the effects of magma degassing and mixing between the three endmembers reconstructed by Caliro et al. (2014). The analytical data of Bocca Grande fumarolic fluids from Chiodini and coworkers (Chiodini et al. 2021 and references therein) are also shown

sample Mi 1-C1 o5 M1 hosted in an olivine crystal of Minopoli 1 scoriae (data from Cannatelli et al. 2007). These volcanic products were emitted during an explosive hydromagmatic eruption that took place 10.3–11.5 ka BP in the northern sector of the Phlegraean Fields caldera. The bulk chemical composition of the considered trachybasalt melt inclusion is reported in Table 6. The temperature of the magma was fixed at 1120 °C, as assumed by Caliro et al. (2014), which is 25 °C lower than the mean homogenization temperature of the Minopoli 1 melt inclusions in olivine, 1145 ± 3 °C.

Magmatic degassing was modeled by means of both: (i) the Papale model for H_2O-CO_2 solubility in magmas (Papale et al 2006), using the calculators available at the web site http://melts.ofm-research.org/CORBA_CTserver/Papale/Papale.php and (ii) the Microsoft Excel code VolatileCalc 2.0 (Newman and Lowenstern 2002), which was downloaded from the web site https://volcanoes.usgs.gov/observatories/yvo/jlowenstern/other/software_jbl.html.

The Papale model is a fully non-ideal multicomponent model which computes the saturation surface in H_2O–CO_2–silicate melts based on the thermodynamic equilibrium condition between gaseous and dissolved volatile components. In the Papale model, the activity coefficients of H_2O and CO_2 dissolved in the melt, describing the deviation from the ideal behavior, are calculated using 24 interaction parameters between volatile components and non-volatile oxide components (SiO_2, Al_2O_3, Fe_2O_3, FeO, MgO, CaO, Na_2O and K_2O), namely 8 for H_2O and 16 for CO_2. These interaction parameters were derived through best-fitting of the about 1100 experimental saturation data for H_2O, CO_2, and $H_2O + CO_2$ in silicate melts of different composition without a-priori assumptions on their

Table 6 Bulk chemical composition of the trachybasalt melt inclusion sample Mi 1-C1 o5 M1 hosted in an olivine crystal of the Minopoli 1 scoriae (from Cannatelli et al. 2007) which was considered for modeling magmatic degassing

SiO_2	TiO_2	Al_2O_3	FeO tot	MnO	MgO	CaO	Na_2O	K_2O	P_2O_5	H_2O
wt%	wt%	wt%	wt%	wt%	wt%	wt%	wt%	wt%	wt%	wt%
47.81	0.66	13.81	6.80	0.12	11.53	7.57	1.50	4.11	0.98	3.89

dependence on pressure and temperature. Therefore, the Papale model is a very effective tool to compute H_2O and CO_2 solubilities for any silicate melt composition. However, the Papale model does not take into account the speciation of H_2O and CO_2 in the melt, being not required by the adopted thermodynamic approach. From this point of view, the Papale model for silicate melts is similar to the Pitzer model for concentrated aqueous solutions.

VolatileCalc, on the one hand, has a major limitation, being calibrated for two silicate melt types only, either rhyolite or basalt.[3] Therefore, only the SiO_2 concentration (and no other non-volatile oxide component) is taken into account by the code. On the other hand, the "degassing path" macro of VolatileCalc is very useful and effective as it computes automatically the melt and gas compositions along a series of depressurization steps of the melt, for open-system and closed-system gas–melt separation, also admitting the presence of a specified percentage of exsolved gas (relative to the total system of melt plus gas) in equilibrium with the melt at initial pressure.

Due to the different characteristics of the Papale model and VolatileCalc, the two models were used here in a synergistic way, in order to maximize the achieved results. First, saturation pressure and CO_2 concentration of the initial melt M1 (for which only the H_2O concentration is known; see Table 6) were calculated by means of both the Papale model and VolatileCalc, assuming that the gas in equilibrium with the initial melt M1 has CO_2 concentration of 60 mol%, in accordance with Caliro et al. (2014). As expected, the results given by the two models are different, with saturation pressure of 4532 bar and concentration of CO_2 dissolved in the melt of 2400 mg/kg for VolatileCalc, versus saturation pressure of 2879 bar and concentration of CO_2 dissolved in the melt of 1446 mg/kg for the Papale model.

Assuming that the Papale model provides results more realistic than VolatileCalc for the equilibrium partitioning of H_2O and CO_2 between a silicate melt and a gas phase, the pressures and concentrations of CO_2 dissolved in the melt calculated by VolatileCalc for different degassing steps were made consistent with the Papale model (with acceptable approximations, see below) by means of the simple relations:

$$P_{sat,Papale,n} = \frac{P_{sat,VolatileCalc,n}}{P_{sat,VolatileCalc,i}} \cdot P_{sat,Papale,i} \quad (21)$$

$$c_{CO_2,Papale,n} = \frac{c_{CO_2,VolatileCalc,n}}{c_{CO_2,VolatileCalc,i}} \cdot c_{CO_2,Papale,i}, \quad (22)$$

in which subscripts n and i refers to any degassing step and the initial melt M1, respectively. Some check calculations carried out with the Papale model indicates that use of this approach causes mean absolute deviations of 4% on the saturation pressure and 5% on the CO_2 concentration, which are acceptable approximations.

Different runs were performed, in order to model open-system degassing, closed-system degassing without gas exsolution, and closed-system degassing with 1, 2.5 and 5% exsolved gas, from the initial composition of magmatic endmember M1 (60 mol% CO_2, 40 mol% H_2O) to the final composition of the magmatic endmember M2 (36 mol% CO_2, 64 mol% H_2O). The concentrations of H_2O and CO_2 in the separated gas phase are displayed as a function of pressure in the graphs of Fig. 23. The pressure of M1 has a unique value, 2879 bar (see above) for all the runs, whereas the pressure of the magmatic endmember M2 depends on the gas–melt separation model, being 1601 bar for open-system degassing, 1560 bar for closed-system degassing without gas exsolution, 1277 bar for closed-system degassing with 1% exsolved gas, 1012 bar for closed-system degassing with 2.5% exsolved gas, and 373 bar for closed-system degassing with 5% exsolved gas.

[3] According to Newman and Lowenstern (2002), the solution model for the basalt–H_2O–CO_2 system can be used to compute melt-gas equilibria for basalt compositions varying from tholeiite to nephelinite because VolatileCalc incorporates the equations of Dixon (1997) describing the dependence of the mole fractions of CO_3^{2-} groups and molecular H_2O on SiO_2 concentration. Nevertheless, there are considerable differences between the VolatileCalc results and the results of the Papale model, for the considered trachybasalt melt.

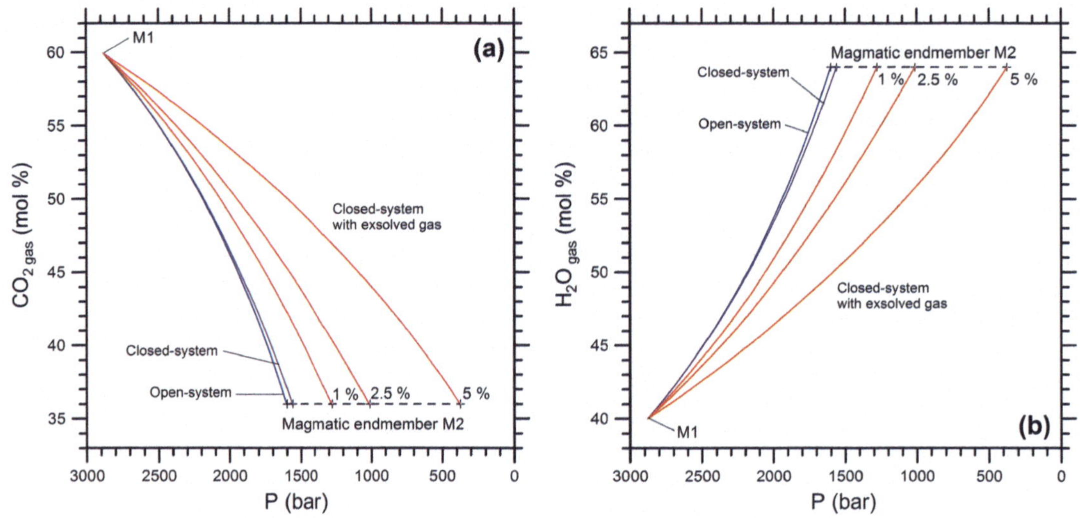

Fig. 23 Concentrations of **a** CO_2 and **b** H_2O in the separated gas phase as a function of pressure for open-system degassing, closed-system degassing without gas exsolution, and closed-system degassing with 1, 2.5 and 5% exsolved gas, for the considered trachybasaltic melt at constant temperature of 1120 °C

The concentrations of H_2O and CO_2 in the melt are displayed as a function of pressure in the graphs of Fig. 24.

The H_2O concentration of the initial melt M1 is 3.89 wt% (Table 6), whereas the computed H_2O concentration of the final melt M2 depends on the gas–melt separation model, being 3.82 wt % for open-system degassing, 3.77 wt% for closed-system degassing without gas exsolution, 3.40 wt% for closed-system degassing with 1% exsolved gas, 3.02 wt% for closed-system degassing with 2.5% exsolved gas, and 1.84 wt % for closed-system degassing with 5% exsolved gas. In other terms, the decrease in the H_2O concentration of the melt caused by degassing is small for the open-system path and the closed-system path without gas exsolution, whereas it becomes more and more important with increasing percentages of exsolved gas.

The CO_2 concentration of the initial melt M1 is 1446 mg/kg (see above), whereas the calculated CO_2 concentration of M2 varies with the gas–melt separation model, being 436 mg/kg for open-system degassing, 424 mg/kg for closed-system degassing without gas exsolution, 339 mg/kg for closed-system degassing with 1% exsolved gas, 263 mg/kg for closed-system degassing with 2.5% exsolved gas, and 92 mg/kg for closed-system degassing with 5% exsolved gas.

In spite of these considerable differences in the final state, the decrease in the CO_2 concentration of the melt triggered by gas separation is always large, since CO_2 is quickly lost by the melt because of its generally low solubility in all silicate liquids.

VolatileCalc computes also the concentrations of molecular H_2O and hydroxyl groups in rhyolitic or basaltic melts, using the model of Stolper and coworkers (Stolper 1982; Silver and Stolper 1985; Silver 1988; Silver et al. 1990). In this model, hydrous silicate melts are assumed to be ideal mixtures of water molecules, hydroxyl groups, and "free" oxygen atoms (i.e., oxygen atoms not associated with hydrogen atoms). The interaction between these three species is described by the homogeneous equilibrium reaction:

$$H_2O_{melt} + O^{2-}_{melt} = 2OH^-_{melt}, \qquad (23)$$

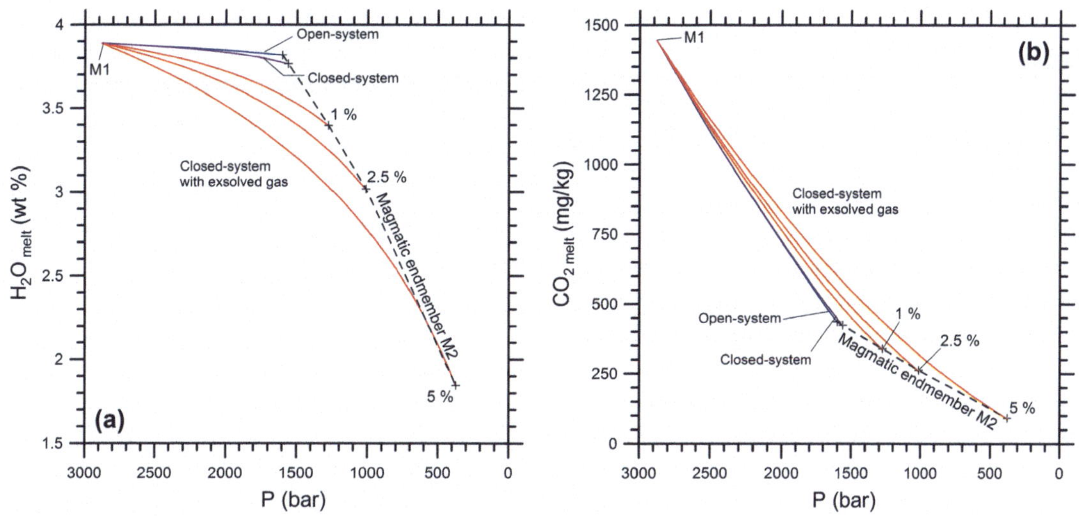

Fig. 24 Concentrations of **a** H_2O and **b** CO_2 dissolved in the considered trachybasaltic melt as a function of pressure for open-system degassing, closed-system degassing without gas exsolution, and closed-system degassing with 1, 2.5 and 5% exsolved gas, at constant temperature of 1120 °C

whose equilibrium constant can be written in terms of mole fractions, $X_{i,m}$, instead of activities, because the activity of each species is equal to its mole fraction in ideal mixtures:

$$K = \frac{(X_{OH^-_m})^2}{X_{H_2O_m} \cdot X_{O^{2-}_m}}. \quad (24)$$

Since water speciation in the melt is of utmost importance for investigating hydrogen isotope partitioning between gaseous H_2O and water dissolved in the melt (see below), the concentrations of molecular H_2O and hydroxyl groups were recalculated adopting, for the equilibrium constant of reaction (23), a value derived from the results of the experimental measurements performed by Fanara et al. (2015) for a trachybasaltic melt of the Phlegraean Fields. These authors performed *inter alias* 25 solubility experiments for this trachybasaltic melt, at pressures of 1000, 2000, 3000, 4000, and 5000 bar and temperature of 1200 °C and reported the concentrations of molecular H_2O and hydroxyl groups measured by near-infrared spectroscopy as well as bulk (total) water concentrations measured by Karl-Fischer titration after thermal dehydration, all expressed in weight units. The corresponding mole fraction of bulk water was obtained using the relation (Silver and Stolper 1985):

$$X_{H_2O,B} = \frac{C_{H_2O}/18.015}{C_{H_2O}/18.015 + (100 - C_{H_2O})/M_O}, \quad (25)$$

in which C_{H_2O} is bulk water concentration in the melt (in wt %) and M_O is defined as follows:

$$M_O = \frac{100}{\sum_i C_i/M_i}, \quad (26)$$

with C_i and M_i indicating the concentration of the i-th oxide component (in wt %), normalized to 100 wt% on a water-free basis, and its molecular weight on a single oxygen basis. The mole fractions of molecular H_2O and hydroxyl groups were computed using relations similar to Eq. (25), but involving the weight percentage of molecular H_2O or hydroxyl groups instead of C_{H_2O}. The mole fraction of "free" oxygen atoms was calculated through the mass balance (Silver et al. 1990):

$$X_{O^{2-}_m} = 1 - X_{H_2O,B} - 0.5 \cdot X_{OH^-_m}. \quad (27)$$

The equilibrium constant of reaction (23) was then computed for each solubility experiment plugging the values of $X_{OH^-_m}$, $X_{H_2O_m}$, and $X_{O^{2-}_m}$ in Eq. (24). The natural logarithm of K was regressed against bulk water concentration in the melt and an enthalpy of 32.5 kJ/mol was adopted to describe dependence of ln K on the absolute temperature reciprocal, thus obtaining the empirical relation:

$$\ln K = 0.00991 \cdot X^2_{H_2O,B} - 0.230 \cdot X_{H_2O,B} - 0.701 - 3913/T(K). \quad (28)$$

Strictly speaking, the considered enthalpy value applies to tholeiitic basalts (Behrens 2020 and references therein), but it is permissible to apply it to trachybasalts as well, also considering that possible uncertainties have limited effects due to the small difference between the temperature of the experiments performed by Fanara et al. (2015), 1200 °C, and the temperature of the isothermal degassing model, 1120 °C.

As already mentioned above, the equilibrium constant defined by Eq. (28) was used to compute the water speciation in the considered trachybasaltic melt during open-system degassing, closed-system degassing without gas exsolution, and closed-system degassing with 1, 2.5 and 5% exsolved gas, at constant temperature of 1120 ° C. In detail, Eq. (24) was rewritten as:

$$K = \frac{(X_{OH^-_m})^2}{(X_{H_2O,B} - 0.5 \cdot X_{OH^-_m}) \cdot (1 - X_{H_2O,B} - 0.5 \cdot X_{OH^-_m})}. \quad (29)$$

considering both Eq. (27) and the following mass balance (Silver et al. 1990):

$$X_{H_2O,m} = X_{H_2O,B} - 0.5 \cdot X_{OH,m}. \quad (30)$$

Then Eq. (29) was rearranged and solved for $X_{OH^-_m}$ using the quadratic formula:

$$X_{OH^-_m} = \frac{-b - \sqrt{b^2 - 4ac}}{2a} \quad (31)$$

with $a = K/4 - 1$, $b = -K/2$, $c = K \cdot (X_{H_2O,B} - X^2_{H_2O,B})$. The computed value of $X_{OH^-_m}$ was then inserted in Eqs. (30) and (27) to calculate $X_{H_2O_m}$, and $X_{O_m{}^{2-}}$, respectively. For completeness, it is convenient to define also the parameter $Y_{H_2O,m}$, indicating the mass fraction of molecular water over the total amount of water dissolved in the melt. Based on Eq. (30), this parameter can be expressed as follows:

$$Y_{H_2O,m} = X_{H_2O_m}/X_{H_2O,B} = 1 - (0.5 \cdot X_{OH_m}/X_{H_2O,B}). \quad (32)$$

The computed mole fractions of molecular H_2O and OH^- groups in the trachybasaltic melt of interest during progressive isothermal degassing at 1120 °C are shown in the diagram of Fig. 25 as function of the bulk H_2O concentration in the melt.

The results of closed-system degassing with 5% exsolved gas distribute over the whole range of bulk H_2O concentration, the outcomes of the open-system degassing model cover the smallest interval of bulk H_2O concentration, whereas the results of closed-system degassing without gas exsolution and closed-system degassing with 1 and 2.5% exsolved gas spread over intermediate ranges of bulk H_2O concentration. However, the computed mole fractions of molecular H_2O and OH groups for all these gas–melt separation models overlap each other, indicating that water speciation in the melt is the same for all the models, as it depends only on bulk H_2O concentration in the melt. Moreover, the mole fraction of molecular water is significantly higher than that of the hydroxyl groups at the initial H_2O concentration, 3.89 wt%, but the former decreases much faster than the latter during progressive degassing so that the mole fraction of molecular H_2O approaches that of the OH groups at the end of closed-system degassing with 5% exsolved gas, causing a concurrent decrease in the $Y_{H_2O,m}$ parameter from 0.813 to 0.696. As shown below, these different variations in water speciation and in the $Y_{H_2O,m}$ parameter, along the distinct degassing paths of interest, determine considerable differences in the partitioning of

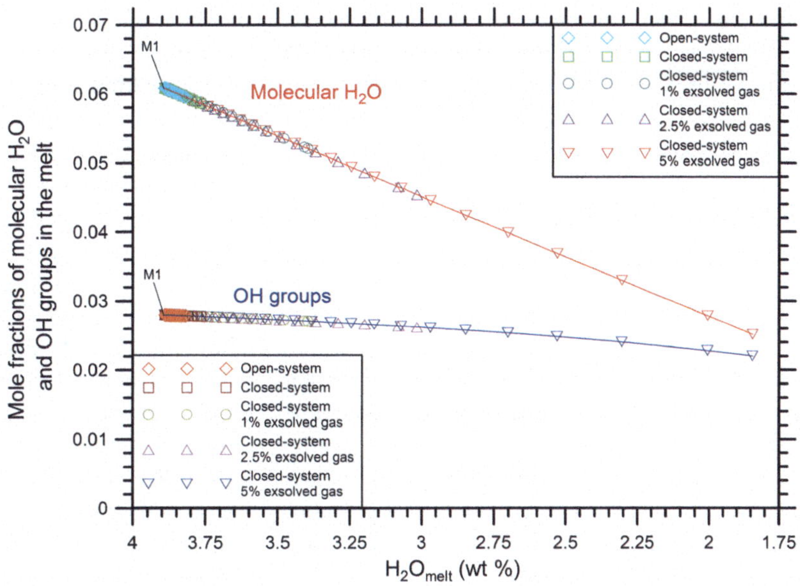

Fig. 25 Concentrations of molecular water and hydroxyl groups in the trachybasaltic magma of interest as a function of bulk H_2O concentration in the melt for open-system degassing, closed-system degassing without gas exsolution, and closed-system degassing with 1, 2.5 and 5% exsolved gas, at constant temperature of 1120 °C

hydrogen isotopes between the melt and the separated gas phase.

The hydrogen isotope fractionation between gaseous water and bulk water dissolved in the melt, $10^3 \cdot \ln \alpha_{D,H_2O,gas-H_2O,B,m}$, is given by the following isotopic balance (Holloway and Blank 1994):

$$10^3 \cdot \ln \alpha_{D,H_2O,gas-H_2O,B,m} = Y_{H_2O,m} \cdot 10^3 \cdot \ln \alpha_{D,H_2O,gas-H_2O,m} + \left(1 - Y_{H_2O,m}\right) \cdot 10^3 \cdot \ln \alpha_{D,H_2O,gas-OH,m} \quad (33)$$

involving the fractionation factors between gaseous water and molecular H_2O in the melt, $10^3 \cdot \ln \alpha_{D,H_2O,gas-H_2O,m}$ as well as between gaseous water and OH groups in the melt, $10^3 \cdot \ln \alpha_{D,H_2O,gas-OH,m}$, the temperature dependence of which is expressed by the equations (see discussion in De Hoog et al. 2009):

$$10^3 \cdot \ln \alpha_{D,H_2O,gas-H_2O,m} = -29.9 + 16.7 \cdot 10^6/T^2 \quad (34)$$

$$10^3 \cdot \ln \alpha_{D,H_2O,gas-OH,m} = 35.0 + 16.7 \cdot 10^6/T^2. \quad (35)$$

Since the degassing of the considered trachybasaltic magma is assumed to take place at constant temperature of 1120 °C, the fractionation factors between exsolved gaseous water and both molecular water and the hydroxyl groups in the melt, given by Eqs. (34) and (35), respectively, are both constant, namely −21.3 ‰ and + 43.6 ‰, respectively. In contrast, the hydrogen isotope fractionation factor between exsolved gaseous water and bulk water dissolved in the melt varies continuously during degassing because the mole fractions of the hydroxyl groups and molecular water as well as the $Y_{H_2O,m}$ parameter are a function of the total amount of water dissolved in the melt. Nevertheless, also $10^3 \cdot \ln \alpha_{D,H_2O,gas-H_2O,B,m}$ exhibits small variations, even for closed-system degassing with 5% exsolved gas, from + 41.8 ‰ to + 42.2 ‰.

For closed-system gas–melt separation, the δD value of bulk water dissolved in the melt after removal of a mass fraction $(1-F_{H_2O})$ of gaseous water is given by the relation (Taylor 1986):

$$\delta D_{H_2O,B,m} = \delta D_{i,H_2O,B,m} - 10^3 \cdot \ln \alpha_{D,H_2O,gas-H_2O,B,m} \cdot \left(1 - F_{H_2O}\right) \quad (36)$$

where $\delta D_{i,H_2O,B,m}$ is the initial δD value of bulk water dissolved in the melt, that is, before degassing begins. For open-system, Rayleigh-type gas–melt separation, in which the gas is immediately removed from the melt, the δD value of bulk water dissolved in the melt varies as a function of F_{H_2O} according to the equation (Taylor 1986):

$$\delta D_{H_2O,B,m} = \delta D_{i,H_2O,B,m} - \left(1000 + \delta D_{i,H_2O,B,m}\right) \cdot \left(1 - F_{H_2O}^{\alpha_{D,H_2O,gas-H_2O,B,m} - 1}\right). \quad (37)$$

Closed-system gas–melt separation is often assumed to describe non-eruptive magmatic degassing, but open-system degassing of the trachybasaltic magma presumably present below the Solfatara volcano was invoked by Caliro et al. (2014) to explain the variations in the gas chemistry of local fumaroles.

Owing to the large fractionations of hydrogen isotopes, the initial δD value of bulk water dissolved in the melt cannot be calculated as a difference of delta values, but must be computed using the equation:

$$\delta D_{i,H_2O,B,m} = \frac{\delta D_{i,H_2O,gas} + 1000}{\alpha_{D,H_2O,gas-H_2O,B,m}} - 1000 \quad (38)$$

in which $\delta D_{i,H_2O,gas}$ is assumed equal to −20 ‰, following Caliro et al. (2007), who considered the groundbreaking results of Giggenbach (1992b). For the same token, the δD value of the gaseous water during the degassing process is obtained utilizing the corresponding relation:

$$\delta D_{H_2O,gas} = \alpha_{D,H_2O,gas-H_2O,B,m} \cdot \left(\delta D_{H_2O,B,m} + 1000\right) - 1000 \quad (39)$$

The computed δD values of both exsolved gaseous water and bulk water dissolved in the melt along different isothermal degassing paths of the considered trachybasaltic magma are contrasted with the bulk water concentration in the melt in Fig. 26. Again, although the considered gas–melt separation models span distinct ranges of bulk water concentration in the melt, the δD values of exsolved gaseous water for the different degassing models are superimposed onto each other and the δD values of bulk water dissolved in the melt for the distinct degassing simulations overlap. This similarity in the results of the different degassing models is not surprising because the gas–melt fractionation of hydrogen isotopes depends on water speciation in the melt which, in turn, is a function of its bulk H_2O concentration (see above).

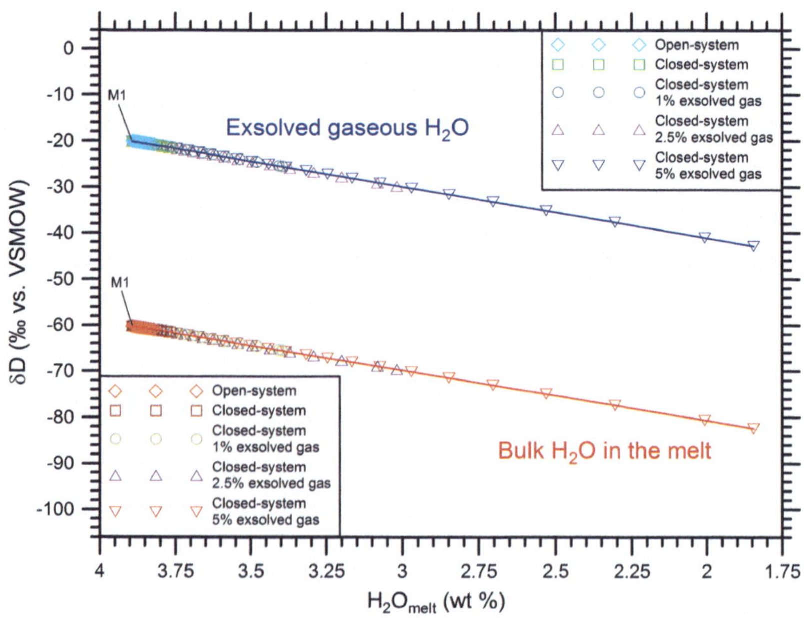

Fig. 26 δD values of exsolved gaseous water and bulk water dissolved in the trachybasaltic magma of interest as a function of bulk H_2O concentration in the melt, for open-system degassing, closed-system degassing without gas exsolution, and closed-system degassing with 1, 2.5 and 5% exsolved gas, at constant temperature of 1120 °C

Moreover, the δD values of exsolved gaseous water exhibits a small decrease during open-system degassing and closed-system degassing without gas exsolution, −0.8 and −1.4 ‰ units, respectively, whereas it experiences a progressively higher drop during closed-system degassing with 1, 2.5, and 5% exsolved gas, namely −5.5, −9.7, and −22.8 ‰ units, respectively. Similarly, the δD value of bulk water dissolved in the melt undergoes a small reduction during open-system degassing and closed-system degassing without gas exsolution, −0.8 and −1.3 ‰ units, respectively, whereas it suffers a gradually higher descent during closed-system degassing with 1, 2.5, and 5% exsolved gas, namely −5.3, −9.4, and −22.2 ‰ units, respectively. Summing up, a significant decrease in the δD values of both exsolved gaseous water and bulk water dissolved in the melt requires the presence of exsolved gas in equilibrium with the melt, otherwise the drop in the δD values of both exsolved gaseous water and bulk water dissolved in the melt is insignificant for both closed-system and open-system gas–melt separation.

It has been known since a long time that loss of gases from granites (Nabelek et al. 1983) and rhyolites (Taylor et al. 1983) at magmatic temperature depletes gradually the hydrous silicate melt in water and deuterium without significantly affecting the oxygen isotope composition. More recently, Mandeville et al. (2009) investigating the rhyodacitic and andesitic magmas of Mount Mazama–Crater Lake (Oregon), asserted that "*complete degassing of the volatiles H_2O and CO_2 produces no change in $\delta^{18}O$ of the melt (glass) due to the overwhelming concentration of oxygen in the silicate magma*".

To get mileage out of these qualitative considerations, modeling of the degassing process was attempted to evaluate the effects of magma degassing on gas–melt partitioning of oxygen isotopes, although it is impossible to take into account: (i) the fractionation factor between the exsolved gaseous water and the hydroxyl groups in the melt as well as the fractionation factor between the exsolved gaseous water and molecular water of the melt, as done for the hydrogen isotopes (see above); (ii) the fractionation factor between exsolved gaseous CO_2 and the carbonate groups in the melt as well as the fractionation factor between exsolved gaseous CO_2 and the molecular CO_2 of the melt.

However, it is possible to use suitable values of the H_2O-melt and CO_2-melt fractionation factors and relations similar to Eqs. (36) and (37), for closed- and open-system gas–melt separation, respectively. Thus, for closed-system gas–melt separation, the $\delta^{18}O$ value of the melt after removal of a mass fraction ($1-F_{gas}$) of gas ($H_2O + CO_2$) is described by the relation (Taylor 1986):

$$\delta^{18}O_{melt} = \delta^{18}O_{i,melt} - 10^3 \cdot \ln \alpha_{^{18}O,gas-melt} \cdot \left(1 - F_{gas}\right), \tag{40}$$

where $\delta^{18}O_{i,melt}$ is the initial $\delta^{18}O$ value of the melt, that is, before degassing begins. For Rayleigh-type open-system, gas–melt separation, in which the gas is instantaneously removed from the melt, the $\delta^{18}O$ value of the melt varies as a function of F_{gas} according to the equation (Taylor 1986):

$$\delta^{18}O_{melt} = \delta^{18}O_{i,melt} - \left(1000 + \delta^{18}O_{i,melt}\right) \cdot \left(1 - F^{\alpha_{^{18}O,gas-melt} - 1}\right). \tag{41}$$

The H_2O-melt fractionation factor was calculated using the following relation:

$$10^3 \cdot \ln \alpha_{^{18}O,H_2O-melt} = -4.235 \cdot 10^6/T^2 + 7.912 \cdot 10^3/T - 2.54. \tag{42}$$

Equation (42) was obtained by Zhao and Zheng (2003) for a basalt of normalized chemical composition. It describes the temperature dependence of the water-basalt oxygen isotope fractionation factor and can be applied to melts of basaltic composition because of the negligible oxygen isotope fractionation between rock and melt of the same compositions (see discussion in Zhao and Zheng 2003).

The CO_2-melt fractionation factor was computed by means of the equation:

$$10^3 \cdot \ln \alpha_{^{18}O,CO_2-melt} = 3.6635 \cdot 10^6/T^2 - 0.4965 \cdot \left(10^6/T^2\right)^2 + 0.000590 \cdot \left(10^6/T^2\right)^3. \quad (43)$$

Equation (43) was obtained combining the reduced partition function ratios for CO_2 (from Chacko et al. 1991) and that for the basalt melt (which was derived by Appora et al. 2003 based on the experimental outcomes of Muehlenbachs and Kushiro 1974). The result was then divided by two to express it on a single oxygen atom basis.

The H_2O-melt and CO_2-melt oxygen isotope fractionation factors given by Eqs. (42) and (43), respectively, were then merged through the relation:

$$10^3 \cdot \ln \alpha_{^{18}O,gas-melt} = Y_{CO_2,m} \cdot 10^3 \cdot \ln \alpha_{^{18}O,CO_2-melt} + \left(1 - Y_{CO_2,m}\right) \cdot 10^3 \cdot \ln \alpha_{^{18}O,H_2O-melt}, \quad (44)$$

in which $Y_{CO_2,m} = 2 \cdot n_{CO_2,B}/\left(2 \cdot n_{CO_2,B} + n_{H_2O,B}\right)$ with $n_{CO_2,B}$ and $n_{H_2O,B}$ indicating the moles of bulk CO_2 and bulk H_2O dissolved in the melt, respectively.

At the considered temperature of 1120 °C, the H_2O-melt and the CO_2-melt oxygen isotope fractionation factors are 0.957 and 1.756. The range spanned by Y_{CO_2} is: (i) from 0.030 to 0.0093 for open-system gas–melt separation, (ii) from 0.030 to 0.0091 for closed-system gas–melt separation without exsolved gas, (iii) from 0.030 to 0.0081 for closed-system gas–melt separation with 1% exsolved gas, (iv) from 0.030 to 0.0071 for closed-system gas–melt separation with 2.5% exsolved gas, and (v) from 0.030 to 0.0041 for closed-system gas–melt separation with 5% exsolved gas. Therefore, the gas–melt oxygen isotope fractionation factor is not significantly different from the H_2O-melt oxygen isotope fractionation factor, irrespective of the type and extent of gas–melt separation, as it varies (i) from 0.981 to 0.965 for open-system gas–melt separation, (ii) from 0.981 to 0.964 for both closed-system gas–melt separation without exsolved gas and closed-system gas–melt separation with 1% exsolved gas, (iii) from 0.981 to 0.963 for closed-system gas–melt separation with 2.5% exsolved gas, and (iv) from 0.981 to 0.960 for closed-system gas–melt separation with 5% exsolved gas.

Upon computing the $\delta^{18}O$ value of the melt, the following relation, similar to Eq. (39), was utilized to calculate the $\delta^{18}O$ value of the exsolved gas (H_2O + CO_2) throughout the degassing process, including the initial condition:

$$\delta^{18}O_{gas} = \alpha_{^{18}O,gas-melt} \cdot \left(\delta^{18}O_{melt} + 1000\right) - 1000. \quad (45)$$

In addition to the limitations related to the modeling of gas–melt oxygen isotope fractionation during degassing, some uncertainties are also associated with the choice of the initial $\delta^{18}O$ value of the melt. According to Caliro et al. (2007) the magmatic fluids of the Solfatara have an estimated $\delta^{18}O$ value of + 10.5 ± 0.3 ‰ and are in isotopic equilibrium with a melt with $\delta^{18}O$ value of + 8.20 ‰, which is typical of the trachytic volcanic product of the Phlegraean Fields area based on the early studies of Turi and Taylor (1976) and Turi et al. (1991). According to the recent study of Iovine et al. (2018), the $\delta^{18}O$ value of the melt in isotopic equilibrium with the analyzed feldspar, pyroxene, and olivine crystals ranges from +6.95 to +8.80 ‰ for the volcanic products of the Campi Flegrei. Therefore, the intermediate value, + 7.88 ‰, was assumed to be representative of the initial (undegassed) melt in a first series of simulations, whereas the effects of the uncertainty of ± 0.93 ‰ units in the $\delta^{18}O$ value of the initial melt were considered afterwards.

The $\delta^{18}O$ values of both the melt and the gas phase (H_2O + CO_2) separated along distinct isothermal degassing paths of the trachybasaltic magma of interest are plotted against the bulk gas concentration in the melt in Fig. 27. Incidentally,

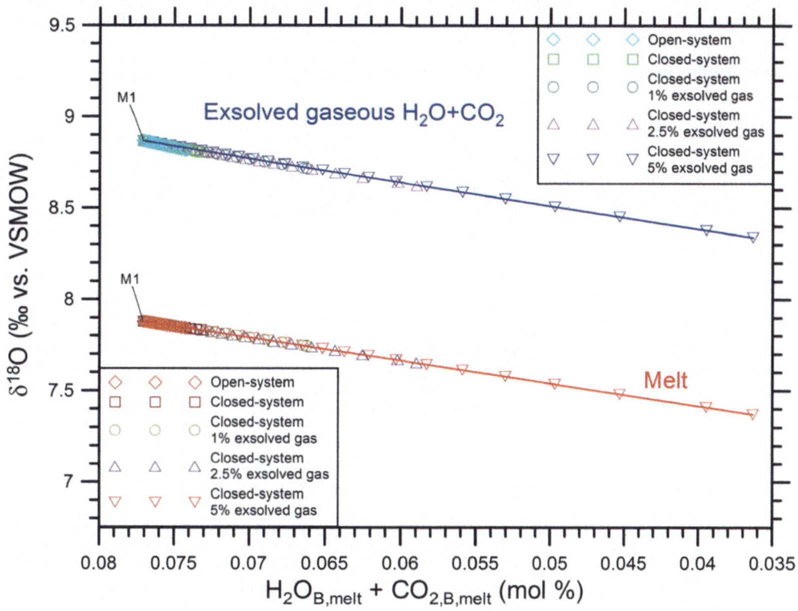

Fig. 27 δ^{18}O values of the trachybasaltic melt and exsolved gas phase ($H_2O + CO_2$) as a function of the bulk gas concentration in the melt, for open-system degassing, closed-system degassing without gas exsolution, and closed-system degassing with 1, 2.5 and 5% exsolved gas, at constant temperature of 1120 °C

the bulk gas concentration in the melt was computed using the following relation:

$$X_{gas,B} = \frac{C_{H_2O}/18.015 + C_{CO_2}/22.005}{C_{H_2O}/18.015 + C_{CO_2}/22.005 + (100 - C_{H_2O} - C_{CO2})/M_O}, \quad (46)$$

which is an extended version of Eq. (25) to include CO_2 in addition to H_2O.

Although the considered gas–melt degassing simulations extend over different intervals of bulk gas concentration in the melt, the δ^{18}O values of the melt for the distinct degassing simulations are superimposed onto each other and the δ^{18}O values of the exsolved gas for the different degassing models overlap onto each other as well or nearly so. This is the same of what was observed above for the δD values of exsolved gaseous water and the bulk water dissolved in the melt.

Furthermore, the δ^{18}O value of the exsolved gas phase experiences a small descent during open-system degassing and closed-system degassing without gas exsolution, −0.053 and −0.065 ‰ units, respectively, whereas it exhibits a gradually greater decrease during closed-system degassing with 1, 2.5, and 5% exsolved gas, namely −0.15, −0.24, and −0.53 ‰ units, respectively. Likewise, the δ^{18}O value of the melt suffers a small drop during open-system degassing and closed-system degassing without gas exsolution, −0.037 and −0.048 ‰ units, respectively, whereas it undergoes an increasingly higher reduction during closed-system degassing with 1, 2.5, and 5% exsolved gas, namely −0.14, −0.23, and −0.51 ‰ units, respectively. All in all, a significant decrease in the δ^{18}O values of both the exsolved gas and melt calls for the occurrence of exsolved gas in equilibrium with the melt, otherwise the drop in the δ^{18}O values of both the exsolved gas and melt is very small for both closed-system and open-system gas–melt separation.

Simulations were repeated for closed-system degassing with 5% exsolved gas only, considering the two extreme δ^{18}O values, +6.95 and +8.80 ‰, for the initial (undegassed) melt instead of the intermediate value, +7.88 ‰, assumed in previous modeling exercises. The results of these simulations, shown in Fig. 28, indicate that the uncertainty of ±0.93 ‰ units in the δ^{18}O value of the initial melt determines exactly the same uncertainty in the δ^{18}O values of the melt and exsolved gas phase ($H_2O + CO_2$) along the whole melt degassing path.

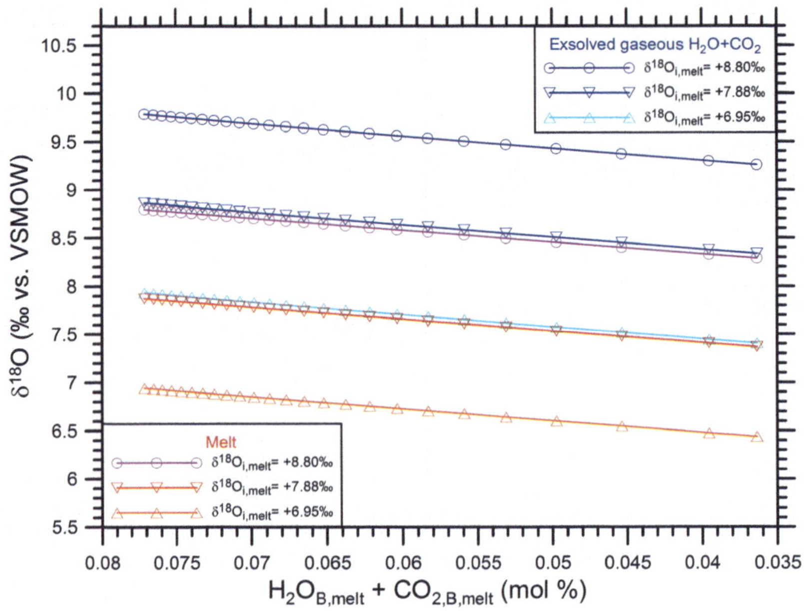

Fig. 28 $\delta^{18}O$ values of the trachybasaltic melt and exsolved gas phase ($H_2O + CO_2$) as a function of the bulk gas concentration in the melt, for closed-system degassing with 5% exsolved gas, at constant temperature of 1120 °C and different $\delta^{18}O$ value of the initial melt (see legend)

All in all, in spite of the considerable uncertainties in the computed $\delta^{18}O$ values for the gases exsolved from the magma, they can be merged with the corresponding δD values, to obtain a suitable term of comparison for the analytical data of the Solfatara fluids, as discussed in Sect. 10.4.

10.3 Recycling of Condensed Steam

Entrainment of secondary steam (i.e., steam produced through complete evaporation of down-flowing condensed steam) into the up-flowing fumarolic fluids appears to be a likely process, by analogy with what was recognized in the vapor-dominated geothermal reservoirs of Larderello in Italy, The Geysers in California (USA), Matsukawa in Japan, and both Darajat and Kamojang in Indonesia. A short digression on this topic is needed.

The vapor-dominated geothermal reservoirs are characterized by very small vertical gradients of temperatures and pressures in their natural (pre-exploitation) state. In fact, temperatures of 240 ± 5 °C are found over thicknesses of several hundred meters to 2 km, while pressures are equal to the water saturation pressures at the corresponding temperatures within measurement error. According to White et al. (1971) and Truesdell and White (1973), the following two important consequences can be drawn from these observations: (1) The vapor-dominated geothermal reservoir must be sealed laterally, otherwise it would be inundated by groundwaters coming from the neighboring aquifers with pressure profile following the hydrostatic gradient. (2) Liquid water must be present all over the vapor-dominated reservoir to explain the vapor pressure–temperature relation for saturated water.

White et al. (1971) and Truesdell and White (1973) developed a conceptual model of vapor-dominated systems explaining one of the most perplexing features of vapor-dominated reservoirs, that is, the occurrence of a considerable upward heat flux, in spite of the very small vertical temperature gradient. These authors invoked a mechanism known as "heat pipe" (Eastman, 1968). Accordingly, in the central zone of a vapor-dominated reservoir, both steam and liquid water are mobile, in that the steam flows upward, condenses at shallow depth, and the steam condensate flows downward. Since the latent heat released in steam condensation is very large, the vapor–liquid counterflow can sustain great heat fluxes with small to very small net mass transport.

The conceptual model developed by White and coworkers provides a qualitative description of how vapor-dominated systems work. Nevertheless, some aspects were not clarified, such as the amount and distribution of water stored in the geothermal reservoir. This is a feature of great practical importance for the use of geothermal resources, which is directly linked to the fluid reserves in place. Although liquid water has been directly found only in the condensation zones at the reservoir margins, mass balance considerations suggest that very high quantities of liquid water have to be present, under natural-state conditions in vapor-dominated reservoirs, to account for the mass of the produced fluids, otherwise an unreasonably large reservoir thickness would be needed assuming that the extracted fluids were stored in the form of steam (e.g., James 1968; Pruess 1985). Furthermore, gas chemistry indicates that up to 99% of the fluids extracted at The Geysers originated from boiling of the liquid phase hosted in the geothermal reservoir (D'Amore and Truesdell 1979).

Finally, based on numerical modeling studies, Pruess (1985) proposed that vapor-dominated geothermal reservoirs are characterized by dual permeability, where a high-permeability network of interconnected fractures with small volume provides pressure control in the system, while most of the fluid reserves are hosted in liquid form in a porous low-permeability rock matrix.

Some features of the Solfatara magmatic-hydrothermal system appear to be similar to those of vapor-dominated geothermal reservoirs, such as the occurrence of a heat-pipe mechanism at shallow depths. In fact, adopting the approach of Chiodini et al. (2001), the diffuse CO_2 flux from soil measured in 1998–2013 (Cardellini et al 2017) together with the H_2O/CO_2 weight ratio of the Bocca Grande fumarolic fluids in the same period suggest that a considerable amount of steam, 1860 ± 530 t/d on average, is condensed and flows downward. Then, part of it is evaporated and entrained in the upward moving flow of fluids below the Solfatara.

The analytical δD and $\delta^{18}O$ values of fumarolic steam which are available for 81 samples from several studies (Nuti et al. 1985; Bolognesi et al. 1986; Panichi and Volpi 1999; Chiodini et al. 1996; Caliro et al. 2007) are representative of the steam condensate produced through total condensation, whereas the δD and $\delta^{18}O$ values of condensed steam span a range extending from the condition of incipient (equilibrium) condensation to the condition of total condensation, depending on the extent of the steam condensation process. Therefore, it is instructive to calculate the isotope composition of the initial steam condensate. The exchange of oxygen isotopes between CO_2 and H_2O can be neglected in these calculations because most CO_2 remains in the gas phase and, consequently, Y_{CO_2} has a very low value, 0.0004. This is because condensed steam, at an assumed maximum P_{CO_2} value of 1 bar and 100 °C, has a computed CO_2 concentration of 11.1 mmol/kg and, consequently, $Y_{CO_2} = 2 \cdot 0.0111/(2 \cdot 0.0111 + 1000/18.015) = 0.0004$.[4]

At the very beginning of the condensation process, the δD and $\delta^{18}O$ values of condensed steam, $\delta_{CD,i}$, are related to the isotope values of a given fumarolic steam sample, δ_{FS}, by the simple relation:

$$\delta_{CDi} = 1000 \ln \alpha_{L-V} + \delta_{FS} \qquad (47)$$

where 1000 ln $\alpha_{L\text{-}V}$ is the liquid–vapor equilibrium fractionation factor of pure water at the condensation temperature. Assuming that steam condensation occurs at 100 °C, 1000 ln $\alpha_{L\text{-}V}$ is equal to +27.9 ‰ for deuterium and +5.08 ‰ for oxygen-18 (Horita and Wesolowski 1994). For oxygen-18:

(i) First, the $\delta^{18}O$ value of gaseous water at 100 °C is computed using Eq. (18) solved for $\delta^{18}O_{H_2O}$, considering the $CO_{2(g)}$-$H_2O_{(g)}$

[4] Furthermore, condensed steam, at P_{CO_2} of 1 bar and 100 °C, has negligible concentrations of HCO_3^- and CO_3^{2-} ions, 0.0678 and 8.54E-8 mmol/kg, respectively, since pH is 4.173. These results were obtained by using PHREEQC.

fractionation factor at 100 °C, + 33.21‰, which is calculated by means of Eq. (19). (ii) Second, inserting the computed $\delta^{18}O$ values of gaseous water at 100 °C into Eq. (47), it turns out that the $\delta^{18}O_{H_2O}$ values of the incipient steam condensates vary from −0.71 to +3.12 ‰, with a mean of +1.63‰, a median of + 1.74‰, and a standard deviation of 0.76‰.

For deuterium, the analytical δD values of fumarolic steam is directly introduced into Eq. (47), obtaining for the incipient steam condensates a range from −6.90 to +5.79 ‰, a mean of +0.31‰, a median of +0.20‰, and a standard deviation of 2.45‰.

10.4 Interpretation of the δD and $\delta^{18}O$ Values Available for the Solfatara Fumarolic Fluids

The analytical δD values of the 81 available fumarolic samples from several studies (Nuti et al. 1985; Bolognesi et al. 1986; Panichi and Volpi 1999; Chiodini et al. 1996; Caliro et al. 2007) and the corresponding $\delta^{18}O_{H_2O+CO_2}$ values computed by means of Eq. (18) at outlet temperature (see Sect. 10.1) are contrasted in the correlation diagram of Fig. 29 also showing:

(i) the δD and $\delta^{18}O_{H_2O+CO_2}$ values of magmatic fluids generated through closed-system degassing of a magma with $\delta^{18}O$ values of +6.95, +7.88, and +8.80 ‰, assuming the presence of 5% exsolved gas and a constant temperature of 1120 °C, (see Sect. 10.2), and
(ii) the δD and $\delta^{18}O$ values of condensed steam produced through both incipient and total condensation (see Sect. 10.3).

The local meteoric water line, with slope 8 and deuterium excess +15, and the isotope values of local meteoric water, namely $\delta D = -33.8$ ‰ and $\delta^{18}O = -6.1$ ‰ (Valentino et al. 1999 and references therein), are also represented in the correlation diagram of Fig. 29.

The diagram clearly shows that the fumarolic fluids are positioned close to the magma degassing lines and far from the condensed steam area and the point representative of the local meteoric water. Although the uncertainties on the isotope values of magmatic fluids and the large possible variations in the isotope values of condensed steam (and consequently of secondary steam, that

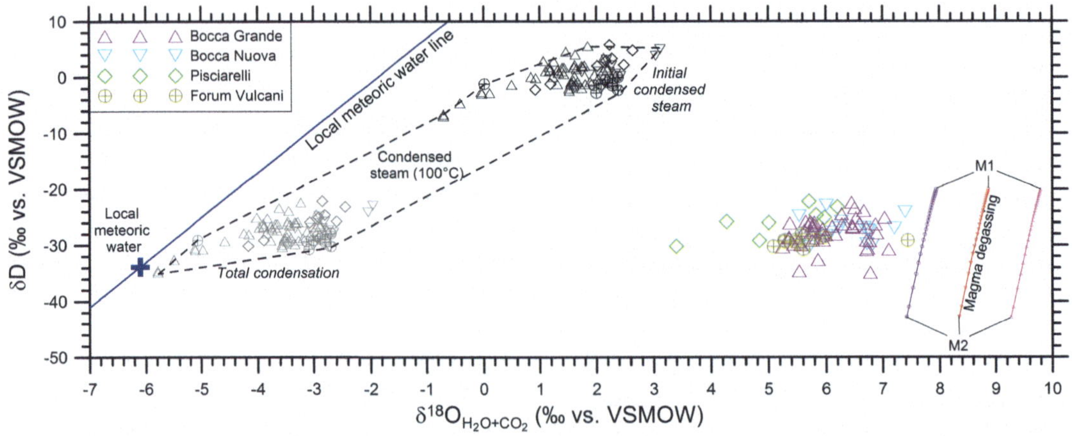

Fig. 29 Correlation diagram of the δD versus $\delta^{18}O_{H_2O+CO_2}$ values for the 81 available samples of Solfatara fumarolic fluids from several studies (see text), also showing the isotope values of fluids produced through magma degassing, steam condensed at 100 °C (black and grey symbols refer to incipient and total condensation, respectively, whereas the symbol shape identifies the fumarolic vent, see legend), and local meteoric water, as well as the local meteoric water line

is, the steam generated through total evaporation of down-flowing condensed steam) do not allow one to compute the relative proportions of the different endmembers involved in the Solfatara fumarolic fluids, there is little doubt that the water released by magmatic degassing prevails by far over the water coming from other sources. This conclusion might seem at variance with the results of the degassing model of Caliro et al. (2014), which predicts dominant contributions of the hydrothermal component and subordinate contributions of the magmatic components. However, this discrepancy is not surprising because the degassing model of Caliro et al. (2014) is constrained by the mass balances involving gas components (i.e., N_2, CO_2, and He) and, therefore, it provides indications on the provenance of these gases, whereas previous conclusions are based on the δD values of H_2O and the $\delta^{18}O$ values of H_2O (and CO_2) and, consequently, they refer to the provenance of H_2O (and CO_2).

References

Allard P, Maiorani A, Tedesco D, Cortecci G, Turi B (1991) Isotopic study of the origin of sulfur and carbon in Solfatara fumaroles, Campi Flegrei caldera. J Volcanol Geoth Res 48(1–2):139–159

Appora I, Eiler JM, Matthews A, Stolper EM (2003) Experimental determination of oxygen isotope fractionations between CO_2 vapor and soda-melilite melt. Geochim Cosmochim Acta 67(3):459–471

Behrens H (2020) Water speciation in oxide glasses and melts. Chem Geol 558:119850

Bellucci F (1994) Nuove conoscenze stratigrafiche sui depositi vulcanici del sottosuolo del settore meridionale della Piana Campana. Bollettino Della Società Geologica Italiana 113(2):395–420

Berndt ME, Allen DE, Seyfried WE Jr (1996) Reduction of CO2 during serpentinization of olivine at 300°C and 500 bar. Geology 24(4):351–354

Bolognesi L, Noto P, Nuti S (1986) Studio chimico ed isotopico della solfatara di Pozzuoli: ipotesi sull'origine e sulle temperature profonde dei fluidi. Rendiconti della Società Italiana di Mineralogia e Petrologia 41 (2):281–295

Boucher C, Marty B, Zimmermann L, Langenfelds R (2018) Atmospheric helium isotopic ratio from 1910 to 2016 recorded in stainless steel containers. Geochem Perspect Lett 6:23–27

Brombach T, Caliro S, Chiodini G, Fiebig J, Hunziker JC, Raco B (2003) Geochemical evidence for mixing of magmatic fluids with seawater, Nisyros hydrothermal system Greece. Bull Volcanol 65(7):505–516

Caliro S, Chiodini G, Moretti R, Avino R, Granieri D, Russo M, Fiebig J (2007) The origin of the fumaroles of La Solfatara (Campi Flegrei, south Italy). Geochim Cosmochim Acta 71(12):3040–3055

Caliro S, Chiodini G, Paonita A (2014) Geochemical evidences of magma dynamics at Campi Flegrei (Italy). Geochim Cosmochim Acta 132:1–15

Cannatelli C, Lima A, Bodnar RJ, De Vivo B, Webster JD, Fedele L (2007) Geochemistry of melt inclusions from the Fondo Riccio and Minopoli 1 eruptions at Campi Flegrei (Italy). Chem Geol 237(3–4):418–432

Carapezza M, Nuccio M, Valenza M (1984) Geochemical surveillance of the Solfatara of Pozzuoli (Phlegrean Fields) during 1983. Bull Volcanol 47(2):303–311

Cardellini C, Chiodini G, Frondini F, Avino R, Bagnato E, Caliro S, Lelli M, Rosiello A (2017) Monitoring diffuse volcanic degassing during volcanic unrests: the case of Campi Flegrei (Italy). Sci Rep 7(1):6757. https://doi.org/10.1038/s41598-017-06941-2

Chacko T, Mayeda TK, Clayton RN, Goldsmith JR (1991) Oxygen and carbon isotope fractionations between CO_2 and calcite. Geochim Cosmochim Acta 55(10):2867–2882

Chelini W, Sbrana A (1987) Subsurface geology. In: Rosi M, Sbrana A (eds) Phlegrean Fields, Progetto Finalizzato 'Geodinamica', Monografie Finali, vol 9. Quaderni de "La Ricerca Scientifica", AGIP-ENEL Joint-Venture for Geothermal Research and Progetto Finalizzato 'Geodinamica'–CNR 9 Consiglio Nazionale delle Ricerche Roma, pp 94-103

Cheminée J, Letolle R, Olive P (1968) Premiéres données isotopiques sur des fumerolles de volcans italiens. Bull Volcanol 32(3):467–475

Chiodini G (2009) CO_2/CH_4 ratio in fumaroles a powerful tool to detect magma degassing episodes atn quiescent volcanoes. Geophys Res Lett 36(2):L02302. https://doi.org/10.1029/2008GL036347

Chiodini G, Marini L (1998) Hydrothermal gas equilibria: The H_2O-H_2-CO_2-CO-CH_4 system. Geochim Cosmochim Acta 62:2673–2687

Chiodini G, Cioni R, Magro G, Marini L, Panichi C, Raco B, Russo M (1996) Gas and water geochemistry: chemical and isotopic variations of Bocca Grande fumarole (Solfatara volcano, Phlegrean Fields). Acta Vulcanol 8:228–232

Chiodini G, Allard P, Caliro S, Parello F (2000) ^{18}O exchange between steam and carbon dioxide in volcanic and hydrothermal gases: implications for the source of water. Geochim Cosmochim Acta 64 (14):2479–2488

Chiodini G, Frondini F, Cardellini C, Granieri D, Marini L, Ventura G (2001) CO_2 degassing and energy release at Solfatara volcano, Campi Flegrei, Italy. J Geophys Res 106:16213–16222

Chiodini G, Caliro S, Cardellini C, Granieri D, Avino R, Baldini A, Donnini M, Minopoli C (2010) Long-term variations of the Campi Flegrei, Italy, volcanic system as revealed by the monitoring of hydrothermal activity. J Geophys Res Solid Earth 115(B3):B03205. https://doi.org/10.1029/2008JB006258

Chiodini G, Vandemeulebrouck J, Caliro S, D'Auria L, De Martino P, Mangiacapra A, Petrillo Z (2015) Evidence of thermal-driven processes triggering the 2005–2014 unrest at Campi Flegrei caldera. Earth Planet Sci Lett 414:58–67

Chiodini G, Avino R, Caliro S, Minopoli C (2011) Temperature and pressure gas geoindicators at the Solfatara fumaroles (Campi Flegrei). Ann Geophys 54 (2). https://doi.org/10.4401/ag-5002

Chiodini G, Caliro S, Avino R, Bini G, Giudicepietro F, De Cesare W, Ricciolino P, Aiuppa A, Cardellini C, Petrillo Z, Selva J, Siniscalchi A, Tripaldi S (2021) Hydrothermal pressure-temperature control on CO_2 emissions and seismicity at Campi Flegrei (Italy). J Volcanol Geoth Res 414:107245

Cioni R, Corazza E, Marini L (1984) The gas/steam ratio as indicator of heat transfer at the Solfatara fumaroles, Phlegraean Fields (Italy). Bulletin of Volcanology 47 (2):295–302.

Cioni R, Corazza E, Fratta M, Guidi M, Magro G, Marini L (1989) Geochemical precursors at Solfatara Volcano, Pozzuoli (Italy). In: Latter JH (ed) Volcanic hazards. IAVCEI proceedings in volcanology, vol 1. Springer, Berlin, Heidelberg, pp 384–398

D'Amore F, Truesdell AH (1979) Models for steam chemistry at Larderello and The Geysers. In: Proceedings, 5th annual workshop on geothermal reservoir engineering, Stanford University, Stanford CA, pp 283–297

Davis JC (2002) Statistics and data analysis in geology, 3rd edn. Wiley, New York, p 656

De Hoog JC, Taylor BE, Van Bergen MJ (2009) Hydrogen-isotope systematics in degassing basaltic magma and application to Indonesian arc basalts. Chem Geol 266(3–4):256–266

Dixon JE (1997) Degassing of alkalic basalts. Am Miner 82(3–4):368–378

Eastman GY (1968) The heat pipe. Scientific American 218(5):38–46.

Fanara S, Botcharnikov RE, Palladino DM, Adams F, Buddensieck J, Mulch A, Behrens H (2015) Volatiles in magmas related to the Campanian Ignimbrite eruption: experiments versus natural findings. Am Mineral 100(10):2284–2297

Faure G (1986) Principles of isotope geology. Wiley, New York, p 589

Fiebig J, Tassi F, D'Alessandro W, Vaselli O, Woodland AB (2013) Carbon-bearing gas geothermometers for volcanic-hydrothermal systems. Chem Geol 351:66–75

Giggenbach WF (1992b) Isotopic shifts in waters from geothermal and volcanic systems along convergent plate boundaries and their origin. Earth Planet Sci Lett 113(4):495–510

Giggenbach WF (1997a) Relative importance of thermodynamic and kinetic processes in governing the chemical and isotopic composition of carbon gases in high-heatflow sedimentary basins. Geochim Cosmochim Acta 61:3763–3785

Giggenbach WF (1997b) The origin and evolution of fluids in magmatic-hydrothermal systems. In: Barnes HL (ed) Geochemistry of hydrothermal ore deposits, 3d edn. Wiley, pp 737–796

Giggenbach WF, Goguel RL (1989) Collection and analysis of geothermal and volcanic water and gas discharges. Unpublished report, Chemistry Division, DSIR-Petone, New Zealand, p 81

Giggenbach WF, Poreda RJ (1993) Helium isotopic and chemical composition of gases from volcanic-hydrothermal systems in the Philippines. Geothermics 22(5–6):369–380

Giggenbach WF (1979) The chemistry of geothermal gases. Lecture Notes, DSIR, Petone

Giggenbach WF (1992a) The composition of gases in geothermal and volcanic systems as a function of tectonic setting. In: Kharaka YK, Maest AS (eds) Proceedings of the 7th international symposium on water-rock interactions, WRI-7, Park City, Utah, 13–18 July 1992, pp 873–878

Holloway JR, Blank JG (1994) Application of experimental results to C-O-H species in natural melts. In: Carroll MR, Holloway JR (eds) Volatiles in magmas (Rev Mineral 30:281–330)

Horita J (2001) Carbon isotope exchange in the system CO_2-CH_4 at elevated temperatures. Geochim Cosmochim Acta 65(12):1907–1919

Horita J, Wesolowski DJ (1994) Liquid-vapor fractionation of oxygen and hydrogen isotopes of water from the freezing to the critical temperature. Geochim Cosmochim Acta 58(16):3425–3437

Iacono-Marziano G, Paonita A, Rizzo A, Scaillet B, Gaillard F (2010) Noble gas solubilities in silicate melts: new experimental results and a comprehensive model of the effects of liquid composition, temperature and pressure. Chem Geol 279(3–4):145–157

Iovine RS, Mazzeo FC, Wörner G, Pelullo C, Cirillo G, Arienzo I, Pack A, D'Antonio M (2018) Coupled $\delta^{18}O$-$\delta^{17}O$ and $^{87}Sr/^{86}Sr$ isotope compositions suggest a radiogenic and ^{18}O-enriched magma source for Neapolitan volcanoes (Southern Italy). Lithos 316:199–211

James R (1968) Wairakei and Larderello: geothermal power systems compared. NZ J Sci 11(4):706–719

Kavouridis T, Kuris D, Leonis C, Liberopoulou V, Leontiadis J, Panichi C, La Ruffa G, Caprai A (1999) Isotope and chemical studies for a geothermal assessment of the island of Nisyros (Greece). Geothermics 28:219–239

Kelley S (2002) Excess argon in K-Ar and Ar-Ar geochronology. Chem Geol 188(1–2):1–22

La Ruffa G, Panichi C, Kavouridis T, Liberopoulou V, Leontiadis J, Caprai A (1999) Isotope and chemical assessment of geothermal potential of Kos Island, Greece. Geothermics 28:205–217

Lee JY, Marti K, Severinghaus JP, Kawamura K, Yoo HS, Lee JB, Kim JS (2006) A redetermination of the isotopic abundances of atmospheric Ar. Geochim Cosmochim Acta 70(17):4507–4512

Mandeville CW, Webster JD, Tappen C, Taylor BE, Timbal A, Sasaki A, Hauri E, Bacon CR (2009) Stable isotope and petrologic evidence for open-system degassing during the climactic and pre-climactic eruptions of Mt. Mazama, Crater Lake, Oregon. Geochimica et Cosmochimica Acta 73 (10):2978–3012

Marini L, Fiebig J (2005) Fluid geochemistry of the magmatic-hydrothermal system of Nisyros (Greece). In: Hunziker JC, Marini L (eds) The geology, geochemistry and evolution of nisyros volcano (Greece). Implications for the volcanic hazards. Memoires de Géologie (Lausanne), 44, pp 121–163

Marini L, Moretti R, Accornero M (2011) Sulfur isotopes in magmatic-hydrothermal systems, melts, and magmas. In: Behrens H, Webster JD (eds) Sulfur in magmas and melts: its importance for natural and technical processes (Rev Mineral Geochem 73 (1):423–492)

Marini L (2007) Geological sequestration of carbon dioxide. In: thermodynamics, kinetics, and reaction path modeling, vol 11, 453 p. Developments in Geochemistry. Elsevier, Amsterdam

Martelli M, Nuccio PM, Stuart FM, Burgess R, Ellam RM, Italiano F (2004) Helium-strontium isotope constraints on mantle evolution beneath the Roman Comagmatic Province Italy. Earth Planet Sci Lett 224(3–4):295–308

Marty B, Trull T, Lussiez P, Basile I, Tanguy JC (1994) He, Ar, O, Sr and Nd isotope constraints on the origin and evolution of Mount Etna magmatism. Earth Planet Sci Lett 126(1–3):23–39

Matsuo S, Suzuki M, Mizutani Y (1978) Nitrogen to argon ratio in volcanic gases. Adv Earth Planet Sci 3:17–25

McCollom TM, Seewald JS (2001) A reassessment of the potential for reduction of dissolved CO_2 to hydrocarbons during serpentinization of olivine. Geochim Cosmochim Acta 65(21):3769–3778

Meija J, Coplen TB, Berglund M, Brand WA, De Bièvre P, Gröning M, Holden NE, Irrgeher J, Loss RD, Walczyk T, Prohaska T (2016) Isotopic compositions of the elements 2013 (IUPAC Technical Report). Pure Appl Chem 88(3):293–306

Mohr PJ, Newell DB, Taylor BN, Tiesinga E (2018) Data and analysis for the CODATA 2017 special fundamental constants adjustment. Metrologia 55(1):125

Muehlenbachs K, Kushiro I (1974) Oxygen isotope exchange and equilibrium of silicates with CO_2 and O_2. Carnegie Inst Wash Yearb 73:232–240

Nabelek PI, O'Neil JR, Papike JJ (1983) Vapor phase exsolution as a controlling factor in hydrogen isotope variation in granitic rocks: the Notch Peak granitic stock, Utah. Earth Planet Sci Lett 66:137–150

Nier AO (1950) A redetermination of the relative abundances of the isotopes of carbon, nitrogen, oxygen, argon, and potassium. Phys Rev 77:789–793

Newman S, Lowenstern JB (2002) VolatileCalc: a silicate melt–H2O–CO2 solution model written in Visual Basic for excel. Comput Geosci 28(5):597–604

Nuccio PM, Paonita A (2001) Magmatic degassing of multicomponent vapors and assessment of magma depth: application to Vulcano Island (Italy). Earth Planet Sci Lett 193(3–4):467–481

Nuti S, Caprai A, Noto P (1985) Hypothesis on the origin of steam and on the deep temperatures of the fluids of Pozzuoli Solfatara (Campania, Italy). Geoth Resour Counc Trans 1985:26–30

Panichi C, Tongiorgi E (1975) Carbon isotopic composition of CO2 from springs, fumaroles, mofettes and travertines of Central and Southern Italy: a preliminary prospection method of a geothermal area. In: Proceedings of the Second united nations symposium on the development and use of geothermal resources, San Francisco, pp 815-825

Panichi C, Volpi G (1999) Hydrogen, oxygen and carbon isotope ratios of Solfatara fumaroles (Phlegrean Fields, Italy): further insight into source processes. J Volcanol Geoth Res 91(2–4):321–328

Paonita A, Caracausi A, Iacono-Marziano G, Martelli M, Rizzo A (2012) Geochemical evidence for mixing between fluids exsolved at different depths in the magmatic system of Mt. Etna (Italy). Geochim Cosmochim Acta 84:380–394

Papale P, Moretti R, Barbato D (2006) The compositional dependence of the saturation surface of $H_2O + CO_2$ fluids in silicate melts. Chem Geol 229:78–95

Principe C, Gogichaishvili A, Arrighi S, Devidze M, La Felice S, Paolillo A, Giordano D, Morales J (2018) Archaeomagnetic dating of Copper Age furnaces at Croce di Papa village and relations on Vesuvius and Phlegraean Fields volcanic activity. J Volcanol Geoth Res 349:217–229

Pruess K (1985) A quantitative model of vapor dominated geothermal reservoirs as heat pipes in fractured porous rock. Lawrence Berkeley Laboratory Report LBL-19366, presented at the geothermal resources council 1985 international symposium on geothermal energy, Kailua Kona, HI, 26–30 Aug 1985

Richet P, Bottinga Y, Javoy M (1977) A review of hydrogen, carbon, nitrogen, oxygen, sulphur, and chlorine stable isotope fractionation among gaseous molecules. Annu Rev Earth Planet Sci 5(1):65–110

Robinson BW (1987) Sulphur and sulphate-oxygen isotopes in New Zealand geothermal systems and volcanic discharges. In: Proceedings of an advisory group meeting, studies on sulphur isotope variations in nature. IAEA, Vienna, 17–20 June 1985, pp 31–48

Sano Y, Takahata N, Nishio Y, Fischer TP, Williams SN (2001) Volcanic flux of nitrogen from the Earth. Chem Geol 171(3–4):263–271

Shaw AM, Hilton DR, Fischer TP, Walker JA, De Leeuw GAM (2006) Helium isotope variations in mineral separates from Costa Rica and Nicaragua: assessing crustal contributions, timescale variations and diffusion-related mechanisms. Chem Geol 230(1–2):124–139

Signorelli S, Buccianti A, Martini M, Piccardi G (1998) Arsenic in fumarolic gases of Vulcano (Aeolian Islands, Italy) from 1978 to 1993: geochemical evidence from multivariate analysis. Geochem J 32 (6):367–382

Silver L, Stolper E (1985) A thermodynamic model for hydrous silicate melts. J Geol 93(2):161–177

Silver LA, Ihinger PD, Stolper E (1990) The influence of bulk composition on the speciation of water in silicate glasses. Contrib Miner Petrol 104(2):142–162

Silver LA (1988) Water in silicate glasses. PhD Thesis, California Institute of Technology, Pasadena, California, 310p

Steiger RH, Jäger E (1977) Subcommission on geochronology: convention on the use of decay constants in geo- and cosmochronology. Earth Planet Sci Lett 36:359–362

Stolper E (1982) The speciation of water in silicate melts. Geochim Cosmochim Acta 46(12):2609–2620

Tassi F, Fiebig J, Vaselli O, Nocentini M (2012) Origins of methane discharging from volcanic-hydrothermal, geothermal and cold emissions in Italy. Chem Geol 310:36–48

Taylor BE, Eichelberger JC, Westrich HR (1983) Hydrogen isotopic evidence of rhyolitic magma degassing during shallow intrusion and eruption. Nature 306 (5943):541–545

Taylor BE (1986) Magmatic volatiles: isotopic variation of C, H, and S. In: Valley JW, Taylor HP, O'Neil (eds) Stable isotopes in high temperature geological processes (Rev Mineral 16:185–225)

Tedesco D, Scarsi P (1999) Chemical (He, H_2, CH_4, Ne, Ar, N_2) and isotopic (He, Ne, Ar, C) variations at the Solfatara crater (southern Italy): mixing of different sources in relation to seismic activity. Earth Planet Sci Lett 171(3):465–480

Tedesco D, Allard P, Sano Y, Wakita H, Pece R (1990) Helium-3 in subaerial and submarine fumaroles of Campi Flegrei caldera Italy. Geochim Cosmochim Acta 54(4):1105–1116

Truesdell AH, White DE (1973) Production of superheated steam from vapor-dominated geothermal reservoirs. Geothermics 2(3–4):154–173

Turi B, Taylor HP Jr, Ferrara G (1991) Comparisons of $^{18}O/^{16}O$ and $^{87}Sr/^{86}Sr$ in volcanic rocks from the Pontine Islands, M. Ernici, and Campania with other areas in Italy. In: Taylor HP Jr

Turi B, Taylor HP (1976) Oxygen isotope studies of potassic volcanic rocks of the Roman Province, Central Italy. Contrib Miner Petrol 55(1):1–31

Valentino GM, Cortecci G, Franco E, Stanzione D (1999) Chemical and isotopic compositions of minerals and waters from the Campi Flegrei volcanic system, Naples, Italy. J Volcanol Geoth Res 91(2–4):329–344

White DE, Muffler LJP, Truesdell AH (1971) Vapor-dominated hydrothermal systems compared with hot-water systems. Econ Geol 66(1):75–97

Zhao ZF, Zheng YF (2003) Calculation of oxygen isotope fractionation in magmatic rocks. Chem Geol 193(1–2):59–80

The Geoindicators Involving H_2O, CO_2, CO, CH_4 and H_2

Abstract

First we reviewed the five distinct reactions involving H_2O, CO_2, H_2, CH_4, and CO, namely: $CO_2 + H_2 = CO + H_2O$, indicated by the acronym RWG; $CO + \frac{1}{2} H_2O = \frac{3}{4} CO_2 + \frac{1}{4} CH_4$, indicated by the acronym CCC; $CO + H_2 = \frac{1}{2} CH_4 + \frac{1}{2} CO_2$, indicated by the acronym SS1; $CO + 3 H_2 = CH_4 + H_2O$, indicated by the acronym SS3; $CO_2 + 4 H_2 = CH_4 + 2 H_2O$, indicated by the acronym SS4. Starting from the axiom that these reactions actually take place in natural magmatic-hydrothermal systems, it is appropriate to work with the redox-independent, multi-component geothermometers derived from the log K of these five reactions, rather than with the redox-dependent geothermometers involving the separate H_2/H_2O, CO/CO_2, and CH_4/CO_2 concentration ratios. In this chapter, geothermometric calculations are performed under the hypothesis of ideal gas behavior for all gas species to simplify the discussion. Assuming attainment of equilibrium among all gas species and that all measured concentrations are equal to those present in the gas equilibration zone, the computed full equilibrium temperature exhibits limited changes with time, with most values varying between 210 and 250°C, whereas total fluid pressure, H_2O partial pressure, and CO_2 partial pressure show a practically continuous decrease with time, indicating a gradual depressurization of the hydrothermal system, with total fluid pressure even lower than 1 bar in the last years of monitoring. Since these results are at variance with the observed expansion of the area affected by soil degassing of deeply derived CO_2 from 1998 to 2020, the full equilibrium hypothesis is considered unrealistic. Alternatively, five distinct equilibrium temperatures and corresponding H_2O partial pressure, CO_2 partial pressure, and total fluid pressure values are calculated, for each sample, hypothesizing that the RWG, CCC, SS1, SS3, and SS4 reactions achieve the equilibrium condition in a pure saturated vapor phase, but independently of each other. These five different equilibrium temperatures reflect the disequilibrium between CO and CH_4 and the variable influence of these two gas species on each reaction. Only the RWG and SS4 equilibrium temperatures are probably meaningful because the RWG reaction includes CO but not CH_4 and the SS4 reaction involves CH_4 but not CO, whereas the other three reactions comprise both CO and CH_4. The RWG equilibrium temperature shows limited changes with time,

Supplementary Information The online version contains supplementary material available at (https://doi.org/10.1007/978-3-030-98471-7_6).

L. Marini et al., *The Solfatara Magmatic-Hydrothermal System*, Advances in Volcanology,
https://doi.org/10.1007/978-3-030-98471-7_6

generally varying between 210 and 250°C, similar to the full equilibrium temperature. In contrast, the SS4 equilibrium temperature exhibits a gradual and marked increase with time from 224°C in 1984 to 440-470°C in 2020. The RWG total fluid pressure is low, being usually in the range 6.5 - 35.9 bar. In contrast, the SS4 total fluid pressure shows a progressive and remarkable increment with time from 20.2-29.6 bar in 1983-1984 to 255-401 bar in 2020. Nevertheless, the RWG equilibrium temperature and fluid pressure did not depart appreciably from the SS4 equilibrium temperature and fluid pressure, respectively, until the end of the 1982-1985 uplift episode, evidently due to full equilibration of gas species, including CO and CH_4, in a pure saturated vapor phase. The attainment of this conditions was possible because the shallow aquifer behaved as a closed system, in the period 1983-1985, with fluid residence time long enough for the attainment of full equilibrium among gas species. In contrast, the differences between the SS4 and RWG equilibrium temperatures and fluid pressures augmented progressively after 1985, reflecting the increasing degree of disequilibrium between CO and CH_4. This different gas behavior indicates that the fluid residence time in the shallow aquifer was enough for CO equilibration but it was too short for CH_4 equilibration. Further considerations are given on the equilibrium CO concentration at the SS4 temperature, the equilibrium CH_4 concentration at the RWG temperature, the equilibrium H_2 concentration at the CCC temperature, and the effects of steam condensation controlled by isenthalpic decompression.■■■

To provide a reference to the reader, several geothermometric and geobarometric functions discussed in this chapter are reported as electronic supplementary material (see file Chapter_5_6.XLS).

1 Theoretical Foundations of the Geoindicators Involving H_2O, CO_2, CO, CH_4 and H_2

Several studies have postulated that the partial pressures of CO_2 and H_2 (and H_2S as well) of hydrothermal fluids are controlled by mineral-solution equilibria (e.g., Giggenbach 1981; Arnórsson et al. 1983; Arnórsson 1985; Arnórsson and Gunnlaugsson 1985; Stefánsson and Arnórsson 2002). Other investigations have proposed to use the homogeneous equilibrium reactions involving gaseous H_2O, CO_2, CO, CH_4 and H_2 as theoretical basis for the derivation of gas geothermometers and P_{CO_2}—indicators (e.g., Hulston and McCabe 1962; Tonani 1973; D'Amore and Nuti 1977; Giggenbach 1980; D'Amore and Celati 1983; Nehring and D'Amore 1984; Bertrami et al. 1985; Chiodini and Cioni 1989; Saracco and D'Amore 1989; Chiodini et al. 1993a; Giggenbach 1993; Chiodini and Marini 1998).

In particular, Chiodini and Marini (1998) have underscored that it is possible to write five reactions involving H_2O, CO_2, CO, CH_4 and H_2, each including four and excluding one of these five gas constituents. These reactions are written below setting equal to one the stoichiometric coefficient of CO in (1)–(4) and of CO_2 in (5), for convenience.

$$CO_2 + H_2 = CO + H_2O \qquad (1)$$

$$CO + \frac{1}{2}H_2O = \frac{3}{4}CO_2 + \frac{1}{4}CH_4 \qquad (2)$$

$$CO + H_2 = \frac{1}{2}CH_4 + \frac{1}{2}CO_2 \qquad (3)$$

$$CO + 3H_2 = CH_4 + H_2O \qquad (4)$$

$$CO_2 + 4H_2 = CH_4 + 2H_2O. \qquad (5)$$

Only two of the five reactions (1)–(5) are mutually independent, as all the five gas constituents are involved in any two of these five reactions. Nevertheless, it is advisable to consider all these five reactions for the reasons that will become evident from the following discussion.

As already shown by Chiodini and Marini (1998), reactions (1)–(5) can be obtained by summing the following reactions and eliminating O_2:

$$H_2O = H_2 + \frac{1}{2}O_2 \quad (6)$$

$$CO_2 = CO + \frac{1}{2}O_2 \quad (7)$$

$$CO_2 + 2H_2O = CH_4 + 2O_2. \quad (8)$$

Reactions (6), (7), and (8) represent the theoretical basis of redox-dependent geothermometers. These geothermometers have the advantage of considering separately the single concentration ratios X_{H_2}/X_{H_2O}, X_{CO}/X_{CO_2}, and X_{CH_4}/X_{CO_2}. However, they have disadvantage of relying on a suitable redox buffer, (e.g., those of D'Amore and Panichi 1980; D'Amore and Gianelli 1984; the hydrothermal redox buffer of Giggenbach 1987; and the Campanian redox buffer of Chiodini and Marini 1998) which must necessarily be specified in order to use these geothermometers. Unfortunately, there is no certainty on the effectiveness of these redox buffers, in general, and for the Solfatara magmatic-hydrothermal system, in particular. Even the Campanian redox buffer might be inadequate, although it has been specifically calibrated for the hydrothermal systems of Vesuvius, Ischia, and Solfatara, using the data acquired in 1983–1996 for the latter system. In fact, it is likely that redox conditions experienced significant changes with time in the zones of the Solfatara magmatic-hydrothermal system where different gas species equilibrate, due to the increasing input of magmatic fluids with time, as shown by Chiodini and coworkers (e.g., Chiodini et al. 2010, 2011, 2015, 2021; Cardellini et al. 2017). We will return on this point in chapter "The Redox Potential and Sulfur Gas Species".

Reactions (1)–(5) constitute the theoretical foundations of the redox-independent, multi-component geothermometers involving sums of log concentration ratios, such as $\log(X_{CO}/X_{CO_2}) - \log(X_{H_2}/X_{H_2O})$ and $3 \cdot \log(X_{CO}/X_{CO_2}) + \log(X_{CO}/X_{CH_4})$. The major disadvantage of these geothermometers is that there is no certainty on the attainment of equilibrium among all the reactants at the same temperature, pressure, and redox conditions, as pointed out by Giggenbach (1991).

Nevertheless, this issue can be mitigated considering (i) the geochemical characteristics of the Solfatara fumarolic fluids and (ii) the geothermometric indications provided the five relevant reactions (1)–(5), as discussed below and in Sect. 2. Although the first point was already discussed in chapter "Chemical and Isotopic Characteristics of Solfatara-Pisciarelli Fumarolic Fluids", we recall that H_2O and CO_2 are by far the major components of these fluids, whereas the concentrations of the other gas constituents of interest are as follows, based on the analytical data of Chiodini and coworkers (e.g., Chiodini et al. 2011, 2015, 2021):

(i) Hydrogen varies from 146 to 1230 μmol/mol, with an average of 446 μmol/mol, a median of 431 μmol/mol, and a standard deviation of 196 μmol/mol, at Bocca Grande, whereas H_2 varies from 237 to 1402 μmol/mol, with an average of 501 μmol/mol, a median of 456 μmol/mol, and a standard deviation of 193 μmol/mol, at Bocca Nuova.

(ii) Methane ranges from 7.2 to 41.4 μmol/mol, with an average of 18.3 μmol/mol, a median of 17.3 μmol/mol, and a standard deviation of 5.6 μmol/mol at Bocca Grande, while CH_4 ranges from 5.3 to 31.4 μmol/mol, with an average of 14.7 μmol/mol, a median of 14.0 μmol/mol, and a standard deviation of 4.8 μmol/mol at Bocca Nuova.

(iii) Carbon monoxide oscillates from 0.104 to 3.35 μmol/mol, with an average of 0.949 μmol/mol, a median of 0.685 μmol/mol, and a standard deviation of 0.659 μmol/mol at Bocca Grande, whereas CO oscillates from 0.170 to 3.78 μmol/mol, with an average of 1.04 μmol/mol, a median of 0.829 μmol/mol, and a standard deviation of 0.677 μmol/mol at Bocca Nuova.

Based on these different concentrations and assuming attainment of homogeneous gas equilibration through the five reactions (1)–(5), it is evident that there is plenty of H_2O and CO_2 and enough H_2 for the readjustment of CO and CH_4 to the equilibrium values at the temperature and pressure conditions of interest, whereas there is a lack of CH_4 and CO for the re-equilibration of H_2. However, these are necessary but not sufficient conditions for the attainment of chemical equilibrium. In other words, it is not certain that CH_4 and CO are doomed to equilibrate, because the achievement of this condition could be prevented by kinetic constraints, especially for CH_4 at relatively low temperatures (see chapter "The Chemical Kinetics of the Reactions Involving H_2O, CO_2, CO, CH_4 and H_2").

Summing up and taking into account all the pros and cons, in the case of Solfatara fluids, it is probably convenient to work with the redox-independent, multi-component geothermometers which are based on reactions (1)–(5). In this chapter, geothermometric calculations will be performed under the hypothesis of ideal gas behavior for all gas species, to simplify the discussion, although considerable deviations from ideality occur at pressures higher than some tens of bar corresponding to temperatures higher than 230–250 °C, as shown in chapter "The Expansion (Decompression) of the Solfatara Fumarolic Fluids". The effects caused by deviations from ideality will be considered in chapter "Calibration and Application of the RWG and SS4 Indicators for the Possible Expansion Paths of Solfatara Fluids".

As regards the nomenclatural aspect, each of these five reactions could be defined by the missing gas species, which is CH_4 in reaction (1), H_2 in reaction (2), H_2O in reaction (3), CO_2 in reaction (4), and CO in reaction (5). However, things change for reactions (2) and (3) written for coexistence of vapor and liquid water (see below), because both H_2 and H_2O are missing in reaction (16), which is the changed form of reaction (2), while there is no missing gas species in reaction (17), which the modified version of reaction (3). A proper definition was suggested earlier for some of these reactions: reaction (1) is known as the reverse water–gas shift reaction, whereas the so-called Sabatier-Senderens reductions comprise reactions (3), (4), (5) and other similar processes. This name was adopted in honor of Paul Sabatier and Jean-Baptiste Senderens who, in 1897, discovered that methane is formed when a mixture of CO or CO_2 and H_2 is passed over finely divided nickel at 300 °C (Ballentyne and Lovett 1980). The Sabatier-Senderens reductions are somewhat different from the Fischer–Tropsch process which is aimed at synthesize gasoline rather than methane (Ballentyne and Lovett 1980). Therefore Sabatier-Senderens is the most proper term for reactions (3), (4), (5) and other similar processes, although they were often called Fischer–Tropsch reactions in the scientific literature. All in all, the following acronyms were adopted: RWG for reaction (1) from reverse water–gas shift reaction, CCC for reaction (2) following Chiodini and Marini (1998), SS1, SS3, and SS4 for reactions (3), (4), and (5), from Sabatier-Senderens reduction, with stoichiometric coefficient of H_2 equal to 1, 3, and 4, respectively.

Reported in Table 1 are the values of the decimal logarithm of the thermodynamic equilibrium constant of the five reactions (1)–(5) calculated by means of SUPCRT92 (Johnson et al. 1992) at temperatures varying from 100 to 350 °C, every 10 °C, and total pressure of 1 bar, corresponding to the saturation pressure at 100 °C. The same log K values are obtained for a total pressure of 165 bar, which is the saturation pressure at 350 °C. This means that the log K values reported in Table 1 apply not only at total pressure of 1 bar but also at higher pressures, up to 165 bar at least.

The log K values of the five reactions (1)–(5) are reproduced with sufficient approximation by the following polynomial relations (T in Kelvin):

$$\log K_{RWG} = 23{,}755{,}011/T^3 - 198{,}131.8/T^2 - 1563.868/T + 1.600396$$
$$\text{with } K_{RWG} = \frac{P_{CO} \cdot P_{H_2O}}{P_{CO_2} \cdot P_{H_2}} = \frac{X_{CO} \cdot X_{H_2O}}{X_{CO_2} \cdot X_{H_2}} \quad (9)$$

Table 1 Decimal logarithms of the thermodynamic equilibrium constant of the five reactions (1)–(5) at 1 bar total pressure as a function of temperature

P (bar)	T (°C)	log K_{RWG}	log K_{CCC}	log K_{SS1}	log K_{SS3}	log K_{SS4}
1	100	−3.5564	7.0673	10.5781	17.5998	14.0434
1	110	−3.4085	6.7646	10.1208	16.8330	13.4245
1	120	−3.2683	6.4772	9.6861	16.1038	12.8356
1	130	−3.1352	6.2038	9.2724	15.4095	12.2744
1	140	−3.0087	5.9434	8.8782	14.7477	11.7390
1	150	−2.8884	5.6952	8.5021	14.1159	11.2275
1	160	−2.7738	5.4584	8.1430	13.5122	10.7384
1	170	−2.6645	5.2321	7.7996	12.9347	10.2702
1	180	−2.5603	5.0157	7.4711	12.3818	9.8215
1	190	−2.4608	4.8085	7.1563	11.8518	9.3910
1	200	−2.3656	4.6101	6.8545	11.3433	8.9777
1	210	−2.2746	4.4197	6.5648	10.8551	8.5804
1	220	−2.1875	4.2370	6.2866	10.3858	8.1984
1	230	−2.1039	4.0616	6.0192	9.9345	7.8306
1	240	−2.0238	3.8929	5.7619	9.5000	7.4762
1	250	−1.9470	3.7306	5.5142	9.0815	7.1346
1	260	−1.8731	3.5744	5.2756	8.6781	6.8049
1	270	−1.8022	3.4239	5.0455	8.2889	6.4867
1	280	−1.7340	3.2788	4.8236	7.9132	6.1792
1	290	−1.6683	3.1388	4.6093	7.5503	5.8820
1	300	−1.6051	3.0037	4.4023	7.1995	5.5944
1	310	−1.5442	2.8732	4.2022	6.8603	5.3161
1	320	−1.4854	2.7471	4.0087	6.5320	5.0465
1	330	−1.4288	2.6251	3.8215	6.2141	4.7854
1	340	−1.3741	2.5072	3.6402	5.9063	4.5321
1	350	−1.3214	2.3930	3.4646	5.6078	4.2865

$$\log K_{CCC} = -18{,}133.23/T^2 + 4{,}425.884/T - 4.663045$$
$$\text{with } K_{CCC} = \frac{P_{CO_2}^{0.75} \cdot P_{CH_4}^{0.25}}{P_{CO} \cdot P_{H_2O}^{0.5}} = \frac{X_{CO_2}^{0.75} \cdot X_{CH_4}^{0.25}}{X_{CO} \cdot X_{H_2O}^{0.5} \cdot P_{tot}^{0.5}} = \frac{X_{CO_2}^{0.75} \cdot X_{CH_4}^{0.25}}{X_{CO} \cdot P_{H_2O}^{0.5}} \quad (10)$$

$$\log K_{SS1} = -82{,}793.77/T^2 + 6{,}969.929/T - 7.506592$$
$$\text{with } K_{SS1} = \frac{P_{CO_2}^{0.5} \cdot P_{CH_4}^{0.5}}{P_{CO} \cdot P_{H_2}} = \frac{X_{CO_2}^{0.5} \cdot X_{CH_4}^{0.5}}{X_{CO} \cdot X_{H_2} \cdot P_{tot}} = \frac{X_{CO_2}^{0.5} \cdot X_{CH_4}^{0.5} \cdot X_{H_2O}}{X_{CO} \cdot X_{H_2} \cdot P_{H_2O}} \quad (11)$$

$$\log K_{SS3} = -212{,}313.4/T^2 + 12{,}058.74/T - 13.19430$$
$$\text{with } K_{SS3} = \frac{P_{CH_4} \cdot P_{H_2O}}{P_{CO} \cdot P^3_{H_2}} = \frac{X_{CH_4} \cdot X_{H_2O}}{X_{CO} \cdot X^3_{H_2} \cdot P^2_{tot}} = \frac{X_{CH_4} \cdot X^3_{H_2O}}{X_{CO} \cdot X^3_{H_2} \cdot P^2_{H_2O}} \quad (12)$$

$$\log K_{SS} = -258{,}773.6/T^2 + 10{,}176.60/T - 11.37445$$
$$\text{with } K_{SS4} = \frac{P_{CH_4} \cdot P^2_{H_2O}}{P_{CO_2} \cdot P^4_{H_2}} = \frac{X_{CH_4} \cdot X^2_{H_2O}}{X_{CO_2} \cdot X^4_{H_2} \cdot P^2_{tot}} = \frac{X_{CH_4} \cdot X^4_{H_2O}}{X_{CO_2} \cdot X^4_{H_2} \cdot P^2_{H_2O}} \quad (13)$$

Since there is the same number of gas species in the left and right hand side of reaction (1), K_{RWG} does not involve total fluid pressure, P_{tot}. In contrast, since the number of gas species in the left and right hand side of reactions (2)–(5) is different, P_{tot} is involved in K_{CCC}, K_{SS1}, K_{SS3}, and K_{SS4}. In these K's in which P_{tot} is written explicitly, the sum of the exponents of the mole fractions appearing in the numerator is different from the sum of the exponents of the mole fractions appearing in the denominator, which is an undesirable situation. Therefore, it is advisable to substitute the P_{H_2O}/X_{H_2O} ratio for P_{tot}, obtaining K's in which the sum of the exponents of the mole fractions appearing in the numerator is equal to the sum of the exponents of the mole fractions appearing in the denominator. Using this trick, the mole fractions of gas species can be used to compute $K_{CCC} \cdot P^{0.5}_{H_2O}$, $K_{SS1} \cdot P_{H_2O}$, $K_{SS3} \cdot P^2_{H_2O}$, and $K_{SS4} \cdot P^2_{H_2O}$. Hence, the next step is to estimate P_{H_2O}.

It was shown in Sect. 4.1 in chapter "The Expansion (Decompression) of the Solfatara Fumarolic Fluids" that, based on the hypothesis of isenthalpic flow, the Bocca Grande and Bocca Nuova fluids are expected to be at saturation conditions (coexistence of water vapor and liquid) in a large temperature interval, possibly extending from the lower saturation temperature (186.1–199.7 °C, with a mean of 192.8 °C, for Bocca Grande and 160.0–167.0 °C, with a mean of 163.4 °C, for Bocca Nuova) to the upper saturation temperature (289.4–299.8 °C, with a mean of 295.8 °C, for Bocca Grande and 307.5–311.6 °C, with a mean of 310.5 °C, for Bocca Nuova). Moreover, the lower saturation temperature is not much higher than the corresponding outlet temperature, namely 31.4 °C on average for the Bocca Grande fluids and 18.0 °C on average for the Bocca Nuova fluids. Therefore, although the Bocca Grande and Bocca Nuova vents discharge superheated fluids, it is reasonable to assume that these fluids are actually at saturation conditions along a significant portion of their upflow path upstream of the discharge points at the surface.

At saturation, the partial pressure of H_2O can be calculated as a function of both the temperature and CO_2 mole fraction by means of Eq. (116) in chapter "The Expansion (Decompression) of the Solfatara Fumarolic Fluids" for Bocca Grande fluids and Eq. (117) in chapter "The Expansion (Decompression) of the Solfatara Fumarolic Fluids" for Bocca Nuova fluids with an uncertainty on P_{H_2O} up to 3.0 and 5.5% (absolute value), respectively. Adopting a general approach, at saturation, water partial pressure is approximated by the following relation (temperature in Kelvin, P_{H_2O} in bar; Giggenbach 1980):

$$\log P_{H_2O} = 5.51 - 2048/T \quad (14)$$

which is valid from 100 to 360 °C, or by Eq. (134) in chapter "The Expansion (Decompression) of the Solfatara Fumarolic Fluids", which can be simplified as follows (temperature in Kelvin, P_{H_2O} in bar; Giggenbach 1987):

$$\log P_{H_2O} = 4.90 - 1820/T \quad (15)$$

assuming $P_{H_2O} = f_{H_2O}$. Equation (15) can be used up to 600 °C. Therefore, at saturation conditions, Eqs. (116) in chapter "The Expansion (Decompression) of the Solfatara Fumarolic Fluids" or (117) in chapter "The Expansion (Decompression) of the Solfatara Fumarolic

Table 2 Decimal logarithms of the thermodynamic equilibrium constant of the four reactions (16), (17), (18) and (19) at saturation conditions (coexistence of water liquid and vapor) as a function of temperature and water pressure. The log K of reaction (1) is also reported

T (°C)	P (bar)	$\log K_{RWG}$	$\log K_{CCC^*}$	$\log K_{SS1^*}$	$\log K_{SS3^*}$	$\log K_{SS4^*}$
100	1.013	−3.5564	7.0701	10.5837	17.6111	14.0547
125	2.32	−3.2009	6.5183	9.8357	16.4706	13.2697
150	4.757	−2.8884	6.0268	9.1652	15.4421	12.5537
175	8.918	−2.6118	5.5859	8.5600	14.5082	11.8964
200	15.536	−2.3656	5.1880	8.0104	13.6551	11.2894
225	25.479	−2.1453	4.8269	7.5085	12.8717	10.7264
250	39.736	−1.9470	4.4975	7.0480	12.1490	10.2021
275	59.431	−1.7677	4.1957	6.6237	11.4797	9.7119
300	85.838	−1.6051	3.9182	6.2312	10.8574	9.2523
325	120.458	−1.4569	3.6619	5.8669	10.2770	8.8201
350	165.211	−1.3214	3.4245	5.5277	9.7340	8.4126

Fluids" or (14) or (15) can be used to calculate P_{H_2O}, which can then be inserted in the expressions of K_{CCC}, K_{SS1}, K_{SS3}, and K_{SS4} related to Eqs. (10)–(13). In this way, the mole fraction products can be directly linked to temperature.

Alternatively, the equilibrium reaction between water liquid and vapor can be summed to reactions (2)–(5), thus obtaining the following pressure-independent reactions:

$$CO + \frac{1}{2}H_2O_{(L)} = \frac{3}{4}CO_2 + \frac{1}{4}CH_4 \quad (16)$$

$$CO + H_2 + H_2O_{(L)} = \frac{1}{2}CH_4 + \frac{1}{2}CO_2 + H_2O \quad (17)$$

$$CO + 3H_2 + 2H_2O_{(L)} = CH_4 + 3H_2O \quad (18)$$

$$CO_2 + 4H_2 + 2H_2O_{(L)} = CH_4 + 4H_2O \quad (19)$$

in which water liquid is indicated by H_2O with subscript L and water vapor by H_2O without subscript. Since the same number of gas species is present in the left and right hand side of these reactions, they are pressure independent and the mole fractions of gas species can be directly used to compute their K values. The activity of liquid water can be assumed to be equal to one, to a first approximation. This manipulation is not necessary for reaction (1) since it is pressure independent, as already noted above. Reactions (16)–(19) are indicated by the same acronyms of the corresponding reactions (2)–(5) but adding an asterisk to indicate the coexistence of vapor and liquid water. The thermodynamic equilibrium constant of the four reactions (16) or CCC*, (17) or SS1*, (18) or SS3*, and (19) or SS4*, was computed by means of SUPCRT92 (Johnson et al. 1992) obtaining the log K values reported in Table 2 together with those of reaction (1) or RWG.

The following geothermometric relations were then obtained through polynomial fitting of the log K values reported in Table 2:

$$\begin{aligned} T_{RWG} = {} & 1.7215 \cdot \log K_{RWG}^4 + 23.359 \cdot \log K_{RWG}^3 \\ & + 133.51 \cdot \log K_{RWG}^2 + 437.95 \cdot \log K_{RWG} \\ & + 744.18 \\ \text{with } & \log K_{RWG} = \log(X_{CO}/X_{CO_2}) \\ & - \log(X_{H_2}/X_{H_2O}) \end{aligned} \quad (20)$$

$$\begin{aligned} T_{CCC^*} = {} & 0.13347 \cdot \log K_{CCC^*}^4 - 3.8787 \cdot \log K_{CCC^*}^3 \\ & + 47.255 \cdot \log K_{CCC^*}^2 - 317.66 \cdot \log K_{CCC^*} \\ & + 1021.05 \\ \text{with } & \log K_{CCC^*} = -3/4 \cdot \log(X_{CO}/X_{CO_2}) \\ & - 1/4 \cdot \log(X_{CO}/X_{CH_4}) \end{aligned} \quad (21)$$

$$T_{SS1^*} = 0.027401 \cdot \log K_{SS1^*}^4 - 1.2248 \cdot \log K_{SS1^*}^3 + 23.003 \cdot \log K_{SS1^*}^2 - 236.72 \cdot \log K_{SS1^*} + 1{,}136.92$$
$$\text{with} \log K_{SS1^*} = -1/2 \cdot \log(X_{CO}/X_{CO_2}) - 1/2 \cdot \log(X_{CO}/X_{CH_4}) - \log(X_{H_2}/X_{H_2O}) \quad (22)$$

$$T_{SS3^*} = 0.0035075 \cdot \log K_{SS3^*}^4 - 0.26842 \cdot \log K_{SS3^*}^3 + 8.6671 \cdot \log K_{SS3^*}^2 - 152.88 \cdot \log K_{SS3^*} + 1{,}232.97$$
$$\text{with } \log K_{SS3^*} = -\log(X_{CO}/X_{CH_4}) - 3 \cdot \log(X_{H_2}/X_{H_2O}) \quad (23)$$

$$T_{SS4^*} = 0.0082655 \cdot \log K_{SS4^*}^4 - 0.52206 \cdot \log K_{SS4^*}^3 + 14.110 \cdot \log K_{SS4^*}^2 - 209.35 \cdot \log K_{SS4^*} + 1{,}381.98$$
$$\text{with } \log K_{SS4^*} = \log(X_{CH_4}/X_{CO_2}) - 4 \cdot \log(X_{H_2}/X_{H_2O}) \quad (24)$$

Equations (20)–(24) reproduce the nominal temperatures with an approximation smaller than 0.1 °C. To be recalled that Eqs. (21)–(24) apply to saturated vapors only, while Eqs. (10)–(13), as well as (9) and (20), apply to both saturated and superheated (dry) vapors.

2 Application of the Geoindicators Involving H_2O, CO_2, CO, CH_4 and H_2 to the Solfatara-Pisciarelli Fluids: Preliminary Considerations

In previous works on the H_2O–CO_2–CO–CH_4–H_2 system (e.g., Bertrami et al. 1985; Chiodini and Cioni 1989; Cioni and Marini 1990; Marini et al. 1991; Chiodini et al. 1993a; Chiodini and Marini 1998), emphasis was placed on carbon monoxide and the RWG reaction was considered together with the CCC reaction as theoretical basis for the development of graphical geoindicators. This choice was largely dictated by the fast re-equilibration kinetics of carbon monoxide[1] and the limited contribution of CH_4 to the log K of the CCC reaction and is reasonable as long as one wants to evaluate the temperature of last gas equilibration. However, looking at the RWG, CCC, SS1, SS3, and SS4 reactions from a different point of view, it could be better to focus on the RWG and SS4 reactions, in which the species not considered are CH_4 and CO, respectively. In this way, fast-reacting CO and slow-reacting CH_4 are taken into account separately.

Changing the point of view, and considering the conversion of CO_2 to CH_4, via CO as intermediate, it would be better to focus on the RWG reaction, which is representative of the first step, whereas the second step could be described by the CCC and/or SS1 and/or SS3 reactions. Based on the stoichiometry of these reactions, it follows that:

(i) The CCC reaction is expected to occur if H_2 is not available. Through this reaction, in fact, three quarters of CO are converted to CO_2 and one quarter of CO is converted to CH_4.
(ii) The SS1 reaction is expected to take place if H_2 is available in quantities similar to CO. Through this reaction, in fact, half CO is converted to CO_2 and half CO is converted to CH_4.
(iii) The SS3 reaction is expected to happen if H_2 is available in much greater quantities than CO. Through this reaction, in fact, all CO is converted to CH_4.

Regardless of which is the second step [either the CCC or SS1 or SS3 reaction], the SS4 reaction is representative of the complete CO_2 to CH_4 conversion process, since it is obtained by adding the RWG reaction to any one of the other three reactions.

Based on previous discussion, to inspect the available data for the Solfatara-Pisciarelli fumaroles all together, it is advisable to consider different plots, although only two of the five RWG, CCC, SS1, SS3, and SS4 reactions are

[1] For instance, CO was found to re-equilibrate almost as fast as H_2 at Vulcano crater (Chiodini et al. 1993b).

mutually independent, as already noted in Sect. 1. The log concentration ratio $\log(X_{CO}/X_{CO_2}) - \log(X_{H_2}/X_{H_2O})$, corresponding to the log K of the RWG reaction, is reported on the abscissa axis of these plots, whereas the following sums of log concentration ratios are alternatively represented on the ordinate axis:

(i) $\log(X_{CH_4}/X_{CO_2}) - 4 \cdot \log(X_{H_2}/X_{H_2O})$, corresponding to $\log K_{SS4} + 2 \cdot \log P_{H_2O}$ (Fig. 1),
(ii) $\log(X_{CH_4}/X_{CO}) - 3 \cdot \log(X_{H_2}/X_{H_2O})$, corresponding to $\log K_{SS3} + 2 \cdot \log P_{H_2O}$ (Fig. 2),
(iii) $0.5 \cdot \log(X_{CH_4}/X_{CO}) - 0.5 \cdot \log(X_{CO}/X_{CO_2}) - \log(X_{H_2}/X_{H_2O})$, corresponding to $\log K_{SS1} + \log P_{H_2O}$ (Fig. 3),
(iv) $0.25 \cdot \log(X_{CH_4}/X_{CO}) - 0.75 \cdot \log(X_{CO}/X_{CO_2})$, corresponding to $\log K_{CCC} + 0.5 \cdot \log P_{H_2O}$ (Fig. 4).

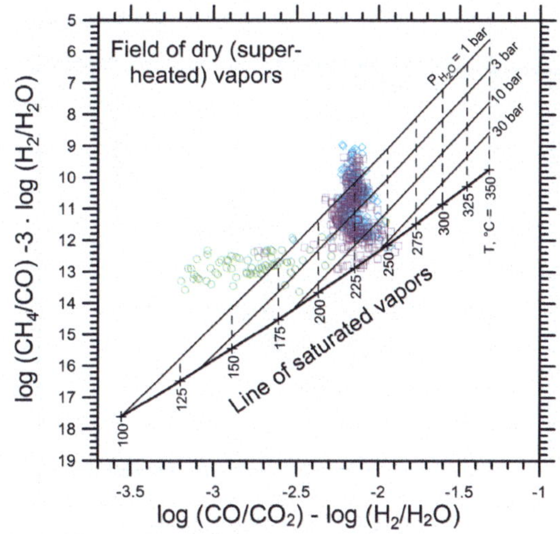

Fig. 2 Plot of log (CO/CO_2) – log (H_2/H_2O) versus log (CH_4/CO) – 3 · log (H_2/H_2O) showing the theoretical grid for gas equilibration in saturated and superheated vapors as well as the data of the Solfatara-Pisciarelli fumarolic fluids of Chiodini and coworkers (e.g., Chiodini et al. 2011, 2015, 2016, 2021). Symbols as in Fig. 1

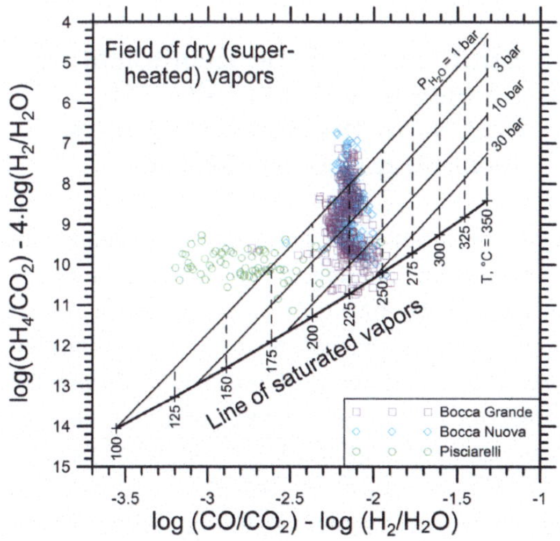

Fig. 1 Plot of log (CO/CO_2) – log (H_2/H_2O) versus log (CH_4/CO_2) – 4 · log (H_2/H_2O) showing the theoretical grid for gas equilibration in saturated and superheated vapors as well as the data of the Solfatara-Pisciarelli fumarolic fluids of Chiodini and coworkers (e.g., Chiodini et al. 2011, 2015, 2016, 2021)

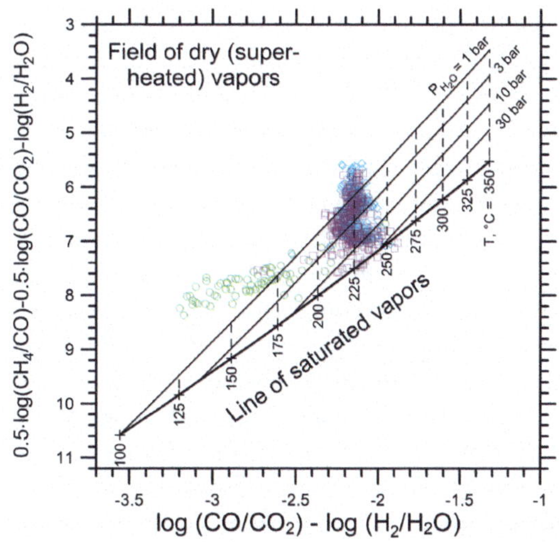

Fig. 3 Plot of log (CO/CO_2) – log (H_2/H_2O) versus 0.5 · log (CH_4/CO) – 0.5 · log (CO/CO_2) – log (H_2/H_2O) showing the theoretical grid for gas equilibration in saturated and superheated vapors as well as the data of the Solfatara-Pisciarelli fumarolic fluids of Chiodini and coworkers (e.g., Chiodini et al. 2011, 2015, 2016, 2021). Symbols as in Fig. 1

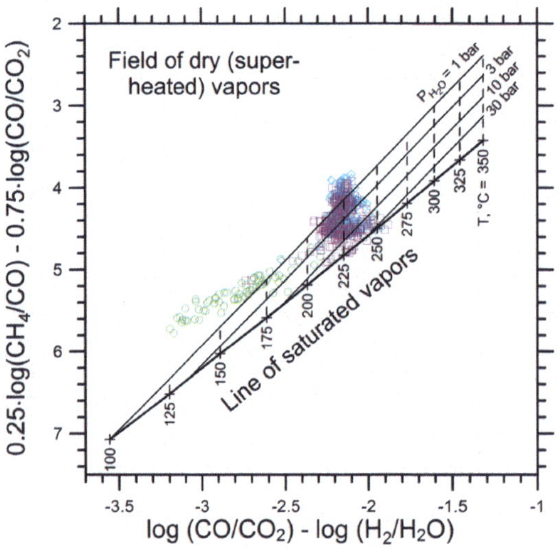

Fig. 4 Plot of log (CO/CO_2) – log (H_2/H_2O) versus 0.25 · log (CH_4/CO) – 0.75 · log (CO/CO_2) showing the theoretical grid for gas equilibration in saturated and superheated vapors as well as the data of the Solfatara-Pisciarelli fumarolic fluids of Chiodini and coworkers (e.g., Chiodini et al. 2011, 2015, 2016, 2021). Symbols as in Fig. 1

Only a portion of these diagrams is shown in Figs. 1, 2, 3 and 4 representing a zoomed view of the area occupied by the data of interest. These four diagrams are very similar to each other for both the wedge-shaped theoretical grid and the spread of sample points. In fact: (i) the theoretical grid is delimited by the isobar corresponding to P_{H_2O} of 1 bar, at the top, and the line of saturated vapors, at the bottom, with the other isobars, corresponding to P_{H_2O} of 3, 10 and 30 bars occupying intermediate positions, (ii) the isotherms are vertical being controlled by the log K of the RWG reaction and (iii) the Bocca Grande and Bocca Nuova samples group together in a single cluster, extending from the line of saturated vapors to beyond (above) the isobar corresponding to P_{H_2O} of 1 bar, with most samples in the field of the dry vapors, whereas most Pisciarelli samples form a linear trend to the left of the Bocca Grande and Bocca Nuova samples.

The spread of the Pisciarelli samples is probably due to water loss through steam condensation occurring close to the surface. In fact, the Pisciarelli vents discharge either saturated steam or slightly superheated steam, with outlet temperature ranging from 90.5 to 105.6 °C, a mean outlet temperature of 97.1 °C, and a median outlet temperature of 96.0 °C, as noted in Sect. 1 in chapter "The Expansion (Decompression) of the Solfatara Fumarolic Fluids".

Although the Bocca Grande and Bocca Nuova fluids are superheated at outlet conditions, occurrence of steam condensation and separation of condensed steam might have affected also these fluids along their upflow path, namely between the upper and lower saturation temperatures (see Sects. 1 and 4.1 in chapter "The Expansion (Decompression) of the Solfatara Fumarolic Fluids"), assuming isenthalpic decompression. This process is discussed in Sect. 8.

Based on Figs. 1, 2, 3 and 4 and related equations (see Sect. 1), different equilibrium temperatures can be obtained for the Solfatara fumarolic fluids, depending on the adopted interpretative approach, as discussed in the next sections.

3 Application of the Geoindicators Involving H_2O, CO_2, CO, CH_4 and H_2 to the Solfatara Fluids Assuming Full Equilibrium

Assuming both full equilibrium (i.e., attainment of equilibrium among all gas species) and that the measured concentrations of all gas species are equal to those present in the gas equilibration zone (i.e., occurrence of no change during the transit of the fluid from the gas equilibration zone to the surface discharge), the following calculations are performed for each Bocca Grande and Bocca Nuova sample: (i) the equilibrium temperature is computed by means of Eq. (20); (ii) the P_{H_2O} is calculated using any of the relations (10)–(13), all providing the same result; (iii) the P_{CO_2} is calculated by multiplying P_{H_2O} times the X_{CO_2}/X_{H_2O} ratio; (iv) the total fluid pressure, P_{tot} is assumed to be given by the sum $P_{H_2O} + P_{CO_2}$. The first two steps are equivalent to read the equilibrium temperature and P_{H_2O} on the grid of any of the four plots of Figs. 1, 2, 3 and 4, in correspondence of each sample.

Table 3 Main statistical parameters of the equilibrium temperature, P_{H_2O}, P_{CO_2}, and total fluid pressure calculated, assuming full equilibrium among gas species, for the 330 samples of Bocca Grande (BG) and the 294 samples of Bocca Nuova (BN) fumarolic fluids based on the data of Chiodini and coworkers (e.g., Chiodini et al. 2011, 2015, 2016, 2021)

Vent	Parameter	T (°C)	P_{H_2O} (bar)	P_{CO_2} (bar)	P_{tot} (bar)
BG	Mean	223.3	5.01	1.10	6.11
BG	Median	223.2	2.08	0.59	2.61
BG	Min	163.6	0.25	0.07	0.36
BG	Max	259.0	57.6	10.5	68.1
BG	Std. Dev	9.9	6.95	1.38	8.31
BN	Mean	225.8	3.77	0.79	4.56
BN	Median	225.6	2.18	0.58	2.80
BN	Min	183.8	0.16	0.07	0.22
BN	Max	252.8	40.3	6.72	47.0
BN	Std. Dev	7.0	4.40	0.70	5.10

The main statistical parameters of the equilibrium temperature, P_{H_2O}, P_{CO_2}, and P_{tot} thus obtained for the Bocca Grande and Bocca Nuova fluids are listed in Table 3. The 330 fluid samples from Bocca Grande appear to have attained full equilibrium at a mean temperature of 223 ± 10 °C (1σ) and a mean total fluid pressure of 6.1 ± 8.3 bar, whereas the 294 samples of Bocca Nuova appear to have achieved full equilibrium at an average temperature of 226 ± 7 °C (1σ) and a mean total fluid pressure of 4.6 ± 5.1 bar, usually under superheated conditions and uncommonly under saturated conditions.

As shown by the chronogram of Fig. 5, the temperature of full equilibrium exhibits limited changes with time, with most values varying

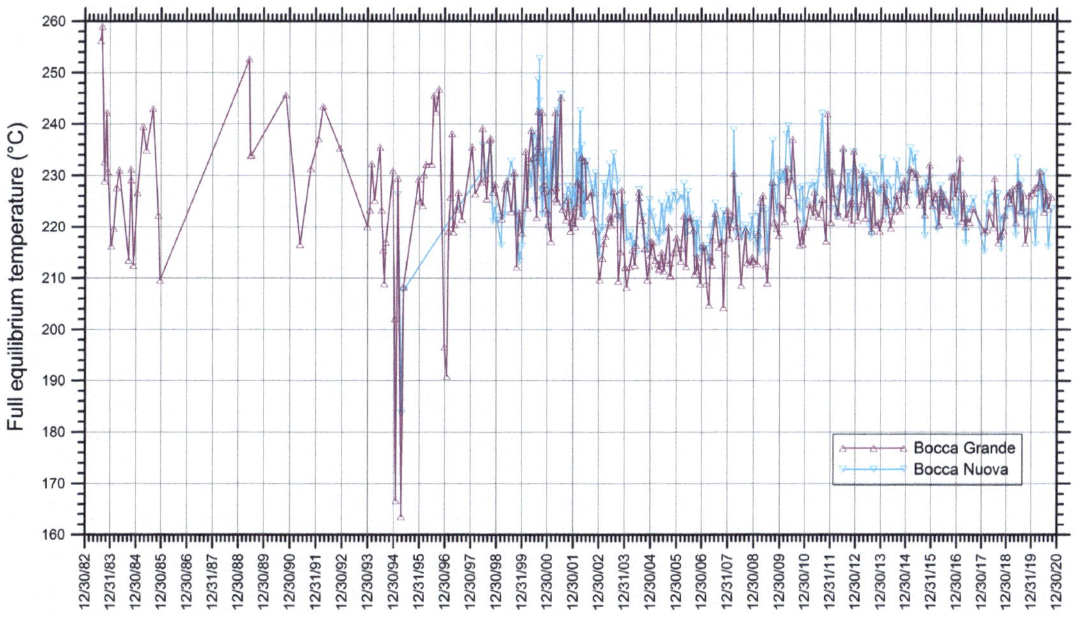

Fig. 5 Chronogram of the full equilibrium temperature indicated by the RWG reaction, for Bocca Grande and Bocca Nuova fumarolic fluids based on the data of Chiodini and coworkers (e.g., Chiodini et al. 2011, 2015, 2016, 2021)

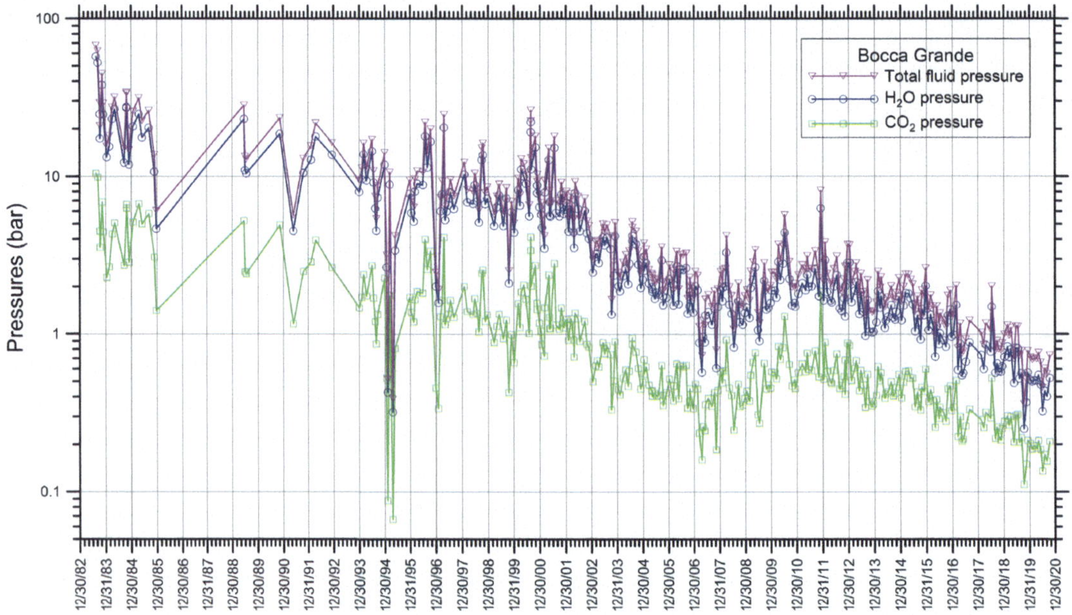

Fig. 6 Chronogram of total fluid pressure, P_{H_2O}, and P_{CO_2}, based on the hypothesis of full equilibrium among all gas species for the Bocca Grande fumarolic fluids, based on the data of Chiodini and coworkers (e.g., Chiodini et al. 2011, 2015, 2016, 2021)

between 210 and 250 °C, apart from few exceptions, such as the low equilibrium temperature values of 166.7 °C on 4 February 1995 and 163.6 °C on 22 April 1995, which are not very different from the corresponding outlet temperatures, 162 and 164 °C, respectively, for Bocca Grande, as well as the relatively low equilibrium temperature of 183.7 °C on 28 April 1995, for Bocca Nuova. These anomalously low full equilibrium temperatures reflect the anomalously low CO concentrations, 0.11 μmol/mol on 4 February 1995 and 0.104 μmol/mol on 22 April 1995, for Bocca Grande, and 0.17 μmol/mol on 28 April 1995, for Bocca Nuova.

Total fluid pressure, P_{H_2O}, and P_{CO_2} show a practically continuous decrease with time for both Bocca Grande (Fig. 6) and Bocca Nuova (Fig. 7), indicating a gradual depressurization of the hydrothermal system.

This phenomenon, however, is at variance with the expansion of the area affected by diffuse soil degassing of deeply derived CO_2 observed from 1998 to 2016 (Chiodini et al. 2010, 2021; Cardellini et al. 2017; see chapter "Soil CO_2 Diffuse Degassing and Thermal Energy Release as Indicators of Volcanic Unrest in the Solfatara-Pisciarelli Area" for further details), while total fluid pressures lower than 1 bar, as in the last years of monitoring, are unrealistic, being too low.

Moreover, the occurrence of superheated (dry) conditions along the whole flowpath of the uprising Solfatara fluids is expected to allow the transport of strongly acidic gases (e.g., SO_2, HCl and HF) like at White Island (Giggenbach 1987) and Vulcano Island (Chiodini et al. 1993b) and of neutral species (e.g., H_3BO_3 and SiO_2) like in the Larderello geothermal field (Andreussi et al. 1994). Since all these gas species are virtually absent or not reported in the fumarolic fluids discharged so far from the Solfatara fumaroles, it is likely that superheated conditions are only present along the final part of the flowpath of uprising fluids, downstream of the zone where the local temperature is equal to the lower saturation temperature, that is close or relatively close to the surface.

Last but not least, attainment of gas equilibrium in a superheated vapor phase contrasts with the presence of saturation conditions (coexistence of water vapor and liquid) along most of the upflow

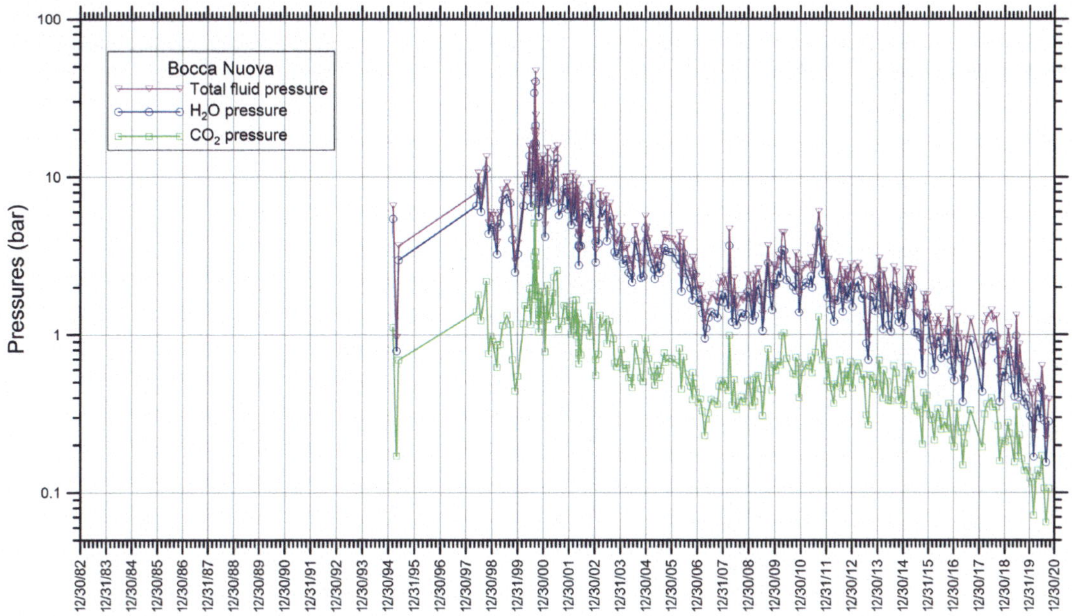

Fig. 7 Chronogram of total fluid pressure, P_{H_2O}, and P_{CO_2}, based on the hypothesis of equilibration among all gas species for the Bocca Nuova fumarolic fluids, based on the data of Chiodini and coworkers (e.g., Chiodini et al. 2011, 2015, 2016, 2021)

path of the Bocca Grande and Bocca Nuova fluids, as indicated by the measured outlet temperatures and the hypothesis of isenthalpic flow (see chapter "The Expansion (Decompression) of the Solfatara Fumarolic Fluids"). For all these reasons, the results of the full equilibrium approach are considered to be scarcely realistic, at least for the samples collected after the end of the 1982-1985 uplift episode.

4 Application of the Geoindicators Involving H_2O, CO_2, CO, CH_4 and H_2 to the Solfatara Fluids Assuming Independent Equilibrium in a Pure Saturated Vapor Phase

In this approach, the RWG, CCC, SS1, SS3, and SS4 reactions are assumed to attain the equilibrium condition in a pure saturated vapor phase, but independently of each other. Moreover, the measured concentrations of all gas species are assumed to be equal to those present in the gas equilibration zone also in this approach. The following calculations are performed for each Bocca Grande and Bocca Nuova sample: (i) The equilibrium temperature of each reaction (RWG, CCC, SS1, SS3, and SS4) is computed by means of Eqs. (20)–(24). (ii) The P_{H_2O} of each reaction is calculated using Eq. (14). (iii) The P_{CO_2} is calculated by multiplying the P_{H_2O} value of each reaction times the X_{CO_2}/X_{H_2O} ratio. (iv) For each reaction, the total fluid pressure is assumed to be given by the sum $P_{H_2O} + P_{CO_2}$.

Alternatively, to obtain the equilibrium temperature of the RWG reaction and the corresponding P_{H_2O} value, each sample is projected onto the line of saturated vapors, by moving it vertically (and downward in most cases), in any of the four plots of Figs. 1, 2, 3 and 4. To obtain the equilibrium temperatures of the CCC, SS1, SS3, and SS4 reactions and the corresponding P_{H_2O} values, each sample is projected onto the line of saturated vapors, by shifting it horizontally (and to the right, in most cases), in the four plots of Figs. 1, 2, 3 and 4 The main statistical parameters of the five equilibrium temperatures

Table 4 Main statistical parameters of the equilibrium temperatures calculated for the Bocca Grande (BG) and Bocca Nuova (BN) fumarolic fluids, assuming that the RWG, CCC, SS1, SS3, and SS4 reactions attain equilibrium in a pure saturated vapor phase independently of each other, based on the data of Chiodini and coworkers (e.g., Chiodini et al. 2011, 2015, 2016, 2021)

Vent	Parameter	T(RWG)	T_(CCC)	T_(SS1)	T_(SS3)	T_(SS4)
		°C	°C	°C	°C	°C
BG	Mean	223.3	258.4	275.1	290.9	312.1
BG	Median	223.2	256.1	270.6	287.5	308.1
BG	Min	163.6	203.8	221.7	224.4	224.0
BG	Max	259.0	300.1	335.8	372.4	440.1
BG	Std. Dev	9.9	17.1	24.8	32.6	51.1
BN	Mean	225.8	264.8	283.4	301.1	332.6
BN	Median	225.6	263.0	284.0	302.3	333.3
BN	Min	183.8	220.7	238.8	248.0	247.6
BN	Max	252.8	308.2	347.3	389.1	470.1
BN	Std. Dev	7.0	15.6	23.7	31.8	47.1

thus obtained for the Bocca Grande and Bocca Nuova fluids are listed in Table 4.

Inspection of Table 4 shows that the RWG, CCC, SS1, SS3, and SS4 reactions indicate progressively higher equilibrium temperatures for both fumarolic vents. This fact is due to the increasing weight of CH_4 and the decreasing weight of CO on each reaction along the sequence RWG → CCC → SS1 → SS3 → SS4, as pointed out by the $|\nu_{CH_4}|/(|\nu_{CH_4}| + |\nu_{CO}|)$ ratio (with ν_i indicating the stoichiometric coefficient of the i-th gas species in the reaction), which is 0 for the RWG reaction, 0.20 for the CCC reaction, 0.33 for the SS1 reaction, 0.50 for the SS3 reaction, and 1 for the SS4 reaction. In other terms, the different equilibrium temperatures indicated by the five considered reactions reflect the lack of equilibrium between CO and CH_4 and the distinct influence of these two gas species on each reaction. Consequently, only the two extreme equilibrium temperatures given by the RWG and SS4 reactions are probably meaningful because the RWG reaction includes CO but not CH_4 and the SS4 reaction involves CH_4 but not CO. In contrast, the intermediate equilibrium temperatures given by the CCC, SS1, and SS3 reactions are probably meaningless because they are distinct combinations of the two equilibrium temperatures given by the RWG and SS4 reactions.

The spread of the equilibrium temperatures in the case under consideration is completely different from the full equilibrium case (Sect. 3), in which the five considered reactions give the same full equilibrium temperature for any sample. However, that the full equilibrium temperatures calculated in Sect. 3 are equal to the RWG temperatures computed in this section. Nevertheless, the full equilibrium P_{H_2O}, P_{CO_2}, and total fluid pressure values obtained in Sect. 3 are generally lower than the corresponding values for the case under consideration, even those obtained for the RWG reaction (see below).

Table 4 also shows that the main statistical parameters computed for Bocca Nuova are generally higher than those calculated for Bocca Grande, due to the different time interval of monitoring, from June 1983 to October 2020, for Bocca Grande, and from March 1995 to October 2020, for Bocca Nuova.

The time changes of the equilibrium temperatures indicated by the five reactions of interest are shown in the chronograms of Fig. 8 for Bocca Grande and Fig. 9 for Bocca Nuova, although only the equilibrium temperatures given by the RWG and SS4 reactions are meaningful, as noted above. The outlet temperature, the lower saturation temperature and the upper saturation

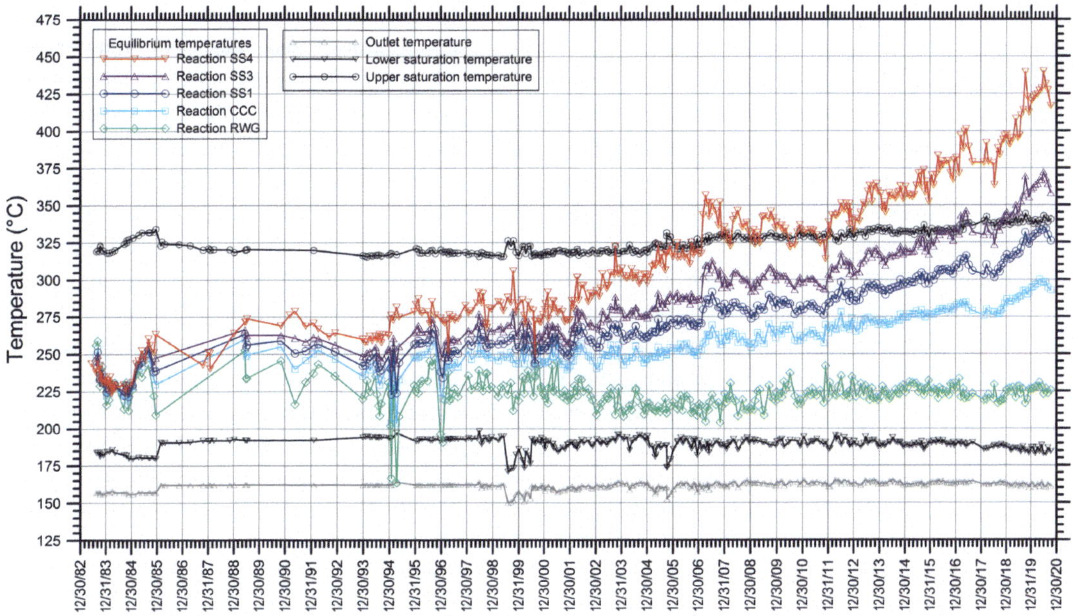

Fig. 8 Chronograms of the equilibrium temperatures indicated by the RWG, CCC, SS1, SS3, and SS4 reactions for the Bocca Grande fumarolic fluids, based on the data of Chiodini and coworkers (e.g., Chiodini et al. 2011, 2015, 2016, 2021). Also shown are the outlet temperature, the lower saturation temperature, and the upper saturation temperature

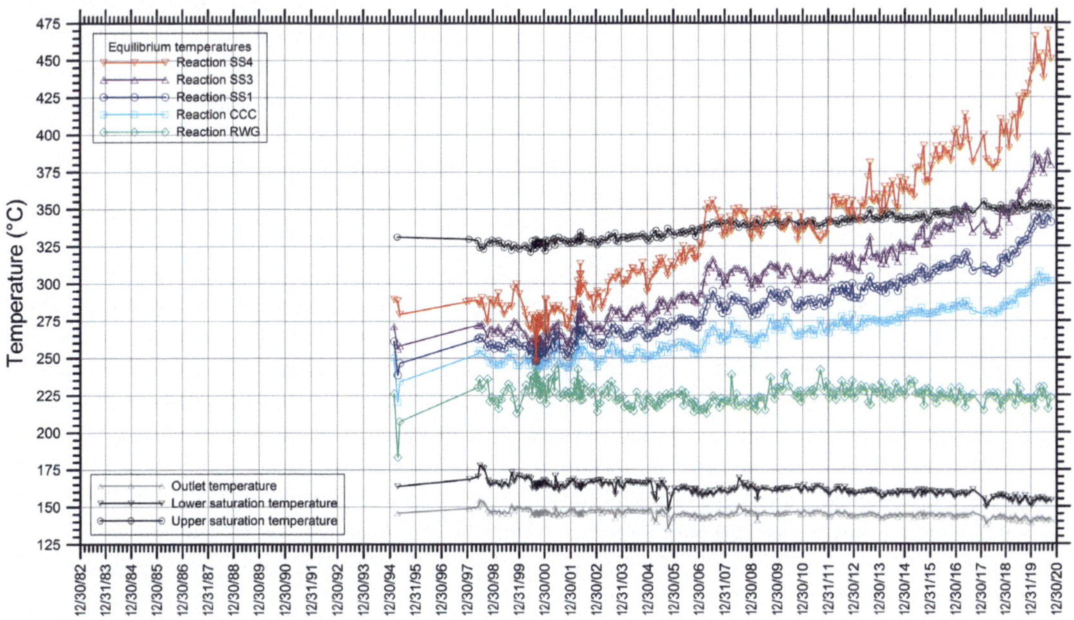

Fig. 9 Chronograms of the equilibrium temperatures indicated by the RWG, CCC, SS1, SS3, and SS4 reactions for the Bocca Nuova fumarolic fluids, based on the data of Chiodini and coworkers (e.g., Chiodini et al. 2011, 2015, 2016, 2021). Also shown are the outlet temperature, the lower saturation temperature, and the upper saturation temperature

temperature are also represented in these two chronograms for comparison.

Inspection of these two figures shows what follows:

(i) Since the equilibrium temperature indicated by the RWG reaction is always higher than the lower saturation temperature and always lower than the upper saturation temperature, for both fumaroles, the hypothesis of gas equilibration in a pure saturated vapor phase is consistent with the assumption of isenthalpic fluid expansion. For both fumaroles, the equilibrium temperature given by the SS4 reaction is always higher than the lower saturation temperature, it is lower than the upper saturation temperature before 2007, it is close to the upper saturation temperature in 2007–2011, but it is higher than the upper saturation temperature from 2012 to 2020. Therefore, the hypothesis of gas equilibration in a single saturated vapor phase is consistent with the assumption of isenthalpic fluid expansion before 2012, whereas there is a conflict between these two hypotheses in the period 2012–2020. This conflict can be solved assuming that fluid decompression deviated from the isenthalpic condition in 2012–2020. The other three equilibrium temperatures, corresponding to the CCC, SS1, and SS3 reactions are intermediate between the RWG and SS4 temperatures, but probably meaningless, as already noted above.

(ii) The RWG equilibrium temperature exhibits limited changes with time mostly fluctuating between 210 and 250 °C, for both fumaroles, similar to what was observed in Sect. 3 for the full equilibrium case. In contrast, the SS4 equilibrium temperature shows a remarkable and continuous increase with time (apart from short-term fluctuations) from 224 °C in March 1984 to 440 °C in July 2020 for Bocca Grande and from 248 °C in August 2000 to 470 °C in September 2020 for Bocca Nuova. Moreover, the SS1 temperature grows faster than the CCC temperature, and the SS3 temperature increments faster than the SS1 temperature but less steeply than the SS4 temperature.

(iii) For Bocca Grande, the equilibrium temperatures indicated by all five reactions exhibit small differences until 1985, indicating the attainment of full equilibrium, in a pure saturated vapor phase, among all gas species, including CO and CH_4. In contrast, after 1985, the discrepancies between the different equilibrium temperatures increase gradually, indicating a growing degree of disequilibrium between CO and CH_4. Interestingly, this shift from equilibrium to disequilibrium, between CH_4 and CO, corresponds to the end of the 1982–1985 uplift episode and related seismicity (e.g., Chiodini et al. 2010 and references therein). Carbon monoxide is always in disequilibrium with methane in Bocca Nuova fluids, since geochemical monitoring of this fumarole began in March 1995.

(iv) In spite of this usual CO–CH_4 disequilibrium, the equilibrium temperatures indicated by all five reactions are concordant or relatively concordant for the samples collected from Bocca Grande on 18 April 1997 and 22 and 23 August 2000 and from Bocca Nuova on 23 August 2000 and 13 September 2000. It must be underscored that these samples with concordant equilibrium temperatures have H_2O concentration somewhat higher and H_2 concentration fairly lower than those of most samples, but the reason (or reasons) of these differences are unclear.

At any given time, the RWG and SS4 equilibrium temperatures for Bocca Nuova are generally higher than the corresponding equilibrium temperatures for Bocca Grande, but differences are usually lower than 10–20 °C, as shown by the chronograms of Fig. 10. This generally good correspondence between the equilibrium temperatures of the two fumaroles reflects the

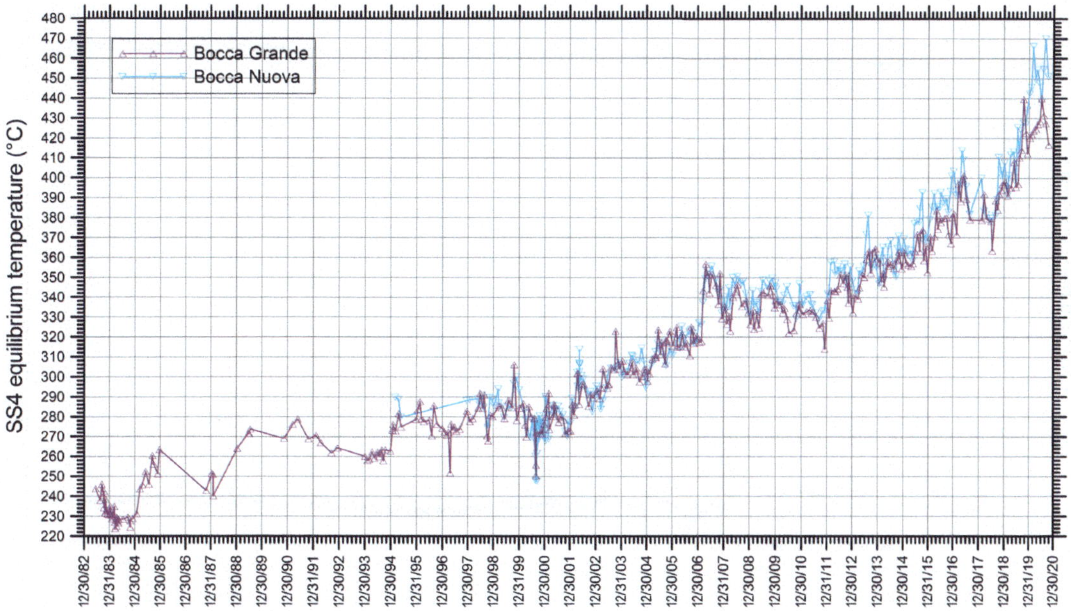

Fig. 10 Chronograms of the equilibrium temperature indicated by the SS4 reaction for Bocca Grande and Bocca Nuova fumarolic fluids, based on the data of Chiodini and coworkers (e.g., Chiodini et al. 2011, 2015, 2016, 2021)

compositional similarity of the fumarolic fluids discharged from the two vents.

The time changes of the total fluid pressures indicated by the RWG and SS4 reactions are shown in the chronograms of Fig. 11 for Bocca Grande and Fig. 12 for Bocca Nuova, whereas the total fluid pressures given by the reactions CCC, SS1, and SS3 were considered meaningless and, therefore, disregarded. The following inferences can be drawn from these two figures:

(i) The RWG total fluid pressure shows relatively small changes with time, oscillating between 6.5 and 35.9 bar, around a mean of 21.9 bar and a median of 22.0 bar, with a standard deviation of 3.6 bar for Bocca Grande in the 1983–2020 period and between 10.0 and 32.1 bar, around a mean of 22.7 bar and a median of 22.9 bar, with a standard deviation of 2.6 bar for Bocca Nuova in the 1995–2020 period. In contrast, the SS4 total fluid pressure exhibits a gradual and pronounced increase with time from 20.2–29.6 bar in 1983–1984 to 255–316 bar in 2020 for Bocca Grande and from 45.0–56.8 bar in 1995–1998 to 301–401 bar in 2020 for Bocca Nuova, apart from short-term fluctuations.

(ii) Similar to the equilibrium temperatures, also the total fluid pressures given by the RGW and SS4 reactions show negligible deviations until 1985, for Bocca Grande, whereas the gradually growing CO–CH_4 disequilibrium induced progressively increasing discrepancies between the two chronograms, after 1985, that is, after the cessation of the 1982–1985 uplift episode and related seismicity (e.g., Chiodini et al. 2010 and references therein).

(iii) In spite of these increasing discrepancies, the total fluid pressures given by the two reactions are concordant or relatively concordant for the few samples with concordant equilibrium temperatures, which were collected from Bocca Grande on 18 April 1997 and 22 and 23 August 2000 and from Bocca Nuova on 23 August 2000 and 13 September 2000.

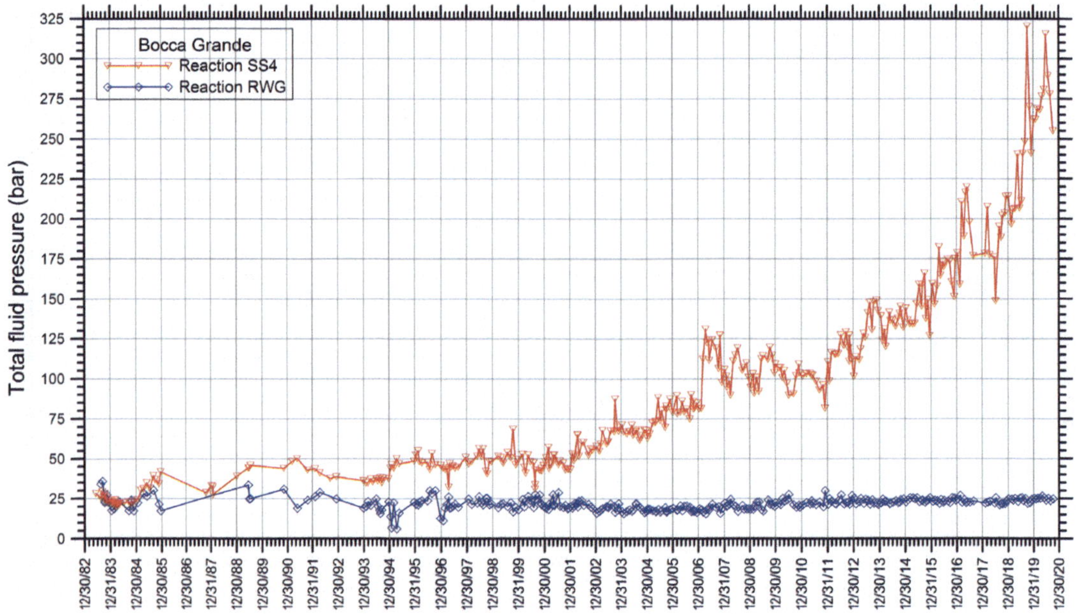

Fig. 11 Chronograms of the total fluid pressure indicated by the RWG and SS4 reactions for the Bocca Grande fumarolic fluids, based on the data of Chiodini and coworkers (e.g., Chiodini et al. 2011, 2015, 2016, 2021)

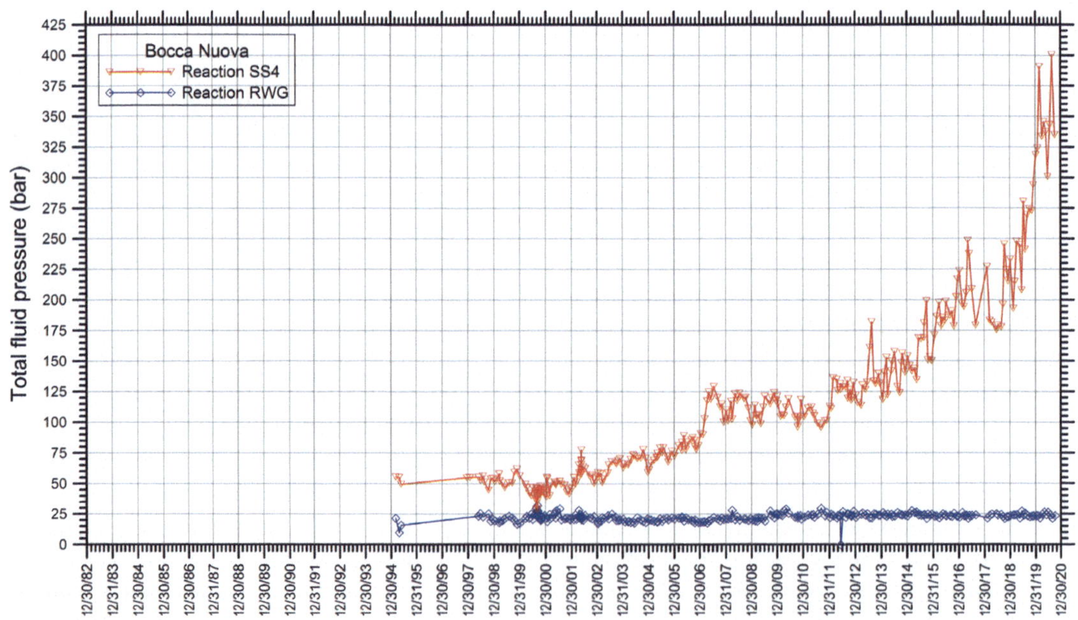

Fig. 12 Chronograms of the total fluid pressure indicated by the RWG and SS4 reactions for the Bocca Nuova fumarolic fluids, based on the data of Chiodini and coworkers (e.g., Chiodini et al. 2011, 2015, 2016, 2021)

Summing up, the RWG and SS4 reactions provide two strongly contrasting scenarios. In fact, the equilibrium temperature and total fluid pressure given by the CO-controlled RWG reaction are low and exhibit small temporal variations, while the equilibrium temperature and total fluid pressure given by the CH_4-controlled SS4 reaction show a considerable and progressive increment with time, due to the gradually growing CO–CH_4 disequilibrium.

5 Comparison of CO Concentrations Consistent with the SS4 Equilibrium Temperature and Analytical Values

Since carbon monoxide is not involved in the SS4 reaction, this reaction is adopted to calculate the equilibrium temperature, assuming attainment of equilibrium in a pure saturated vapor phase. Then, the SS4 equilibrium temperature is inserted into Eq. (9), together with the analytical values of X_{H_2O}, X_{CO_2}, and X_{H_2}, and Eq. (9) is solved for X_{CO}. As expected, by using this computed equilibrium X_{CO} instead of the measured X_{CO}, the RWG, CCC, SS1, and SS3 reactions indicate the same temperature, P_{H_2O}, P_{CO_2}, and total fluid pressure values given by the SS4 reaction, assuming again attainment of equilibrium in a pure saturated vapor phase. The following observations can be drawn from the chronograms of the equilibrium and analytical CO concentrations for Bocca Grande (Fig. 13) and Bocca Nuova (Fig. 14):

(i) The equilibrium CO concentration of Bocca Grande exhibits a gradual but marked increase with time from 0.22–0.51 μmol/mol in June 1983–January 1985 to 40.9–62.1 μmol/mol in 2020, with a mean of 6.59 μmol/mol, a median of 2.78 μmol/mol and a standard deviation of 9.94 μmol/mol. The equilibrium CO concentration of Bocca Nuova shows an even more pronounced increment from 1.1–1.2 μmol/mol in March–May 1995 to 50.0–92.0 μmol/mol in 2020.

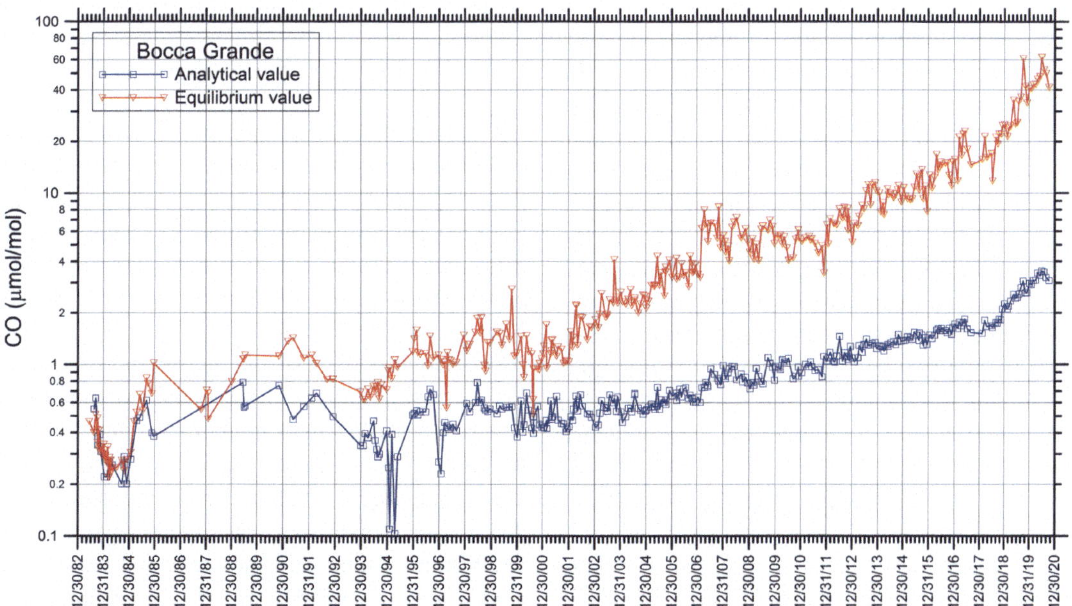

Fig. 13 Chronogram of CO concentration (in log-scale) showing both the analytical value and the equilibrium value indicated by the SS4 reaction in a pure saturated vapor phase for the Bocca Grande fumarolic fluids, based on the data of Chiodini and coworkers (e.g., Chiodini et al. 2011, 2015, 2016, 2021)

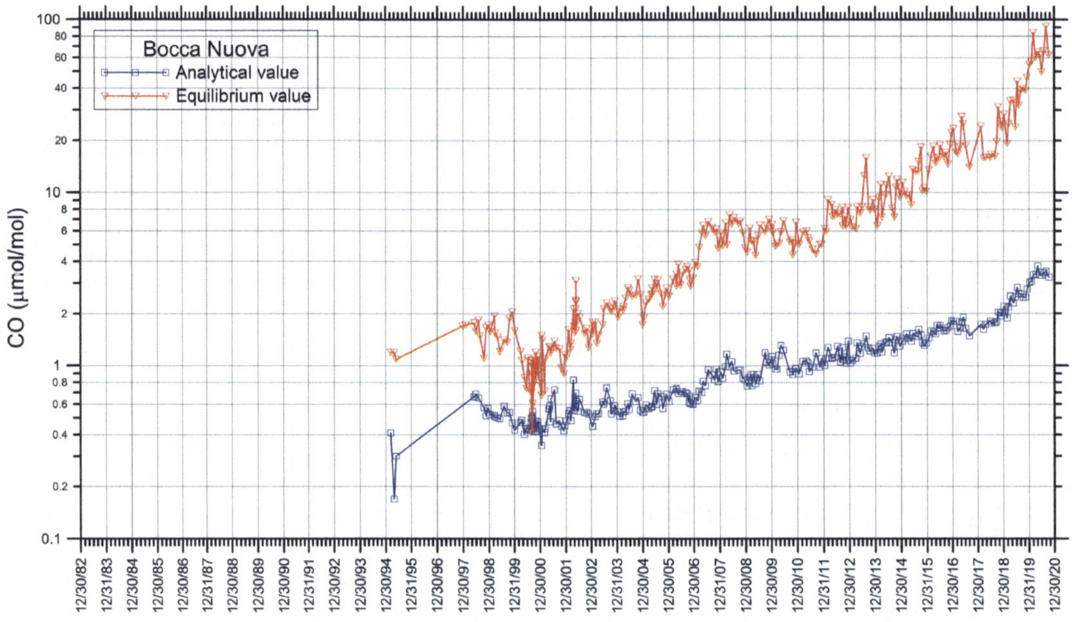

Fig. 14 Chronogram of CO concentration (in log-scale) showing both the analytical value and the equilibrium value indicated by the SS4 reaction in a pure saturated vapor phase for the Bocca Nuova fumarolic fluids, based on the data of Chiodini and coworkers (e.g., Chiodini et al. 2011, 2015, 2016, 2021)

(ii) The equilibrium CO concentration of Bocca Grande agrees with the analytical value, with absolute deviations lower than 40%, from June 1983 to September 1985. After this date, corresponding with the termination of the 1982–1985 uplift episode and related seismicity (e.g., Chiodini et al. 2010 and references therein), the equilibrium CO concentration increases quicker than the analytical value and the deviations between the two chronograms become progressively greater with time.

(iii) Despite these increasing discrepancies with time, the equilibrium CO concentration agrees with the analytical CO value within 40%, for the samples collected from Bocca Grande on 18 April 1997 and 22 and 23 August 2000 and from Bocca Nuova on 23 August 2000 and 13 September 2000. Not surprisingly, these are the same samples with concordant equilibrium temperatures and total fluid pressures (see Sect. 4).

Further information is provided by the chronogram of the percentage of reacted CO, which is given by the simple relation:

$$100 \cdot \left|X_{CO,eq} - X_{CO,meas}\right| / X_{CO,eq} \qquad (25)$$

with the pedices ‘eq’ and ‘meas’ standing for equilibrium and measured, respectively (Fig. 15). The absolute value is adopted in Eq. (25) because equilibrium CO is lower than measured CO for six Bocca Grande samples collected in 1983–1984 and the two Bocca Nuova samples obtained in August–September 2000. However, these are exceptions, because $X_{CO,eq} > X_{CO,meas}$ for all the other 667 fumarolic gas samples. The percentage of reacted CO for the Bocca Grande fluids was usually below 30% until September 1985, which is not surprising due to the satisfactory agreement between the equilibrium values and the analytical values, as noted above. Afterwards, the percentage of reacted CO has experienced a considerable increase with time, for both the Bocca Grande

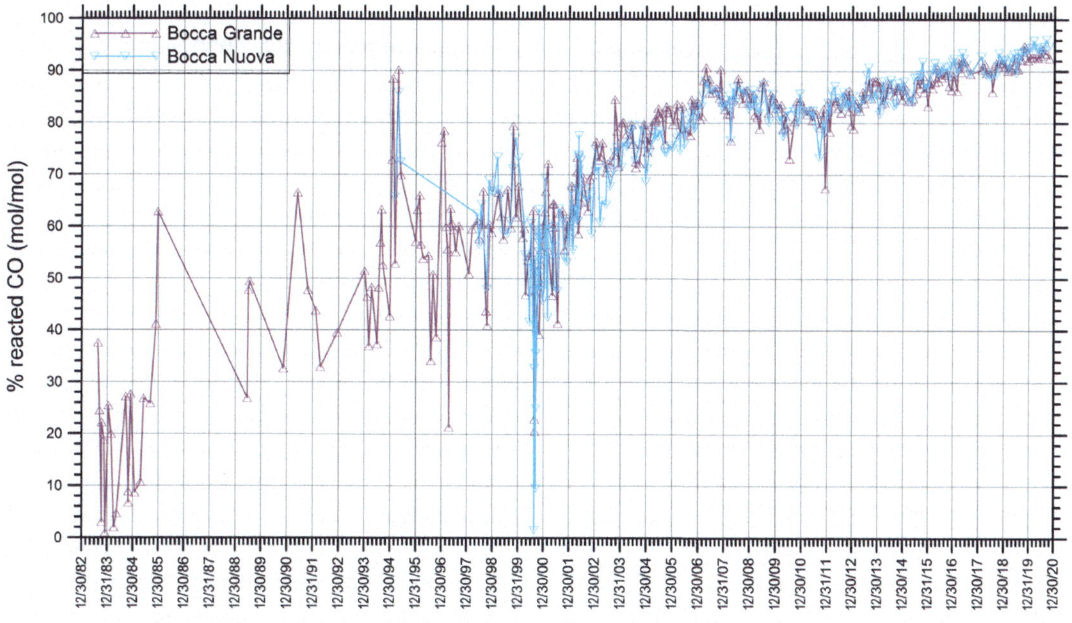

Fig. 15 Chronogram of the percentage of reacted CO for both Bocca Nuova and Bocca Grande, based on the equilibrium CO concentration indicated by the SS4 reaction in a pure saturated vapor phase and the analytical CO concentration (data from Chiodini and coworkers, e.g., Chiodini et al. 2011, 2015, 2016, 2021)

and Bocca Nuova fluids, varying from 60 to 80% in the years 2002–2006, generally oscillating from 80 to 90% in the years 2007–2015, and often exceeding the 90% threshold afterwards. These high percentages of reacted CO imply a fast kinetics for the reactions converting CO in CO_2 and/or CH_4, that is the reverse of reaction (1) and reactions (2), (3), and (4). This topic will be resumed in chapter "The Chemical Kinetics of the Reactions Involving H_2O, CO_2, CO, CH_4 and H_2".

6 Comparison of CH_4 Concentrations Consistent with the RWG Equilibrium Temperature and Analytical Values

According to Fiebig et al. (2007), the CH_4 concentrations and $\delta^{13}C$ values of geothermal fluids emitted from the Mediterranean volcanoes Vesuvius and Ischia, in Italy, and Nisyros, in Greece, are consistent with the achievement of chemical and isotopic equilibrium between CH_4 and CO_2, implying the occurrence of CO_2 methanation. However, these inferences were questioned by Fiebig et al. (2019) who reassessed the CO_2 and CH_4 isotopic data of global volcanic–hydrothermal fluids, showed that there is a fortuitous correspondence between the temperatures indicated by CO_2-CH_4 carbon isotope fractionation and actual subsurface temperatures, and concluded that the attainment of chemical and isotopic equilibrium between CO_2 and CH_4 in geothermal discharges cannot be definitively proven.

Starting from these preliminary considerations, to elucidate the doubts concerning CH_4 behavior in the Solfatara fluids, it is useful to assume attainment of gas equilibration in a pure saturated vapor phase at the methane-independent RWG temperature and to compute the equilibrium CH_4 concentration inserting the measured values of X_{H_2O}, X_{CO_2}, and X_{H_2}, and the RWG temperature into Eq. (13), which is then solved for X_{CH_4}. As expected, utilizing the calculated equilibrium X_{CH_4} instead of the analytical X_{CH_4}, the CCC, SS1, SS3, and SS4 reactions provide the same temperature, P_{H_2O}, P_{CO_2}, and total fluid pressure values given by the RWG reaction. To be noted that Eqs. (10), (11), and (12) could be used to compute the equilibrium CH_4 concentration instead of Eq. (13), provided that the RWG temperature and the mole fractions of relevant gas species are involved in calculations.

As shown by the chronogram of Fig. 16, the computed equilibrium CH_4 concentrations are in satisfactory agreement with the analytical values of Bocca Grande fluids, for the period June 1983 to September 1985, with absolute deviations of the log-concentrations generally lower than 30%. After September 1985, corresponding with the termination of the 1982–1985 uplift episode and related seismicity (e.g., Chiodini et al. 2010 and references therein), the equilibrium CH_4 concentrations deviate more and more from the measured CH_4 concentrations of Bocca Grande fluids. In fact, the equilibrium CH_4 concentration of Bocca Grande fumarolic fluids experiences an increase of three–four orders of magnitude with time, from 8.5–53.1 μmol/mol, in June 1983–September 1985, to 42,800–184,000 μmol/mol, in October 2019–October 2020, with an average of 6800 μmol/mol, a median of 1510 μmol/mol, and a standard deviation of 19,900 μmol/mol, whereas the analytical CH_4 concentrations oscillate between 7.21 and 41.4 μmol/mol, around the mean of 18.3 μmol/mol and the median of 17.3 μmol/mol, with a standard deviation of 5.61 μmol/mol.

Although the chronogram of Bocca Nuova begins in 1995 (Fig. 17), it is very similar to that of Bocca Grande. In fact, the analytical CH_4 concentration does not show significant changes with time, varying between 5.29 and 31.4 μmol/mol, around a mean of 14.7 μmol/mol and a median of 14.0 μmol/mol, with a standard deviation of 4.80 μmol/mol. In contrast, the equilibrium CH_4 concentration increases substantially with time, from a minimum of 11.8 μmol/mol to a maximum of 391,000 μmol/mol, with an average of 9490 μmol/mol, a median of 1300 μmol/mol, and a standard deviation of 34,400 μmol/mol.

In spite of these increasingly large differences between equilibrium and analytical CH_4

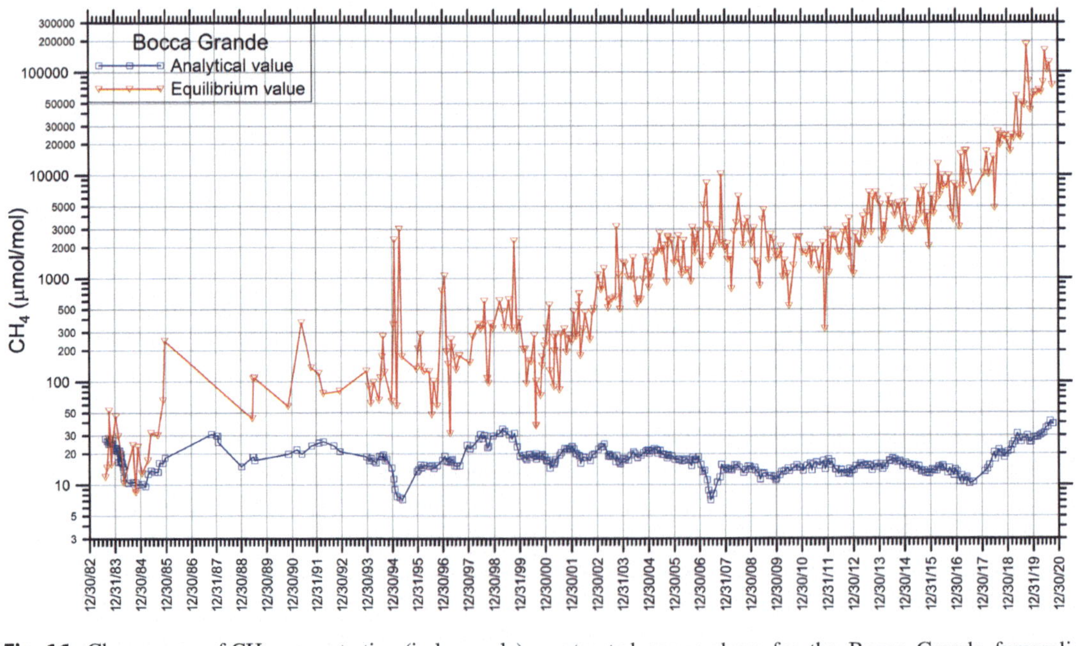

Fig. 16 Chronogram of CH_4 concentration (in log-scale) showing both the analytical value and the equilibrium value indicated by the RWG reaction in a pure saturated vapor phase for the Bocca Grande fumarolic fluids, based on the data of Chiodini and coworkers (e.g., Chiodini et al. 2011, 2015, 2016, 2021)

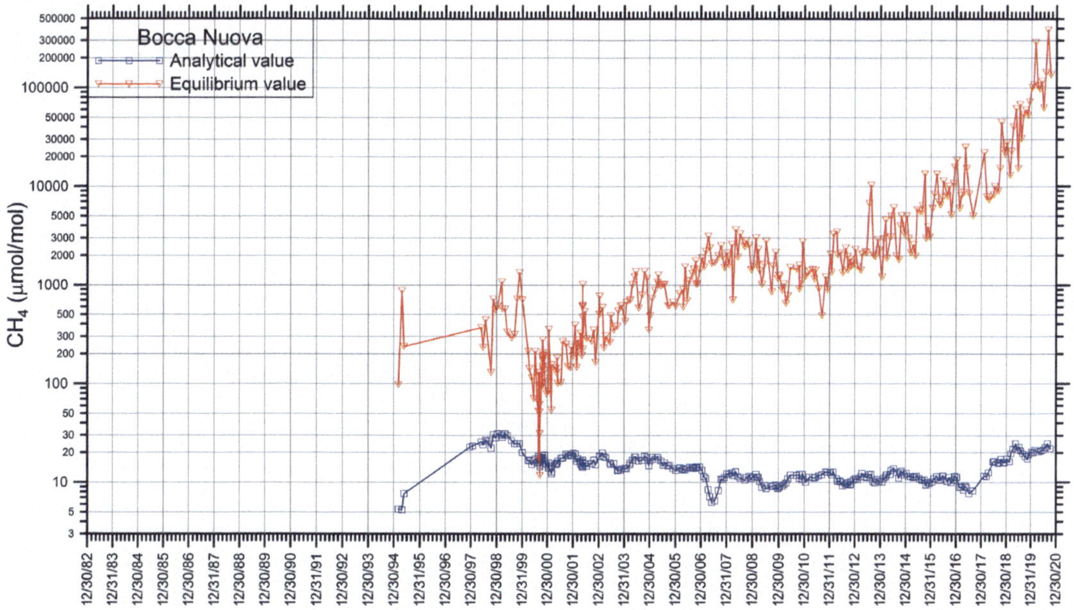

Fig. 17 Chronogram of CH_4 concentration (in log-scale) showing both the analytical value and the equilibrium value indicated by the RWG reaction in a pure saturated vapor phase for the Bocca Nuova fumarolic fluids, based on the data of Chiodini and coworkers (e.g., Chiodini et al. 2011, 2015, 2016, 2021)

concentrations with time, the equilibrium log-concentration agrees with the analytical value within 17–19%, for the samples collected from Bocca Grande on 18 April 1997 and 22 and 23 August 2000 and the agreement is even better for the Bocca Nuova sample collected on 23 August 2000, with a deviation of 1% only, and a sample collected at Bocca Nuova on 13 September 2000, with a deviation of 10%. Not surprisingly, these samples have also concordant equilibrium temperatures and total fluid pressures (see Sect. 4).

The amount of methane to be produced through conversion of CO and CO_2, according to reactions (2)–(5), for attainment of the CH_4 equilibrium concentration, is increasingly higher with time. Although the availability of CO_2 is not a limiting factor, the low reactivity of CH_4 appears to be an impassable obstacle for its equilibration, at least after 1985. Nevertheless, the satisfactory agreement between equilibrium CH_4 concentrations and analytical values of Bocca Grande fluids, in the period 1983–1985, suggests that the zone where the RGW reaction equilibrates (or, in other terms, CO equilibrates) acted as a closed system, with residence times of the gases long enough for CH_4 equilibration, even though the achievement of this condition was favored by the low CH_4 equilibrium concentrations of that period. Subsequently, it appears that the zone where the RGW reaction equilibrates behaved as an open system, being crossed by a gas flow coming from below, with gas residence times too short for the equilibration of methane, also due to the progressively higher CH_4 equilibrium concentrations with time.

7 Comparison of H_2 Concentrations Consistent with the CCC Equilibrium Temperature and Analytical Values

Hydrogen reacts quickly to changes in temperature and redox conditions (Giggenbach 1987; Chiodini et al. 1993b) and most multi-component geothermometers involving CH_4 and NH_3 are actually hydrogen geothermometers (Taran 1986). The fast reactivity of hydrogen has been

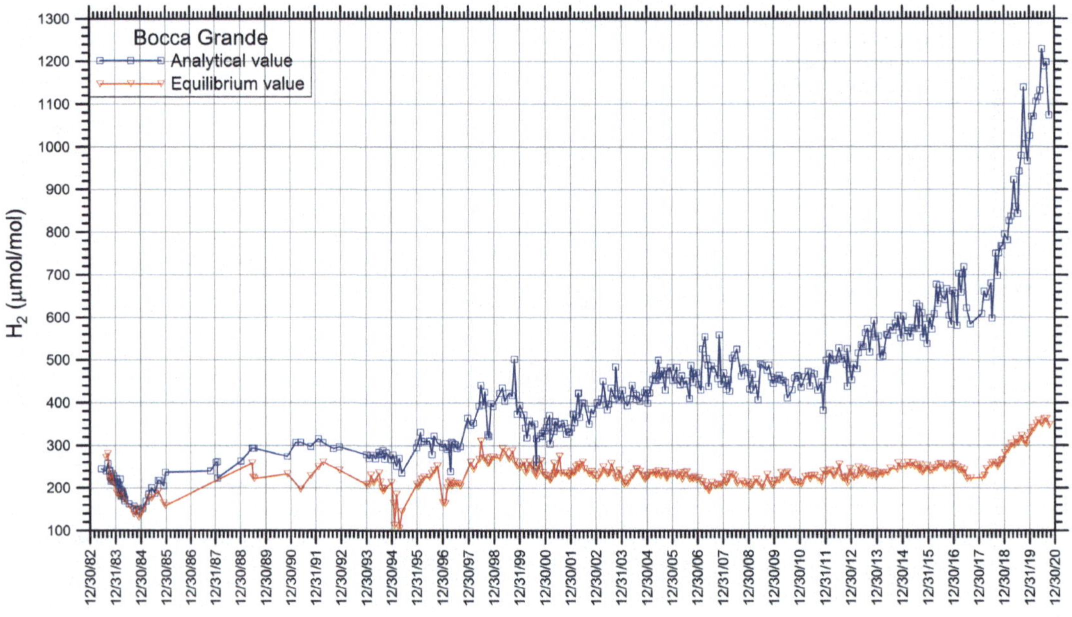

Fig. 18 Chronogram of the H_2 concentration (in log-scale) showing both the analytical value and the equilibrium value indicated by the CCC reaction in a pure saturated vapor phase for the Bocca Grande fumarolic fluids, based on the data of Chiodini and coworkers (e.g., Chiodini et al. 2011, 2015, 2016, 2021)

utilized to formulate the H_2–Ar geothermometer (Giggenbach and Goguel 1989) and the H_2–N_2 geothermometer (Brombach et al. 2003). For the same reason, it is interesting to evaluate the possible changes of hydrogen in the Solfatara fumarolic fluids.

Since hydrogen is not included in the CCC reaction, the equilibrium temperature given by this reaction, hypothesizing attainment of equilibrium in a pure saturated vapor phase, is introduced into Eq. (9), together with the measured values of X_{H_2O}, X_{CO_2}, and X_{CO}, and Eq. (9) is solved for X_{H_2}. Then, these computed X_{H_2} values are compared with the corresponding analytical values, to test the meaningfulness of the CCC temperatures, which was already questioned in Sect. 4. Not surprisingly, by using the computed equilibrium X_{H_2} instead of the analytical X_{H_2}, the RWG, SS1, SS3, and SS4 reactions provide the same temperature, P_{H_2O}, P_{CO_2}, and total fluid pressure values given by the CCC reaction, assuming again attainment of equilibrium in a pure saturated vapor phase.

The chronogram of the equilibrium and analytical H_2 concentrations for Bocca Grande (Fig. 18) shows that the equilibrium H_2 concentrations of these fluids are in excellent agreement with the measured values, with absolute deviations lower than 14%, from June 1983 to September 1985. After this date, corresponding with the cessation of the 1982–1985 uplift episode and related seismicity (e.g., Chiodini et al. 2010 and references therein), the analytical H_2 concentrations of Bocca Grande experienced a considerable increase with time, from 146–258 μmol/mol in 1983–1985 to 967–1230 μmol/mol in October 2019–October 2020. Similarly, the analytical H_2 concentrations of Bocca Nuova increased from 255–272 μmol/mol in 1995 to 988–1402 μmol/mol in October 2019–October 2020 (Fig. 19).

In contrast, the equilibrium H_2 concentrations exhibited limited oscillations, from 106 to 275 μmol/mol, around an average of 236 μmol/mol and a median of 234 μmol/mol, with a standard deviation of 36.0 μmol/mol at

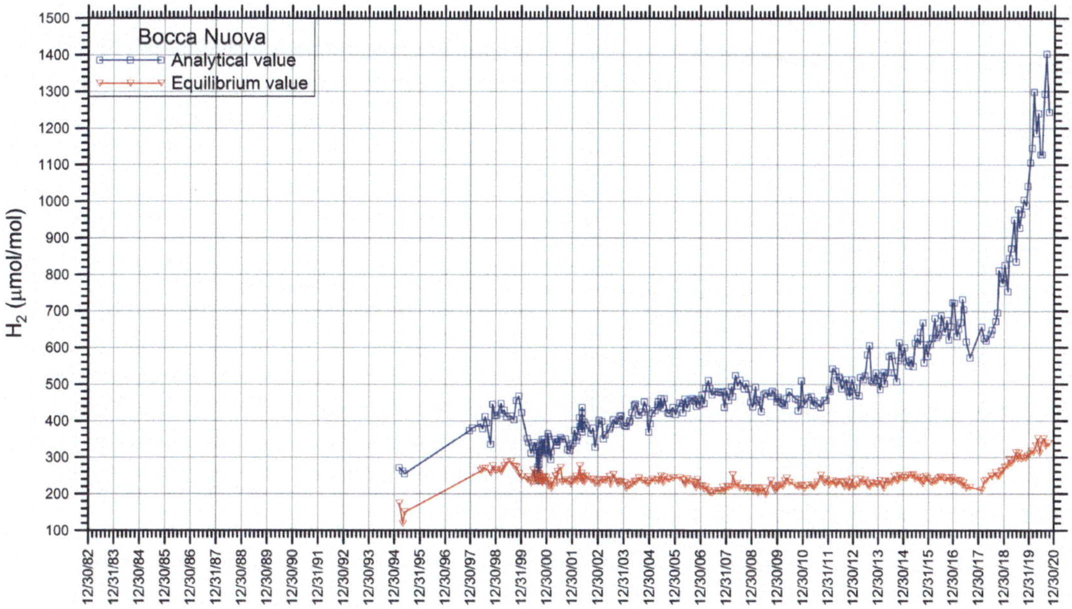

Fig. 19 Chronogram of the H_2 concentration (in log-scale) showing both the analytical value and the equilibrium value indicated by the CCC reaction in a pure saturated vapor phase for the Bocca Nuova fumarolic fluids, based on the data of Chiodini and coworkers (e.g., Chiodini et al. 2011, 2015, 2016, 2021)

Bocca Grande, and from 120 to 354 µmol/mol, around an average of 241 µmol/mol and a median of 237 µmol/mol, with a standard deviation of 27.2 µmol/mol at Bocca Nuova. Consequently, the absolute deviations between analytical and equilibrium H_2 concentration increase with time up to 73% for Bocca Grande and 76% for Bocca Nuova.

In order to equilibrate H_2 at the temperatures indicated by the CCC reaction for a pure saturated vapor phase, it would be necessary to convert considerable amounts of H_2, according to reactions (1), (3), (4), and (5). Reaction (1), however, is likely to proceed in the opposite direction, consuming CO (see Sect. 5), whereas reactions (3), (4), and (5) are unlikely to take place because they would produce considerable amounts of CH_4, in contrast with what was observed in Sect. 6. This matter will be reconsidered in chapter "The Chemical Kinetics of the Reactions Involving H_2O, CO_2, CO, CH_4 and H_2" Fig. 18.

8 Effects of Steam Condensation

As shown in Sect. 4.1 in chapter "The Expansion (Decompression) of the Solfatara Fumarolic Fluids", the maximum mass fraction of condensed steam (in % units), possibly produced during the isenthalpic decompression of Solfatara fluids, is the complement to 100 of the lowest mass steam fraction, y%, which is given by Eqs. (114) and (115) in chapter "The Expansion (Decompression) of the Solfatara Fumarolic Fluids", for Bocca Grande and Bocca Nuova, respectively. Since the lowest mass steam fraction is a function of the CO_2 mol fraction only, as dictated by Eqs. (114) and (115) in chapter "The Expansion (Decompression) of the Solfatara Fumarolic Fluids", also the maximum mass fraction of condensed steam depends on the CO_2 mol fraction only, as shown by the correlation diagram of Fig. 20. For Bocca Grande, the maximum mass fraction of condensed steam

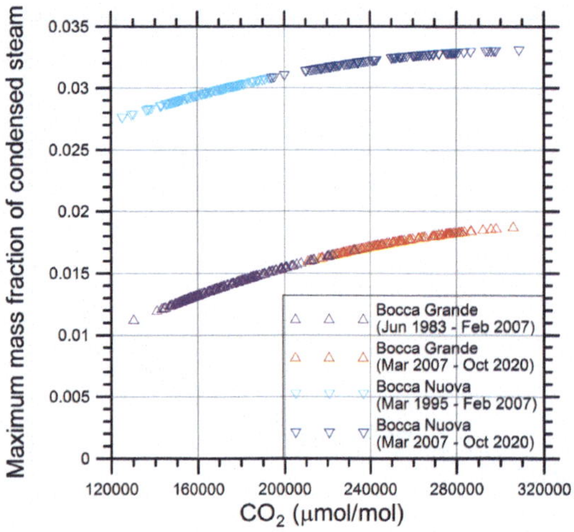

Fig. 20 Correlation diagram between the maximum mass fraction of condensed steam possibly produced during the isenthalpic decompression of Solfatara fluids and the analytical CO_2 concentration for the fumarolic fluids of Bocca Grande and Bocca Nuova, based on the data of Chiodini and coworkers (e.g., Chiodini et al. 2011, 2015, 2021)

varies between 0.0114 and 0.0189, with a mean of 0.0154, a median of 0.0152, and a standard deviation of 0.00195 whereas, for Bocca Nuova, it is about twice as large, oscillating between 0.0275 and 0.0330, with a mean of 0.0309, a median of 0.0314, and a standard deviation of 0.00147.

Owing to the dependence of the maximum mass fraction of condensed steam on the CO_2 mol fraction, the chronograms of these two variables have the same look, even though values are different. In particular, the chronogram of the maximum mass fraction of condensed steam for Bocca Grande (Fig. 21) is characterized by three ups and downs before 2000, and by a continuous increase afterwards, with an higher slope in the period 2000–2006. The chronogram of the maximum mass fraction of condensed steam for Bocca Nuova is very similar to that of Bocca Grande but, since geochemical monitoring began later, it shows only a decrease in 1995–2000 followed by an increase afterwards, again with a change in slope in 2007.

The mol fraction of H_2O in the initial gas mixtures of interest, that is, before the occurrence of steam condensation, $X_{H_2O,Vo}$, was calculated by means of the simple relation:

$$X_{H_2O,Vo} = X_{H_2O,V} \cdot \frac{100}{y(\%)} \tag{26}$$

where $X_{H_2O,V}$ is the mol fraction of H_2O in the remaining vapor, which is assumed to be equal to the analytical value. The concentrations of the other gas species in the initial vapor (subscript Vo) were computed considering their partitioning between the steam condensate (subscript L) and the remaining vapor (subscript V) as dictated by the mass balance:

$$X_{i,Vo} = X_{i,V} \cdot y + X_{i,L} \cdot (1 - y) \tag{27}$$

and the vapor–liquid distribution coefficient:

$$B_i = X_{i,V} / X_{i,L} \tag{28}$$

at the mean condensation temperature, which was computed by averaging the upper and lower saturation temperatures. Inserting Eq. (28) in Eq. (27) and rearranging leads to:

$$X_{i,Vo} = X_{i,V} \cdot \left(y + \frac{1 - y}{B_i} \right) \tag{29}$$

which is the needed relation. The vapor–liquid distribution coefficients were taken from Giggenbach (1980) for CO_2, H_2S, N_2, CH_4, and H_2 and from Bertrami et al. (1985) for CO, whereas those of He and Ar were derived from the Henry's constant, $K_{H,i}$, calculated by running SUPCRT92 (Johnson et al. 1992), by means of the relation (from Giggenbach 1980):

$$B_i = \frac{K_{H,i} \cdot V_S}{R \cdot T} \tag{30}$$

using consistent units. For instance, R = 0.0831446 (L bar)/(mol K), with $K_{H,i}$ in (bar kg)/mol, V_S in L/kg, and T in K. In this

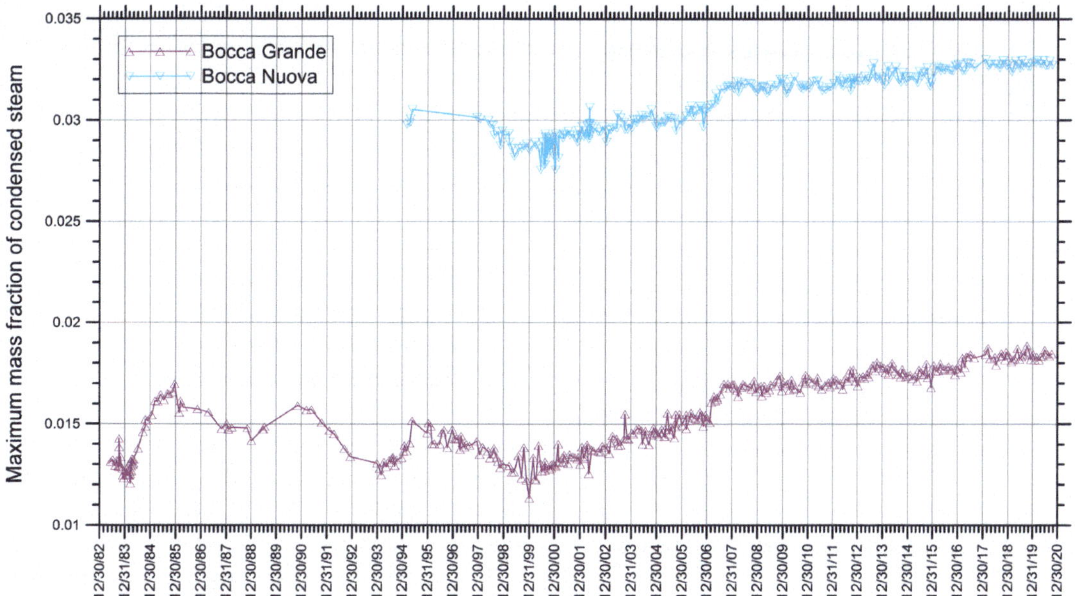

Fig. 21 Chronograms of the maximum mass fraction of condensed steam for Bocca Grande and Bocca Nuova, based on the data of Chiodini and coworkers (e.g., Chiodini et al. 2011, 2015, 2016, 2021)

way, the following two equations were obtained (T in °C):

$$\log B_{He} = 6.3769 - 0.01456 \cdot T \quad (31)$$

$$\log B_{Ar} = 6.1639 - 0.01408 \cdot T \quad (32)$$

The concentrations in the initial vapor computed by means of Eq. (29) were then readjusted in order to make their sum equal to the sum of the analytical concentrations. For Bocca Grande fluids, the average absolute deviation of the readjusted concentrations in the initial vapor from the corresponding analytical values is 0.65% (range 0.30–1.17%) for H_2O, while it varies from 2.40 to 2.49% (minimum 1.94–1.96%; maximum 2.61–2.65%) for the other gas species. For Bocca Nuova fluids, the mean absolute deviation of the readjusted concentrations in the initial vapor from the corresponding analytical values is 1.29% (range 0.69–2.06%) for H_2O, whereas it varies from 4.81 to 4.86% (minimum 4.51–4.57%; maximum 4.88–4.93%) for the other gas species.

The difference between the RWG temperatures computed using the readjusted concentrations in the initial vapor and those calculated utilizing the analytical concentrations has average and median of 1.6 °C, range of 1.1–2.0 °C, and standard deviation of 0.2 °C, for Bocca Grande, and mean and median of 3.3 °C, range of 2.7 to 3.6 °C, and standard deviation of 0.2 °C, for Bocca Nuova.

The difference between the SS4 temperatures computed using the readjusted concentrations in the initial vapor and those calculated utilizing the analytical concentrations has average of −3.2 °C, median of −2.9 °C, range of −5.0 to −1.9 °C, and standard deviation of 0.8 °C, for Bocca Grande, and mean of −6.6 °C, median of −6.7 °C, range of −9.3 to −4.8 °C, and standard deviation of 1.0 °C, for Bocca Nuova. These uncertainties in the RWG and SS4 temperatures caused by steam condensation during isenthalpic decompression are much lower than the uncertainties brought about by the lack of provision for departures from the perfect gas behavior (see chapter "Calibration and Application of the RWG and SS4 Indicators for the Possible Expansion Paths of Solfatara Fluids").

9 Summary of the Results Provided by the Geoindicators Involving H_2O, CO_2, CO, CH_4 and H_2 for the Solfatara Fluids

As shown in previous sections of this chapter, two distinct approaches to geothermometry were adopted. One is based on the attainment of equilibrium among all gas species, that is, on the full equilibrium hypothesis, while the other is based on the assumption that the RWG, CCC, SS1, SS3, and SS4 reactions attain the equilibrium condition in a pure saturated vapor phase, but independently of each other.

9.1 The Full Equilibrium Approach

The full equilibrium approach gives, on average, equilibrium temperatures of 223.3 ± 9.9 °C for Bocca Grande and 225.8 ± 7.0 °C for Bocca Nuova, P_{H_2O} values of 5.01 ± 6.95 bar for Bocca Grande and 3.77 ± 4.40 bar for Bocca Nuova, P_{CO_2} values of 1.10 ± 1.38 bar for Bocca Grande and 0.79 ± 0.70 bar for Bocca Nuova, and total fluid pressure values of 6.11 ± 8.31 bar for Bocca Grande and 4.56 ± 5.10 bar for Bocca Nuova. Time changes of total fluids pressure are scarcely realistic because they indicate a gradual depressurization of the hydrothermal system, which is at variance with the expansion of the area affected by soil degassing of deeply derived CO_2 observed from 1998 to 2020 (Chiodini et al. 2010, 2021; Cardellini et al. 2017). Furthermore, total fluid pressures lower than 1 bar, as in the last years of monitoring, are unrealistic, being too low.

In most cases, full equilibrium temperatures and pressures are consistent with the attainment of gas equilibration in a superheated vapor phase and the presence of dry conditions along the whole flowpath of the uprising fluids. However, this outcome contrasts with the probable presence of saturation conditions (coexistence of water vapor and liquid) along most of the upflow path of the Bocca Grande and Bocca Nuova fluids, as indicated by measured outlet temperatures and the hypothesis of isenthalpic flow. Moreover, the transport of strongly acidic gases (e.g., SO_2, HCl and HF) and other neutral species (e.g., H_3BO_3 and SiO_2) is expected to take place under dry conditions, but is at variance with available analytical data.

9.2 The Approach Based on the Independent Equilibrium in a Pure Saturated Vapor Phase

Alternatively, five different equilibrium temperatures and related P_{H_2O}, P_{CO_2}, and total fluid pressure values are obtained, for each gas mixture, by assuming that the RWG, CCC, SS1, SS3, and SS4 reactions attain the equilibrium condition in a pure saturated vapor phase, but independently of each other. These five different equilibrium temperatures reflect the lack of equilibrium between CO and CH_4 and the distinct influence of these two gas species on each reaction. Only the two extreme equilibrium temperatures given by the RWG and SS4 reactions are probably meaningful because the RWG reaction includes CO but not CH_4 and the SS4 reaction involves CH_4 but not CO. In contrast, the intermediate equilibrium temperatures given by the CCC, SS1, and SS3 reactions are probably meaningless because they are distinct combinations of the two equilibrium temperatures given by the RWG and SS4 reactions.

The RWG equilibrium temperature is equal to the full equilibrium temperature and exhibits limited changes with time mostly fluctuating between 210 and 250 °C. In contrast, the SS4 equilibrium temperature shows a progressive but remarkable increase with time from 224 °C in March 1984 to 440 °C in July 2020 for Bocca Grande and from 248 °C in August 2000 to 470 °C in September 2020 for Bocca Nuova.

The RWG total fluid pressure shows small changes with time mostly oscillating between 6.5 and 35.9 bar for Bocca Grande through the 1983–2020 period. In contrast, the SS4 total fluid pressure exhibits a gradual but pronounced increase with time from 20.2–29.6 bar in 1983–1984 to 255–316 bar in 2020 for Bocca Grande and from 45.0–56.8 bar in 1995–1998 to 301–401 bar in 2020 for Bocca Nuova, apart from short-term fluctuations.

The RWG equilibrium temperature and fluid pressure do not deviate significantly from the SS4 equilibrium temperature and fluid pressure, respectively, until 1985, due to attainment of full equilibrium among relevant gas species, including CO and CH_4, in a pure saturated vapor phase. In contrast, the differences between the SS4 and RWG equilibrium temperatures and fluid pressures increase progressively after 1985, due to the growing degree of disequilibrium between CO and CH_4. Interestingly, this shift from equilibrium to disequilibrium between CH_4 and CO corresponds to the end of the 1982–1985 uplift episode and related seismicity (e.g., Chiodini et al. 2010 and references therein). Nevertheless, the RWG and SS4 equilibrium temperatures and fluid pressures are concordant or relatively so for the samples collected from Bocca Grande on 18 April 1997 and 22 and 23 August 2000 and from Bocca Nuova on 23 August 2000 and 13 September 2000.

Summing up, the two RWG and SS4 reactions provide two strongly contrasting scenarios. In fact, the RWG reaction exhibits a substantial lack of temporal variations in both the equilibrium temperature and total fluid pressure, while time changes are very pronounced for the equilibrium temperature and total fluid pressure given by the SS4 reaction.

9.3 Considerations on Carbon Monoxide

The recalculated (equilibrium) CO concentration consistent with the SS4 equilibrium temperature exhibits a gradual but marked increase with time from 0.22–0.51 μmol/mol in June 1983–January 1985 to 40.9–62.1 μmol/mol in 2020, for Bocca Grande, and from 1.1–1.2 μmol/mol in March–May 1995 to 50.0–92.0 μmol/mol in 2020, for Bocca Nuova. The equilibrium CO concentration agrees with the analytical value from June 1983 to September 1985, whereas after this date, the equilibrium CO concentration becomes greater and greater than the analytical value, which oscillates from 0.10 to 3.53 μmol/mol, at Bocca Grande, and from 0.17 to 3.78 μmol/mol, at Bocca Nuova. The considerable differences between the equilibrium CO concentrations and the analytical values imply that the reactions converting CO in CO_2 and/or CH_4 proceed quickly (see chapter "The Chemical Kinetics of the Reactions Involving H_2O, CO_2, CO, CH_4 and H_2").

9.4 Considerations on Methane

The recalculated (equilibrium) CH_4 concentration consistent with the RWG equilibrium temperature experiences an increase of three to four orders of magnitude with time, from 8.5–53.1 μmol/mol, in June 1983–September1985, to 42,800–184,000 μmol/mol, in October 2019–October 2020, at Bocca Grande, and from a minimum of 11.8 μmol/mol to a maximum of 391,000 μmol/mol at Bocca Nuova, In contrast, the analytical CH_4 concentration oscillates between 7.21 and 41.4 μmol/mol, at Bocca Grande, and from 5.29 and 31.4 μmol/mol, at Bocca Nuova. This implies that the amounts of methane to be produced through conversion of CO and CO_2 for attaining the equilibrium concentrations at the RWG temperature are increasingly higher with time and are probably unrealistic due to the low CH_4 reactivity. The hydrothermal aquifer zone where the RGW reaction equilibrates (or, in other terms, CO equilibrates) behaved as a closed system, in the period 1983–1985, with residence times of the gases long enough for CH_4 equilibration, even though the achievement of this condition was favored by the low CH_4 equilibrium concentrations of that period. Afterwards, the hydrothermal aquifer zone of interest acted as an open system, with gas residence times too short for the equilibration of methane also due to the progressively higher CH_4 equilibrium concentrations with time.

9.5 Considerations on Hydrogen

Although the CCC equilibrium temperature is probably meaningless, being a combination of the RWG and SS4 equilibrium temperatures, it was used to recalculate the equilibrium concentration of H_2, since this gas species is not involved in the CCC reaction. It turns out that the equilibrium H_2 concentration exhibits limited oscillations, from 106 to 275 μmol/mol at Bocca Grande and from 120 to 354 μmol/mol at Bocca Nuova, whereas the analytical H_2 concentration experienced a considerable increase with time, from 146–258 μmol/mol in 1983–1985 to 967–1230 μmol/mol in October 2019–October 2020 at Bocca Grande and from 255–272 μmol/mol in 1995 to 988–1402 μmol/mol in October 2019–October 2020 at Bocca Nuova.

In order to equilibrate H_2 at the temperatures indicated by the CCC reaction for a pure saturated vapor phase, it would be necessary to convert considerable amounts of H_2, through reactions that are unlikely to occur or are likely to take place in the opposite direction.

9.6 Considerations on Steam Condensation

Assuming isenthalpic decompression of Solfatara fluids, the maximum mass fraction of condensed steam is a function of the CO_2 mol fraction only, as dictated by Eqs. (114) and (115) in chapter "The Expansion (Decompression) of the Solfatara Fumarolic Fluids", and results to vary between 0.0114 and 0.0189, with a mean of 0.0154, a median of 0.0152, and a standard deviation of 0.00195, for Bocca Grande, whereas it is about twice as large, oscillating between 0.0275 and 0.0330, with a mean of 0.0309, a median of 0.0314, and a standard deviation of 0.00147, for Bocca Nuova.

Since the maximum mass fraction of condensed steam is relatively small, the computed concentrations in the initial vapor (before steam condensation) do not deviate significantly from the analytical values, with average absolute deviation of 0.65% for H_2O and 2.40 to 2.49% for the other gas species, at Bocca Grande, and 1.29% for H_2O and 4.81 to 4.86% for the other gas species, at Bocca Nuova.

On average, (1) the RWG temperatures computed using the concentrations in the initial vapor differ from those calculated utilizing the analytical concentrations by +1.6 °C for Bocca Grande and +3.3 °C for Bocca Nuova, and (2) the SS4 temperatures computed using the concentrations in the initial vapor diverge from those calculated utilizing the analytical concentrations by −3.2 °C for Bocca Grande and −6.6 °C for Bocca Nuova. Steam condensation causes uncertainties on the RWG and SS4 temperatures which are much lower than those induced by the lack of provision for deviations from the ideal gas behavior (see chapter "Calibration and Application of the RWG and SS4 Indicators for the Possible Expansion Paths of Solfatara Fluids").

References

Andreussi P, Corsi R, Guidi M, Marini L (1994) Theoretical prediction of physical and chemical characteristics of the "first drop" of condensate from superheated geothermal steam. Implications for corrosion and scaling in turbines. Geothermics 23:233–255

Arnórsson S (1985) Gas pressures in geothermal systems. Chem Geol 49(1–3):319–328

Arnórsson S, Gunnlaugsson E, Svavarsson (1983) The chemistry of geothermal waters in Iceland. III. Chemical geothermometry in geothermal investigations. Geochim Cosmochim Acta 47:567–577

Arnórsson S, Gunnlaugsson E (1985) New gas geothermometers for geothermal exploration - Calibration and application. Geochim Cosmochim Acta 49:1307–1325

Ballentyne DWG, Lovett DR (1980) A dictionary of named effects and laws in chemistry, physics and mathematics. Chapman & Hall, 360 pp

Bertrami R, Cioni R, Corazza E, D'Amore F, Marini L (1985) Carbon monoxide in geothermal gases. Reservoir temperature calculations at Larderello (Italy). Geotherm Res Counc Trans 9:299–303

Brombach T, Caliro S, Chiodini G, Fiebig J, Hunziker JC, Raco B (2003) Geochemical evidence for mixing of magmatic fluids with seawater, Nisyros hydrothermal system, Greece. Bull Volcanol 65(7):505–516

Cardellini C, Chiodini G, Frondini F, Avino R, Bagnato E, Caliro S, Lelli M, Rosiello A (2017) Monitoring diffuse volcanic degassing during volcanic unrests: the case of Campi Flegrei (Italy). Sci Rep 7(1):6757. https://doi.org/10.1038/s41598-017-06941-2

Chiodini G, Avino R, Caliro S, Minopoli C (2011) Temperature and pressure gas geoindicators at the Solfatara fumaroles (Campi Flegrei). Ann Geophys 54 (2). https://doi.org/10.4401/ag-5002

Chiodini G, Cioni R (1989) Gas geobarometry for hydrothermal systems and its application to some Italian geothermal areas. Appl Geochem 4:465–472

Chiodini G, Caliro S, Cardellini C, Granieri D, Avino R, Baldini A, Donnini M, Minopoli C (2010) Long-term variations of the Campi Flegrei, Italy, volcanic system as revealed by the monitoring of hydrothermal activity. J Geophys Res Solid Earth 115(B3):B03205. https://doi.org/10.1029/2008JB006258

Chiodini G, Caliro S, Avino R, Bini G, Giudicepietro F, De Cesare W, Ricciolino P, Aiuppa A, Cardellini C, Petrillo Z, Selva J, Siniscalchi A, Tripaldi S (2021) Hydrothermal pressure-temperature control on CO_2 emissions and seismicity at Campi Flegrei (Italy). J Volcanol Geotherm Res 414:107245

Chiodini G, Cioni R, Leonis C, Marini L, Raco B (1993a) Fluid geochemistry of Nisyros island, Dodecanese, Greece. J Volcanol Geotherm Res 56:95–112

Chiodini G, Paonita A, Aiuppa A, Costa A, Caliro S, De Martino P, Acocella V, Vandemeulebrouck J (2016) Magmas near the critical degassing pressure drive volcanic unrest towards a critical state. Nat Commun 7 (1):1–9

Chiodini G, Cioni R, Marini L (1993b) Reactions governing the chemistry of crater fumaroles from Vulcano Island, Italy, and implications for volcanic surveillance. Appl Geochem 8:357–371

Chiodini G, Marini L (1998) Hydrothermal gas equilibria: the H_2O–H_2–CO_2–CO–CH_4 system. Geochim Cosmochim Acta 62:2673–2687

Chiodini G, Vandemeulebrouck J, Caliro S, D'Auria L, De Martino P, Mangiacapra A, Petrillo Z (2015) Evidence of thermal-driven processes triggering the 2005–2014 unrest at Campi Flegrei caldera. Earth Planet Sci Lett 414:58–67

Cioni R, Marini L (1990) The determination of deep temperatures by means of the CO–CO_2-H_2–H_2O geothermometer: an example using fumaroles in the Campi Flegrei, Italy. A comment. Bull Volcanol 53 (1):67–69

D'Amore F, Celati R (1983) Methodology for calculating steam quality in geothermal reservoir. Geothermics 12:129–140

D'Amore F, Gianelli G (1984) Mineral assemblages and oxygen and sulphur fugacities in natural water-rock interaction processes. Geochim Cosmochim Acta 48:847–857

D'Amore F, Nuti S (1977) Notes on the chemistry of geothermal gases. Geothermics 6:39–45

D'Amore F, Panichi C (1980) Evaluation of deep temperature of hydrothermal systems by a new gas-geothermometer. Geochim Cosmochim Acta 44:549–556

Fiebig J, Woodland AB, Spangenberg J, Oschmann W (2007) Natural evidence for rapid abiogenic hydrothermal generation of CH_4. Geochim Cosmochim Acta 71(12):3028–3039

Fiebig J, Stefánsson A, Ricci A, Tassi F, Viveiros F, Silva S, Lopez TM, Schreiber C, Hofmann S, Mountain BW (2019) Abiogenesis not required to explain the origin of volcanic-hydrothermal hydrocarbons. Geochem Perspect Lett 11:23–27

Giggenbach WF (1980) Geothermal gas equilibria. Geochim Cosmochim Acta 44:2021–2032

Giggenbach WF (1981) Geothermal mineral equilibria. Geochim Cosmochim Acta 45(3):393–410

Giggenbach WF (1987) Redox processes governing the chemistry of fumarolic gas discharges from White Island, New Zeland. Appl Geochem 2:143–161

Giggenbach WF (1991) Chemical techniques in geothermal exploration. In: D'Amore F, co-ordinator (eds) Application of geochemistry in geothermal reservoir development. UNITAR, pp 119–144

Giggenbach WF (1993) Redox control of gas compositions in Philippine volcanic-hydrothermal systems. Geothermics 22:575-587.

Giggenbach WF, Goguel RL (1989) Collection and analysis of geothermal and volcanic water and gas discharges. Unpublished report, Chemistry Division, DSIR-Petone, New Zealand, 81 pp

Hulston JR, McCabe WJ (1962) Mass spectrometer measurements in the thermal areas of New Zealand: Part 1. Carbon dioxide and residual gas analyses. Geochim Cosmochim Acta 26(3):383–397

Johnson JW, Oelkers EH, Helgeson HC (1992) SUPCRT92: a software package for calculating the standard molal thermodynamic properties of minerals, gases, aqueous species, and reactions from 1 to 5000 bar and 0 to 1000 °C. Comput Geosci 18(7):899–947

Marini L, Agostini A, Cioni R, Guidi M, Leon O (1991) Guagua Pichincha volcano, Ecuador: fluid geochemistry in volcanic surveillance. J Volcanol Geotherm Res 46:21–35

Nehring NL, D'Amore F (1984) Gas chemistry and thermometry of the Cerro Prieto, Mexico, geothermal field. Geothermics 13:75–89

Saracco L, D'Amore F (1989) CO2B: a computer program for applying a gas geothermometer to geothermal systems. Comput Geosci 15(7):1053–1065

Stefánsson A, Arnórsson S (2002) Gas pressures and redox reactions in geothermal fluids in Iceland. Chem Geol 190(1–4):251–271

Taran YA (1986) Gas geothermometers for hydrothermal systems. Geochem Int 23(7):111–126

Tonani F (1973) Equilibria that control the hydrogen content of geothermal gases. Unpublished Phillips Petroleum Co. report

The Chemical Kinetics of the Reactions Involving H_2O, CO_2, CO, CH_4 and H_2

Abstract

To contextualize the kinetics of the reactions involving H_2O, CO_2, CO, CH_4 and H_2, first of all, we reviewed (i) the groundbreaking researches performed by Paul Sabatier and coworkers on the conversion of CO_2 and CO to CH_4, (ii) the investigations performed by geochemists on the methanation of CO_2 and CO, aimed at clarifying the origin of the so-called abiotic methane, and (iii) the kinetic studies carried out by chemical engineers on the hydrogenation of CO_2 and CO to CH_4 as well as on the steam reforming of methane coupled with the water-gas shift reaction. Then, adopting the kinetic model of Xu and Froment (1989a), we implemented two reaction path models, one involving the three reactions RWG-WG, SS3-RSS3, and SS4-RSS4, the other comprising the two reactions RWG-WG and SS3-RSS3 only. Both models were used to simulate the heating of fourteen suitably selected samples of Solfatara fluids from the outlet temperature up to a maximum temperature higher than the SS4-RSS4 inversion temperature, for the 3-reactions model, or the SS3-RSS3 inversion temperature, for the 2-reactions model, by a few tens of degrees. In the simulations of both types, H_2O and CO_2 concentrations undergo nil to negligible variations, H_2 and CH_4 concentrations experience variable changes depending on the maximum temperature, whereas CO concentration suffers significant variations, passing through a minimum value, irrespective of the maximum temperature. In the 3-reactions model, reactions RSS3 and RSS4 invert at the same temperature, which is the full equilibrium temperature of the H_2O-CO_2-H_2-CH_4-CO system. The full equilibrium temperature increases with time, from 235°C in 1983 to 342-347°C in 2020. The inversion temperature of reaction WG is representative of the partial equilibrium condition among H_2O, CO_2, H_2, and CO and corresponds to the minimum attained by the CO concentration. Among the fourteen selected samples of Solfatara fluids: (i) there are small differences, 1 to 14°C, between the full equilibrium temperature from the heating simulations and the SS4 temperature calculated from the analytical data (see section "The Geoindicators Involving H_2O, CO_2, CO, CH_4 and H_2"), for the samples collected from 1983 to 2006, but (ii) the samples collected from 2012 to 2020 have full equilibrium temperature from 311 to 347°C, which is lower than the SS4 temperature (340 to 470°C) by 29 to 123°C. In the 2-reactions model, the SS3-RSS3

Supplementary Information The online version contains supplementary material available at (https://doi.org/10.1007/978-3-030-98471-7_7).

L. Marini et al., *The Solfatara Magmatic-Hydrothermal System*, Advances in Volcanology,
https://doi.org/10.1007/978-3-030-98471-7_7

inversion temperature increases with time from 219°C in 1983 to 432-460°C in 2020, whereas the WG-RWG inversion temperature marks again the minimum CO concentration. For the fourteen selected samples of Solfatara fluids, there is a small difference, 7-13°C only, between the SS3-RSS3 inversion temperature and the SS4 temperature computed from the analytical data. Irrespective of the considered deep temperature, there is no doubt that the Solfatara magmatic-hydrothermal system experienced a gradual heating from 1983 to 2020. ■■■

The kinetics of relevant reactions involving H_2O, CO_2, CO, CH_4 and H_2 is consider in this chapter with the aim to quantify its effects on the results of the geothermometric functions already discussed in chapter "The Geoindicators Involving H_2O, CO_2, CO, CH_4 and H_2" from a purely thermodynamic point of view, especially on the RWG and SS4 equilibrium temperatures.

A good starting point is represented by the groundbreaking researches performed by Paul Sabatier and coworkers, during the last years of the nineteenth century and the first decades of the twentieth century, on the hydrogenation of CO_2 and CO to CH_4, which are summarized in Sect. 1. Afterwards, the catalysts promoting these processes have been the subject of a large number of investigations.

Geochemists studied the methanation of CO_2 and CO with the main aim to elucidate the origin of the so-called abiotic methane, which is derived from inorganic starting material, often using the isotopes of C and H as tracers. A short review of these geochemical investigations is given in Sect. 2.

A considerable number of researches were performed by chemical engineers for practical purposes and were addressed to understand the reaction mechanisms of the conversion of CO_2 and CO to CH_4 and to derive the related kinetic laws which are used to manage production processes in industrial plants. This subject is covered in Sect. 3. Owing to the lack of the kinetic data for the inverse reactions, it was necessary to examine also the kinetics of steam reforming, which is discussed in Sect. 4.

Finally, kinetics considerations are applied to the Solfatara fluids in Sect. 5, the processes controlling H_2O, CO_2, CO, CH_4, H_2 and redox conditions are discussed, practical implications are drawn, and some suggestions on the work to be done in future studies are given.

1 The Hydrogenation of CO_2 and CO to CH_4: A Synthesis of the Achievements of Paul Sabatier

Paul Sabatier was awarded the Nobel Prize in Chemistry in 1912 for his method of hydrogenating organic compounds in the presence of finely divided metals (Nobel Foundation 1966). Paul Sabatier wrote the classical book "*La Catalyse en Chimie Organique*" that was first published in 1913. Here, reference is made to the English translation of this Sabatier's book, which was edited by Emmet E. Reid and published in 1922 (Sabatier 1922).

Since 1897, Paul Sabatier performed a series of studies, initially with Jean-Baptiste Senderens (1897–1905) and later on with Alfonse Mailhe (1904–1908), and Marcel Murat (1912–1914) leading to establishing a general method of direct hydrogenation of volatile organic compounds, based on the use of finely divided catalytic metals and extending it to a large number of applications.

As early as 1902, the general method of direct hydrogenation of volatile organic compounds was adopted in many laboratories, based in France and other countries, and numerous chemists contributed, together with Sabatier and coworkers, to expand its application. The process consists in passing the vapors of the substance of interest mixed with hydrogen over a layer of the catalyst, either platinum black or nickel, cobalt, iron, or copper reduced from the oxides in the same tube in which the hydrogenation is to be performed, maintained at a suitable temperature, sometimes ambient temperature but more frequently at values between 150 and 200 °C. Among the five metals mentioned above, the most active is nickel.

In addition to the hydrogenation of volatile organic compounds, Sabatier and coworkers investigated also the hydrogenation of carbon monoxide and carbon dioxide. They found that the hydrogenation of carbon monoxide catalyzed by nickel is a simple method for the synthesis of methane. This process is described by the reaction which was called SS3 in Sect. 1 in chapter "The Geoindicators Involving H_2O, CO_2, CO, CH_4 and H_2" and is re-written here below for convenience:

$$CO + 3H_2 = CH_4 + H_2O. \quad (1)$$

The SS3 reaction begins close to 180–200 °C and proceeds quickly at 230–250 °C. If the H_2:CO concentration ratio is 3:1, the SS3 reaction is virtually complete and the product gas is nearly pure methane.

At temperatures >250 °C, nickel catalyzes the decomposition of carbon monoxide into carbon and carbon dioxide, which is then partially hydrogenated. For instance:

(i) At 380 °C and H_2:CO concentration ratio 3:1, the reaction proceeds quickly producing a gas mixture constituted by 67.9 vol% CH_4, 21.6% H_2, and 10.5% CO_2,
(ii) At the same temperature and H_2:CO concentration ratio 1:1, the obtained gas mixture is formed by 52.6 vol% CO_2, 39.8 vol% CH_4, and 7 vol% H_2, in partial agreement with the reaction which was named SS1 in Sect. 1 in chapter "The Geoindicators Involving H_2O, CO_2, CO, CH_4 and H_2" and is written again here below:

$$CO + H_2 = 1/2\,CH_4 + 1/2\,CO_2. \quad (2)$$

In other terms, increasing the proportion of carbon monoxide, much H_2 passes through and the percentage of CO_2 becomes very large.

As pointed out by Sabatier, also CO_2 is readily hydrogenated over nickel to produce methane according to the reaction which was termed SS4 in Sect. 1 in chapter "The Geoindicators Involving H_2O, CO_2, CO, CH_4 and H_2" and is re-written here below:

$$CO_2 + 4H_2 = CH_4 + 2H_2O. \quad (3)$$

The SS4 reaction begins at a temperature close to 230 °C and proceeds quickly above 300 °C without any significant impediments up to 400 °C. With gas mixtures containing an excess of H_2 with respect to the H_2:CO_2 concentration ratio 4:1, CO_2 disappears almost completely, especially at high temperature. For instance, a gas mixture made up of 82% H_2 and 18% CO_2 reacts producing a gas containing 17.2% CO_2 at 258 °C and 0.5% CO_2 at 283 °C. The practical implication is that, at 300 °C and with an excess of H_2, the SS4 reaction is a useful way for preparing pure CH_4, for instance, if liquid air is used for condensing methane leaving hydrogen as gas.

Also cobalt catalyzes the SS3 reaction, which begins at 270 °C, is rapid at 300 °C, but is opposed by the decomposition of CO in C and CO_2 even more than the nickel-catalyzed process. Also the hydrogenation of CO_2 is catalyzed by cobalt at temperature of 300 °C or higher, is quick at 360 °C and even more so at 400 °C and occurs without any obstacles.

In contrast, platinum, iron and copper do not catalyze the direct hydrogenation of CO and CO_2 into CH_4, according to the SS3 and SS4 reactions. However, between 350 and 400 °C, copper begins to promote the reaction which was called RWG in Sect. 1 in chapter "The Geoindicators Involving H_2O, CO_2, CO, CH_4 and H_2" and is re-written here below:

$$CO_2 + H_2 = CO + H_2O. \quad (4)$$

The RWG reaction becomes important at 420–450 °C.

Sabatier and coworkers recognized that the production of methane by the direct hydrogenation of carbon dioxide over nickel can be utilized for the commercial production of a gas rich in methane which can be used either for heating or for lighting because of its high calorific and illuminating power. They also recognized that the preparation of hydrogen is too expensive to be used for the production of illuminating gas. Therefore, they suggested different methods employing a cheap commercial gas such as: (i) the water gas, that is, a mixture of CO and H_2 obtained by reacting steam and red-hot carbon or coke and (ii) the Riché gas, that is, a mixture of CO, H_2, CH_4 and CO_2 prepared by heating woody or cellulose materials.

2 The Hydrogenation of CO_2 and CO to CH_4: The Geochemical Studies

The hydrogenation of CO_2 and CO to CH_4 has been the subject of several geochemical studies mainly aimed at understanding the origin of abiotic methane. The isotopes of C and H were employed as tracers in some of these geochemical investigations.

High H_2 concentrations were reported for the hydrothermal aqueous solutions circulating through ultramafic rocks below the seafloor (e.g., Charlou and Donval 1993; Kelley and Früh-Green 1999) as well as for the gas discharges from continental serpentinization areas (e.g., Etiope and Whiticar 2019 and references therein), suggesting that CH_4 and light hydrocarbons present in these fluids might have an abiotic origin. Starting from these premises, ultramafic rocks and serpentinites have been the subject of several investigations on abiotic hydrocarbon formation (see below) although serpentinization is not the only fluid-rock reaction that can produce sufficiently reducing conditions to promote H_2 production and consequent CO_2 hydrogenation to CH_4 and light hydrocarbons. In fact, many other rock types and reactions may favor these processes (e.g., Potter et al. 2004; Sherwood Lollar et al. 2007).

The first laboratory experiments to investigate the production of hydrogen and methane during serpentinization of olivine and pyroxenes were probably performed by Janecky and Seyfried (1986). Dunite, lherzolite, and harzburgite were reacted with seawater in the gold-titanium reaction cells of Seyfried et al. (1979) at 500 bar pressure and temperature of 300 °C, apart from a single experiment carried out at 200 °C. However, this research was chiefly focused on the changes in seawater chemistry and the characterization of solid product phases, while gases were analyzed in one experiment only. Hydrogen was found to be a major component and methane was also detected but at lower concentration.

Adopting the same experimental approach, Berndt et al. (1996) reacted olivine (Fo88) with a CO_2-bearing NaCl fluid, comprising 0.5 mol/kg dissolved NaCl and 8.9 mmol/kg dissolved $NaHCO_3$, at 300 °C and 500 bar. Based on the obtained results, Berndt et al. (1996) concluded that conversion of divalent iron in olivine to trivalent iron in magnetite during olivine serpentinization leads to production of H_2 and conversion of dissolved CO_2 to a variety of hydrocarbon gases (mainly methane, ethane, and propane) and a carbonaceous phase. They also recognized that the product hydrocarbon gases fit a Schulz-Flory distribution, consistent with the catalysis by mineral reactants or products, possibly magnetite.

The role of reduction of dissolved inorganic carbon to hydrocarbons catalyzed by typical hydrothermal minerals was soon questioned by McCollom and Seewald (2001) who performed an experimental study similar to that of Berndt et al. (1996), reacting dissolved CO_2 and olivine at 300 °C and 350 bar, but employing ^{13}C-labelled reactants to trace the source of carbon in the hydrocarbon products. It turned out that the

^{13}C label was found only in a small fraction (2–15%) of the produced CH_4 whereas none of the C_2H_6 and C_3H_8 was labeled, suggesting that most hydrocarbons were generated from reduced carbon sources already present in the olivine at the beginning of the experiment rather than through reduction of dissolved CO_2.

Although magnetite was generated in these experiments, as typically occurs in serpentinization, the nil to negligible extent of CO_2 hydrogenation suggested that the role of magnetite as effective catalyst for this process is questionable. Therefore, other researches were carried out to reconsider the role of magnetite and to assess the possible catalytic role of other minerals occurring in serpentinites including chromite [$FeCr_2O_4$], pentlandite [$(Fe, Ni)_9S_8$], awaruite [Ni_2Fe to Ni_3Fe] and other Ni–Fe alloys.

The catalytic effect of Ni–Fe alloys was investigated by Horita and Berndt (1999) who performed two-steps experiments under conditions similar to those of the serpentinization of ultramafic rocks. First, a mixture of magnetite and Ni–Fe alloy was produced at 400 °C and 500 bar in an H_2-rich aqueous solution inside a closed, flexible gold-cell hydrothermal apparatus. Then, a $NaHCO_3$ solution was injected in the vessel, which was kept at pressure of 500 bar and temperature of 200 or 300 or 400 °C for different time intervals, monitoring the concentrations of dissolved inorganic carbonate (DIC), CH_4, and H_2. The isotopic values of DIC and CH_4 were also monitored in two runs. Very high H_2 concentrations (173–297 mmol/kg) were measured. The extent of CO_2 to CH_4 conversion increased with the amounts of Ni–Fe alloy present in the vessel, while no CH_4 was produced in a control experiment without the Ni–Fe alloy, proving that this alloy acts as catalyzer of CO_2 hydrogenation. Conversion of CO_2 to CH_4 was (i) almost complete in the 300 °C experiments in two weeks or less, (ii) >40% in the 200 °C experiment after 3 months, (iii) highly variable in the 400 °C experiments, possibly due to passivation of the catalyst at this high temperature. Although the Ni–Fe alloy employed by Horita and Berndt (1999) in their experiments was considerably enriched in Ni with respect to the awaruite occurring in natural serpentinites, it is reasonable to assume that natural awaruite have comparable catalytic properties. Thus, it was proven, for the first time, that naturally occurring minerals were able to catalyze the conversion of CO_2 to CH_4 overcoming the kinetic inhibitions.

The other experiments performed so far to understand the potential role of magnetite, chromite, and pentlandite as catalysts of CO_2 reduction to CH_4 (McCollom and Seewald 2001; Foustoukos and Seyfried 2004; Fu et al. 2007; Ji et al. 2008) did not provide conclusive evidence neither for nor against.

The $\delta^{13}C$ values of product CH_4 in the experiments of Horita and Berndt (1999) resulted to be as low as those of microbial methane, and $CH_4/(C_2H_6 + C_3H_8)$ ratios were similarly high, which was in contrast to the axiom that only methanogenic microorganisms were able to produce nearly pure CH_4 in natural environments (e.g., Whiticar 1990).

As underscored by Etiope and Whiticar (2019), the experimental works described above and similar investigations performed with liquid phases provided ambiguous and sometimes conflicting results. In contrast, high degrees of CO_2 to CH_4 conversion were obtained in experiments performed employing a gas phase with high H_2/CO_2 ratio and a suitable catalyst, even at low temperatures. Details are given below for some of these experimental researches (Taran et al. 2007, 2010; Medina et al. 2000; Zhang et al. 2013a; Etiope and Ionescu 2015).

Taran et al. (2007) investigated the isotopic effects of the catalytic reaction between CO and H_2 and accessory reactions injecting a gas mixture made up of $5N_2 + 3H_2 + 2CO$ into a 100 ml stainless steel fixed-bed reaction vessel filled with a mixture of reduced magnetite catalyst and an approximately equal amount of an Al–silicate promoter (ZSM-5 synthetic zeolite). In two runs a mixed Fe + Co-catalyst was used to minimize the fraction of unsaturated hydrocarbons. Seven runs of duration varying from 56 to 96 h were conducted at temperatures from 254 to 316 °C and pressure of 30 atm. The $\delta^{13}C$ values were measured for CO, CO_2, the C_1 to C_4 hydrocarbons, and the "oil" fraction, while the δ^2H values were determined for the produced water and the "oil" fraction. An unambiguous inverse distribution of $\delta^{13}C$ values was not observed for the light hydrocarbons produced under static, closed-system conditions and under dynamic, open-system conditions. An inverse C isotope trend was only found for the light hydrocarbons produced at low conversion of CO under dynamic, open-system conditions. This means that stable carbon isotopes do not allow one to distinguish the hydrocarbons produced through abiotic synthesis from those formed by degradation of organic matter.

Taran et al. (2010) injected a gas mixture made up of $3H_2 + CO_2$ and 10 vol% of added N_2 in a 100 ml stainless steel vertical flow-through fixed-bed reaction vessel filled with a Fe or Co catalyst. Experiments were carried out at 350 °C and 100 bar with the Fe catalyst and at 245 °C and 100 bar with the Co catalyst. The gaseous hydrocarbon products resulted to be dominated by CH_4, whereas the CH_4/C_2 ratio was 55–85 with the Co catalyst and 4.3–5.7 with the Fe catalyst. The measured carbon isotopic composition, in general, exhibits a positive trend (namely the enrichment in ^{13}C with the increase in the chain length) with the main isotope difference between CH_4 and C_2H_6. The apparent isotope fractionation between CO_2 and hydrocarbons is much greater than the equilibrium value and is chiefly controlled by the catalyst composition and/or the experiment temperature. The H isotopic composition of the products shows a positive trend (namely the enrichment in ^2H with the increase in the chain length), similar to that of thermogenic hydrocarbons, that is, hydrocarbons produced by thermal decomposition of organic matter contained in sediments. Altogether, the results of Taran et al. (2007, 2010) show that (i) the carbon source (CO or CO_2) has little influence on the chemical and isotopic composition of product hydrocarbons under open-system conditions over the Fe catalyst and (ii) the gaseous products are much more similar to thermogenic natural hydrocarbons than to CH_4-rich gases from ultramafic-hosted hydrothermal systems.

Medina et al. (2000) investigated CO_2 hydrogenation in the presence of Fe/SiO_2 and Ni/SiO_2 catalysts at temperatures of 180–200 °C. Carbon dioxide and H_2 gases were introduced into the reactor in a stoichiometric ratio of 1:4 at total pressure of 1 atm. Methane was the only product in the Ni-catalyzed CO_2 hydrogenation, whereas small amounts of C_2H_6 and C_3H_8 were found in addition to CH_4 in the Fe-catalyzed process.

Zhang et al. (2013a) studied the synthesis of hydrocarbon gases, at 400 °C and 500 bar for 2–60 h, in a gold-tube system with montmorillonite K-10 loaded with Fe^{3+} and Ni^{3+} as catalysts, using as reactants hydrogen gas and four different carbon sources, namely graphite, a 20% Na_2CO_3 solution, and two types of CO_2 gases with different isotopic compositions. The experimental results showed that: (i) the hydrogenation of dissolved carbonate could hardly occur, (ii) the reaction between gaseous CO_2 and H_2 gas was easier than that between graphite and H_2 gas, and (iii) the ^{13}C-depleted CO_2 is more reactive than the ^{13}C-enriched CO_2.

Etiope and Ionescu (2015) performed batch experiments of CO_2 hydrogenation in borosilicate bottles filled with H_2 and CO_2, in a stoichiometric ratio of 10:1, mixed with Ar. Experiments were carried out at temperatures of 20–25, 50, and 90 °C, with Ru catalysts, for 0.9–122 days. The experimental outcomes proved that abiotic CH_4, accompanied by minor amounts of C_2–C_6 hydrocarbons, can be generated at temperatures below 100 °C in the presence of ruthenium catalysts, whereas CH_4 production does not occur with Fe and Ni oxide catalysts, under the same conditions. The $\Delta^{13}C_{CO_2-CH_4}$ value decreases over time and with increasing temperature and the final $\delta^{13}C_{CH_4}$ value approaches the initial $\delta^{13}C_{CO_2}$ value.

Further indications are provided by the concentrations and $\delta^{13}C$ values of CO_2 and CH_4 in the terrestrial hydrothermal gases discharged from the Mediterranean volcanoes of Vesuvius and Ischia in Italy and Nisyros in Greece, that appear to be consistent with equilibrium with CO_2, implying the occurrence of hydrogenation of CO_2 to CH_4, according to Fiebig et al. (2007).

However, the same topic was reconsidered by Fiebig et al. (2019) who concluded that δ^2H and $\delta^{13}C$ values of CH_4 and the $\delta^{13}C$ values of the C_2 to C_4 n-alkanes from volcanic-hydrothermal emissions exhibit isotopic trends that are consistent with the occurrence of organic matter degradation under relatively high temperatures and open system conditions. Therefore, there is no need to invoke any significant contribution from an abiotic source. Similar conclusions were also reached by Reeves and Fiebig (2020).

On the one hand, understanding whether hydrocarbon (and in particular CH_4) are produced from biotic or abiotic sources is a matter of scientific interest and, in our case, is a useful information for reconstructing the conceptual model of the Solfatara magmatic-hydrothermal system. On the other hand, this information is irrelevant for using the chemical geoindicators involving H_2O, CO_2, CO, CH_4 and H_2 at the Solfatara (see Sect. 5) which are based on the chemical equilibrium hypothesis.

Thus, the question relevant for the use of geothermometers and geobarometers is whether in natural magmatic-hydrothermal systems, and specifically in the Solfatara magmatic-hydrothermal system, there are effective catalysts which are able to promote and speed up the reactions involving H_2O, CO_2, CO, CH_4 and H_2 and, possibly, other relevant reactions (e.g., the thermal decomposition of organic matter contained in sediments or in other rocks present under the volcanic products or interstratified with them), so as to allow CO_2 and CH_4 to achieve or approach the chemical equilibrium condition in a relatively short time.

The pertinent experimental investigations performed so far with liquid aqueous solutions provided ambiguous and sometimes conflicting results. Nevertheless, it has been proven that Ni–Fe alloys effectively catalyze CO_2 hydrogenation (Horita and Berndt 1999), even if the Ni–Fe alloys used in the experiments are considerably enriched in nickel compared to the awaruite present in natural serpentinites. Other catalysts might be identified in future studies.

In contrast, high degrees of CO_2 to CH_4 conversion were achieved in experiments performed employing a gas phase with high H_2/CO_2 ratio and a suitable catalyst, even at low temperatures (Etiope and Whiticar 2019). Even if unanswered questions remain, for example how these catalysts are reactivated in the natural systems of interest, the knowledge gathered so far by geochemical studies (see above) authorize us to apply suitable kinetic laws to the Solfatara fluids. Accepting this *modus operandi*, the next step is to review the studies carried out by chemical engineers to manage pertinent production processes in industrial plants, focusing on the rates and reaction mechanisms of relevant processes, first of all hydrogenation of CO_2 and CO to CH_4. This topic is the subject of the next section.

3 The Hydrogenation of CO_2 and CO to CH_4: Reaction Path, Reaction Mechanisms, and Rate Equations

In the pertinent scientific literature there is a lack of consensus on how the hydrogenation of carbon dioxide to methane proceeds (e.g., Vlasenko and Yuzefovich 1969; Darensbourg et al. 1985; Wang et al. 2011). In fact, the two following reaction paths were proposed:

(i) The conversion of CO_2 to CO via the RWG reaction, followed by CO methanation via the SS3 reaction (e.g., Falconer and Zağli

1980; Weatherbee and Bartholomew 1982; Peebles et al. 1983; Marwood et al. 1997; Lapidus et al. 2007).

(ii) The direct hydrogenation of CO_2 to methane via the SS4 reaction, that is through a direct mechanism which does not include the RWG reaction and CO methanation as intermediate steps (e.g., Fujita et al. 1987; Schild et al. 1990).

As pointed out by Weatherbee and Bartholomew (1982), the studies performed by Dalmon and Martin (1979) and Falconer and Zağli (1980), provide evidence in favor of the first of these two reaction paths, as confirmed later on by Weatherbee and Bartholomew (1984). If so, the rate of CO_2 methanation is controlled by the slowest reaction between the RWG and the SS3.

In terms of reaction mechanisms, there is a general agreement on the involvement of the following 13 distinct steps in CO methanation as underscored by Sughrue and Bartholomew (1982[1]; the numbers used to identify these steps are also used to define the rate constants and adsorption constants in the following discussion):

(1) Adsorption of H_2 onto the catalyst surface.
(2) Adsorption of CO onto the catalyst surface.
(3) Dissociation of adsorbed CO to adsorbed O and C atoms.
(4) to (7) Step-by-step hydrogenation of the adsorbed C atom to produce CH_4.
(8) Desorption of CH_4.
(9) and (10) Two-step hydrogenation of the adsorbed O atom to generate H_2O.
(11) Desorption of H_2O.
(12) Reaction between adsorbed CO and adsorbed O to form CO_2.
(13) Desorption of CO_2.

Nevertheless, there is no certainty on the rate-determining step and, therefore, on the rate equation. In fact, assuming in separate attempts that each one of these reaction steps is the rate-determining step leads to distinct Langmuir–Hinshelwood rate equations[2] describing CO methanation. For instance:

(i) If the CO adsorption is the slowest step, the rate expression is:

$$r_{CH_4} = \frac{k_{CO} \cdot P_{CO}}{1 + K_{H_2}^{0.5} \cdot P_{H_2}^{0.5}} \tag{5}$$

(ii) If the H_2 adsorption is the slowest step, the rate equation is:

$$r_{CH_4} = \frac{k_{H_2} \cdot P_{H_2}}{(1 + K_{CO} \cdot P_{CO})^2} \tag{6}$$

(iii) If the dissociation of adsorbed CO to adsorbed O and C atoms is the slowest step, the rate relation is:

$$r_{CH_4} = \frac{k_3 \cdot K_{CO} \cdot P_{CO}}{\left(1 + K_{CO} \cdot P_{CO} + K_{H_2}^{0.5} \cdot P_{H_2}^{0.5}\right)^2} \tag{7}$$

(iv, a) If the hydrogenation of the adsorbed C atom to CH is the slowest step, the rate expression is:

$$r_{CH_4} = \frac{k_4 \cdot k_9 \cdot K_{H_2} \cdot \left(\frac{P_{H_2}}{k_{-3}}\right)}{\left(1 + K_{H_2}^{0.5} \cdot P_{H_2}^{0.5} + K_{CO} \cdot P_{CO} + \frac{k_9 \cdot K_{H_2}^{0.5} \cdot P_{H_2}^{0.5}}{k_{-3}}\right)^2} \tag{8}$$

[1] Sughrue and Bartholomew (1982) investigated the kinetics of CO methanation performing several kinetic tests at 200 to 350 °C and total pressure of 6.9 bar, with N_2 making up 95 to 97% of the feed gas, CO partial pressure ranging from 0.05 to 0.12 bar, and H_2 partial pressure varying from 0.12 to 0.30 bar. The H_2/CO ratio was kept close to 3, which is the stoichiometric ratio of the SS3 reaction. The partial pressure of CO was kept at sufficiently low values to preclude the deactivation of the catalyst due to carbon deposition (Bartholomew et al. 1980) and the formation of nickel carbonyl (Shen et al. 1981). A monolithic nickel catalyst with a coating of 5% Ni/Al_2O_3 was used in most experiments.

[2] The Langmuir–Hinshelwood mechanism describes bimolecular chemical reactions on solid surfaces, as proposed by Irving Langmuir in 1921 and further elaborated by Cyril Hinshelwood in 1926. According to this model, two atoms or molecules are adsorbed on a surface, and at least one diffuses on the surface until both are close enough to interact. This result, in turn, implies that that in the synthesis of methane from carbon monoxide and hydrogen on nickel catalysts, the reactants interact in the adsorbed layer.

(iv, b) If the hydrogenation of CH to CH_2 is the slowest step, the rate equation is:

$$r_{CH_4} = \frac{k_5 \cdot k_4 \cdot k_{12} \cdot K_{CO} \cdot P_{CO} \cdot K_{H_2} \cdot P_{H_2}}{\left(1 + K_{H_2}^{0.5} \cdot P_{H_2}^{0.5} + K_{CO} \cdot P_{CO} + \frac{k_{12}}{k_{-3}} \cdot K_{CO} \cdot P_{CO} + \frac{k_4 \cdot k_{12}}{k_{-3}} \cdot K_{CO} \cdot K_{H_2}^{0.5} \cdot P_{CO} \cdot P_{H_2}^{0.5}\right)^2}. \quad (9)$$

However, none of these relations explains adequately the available kinetic data (see above), suggesting that the rate of the process might be controlled by a combination of elementary steps rather than a single step. This approach was adopted by Gardner and Bartholomew (1981), who postulated a dual-step mechanism in which both CO dissociation and carbon hydrogenation are assumed to be rate-controlling and in balance with each other. The rate expression based on this dual mechanism is:

$$r_{CH_4} = \frac{k_3 \cdot K_{CO} \cdot P_{CO}}{\left(1 + K_{CO} \cdot P_{CO} + K_{H_2}^{0.5} \cdot P_{H_2}^{0.5} + \frac{k_3 \cdot K_{CO} \cdot P_{CO}}{k_4 \cdot K_{H_2}^{0.5} \cdot P_{H_2}^{0.5}}\right)^2}. \quad (10)$$

However, this relation does not explain the changes in the H_2 reaction order with temperature. According to Sughrue and Bartholomew (1982), the combination of Eqs. (6) and (8) in a series provides a suitable explanation of the available rate data and the corresponding rate expression is relatively simple, as it involves only four constants:

$$r_{CH_4} = \frac{k_1 \cdot k_2 \cdot P_{H_2}}{k_1 \cdot \left(1 + K_{H_2}^{0.5} \cdot P_{H_2}^{0.5} + K_{CO} \cdot P_{CO}\right)^2 + k_2 \cdot (1 + K_{CO} \cdot P_{CO})^2} \quad (11)$$

where $k_1 = k_{H_2}$ and $k_2 = (k_4 \cdot k_9 \cdot K_{H_2})/k_{-3}$. The rate computed using Eq. (11) is expressed in units of molecules sites^{-1} s^{-1}, whereas the partial pressures of H_2 and CO are in kPa. Sughrue and Bartholomew (1982) set the activation energy for k_2 equal to 70 kJ mol^{-1} based on literature data (e.g., McCarty and Wise 1979; Van Ho and Harriott 1980; Gardner and Bartholomew 1981) and performed a non-linear least squares fit of all the data from their study to evaluate the rate constants and adsorption constants in Eq. (11) as a function of temperature. The following relations were thus obtained:

$$k_1 = 3.657 \cdot 10^{11} \cdot \exp\left(\frac{-143.7}{RT}\right) \quad (12)$$

$$k_2 = 5.735 \cdot 10^{3} \cdot \exp\left(\frac{-70}{RT}\right) \quad (13)$$

$$K_{H_2} = 9.706 \cdot 10^{-12} \cdot \exp\left(\frac{92}{RT}\right) \quad (14)$$

$$K_{CO} = 4.124 \cdot 10^{-9} \cdot \exp\left(\frac{68.5}{RT}\right) \quad (15)$$

in which $R = 8.31446 \cdot 10^{-3}$ kJ K^{-1} mol^{-1}. Equation (11) is consistent with the presence of three rate-controlling mechanisms of CO methanation on nickel catalysts, depending on temperature. In detail, the rate-controlling steps are: (i) H_2 adsorption, at low temperatures (200–250 °C), (ii) both CO dissociation and atomic C hydrogenation, which are in balance, at intermediate temperatures (250–300 °C), and (iii) C hydrogenation, at comparatively high reaction temperatures (>300 °C).

Carbon dioxide methanation probably proceeds via dissociative adsorption of both H_2, with production of two H atoms, and CO_2, with generation of CO and atomic O. The ensuing implication is that, after CO_2 adsorption onto the catalyst surface and dissociation to CO and atomic oxygen, CO_2 methanation actually proceeds via the same reaction steps as CO methanation.

Weatherbee and Bartholomew (1982) acquired rate data for the hydrogenation of CO_2 on silica-supported Ni at total pressure of 1.4 to 1.75 bar and temperatures of 227 to 327 °C as a function of H_2 and CO_2 partial pressures. The experimental kinetic data were interpreted through an approach similar to that used by Sughrue and Bartholomew (1982) for CO methanation and the following Langmuir–Hinshelwood rate equation was finally obtained:

$$r_{CH_4} = \frac{\left(\frac{K_1 \cdot K_{10} \cdot k_{11} \cdot K_2 \cdot k_4}{2}\right)^{1/2} \cdot \Lambda^2 \cdot P_{CO_2}^{1/2} \cdot P_{H_2}^{1/2}}{\left(1 + \left(\frac{2 \cdot K_2 \cdot k_4}{K_1 \cdot K_{10} \cdot k_{11}}\right)^{1/2} \frac{P_{CO_2}^{1/2}}{P_{H_2}^{1/2}} + \left(\frac{K_1 \cdot K_{10} \cdot k_{11} \cdot K_2}{2 \cdot k_4}\right)^{1/2} \cdot P_{CO_2}^{1/2} \cdot P_{H_2}^{1/2} + \frac{P_{CO}}{K_3}\right)^2} \quad (16)$$

in which r_{CH_4} has units of molecules of methane produced per site per second whereas the total number of available surface sites, Λ, is equal to one being normalized to the total number of surface sites determined by H_2 adsorption. The following polynomials (T in K) are obtained by fitting the values reported by Weatherbee and Bartholomew (1982) at temperatures of 500, 525, 550, 575, and 600 K:

$$\log(K_1 \cdot K_{10} \cdot k_{11}) = -99{,}434/T^2 - 1647/T + 4.524 \quad (17)$$

$$\log K_2 = -8{,}505{,}142/T^2 + 29{,}376/T - 26.517 \quad (18)$$

$$\log k_4 = 3{,}724{,}109/T^2 - 18{,}556/T + 22.948 \quad (19)$$

$$\log K_3 = -4{,}630{,}956/T^2 + 12{,}731/T - 8.302 \quad (20)$$

The activation energy for CO dissociation, determined from the temperature dependence of the Langmuir–Hinshelwood reaction rate constant, has the constant value of 94 kJ/mol over the whole range of investigated temperatures (500–600 K).

Many other investigations were performed to describe the kinetics of the hydrogenation of CO_2 and CO to CH_4. One of the most recent and thorough studies was carried out by Kopyscinski (2010) who investigated CO methanation through kinetic experiments carried out at five different temperatures from 280 to 360 °C, total pressure up to 2 bar (absolute), adopting a reactant gas mixture containing 10 to 14 vol% CO and 50 to 72 vol% H_2 in diluent Ar. The inhibiting effect of the products (CH_4, H_2O, and CO_2) was investigated and turned out to be important for H_2O, whereas CH_4 and CO_2 do not have appreciable effects on the reaction rate.

To find a suitable rate equation of CO methanation, Kopyscinski (2010) considered sixteen different kinetic models differing from each other for the rate-limiting step. The reverse reaction was not included in the rate equation because it was considered to have nil to negligible importance below 380 °C. Kopyscinski (2010) derived also two rate equations for the water–gas shift reaction including the reverse reaction, due to its small equilibrium constant in the considered temperature interval.

Since the three models called 12b, 14b, and 6b were found to provide an equally good description of the experimental data, Kopyscinski (2010) recognized that it is difficult to select the best model. These three models assume the rate-limiting step to be the reaction of the surface carbon species (i.e., C*, CH*, or COH*) with an adsorbed hydrogen (H*), and water to be adsorbed as a hydroxyl species (OH*).

The rate equations for the methanation, r_1, and the water–gas shift reaction, r_2, for model 12b are:

$$r_1 = \frac{k_1 \cdot K_C \cdot P_{CO}^{0.5} \cdot P_{H_2}^{0.5}}{\left(1 + K_C \cdot P_{CO} + K_{OH} \cdot P_{H_2O} \cdot P_{H_2}^{-0.5}\right)^2} \quad (21)$$

$$r_2 = \frac{k_2 \cdot \left(K_\alpha \cdot P_{CO} \cdot P_{H_2O} - \frac{P_{CO_2} \cdot P_{H_2}}{K_{eq}}\right)}{P_{H_2}^{0.5} \cdot \left(1 + K_C \cdot P_{CO} + K_{OH} \cdot P_{H_2O} \cdot P_{H_2}^{-0.5}\right)^2} \quad (22)$$

in which: $\ln k_1 = 0.15$, $\ln K_{OH} = -0.41$, $\ln K_C = 0.57$, $\ln k_2 = 2.0$, and $\ln K_\alpha = -1.07$ at the reference temperature $T_r = 598.15$ K (325 °C); $E_{A1}/(R \cdot T_r) = 14.9$, $\Delta H_{OH}/(R \cdot T_r) = -14.6$, $\Delta H_C/(R \cdot T_r) = -12.3$, $E_{A2}/(R \cdot T_r) = 32.5$, and $\Delta H_\alpha/(R \cdot T_r) = -1.3$. In Eqs. (21) to (26), the partial pressures are expressed in bar and the reaction rates in mol s^{-1} kg^{-1} catalyst.

The rate equations for the methanation, r_1, and the water–gas shift reaction, r_2, for model 14b are:

$$r_1 = \frac{k_1 \cdot K_{CH} \cdot P_{CO}^{0.5} \cdot P_{H_2}}{\left(1 + K_{CH} \cdot P_{CO}^{0.5} \cdot P_{H_2}^{0.5} + K_{OH} \cdot P_{H_2O} \cdot P_{H_2}^{-0.5}\right)^2} \quad (23)$$

$$r_2 = \frac{k_2 \cdot \left(K_\alpha \cdot P_{CO} \cdot P_{H_2O} - \frac{P_{CO_2} \cdot P_{H_2}}{K_{eq}}\right)}{P_{H_2}^{0.5} \cdot \left(1 + K_{CH} \cdot P_{CO}^{0.5} \cdot P_{H_2}^{0.5} + K_{OH} \cdot P_{H_2O} \cdot P_{H_2}^{-0.5}\right)^2} \quad (24)$$

in which: $\ln k_1 = 0.08$, $\ln K_{OH} = -0.40$, $\ln K_{CH} = 0.93$, $\ln k_2 = 2.12$, and $\ln K_\alpha = -1.02$ at the reference temperature $T_r = 598.15$ K (325 °C); $E_{A1}/(R \cdot T_r) = 12.7$, $\Delta H_{OH}/(R \cdot T_r) = -17.6$, $\Delta H_{CH}/(R \cdot T_r) = -10.2$, $E_{A2}/(R \cdot T_r) = 31.3$, and $\Delta H_\alpha/(R \cdot T_r) = -0.35$.

The rate equations for the methanation, r_1, and the water–gas shift reaction, r_2, for model 6b are:

$$r_1 = \frac{k_1 \cdot K_{COH} \cdot P_{CO} \cdot P_{H_2}}{\left(1 + K_{COH} \cdot P_{CO} \cdot P_{H_2}^{0.5} + K_{OH} \cdot P_{H_2O} \cdot P_{H_2}^{-0.5}\right)^2} \quad (25)$$

$$r_2 = \frac{k_2 \cdot \left(K_\alpha \cdot P_{CO} \cdot P_{H_2O} - \frac{P_{CO_2} \cdot P_{H_2}}{K_{eq}}\right)}{P_{H_2}^{0.5} \cdot \left(1 + K_{COH} \cdot P_{CO} \cdot P_{H_2}^{0.5} + K_{OH} \cdot P_{H_2O} \cdot P_{H_2}^{-0.5}\right)^2} \quad (26)$$

in which: $\ln k_1 = 0.042$, $\ln K_{OH} = -0.53$, $\ln K_{COH} = 2.58$, $\ln k_2 = 2.46$, and $\ln K_\alpha = -1.04$ at the reference temperature $T_r = 598.15$ K (325 ° C); $E_{A1}/(R \cdot T_r) = 11.8$, $\Delta H_{OH}/(R \cdot T_r) = -19.8$, $\Delta H_{COH}/(R \cdot T_r) = 1.3$, $E_{A2}/(R \cdot T_r) = 33.0$, and $\Delta H_\alpha/(R \cdot T_r) = 3.0$.

For model 12b, the activation energy of the methanation reaction and the water–gas shift reaction are 74.1 kJ mol^{-1} and 161.6 kJ mol^{-1}, respectively. The first value is close to the activation energy estimated under comparable experimental conditions by Gardner and Bartholomew (1981), 72–78 kJ mol^{-1}, McCarty and Wise (1979), 71 kJ mol^{-1}, and Hayes et al. (1985), 78 kJ mol^{-1}. It is difficult to compare the value of the activation energy for the water–gas shift reaction with literature values, as it is a lumped parameter containing the activation energy and the heat of adsorption of hydrogen and carbon dioxide.

The rate equations and the reaction mechanisms for the methanation of CO and CO_2 are certainly interesting to understand how the kinetics of these reactions has been described in the chemical engineering literature, but the application of the rate equations reported above to the Solfatara fluids is hampered by the lack of the kinetic data for the inverse reactions. Therefore, it is necessary to examine the kinetics of the steam reforming of methane accompanied by the water–gas shift reaction, which is the subject of the next section.

4 The Kinetics of the Steam Reforming of Methane Accompanied by the Water–Gas Shift Reaction

A very brief description of the steam reforming of methane and dry reforming of methane is given here, whereas the reader is referred to Van Hook (1980), Rostrup-Nielsen (1984), Wang et al. (1996), and Armor (1999) for more details on these processes. The steam reforming of methane is the process whereby methane reacts with water vapor in the presence of a catalyst producing syngas, namely a mixture of hydrogen and carbon monoxide with a H_2/CO ratio of 3:1. The steam reforming of methane is represented by the following reaction:

$$CH_4 + H_2O \rightarrow CO + 3H_2, \quad (27)$$

which is the reverse of the SS3 reaction. The steam reforming of methane is an endothermic process and, therefore, it requires very high temperatures, which make this process highly energy-consuming. Therefore, alternative processes of methane reforming were investigated and implemented, including the so called dry reforming of methane whereby methane reacts with carbon dioxide in the presence of a catalyst, producing a syngas with a H_2/CO ratio of 1:1. The dry reforming of methane is described by the following reaction:

$$CH_4 + CO_2 \rightarrow 2CO + 2H_2, \quad (28)$$

which is the reverse of the SS1 reaction. Also the dry reforming of methane is endothermic and consumes a great amount of energy. Another problem of this process is the significant production of elemental carbon which is deposited on the surface of the catalyst, undermining its role in the reaction and reducing its life.

Apart from osmium, the other group VIII metals (namely Fe, Co, Ni, Ru, Rh, Pd, Ir, and Pt) are more or less effective for the catalysis of methane reforming processes, but the most common choice is nickel because of its low cost and good performance. Besides, several different metal oxides (e.g., α- and γ-Al_2O_3, TiO_2, SiO_2, MgO, ZrO_2 and La_2O_3) can be employed as catalyst supports, but the most frequently utilized is α-Al_2O_3, a relatively cheap substance with good thermal stability, which is a crucial requirement since the process is performed at high temperatures.

The carbon monoxide produced through the steam reforming of methane, (27), and the dry reforming of methane, (28), can be further converted into H_2 and CO_2 by reaction with steam, that is through the water–gas shift reaction:

$$CO + H_2O \rightarrow CO_2 + H_2, \quad (29)$$

which is the reverse of the RWG reaction. In contrast to reactions (27) and (28), reaction (29) is mildly exothermic and, therefore, it is favored by relatively low temperatures. The water–gas shift reaction is catalyzed by both high-temperature catalysts, operating in the temperature range 310–450 °C and typically constituted by mixtures of Fe_2O_3 and Cr_2O_3, and low-temperature catalysts, working at 200–250 °C and usually made up of mixtures of ZnO, CuO, Cr_2O_3 and Al_2O_3, although the oxides of other metals also play a catalytic role (Smith et al. 2010).

Taken together, both reactions (27) and (29) and reactions (28) and (29) constitute the overall reaction (30) that describes the conversion of methane into carbon dioxide and hydrogen (e.g., Le Valley et al. 2014):

$$CH_4 + 2H_2O \rightarrow CO_2 + 4H_2, \quad (30)$$

and is the reverse of the SS4 reaction.

The first thorough investigation of the catalytic reaction between steam and methane is probably that of Neumann and Jacob (1924), but the first extensive study was performed by Akers and Camp (1955). Van Hook (1980) provided a good review of the kinetic studies carried out until 1973 in which nickel catalysts were employed and large intervals of temperature (260–925 °C) and pressure (0.01–50 bar) were covered. Afterwards, the kinetics of the steam reforming of methane and related reactions was investigated by several studies (e.g., Allen et al. 1975; Quach and Rouleau 1975; Münster and Grabke 1981; De Deken et al. 1982; Xu and Froment 1989a, b), among which the one performed by Xu and Froment (1989a) is of particular importance, since its reaction kinetic model is considered to be the most general in form and has been extensively applied in other works (e.g., Ding and Alpay 2000; Hou and Hughes 2001; Seo et al. 2006; Kang and Lee 2013; Parlikkad et al. 2013; Zhang et al. 2013b).

Xu and Froment (1989a) derived rate equations for the steam reforming of methane, accompanied by the water–gas shift reaction, on a catalyst containing 15.2% nickel supported on magnesium spinel, based on the results of experimental tests performed under the following conditions, where W is the weight of the catalyst in kg and $F^o_{CH_4}$ and $F^o_{CO_2}$ are the molar feed rates of CH_4 and CO_2, in kmol/h:

(i) $W/F^o_{CH_4} = 0.2 kg_{cat} \cdot h/kmol_{CH_4}$, molar H_2O/CH_4 ratio 3.0–5.0, molar H_2/CH_4 ratio 1.25, temperature 500–575 °C, and total gas pressure 3–15 bar for the 220 steam reforming runs;

(ii) $W/F^o_{CO_2} = 0.25 kg_{cat} \cdot h/kmol_{CO_2}$, molar H_2/CO_2 ratio 0.5–1.0, temperature of 300–400 °C and total gas pressure of 3–10 bar for the 60 runs of the reverse water–gas shift and reverse methanation reactions.

The thermodynamic analysis of the obtained results showed that the process of interest can be described on the basis of the reactions (27), indicated by the acronym RSS3 being the reverse of the SS3 reaction, (29), indicated by the acronym WG, and (30) indicated by the acronym RSS4 being the reverse of the SS4 reaction. The corresponding intrinsic rate equations are as follows, in which the rates have units of kmol kg_{cat}^{-1} h^{-1}:

$$r_{RSS3} = \frac{\frac{k_{RSS3}}{P_{H_2}^{2.5}} \cdot \left(P_{CH_4} \cdot P_{H_2O} - \frac{P_{H_2}^3 \cdot P_{CO}}{K_{RSS3}}\right)}{(DEN)^2} \quad (31)$$

$$r_{WG} = \frac{\frac{k_{WG}}{P_{H_2}} \cdot \left(P_{CO} \cdot P_{H_2O} - \frac{P_{H_2} \cdot P_{CO_2}}{K_{WG}}\right)}{(DEN)^2} \quad (32)$$

$$r_{RSS4} = \frac{\frac{k_{RSS4}}{P_{H_2}^{3.5}} \cdot \left(P_{CH_4} \cdot P_{H_2O}^2 - \frac{P_{H_2}^4 \cdot P_{CO_2}}{K_{RSS4}}\right)}{(DEN)^2} \quad (33)$$

in which:

$$DEN = 1 + K_{CO} \cdot P_{CO} + K_{H_2} \cdot P_{H_2} + K_{CH_4} \cdot P_{CH_4} + K_{H_2O} \cdot P_{H_2O} \cdot P_{H_2}^{-1} \quad (34)$$

$$k_{RSS3} = 4.225 \cdot 10^{15} \cdot \exp\left(\frac{-240.1}{RT}\right) \quad (35)$$

$$k_{WG} = 1.955 \cdot 10^{6} \cdot \exp\left(\frac{-67.13}{RT}\right) \quad (36)$$

$$k_{RSS4} = 1.020 \cdot 10^{15} \cdot \exp\left(\frac{-243.9}{RT}\right) \quad (37)$$

$$K_{CO} = 8.23 \cdot 10^{-5} \cdot \exp\left(\frac{70.61}{RT}\right) \quad (38)$$

$$K_{H_2} = 6.12 \cdot 10^{-9} \cdot \exp\left(\frac{82.90}{RT}\right) \quad (39)$$

$$K_{CH_4} = 6.65 \cdot 10^{-4} \cdot \exp\left(\frac{38.28}{RT}\right) \quad (40)$$

$$K_{H_2O} = 1.77 \cdot 10^{5} \cdot \exp\left(\frac{-88.68}{RT}\right) \quad (41)$$

with R = 8.31446 $\cdot 10^{-3}$ kJ K^{-1} mol^{-1} and T in K.

As already noted above, the rate equations of Xu and Froment (1989a) were calibrated at temperatures of 500–575 °C for the steam reforming reactions and 300–400 °C for the reverse water–gas shift and reverse methanation reactions. These temperatures bracket a part of the range which is expected to be present in the Solfatara magmatic-hydrothermal system (see Sect. 5). Nevertheless, the rate equations of Xu and Froment (1989a) were already used outside the calibration range in previous studies, namely, at temperatures of: (i) 428–467.5 °C by Ding and Alpay (2000); (ii) 475–550 °C for the steam reforming reactions and 325–400 °C for the reverse water–gas shift and reverse methanation reactions by Hou and Hughes (2001); (iii) 100–640 °C for the steam reforming reactions and 190–616 °C for the reverse water–gas shift and reverse methanation reactions by Seo et al. (2006); (iv) 275 °C by Kang and Lee (2013); (v) 250–360 °C by Zhang et al. (2013b). Therefore, it is permissible to apply the rate equations of Xu and Froment (1989a) also at the Solfatara magmatic-hydrothermal system, which represents the subject of the next section.

5 Application of the Reaction Path Model Approach to the Solfatara Fluids: Results and Limitations

It was shown in Sect. 2 that nickel–iron alloys similar to natural awaruite but enriched in nickel act as catalysts of CO_2 methanation. Therefore, it is reasonable to assume the possible presence of suitable catalysts promoting the CO_2–CH_4 conversion and/or the reverse reaction in natural magmatic-hydrothermal systems, such as that of Solfatara, although there are many unclear points on these natural catalysts, such as their mineralogical, chemical and physical characteristics, amount and distribution in the system of interest, and how they are reactivated.

It was discussed in Sect. 3, that the most likely reaction path of CO_2 methanation involves the conversion of CO_2 to CO via the RWG reaction, followed by CO methanation via the SS3 reaction. If so, the slowest reaction between RWG and SS3 is expected to control the rate of CO_2–CH_4 conversion and/or the reverse reaction and the SS4 reaction can be disregarded. Alternatively, the three reactions WG, RSS3, and RSS4 might co-occur, as assumed by the kinetic model of Xu and Froment (1989a). Therefore, two distinct reaction path models were set up, one comprising the RWG, SS3, and SS4 reactions, the other including the RWG and SS3 reactions only. The reverse reactions, namely the WG, RSS3, and RSS4 reaction, are also taken into account in both models. As shown in Sect. 4, the kinetics of these reactions is described by the rate equations proposed by Xu and Froment (1989a), representing the foundations on which the two reaction path models of interest were built.

5.1 Application of the Reaction Path Model Involving the Reactions WG, RSS3, and RSS4 to the Solfatara Fluids

Assuming the co-occurrence of the three reactions WG, RSS3, and RSS4, as postulated by the kinetic model of Xu and Froment (1989a), and taking into account the stoichiometric coefficients of the gas constituents involved in these three reactions, it is possible to write the following equations:

$$n_{H_2O,t+\Delta t} = n_{H_2O,t} - r_{WG} \cdot W \cdot \Delta t - r_{RSS3} \cdot W \cdot \Delta t - 2 \cdot r_{RSS4} \cdot W \cdot \Delta t \quad (42)$$

$$n_{H_2,t+\Delta t} = n_{H_2,t} + r_{WG} \cdot W \cdot \Delta t + 3 \cdot r_{RSS3} \cdot W \cdot \Delta t + 4 \cdot r_{RSS4} \cdot W \cdot \Delta t \quad (43)$$

$$n_{CO_2,t+\Delta t} = n_{CO_2,t} + r_{WG} \cdot W \cdot \Delta t + r_{RSS4} \cdot W \cdot \Delta t \quad (44)$$

$$n_{CH_4,t+\Delta t} = n_{CH_4,t} - r_{RSS3} \cdot W \cdot \Delta t - r_{RSS4} \cdot W \cdot \Delta t \quad (45)$$

$$n_{CO,t+\Delta t} = n_{CO,t} - r_{WG} \cdot W \cdot \Delta t + r_{RSS3} \cdot W \cdot \Delta t, \quad (46)$$

where $n_{i,t+\Delta t}$ and $n_{i,t}$ are the moles of the i-th gas constituent at time t + Δt and t, respectively, r_{RSS3}, r_{WG}, and r_{RSS4} are the rates of reactions RSS3, WG, and RSS4 (in kmol kg_{cat}^{-1} h^{-1}), which are given by Eqs. (31), (32), and (33), respectively, W is the weight of the catalyst in kg, and Δt is the considered time step in hours. Since W is unknown, it is necessarily set to 1, and rates and times are relative rather than absolute.

In spite of this limitation, the heating simulation was performed for twelve samples of Bocca Grande and two samples of Bocca Nuova (data from Chiodini et al. 2021 and references therein), which are characterized by different chemistry and SS4 temperature (see Table 1). The heating simulation basically consists in solving Eqs. (42)–(46) for a series of time steps, each one corresponding to an increase of 1 °C in temperature. The heating simulation begins at the outlet temperature and ends at a maximum temperature which is higher than the inversion temperature of the RSS3 and RSS4 reactions by a few tens of degrees. At the outlet temperature, H_2O partial pressure was assumed to be equal to 1.013 bar times the measured H_2O mole fraction. Between the outlet temperature and the lower saturation temperature (LST), H_2O partial pressure was assumed to increase linearly with temperature, which is a reasonable approximation. From the lower saturation temperature to the maximum considered temperature, H_2O partial pressure was assumed to be fixed by Eq. (14) in chapter "The Geoindicators Involving H_2O, CO_2, CO, CH_4 and H_2" due to the coexistence of liquid water and steam.[3] The partial pressures of the other gas constituents, at each time/temperature step, were computed by means of the simple relation:

$$P_i = P_{H_2O} \cdot (X_i / X_{H_2O}). \qquad (47)$$

At each time/temperature step, the kinetic constants were calculated by means of Eqs. (35), (36), and (37), the absorption constants were computed by using Eqs. (38) to (41), and partial pressures were inserted in Eqs. (31), (32), (33) and (34), to obtain the rates of reactions RSS3, WG, and RSS4.[4] Then, using Eqs. (42) to (46), the gas composition was obtained at each time/temperature step. Finally, the gas composition at each time/temperature step was used to calculate the RWG, CCC, SS1, SS3, and SS4 temperatures, although only the RWG and SS4 temperatures are probably meaningful and the other three temperatures are probably meaningless as already noted in Sect. 4 in chapter "The Geoindicators Involving H_2O, CO_2, CO, CH_4 and H_2". As an example, the heating simulation for the sample collected on November 30, 1983 is reported as electronic supplementary material (file ESM Ch.7_Heat BG 30.11.1983 WG+SS3+SS4).

The main results of these heating simulations are summarized in Table 1 and displayed in different graphs. In particular: (i) The evolution of the concentration of each gas constituent during the heating simulation for the fourteen selected fluids is displayed in Figs. 1, 2, 3, 4 and 5. (ii) The variations in the RWG, CCC, SS1, SS3, and SS4 temperatures during the heating simulation for the fourteen selected fluids are shown in Figs. 6a, 7a, 8a, 9a, 10a, 11a, 12a, 13a, 14a, 15a, 16a, 17a, 18a and 19a. (iii) The changes in the decimal logarithm of the absolute rate of the reactions WG/RWG, SS3/RSS3, and SS4/RSS4 are presented in Figs. 6b, 7b, 8b, 9b, 10b, 11b, 12b, 13b, 14b, 15b, 16b, 17b, 18b and 19b.

Heating is the process simulated for the fourteen selected gaseous mixtures since available data refer to the surface discharge point. However, in the natural system, the opposite process obviously takes place, namely, cooling of the gaseous mixtures from the SS4-RSS4 inversion temperature to the outlet temperature. Therefore, the effects of the cooling process are considered in the following discussion. For all fourteen selected samples, the concentrations of H_2O and CO_2 undergo nil to negligible variations during cooling (Figs. 1 and 2). This constancy or near-constancy of H_2O and CO_2 concentrations is not surprising because they are by far the two major constituents of Solfatara fluids.

In contrast, the concentrations of H_2 and CH_4 experience significant changes during cooling, at least at temperatures higher than 220–230 °C (Figs. 3 and 4), while CO concentration undergoes important variations in the whole

[3] This hypothesis is realistic along the whole saturation decompression path, whereas it is realistic from the lower saturation temperature to the upper saturation temperature (UST) along the isenthalpic expansion path.

[4] To be noted that the thermodynamic equilibrium constants of the three considered reactions, K_{RSS3}, K_{WG}, and K_{RSS4}, appearing in Eq. (31), (32), and (33), are calculated using relations linear in the absolute temperature reciprocal instead of the polynomials presented in chapter "The Geoindicators Involving H_2O, CO_2, CO, CH_4 and H_2".

Table 1 Salient characteristics of the twelve samples of Bocca Grande and the two samples of Bocca Nuova (data from Chiodini et al. 2021 and references therein), whose heating was simulated considering the occurrence of the three reactions WG, RSS3, and RSS4, from the outlet temperature up to a maximum temperature higher than the equilibrium temperature (i.e., SS4-RSS4 inversion temperature) by a few tens of degrees

Fum	Date	Outlet T	H_2O	CO_2	CH_4	H_2	CO	LST	UST	T RWG, Sect. 4 in chapter "The Geoindicators Involving H_2O, CO_2, CO, CH_4 and H_2"	T SS4, Sect. 4 in chapter "The Geoindicators Involving H_2O, CO_2, CO, CH_4 and H_2"	CO, T SS4, Sect. 5 in chapter "The Geoindicators Involving H_2O, CO_2, CO, CH_4 and H_2"	Tinv, WG, heat	Tinv, RSS3, heat	Tinv, RSS4 = Teq, heat	Eq. CH_4, heat	Eq. H2, heat	Eq. CO, heat
		°C	μmol/mol					°C				μmol/mol	°C			μmol/mol		
BG	30/11/1983	156	846,000	151,900	25.1	216	0.310	197	292	231	231	0.313	195	234	235	25.4	215	0.321
BG	13/10/1983	156	828,000	169,000	27.2	241	0.380	195	294	229	243	0.489	196	246	246	27.9	238	0.496
BG	10/08/1994	162	837,300	160,500	18.8	281	0.320	196	293	215	262	0.744	196	264	265	20.6	274	0.745
BG	15/06/2000	153	841,500	156,300	18.6	347	0.544	197	293	234	280	1.19	201	281	281	22.3	331	1.13
BG	23/03/2004	163	812,900	184,600	17.4	394	0.498	194	295	212	301	2.26	198	295	295	25.5	360	1.86
BG	14/12/2006	157.2	799,100	198,200	18.9	472	0.608	193	296	209	321	3.86	197	306	307	33.5	412	2.77
BG	13/12/2012	162.3	762,000	236,000	13.8	480	1.28	190	298	235	340	6.30	207	311	311	34.8	394	3.50
BG	17/12/2013	162.5	739,000	259,000	14.9	554	1.24	189	299	219	360	10.4	204	317	317	47.0	422	4.62
BG	26/09/2016	163.3	745,087	251,932	14.6	668	1.54	189	299	222	380	15.2	207	325	325	59.3	485	5.71
BG	12/05/2017	163.8	720,275	276,720	11.7	703	1.76	188	299	220	400	22.2	207	326	327	66.7	477	6.55
BG	09/01/2020	162.8	713,802	281,537	28.9	1027	2.99	187	300	227	420	40.9	214	337	338	132	609	9.91
BG	01/07/2020	160.6	702,044	292,171	35.7	1230	3.53	187	300	223	440	62.1	214	342	342	176	662	12.0
BN	03/08/2020	140.6	725,701	269,675	22.9	1293	3.45	188	299	225	455	66.1	216	346	347	166	712	12.2
BN	01/09/2020	141.9	701,653	293,419	25.1	1402	3.53	187	300	216	470	92.0	211	347	347	197	705	13.6

temperature range of interest (Fig. 5). The following considerations can be drawn on the evolution of H_2 concentration during cooling of Solfatara fluids:

(i) For the fluids initially at temperatures greater than 300 °C, H_2 concentration increases considerably upon cooling to 220–230 °C, while it experiences nil to negligible changes upon further cooling to outlet temperature (140.6–163.8 °C). For instance, the Bocca Nuova fluid sampled on the 1st September 2020 has an initial H_2 concentration of 705 µmol/mol at 347 °C, which rises to 1391 µmol/mol at 230 °C and 1398 µmol/mol at 220 °C, attaining the final value of 1402 µmol/mol at the outlet temperature of 141.9 °C. The increase in H_2 concentration of this fluid upon cooling from 347 to 230 °C is controlled by the relatively high rate of reaction SS4 in this temperature range (Fig. 19b).

(ii) For the fluids initially at temperatures lower than 300 °C, H_2 concentration remains practically constant during cooling. For example, the Bocca Grande sample collected on the 30th November 1983 has an initial H_2 concentration of 215 µmol/mol at 235 °C, the same value at 230 and 220 °C, and a final value of 216 µmol/mol at the outlet temperature of 156 °C. This constancy of H_2 concentration is explained by the relatively low rates of the three reactions WG/RWG, SS3, and SS4 in the temperature range of interest (Fig. 6b).

Methane evolution during the cooling of Solfatara fluids is the opposite of that of hydrogen, as shown by Fig. 4, which is a sort of mirror image of Fig. 3. In detail:

(a) For the fluids at initial temperature higher than 300 °C, CH_4 concentration decreases substantially upon cooling to 220–230 °C, while it undergoes nil to negligible variations upon further cooling to outlet temperature (140.6–163.8 °C). For instance, the Bocca Nuova fluid sampled on the 1st September 2020 has an initial CH_4 concentration of 197 µmol/mol at 347 °C, which reduces to 27.8 µmol/mol at 230 °C and to 26.4 µmol/mol at 220 °C, reaching the final value of 25.1 µmol/mol at the outlet temperature of 141.9 °C. Again, the decrease in CH_4 concentration upon cooling from 347 to 230 °C is mainly due to the relatively high rate of reaction SS4 in this temperature range (Fig. 19b).

(b) For the fluids at initial temperature lower than 300 °C, CH_4 exhibits small to very small concentration changes during cooling. For example, the Bocca Grande sample collected on the 30th November 1983 has CH_4 concentration of 25.4 µmol/mol at both 235 °C (initial temperature) and 230 °C, 25.3 µmol/mol at 220 °C, and 25.1 µmol/mol at the outlet temperature of 156 °C. Again, this quasi-constancy of CH_4 concentration is ascribable to the relatively low rates of the three reactions WG, RSS3, and RSS4 at temperatures $\leq$235 °C (Fig. 6b).

As shown by Fig. 5, carbon monoxide evolution during the cooling of Solfatara fluids is different from those of hydrogen and methane, as CO concentration experiences a decrease from the initial temperature (347–235 °C) to the WG-RWG inversion temperature (195–216 °C), where CO concentration attains the minimum value, followed by an increase upon further cooling to outlet temperature (140.6–163.8 °C). In particular:

(1) The Bocca Nuova sample obtained on the 1st September 2020 has an initial CO concentration of 13.6 µmol/mol at 347 °C, which reduces to the minimum value of 3.16 µmol/mol upon cooling to 211 °C (the WG-RWG inversion temperature), and increases upon further cooling attaining the final value of 3.53 µmol/mol at the outlet temperature of 141.9 °C.

(2) The Bocca Grande sample collected on the 30th November 1983 has an initial CO concentration of 0.321 µmol/mol at 235 °C, which decreases to the minimum value of 0.152 µmol/mol upon cooling to 195 °C (the WG-RWG inversion temperature), and rises

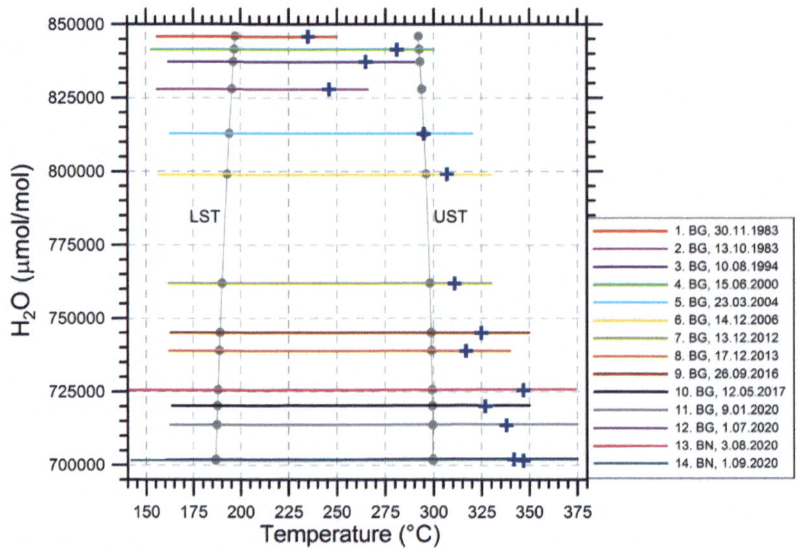

Fig. 1 Water concentration of the fourteen selected Solfatara fluids (data from Chiodini et al. 2021 and references therein) during the heating simulations involving the three reactions WG, RSS3, and RSS4 (curves, see legend). The blue cross indicates the SS4-RSS4 inversion temperature. Also shown are the upper and lower saturation temperatures (grey lines with closed circles)

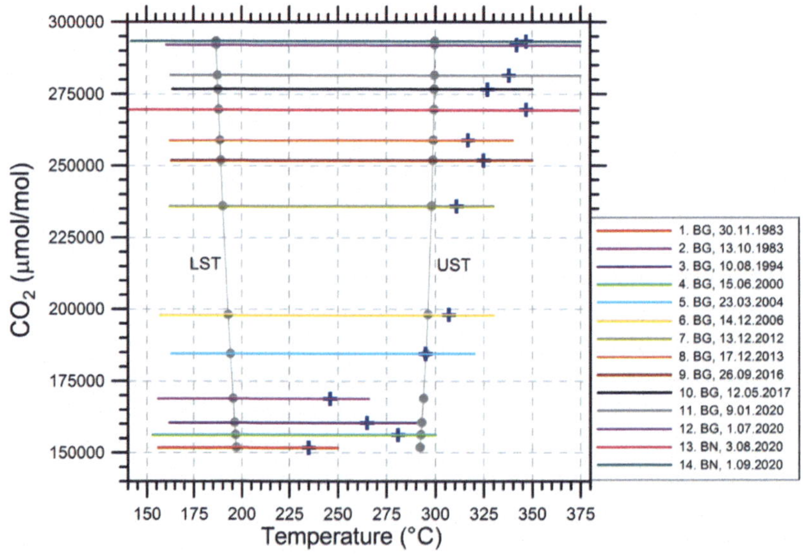

Fig. 2 Carbon dioxide concentration of the fourteen selected Solfatara fluids (data from Chiodini et al. 2021 and references therein) during the heating simulations involving the three reactions WG, RSS3, and RSS4 (curves, see legend). The blue cross indicates the SS4-RSS4 inversion temperature. Also shown are the upper and lower saturation temperatures (grey lines with closed circles)

upon further cooling achieving the final value of 0.310 µmol/mol at the outlet temperature of 156 °C.

In both samples, the changes in CO concentration are mainly controlled by the WG reaction, above the WG-RWG inversion temperature, and

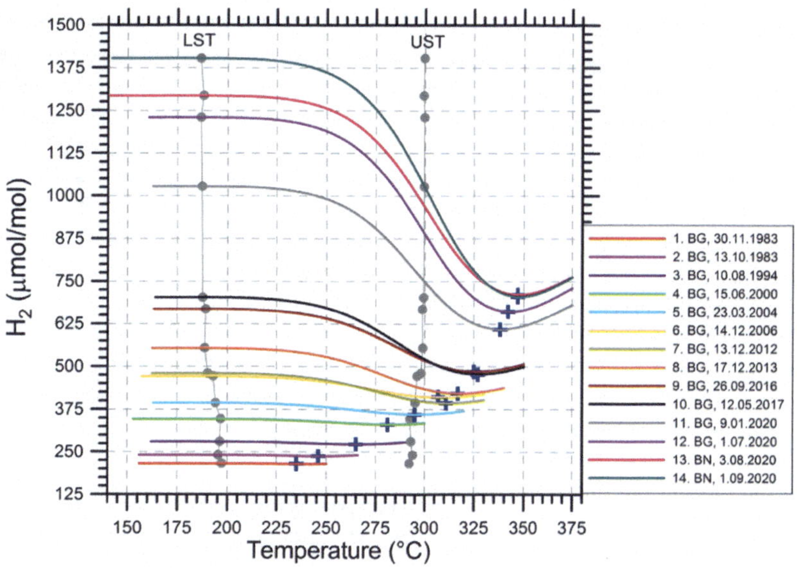

Fig. 3 Hydrogen concentration of the fourteen selected Solfatara fluids (data from Chiodini et al. 2021 and references therein) during the heating simulations involving the three reactions WG, RSS3, and RSS4 (curves, see legend). The blue cross indicates the SS4-RSS4 inversion temperature. Also shown are the upper and lower saturation temperatures (grey lines with closed circles)

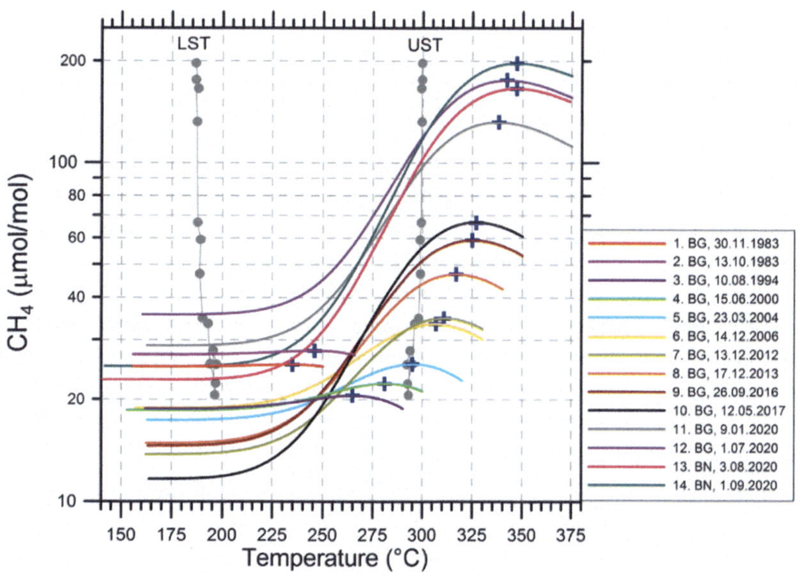

Fig. 4 Methane concentration of the fourteen selected Solfatara fluids (data from Chiodini et al. 2021 and references therein) during the heating simulations involving the three reactions WG, RSS3, and RSS4 (curves, see legend). The blue cross indicates the SS4-RSS4 inversion temperature. Also shown are the upper and lower saturation temperatures (grey lines with closed circles)

by the RWG reaction below the WG-RWG inversion temperature. In fact, both reactions proceed faster than the SS3 reaction at all temperatures (Figs. 6b and 19b). Inspection of Figs. 7b, 8b, 9b, 10b, 11b, 12b, 13b, 14b, 15b, 16b, 17b and 18b suggests that previous considerations on the reactions controlling the concentrations of H_2, CH_4, and CO during the upflow of the gaseous mixtures collected at Bocca Grande on the 30th November 1983 and at Bocca Nuova on the 1st September 2020 can be extended to the other twelve Solfatara fluid samples of interest.

Furthermore, the diagrams of Figs. 6b, 7b, 8b, 9b, 10b, 11b, 12b, 13b, 14b, 15b, 16b, 17b, 18b and 19b show that the SS3-RSS3 and SS4-RSS4 inversion temperatures differ by 1 °C or less. Since the time step of 1 °C was adopted in the heating simulations, it can be concluded that reactions RSS3 and RSS4 invert at the same

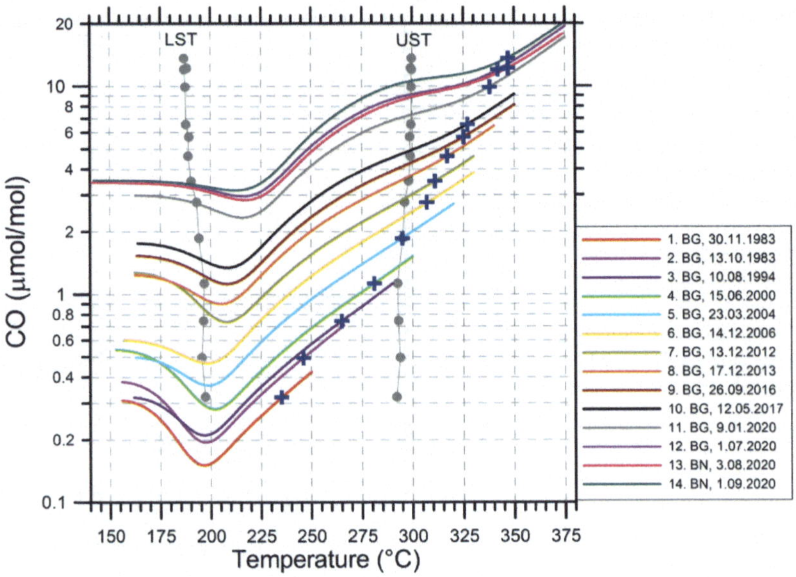

Fig. 5 Carbon monoxide concentration of the fourteen selected Solfatara fluids (data from Chiodini et al. 2021 and references therein) during the heating simulations involving the three reactions WG, RSS3, and RSS4 (curves, see legend). The blue cross indicates the SS4-RSS4 inversion temperature. Also shown are the upper and lower saturation temperatures (grey lines with closed circles)

temperature. In addition, since all five gas constituents of interest are involved in reactions RSS3 and RSS4, the inversion temperature of these two reactions is representative of the full equilibrium temperature for the H_2O–CO_2–H_2–CH_4–CO system. On the other hand, the inversion temperature of reaction WG, which involves H_2O, CO_2, H_2, and CO (but not CH_4), is representative of the partial equilibrium condition among these four gas constituents and corresponds to the minimum CO concentration as already noted above.

The evolution of the RWG, CCC, SS1, SS3, and SS4 reaction temperatures during cooling is shown in Figs. 6a, 7a, 8a, 9a, 10a, 11a, 12a, 13a, 14a, 15a, 16a, 17a, 18a and 19a for the fourteen Solfatara fluids under considerations. In all these diagrams, the five reaction temperatures converge at the full equilibrium temperature, which increases with time, from 235 °C on 30th November 1983 to 342 °C on 1st July 2020 for Bocca Grande and reaches the maximum value of 347 °C on 3rd August 2020 and on 1st September 2020 for Bocca Nuova (see Table 1). In the diagrams of Figs. 6a, 7a, 8a, 9a, 10a, 11a, 12a, 13a and 14a, the five reaction temperatures converge also at temperatures between the outlet value and 183 °C, whereas in the diagrams of Figs. 15a, 16a, 17a, 18a and 19a, the five reaction temperatures are expected to converge at temperatures lower than the outlet value. However, the convergence or the tendency to convergence of reaction temperatures at these low temperature values is coincidental because it is not corroborated by the zeroing of reaction rates.

5.2 Application of the Reaction Path Model Involving Reactions WG and RSS3 to the Solfatara Fluids

Assuming that only reactions WG and RSS3 occur and considering the stoichiometric coefficients of the gas species participating to these two reactions, Eqs. (42)–(46), can be rewritten as follow:

$$n_{H_2O,t+\Delta t} = n_{H_2O,t} - r_{WG} \cdot W \cdot \Delta t - r_{RSS3} \cdot W \cdot \Delta t \tag{48}$$

$$n_{H_2,t+\Delta t} = n_{H_2,t} + r_{WG} \cdot W \cdot \Delta t + 3 \cdot r_{RSS3} \cdot W \cdot \Delta t \tag{49}$$

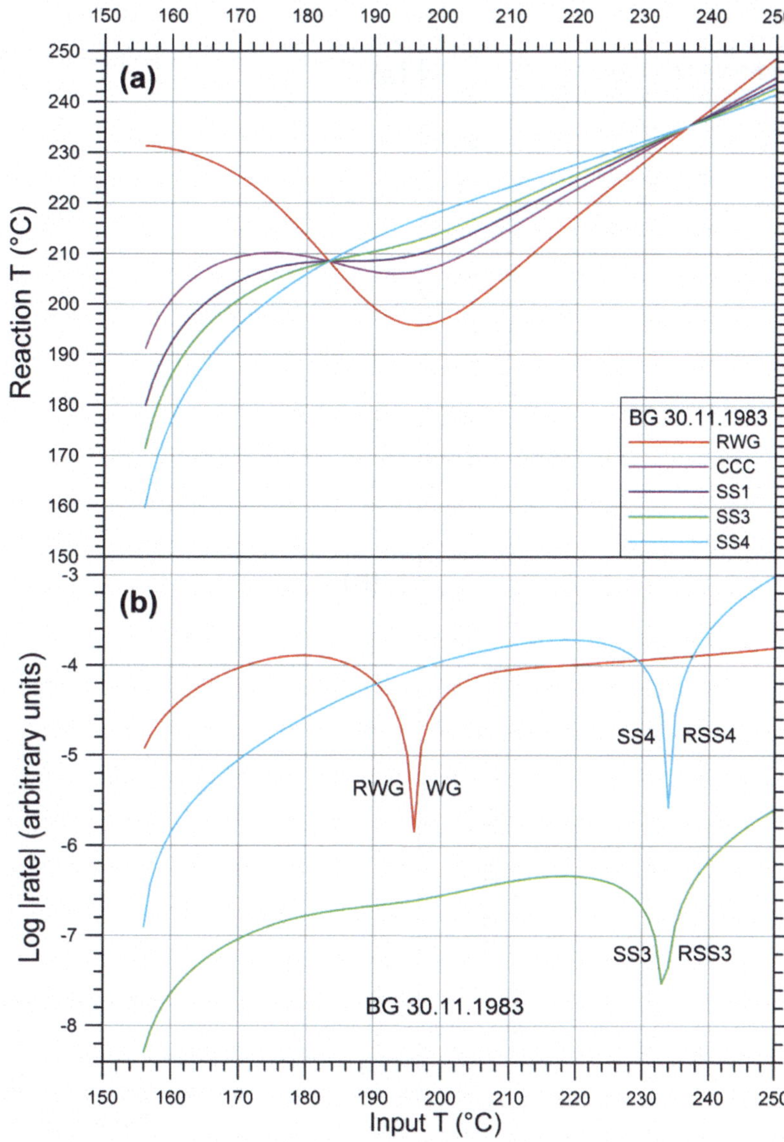

Fig. 6 a RWG, CCC, SS1, SS3, and SS4 temperatures and **b** log |rate| of reactions WG, RSS3, and RSS4 during the heating simulation involving these three reactions for the sample collected on 30.11.1983 from Bocca Grande (data from Chiodini et al. 2021 and references therein)

$$n_{CO_2,t+\Delta t} = n_{CO_2,t} + r_{WG} \cdot W \cdot \Delta t \tag{50}$$

$$n_{CH_4,t+\Delta t} = n_{CH_4,t} - r_{RSS3} \cdot W \cdot \Delta t \tag{51}$$

$$n_{CO,t+\Delta t} = n_{CO,t} - r_{WG} \cdot W \cdot \Delta t + r_{RSS3} \cdot W \cdot \Delta t. \tag{52}$$

The heating of the fourteen selected samples (data from Chiodini et al. 2021 and references therein) was simulated again using Eqs. (48)–(52) instead of Eqs. (42)–(46), but adopting the same approach of Sect. 5.1, apart from a minor adjustment concerning the calculation of the H_2O partial pressure. In fact, some preliminary calculations indicated that the SS3/RSS3 reaction inverts at temperatures higher than 360 °C, which is the upper validity limit of Eq. (14) in chapter "The Geoindicators Involving H_2O, CO_2, CO, CH_4 and H_2". Therefore, in these new heating simulations, the H_2O partial pressure was assumed to be fixed by Eq. (15) in chapter "The Geoindicators Involving H_2O, CO_2, CO, CH_4

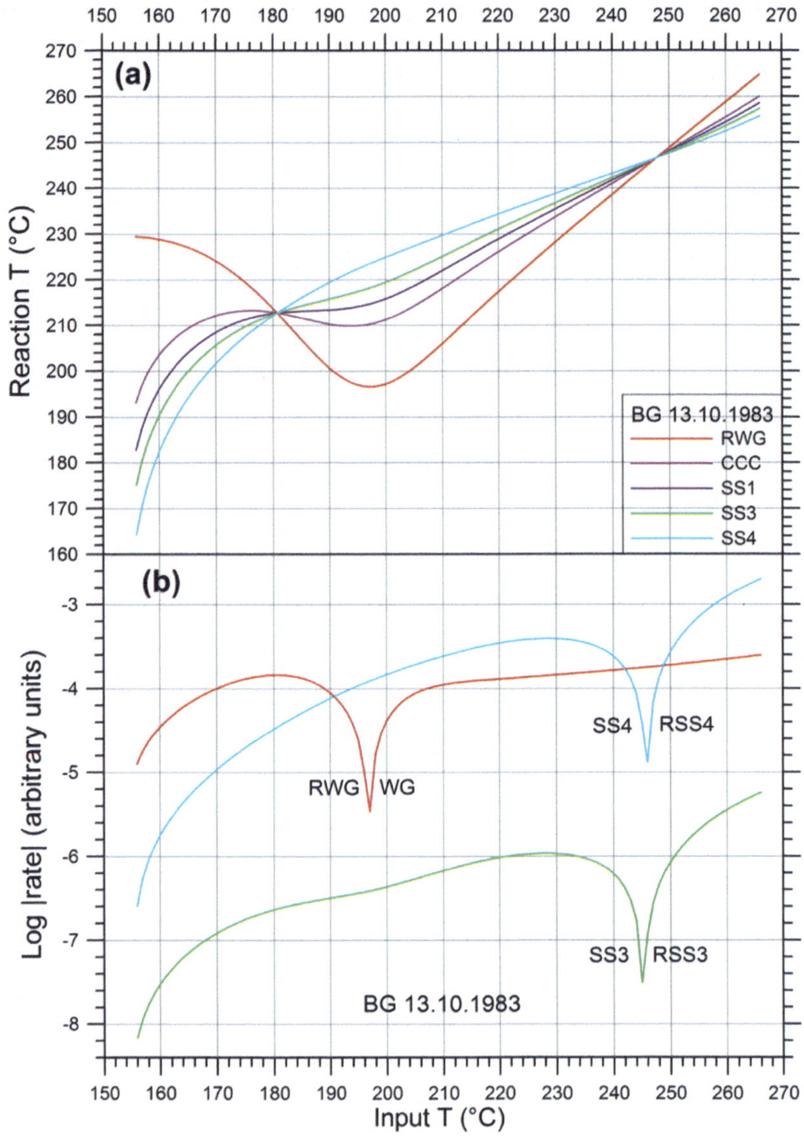

Fig. 7 **a** RWG, CCC, SS1, SS3, and SS4 temperatures and **b** log |rate| of reactions WG, RSS3, and RSS4 during the heating simulation involving these three reactions for the sample collected on 13.10.1983 from Bocca Grande (data from Chiodini et al. 2021 and references therein)

and H_2", as it can be used up to 600 °C, instead of Eq. (14) in chapter "The Geoindicators Involving H_2O, CO_2, CO, CH_4 and H_2". Since Eq. (15) in chapter "The Geoindicators Involving H_2O, CO_2, CO, CH_4 and H_2" gives P_{H_2O} values significantly lower than those given by Eq. (14) chapter "The Geoindicators Involving H_2O, CO_2, CO, CH_4 and H_2" at the same temperature, the temperatures calculated during the heating simulations involving Eq. (15) in chapter "The Geoindicators Involving H_2O, CO_2, CO, CH_4 and H_2" are some tens of degrees lower than those obtained considering Eq. (14) in chapter "The Geoindicators Involving H_2O, CO_2, CO, CH_4 and H_2". As an example, the heating simulation for the sample collected on November 30, 1983 is reported as electronic supplementary material (file ESM Ch.7_Heat BN 1.09.2020 WG +SS3).

The key outcomes of these new heating simulations are summarized in Table 2 and are shown in different diagrams. In particular: (i) The evolution of the concentration of each gas species during the heating simulation for the

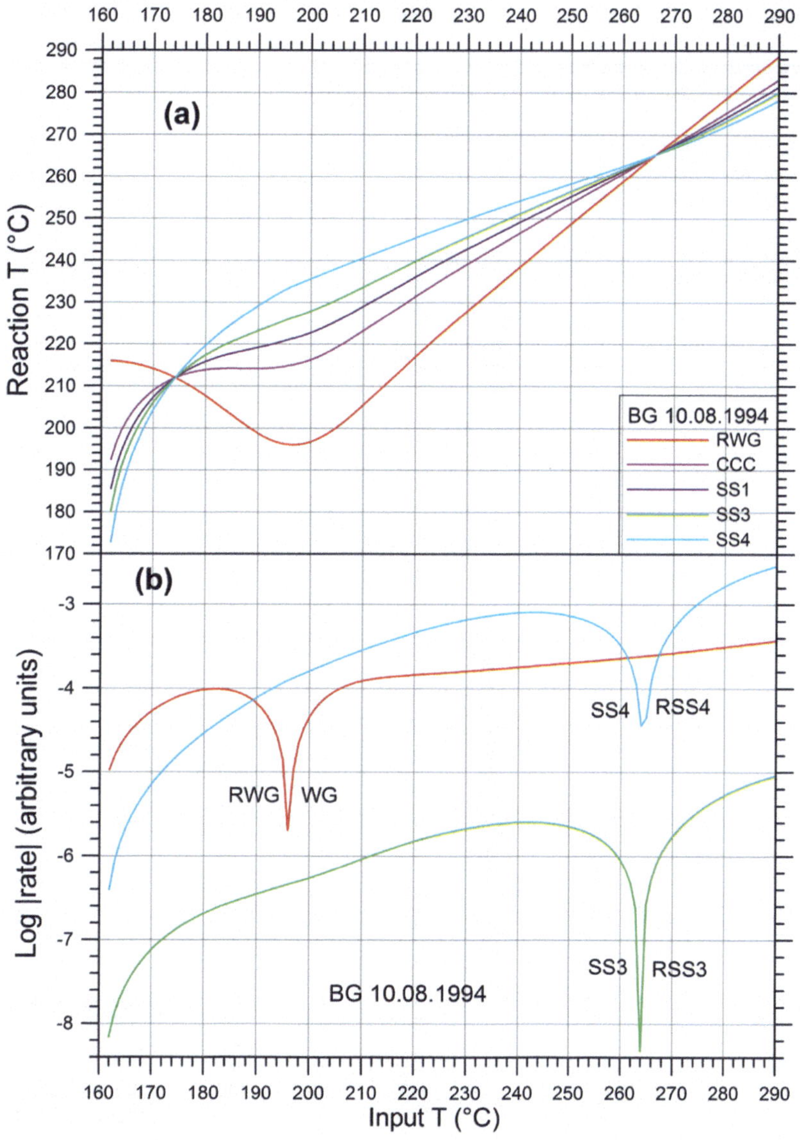

Fig. 8 **a** RWG, CCC, SS1, SS3, and SS4 temperatures and **b** log |rate| of reactions WG, RSS3, and RSS4 during the heating simulation involving these three reactions for the sample collected on 10.08.1994 from Bocca Grande (data from Chiodini et al. 2021 and references therein)

fourteen selected samples is presented in Figs. 20, 21, 22, 23 and 24. (ii) The variations in the RWG, CCC, SS1, SS3, and SS4 temperatures during the heating simulation for the fourteen selected fluids are displayed in Figs. 25a, 26a, 27a, 28a, 29a, 30a, 31a, 32a, 33a, 34a, 35a, 36a, 37a and 38a. (iii) The changes in the decimal logarithm of the absolute rate of the reactions WG/RWG and SS3/RSS3 are shown in Figs. 25b, 26b, 27b, 28b, 29b, 30b, 31b, 32b, 33b, 34b, 35b, 36b, 37b and 38b.

Again, the effects of cooling (i.e., the process occurring in the natural system) are considered in the following discussion although the opposite process of heating was simulated for Solfatara fluids. Not surprisingly, the concentrations the two major constituents, H_2O and CO_2, of all the fourteen selected samples experience nil to negligible changes during the cooling of Solfatara fluids also according to these new simulations involving the WG and RSS3 reactions only (Figs. 20 and 21).

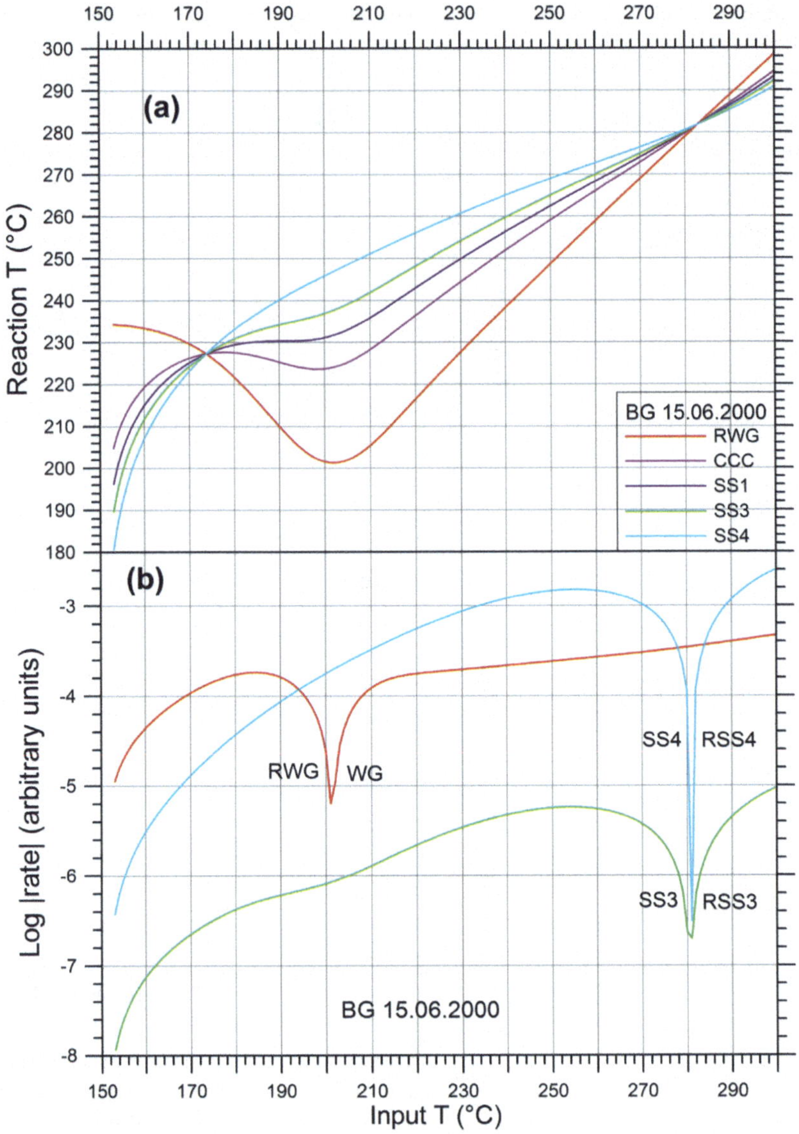

Fig. 9 a RWG, CCC, SS1, SS3, and SS4 temperatures and **b** log |rate| of reactions WG, RSS3, and RSS4 during the heating simulation involving these three reactions for the sample collected on 15/06/2000 from Bocca Grande (data from Chiodini et al. 2021 and references therein)

Differently from the major gas constituents, the concentrations of H_2 and CH_4 undergo substantial variations during cooling of Solfatara fluids, at least at temperatures higher than 250 °C approximately and only for the samples with initial temperature (corresponding to the SS3-RSS3 inversion temperature) higher than 400 °C (Figs. 22 and 23), whereas CO concentration experiences significant changes at all temperatures (Fig. 24). The following further considerations can be drawn on the evolution of hydrogen concentration during cooling of Solfatara fluids:

(i) For the fluids with initial temperature higher than 400 °C, H_2 concentration increases substantially upon cooling to ca. 250 °C, whereas it undergoes nil to negligible variations upon further cooling to outlet temperature (140.6–163.8 °C). For example, the Bocca Nuova fluid collected on the 1st September 2020 has initial H_2 concentration of 1301 μmol/mol at 460 °C, which increments to 1387 μmol/mol at 300 °C and to 1400 μmol/mol at 250 °C, finally attaining 1402 μmol/mol at the outlet temperature of

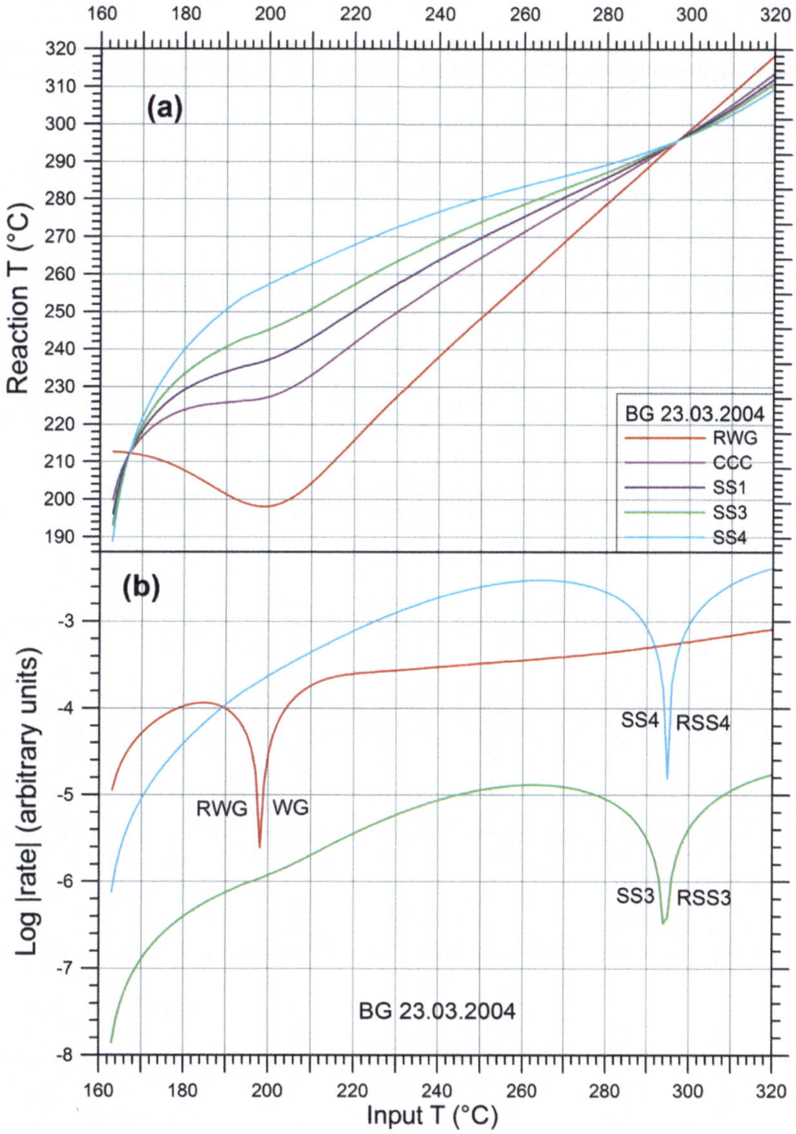

Fig. 10 **a** RWG, CCC, SS1, SS3, and SS4 temperatures and **b** log |rate| of reactions WG, RSS3, and RSS4 during the heating simulation involving these three reactions for the sample collected on 23.03.2004 from Bocca Grande (data from Chiodini et al. 2021 and references therein)

141.9 °C. The increase in H_2 concentration from 460 to 250 °C is controlled by the high rate of reaction WG in this temperature interval (Fig. 38b).

(ii) For the fluids initially at temperatures lower than 400 °C, H_2 concentration does not change appreciably upon cooling. For example, the Bocca Grande sample collected on the 30th November 1983 has an initial H_2 concentration of 216 μmol/mol at 219 °C, which does not change during cooling to the outlet temperature of 156 °C. This constancy of H_2 concentration is explained by the low rate of reaction WG and the even lower rate of reaction RSS3 in the entire temperature interval of interest (Fig. 25b).

Again, the evolution of methane concentration upon cooling of Solfatara fluids is the contrary of the evolution of hydrogen concentration, as shown by Fig. 23, which is, to some extent, a

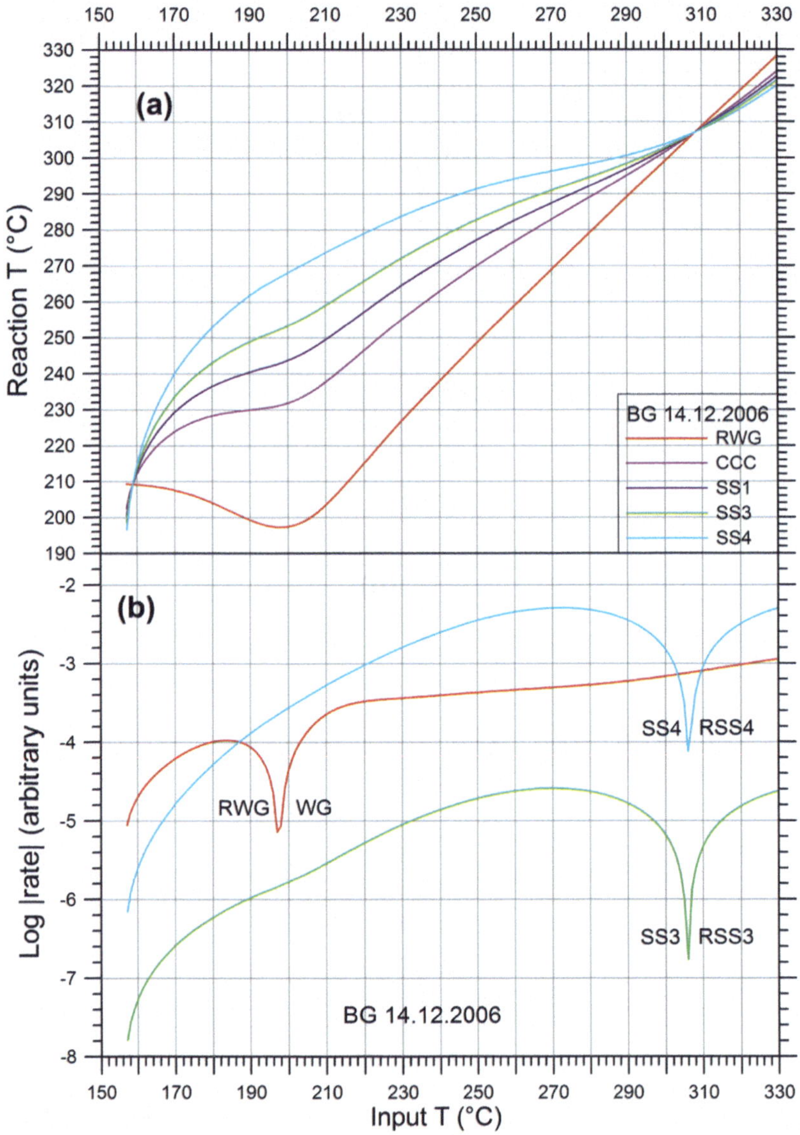

Fig. 11 a RWG, CCC, SS1, SS3, and SS4 temperatures and **b** log |rate| of reactions WG, RSS3, and RSS4 during the heating simulation involving these three reactions for the sample collected on 14.12.2006 from Bocca Grande (data from Chiodini et al. 2021 and references therein)

mirror image of Fig. 22. Based on Fig. 23, the following further details can be obtained on the evolution of methane concentration during cooling of Solfatara fluids:

(a) For the samples initially at temperature higher than 400 °C, CH_4 concentration decreases appreciably upon cooling to 250 °C, while it experiences nil to negligible changes upon further cooling to outlet temperature (140.6–163.8 °C). For instance, the Bocca Nuova fluid sampled on the 1st September 2020 has initial CH_4 concentration of 31.3 µmol/mol at 460 °C, which decreases to 25.2 µmol/mol at 250 °C, reaching the final value of 25.1 µmol/mol at the outlet temperature of 141.9 °C. Again, the decrement in CH_4 concentration from 460 to 250 °C is mainly due to the high rate of reaction WG in this temperature range (Fig. 38b).

(b) For the fluids with initial temperature lower than 400 °C, CH_4 suffers little concentration changes, if any, during cooling. For example, the Bocca Grande sample collected on the

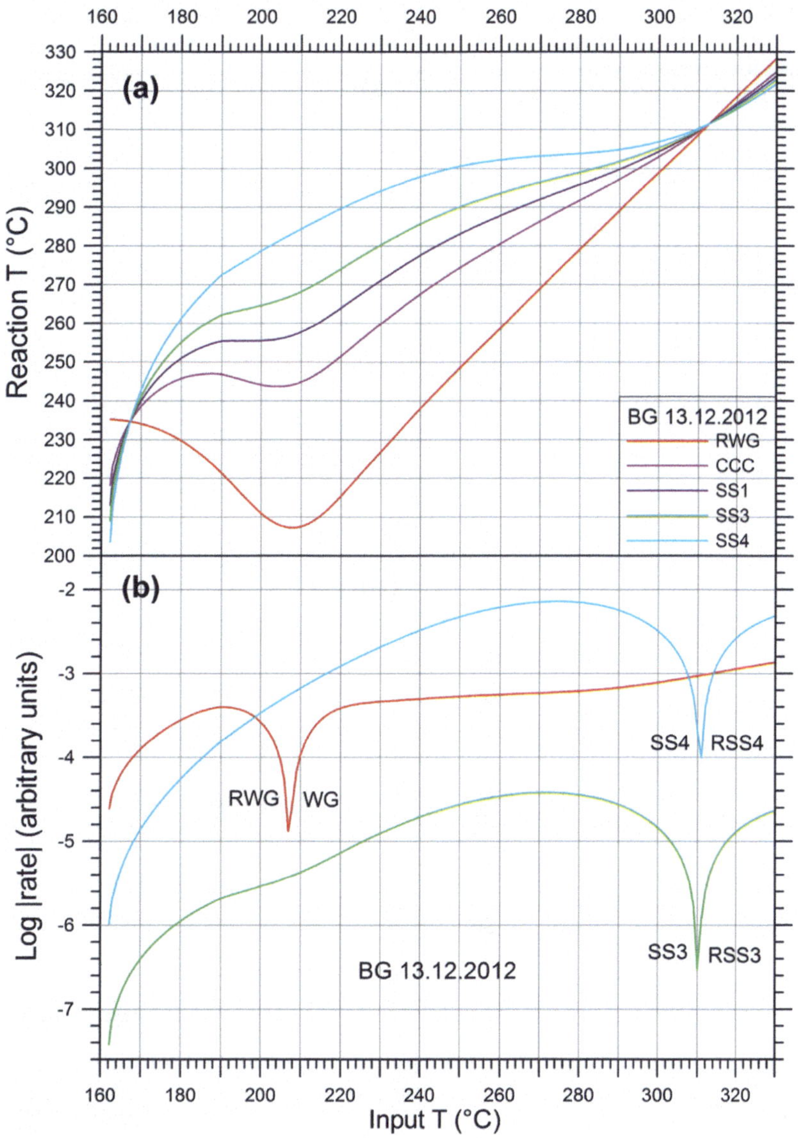

Fig. 12 **a** RWG, CCC, SS1, SS3, and SS4 temperatures and **b** log |rate| of reactions WG, RSS3, and RSS4 during the heating simulation involving these three reactions for the sample collected on 13.12.2012 from Bocca Grande (data from Chiodini et al. 2021 and references therein)

30th November 1983 has a constant CH_4 concentration of 25.1 μmol/mol during the whole cooling process from the initial temperature of 219 °C to the final (outlet) temperature of 156 °C. Again, this constancy of CH_4 concentration is explained by the low rate of reaction WG and the even lower rate of reaction RSS3 in the whole temperature range of interest (Fig. 25b).

The evolution of carbon monoxide concentration during cooling of Solfatara fluids differs from the evolution of hydrogen and methane concentration, in that CO first decreases from the initial temperature (460 to 219 °C) to the temperature where reaction WG inverts (197–216 °C) and then it increases upon cooling to outlet temperature (140.6–163.8 °C). In detail:

(1) For the sample collected on the 1st September 2020 at Bocca Nuova, CO concentration decreases from the initial value, 79.5 μmol/mol at 460 °C, to 3.18 μmol/mol at 211 °C, the inversion temperature of reaction WG, while it increases upon further

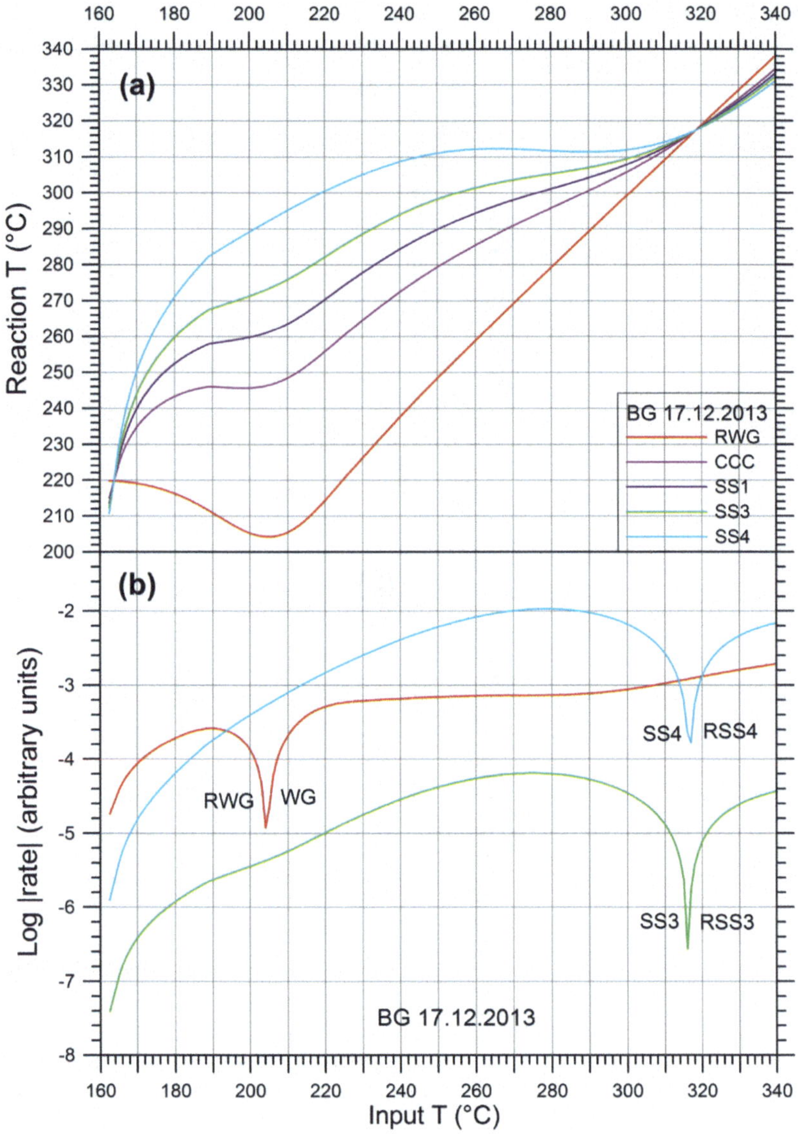

Fig. 13 a RWG, CCC, SS1, SS3, and SS4 temperatures and **b** log |rate| of reactions WG, RSS3, and RSS4 during the heating simulation involving these three reactions for the sample collected on 17.12.2013 from Bocca Grande (data from Chiodini et al. 2021 and references therein)

cooling reaching the final value of 3.53 μmol/mol at the outlet temperature of 141.9 °C.

(2) For the fluid sampled on the 30th November 1983 at Bocca Grande, CO concentration decrements from the initial value, 0.228 μmol/mol at 219 °C, to 0.160 μmol/mol at 197 °C, the WG-RWG inversion temperature, whereas it grows upon further cooling reaching the final value of 0.310 μmol/mol at the outlet temperature of 156 °C.

In both fluids, the variations in CO concentration are governed by the WG reaction, above the WG-RWG inversion temperature, and the RWG reaction below the WG-RWG inversion temperature, whereas the SS3 reaction has a much lower rate at all temperatures (Figs. 25b, 26b, 27b, 28b, 29b, 30b, 31b, 32b, 33b, 34b, 35b, 36b, 37b and 38b).

Examination of Figs. 25b, 26b, 27b, 28b, 29b, 30b, 31b, 32b, 33b, 34b, 35b, 36b, 37b, and 38b indicates that the WG reaction above the WG-RWG inversion temperature and the RWG

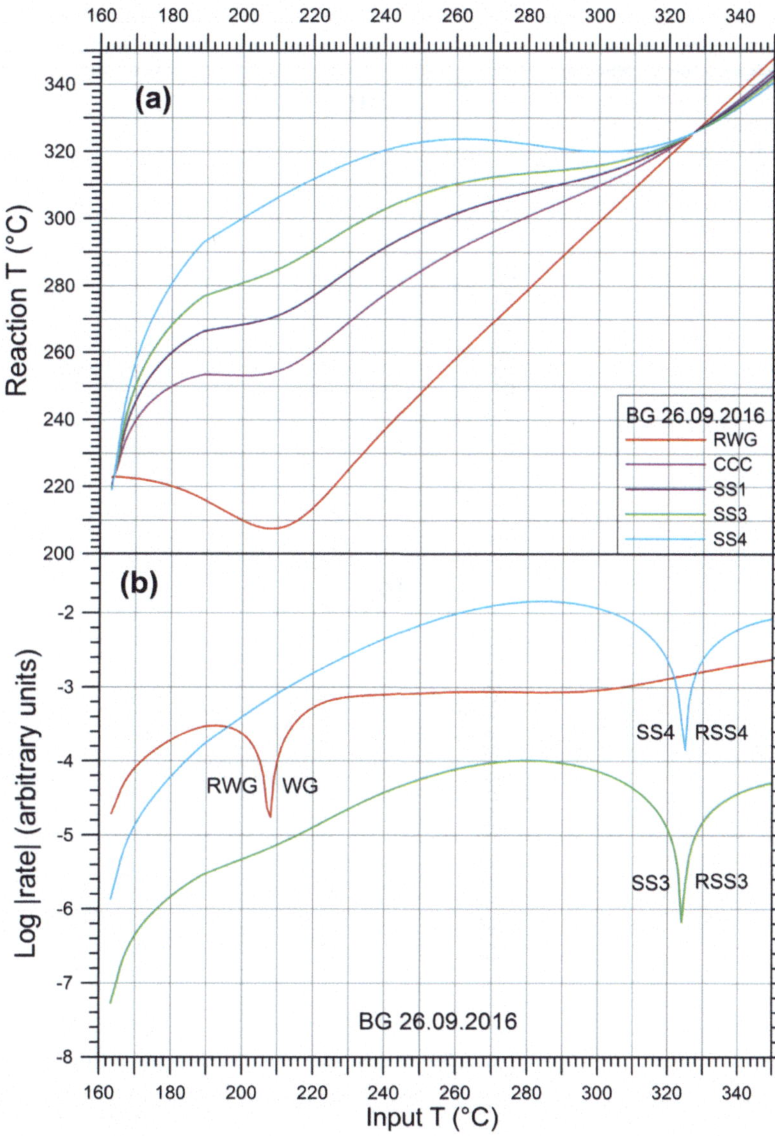

Fig. 14 **a** RWG, CCC, SS1, SS3, and SS4 temperatures and **b** log |rate| of reactions WG, RSS3, and RSS4 during the heating simulation involving these three reactions for the sample collected on 26.09.2016 from Bocca Grande (data from Chiodini et al. 2021 and references therein)

reaction below the WG-RWG inversion temperature proceed with rates significantly greater than those of reaction SS3 and, consequently, they govern the concentrations of H_2, CH_4, and CO upon cooling, not only in the fluid samples collected at Bocca Grande on the 30th November 1983 and at Bocca Nuova on the 1st September 2020, but also in the other twelve selected samples of Solfatara fluids.

Returning to consider the diagram of Fig. 24, it can be noted that the blue crosses representative of the CO concentration at the SS3-RSS3 inversion temperature for the fourteen selected samples are positioned along the trend delineated by the SS4 temperature (computed from the analytical data) and the related equilibrium CO concentration of all the Bocca Grande and Bocca Nuova fluids. In fact, the difference between the SS4 temperature (computed from the analytical data) and the SS3-RSS3 inversion temperature varies from 7 to 13 °C for the fourteen selected samples, whereas the absolute value of the percent deviation of the equilibrium CO concentration at the SS4 temperature from the equilibrium

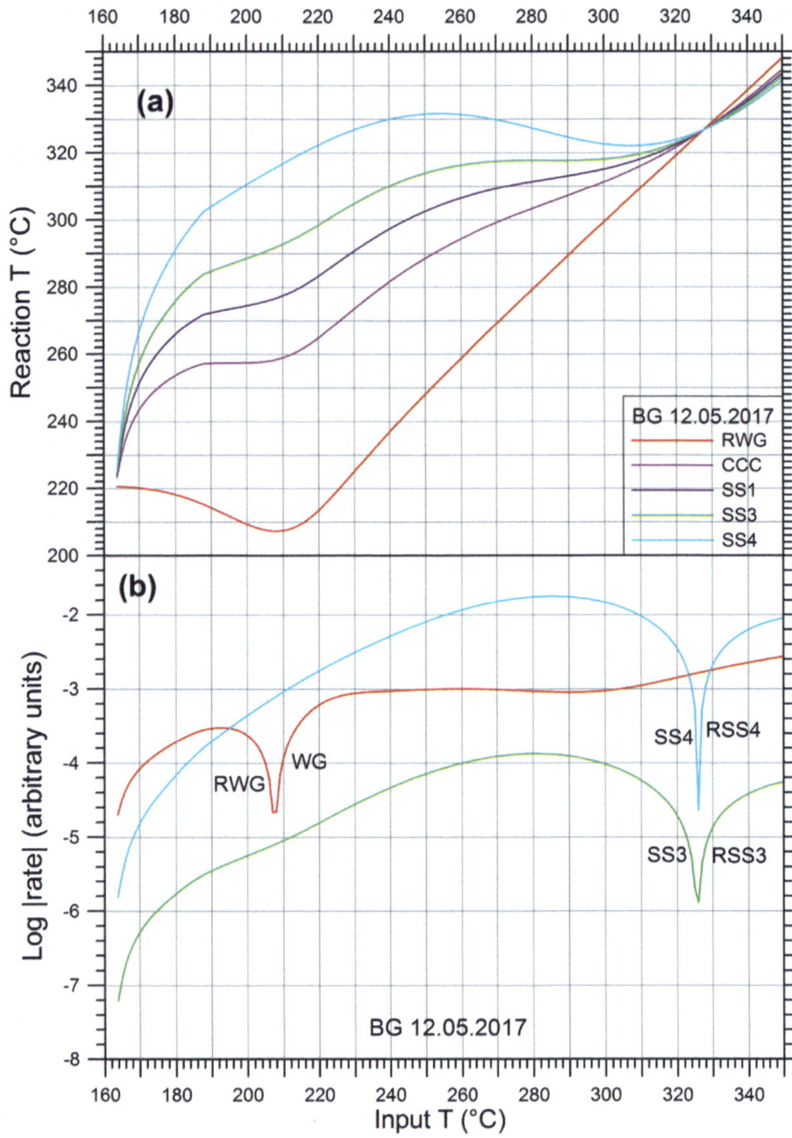

Fig. 15 **a** RWG, CCC, SS1, SS3, and SS4 temperatures and **b** log |rate| of reactions WG, RSS3, and RSS4 during the heating simulation involving these three reactions for the sample collected on 12.05.2017 from Bocca Grande (data from Chiodini et al. 2021 and references therein)

CO concentration at the SS3-RSS3 inversion temperature ranges from 11 to 37% for the fourteen selected samples.

As shown in Table 2 and in the diagrams of Figs. 25b, 26b, 27b, 28b, 29b, 30b, 31b, 32b, 33b, 34b, 35b, 36b, 37b and 38b the SS3-RSS3 inversion temperature is 21–249 °C higher than the WG-RWG inversion temperature. The two inversion temperatures are representative of two distinct partial equilibrium states, involving H_2O, CH_4, H_2, and CO (but not CO_2) and H_2O, CO_2, H_2, and CO (but not CH_4), respectively. Furthermore:

(i) As displayed by Figs. 25b, 26b, 27b, 28b, 29b, 30b, 31b, 32b, 33b, 34b, 35b, 36b, 37b and 38b, carbon monoxide concentration attains the minimum value at the WG-RWG inversion temperature, similar to what was observed in Sect. 5.1.

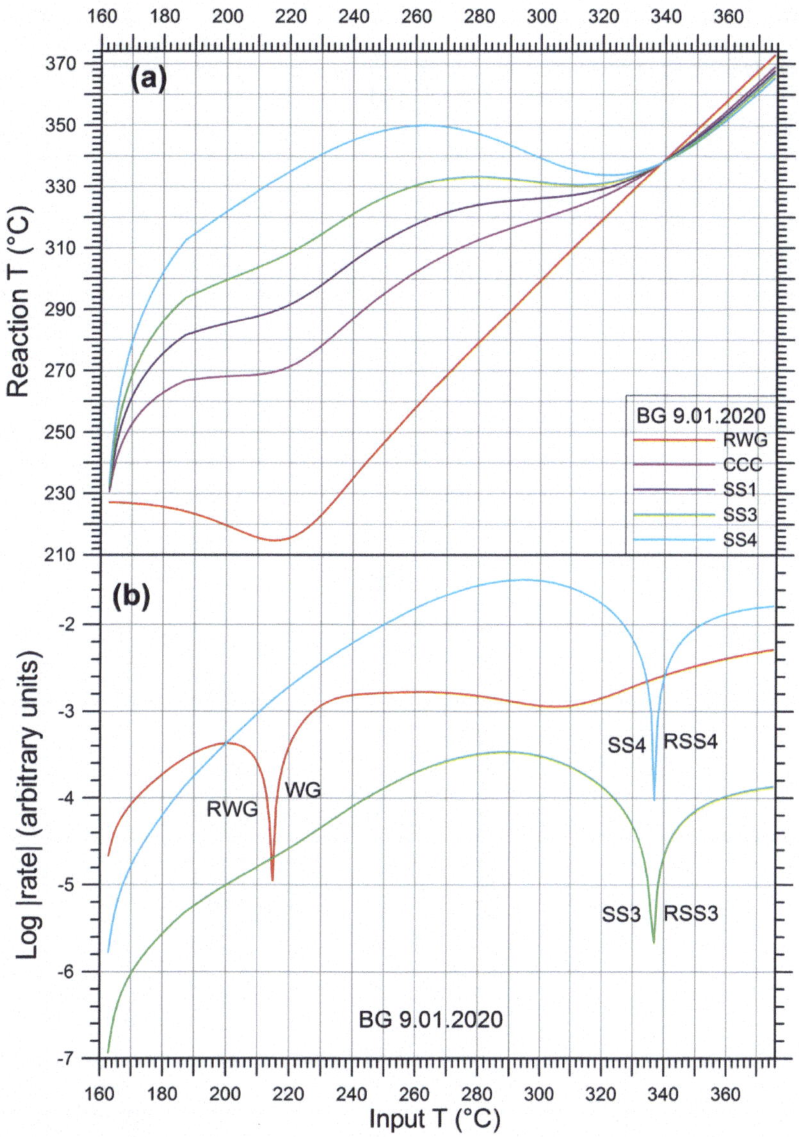

Fig. 16 **a** RWG, CCC, SS1, SS3, and SS4 temperatures and **b** log |rate| of reactions WG, RSS3, and RSS4 during the heating simulation involving these three reactions for the sample collected on 9.01.2020 from Bocca Grande (data from Chiodini et al. 2021 and references therein)

(ii) The RWG, CCC, SS1, SS3, and SS4 reaction temperatures converge at the SS3-RSS3 inversion temperature, which increases with time, from 219 °C on 30th November 1983 to 432 °C on 1st July 2020 for Bocca Grande and reaches even higher values for Bocca Nuova, namely, 447 and 460 °C on 3rd August 2020 and 1st September 2020, respectively, (see Figs. 25a, 26a, 27a, 28a, 29a, 30a, 31a, 32a, 33a, 34a, 35a, 36a, 37a and 38a and Table 2). The convergence of these five reaction temperatures was already observed in Sect. 5.1, but it occurred at the full equilibrium temperature for the H_2O–CO_2–H_2–CH_4–CO system.

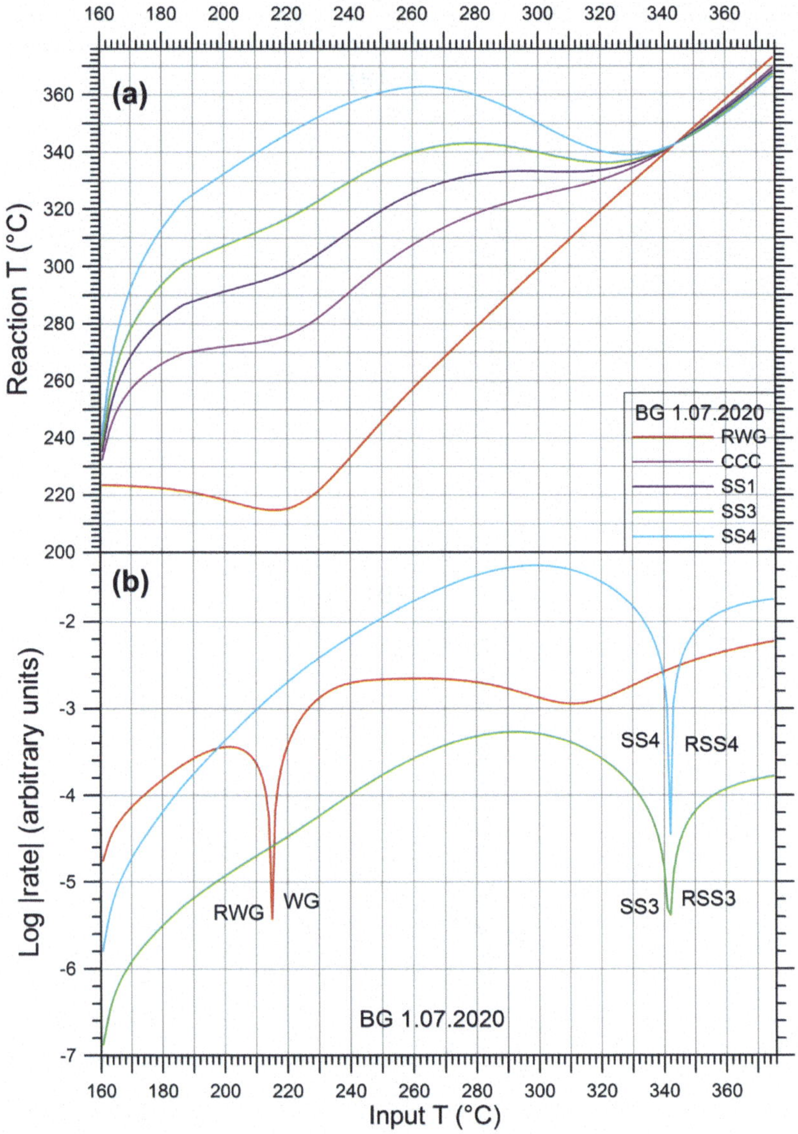

Fig. 17 **a** RWG, CCC, SS1, SS3, and SS4 temperatures and **b** log |rate| of reactions WG, RSS3, and RSS4 during the heating simulation involving these three reactions for the sample collected on 1.07.2020 from Bocca Grande (data from Chiodini et al. 2021 and references therein)

In the plots of Figs. 25a, 26a, 27a, 28a, 29a, 30a, 31a, 32a, 33a, the RWG, CCC, SS1, SS3, and SS4 reaction temperatures converge also at temperatures between the outlet value and 188 °C, while in the graphs of Figs. 34a, 35a, 36a, 37a and 38a, the RWG, CCC, SS1, SS3, and SS4 reaction temperatures are expected to converge at temperatures below the outlet value. However, this convergence or tendency to convergence of RWG, CCC, SS1, SS3, and SS4 reaction temperatures at low temperature values is accidental being not substantiated by the zeroing of reaction rates.

5.3 Final Considerations on the Application of Reaction Path Modeling to Solfatara Fluids

Both (1) the SS3-RSS3 inversion temperature from the heating simulations involving the reactions WG and RSS3 and (2) the full equilibrium temperature from the heating simulations involving the reactions WG, RSS3, and RSS4, for the twelve selected Bocca Grande samples (data from Chiodini et al. 2021 and references

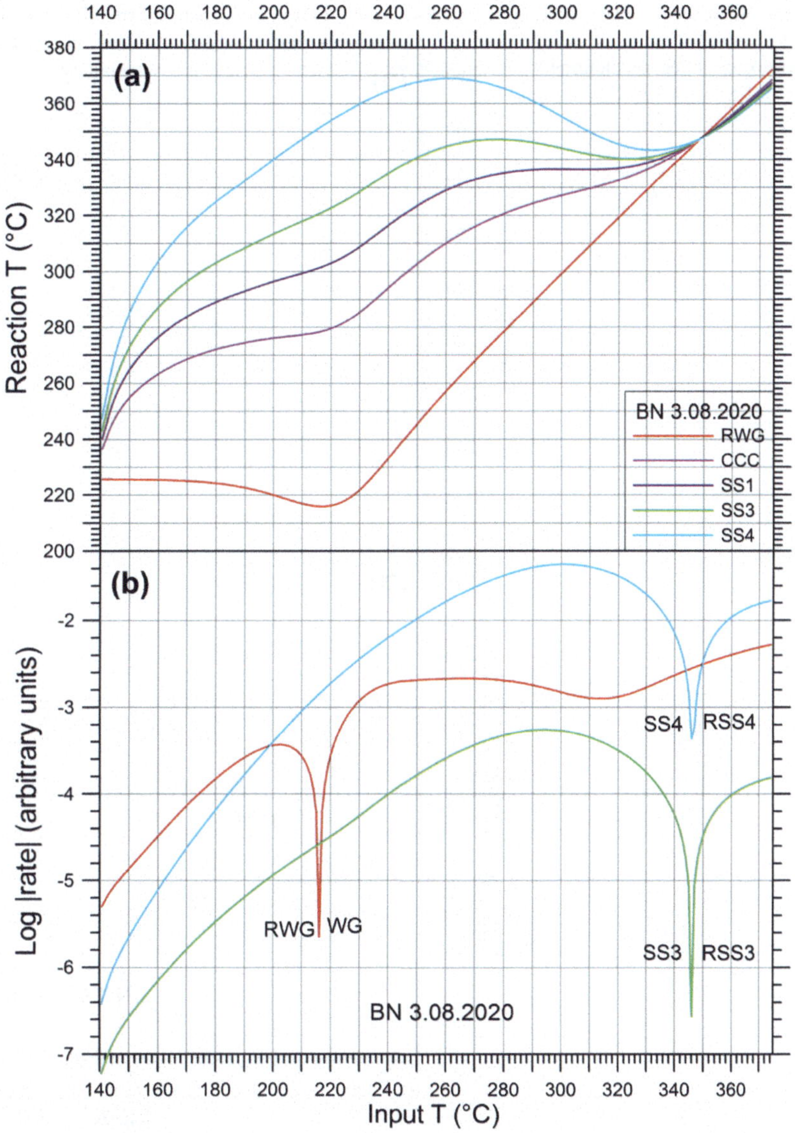

Fig. 18 **a** RWG, CCC, SS1, SS3, and SS4 temperatures and **b** log |rate| of reactions WG, RSS3, and RSS4 during the heating simulation involving these three reactions for the sample collected on 3.08.2020 from Bocca Nuova (data from Chiodini et al. 2021 and references therein)

therein), are reported in the chronogram of Fig. 39 also showing the SS4 temperature computed from the analytical data for all the Bocca Grande samples. The two samples obtained at Bocca Nuova on 3/08/2020 and 1/09/2020 (data from Chiodini et al. 2021 and references therein) are not shown in this figure.

As anticipated in previous section, there are limited differences, 7–13 °C, between the SS4 temperature computed from the analytical data and the SS3-RSS3 inversion temperature from the heating simulations involving the reactions WG and RSS3, for the fourteen selected Solfatara samples. Therefore, it can be inferred that the occurrence of these two reactions during the cooling of Solfatara fluids causes moderate re-equilibration effects.

There are small differences, 1–14 °C, also between the SS4 temperature calculated from the analytical data and the full equilibrium

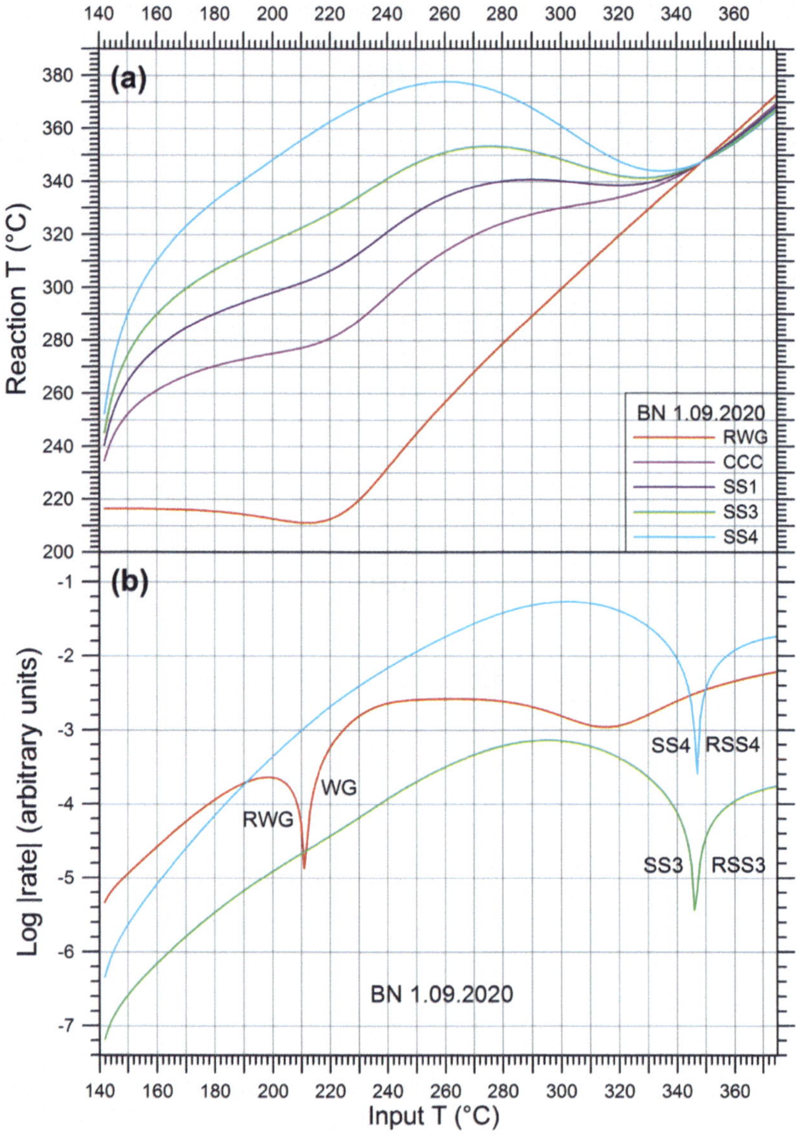

Fig. 19 **a** RWG, CCC, SS1, SS3, and SS4 temperatures and **b** log |rate| of reactions WG, RSS3, and RSS4 during the heating simulation involving these three reactions for the sample collected on 1.09.2020 from Bocca Nuova (data from Chiodini et al. 2021 and references therein)

temperature from the heating simulations involving the reactions WG, RSS3, and RSS4, for the selected samples with SS4 temperature varying from 231 to 321 °C, which were collected from 1983 to 2006. In contrast, the samples collected from 2012 to 2020 have SS4 temperature in the range 340–470 °C and significantly lower full equilibrium temperature, in the interval 311–347 °C, with differences of 29–123 °C. Therefore, it can be deduced that the occurrence of the three reactions WG, RSS3, and RSS4 during the cooling of Solfatara fluids determines limited re-equilibration effects if the initial temperatures are lower than 305–320 °C, whereas re-equilibration effects are more and more important for increasingly higher initial temperatures.

Irrespective of the considered deep temperature (the SS4 temperature computed from the analytical data or the SS3–RSS3 inversion

Table 2 Prominent characteristics of the twelve samples of Bocca Grande and the two samples of Bocca Nuova (data from Chiodini et al. 2021 and references therein), whose heating was simulated allowing the occurrence of reactions WG and RSS3 only, from the outlet temperature up to a maximum temperature higher than the inversion temperature of the RSS3 reaction by a few tens of degrees

Fum	Date	Outlet T	H_2O	CO_2	CH_4	H_2	CO	LST	UST	T RWG, Sect. 4 in chapter "The Geoindicators Involving H_2O, CO_2, CO, CH_4 and H_2"	T SS4, Sect. 4 in chapter "The Geoindicators Involving H_2O, CO_2, CO, CH_4 and H_2"	CO, T SS4, Sect. 5 in chapter "The Geoindicators Involving H_2O, CO_2, CO, CH_4 and H_2"	Tinv, WG heating	Tinv, RSS3 heating	Eq. CH_4, heating	Eq. H2, heating	Eq. CO, heating
		°C	µmol/mol					°C				µmol/mol	°C		µmol/mol		
BG	30/11/1983	156	846,000	151,900	25.1	216	0.310	197	292	231	231	0.313	198	219	25.1	216	0.228
BG	13/10/1983	156	828,000	169,000	27.2	241	0.380	195	294	229	243	0.489	199	230	27.2	241	0.365
BG	10/08/1994	162	837,300	160,500	18.8	281	0.320	196	293	215	262	0.744	197	249	18.8	281	0.574
BG	15/06/2000	153	841,500	156,300	18.6	347	0.544	197	293	234	280	1.19	203	268	18.6	347	0.954
BG	23/03/2004	163	812,900	184,600	17.4	394	0.498	194	295	212	301	2.26	199	289	17.4	393	1.84
BG	14/12/2006	157.2	799,100	198,200	18.9	472	0.608	193	296	209	321	3.86	198	309	18.9	469	3.22
BG	13/12/2012	162.3	762,000	236,000	13.8	480	1.28	190	298	235	340	6.30	209	329	13.9	476	5.34
BG	17/12/2013	162.5	739,000	259,000	14.9	554	1.24	189	299	219	360	10.4	205	349	15.1	545	8.9
BG	26/09/2016	163.3	745,087	251,932	14.6	668	1.54	189	299	222	380	15.2	208	371	15.0	655	13.4
BG	12/05/2017	163.8	720,275	276,720	11.7	703	1.76	188	299	220	400	22.2	208	391	12.3	682	19.7
BG	09/01/2020	162.8	713,802	281,537	28.9	1027	2.99	187	300	227	420	40.9	215	413	30.9	985	36.8
BG	01/07/2020	160.6	702,044	292,171	35.7	1230	3.53	187	300	223	440	62.1	215	432	39.6	1162	55.1
BN	03/08/2020	140.6	725,701	269,675	22.9	1293	3.45	188	299	225	455	66.1	216	447	27.0	1222	59.0
BN	01/09/2020	141.9	701,653	293,419	25.1	1402	3.53	187	300	216	470	92.0	211	460	31.3	1301	79.5

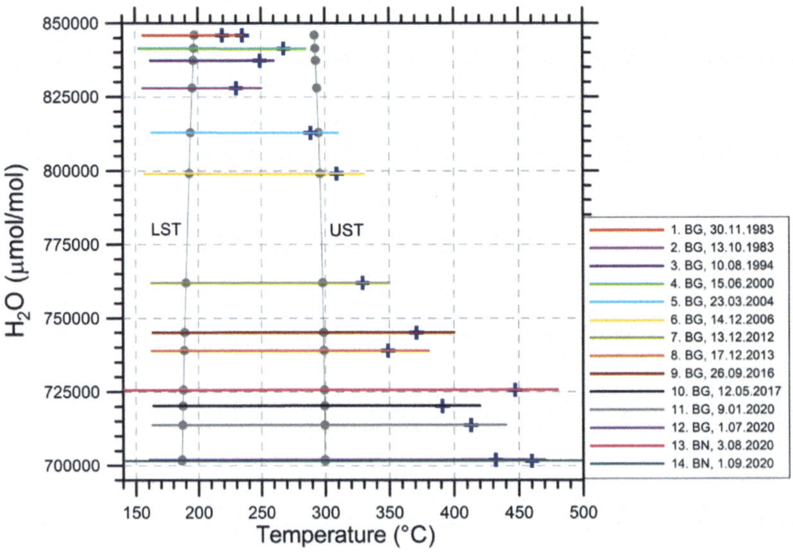

Fig. 20 Water concentration of the fourteen selected Solfatara samples (data from Chiodini et al. 2021 and references therein) during the heating simulation involving reactions WG and RSS3 only (curves, see legend). The blue cross indicates the SS3-RSS3 inversion temperature. Also shown are the upper and lower saturation temperatures (grey lines with closed circles)

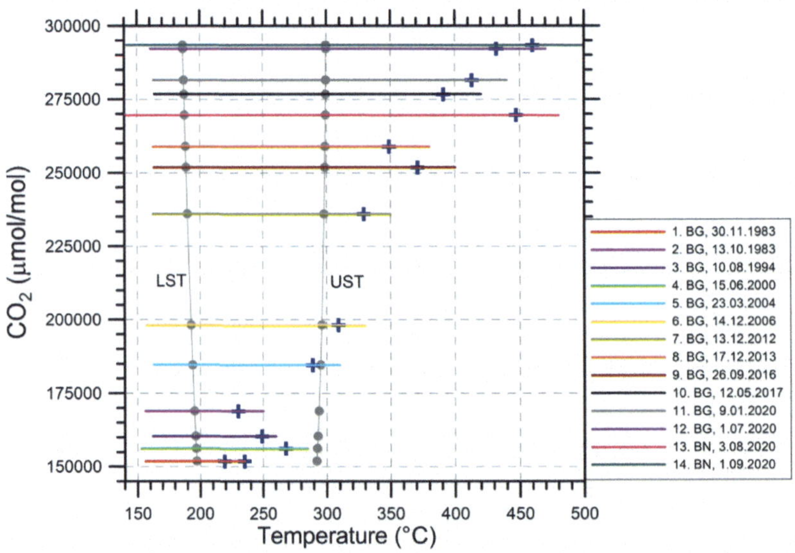

Fig. 21 Carbon dioxide concentration of the fourteen selected Solfatara samples (data from Chiodini et al. 2021 and references therein) during the heating simulation involving reactions WG and RSS3 only (curves, see legend). The blue cross indicates the SS3-RSS3 inversion temperature. Also shown are the upper and lower saturation temperatures (grey lines with closed circles)

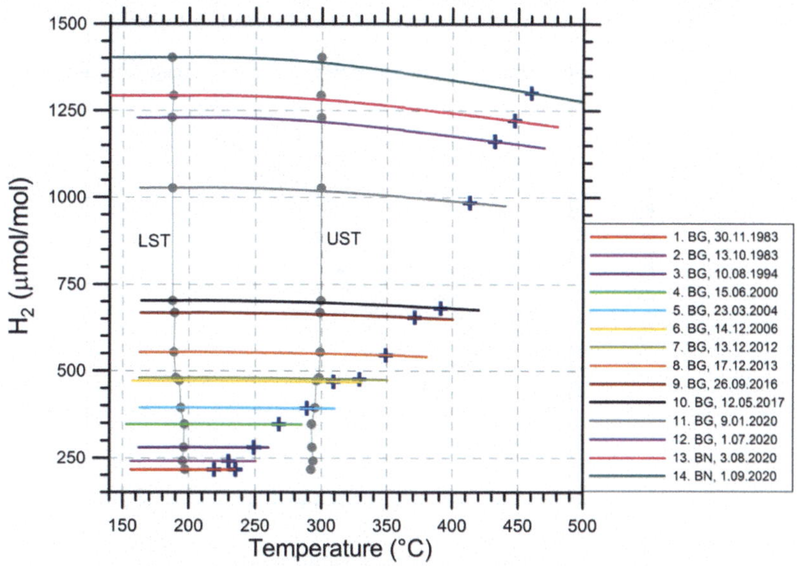

Fig. 22 Hydrogen concentration of the fourteen selected Solfatara samples (data from Chiodini et al. 2021 and references therein) during the heating simulation involving reactions WG and RSS3 only (curves, see legend). The blue cross indicates the SS3-RSS3 inversion temperature. Also shown are the upper and lower saturation temperatures (grey lines with closed circles)

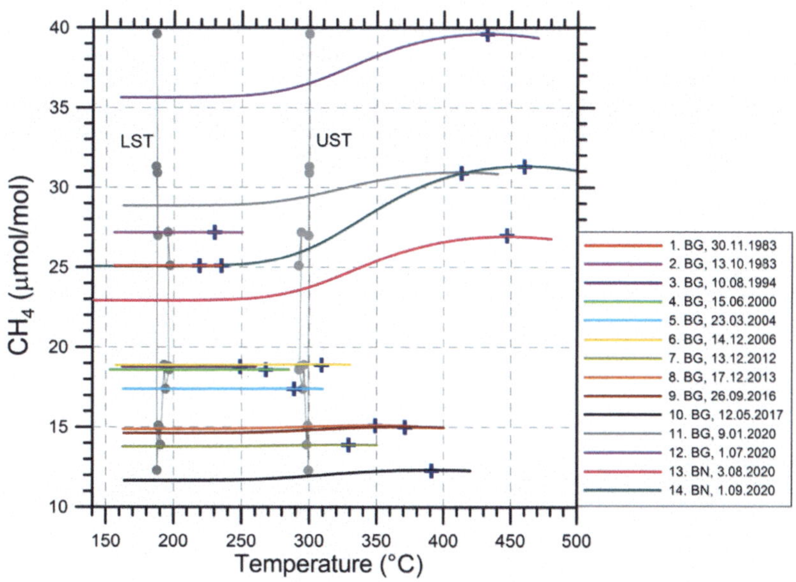

Fig. 23 Methane concentration of the fourteen selected Solfatara samples (data from Chiodini et al. 2021 and references therein) during the heating simulation involving reactions WG and RSS3 only (curves, see legend). The blue cross indicates the SS3-RSS3 inversion temperature. Also shown are the upper and lower saturation temperatures (grey lines with closed circles)

temperature from the heating simulations involving the reactions WG and RSS3 or the full equilibrium temperature from the heating simulations involving the reactions WG, RSS3, and RSS4), it can be concluded that over the years, from 1983 to 2020, the fluids released from Solfatara fumaroles indicate the occurrence of a gradual heating of the related magmatic-hydrothermal system, with the provenance of the fluids either from progressively hotter and deeper aquifer zones or from the same aquifer zone (at least after 1985) which experienced a gradual pressurization and heating with time.

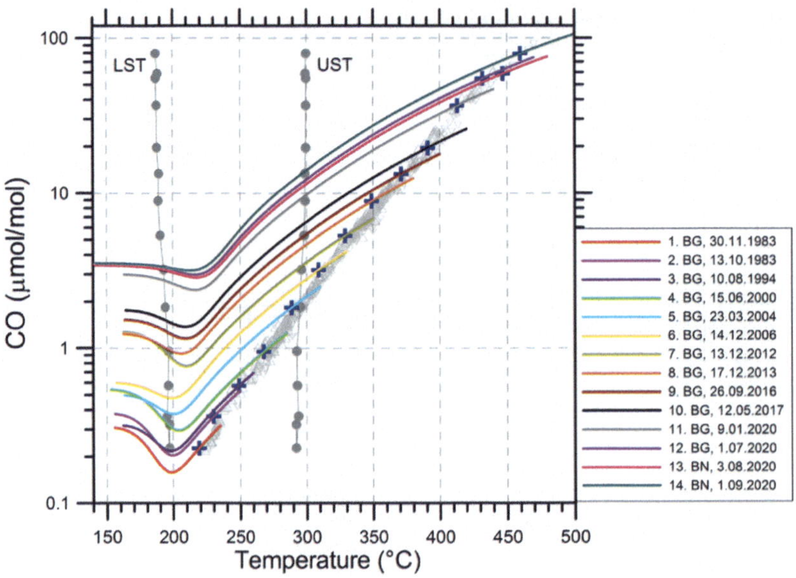

Fig. 24 Carbon monoxide concentration of the fourteen selected Solfatara samples (data from Chiodini et al. 2021 and references therein) during the heating simulation involving reactions WG and RSS3 only (curves, see legend). The blue cross indicates the SS3-RSS3 inversion temperature, whereas the grey lines with closed circles identify the upper and lower saturation temperatures. Also shown is the SS4 temperature computed from the analytical data and the related equilibrium CO concentration (Sects. 4 and 5 in chapter "The Geoindicators Involving H_2O, CO_2, CO, CH_4 and H_2") for all the samples of Bocca Grande (grey triangles) and Bocca Nuova (grey inverted triangles)

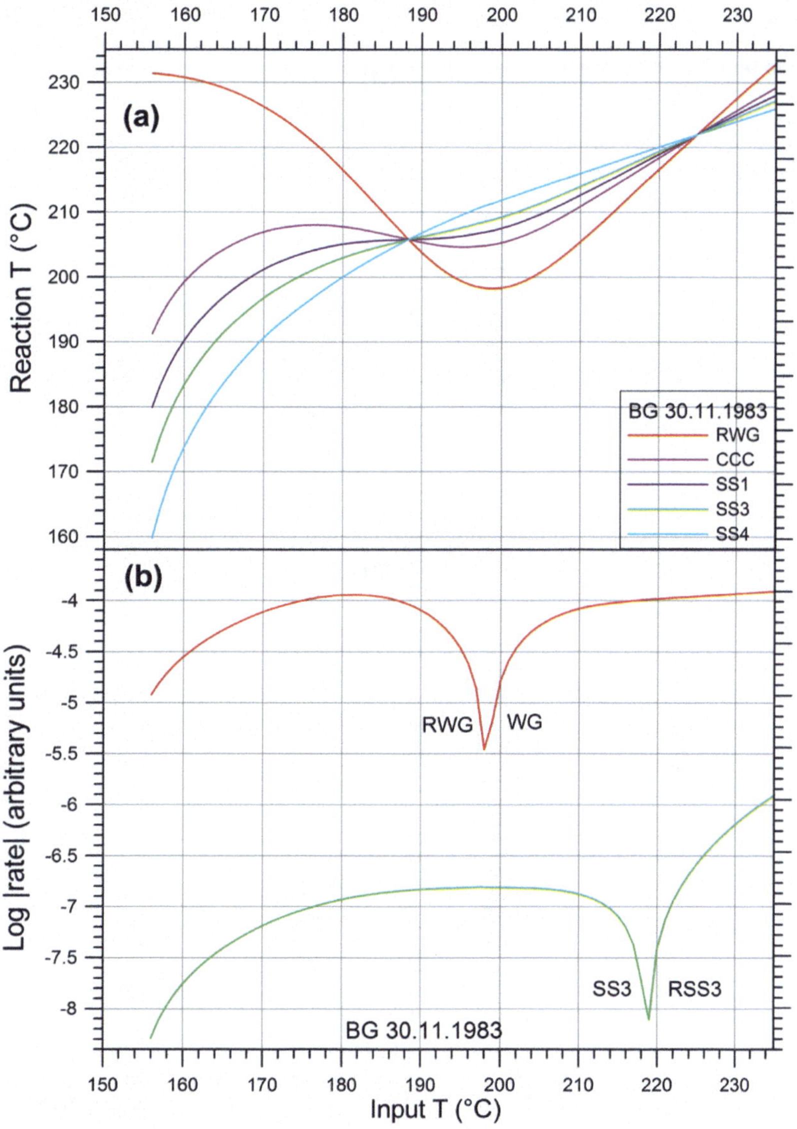

Fig. 25 **a** RWG, CCC, SS1, SS3, and SS4 temperatures and **b** log |rate| of reactions WG and RSS3 during the heating simulation involving these two reactions for the sample collected on 30.11.1983 from Bocca Grande (data from Chiodini et al. 2021 and references therein)

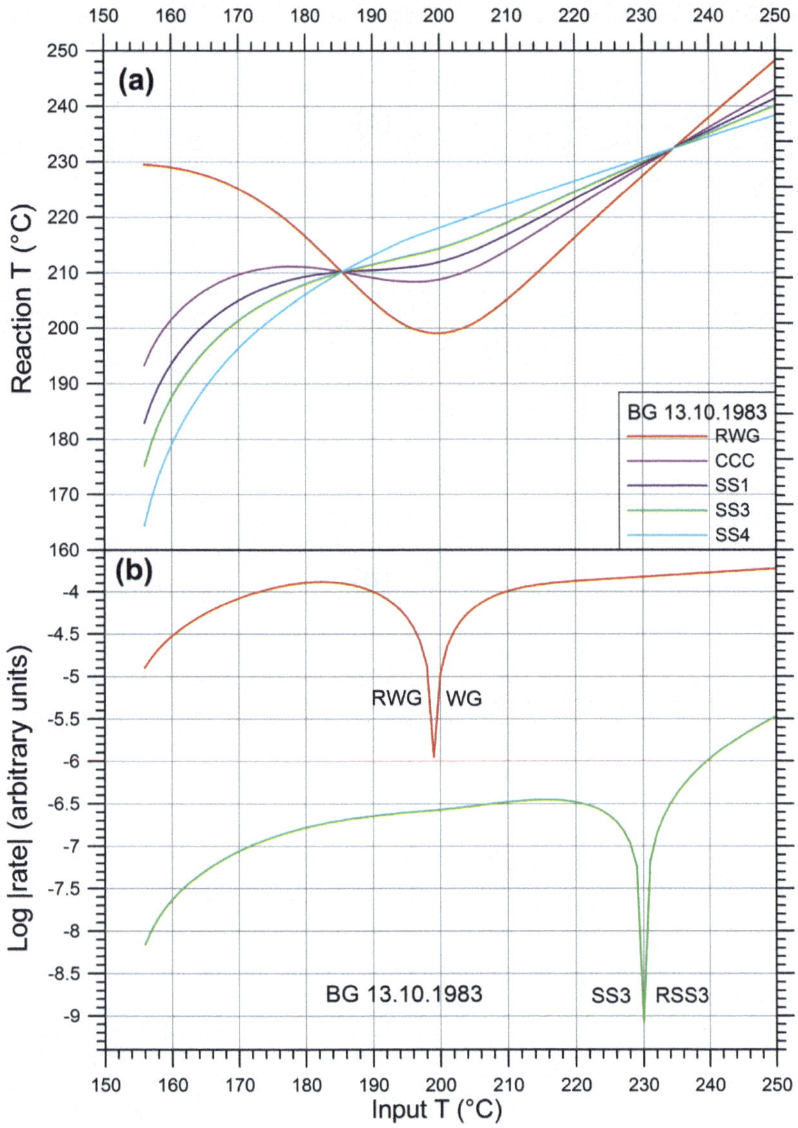

Fig. 26 **a** RWG, CCC, SS1, SS3, and SS4 temperatures and **b** log |rate| of reactions WG and RSS3 during the heating simulation involving these two reactions for the sample collected on 13.10.1983 from Bocca Grande (data from Chiodini et al. 2021 and references therein)

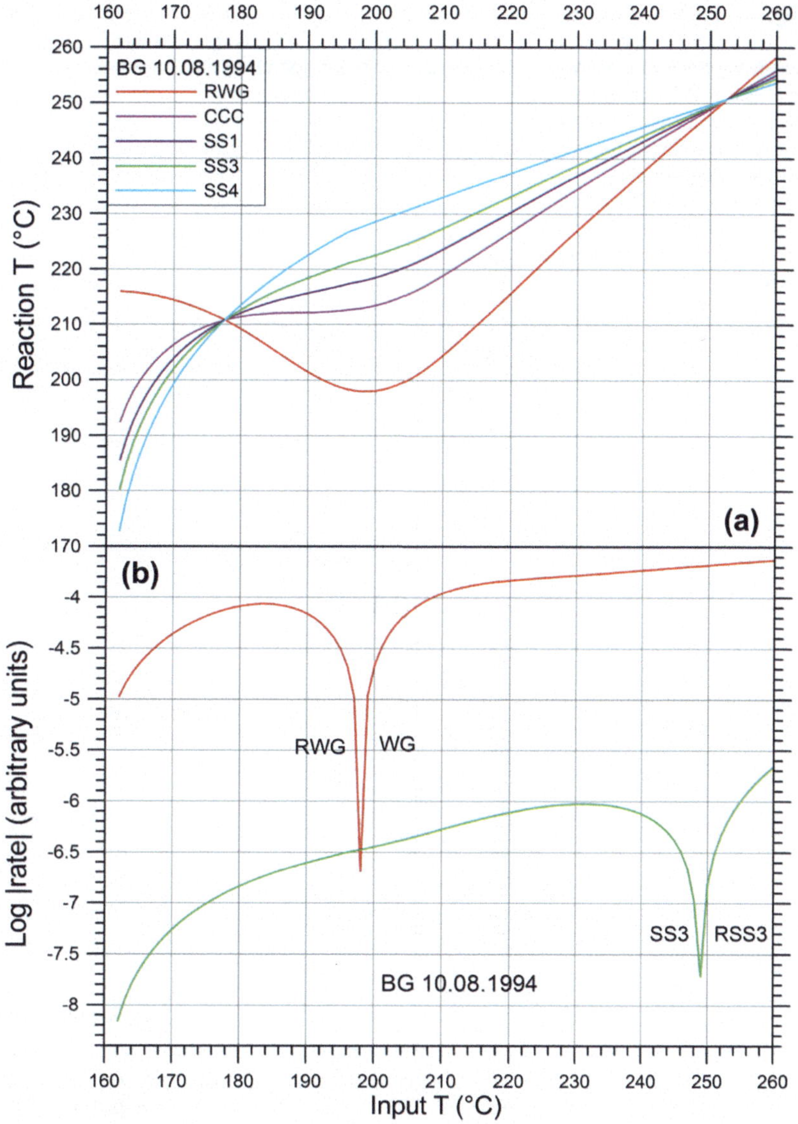

Fig. 27 **a** RWG, CCC, SS1, SS3, and SS4 temperatures and **b** log |rate| of reactions WG and RSS3 during the heating simulation involving these two reactions for the sample collected on 10.08.1994 from Bocca Grande (data from Chiodini et al. 2021 and references therein)

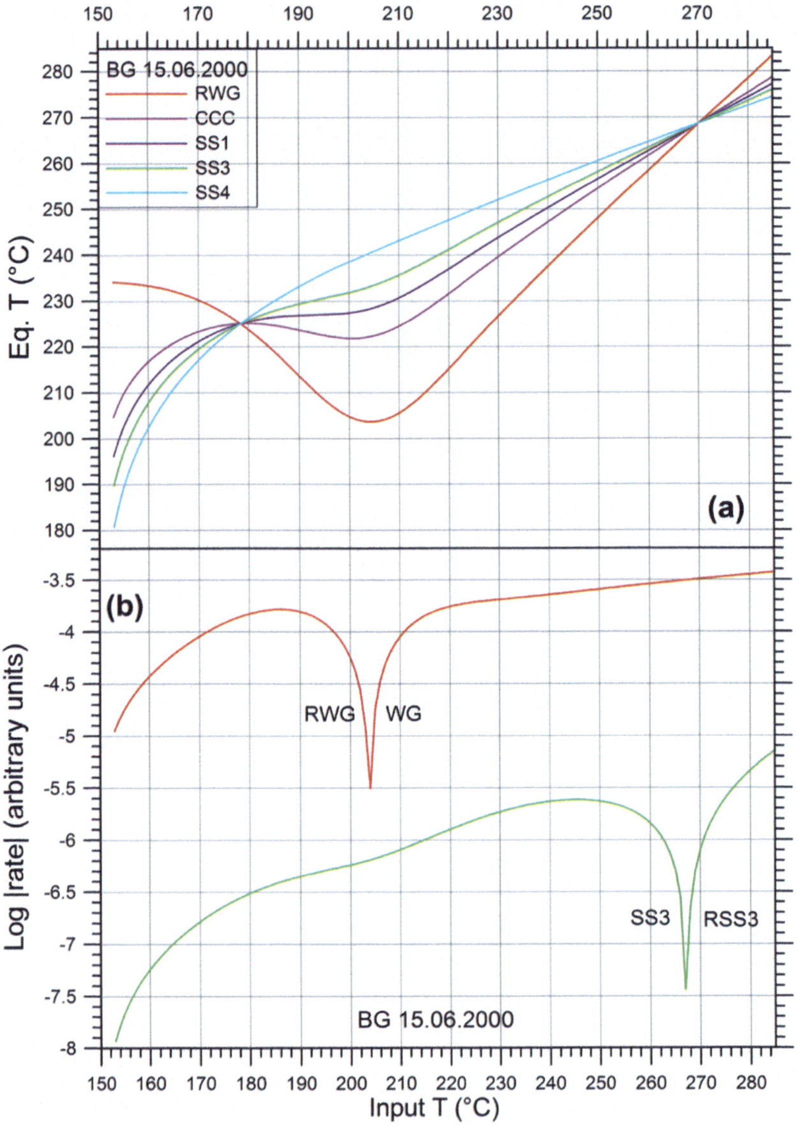

Fig. 28 **a** RWG, CCC, SS1, SS3, and SS4 temperatures and **b** log |rate| of reactions WG and RSS3 during the heating simulation involving these two reactions for the sample collected on 15.06.2000 from Bocca Grande (data from Chiodini et al. 2021 and references therein)

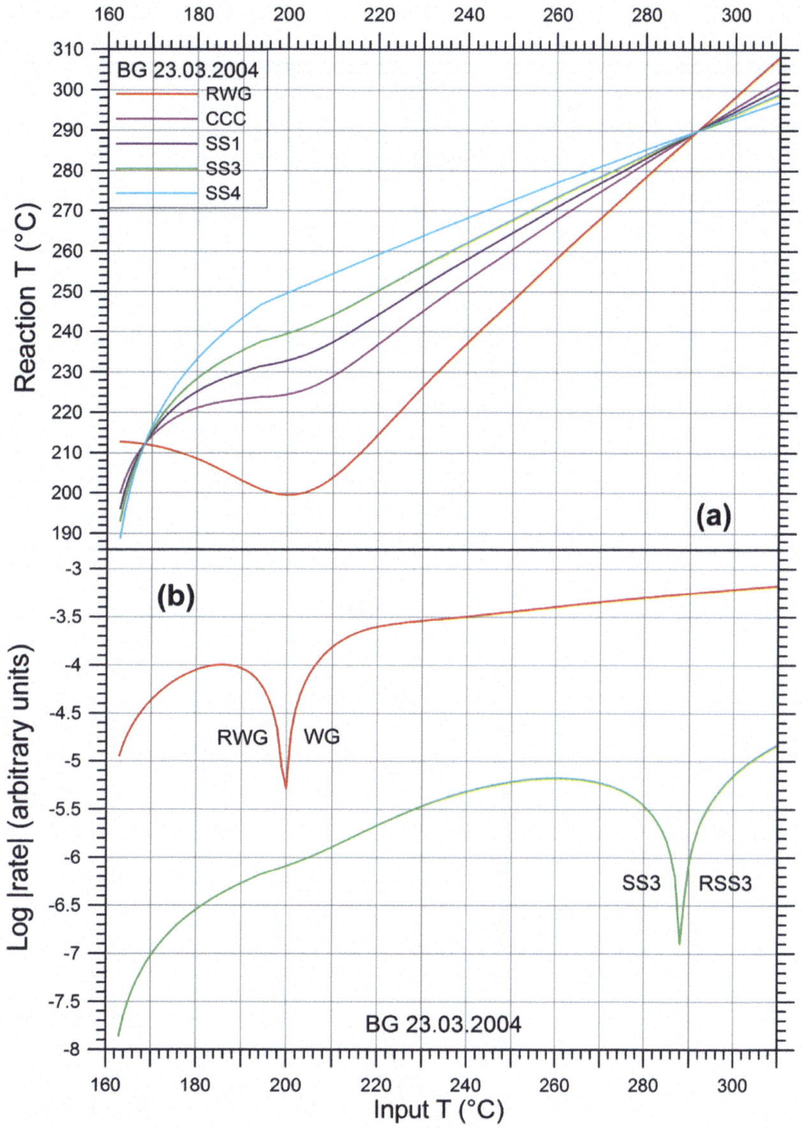

Fig. 29 **a** RWG, CCC, SS1, SS3, and SS4 temperatures and **b** log |rate| of reactions WG and RSS3 during the heating simulation involving these two reactions for the sample collected on 23.03.2004 from Bocca Grande (data from Chiodini et al. 2021 and references therein)

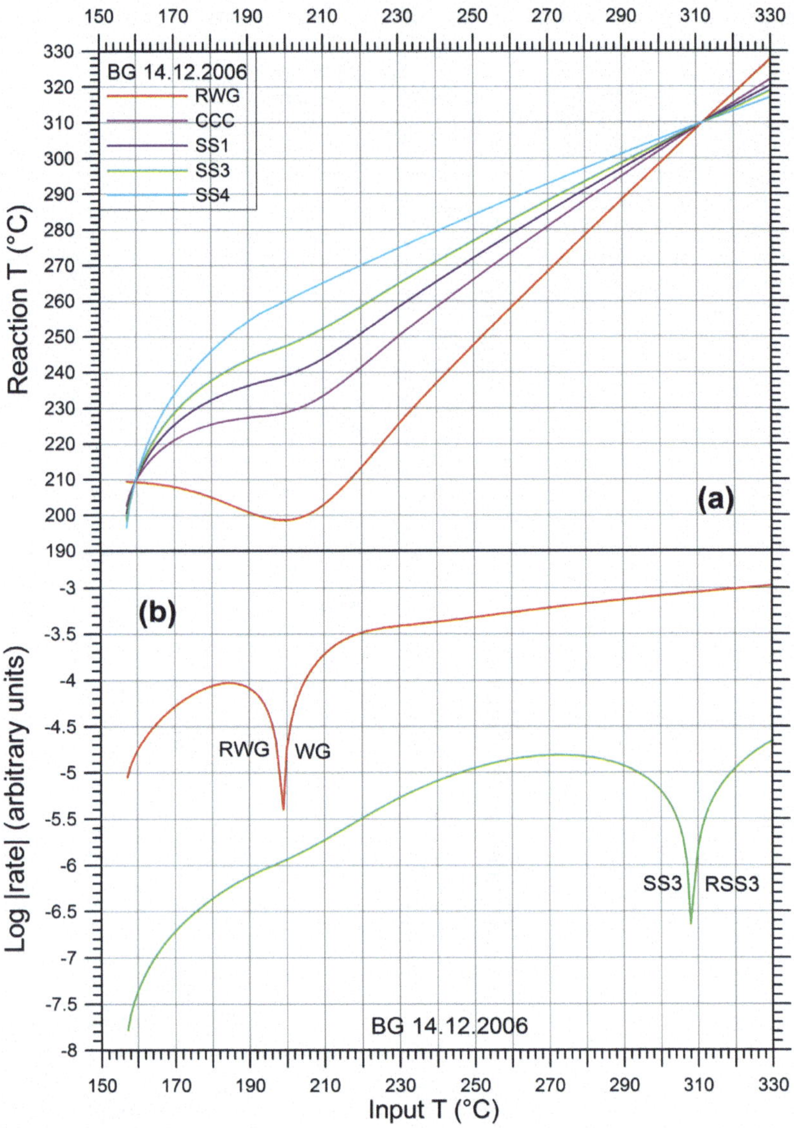

Fig. 30 **a** RWG, CCC, SS1, SS3, and SS4 temperatures and **b** log |rate| of reactions WG and RSS3 during the heating simulation involving these two reactions for the sample collected on 14.12.2006 from Bocca Grande (data from Chiodini et al. 2021 and references therein)

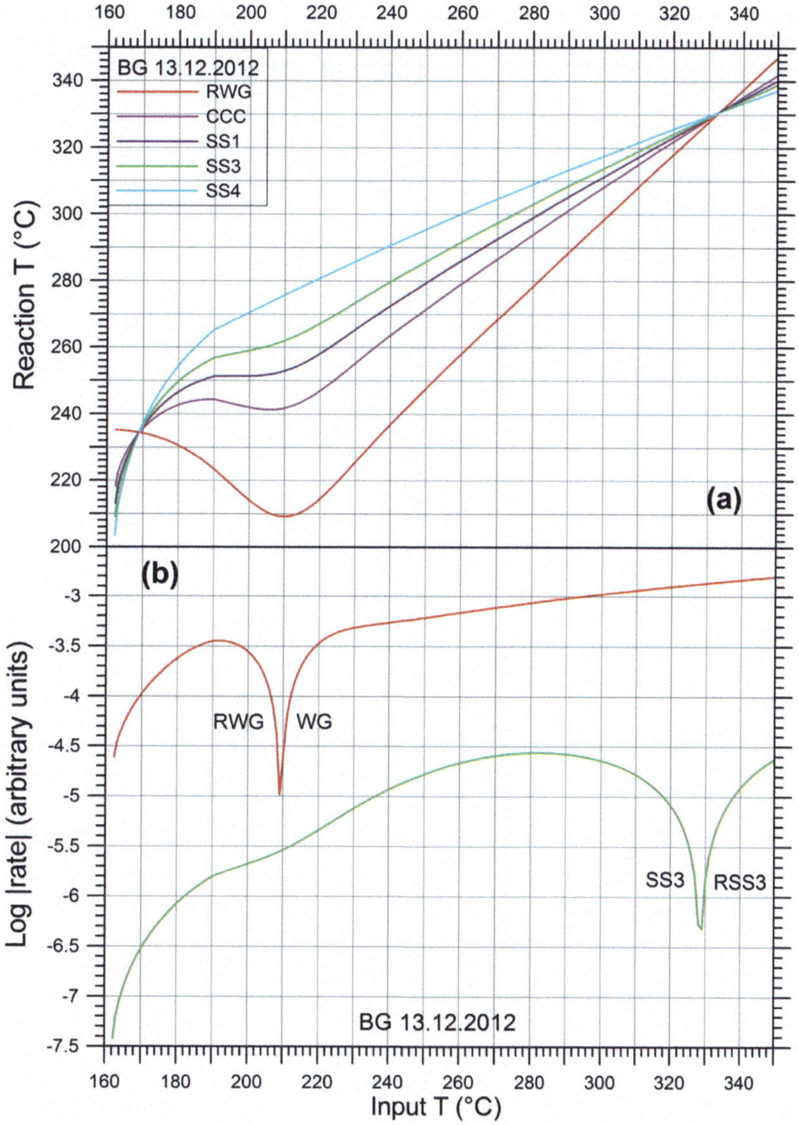

Fig. 31 **a** RWG, CCC, SS1, SS3, and SS4 temperatures and **b** log |rate| of reactions WG and RSS3 during the heating simulation involving these two reactions for the sample collected on 13.12.2012 from Bocca Grande (data from Chiodini et al. 2021 and references therein)

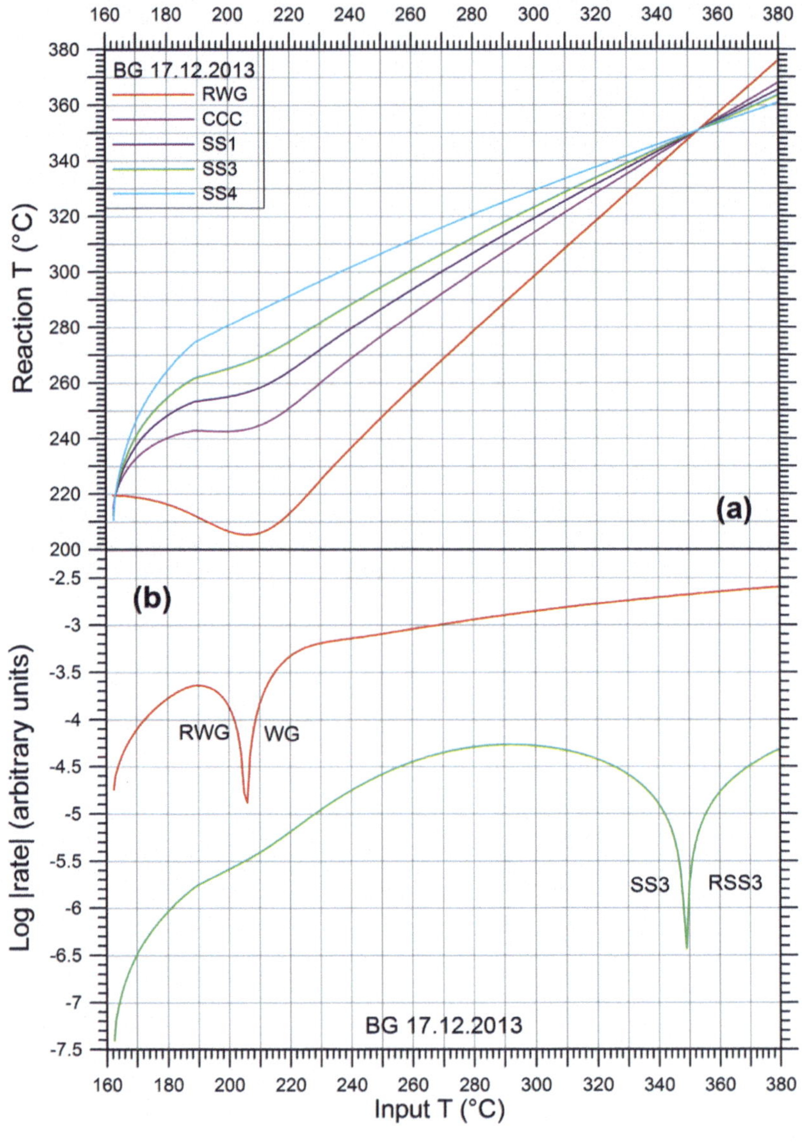

Fig. 32 **a** RWG, CCC, SS1, SS3, and SS4 temperatures and **b** log |rate| of reactions WG and RSS3 during the heating simulation involving these two reactions for the sample collected on 17.12.2013 from Bocca Grande (data from Chiodini et al. 2021 and references therein)

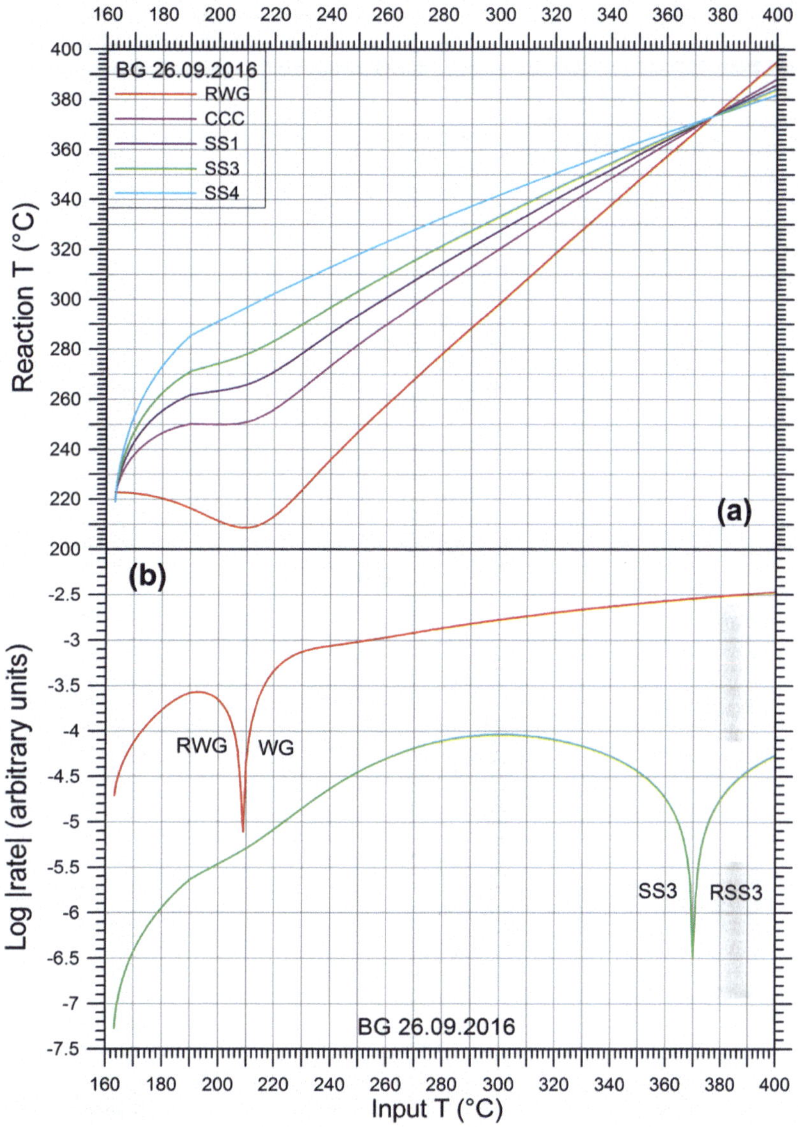

Fig. 33 **a** RWG, CCC, SS1, SS3, and SS4 temperatures and **b** log |rate| of reactions WG and RSS3 during the heating simulation involving these two reactions for the sample collected on 26.09.2016 from Bocca Grande (data from Chiodini et al. 2021 and references therein)

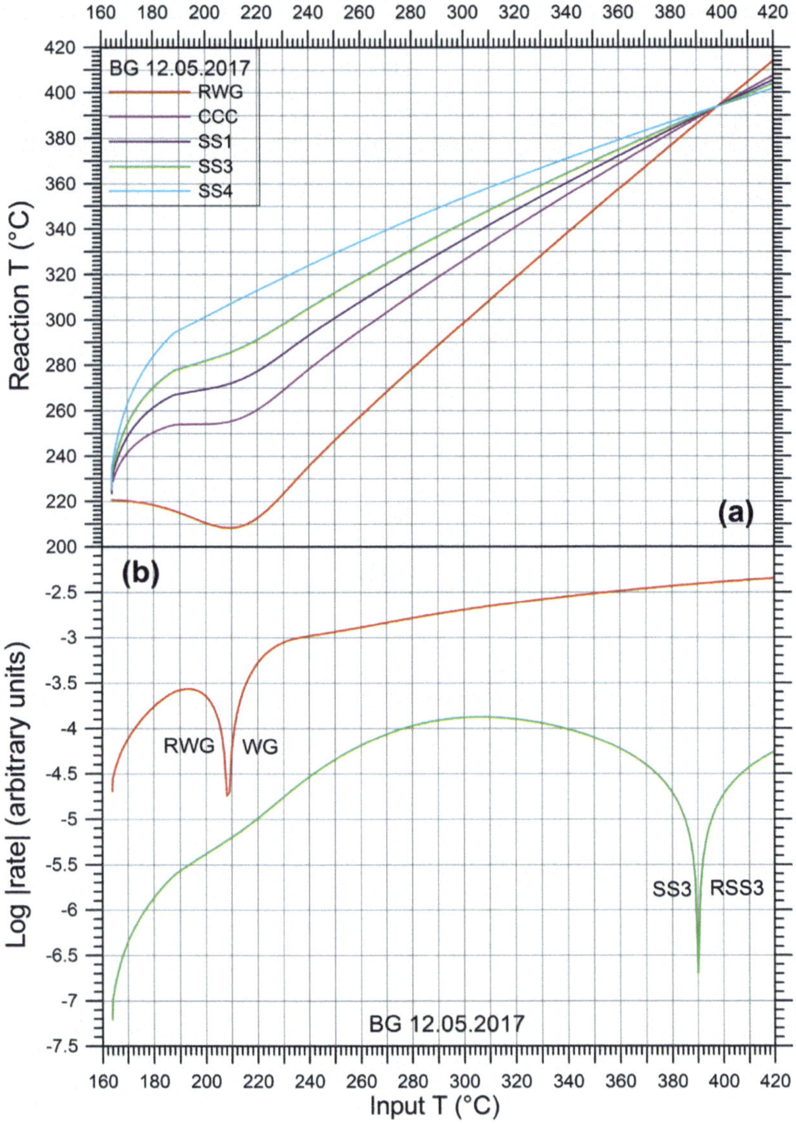

Fig. 34 **a** RWG, CCC, SS1, SS3, and SS4 temperatures and **b** log |rate| of reactions WG and RSS3 during the heating simulation involving these two reactions for the sample collected on 12.05.2017 from Bocca Grande (data from Chiodini et al. 2021 and references therein)

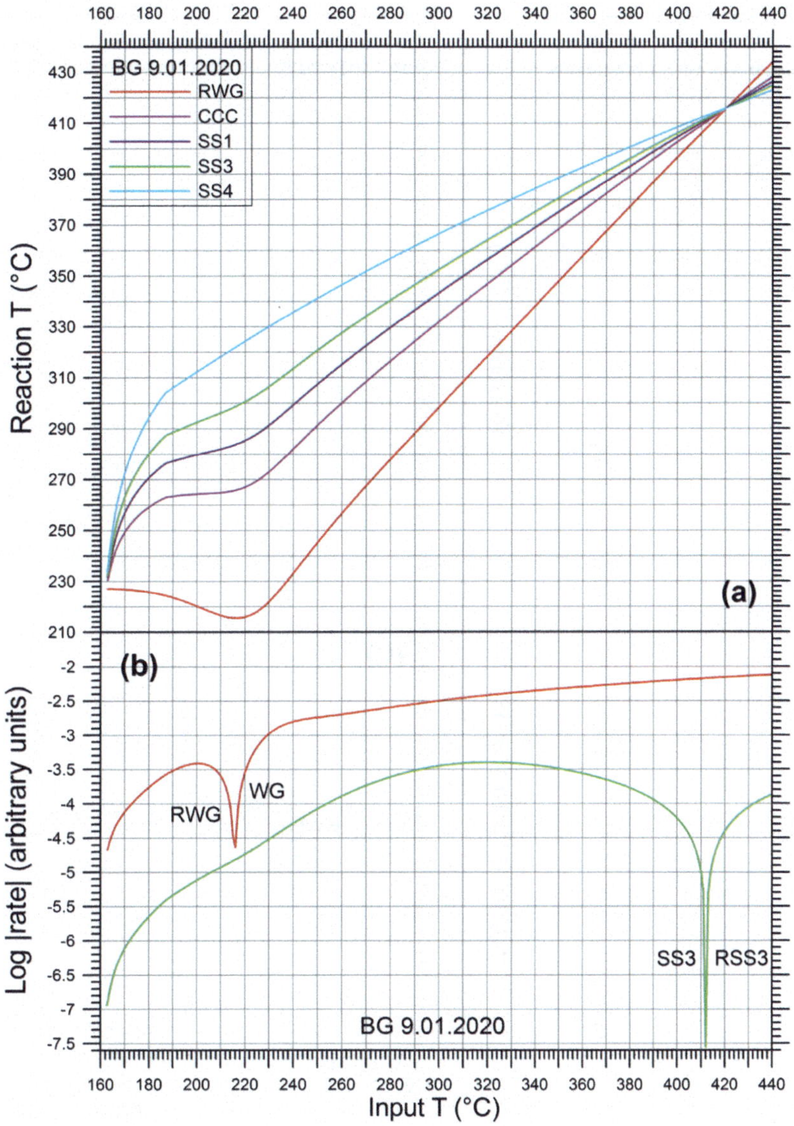

Fig. 35 **a** RWG, CCC, SS1, SS3, and SS4 temperatures and **b** log |rate| of reactions WG and RSS3 during the heating simulation involving these two reactions for the sample collected on 9.01.2020 from Bocca Grande (data from Chiodini et al. 2021 and references therein)

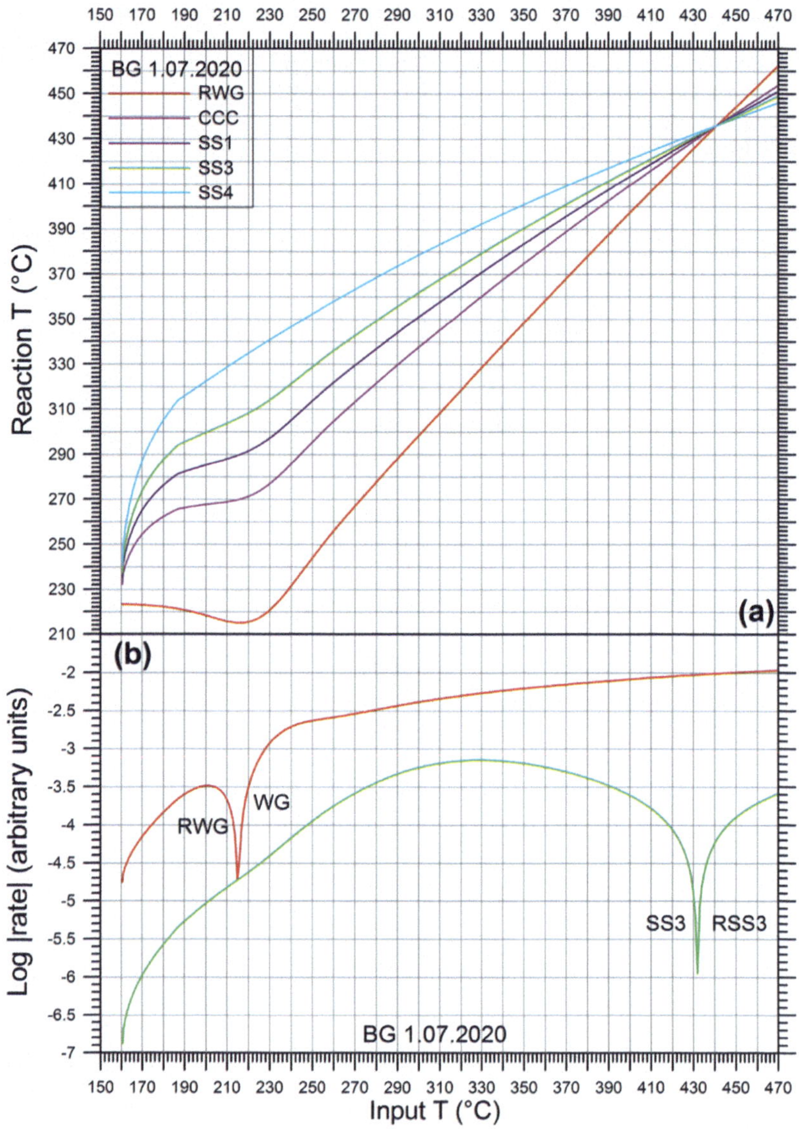

Fig. 36 **a** RWG, CCC, SS1, SS3, and SS4 temperatures and **b** log |rate| of reactions WG and RSS3 during the heating simulation involving these two reactions for the sample collected on 1.07.2020 from Bocca Grande (data from Chiodini et al. 2021 and references therein)

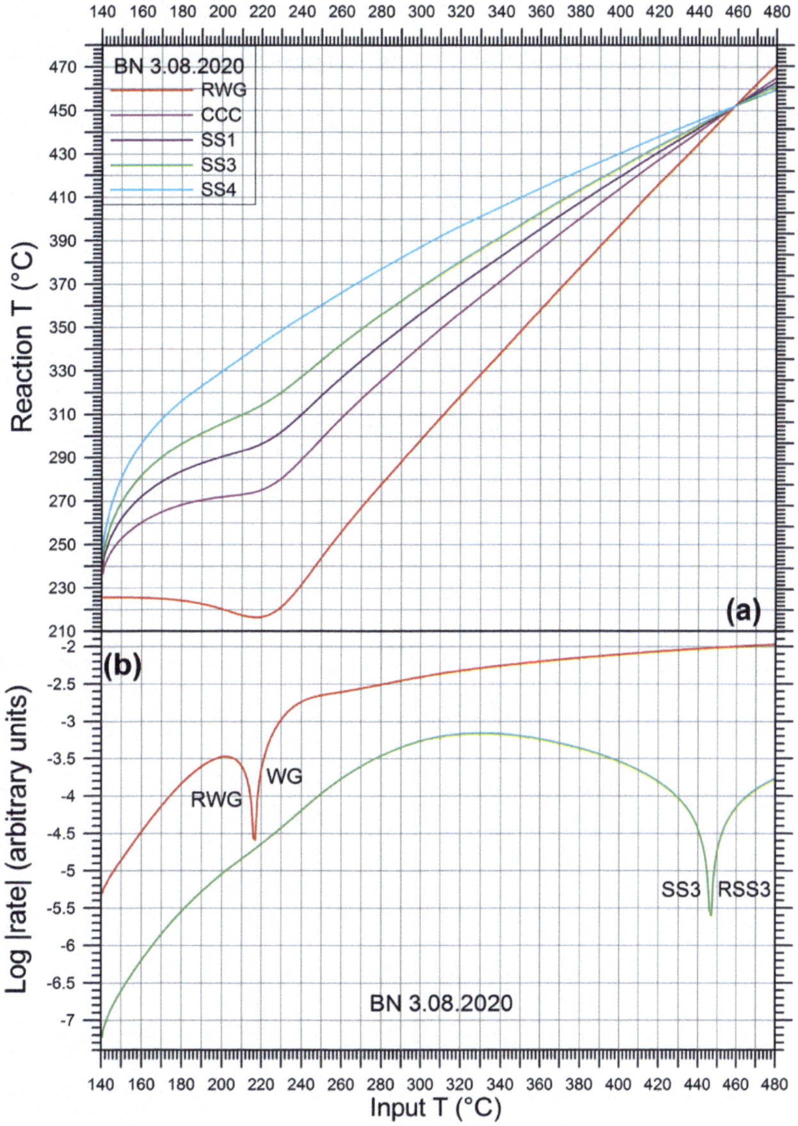

Fig. 37 **a** RWG, CCC, SS1, SS3, and SS4 temperatures and **b** log |rate| of reactions WG and RSS3 during the heating simulation involving these two reactions for the sample collected on 3.08.2020 from Bocca Nuova (data from Chiodini et al. 2021 and references therein)

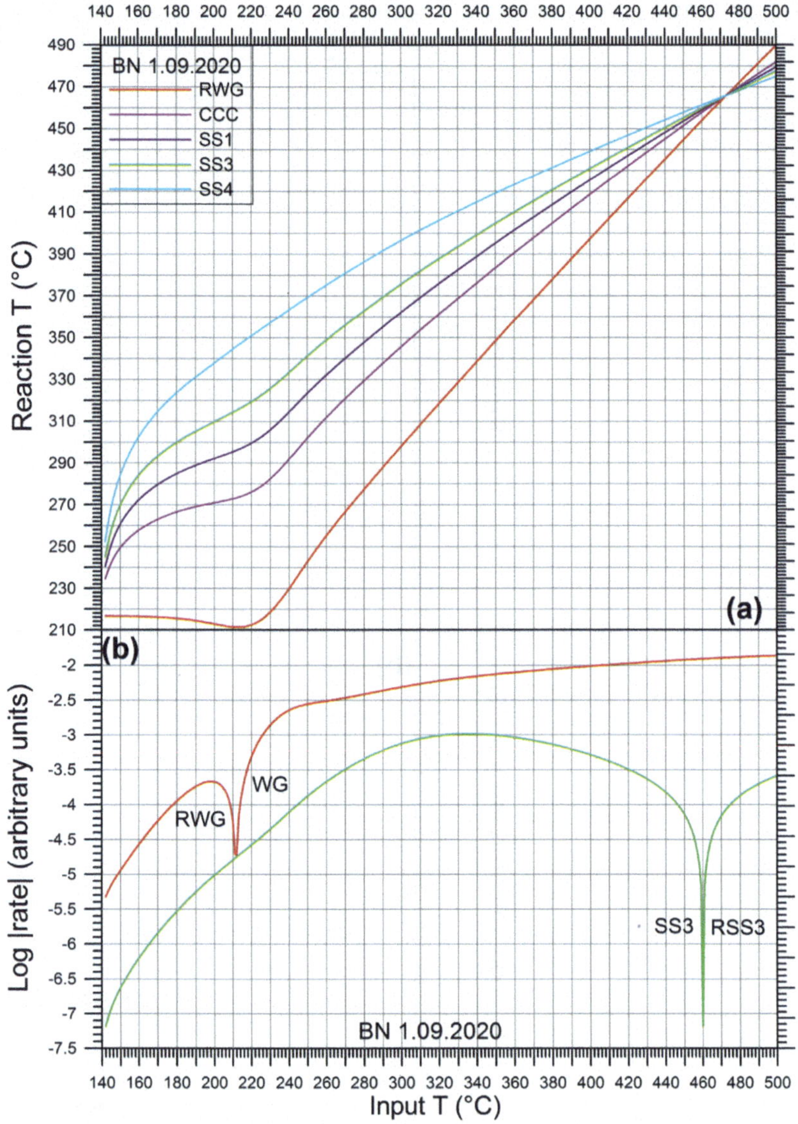

Fig. 38 **a** RWG, CCC, SS1, SS3, and SS4 temperatures and **b** log |rate| of reactions WG and RSS3 during the heating simulation involving these two reactions for the sample collected on 1.09.2020 from Bocca Nuova (data from Chiodini et al. 2021 and references therein)

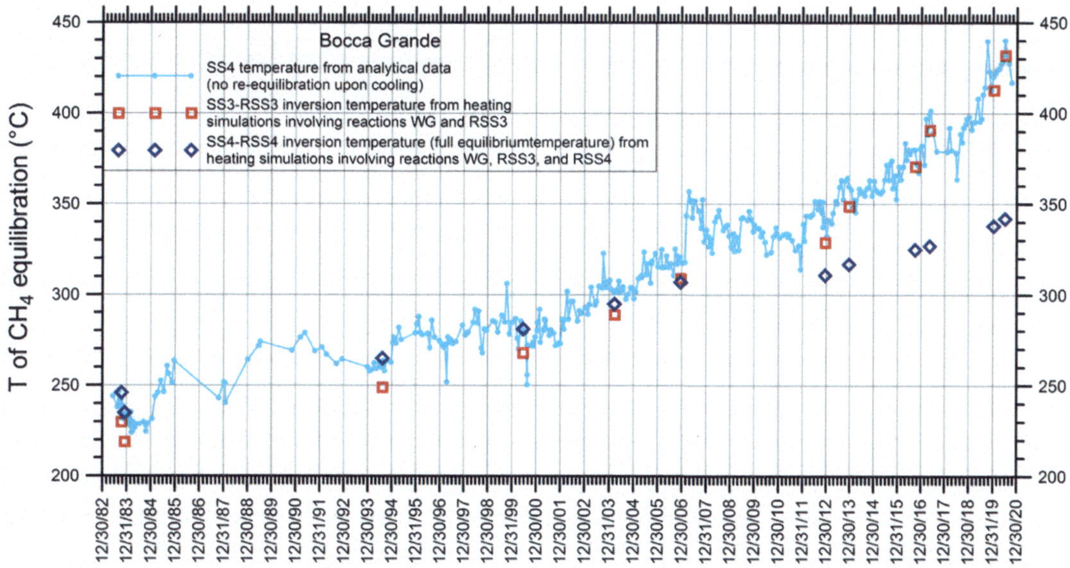

Fig. 39 Chronogram of the SS4 temperature computed from the analytical data for all the Bocca Grande samples (data from Chiodini et al. 2021 and references therein) also showing both (1) the SS3-RSS3 inversion temperature from the heating simulations involving the reactions WG and RSS3 and (2) the full equilibrium temperature from the heating simulations involving the reactions WG, RSS3, and RSS4, for the twelve selected Bocca Grande samples

For what concerns the work to be done in future, it might be useful (i) to implement heating simulations in which the deviations from the ideal gas behavior are taken into account and (ii) run the heating simulations, especially those involving the three reactions WG, RSS3, and RSS4, for all the Bocca Grande and Bocca Nuova samples.

References

Akers WW, Camp DP (1955) Kinetics of the methane-steam reaction. AIChE J 1(4):471–475

Allen DW, Gerhard ER, Likins Jr MR (1975) Kinetics of the methane-steam reaction. Ind Eng Chem Process Des Dev 14(3):256–259

Armor JN (1999) The multiple roles for catalysis in the production of H_2. Appl Catal A 176(2):159–176

Bartholomew CH, Weatherbee GD, Jarvi GA (1980) Effects of carbon deposits on the specific activity of nickel and nickel bimetallic catalysts. Chem Eng Commun 5(1–4):125–134

Berndt ME, Allen DE, Seyfried Jr WE (1996) Reduction of CO_2 during serpentinization of olivine at 300 °C and 500 bar. Geology 24(4):351–354

Charlou JL, Donval JP (1993) Hydrothermal methane venting between 12° N and 26° N along the Mid-Atlantic Ridge. J Geophys Res Solid Earth 98 (B6):9625–9642

Chiodini G, Caliro S, Avino R, Bini G, Giudicepietro F, De Cesare W, Ricciolino P, Aiuppa A, Cardellini C, Petrillo Z, Selva J, Siniscalchi A, Tripaldi S (2021) Hydrothermal pressure-temperature control on CO_2 emissions and seismicity at Campi Flegrei (Italy). J Volcanol Geotherm Res 414:107245

Dalmon JA, Martin GA (1979) Intermediates in CO and CO_2 hydrogenation over Ni catalysts. J Chem Soc Faraday Trans 1: Phys Chem Condens Phases 75:1011–1015

Darensbourg DJ, Bauch CG, Ovalles C (1985) Mechanistic aspects of catalytic carbon dioxide methanation. Rev Inorg Chem 7(4):315–340

De Deken JC, Devos EF, Froment GF (1982) Steam reforming of natural gas: intrinsic kinetics, diffusional influences, and reactor design. In: Wei J, Georgakis C (eds) Chemical reaction engineering, ACS symposium series, vol 196, Chapter 16, pp 181–197

Ding Y, Alpay E (2000) Adsorption-enhanced steam–methane reforming. Chem Eng Sci 55(18):3929–3940

Etiope G, Ionescu A (2015) Low-temperature catalytic CO_2 hydrogenation with geological quantities of ruthenium: a possible abiotic CH_4 source in chromitite-rich serpentinized rocks. Geofluids 15 (3):438–452

Etiope G, Whiticar MJ (2019) Abiotic methane in continental ultramafic rock systems: towards a genetic model. Appl Geochem 102:139–152

Falconer JL, Zağli AE (1980) Adsorption and methanation of carbon dioxide on a nickel/silica catalyst. J Catal 62(2):280–285

Fiebig J, Woodland AB, Spangenberg J, Oschmann W (2007) Natural evidence for rapid abiogenic hydrothermal generation of CH_4. Geochim Cosmochim Acta 71(12):3028–3039

Fiebig J, Stefánsson A, Ricci A, Tassi F, Viveiros F, Silva S, Lopez TM, Schreiber C, Hofmann S, Mountain BW (2019) Abiogenesis not required to explain the origin of volcanic-hydrothermal hydrocarbons. Geochem Perspect Lett 11:23–27

Foustoukos DI, Seyfried WE (2004) Hydrocarbons in hydrothermal vent fluids: the role of chromium-bearing catalysts. Science 304(5673):1002–1005

Fu Q, Lollar BS, Horita J, Lacrampe-Couloume G, Seyfried Jr WE (2007) Abiotic formation of hydrocarbons under hydrothermal conditions: constraints from chemical and isotope data. Geochim Cosmochim Acta 71(8):1982–1998

Fujita S, Terunuma H, Kobayashi H, Takezawa N (1987) Methanation of carbon monoxide and carbon dioxide over nickel catalyst under the transient state. React Kinet Catal Lett 33(1):179–184

Gardner DC, Bartholomew CH (1981) Gravimetric study of adsorbed intermediates in methanation of carbon monoxide. Ind Eng Chem Fundam 20(3):229–234

Hayes RE, Thomas WJ, Hayes KE (1985) A study of the nickel-catalyzed methanation reaction. J Catal 92:312–326

Horita J, Berndt ME (1999) Abiogenic methane formation and isotopic fractionation under hydrothermal conditions. Science 285(5430):1055–1057

Hou K, Hughes R (2001) The kinetics of methane steam reforming over a Ni/α-Al_2O catalyst. Chem Eng J 82 (1–3):311–328

Janecky DR, Seyfried Jr WE (1986) Hydrothermal serpentinization of peridotite within the oceanic crust: experimental investigations of mineralogy and major element chemistry. Geochim Cosmochim Acta 50 (7):1357-1378

Ji F, Zhou H, Yang Q (2008) The abiotic formation of hydrocarbons from dissolved CO_2 under hydrothermal conditions with cobalt-bearing magnetite. Orig Life Evol Biosph 38(2):117–125

Kang WR, Lee KB (2013) Effect of operating parameters on methanation reaction for the production of synthetic natural gas. Korean J Chem Eng 30(7):1386–1394

Kelley DS, Früh-Green GL (1999) Abiogenic methane in deep-seated mid-ocean ridge environments: insights from stable isotope analyses. J Geophys Res Solid Earth 104(B5):10439–10460

Kopyscinski J (2010) Production of synthetic natural gas in a fluidized bed reactor: understanding the hydrodynamic, mass transfer, and kinetic effects. Doctoral dissertation, ETH Zurich, 233 pp

Lapidus AL, Gaidai NA, Nekrasov NV, Tishkova LA, Agafonov YA, Myshenkova TN (2007) The mechanism of carbon dioxide hydrogenation on copper and nickel catalysts. Pet Chem 47(2):75–82

Le Valley TL, Richard AR, Fan M (2014) The progress in water gas shift and steam reforming hydrogen production technologies–a review. Int J Hydrogen Energy 39(30):16983–17000

Marwood M, Doepper R, Renken A (1997) In-situ surface and gas phase analysis for kinetic studies under transient conditions. The catalytic hydrogenation of CO_2. Appl Catal A 151:223–246

McCarty JG, Wise H (1979) Hydrogenation of surface carbon on alumina-supported nickel. J Catal 57 (3):406–416

McCollom TM, Seewald JS (2001) A reassessment of the potential for reduction of dissolved CO_2 to hydrocarbons during serpentinization of olivine. Geochim Cosmochim Acta 65(21):3769–3778

Medina JC, Butala SJ, Bartholomew CH, Lee ML (2000) Low temperature iron-and nickel-catalyzed reactions leading to coalbed gas formation. Geochim Cosmochim Acta 64(4):643–649

Münster P, Grabke HJ (1981) Kinetics of the steam reforming of methane with iron, nickel, and iron-nickel alloys as catalysts. J Catal 72(2):279–287

Neumann B, Jacob K (1924) Die Gleichgewichtsverhältnisse bei der Methanbildung aus Kohlenoxyd und Wasserstoff, bzw. Kohlendioxyd und Wasserstoff. Zeitschrift für Elektrochemie und angewandte physikalische Chemie 30(23–24):557–576

Nobel Foundation (1966) Nobel lectures, chemistry 1901–1921. Elsevier, Amsterdam

Parlikkad NR, Chambrey S, Fongarland P, Fatah N, Khodakov A, Capela S, Guerrini O (2013) Modeling of fixed bed methanation reactor for syngas production: operating window and performance characteristics. Fuel 107:254–260

Peebles DE, Goodman DW, White JM (1983) Methanation of carbon dioxide on nickel (100) and the effects of surface modifiers. J Phys Chem 87(22):4378–4387

Potter J, Rankin AH, Treloar PJ (2004) Abiogenic Fischer-Tropsch synthesis of hydrocarbons in alkaline igneous rocks; fluid inclusion, textural and isotopic evidence from the Lovozero complex, NW Russia. Lithos 75(3–4):311–330

Quach TQP, Rouleau D (1975) Kinetics of the methane–steam reaction over nickel catalyst in a continuous stirred tank reactor. J Appl Chem Biotech 25(6):445–459

Reeves EP, Fiebig J (2020) Abiotic synthesis of methane and organic compounds in Earth's lithosphere. Elements 16(1):25–31

Rostrup-Nielsen JR (1984) Catalytic steam reforming. In: Anderson JR, Boudart M (eds) Catalysis, vol 5. Springer, Berlin, pp 1–117

Sabatier P (1922) Catalysis in organic chemistry. D. Van Nostrand Company, 466 pp

Schild C, Wokaun A, Baiker A (1990) On the mechanism of CO and CO_2 hydrogenation reactions on zirconia-supported catalysts: a diffuse reflectance FTIR study: Part II. Surface species on copper/zirconia catalysts: implications for methanol synthesis selectivity. J Mol Catal 63(2):243–254

Seo YS, Seo DJ, Seo YT, Yoon WL (2006) Investigation of the characteristics of a compact steam reformer integrated with a water-gas shift reactor. J Power Sources 161(2):1208–1216

Seyfried WE, Gordon PC, Dickson FW (1979) A new reaction cell for hydrothermal solution equipment. Am Miner 64(5–6):646–649

Shen WM, Dumesic JA, Hill Jr CG (1981) Criteria for stable Ni particle size under methanation reaction conditions: Nickel transport and particle size growth via nickel carbonyl. Journal of Catalysis, 68(1):152–165.

Sherwood Lollar B, Voglesonger K, Lin L-H, Lacrampe-Couloume G, Telling J, Abrajano TA, Onstott TC, Pratt LM (2007) Hydrogeologic controls on episodic H_2 release from Precambrian fractured rocks—energy for deep subsurface life on Earth and Mars. Astrobiology 7(6):971–986

Smith RJB, Loganathan M, Shantha MS (2010) A review of the water gas shift reaction kinetics. Int J Chem React Eng 8(1). https://doi.org/10.2202/1542-6580.2238

Sughrue EL, Bartholomew CH (1982) Kinetics of carbon monoxide methanation on nickel monolithic catalysts. Appl Catal 2(4–5):239–256

Taran YA, Kliger GA, Sevastianov VS (2007) Carbon isotope effects in the open-system Fischer-Tropsch synthesis. Geochim Cosmochim Acta 71(18):4474–4487

Taran YA, Kliger GA, Cienfuegos E, Shuykin AN (2010) Carbon and hydrogen isotopic compositions of products of open-system catalytic hydrogenation of CO_2: implications for abiogenic hydrocarbons in Earth's crust. Geochim Cosmochim Acta 74(21):6112–6125

Van Ho S, Harriott P (1980) The kinetics of methanation on nickel catalysts. J Catal 64(2):272–283

Van Hook JP (1980) Methane-steam reforming. Catal Rev Sci Eng 21(1):1–51

Vlasenko VM, Yuzefovich GE (1969) Mechanism of the catalytic hydrogenation of oxides of carbon. Russ Chem Rev 38(9):728–739

Wang S, Lu GQ, Millar GJ (1996) Carbon dioxide reforming of methane to produce synthesis gas over metal-supported catalysts: state of the art. Energy Fuels 10(4):896–904

Wang W, Wang S, Ma X, Gong J (2011) Recent advances in catalytic hydrogenation of carbon dioxide. Chem Soc Rev 40(7):3369–4260

Weatherbee GD, Bartholomew CH (1982) Hydrogenation of CO_2 on group VIII metals: II. Kinetics and mechanism of CO_2 hydrogenation on nickel. J Catal 77(2):460–472

Weatherbee GD, Bartholomew CH (1984) Hydrogenation of CO_2 on group VIII metals: IV. Specific activities and selectivities of silica-supported Co, Fe, and Ru. J Catal 87(2):352–362

Whiticar MJ (1990) A geochemial perspective of natural gas and atmospheric methane. Org Geochem 16(1–3):531–547

Xu J, Froment GF (1989a) Methane steam reforming, methanation and water-gas shift: I. Intrinsic kinetics. AIChE J 35(1):88–96

Xu J, Froment GF (1989b) Methane steam reforming: II. Diffusional limitations and reactor simulation. AIChE J 35(1):97–103

Zhang S, Mi J, He K (2013a) Synthesis of hydrocarbon gases from four different carbon sources and hydrogen gas using a gold-tube system by Fischer–Tropsch method. Chem Geol 349:27–35

Zhang J, Fatah N, Capela S, Kara Y, Guerrini O, Khodakov AY (2013b) Kinetic investigation of carbon monoxide hydrogenation under realistic conditions of methanation of biomass derived syngas. Fuel 111:845–854

Calibration and Application of the RWG and SS4 Geoindicators for the Possible Expansion Paths of Solfatara Fluids

Abstract

Considering the fugacity coefficients computed by means of the Peng-Robinson EOS, the geoindicators related to the RWG and SS4 reactions have been calibrated for the saturation expansion path involving a brine containing 21 wt% NaCl or 33.5 wt% NaCl and the linear P-T depressurization path, whereas only the geoindicators related to the RWG reaction have been calibrated for the isenthalpic decompression path. The obtained geothermometers and geobarometers allows one to compute the equilibrium temperature and pressure as a function of the CO_2 mole fraction and either the sum of the analytical log ratios α = log (CO/CO_2) - log (H_2/H_2O) for the RWG reaction or the sum of analytical log ratios = log (CH_4/CO_2) - 4 · log (H_2/H_2O) for the SS4 reaction.The RWG geoindicators give low and relatively constant equilibrium temperatures, 215-226°C, and pressures, 16.1-28.2 bar, which are probably representative of the shallow permeable zone where carbon monoxide attains the last equilibrium condition. In contrast, the equilibrium temperatures and total fluid pressures computed by means of the SS4 geoindicators increase significantly with time, namely (i) from 235-267° C, 31-54 bar in 1983–1984 to 576-604°C, 1167-1326 bar in 2020, for Bocca Grande, and up to 605-645°C, 1293-1543 bar in 2020, for Bocca Nuova, assuming saturation with 21 wt % NaCl or (ii) from 222-252°C, 17-29 bar in 1983–1984 to 565-595°C, 1028-1198 bar in 2020, for Bocca Grande, and up to 595-632°C, 1160-1393 bar in 2020 for Bocca Nuova, hypothesizing saturation with 33.5 wt% NaCl. Thus, both the SS4 and RWG reactions equilibrated under the same P-T conditions in 1983-1984, indicating the provenance of fumarolic fluids from a shallow permeable zone, as already noted by Cioni et al. (1984). The increasing differences between the temperatures and pressures calculated by means of the RWG and SS4 geoindicators, in the following years, suggest that fumarolic fluids came from either: (1) progressively deeper zones of the Solfatara magmatic-hydrothermal system, characterized by gradually higher pressures and temperatures, or (2) the same deep permeable zone that underwent a progressive increment in both pressure and temperature with time.

Supplementary Information The online version contains supplementary material available at (https://doi.org/10.1007/978-3-030-98471-7_8).

To help the reader, most geothermometric and geobarometric functions presented in this chapter are reported as electronic supplementary material (see file Chapter 8.XLS). Nevertheless, the

L. Marini et al., *The Solfatara Magmatic-Hydrothermal System*, Advances in Volcanology,
https://doi.org/10.1007/978-3-030-98471-7_8

reader can also write the relations of this chapter to another Microsoft Excel spreadsheet following the indications given below.

1 The Adopted Calibration Approach

It was shown in chapter "The Geondicators Involving H_2O, CO_2, CO, CH_4 and H_2" that the two most appropriate reactions for the geothermometry and geobarometry of Solfatara fluids are the reverse water–gas shift reaction, which is rewritten here below for convenience:

$$CO_2 + H_2 = CO + H_2O \quad (1)$$

and the Sabatier-Senderens reduction with stoichiometric coefficient of H_2 equal to 4, which is also written again for convenience:

$$CO_2 + 4H_2 = CH_4 + 2H_2O. \quad (2)$$

As already mentioned in Sect. 6.1, reactions (1) and (2) are indicated by the acronyms RWG and SS4, respectively. Since the behavior of the gas species involved in reactions (1) and (2) deviate considerably from that of ideal gases at the P,T conditions met by Solfatara fluids along their possible decompression paths (see chapter "The Expansion (Decompression) of the Solfatara Fumarolic Fluids"), it is convenient to write the thermodynamic equilibrium constants of these two reactions in a rigorous way, that is, considering the fugacities of gaseous species instead of their partial pressures.

The thermodynamic equilibrium constant of the RWG reaction is:

$$\log K_{RWG} = \log \frac{f_{CO}}{f_{CO_2}} - \log \frac{f_{H_2}}{f_{H_2O}} \quad (3)$$

Recalling that the fugacity of the i-th gas species, f_i, is equal to the product of its partial pressure, P_i, times its fugacity coefficient, φ_i, Eq. (3) can be rewritten as follows:

$$\log K_{RWG} = \log \frac{P_{CO}}{P_{CO_2}} + \log \frac{\varphi_{CO}}{\varphi_{CO_2}} - \log \frac{P_{H_2}}{P_{H_2O}} - \log \frac{\varphi_{H_2}}{\varphi_{H_2O}} \quad (4)$$

Since the partial pressure of the i-th gas species, P_i, is equal to the product of the total fluid pressure, P_{tot}, times its mole fraction, X_i, Eq. (4) can be simplified in this way:

$$\log K_{RWG} = \log \frac{X_{CO}}{X_{CO_2}} + \log \frac{\varphi_{CO}}{\varphi_{CO_2}} - \log \frac{X_{H_2}}{X_{H_2O}} - \log \frac{\varphi_{H_2}}{\varphi_{H_2O}} \quad (5)$$

It is useful to rearrange Eq. (5) separating the mole fraction ratios, which are obtained from the analysis of the Solfatara fluids, from the other terms, which can be expressed as functions of the temperature, pressure and CO_2 mole fraction. Doing so and posing $\log(X_{CO}/X_{CO_2}) - \log(X_{H_2}/X_{H_2O}) = \alpha$ one obtains:

$$\alpha = \log K_{RWG} - \log \frac{\varphi_{CO}}{\varphi_{CO_2}} + \log \frac{\varphi_{H_2}}{\varphi_{H_2O}}. \quad (6)$$

This is one of the two relations of interest. Adopting the same approach, the thermodynamic equilibrium constant of the SS4 reaction can be written as:

$$\log K_{SS4} = \log \frac{f_{CH_4}}{f_{CO_2}} - 2 \cdot \log \frac{f_{H_2}}{f_{H_2O}} - 2 \cdot \log f_{H_2}. \quad (7)$$

The two fugacity ratios f_{CH_4}/f_{CO_2} and f_{H_2}/f_{H_2O} can be handled as done above for the fugacity ratios f_{CO}/f_{CO_2} and f_{H_2}/f_{H_2O}, whereas the term $-2 \cdot \log f_{H_2}$ requires a different treatment. First, it is possible to write:

$$\log f_{H_2} = \log P_{tot} + \log X_{H_2} + \log \varphi_{H_2} \quad (8)$$

Since $P_{H_2O} = P_{tot} \cdot X_{H_2O}$, it follows that $\log P_{tot} = \log P_{H_2O} - \log X_{H_2O}$. Inserting this expression in Eq. (8) and rearranging, one obtains:

$$\log f_{H_2} = \log P_{H_2O} + \log \frac{X_{H_2}}{X_{H_2O}} + \log \varphi_{H_2}. \quad (9)$$

Since $f_{H_2O} = P_{H_2O} \cdot \varphi_{H_2O}$, it follows that $\log P_{H_2O} = \log f_{H_2O} - \log \varphi_{H_2O}$. Introducing this relation in Eq. (9) and reorganizing, it is possible to write:

$$\log f_{H_2} = \log f_{H_2O} + \log \frac{X_{H_2}}{X_{H_2O}} + \log \frac{\varphi_{H_2}}{\varphi_{H_2O}}. \quad (10)$$

Splitting the $\log f_{H_2O}$ in the sum $\log P_{H_2O} + \log \varphi_{H_2O}$, one obtains:

$$\log f_{H_2} = \log P_{H_2O} + \log \varphi_{H_2O} + \log \frac{X_{H_2}}{X_{H_2O}} + \log \frac{\varphi_{H_2}}{\varphi_{H_2O}} \quad (11)$$

Finally, inserting Eq. (11) in Eq. (7) and rearranging, as done above for the RWG reaction, and posing $\log(X_{CH_4}/X_{CO_2}) - 4 \cdot \log(X_{H_2}/X_{H_2O}) = \beta$, it is possible to get the following relation:

$$\beta = \log K_{SS4} - \log \frac{\varphi_{CH_4}}{\varphi_{CO_2}} + 4 \cdot \log \frac{\varphi_{H_2}}{\varphi_{H_2O}} + 2 \cdot \log P_{H_2O} + 2 \cdot \log \varphi_{H_2O} \quad (12)$$

This is the other sought relation, in which the analytical data are on the left-hand side, whereas the parameters that can be expressed as functions of the temperature, pressure and CO_2 mole fraction are on the right-hand side.

The two relations (6) and (12) can then be calibrated for the expansion paths of interest, considering the fugacity coefficients computed by means of the Peng-Robinson EOS for the isenthalpic decompression path (Sect. 5 in chapter "The Expansion (Decompression) of the Solfatara Fumarolic Fluids"), the saturation expansion path for a brine containing 21 wt% NaCl (Sect. 7.3 in chapter "The Expansion (Decompression) of the Solfatara Fumarolic Fluids") or 33.5 wt% NaCl (Sect. 7.4 in chapter "The Expansion (Decompression) of the Solfatara Fumarolic Fluids"), and the linear P-T expansion path (Sect. 7.5 in chapter "The Expansion (Decompression) of the Solfatara Fumarolic Fluids").

The log K_{RWG} and log K_{SS4} are computed as a function of temperature and pressure using SUPCRT92 (Johnson et al. 1992), as already done in chapter "The Geoindicators Involving H_2O, CO_2, CO, CH_4 and H_2". Finally, the right-hand terms of Eqs. (6) and (12) are fitted to polynomials of the temperature and the CO_2 mole fraction, thus obtaining the geothermometric functions we are looking for.

2 The RWG and SS4 Geoindicators for the Isenthalpic Decompression Path of Solfatara Fluids

As shown by the diagram of Fig. 1a, the temperature and the sum of log ratios $\alpha = \log(X_{CO}/X_{CO_2}) - \log(X_{H_2}/X_{H_2O})$, are connected by a monotonic relation of gradually reducing slope with increasing temperature, which splits in five distinct branches above 280 °C approximately. The diagram also shows that the five α-temperature relations break at temperatures somewhat lower than 610 °C at X_{CO_2} of 0.05, 576 °C at X_{CO_2} of 0.10, 514 °C at X_{CO_2} of 0.20, 460 °C at X_{CO_2} of 0.30, and 414 °C at X_{CO_2} of 0.40, where the different functions become almost vertical, that is, α increases at nearly constant temperature. Therefore, the relation between the dependent variable (1/T to improve linearity) and the predictors (α and X_{CO_2}) was fitted to a series of polynomials, up to the temperatures recalled above, thus obtaining

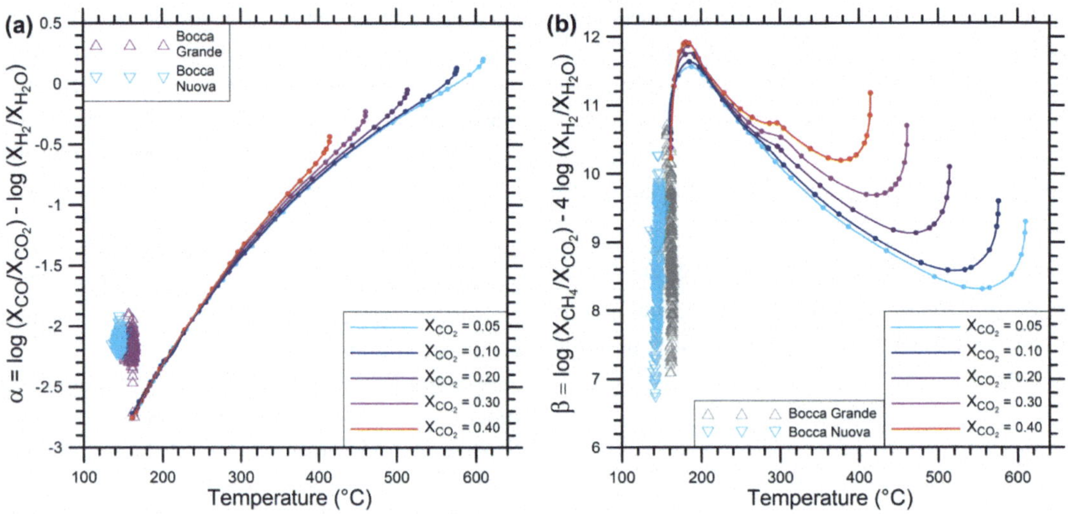

Fig. 1 Temperature dependence of **a** the sum of log ratios $\alpha = \log(X_{CO}/X_{CO_2}) - \log(X_{H_2}/X_{H_2O})$ and **b** the sum of log ratios $\beta = \log(X_{CH_4}/X_{CO_2}) - 4 \cdot \log(X_{H_2}/X_{H_2O})$, at CO_2 mole fraction in the vapor phase of 0.05, 0.10, 0.20, 0.30, and 0.40, for the isenthalpic decompression path of Solfatara fluids

the RWG geothermometric equation for the isenthalpic decompression path of Solfatara fluids.

Calculations are suitably performed using a Microsoft Excel spreadsheet with α in column A and X_{CO_2} in column C. The RWG geothermometric equation is as follows for the first sample with α in cell A2 and X_{CO_2} in cell C2:

$$\begin{aligned} T_{RWG}(^{\circ}C) = 1/((&-0.004117048 * C2 \wedge 4 \\ &+0.004453214 * C2 \wedge 3 \\ &-0.001350117 * C2 \wedge 2 \\ &+0.0004291857 * C2 \\ &+0.001161159) \\ &+(-0.02902604 * C2 \wedge 4 \\ &+0.02732564 * C2 \wedge 3 \\ &-0.008459008 * C2 \wedge 2 \\ &+0.001198208 * C2 \\ &-0.0004203432) * A2 \\ &+(-0.01527073 * C2 \wedge 4 \\ &+0.01436828 * C2 \wedge 3 \\ &-0.004524927 * C2 \wedge 2 \\ &+0.0005820721 * C2 \\ &-0.000009399782) * A2 \wedge 2) - 273.15 \end{aligned} \tag{13}$$

The equal sign and the right-hand side of Eq. (13) are written in cell D2 of the Excel spreadsheet to compute the RWG equilibrium temperature for the first sample. Then, the equation is quickly copied into the adjacent cells of column D dragging the fill handle down, thus computing the RWG equilibrium temperatures for all the available samples.

The absolute error on the RWG temperature computed using Eq. (13) is $1.04\,^{\circ}C \pm 0.89\,^{\circ}C$ (1σ) on average, with minimum of 0.00174 °C and maximum of 4.07 °C. Although Eq. (13) can be used up to 610–414 °C, depending on X_{CO_2}(see above), the computed RWG temperatures are much lower for all the Solfatara fluid samples (see Fig. 1a and Sect. 6.1).

As pointed out by the plot of Fig. 1b, the CO_2 mole fraction has a strong effect on the relation between the sum of log ratios $\beta = \log(X_{CH_4}/X_{CO_2}) - 4 \cdot \log(X_{H_2}/X_{H_2O})$ and the temperature, determining a significant separation of the curves for distinct X_{CO_2} values. Furthermore, at a given X_{CO_2} value, the $\beta - T$ functions are concave downwards at low temperatures (below 200 °C approximately), relatively close to linear at intermediate temperatures (apart from a minor inflection close to 300 °C for $X_{CO_2} \geq 0.20$), and concave upwards at high temperatures (above 300–500 °C, depending on X_{CO_2}). The shape of the $\beta - T$ functions implies what follows:

(i) At X_{CO_2} of 0.05, there are two possible solutions for $10.66 < \beta < 11.57$, a single solution for $9.30 < \beta < 10.66$, two possible solutions for $8.32 < \beta < 9.30$, and no solutions for $\beta < 8.32$.
(ii) At X_{CO_2} of 0.10, there are two possible solutions for $10.61 < \beta < 11.64$, a single solution for $9.60 < \beta < 10.61$, two possible solutions for $8.59 < \beta < 9.60$, and no solutions for $\beta < 8.59$.
(iii) At X_{CO_2} of 0.20, there are two possible solutions for $10.50 < \beta < 11.75$, a single solution for $10.10 < \beta < 10.50$, two possible solutions for $9.14 < \beta < 10.10$, and no solutions for $\beta < 9.14$.
(iv) At X_{CO_2} of 0.30, there are three possible solutions for $10.39 < \beta < 10.71$, two possible solutions for $10.71 < \beta < 11.87$ and $9.69 < \beta < 10.39$, and no solution for $\beta < 9.69$.
(v) At X_{CO_2} of 0.40, there are three possible solutions for $10.24 < \beta < 11.18$, two possible solutions for $11.18 < \beta < 11.93$ and $10.20 < \beta < 10.24$, and no solution for $\beta < 10.20$.

Owing to these uncertainties, no SS4 geothermometric equation was derived for the isenthalpic decompression path of Solfatara fluids.

Utilizing the Microsoft Excel spreadsheet with X_{CO_2} in column C and the RWG equilibrium temperature (in °C) computed by means of Eq. (13) in column D, the following geobarometric equation allows one to compute the total fluid pressure, P_{tot}, for the first sample with X_{CO_2} in cell C2 and T_{RWG} in cell D2:

$$\begin{aligned}P_{tot}(\text{bar}) = 10\text{^}((&-4885985000000000 * C2 \wedge 4 \\ &+ 4237538000000000 * C2 \wedge 3 \\ &- 1261351000000000 * C2 \wedge 2 \\ &+ 180260700000000 * C2 \\ &- 4721313000000) \\ &\quad * (1/(D2 + 273.15)) \wedge 4 \\ &+ (34090430000000 * C2 \wedge 4 \\ &- 29387090000000 * C2 \wedge 3 \\ &+ 8638850000000 * C2 \wedge 2 \\ &- 1227876000000 * C2 \\ &\quad + 30813650000) * (1/(D2 + 273.15)) \wedge 3 \\ &+ (-88604820000 * C2 \wedge 4 \\ &\quad + 75932420000 * C2 \wedge 3 \\ &- 22026330000 * C2 \wedge 2 + 3111652000 * C2 \\ &- 74906530) * (1/(D2 + 273.15)) \wedge 2 \\ &\quad + (101653700 * C2 \wedge 4 \\ &- 86627850 * C2 \wedge 3 + 24773630 * C2 \wedge 2 \\ &- 3476773 * C2 + 78706.79) \\ &\quad * (1/(D2 + 273.15)) \\ &+ (-43428.86 * C2 \wedge 4 + 36816.96 * C2 \wedge 3 \\ &- 10369.28 * C2 \wedge 2 \\ &\quad + 1446.044 * C2 - 27.32279))\end{aligned} \tag{14}$$

The equal sign and the right-hand side of Eq. (14) are written in cell E2 of the Excel spreadsheet to compute the total fluid pressure corresponding to the RWG equilibrium temperature for the first sample. Then, the relation is speedily copied into the adjacent cells of column E dragging the fill handle down, thus computing the total fluid pressure corresponding to the RWG equilibrium temperatures for all the other samples of interest.

The absolute error on the computed P_{tot} is $1.24\% \pm 1.13\%$ (1σ) on average, with minimum of 0.00267%, and maximum of 4.83%. Equation (14) holds true up to total pressures of 700 bar for X_{CO_2} of 0.05, 1000 bar for X_{CO_2} of 0.10, 900 bar for X_{CO_2} of 0.20 and 0.30, and 700 bar for X_{CO_2} of 0.40. Strictly speaking Eqs. (13) and (14) apply to Bocca Grande fluids only, but they can be used for Bocca Nuova fluids as well, accepting somewhat higher uncertainties.

3 The RWG and SS4 Geoindicators for the Saturation Decompression Path of Solfatara Fluids Involving a Brine Containing 21 wt% NaCl and the Related Vapor Phase

The diagram of Fig. 2a, in which the sum of log ratios $\alpha = \log(X_{CO}/X_{CO_2}) - \log(X_{H_2}/X_{H_2O})$ is contrasted with the temperature, shows that these two variables are linked by a monotonic relation of progressively decreasing slope with increasing temperature. Above 350 °C approximately, this relation splits in different branches, each corresponding to a different CO_2 mole fraction. The relation between the dependent variable (1/T to increase linearity) and the predictors (α and X_{CO_2}) was fitted to a series of polynomials, thus deriving the following RWG geothermometric equation for the saturation decompression path of Solfatara fluids involving a brine containing 21 wt% NaCl and the related vapor phase:

$$
\begin{aligned}
T_{RWG}(^{\circ}C) = 1/((&0.0008487619 * C2 \wedge 4 \\
&- 0.001191429 * C2 \wedge 3 \\
&+ 0.0006858667 * C2 \wedge 2 \\
&- 0.0003799214 * C2 \\
&+ 0.001147017) \\
&+ (-0.005137695 * C2 \wedge 4 \\
&+ 0.005929029 * C2 \wedge 3 \\
&- 0.002740698 * C2 \wedge 2 \\
&+ 0.0005322879 * C2 \\
&- 0.00077217) * A2 \\
&+ (-0.009782143 * C2 \wedge 4 \\
&+ 0.01168796 * C2 \wedge 3 \\
&- 0.005585195 * C2 \wedge 2 \\
&+ 0.001486423 * C2 \\
&- 0.0005542923) * A2 \wedge 2 \\
&+ (-0.005205648 * C2 \wedge 4 \\
&+ 0.006267498 * C2 \wedge 3 \\
&- 0.003011277 * C2 \wedge 2
\end{aligned}
$$

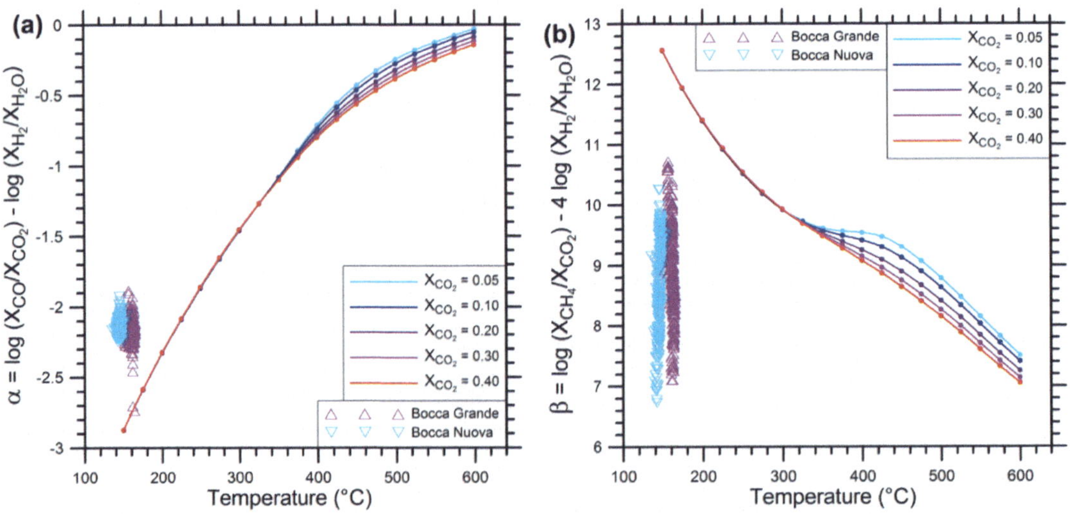

Fig. 2 Temperature dependence of **a** the sum of log ratios $\alpha = \log(X_{CO}/X_{CO_2}) - \log(X_{H_2}/X_{H_2O})$ and **b** the sum of log ratios $\beta = \log(X_{CH_4}/X_{CO_2}) - 4 \cdot \log(X_{H_2}/X_{H_2O})$, at CO_2 mole fraction in the vapor phase of 0.05, 0.10, 0.20, 0.30, and 0.40, for the saturation decompression path of Solfatara fluids involving a brine containing 21 wt% NaCl and the related vapor phase

$$+0.0008427649 * C2 - 0.0002680103) * A2 \wedge 3 + (-0.0008588714 * C2 \wedge 4 + 0.001036726 * C2 \wedge 3 - 0.000498879 * C2 \wedge 2 + 0.0001418067 * C2 - 0.00004093) * A2 \wedge 4) - 273.15 \quad (15)$$

Again, Eq. (15) is written for the first sample with α in cell A2 and X_{CO_2} in cell C2 of the Microsoft Excel spreadsheet. The equal sign and the right-hand side of Eq. (15) are written in cell F2 of the Excel spreadsheet to compute the RWG equilibrium temperature for the first sample. Then, the equation is quickly copied into the adjacent cells of column F dragging the fill handle down, thus computing the RWG equilibrium temperatures for all the available samples.

The absolute error on the RWG temperature calculated by means of Eq. (15) is 1.05 °C ± 0.85 °C (1σ) on average, with minimum of 0.00311 °C and maximum of 4.14 °C. Although Eq. (15) can be applied up to 600 °C, the computed RWG temperatures are much lower for all the Solfatara fluid samples (see Fig. 2a and Sect. 6.1).

The graph of Fig. 2b, in which the sum of log ratios $\beta = \log(X_{CH_4}/X_{CO_2}) - 4 \cdot \log(X_{H_2}/X_{H_2O})$ is plotted against the temperature, points out that these two variables are connected by a monotonic relation, which splits in distinct branches, one for each CO_2 mole fraction, above 340°C approximately. Since the function for CO_2 mole fraction of 0.05 is subhorizontal between 340 and 420°C a large error on the computed SS4 temperature is expected under these conditions. In spite of these complications, the relation between the dependent variable (1/T to increase linearity) and the predictors ((α) and X_{CO_2}) was fitted to a series of polynomials, thus obtaining the following two SS4 geothermometric equations:

$$T_{SS4}(^{\circ}C) = 1/((-249.7937 * C2 \wedge 4 + 260.668 * C2 \wedge 3 - 98.31472 * C2 \wedge 2 + 16.41854 * C2 - 1.147862) + (89.79142 * C2 \wedge 4 - 93.69606 * C2 \wedge 3 + 35.33578 * C2 \wedge 2 - 5.899114 * C2 + 0.4110321) * B2 + (-12.07122 * C2 \wedge 4 + 12.59559 * C2 \wedge 3 - 4.749825 * C2 \wedge 2 + 0.7927105 * C2 - 0.0550276) * B2 \wedge 2 + (0.7193242 * C2 \wedge 4 - 0.7505385 * C2 \wedge 3 + 0.2830093 * C2 \wedge 2 - 0.04721837 * C2 + 0.003268372) * B2 \wedge 3 + (-0.01603161 * C2 \wedge 4 + 0.01672658 * C2 \wedge 3 - 0.006306754 * C2 \wedge 2 + 0.001051957 * C2 - 0.00007262834) * B2 \wedge 4) - 273.15 \quad (16)$$

$$T_{SS4}(^{\circ}C) = 1/((-338.1988 * C2 \wedge 4 + 324.4917 * C2 \wedge 3 - 102.6725 * C2 \wedge 2 + 10.85846 * C2 - 0.03389924) + (164.3052 * C2 \wedge 4 - 157.9805 * C2 \wedge 3 + 50.16642 * C2 \wedge 2 - 5.346373 * C2 + 0.02051445) * B2 + (-29.85781 * C2 \wedge 4 + 28.76961 * C2 \wedge 3 - 9.16867 * C2 \wedge 2 + 0.9845943 * C2 - 0.004369622) * B2 \wedge 2 + (2.405338 * C2 \wedge 4 - 2.322631 * C2 \wedge 3 + 0.7428868 * C2 \wedge 2 - 0.08038292 * C2 + 0.0004033068) * B2 \wedge 3 + (-0.07247987 * C2 \wedge 4 + 0.07013834 * C2 \wedge 3 - 0.02251543 * C2 \wedge 2 + 0.002454832 * C2 - 0.00001364184) * B2 \wedge 4) - 273.15 \quad (17)$$

Equation (16) is valid for $\beta > 9.5$, whereas Eq. (17) holds true for $\beta < 9.5$. Both equations are written for the first sample with β in cell B2 and X_{CO_2} in cell C2 of the Microsoft Excel spreadsheet.

The equal sign and the right-hand side of Eqs. (16) and (17) are written in cells G2 and H2, respectively, of the Excel spreadsheet to compute the two possible SS4 equilibrium temperatures for the first sample. Then the two equations are rapidly copied into the adjacent cells of columns G and H dragging the fill handle down, thus calculating the two possible SS4 equilibrium temperatures for all the samples of interest. Finally, the IF function is used to select the correct SS4 equilibrium temperature for each sample depending on the value of β. For the first sample, the IF function is as follows: T_{SS4} (° C) = IF(B2 > 9.5;G2;H2).

The absolute error on the SS4 temperature calculated using Eqs. (16) and (17) is 2.95 °C ± 2.84 °C (1σ) on average, with minimum of 0.0755 °C, and maximum of 16.5 °C. Large errors occur for β close to 9.5–9.6 and $X_{CO_2} = 0.05$, as expected (see above). Equation (16) can be utilized up to 350–400 °C, depending on the CO_2 mole fraction. Strictly speaking, Eq. (17) can be applied up to 600 °C, but it can be probably used also at temperatures somewhat higher due to the good linearity of the β-T relationship in this region (see Fig. 2b). Actually, a few T_{SS4} values computed for Solfatara fluids resulted to be > 600 °C (see Sect. 6.2).

For the saturation decompression path of Solfatara fluids involving a brine containing 21 wt% NaCl and the related vapor phase, the H_2O partial pressure, P_{H_2O} (in bar), is given by the following relation (T in K), from 150 to 500 °C:

$$\log P_{H_2O} = 5.3323 - \frac{1986.4}{T}, \qquad (18)$$

while, from 500 to 600 °C, P_{H_2O} (in bar), is calculated by the following polynomial function of the absolute temperature reciprocal:

$$\log P_{H_2O} = \frac{-4.2374E + 09}{T^3} + \frac{1.4105E + 07}{T^2} - \frac{1.6926E + 04}{T} + 10.215. \qquad (19)$$

Both Eqs. (18) and (19) are based on the data of Tanger and Pitzer (1989). Total fluid pressure is then obtained by means of Eq. (135 in chapter "The Expansion (Decompression) of the Solfatara Fumarolic Fluids").

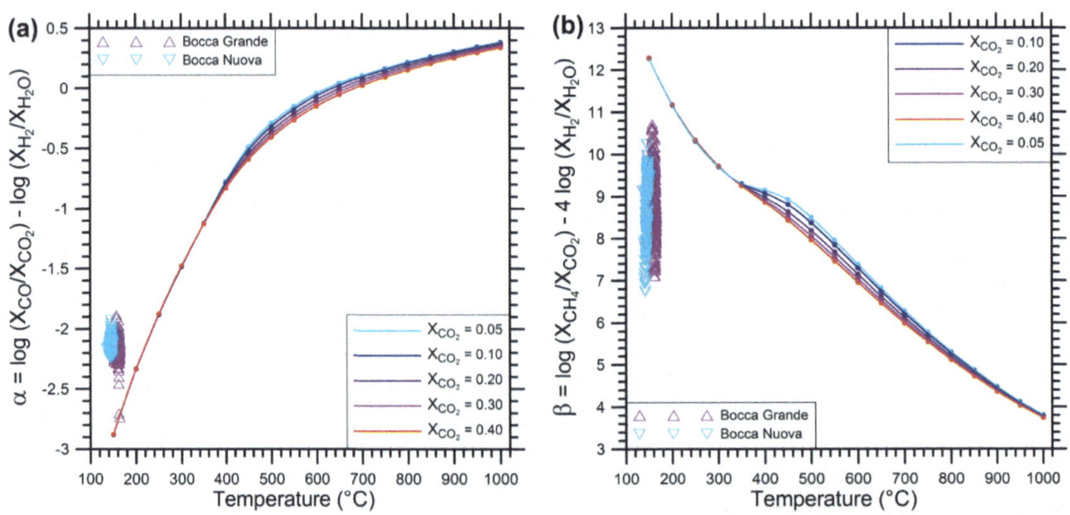

Fig. 3 Temperature dependence of **a** the sum of log ratios $\alpha = \log(X_{CO}/X_{CO_2}) - \log(X_{H_2}/X_{H_2O})$ and **b** the sum of log ratios $\beta = \log(X_{CH_4}/X_{CO_2}) - 4 \cdot \log(X_{H_2}/X_{H_2O})$, at CO_2 mole fraction in the vapor phase of 0.05, 0.10, 0.20, 0.30, and 0.40, for the saturation decompression path of Solfatara fluids, involving a brine containing 33.5 wt% NaCl and the related vapor phase

4 The RWG and SS4 Geoindicators for the Saturation Decompression Path of Solfatara Fluids Involving a Brine Containing 33.5 wt% NaCl and the Related Vapor Phase

The two diagrams of Fig. 3 are similar in shape to the corresponding graphs of Fig. 2, but the curves for the different CO_2 mole fractions extend to a maximum temperature of 1000 °C in Fig. 3, owing to the higher NaCl concentration of the brine, instead of ending at 600 °C like in Fig. 2. Adopting the same approach of the previous sections, the following relations were obtained:

$$
\begin{aligned}
T_{RWG}(^{\circ}C) = {} & 1/((0.0006494286 * C2 \wedge 4 - 0.0009179286 * C2 \wedge 3 + 0.00054225 * C2 \wedge 2 \\
& - 0.0003057114 * C2 + 0.001126855) \\
& + (-0.0008169238 * C2 \wedge 4 + 0.001025307 * C2 \wedge 3 - 0.0005671433 * C2 \wedge 2 \\
& + 0.0001470264 * C2 - 0.0007001285) * A2 \\
& + (-0.002176086 * C2 \wedge 4 + 0.002977836 * C2 \wedge 3 - 0.001699695 * C2 \wedge 2 \\
& + 0.0007351963 * C2 - 0.0004242141) * A2 \wedge 2 \\
& + (-0.001175705 * C2 \wedge 4 + 0.001643155 * C2 \wedge 3 - 0.0009466767 * C2 \wedge 2 \\
& + 0.0004391347 * C2 - 0.0002010126) * A2 \wedge 3 \\
& + (-0.0001894238 * C2 \wedge 4 + 0.0002678188 * C2 \wedge 3 - 0.0001554373 * C2 \wedge 2 \\
& + 0.00007452374 * C2 - 0.00002997456) * A2 \wedge 4) - 273.15
\end{aligned} \tag{20}
$$

$$
\begin{aligned}
T_{SS4}(^{\circ}C) = {} & 1/((-113.5104 * C2 \wedge 4 + 122.3038 * C2 \wedge 3 - 48.28715 * C2 \wedge 2 \\
& + 8.37572 * C2 - 0.4907166) \\
& + (42.69314 * C2 \wedge 4 - 46.00396 * C2 \wedge 3 + 18.16341 * C2 \wedge 2 - 3.146579 * C2 \\
& + 0.1819683) * B2 \\
& + (-6.001729 * C2 \wedge 4 + 6.467653 * C2 \wedge 3 - 2.553687 * C2 \wedge 2 \\
& + 0.441878 * C2 - 0.02521197) * B2 \wedge 2 \\
& + (0.3737274 * C2 \wedge 4 - 0.4027716 * C2 \wedge 3 + 0.1590393 * C2 \wedge 4 \\
& - 0.02748992 * C2 + 0.001551101) * B2 \wedge 3 \\
& + (-0.008697397 * C2 \wedge 4 + 0.009374038 * C2 \wedge 3 - 0.003701702 * C2 \wedge 2 \\
& + 0.0006392084 * C2 - 0.00003570342) * B2 \wedge 4) - 273.15
\end{aligned} \tag{21}
$$

$$
\begin{aligned}
T_{SS4}(^{\circ}C) = {} & 1/((-0.0305901 * C2 \wedge 4 + 0.03179562 * C2 \wedge 3 - 0.01085979 * C2 \wedge 2 \\
& - 0.00006409093 * C2 + 0.001033542) \\
& + (0.02283002 * C2 \wedge 4 - 0.02425956 * C2 \wedge 3 + 0.008619277 * C2 \wedge 2 - 0.0001632477 * C2 \\
& - 0.0003645792) * B2 + (-0.00624905 * C2 \wedge 4 \\
& + 0.006768169 * C2 \wedge 3 - 0.002508957 * C2 \wedge 2 \\
& + 0.0001103516 * C2 + 0.0001271469) * B2 \wedge 2 \\
& + (0.0007407101 * C2 \wedge 4 - 0.0008175034 * C2 \wedge 3 + 0.0003161627 * C2 \wedge 2 \\
& - 0.00002092373 * C2 - 0.00001537445) * B2 \wedge 3 \\
& + (-0.00003206605 * C2 \wedge 4 + 0.0000361006 * C2 \wedge 3 - 0.00001460794 * C2 \wedge 2 \\
& + 0.000001321086 * C2 + 0.0000006881956) * B2 \wedge 4) - 273.15
\end{aligned} \tag{22}
$$

Equation (21) holds true for $\beta > 8.5$, whereas Eq. (22) is valid for $\beta < 8.5$. The three Eqs. (20), (21), and (22) are written for the first sample with α in cell A2, β in cell B2, and X_{CO_2} in cell C2 of the Microsoft Excel spreadsheet. They are written in cells J2, K2, and L2 for the first sample, as explained in previous sections, and quickly applied to all the considered samples. Finally, the right SS4 equilibrium temperature is selected using the IF function as follows: $T_{SS4}(°C) = IF(B2 > 8.5;K2;L2)$.

The absolute error on the RWG temperature computed by using Eq. (20) is $2.33\,°C \pm 1.87\,(1\sigma)$ °C on average, with range 0.0115–7.39 °C. Equation (20) can be applied up to 950 °C, although the computed RWG temperatures are significantly lower for all Solfatara fluid samples (see Sect. 6.1). The absolute error on the SS4 temperature calculated by means of Eqs. (21) and (22) is $1.57\,°C \pm 1.62\,°C\,(1\sigma)$ on average, with range 0.0132–8.83 °C. Equation (21) works up to temperatures of 450 °C, whereas Eq. (22) can be utilized up to 1000 °C, although all the calculated SS4 temperatures are lower than this threshold for all Solfatara fluid samples (see Sect. 6.2).

For the saturation decompression path of Solfatara fluids involving a brine containing 33.5 wt% NaCl and the related vapor phase, the H_2O partial pressure, P_{H_2O} (in bar), is given by the following relation (T in K), from 150 to 550 °C:

$$\log P_{H_2O} = 5.3427 - \frac{2036.5}{T}, \quad (23)$$

whereas from 550 to 1000 °C, P_{H_2O} (in bar), is computed by the following polynomial function of the inverse absolute temperature:

$$\log P_{H_2O} = \frac{-1.0521E+09}{T^3} + \frac{2.3948E+06}{T^2} - \frac{2.7508E+03}{T} + 4.5720. \quad (24)$$

Equations (23) and (24) were both derived from the data of Driesner and Heinrich (2007). Total fluid pressure is then obtained by means of Eq. (135 in chapter "The Expansion (Decompression) of the Solfatara Fumarolic Fluids").

5 The RWG and SS4 Geoindicators for the Linear P–T Expansion Path of Solfatara Fluids

The two graphs of Fig. 4 have shape similar to those of the analogous diagrams of Figs. 2 and 3, apart from a relatively small but annoying difference in Fig. 4b at X_{CO_2} of 0.05 and 0.10. In fact, these two curves are not monotonic, but are instead characterized by the presence of a relative (local) minimum at ~340 °C and a relative (local) maximum at ~400 °C, which were both disregarded to avoid complicating too much the polynomial fitting exercise and the obtained relations. Nevertheless, it must be kept in mind that the computed SS4 equilibrium temperature might be affected by large uncertainties for $\beta > 10$and $X_{CO_2} < 0.15$ approximately. Adopting the same approach of the previous sections, the following equations were derived:

$$\begin{aligned} T_{RWG}(°C) = {} & 1/((0.00363781 * C2 \wedge 4 \\ & - 0.004382643 * C2 \wedge 3 + 0.002230483 \\ & * C2 \wedge 2 \\ & - 0.0008144171 * C2 + 0.001177183) \\ & + (-0.01323543 * C2 \wedge 4 + 0.01495066 \\ & * C2 \wedge 3 \\ & - 0.006531325 * C2 \wedge 2 + 0.001623859 * C2 \\ & - 0.0009366124) * A2 + (-0.01162897 \\ & * C2 \wedge 4 \\ & + 0.01391167 * C2 \wedge 3 - 0.006960909 \\ & * C2 \wedge 2 \\ & + 0.002262881 * C2 - 0.0004639441) \\ & * A2 \wedge 2 \\ & + (0.02993619 * C2 \wedge 4 - 0.03334489 \\ & * C2 \wedge 3 \\ & + 0.01388226 * C2 \wedge 2 - 0.00311747 * C2 \\ & + 0.0005435745) * A2 \wedge 3 + (0.0398447 \\ & * C2 \wedge 4 \\ & - 0.0451618 * C2 \wedge 3 + 0.01972618 * C2 \wedge 2 \\ & - 0.0049412 * C2 + 0.000789153) * A2 \wedge 4 \\ & + (0.01634535 * C2 \wedge 4 - 0.01863088 \\ & * C2 \wedge 3 \\ & + 0.008256445 * C2 \wedge 2 - 0.002122088 * C2 \\ & + 0.000318948) * A2 \wedge 5 + (0.002221106 \\ & * C2 \wedge 4 \end{aligned}$$

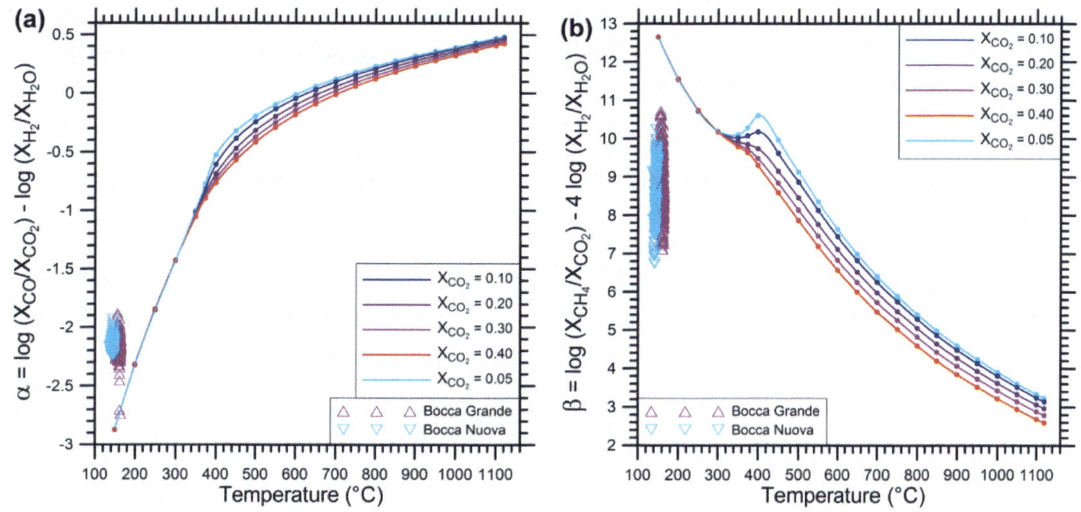

Fig. 4 Temperature dependence of **a** the sum of log ratios $\alpha = \log(X_{CO}/X_{CO_2}) - \log(X_{H_2}/X_{H_2O})$ and **b** the sum of log ratios $\beta = \log(X_{CH_4}/X_{CO_2}) - 4 \cdot \log(X_{H_2}/X_{H_2O})$, at CO_2 mole fraction in the vapor phase of 0.05, 0.10, 0.20, 0.30, and 0.40, for the linear P–T expansion path of Solfatara fluids

$$
\begin{aligned}
&- 0.002539624 * \mathrm{C2} \wedge 3 + 0.001134232 \\
&\quad * \mathrm{C2} \wedge 2 \\
&- 0.0002949376 * \mathrm{C2} + 0.00004246812) \\
&\quad * \mathrm{A2} \wedge 6) - 273.15
\end{aligned} \tag{25}
$$

$$
\begin{aligned}
\mathrm{T_{SS4}}(^{\circ}\mathrm{C}) = {} & \mathrm{ABS}(1/((-38.34438 * \mathrm{C2} \wedge 4 \\
& + 33.6633 * \mathrm{C2} \wedge 3 - 9.928606 * \mathrm{C2} \wedge 2 \\
& + 1.375186 * \mathrm{C2} - 0.1551145) \\
& + (10.07565 * \mathrm{C2} \wedge 4 - 8.835904 * \mathrm{C2} \wedge 3 \\
& + 2.599555 * \mathrm{C2} \wedge 2 - 0.3585454 * \mathrm{C2} \\
& + 0.0403786) * \mathrm{B2} + (-0.8797584 * \mathrm{C2} \wedge 4 \\
& + 0.770722 * \mathrm{C2} \wedge 3 - 0.226224 * \mathrm{C2} \wedge 2 \\
& + 0.03107841 * \mathrm{C2} - 0.003467458) * \mathrm{B2} \wedge 2 \\
& + (0.02552444 * \mathrm{C2} \wedge 4 - 0.02233979 \\
& \quad * \mathrm{C2} \wedge 3 \\
& + 0.006543137 * \mathrm{C2} \wedge 2 \\
& \quad - 0.0008955432 * \mathrm{C2} \\
& + 0.0000995706) * \mathrm{B2} \wedge 3) - 273.15)
\end{aligned} \tag{26}
$$

$$
\begin{aligned}
\mathrm{T_{SS4}}(^{\circ}\mathrm{C}) = {} & 1/((0.03279402 * \mathrm{C2} \wedge 4 \\
& - 0.03662225 * \mathrm{C2} \wedge 3 + 0.01476645 \\
& \quad * \mathrm{C2} \wedge 2 \\
& - 0.002045239 * \mathrm{C2} + 0.000350271) \\
& + (-0.0176815 * \mathrm{C2} \wedge 4 \\
& + 0.0201321 * \mathrm{C2} \wedge 3 - 0.008330952 \\
& \quad * \mathrm{C2} \wedge 2 \\
& + 0.001295434 * \mathrm{C2} \\
& \quad + 0.0001209711) * \mathrm{B2} \\
& + (0.00295229 * \mathrm{C2} \wedge 4 - 0.003429423 \\
& \quad * \mathrm{C2} \wedge 3 \\
& + 0.001468829 * \mathrm{C2} \wedge 2 \\
& \quad - 0.0002381443 * \mathrm{C2} \\
& - 0.000003684129) * \mathrm{B2} \wedge 2 \\
& \quad + (-0.0001525137 * \mathrm{C2} \wedge 4 \\
& + 0.0001811832 * \mathrm{C2} \wedge 3 \\
& \quad - 0.00008119045 * \mathrm{C2} \wedge 2 \\
& + 0.00001427497 * \mathrm{C2} \\
& \quad + 0.0000001456421) * \mathrm{B2} \wedge 3) - 273.15
\end{aligned} \tag{27}
$$

Equation (26) holds true below the critical point of pure water, whereas Eq. (27) is valid above the critical point of pure water. Again, the three relations (25), (26), and (27) make reference to the first sample with α in cell A2, β in cell B2, and X_{CO_2} in cell C2 of the Microsoft Excel spreadsheet. They are written as described in previous sections and rapidly applied to all the available samples. The right SS4 equilibrium temperature is chosen utilizing the IF function as follows:$T_{SS4}(°C) = IF(O2 < 373.946; O2; P2)$.

The absolute error on the RWG temperature given by Eq. (25) is $1.74\,°C \pm 1.58\,°C\ (1\sigma)$ on average, with range 0.00298–6.63 °C. Equation (25) can be used up to 1120 °C, although the computed RWG temperatures are considerably lower for all Solfatara fluid samples (see Sect. 6.1). The absolute error on the SS4 temperature computed by means of Eqs. (26) and (27) is $1.71\,°C \pm 1.20\,°C\ (1\sigma)$ on average, with range 0.00675–4.21 °C. However, the error is higher in the temperature range 300–450 °C for $X_{CO_2} < 0.15$ approximately (see above). Although Eqs. (26) and (27) can be utilized up to 1120 °C, the maximum calculated SS4 temperatures for the Solfatara fluids is significantly lower (see Sect. 6.2).

For the linear P–T decompression path, the H_2O partial pressure, P_{H_2O} (in bar), is given by Eq. (14) up to 340 °C and, accepting a larger uncertainty, up to the critical point of pure H_2O. The total fluid pressure is then computed by means of Eq. (135 in chapter "The Expansion (Decompression) of the Solfatara Fumarolic Fluids").

Above the critical point of pure water, total fluid pressure is computed by means of the following relation, with T2 indicating T_{RWG} or T_{SS4} (in °C):

$$\begin{aligned} P_{tot}(\text{bar}) = {} & (-1.07619 * C2 \wedge 4 + 0.09285714 * C2 \wedge 3 \\ & - 0.3716667 * C2 \wedge 2 - 0.2898571 * C2 \\ & + 3.563117) * T2 + (1152.829 * C2 \wedge 4 \\ & - 57.5119 * C2 \wedge 3 + 403.51 * C2 \wedge 2 \\ & + 325.7244 * C2 - 1111.676) \end{aligned} \tag{28}$$

6 Application of the RWG and SS4 Geoindicators to the Solfatara Fluids

6.1 The RWG Equilibrium Temperature and Total Fluid Pressure

The RWG equilibrium temperatures along different expansion paths are not very diverse from each other with differences of 4.8–8.3 °C on the mean, 1.2–2.6 °C on the minimum, and 8.3–14.6 °C on the maximum for Bocca Grande (see Fig. 5 and Table 1) and with differences of 4.9–8.7 °C on the mean, 2.8–4.0 °C on the minimum, and 7.4–13.4 °C on the maximum for Bocca Nuova (see Fig. 6 and Table 2). The saturation path for pure water without provision for deviations from ideality (see Sect. 4 in chapter "The Geoindicators Involving H_2O, CO_2, CO, CH_4 and H_2") gives the highest RWG equilibrium temperatures, followed by the saturation path for 33.5 wt% NaCl, with the fugacity coefficients of relevant gases computed by means of the Peng-Robinson EOS, whereas the other three approaches provide somewhat lower values. All RWG equilibrium temperatures exhibit small time changes, as they generally oscillate between 210 and 250 °C for both fumaroles, as already noted in Sect. 4 in chapter "The Geoindicators Involving H_2O, CO_2, CO, CH_4 and H_2" for the pure-water saturation path without considering the deviations from the ideal behavior. Furthermore, all computed RWG equilibrium temperatures, apart from few exceptions for Bocca Grande, are situated in the saturation field, being above the lower saturation temperature and below the upper saturation temperature.

Consequently, the two assumptions of (i) gas equilibration in a pure saturated vapor phase and (ii) isenthalpic fluid expansion are consistent with each other.

As expected based on what has been observed above for the RWG equilibrium temperature, also the total fluid pressures indicated by the

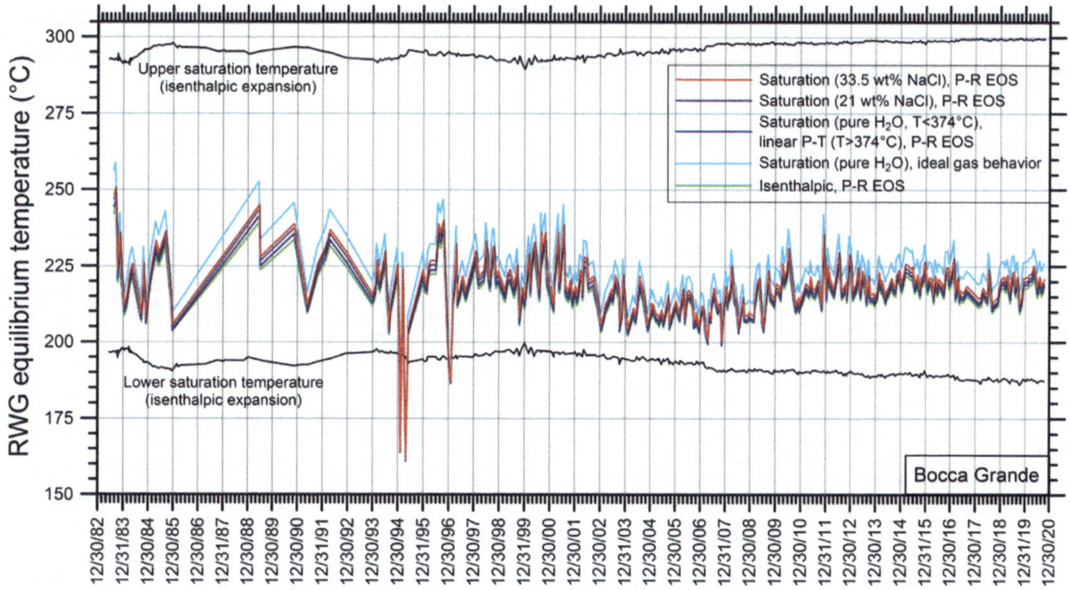

Fig. 5 Chronogram of the temperature indicated by the RWG reaction at equilibrium, along different expansion paths for the Bocca Grande fluids (data from Chiodini et al. 2021 and references therein). The lower and upper saturation temperatures are also shown

Table 1 Statistical parameters of the RWG equilibrium temperature and related total fluid pressure computed for all the Bocca Grande samples collected from 1983 to 2020 (data from Chiodini et al. 2021 and references therein) based on different decompression paths and hypotheses

Expansion path and hypothesis	Parameter	N	Mean	Median	Min	Max	Std. Dev
Saturation, ideal gas behavior	T RWG (°C)	330	223	223	164	259	9.9
Saturation (pure H_2O, T < 374 °C), linear P–T (T > 374 °C), P-R EOS	T RWG (°C)	330	216	216	162	247	8.7
Saturation (21 wt% NaCl), P-R EOS	T RWG (°C)	330	217	217	161	249	9.2
Saturation (33.5 wt% NaCl), P-R EOS	T RWG (°C)	330	219	218	161	251	9.2
Isenthalpic, P-R EOS	T RWG (°C)	330	215	215	162	244	8.4
Saturation, ideal gas behavior	*P RWG* (bar)	*330*	*21.9*	*22.0*	*6.5*	*35.9*	*3.6*
Saturation (pure H_2O, T < 374 °C), linear P–T (T > 374 °C), P-R EOS	*P RWG* (bar)	*330*	*26.9*	*27.0*	*7.8*	*44.2*	*4.5*
Saturation (21 wt% NaCl), P-R EOS	*P RWG* (bar)	*330*	*24.5*	*24.6*	*6.9*	*40.3*	*4.1*
Saturation (33.5 wt% NaCl), P-R EOS	*P RWG* (bar)	*330*	*16.1*	*15.9*	*4.5*	*28.6*	*2.8*
Isenthalpic, P-R EOS	*P RWG* (°C)	*330*	*27.3*	*27.2*	*12.7*	*45.0*	*4.4*

RWG reaction at equilibrium along different decompression paths are comparable to each other with differences of 2.5–5.9 bar on the mean, 0.4–6.1 bar on the minimum, and 4.4–9.1 bar on the maximum for Bocca Grande (see Fig. 7 and Table 1) and with differences of 2.7–6.0 bar on the mean, 0.7–4.1 bar on the minimum, and 4.1–7.9 bar on the maximum for Bocca Nuova (see Fig. 8 and Table 2).

The highest total fluid pressures are obtained for the saturation path for pure water without provision for departures from the ideal gas

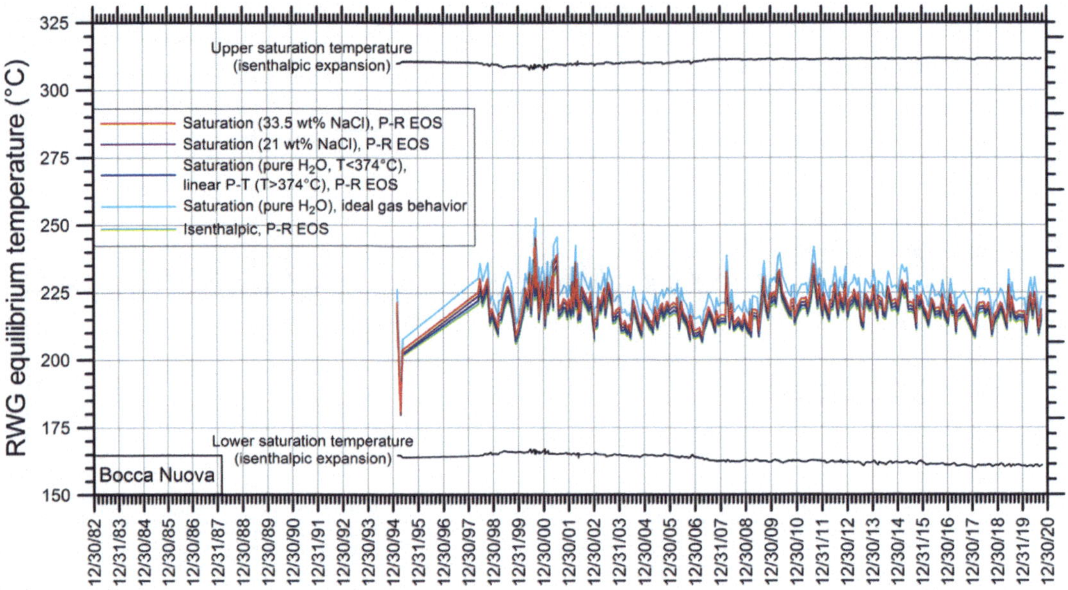

Fig. 6 Chronogram of the temperature indicated by the RWG reaction at equilibrium, along different expansion paths for the Bocca Nuova fluids (data from Chiodini et al. 2021 and references therein). The lower and upper saturation temperatures are also shown

Table 2 Statistical parameters of the RWG equilibrium temperature and related total fluid pressure computed for all the Bocca Nuova samples collected from 1995 to 2020 (data from Chiodini et al. 2021 and references therein) based on different decompression paths and hypotheses

Expansion path and hypothesis	Parameter	N	Mean	Median	Min	Max	Std. Dev
Saturation, ideal gas behavior	T RWG (°C)	294	226	226	184	253	7.0
Saturation (pure H_2O, T < 374 °C), linear P–T (T > 374 °C), P-R EOS	T RWG (°C)	294	218	218	181	241	6.1
Saturation (21 wt% NaCl), P-R EOS	T RWG (°C)	294	220	219	180	244	6.4
Saturation (33.5 wt% NaCl), P-R EOS	T RWG (°C)	294	221	221	181	245	6.4
Isenthalpic, P-R EOS	T RWG (°C)	294	217	217	180	239	5.8
Saturation, ideal gas behavior	*P RWG* (bar)	*294*	*22.7*	*22.9*	*10.0*	*32.1*	*2.6*
Saturation (pure H_2O, T < 374 °C), linear P–T (T > 374 °C), P-R EOS	*P RWG* (bar)	*294*	*27.9*	*28.1*	*12.1*	*39.5*	*3.3*
Saturation (21 wt% NaCl), P-R EOS	*P RWG* (bar)	*294*	*25.4*	*25.5*	*10.7*	*36.2*	*3.0*
Saturation (33.5 wt% NaCl), P-R EOS	*P RWG* (bar)	*294*	*16.7*	*16.6*	*7.2*	*26.0*	*2.1*
Isenthalpic, P-R EOS	*P RWG* (°C)	*294*	*28.2*	*28.3*	*14.1*	*40.0*	*3.3*

behavior. The total fluid pressures for the isenthalpic decompression path are close to those for the saturation for pure water, with provision for deviations from ideality (since RWG temperatures are lower than the critical temperature of pure H_2O), which is not surprising because both comprise the same P,T conditions. Lower total fluid pressures are given by the

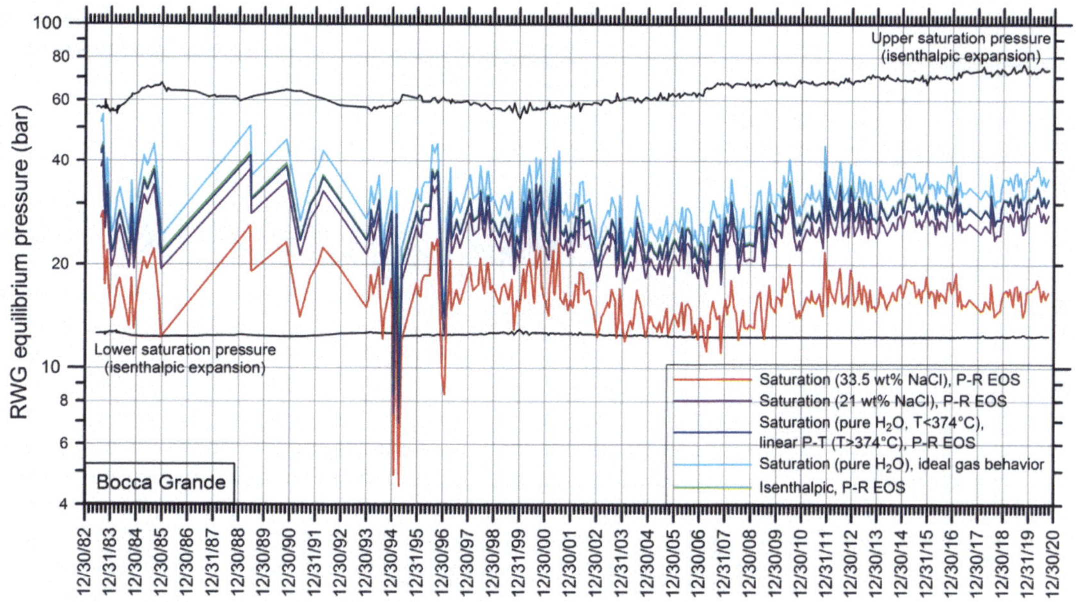

Fig. 7 Chronogram of the total fluid pressure indicated by the RWG reaction at equilibrium, along different expansion paths for the Bocca Grande fluids (data from Chiodini et al. 2021 and references therein). The lower and upper saturation pressures are also shown

saturation path involving a brine containing 21 wt% NaCl and even lower values are computed for the saturation path including a brine with 33.5 wt% NaCl. In other terms, total fluid pressure decreases with increasing salinity of the brine and vice versa.

The total fluid pressure of the RWG reaction at equilibrium experiences limited variations with time, since it usually fluctuates between 15 and 35 bar approximately for both fumaroles, as already noted in Sect. 4 in chapter "The Geoindicators Involving H_2O, CO_2, CO, CH_4 and H_2" for the pure-water saturation path without considering deviations from ideality. All total fluid pressure values given by the RWG reaction at equilibrium, except a few cases concerning Bocca Grande, are positioned in the saturation field, in that they are above the lower saturation pressure and below the upper saturation pressure. Therefore, the two hypotheses of (i) attainment of gas equilibrium in a pure saturated vapor phase and (ii) isenthalpic fluid decompression are coherent with each other, as already noted above.

6.2 The SS4 Equilibrium Temperature and Total Fluid Pressure

The upper and lower saturation temperatures for the isenthalpic depressurization of Solfatara fluids are not reported in the chronograms of Figs. 5 and 6, because no SS4 geothermometric equation was derived for the isenthalpic model due to its intrinsic weaknesses (see Sect. 2).

In contrast to what was observed for the RWG equilibrium temperature (see Sect. 6.1), there are significant differences between the SS4 equilibrium temperatures computed for distinct expansion paths, at least on the average, 72–124 °C for Bocca Grande and 91–143 °C for Bocca Nuova, and on the maximum values, 140–163 °C for Bocca Grande and 142–174 °C for Bocca Nuova, whereas the minimum values exhibit smaller differences, 2–20 °C for Bocca Grande and 6–49 °C for Bocca Nuova (see Figs. 5 and 6 and Tables 3 and 4).

In the whole monitoring period, that is, from 1983 to 2020 for Bocca Grande and from 1995 to 2020 for Bocca Nuova, the lowest SS4

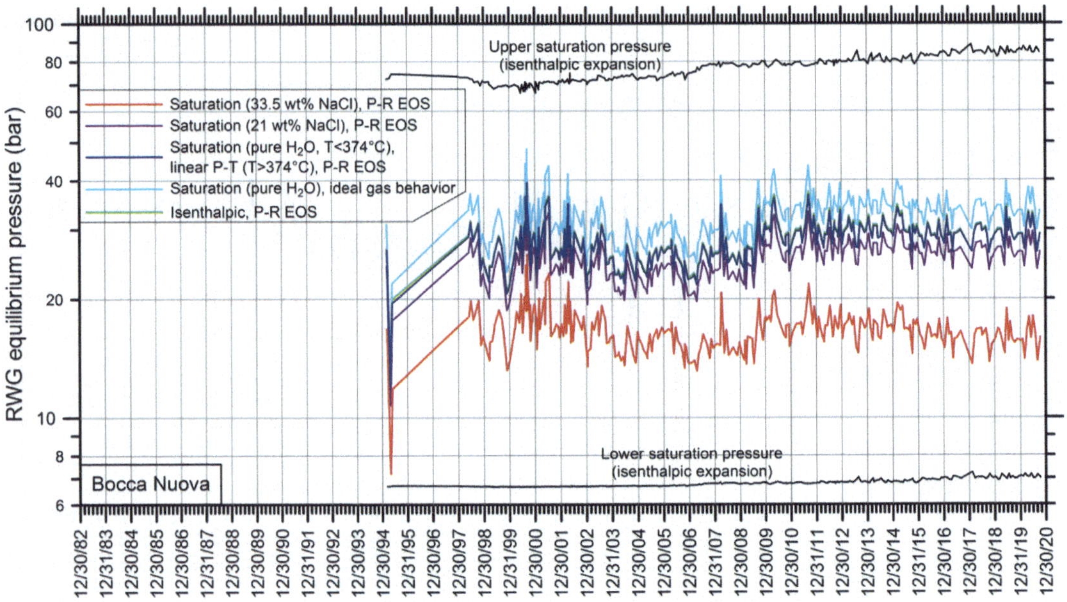

Fig. 8 Chronogram of the total fluid pressure indicated by the RWG reaction at equilibrium, along different expansion paths for the Bocca Nuova fluids (data from Chiodini et al. 2021 and references therein). The lower and upper saturation pressures are also shown

Table 3 Statistical parameters of the SS4 temperature and related total fluid pressure computed for all the Bocca Grande samples collected from 1983 to 2020 (data from Chiodini et al. 2021 and references therein) based on different decompression paths and hypotheses

Expansion path and hypothesis	Parameter	N	Mean	Median	Min	Max	Std. Dev
Saturation, ideal gas behavior	T SS4 (°C)	365	312	308	224	440	51.1
Saturation (pure H_2O, T < 374 °C), linear P–T (T > 374 °C), P-R EOS	T SS4 (°C)	365	436	454	244	580	81.5
Saturation (21 wt% NaCl), P-R EOS	T SS4 (°C)	365	410	428	235	604	101
Saturation (33.5 wt% NaCl), P-R EOS	T SS4 (°C)	365	384	377	222	595	100
Saturation, ideal gas behavior	*P SS4* (bar)	*365*	*159*	*119*	*28.7*	*626*	*121*
Saturation (pure H_2O, T < 374 °C), linear P–T (T > 374 °C), P-R EOS	*P SS4* (bar)	*365*	*518*	*549*	*41.0*	*1022*	*247*
Saturation (21 wt% NaCl), P-R EOS	*P SS4* (bar)	*365*	*435*	*362*	*31.1*	*1334*	*348*
Saturation (33.5 wt% NaCl), P-R EOS	*P SS4* (bar)	*365*	*295*	*163*	*17.0*	*1211*	*297*

equilibrium temperatures are given by the geothermometer for pure water saturation without any provision for deviations from ideality, whereas significantly higher SS4 equilibrium temperatures are computed by means of the three geothermometers involving the fugacity coefficients of gas species. Focusing on the results of these three geothermometers, it can be noted that:

(1) Until 2006, the three SS4 equilibrium temperatures were in the following order (from lowest to highest): (i) saturation with 33.5 wt % NaCl, (ii) saturation with 21 wt% NaCl, and (iii) saturation with pure water (below 374 °C) and linear P–T expansion path (above 374 °C). The differences between the

Table 4 Statistical parameters of the SS4 temperature and related total fluid pressure computed for all the Bocca Nuova samples collected from 1995 to 2020 (data from Chiodini et al. 2021 and references therein) based on different decompression paths and hypotheses

Expansion path and hypothesis	Parameter	N	Mean	Median	Min	Max	Std. Dev
Saturation, ideal gas behavior	T SS4 (°C)	299	333	333	248	470	47.1
Saturation (pure H_2O, T < 374 °C), linear P–T (T > 374 °C), P-R EOS	T SS4 (°C)	299	476	476	297	612	48.4
Saturation (21 wt% NaCl), P-R EOS	T SS4 (°C)	299	452	471	269	645	83.8
Saturation (33.5 wt% NaCl), P-R EOS	T SS4 (°C)	299	424	442	254	632	90.1
Saturation, ideal gas behavior	*P SS4* (bar)	*299*	*201*	*174*	*43.6*	*806*	*143*
Saturation (pure H_2O, T < 374 °C), linear P–T (T > 374 °C), P-R EOS	*P SS4* (bar)	*299*	*634*	*638*	*96.6*	*1132*	*179*
Saturation (21 wt% NaCl), P-R EOS	*P SS4* (bar)	*299*	*555*	*585*	*54.1*	*1543*	*357*
Saturation (33.5 wt% NaCl), P-R EOS	*P SS4* (bar)	*299*	*394*	*312*	*30.1*	*1393*	*330*

three values were ≤ 123 °C for Bocca Grande and ≤ 130 °C for Bocca Nuova.

(2) Afterwards, in the period 2007–2018, the SS4 temperature for saturation with 21 wt% NaCl became the highest, while the differences between the three SS4 temperatures decreased to ≤ 61 °C for Bocca Grande and ≤ 48 °C for Bocca Nuova.

(3) In the last two years (2019 and 2020), the SS4 temperature for the linear P–T expansion path resulted to be the lowest, whereas the differences between the three SS4 temperatures reduced further, to ≤ 24 °C for Bocca Grande and ≤ 33 °C for Bocca Nuova. Hence, in the last two years, the used approach does not influence significantly the computed SS4 temperatures, which are therefore robust results.

The three SS4 equilibrium temperatures given by the geothermometers involving the fugacity coefficients of gas species underwent a considerable increase with time. For instance, the SS4 equilibrium temperature for saturation with 33.5 wt% NaCl of Bocca Grande fluids (Fig. 9) increased from 222 to 252 °C in June 1983–July 1984 to 258–338 °C in 2000–2001, when a marked increase began, with the attainment of values from 402 to 484 °C in 2007. Afterwards there was a moderate decreasing tendency until 2011, when values of 390–441 °C were recorded. Then a striking increment occurred in 2012–2020, with values of 490–520 °C in 2016, 503–544 °C in 2017, 489–537 °C in 2018, 531–593 °C in 2019, and 565–595 °C in 2020.

The SS4 equilibrium temperature for saturation with 21 wt% NaCl of Bocca Grande effluents experienced a similar evolution, from 235 to 267 °C in June 1983–July 1984 to 274–380 °C in 2000–2001, when a considerable increase started, with the achievement of values from 447 to 508 °C in 2007. A limited decrease took place afterwards, until 2011, when values of 431 to 470 °C were reached. Then an impressive increment occurred in 2012–2020, with values of 512–537 °C in 2016, 523–557 °C in 2017, 511–551 °C in 2018, 546–600 °C in 2019, and 576–604 °C in 2020.

The values of the SS4 equilibrium temperature for the linear P–T expansion path of Bocca Grande fluids in the last years are somewhat lower, with 502–535 °C in 2018, 532–577 °C in 2019, and 557–580 °C in 2020.

The SS4 equilibrium temperatures of Bocca Nuova fluids (Fig. 10) compare with or are somewhat higher than those of Bocca Grande effluents. For example, the SS4 equilibrium temperature for saturation with 33.5 wt% NaCl was 510–555 °C in 2018, 532–588 °C in 2019, and 595–632 °C in 2020, the SS4 equilibrium temperature for saturation with 21 wt% NaCl was 528–566 °C in 2018, 547–598 °C in 2019, and

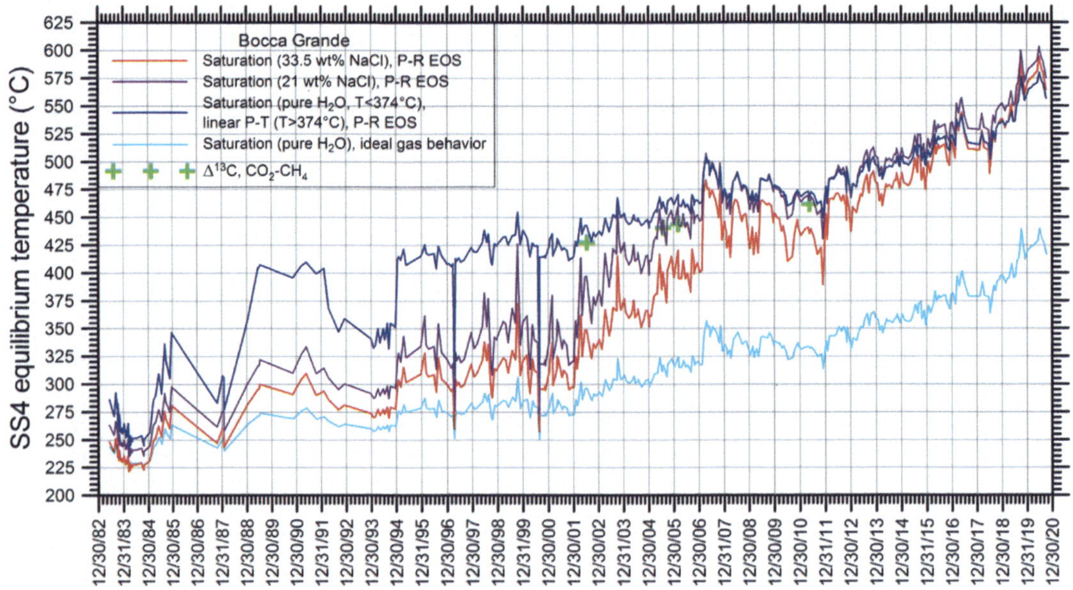

Fig. 9 Chronogram of the temperature indicated by the SS4 reaction at equilibrium, along different expansion paths for the Bocca Grande fluids (data from Chiodini et al. 2021 and references therein). The temperatures indicated by the exchange of C isotopes between CO_2 and CH_4 are also shown

605–645 °C in 2020, and the SS4 equilibrium temperature for the linear P–T expansion path was 515–547 °C in 2018, 536–577 °C in 2019, and 584–612 °C in 2020.

Moreover, it is worth to underscore that the use of geothermometric functions considering the deviations from ideality was not necessary in the 1980s due to the relatively low values of the SS4 temperatures, but it became mandatory afterwards due to their gradual but considerable increase. In the last years of monitoring, the further increase in the SS4 temperatures also led to a progressive reduction of the differences between distinct SS4 temperatures, which are practically independent of the assumed decompression path and, therefore, robust results, as already noted above. Finally, the temperature indicated by the exchange of C isotopes between CO_2 and CH_4 are in satisfactory agreement with the SS4 equilibrium temperatures, apart from the sample collected from Bocca Nuova on September 2nd, 2013, but the available data (4 for Bocca Grande and 5 for Bocca Nuova) are too few to draw sound conclusions.

Similar to the equilibrium temperatures, also the total fluid pressures indicated by the SS4 reaction at equilibrium along distinct expansion paths are different from each other with differences of 135–359 bar on the mean, 396–709 bar on the maximum, and only 2–12 bar on the minimum, for Bocca Grande (see Fig. 11 and Table 3) and 193–433 bar on the mean, 326–737 bar on the maximum, and 10–53 bar on the minimum, for Bocca Nuova (see Fig. 12 and Table 4). The lowest total fluid pressures are indicated by the saturation path for pure water without provision for deviations from ideality, since 2002, and by the saturation path for 33.5% NaCl incorporating the fugacity coefficients of gas species, in previous years. During most of the monitoring period, the highest total fluid pressures are obtained for the expansion path which assumes saturation with pure water below 374 °C, a linear P–T increment above 374 °C, and takes into account the deviations from ideality.

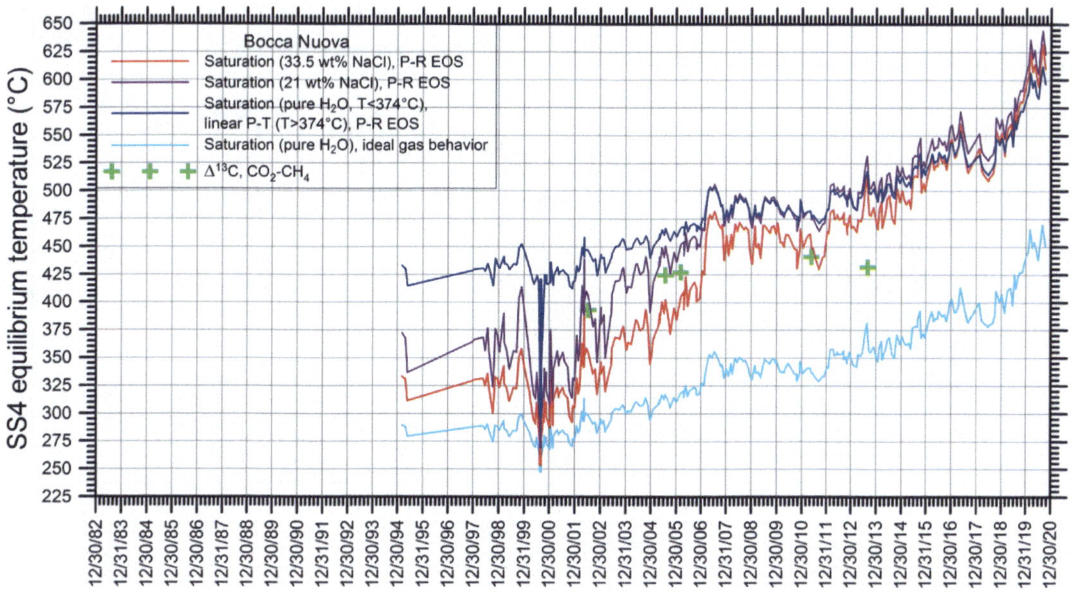

Fig. 10 Chronogram of the temperature indicated by the SS4 reaction at equilibrium, along different expansion paths for the Bocca Nuova fluids (data from Chiodini et al. 2021 and references therein). The temperatures indicated by the exchange of C isotopes between CO_2 and CH_4 are also shown

Nevertheless, in the last years, the total fluid pressures along the linear P,T decompression path are somewhat lower than those calculated for saturation with 33.5 wt% NaCl and those computed for saturation with 21 wt% NaCl for both Bocca Grande (Fig. 11) and Bocca Nuova effluents (Fig. 12).

Focusing on the SS4 total fluid pressures at equilibrium calculated using the geobarometers involving the fugacity coefficients of relevant gases, it can be observed that these pressure values undergo a substantial increment with time for Bocca Grande fluids (Fig. 11). Based on the geobarometer for saturation with 33.5 wt% NaCl, total fluid pressure increased from 17 to 29 bar in June 1983–July 1984 to 32–103 bar in 2000–2001, when a further increase commenced, with the attainment of 211–567 bar in 2007. A moderate decrease in the SS4 total fluid pressure occurred from 2007 to 2011, when values of 187–310 bar were attained. Then a conspicuous increase took place in 2012–2020 with values of 621–776 bar in 2016, 682–913 bar in 2017, 625–888 bar in 2018, 832–1211 bar in 2019, and 1028–1198 bar in 2020.

The total fluid pressure computed using the geobarometer for saturation with 21 wt% NaCl was 31–54 bar in June 1983–July 1984 and 59–183 bar in 2000–2001, when a further increment occurred, with the achievement of 453–757 bar in 2007. A limited decrease in the SS4 total fluid pressure took place from 2007 to 2011, when values of 398–586 bar were attained. Then, a noticeable increment occurred in 2012–2020, with values 812–948 bar in 2016, 865–1066 bar in 2017, 817–1049 bar in 2018, 995–1334 bar in 2019, and 1167–1326 bar in 2020.

Assuming saturation with pure water below 374 °C and a linear P–T increment above 374 °C leads to total fluid pressures of 746–821 bar in 2016, 780–884 bar in 2017, 742–861 bar in 2018, 848–1014 bar in 2019, and 939–1022 bar in 2020.

The SS4 total fluid pressure of Bocca Nuova fluids experienced similar changes from 1995 (when this vent began to be monitored) to 2020

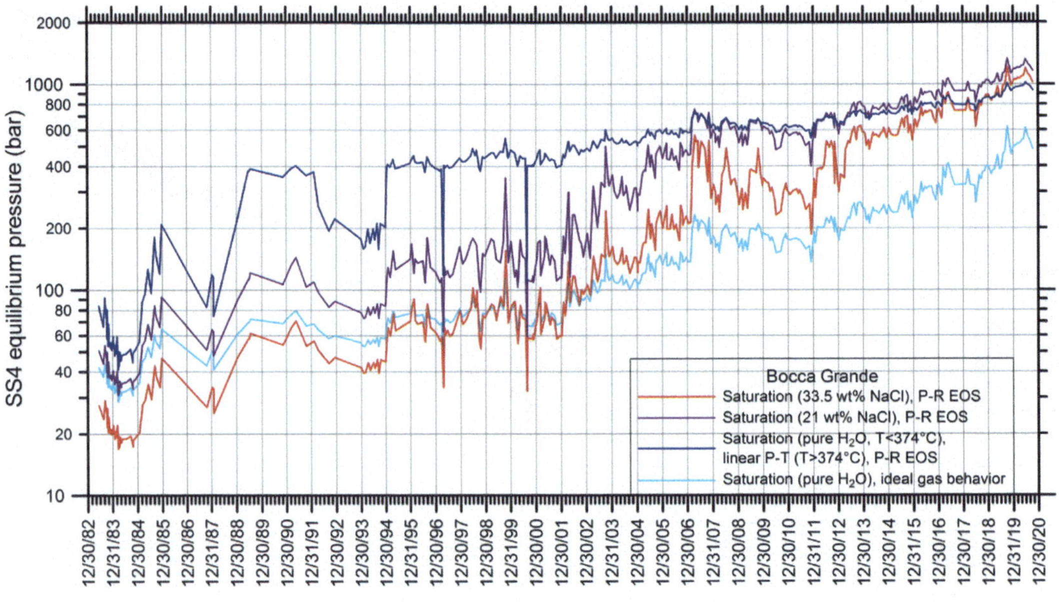

Fig. 11 Chronogram of total fluid pressure (on a logarithmic scale) indicated by the SS4 reaction at equilibrium, along different expansion paths for the Bocca Grande fluids (data from Chiodini et al. 2021 and references therein)

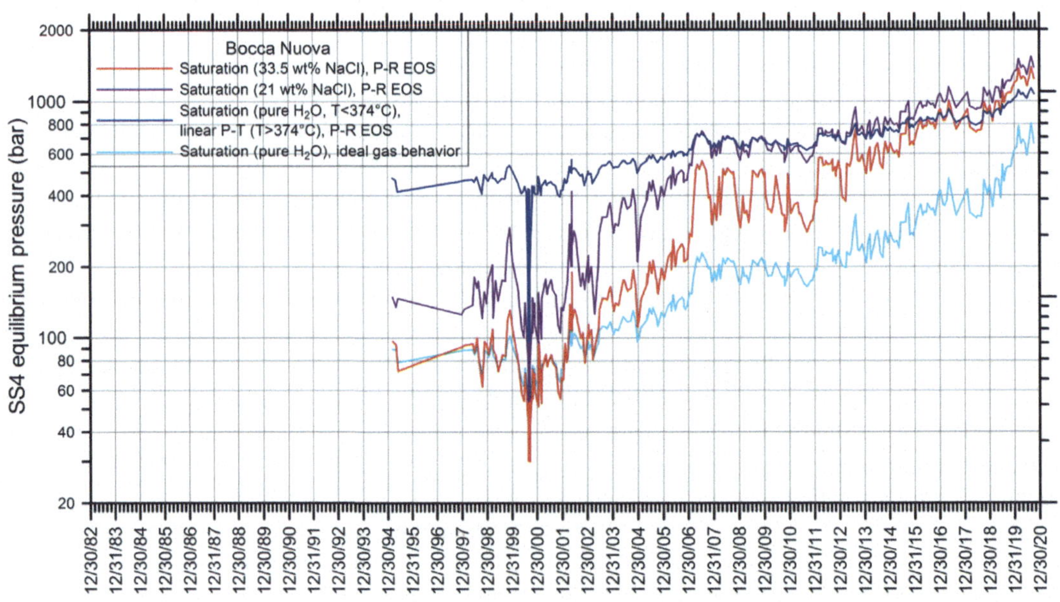

Fig. 12 Chronogram of total fluid pressure (on a logarithmic scale) indicated by the SS4 reaction at equilibrium, along different expansion paths for the Bocca Nuova fluids (data from Chiodini et al. 2021 and references therein)

(Fig. 12). In particular, the values given by the geobarometer for saturation with 21 wt% NaCl are 921 to 1142 bar in 2018, 985 to 1273 bar in 2019, and 1293 to 1543 bar in 2020.

6.3 Final Considerations on the Temperatures and Total Fluid Pressures Indicated by the RWG and SS4 Reactions at Equilibrium

The temperatures and total fluid pressures indicated by the RWG reaction at equilibrium, with mean values of 215–223 °C and 16.1–27.3 bar, for Bocca Grande, and 217–226 °C and 16.7–28.2 bar, for Bocca Nuova, are relatively low and did not experience any important variation with time. Therefore, they are probably representative of a shallow permeable zone, in which carbon monoxide attains the last equilibrium condition. If so (i.e., assuming that CO did not experience any further readjustment downstream of this zone), the depth of the RWG reaction equilibration zone is estimated to be ca. 160 to 280 m below the local water table. Hence, the RWG reaction does not provide any indication on the evolution of the thermodynamic conditions in the deep parts of the Solfatara magmatic-hydrothermal system.

In contrast, the equilibrium temperatures and total fluid pressures indicated by the SS4 reaction, at saturation with 21 wt% NaCl, experienced significant increments with time, from 235–267 °C, 31–54 bar in June 1983–July 1984 to 576–604 °C, 1167–1326 bar in 2020, for Bocca Grande, and up to 605–645 °C, 1293–1543 bar in 2020, for Bocca Nuova. Similar indications are obtained considering the geoindicators calibrated for saturation with 33.5 wt% NaCl, with an increase from 222-252 °C, 17–29 bar in June 1983–July 1984 to 565–595 °C, 1028–1198 bar in 2020, for Bocca Grande, and up to 595–632 °C, 1160–1393 bar in 2020, for Bocca Nuova.

Thus, the SS4 reaction equilibrated under the same P–T conditions of the RWG reaction in 1983–1984, as already noted by Cioni et al. (1984), confirming the provenance of fumarolic fluids from a shallow permeable zone. The increasing discrepancies between the P–T conditions indicated by the RWG and SS4 reactions at equilibrium, in the following years, suggest that fumarolic fluids came from either:

(1) progressively deeper zones of the Solfatara magmatic-hydrothermal system, characterized by gradually higher pressures and temperatures, or
(2) the same deep permeable zone that experienced a progressive increase in both pressure and temperature with time.

These inferences are based on the assumption of constant concentrations of CH_4, H_2, CO_2, and H_2O during the upflow of Solfatara fluids.

In contrast, if CH_4, H_2, CO_2, and H_2O concentrations readjust during the uprise of fumarolic fluids from their source zone to the surface, the SS4 reaction indicates lower P–T conditions, as discussed in chapter "The Chemical Kinetics of the Reactions Involving H_2O, CO_2, CO, CH_4 and H_2". Nevertheless, the qualitative conclusion does not change, in that, from 1983 to 2020, the fluids released from Solfatara fumaroles came from either progressively hotter and deeper zones of the magmatic-hydrothermal system or the same deep zone that underwent a gradual pressurization and heating with time.

References

Chiodini G, Caliro S, Avino R, Bini G, Giudicepietro F, De Cesare W, Ricciolino P, Aiuppa A, Cardellini C, Petrillo Z, Selva J, Siniscalchi A, Tripaldi S (2021) Hydrothermal pressure-temperature control on CO_2 emissions and seismicity at Campi Flegrei (Italy). J Volcanol Geothermal Res 414:107245

Cioni R, Corazza E, Marini L (1984) The gas/steam ratio as indicator of heat transfer at the Solfatara fumaroles, Phlegraean Fields (Italy). Bull Volcanol 47(2):295–302

Driesner T, Heinrich CA (2007) The system H_2O–NaCl. Part I: Correlation formulae for phase relations in temperature–pressure–composition space from 0 to 1000°C, 0 to 5000 bar, and 0 to 1 X_{NaCl}.

Geochimica et Cosmochimica Acta 71 (20):4880–4901

Johnson JW, Oelkers EH, Helgeson HC (1992) SUPCRT92: a software package for calculating the standard molal thermodynamic properties of minerals, gases, aqueous species, and reactions from 1 to 5000 bar and 0 to 1000°C. Comput Geosci 18(7):899–947

Tanger JC IV, Pitzer KS (1989) Thermodynamics of NaCl-H_2O: a new equation of state for the near-critical region and comparisons with other equations for adjoining regions. Geochim Cosmochim Acta 53 (5):973–987

The Redox Potential and Sulfur Gas Species

Abstract

Deviations from ideality were considered to implement an H_2S-H_2 geothermometer and three different H_2S-H_2-H_2O geothermometers based on the heterogeneous gas-solid equilibria involving pyrite-pyrrhotite, pyrite-fayalite-quartz, pyrite-magnetite, and pyrite-hematite. The results obtained by using these geothermometers suggest that the H_2S concentration of Solfatara fluids is probably controlled by other reactions. Therefore, two distinct heterogeneous gas-solid equilibria comprising anhydrite and calcite were utilized to develop a H_2S-CO_2 geothermometer (also involving H_2 and H_2O) and a H_2S-CH_4 geothermometer (also including H_2O), again considering the fugacity coefficients of relevant gas species. The computed H_2S-CO_2 equilibrium temperatures and pressures increase almost continuously with time, from (a) 667±24°C, 1308 ±102 bar, in June 1983–July 1984, to (b) 990 ±30°C, 2958±146 bar in 2019-2020, for Bocca Grande, and 1010±37°C, 3034±183 bar in 2019-2020, for Bocca Nuova. Assuming that gas species equilibrate with anhydrite and calcite in the deepest part of the carbonate layer situated at a depth of 6 - 7 km (to be considered as an educated guess), there was a balance between total fluid pressure and external pressure in June 1983–July 1984. In contrast, a continuous increment in the excess fluid pressure (i.e., the difference between the fluid pressure and the overburden pressure) took place during the last twenty years and this increment was particularly marked in the last 3-4 years, possibly explaining the on-going resurgence of the entire Campi Flegrei caldera. The H_2S-CH_4 geothermometer was used to compute the CH_4 concentration in the zone where H_2S and CO_2 equilibrate with anhydrite and calcite. Methane concentration turned out to be very low, as expected in magmatic fluids, presumably somewhat modified by absorption in deep brines and interaction with carbonate rocks. Deviations from ideality were also considered to calculate the redox potential of Solfatara fluids (at the H_2S-CO_2 equilibrium temperature) which turned out to be similar to those of some melt inclusions from the active Italian volcanoes. Hydrogen sulfide concentrations of Solfatara fluids were also used to evaluate the fugacities of other sulfur species such as S_2, SO_2, and COS.■■■

Supplementary Information The online version contains supplementary material available at (https://doi.org/10.1007/978-3-030-98471-7_9).

Also the geothermometers, geobarometers, and other calculations presented in this chapter are reported as electronic supplementary material (see Microsoft Excel file Chapter 9.XLS).

L. Marini et al., *The Solfatara Magmatic-Hydrothermal System*, Advances in Volcanology,
https://doi.org/10.1007/978-3-030-98471-7_9

Nevertheless, the reader can also write the equations of this chapter to another Excel spreadsheet following the indications given below.

1 Preliminary Considerations

In this chapter, the H_2S concentration of Solfatara fluids is used both for geothermometry and to evaluate the fugacities of other sulfur species, namely S_2, SO_2, and COS. To this purpose, suitable reactions involving not only H_2S, but also H_2O, CO_2, H_2 and CO, are taken into consideration. Therefore, it is necessary to know, at least, the fugacity coefficient of H_2S, which was calculated by means of the Peng-Robinson EOS, at the pressures and temperatures of coexistence of the brine containing 21 wt% NaCl and the related vapor phase. In general, the other possible expansion paths of Solfatara fluids were not considered, to avoid complicating the discussion too much, but it was necessary to take into account the saturation expansion path involving the brine containing 33.5 wt% NaCl for treating the gas–solid reactions involving anhydrite. As already done in Sects. 5, 7.3, 7.4, and 7.5 in chapter "The Expansion (Decompression) of the Solfatara Fumarolic Fluids", calculations were performed for the ternary mixtures H_2O-CO_2-H_2S by means of the Microsoft Excel computer code Prfug.xlsx (Elliott and Lira 2012), for X_{CO_2} equal to 0.05, 0.10, 0.20, 0.30, and 0.40, fixed $X_{H_2S} = 1.35 \cdot 10^{-3}$ (which is the average H_2S mole fraction of the Solfatara fluids sampled by Chiodini and coworkers; Chiodini et al. 2021), and X_{H_2O} computed by difference to unity. Results are listed in Tables 1, 2, 3, 4, 5, together with the fugacity coefficients of H_2O, CO_2, H_2, and CO and other relevant parameters, already discussed in Sect. 7.3 in chapter "The Expansion (Decompression) of the Solfatara Fumarolic Fluids". Since the fugacity coefficients of H_2O and CO_2 were computed by averaging the values for the ternary systems H_2O-CO_2-H_2S, H_2O-CO_2-H_2, and H_2O-CO_2-CO, they might be slightly different (on the fourth or on the third decimal figure) from those listed in Tables 11 to 15 in chapter "The Expansion (Decompression) of the Solfatara Fumarolic Fluids", whereas the values of the fugacity coefficients of H_2 and CO are the same.

As done in Sect. 7.3 in chapter "The Expansion (Decompression) of the Solfatara Fumarolic Fluids", the uncertainties on the computed fugacity coefficient of H_2S were assessed for $X_{CO_2} = 0.20$, at the minimum and maximum temperature and pressure of interest, taking into account alternatively the minimum and maximum concentrations of H_2S determined for Solfatara fluids. Outcomes are compared with those obtained for the average H_2S concentration in Table 6. Inspection of this table suggests that the differences in the computed parameters are acceptable for the purposes of this work.

The online version of SUPCRTBL, which is available at https://models.earth.indiana.edu/supcrtbl.php (last visit on December 4, 2021), was used to investigate the reactions involving elemental sulfur instead of SUPCRT (Johnson et al. 1992) because elemental sulfur is included in SUPCRTBL but not in SUPCRT. SUPCRTBL was implemented by Zimmer et al. (2016) adopting the approach of Holland and Powell (1998) for calculating the standard state thermodynamic properties of mineral end-members. The mineral database of SUPCRTBL is largely based on the internally consistent thermodynamic dataset of Holland and Powell (2011) which, in turn, includes the thermodynamic properties generated by Evans et al (2010) for sulfur-bearing minerals and fluid species.

The thermodynamic properties of the reactions between H_2O, CO_2, H_2, CH_4 and CO calculated by SUPCRT92, which were discussed in chapter "The Geoindicators Involving H_2O, CO_2, CO, CH_4 and H_2", do not differ significantly from those calculated by SUPCRTLB. In particular, the differences between the log K values calculated by means of the two codes vary from a few thousandths to a few hundredths and, therefore, are completely negligible.

Table 1 Compressibility factor, volume, and fugacity of the gas mixture with X_{CO_2} of 0.05 and fugacity coefficients of H_2O, CO_2, H_2S, H_2, and CO in the same gas mixture computed by means of the Peng-Robinson EOS at the P,T conditions fixed by coexistence with a brine containing 21 wt% NaCl

P	T	Z_{mixt}	V_{mixt}	f_{mixt}	φ_{H_2O}	φ_{CO_2}	φ_{H_2S}	φ_{H_2}	φ_{CO}
bar	°C		cm^3/mol	bar					
4.5	150	0.9735	7535.91	4.43	0.9725	1.0035	1.0109	1.0270	1.0206
8.3	175	0.9582	4288.81	7.99	0.9572	1.0071	1.0185	1.0441	1.0345
14.3	200	0.9375	2578.56	13.45	0.9371	1.0132	1.0299	1.0686	1.0552
23.3	225	0.9107	1620.70	21.35	0.9120	1.0234	1.0467	1.1032	1.0852
36.2	250	0.8774	1055.60	32.15	0.8821	1.0396	1.0713	1.1513	1.1282
53.9	275	0.8370	707.12	46.26	0.8478	1.0651	1.1071	1.2181	1.1900
77.7	300	0.7896	484.36	63.92	0.8100	1.1044	1.1594	1.3111	1.2789
109	325	0.7352	336.23	85.43	0.7693	1.1655	1.2373	1.4441	1.4108
148	350	0.6773	237.42	110.62	0.7280	1.2580	1.3515	1.6315	1.6048
196	375	0.6210	170.47	139.89	0.6875	1.3946	1.5163	1.8917	1.8894
255	400	0.5802	127.16	173.82	0.6509	1.5705	1.7238	2.2033	2.2580
326	425	0.5684	101.14	213.75	0.6219	1.7450	1.9229	2.4750	2.6233
407	450	0.5862	86.53	260.54	0.6041	1.8709	2.0556	2.6171	2.8721
497	475	0.6213	77.71	315.22	0.5969	1.9539	2.1306	2.6553	3.0106
592	500	0.6650	72.20	377.44	0.5996	2.0058	2.1644	2.6270	3.0653
689	525	0.7117	68.58	446.68	0.6102	2.0390	2.1743	2.5658	3.0687
783	550	0.7581	66.23	521.63	0.6270	2.0587	2.1685	2.4886	3.0378
878	575	0.8028	64.51	602.87	0.6476	2.0740	2.1586	2.4116	2.9962
972	600	0.8455	63.16	690.62	0.6710	2.0881	2.1487	2.3399	2.9525

Table 2 Compressibility factor, volume, and fugacity of the gas mixture with X_{CO_2} of 0.10 and fugacity coefficients of H_2O, CO_2, H_2S, H_2, and CO in the same gas mixture computed by means of the Peng-Robinson EOS at the P,T conditions fixed by coexistence with a brine containing 21 wt% NaCl

P	T	Z_{mixt}	V_{mixt}	f_{mixt}	φ_{H_2O}	φ_{CO_2}	φ_{H_2S}	φ_{H_2}	φ_{CO}
bar	°C		cm^3/mol	bar					
4.80	150	0.9736	7140.46	4.67	0.9711	1.0025	1.0094	1.0268	1.0202
8.79	175	0.9585	4064.51	8.43	0.9550	1.0055	1.0160	1.0436	1.0339
15.1	200	0.9381	2444.56	14.21	0.9338	1.0108	1.0261	1.0679	1.0542
24.6	225	0.9120	1537.46	22.56	0.9075	1.0200	1.0411	1.1019	1.0837
38.2	250	0.8795	1002.54	34.00	0.8761	1.0348	1.0632	1.1489	1.1259
56.9	275	0.8408	672.99	48.97	0.8403	1.0581	1.0954	1.2136	1.1860
82.0	300	0.7963	462.74	67.77	0.8009	1.0939	1.1419	1.3022	1.2714
115	325	0.7467	323.54	90.79	0.7586	1.1486	1.2099	1.4252	1.3947
156	350	0.6969	231.44	117.97	0.7162	1.2279	1.3050	1.5894	1.5673
207	375	0.6529	169.78	149.91	0.6754	1.3369	1.4322	1.7983	1.8010

(continued)

Table 2 (continued)

P	T	Z_{mixt}	V_{mixt}	f_{mixt}	φ_{H_2O}	φ_{CO_2}	φ_{H_2S}	φ_{H_2}	φ_{CO}
bar	°C		cm^3/mol	bar					
270	400	0.6255	129.87	187.42	0.6395	1.4672	1.5798	2.0256	2.0783
344	425	0.6220	104.85	232.00	0.6117	1.5972	1.7212	2.2201	2.3496
430	450	0.6417	89.74	284.41	0.5951	1.7032	1.8282	2.3363	2.5555
525	475	0.6764	80.16	345.73	0.5891	1.7856	1.9027	2.3846	2.6935
625	500	0.7193	73.98	415.46	0.5929	1.8474	1.9500	2.3826	2.7707
727	525	0.7651	69.85	492.97	0.6045	1.8947	1.9790	2.3516	2.8050
827	550	0.8107	67.10	576.71	0.6224	1.9299	1.9941	2.3036	2.8071
926	575	0.8548	65.07	667.41	0.6442	1.9600	2.0040	2.2522	2.7959
1026	600	0.8969	63.48	765.32	0.6687	1.9878	2.0121	2.2023	2.7791

Table 3 Compressibility factor, volume, and fugacity of the gas mixture with X_{CO_2} of 0.20 and fugacity coefficients of H_2O, CO_2, H_2S, H_2, and CO in the same gas mixture computed by means of the Peng-Robinson EOS at the P,T conditions fixed by coexistence with a brine containing 21 wt% NaCl

P	T	Z_{mixt}	V_{mixt}	f_{mixt}	φ_{H_2O}	φ_{CO_2}	φ_{H_2S}	φ_{H_2}	φ_{CO}
bar	°C		cm^3/mol	bar					
5.40	150	0.9738	6348.11	5.26	0.9680	1.0003	1.0059	1.0264	1.0196
9.88	175	0.9589	3614.56	9.49	0.9501	1.0021	1.0105	1.0429	1.0329
17.0	200	0.9391	2175.26	16.00	0.9267	1.0058	1.0178	1.0667	1.0526
27.6	225	0.9140	1369.72	25.42	0.8977	1.0128	1.0290	1.0997	1.0813
42.9	250	0.8835	895.15	38.36	0.8634	1.0247	1.0459	1.1449	1.1221
64.1	275	0.8481	603.36	55.37	0.8243	1.0440	1.0708	1.2059	1.1794
92.3	300	0.8091	417.96	76.87	0.7819	1.0737	1.1066	1.2868	1.2588
129	325	0.7689	296.11	103.48	0.7372	1.1177	1.1572	1.3930	1.3681
176	350	0.7331	216.42	135.38	0.6936	1.1778	1.2236	1.5224	1.5098
233	375	0.7078	163.60	173.63	0.6535	1.2536	1.3045	1.6686	1.6825
303	400	0.6991	129.02	219.49	0.6202	1.3391	1.3922	1.8111	1.8700
387	425	0.7091	106.26	274.96	0.5957	1.4270	1.4783	1.9306	2.0522
484	450	0.7361	91.51	340.88	0.5822	1.5090	1.5539	2.0107	2.2058
591	475	0.7746	81.60	418.51	0.5791	1.5848	1.6196	2.0565	2.3283
703	500	0.8195	74.92	507.01	0.5856	1.6527	1.6746	2.0734	2.4173
818	525	0.8666	70.33	605.45	0.5998	1.7133	1.7208	2.0706	2.4784
930	550	0.9129	67.16	711.67	0.6202	1.7660	1.7583	2.0541	2.5148
1042	575	0.9575	64.79	826.67	0.6445	1.8145	1.7918	2.0327	2.5386
1154	600	1.0002	62.92	950.82	0.6717	1.8606	1.8231	2.0101	2.5550

Table 4 Compressibility factor, volume, and fugacity of the gas mixture with X_{CO_2} of 0.30 and fugacity coefficients of H_2O, CO_2, H_2S, H_2, and CO in the same gas mixture computed by means of the Peng-Robinson EOS at the P,T conditions fixed by coexistence with a brine containing 21 wt% NaCl

P	T	Z_{mixt}	V_{mixt}	f_{mixt}	φ_{H_2O}	φ_{CO_2}	φ_{H_2S}	φ_{H_2}	φ_{CO}
bar	°C		cm^3/mol	bar					
6.17	150	0.9737	5553.85	6.01	0.9642	0.9979	1.0018	1.0263	1.0190
11.3	175	0.9589	3162.81	10.85	0.9444	0.9983	1.0041	1.0427	1.0321
19.4	200	0.9396	1904.26	18.29	0.9185	1.0003	1.0081	1.0661	1.0516
31.6	225	0.9154	1200.32	29.08	0.8865	1.0051	1.0151	1.0986	1.0798
49.1	250	0.8867	786.11	43.93	0.8488	1.0142	1.0264	1.1424	1.1198
73.2	275	0.8546	532.04	63.54	0.8065	1.0297	1.0439	1.2004	1.1752
105	300	0.8214	371.28	88.51	0.7612	1.0541	1.0696	1.2748	1.2501
148	325	0.7903	266.33	119.75	0.7148	1.0897	1.1055	1.3673	1.3490
201	350	0.7674	198.23	157.76	0.6713	1.1368	1.1512	1.4720	1.4699
266	375	0.7573	153.18	204.19	0.6333	1.1943	1.2051	1.5806	1.6080
347	400	0.7635	123.29	260.92	0.6035	1.2591	1.2636	1.6803	1.7522
443	425	0.7858	103.03	330.72	0.5831	1.3291	1.3248	1.7638	1.8943
553	450	0.8218	89.39	414.77	0.5738	1.4010	1.3857	1.8244	2.0222
675	475	0.8671	79.92	514.75	0.5746	1.4745	1.4466	1.8657	2.1352
804	500	0.9171	73.36	629.49	0.5847	1.5469	1.5055	1.8897	2.2287
935	525	0.9682	68.74	757.64	0.6025	1.6168	1.5616	1.9008	2.3038
1063	550	1.0175	65.50	896.16	0.6264	1.6819	1.6131	1.9016	2.3602
1191	575	1.0646	63.03	1046.40	0.6543	1.7444	1.6625	1.8981	2.4059
1319	600	1.1095	61.08	1208.90	0.6851	1.8051	1.7106	1.8928	2.4448

2 The Redox Potential

As pointed out by Giggenbach (1987), the best parameter to describe the redox potential in the gas equilibration zone is the R_H value, which is equal to the decimal logarithm of the H_2/H_2O fugacity ratio:

$$R_H = \log(f_{H_2}/f_{H_2O}) = \log(X_{H_2}/X_{H_2O}) + \log(\varphi_{H_2}/\varphi_{H_2O}). \quad (1)$$

As shown by Eq. (1), the log-ratio of fugacities can be split in two terms, that is, in the sum of the log-ratio of mole fractions and the log-ratio of fugacity coefficients. To a first approximation, the R_H value is equal to the decimal logarithm of the H_2/H_2O mole fraction ratio in the equilibrium vapour phase, either real or hypothetical, as long as the ratio between the fugacity coefficients of H_2 and H_2O does not diverge significantly from unity. This condition is often adopted without any check, but this practice is permissible as long as temperature and pressure are relatively low, whereas it leaves much to be desired at high temperatures and pressures.

In the case of the Solfatara fluids, it is worth to compute the ratio between the fugacity coefficients of H_2 and H_2O based on a reasonable hypothesis. For instance, assuming that the SS4 reaction equilibrates at saturation with a brine containing 21 wt% NaCl, the $\log(\varphi_{H_2}/\varphi_{H_2O})$ can be computed as a function of the SS4 equilibrium temperature and the CO_2 mole fraction. Calculations are conveniently performed using

Table 5 Compressibility factor, volume, and fugacity of the gas mixture with X_{CO_2} of 0.40 and fugacity coefficients of H_2O, CO_2, H_2S, H_2, and CO in the same gas mixture computed by means of the Peng-Robinson EOS at the P,T conditions fixed by coexistence with a brine containing 21 wt% NaCl

P	T	Z_{mixt}	V_{mixt}	f_{mixt}	φ_{H_2O}	φ_{CO_2}	φ_{H_2S}	φ_{H_2}	φ_{CO}
bar	°C		cm^3/mol	bar					
7.20	150	0.9731	4757.69	7.01	0.9597	0.9950	0.9968	1.0265	1.0188
13.2	175	0.9583	2709.29	12.65	0.9374	0.9939	0.9962	1.0430	1.0318
22.6	200	0.9392	1631.59	21.33	0.9084	0.9941	0.9965	1.0665	1.0513
36.8	225	0.9158	1029.33	33.92	0.8729	0.9965	0.9986	1.0988	1.0796
57.2	250	0.8890	675.54	51.31	0.8315	1.0028	1.0037	1.1419	1.1195
85.4	275	0.8605	459.17	74.36	0.7858	1.0149	1.0135	1.1977	1.1741
123	300	0.8336	322.93	103.94	0.7380	1.0347	1.0294	1.2667	1.2462
172	325	0.8122	234.59	141.39	0.6907	1.0640	1.0527	1.3480	1.3379
234	350	0.8021	177.58	187.71	0.6485	1.1025	1.0832	1.4341	1.4447
311	375	0.8065	139.82	245.35	0.6138	1.1500	1.1204	1.5179	1.5622
404	400	0.8268	114.44	317.15	0.5886	1.2053	1.1636	1.5926	1.6838
517	425	0.8618	96.85	407.14	0.5734	1.2690	1.2134	1.6566	1.8068
645	450	0.9084	84.70	517.21	0.5693	1.3395	1.2692	1.7072	1.9246
787	475	0.9629	76.07	649.90	0.5753	1.4166	1.3309	1.7474	2.0368
938	500	1.0206	69.98	803.78	0.5907	1.4972	1.3962	1.7774	2.1384
1090	525	1.0781	65.61	976.95	0.6136	1.5789	1.4629	1.7990	2.2277
1240	550	1.1327	62.50	1165.04	0.6425	1.6580	1.5282	1.8128	2.3023
1390	575	1.1844	60.11	1369.98	0.6755	1.7361	1.5931	1.8231	2.3682
1539	600	1.2333	58.20	1592.56	0.7117	1.8133	1.6579	1.8317	2.4280

Table 6 Compressibility factor, volume, and fugacity of the gas mixture with CO_2 mole fraction of 0.20 and fugacity coefficients of H_2O, CO_2, and H_2S, in the same gas mixture computed by means of the Peng-Robinson EOS at the minimum and maximum temperature and pressure of interest and minimum, mean, and maximum concentrations of H_2S

X_{H_2O}	X_{H_2S}	T	P	Z_{mixt}	V_{mixt}	f_{mixt}	φ_{H_2O}	φ_{CO_2}	φ_{H_2S}
		°C	bar		cm^3/mol	bar			
0.79940	6.02E-04	150	5.40	0.9738	6348.09	5.26	0.9680	1.0003	1.0060
0.79865	1.35E-03	150	5.40	0.9738	6348.28	5.26	0.9680	1.0003	1.0059
0.79499	5.01E-03	150	5.40	0.9740	6349.22	5.26	0.9680	1.0001	1.0057
0.79940	6.02E-04	600	1154	1.0002	62.92	950.80	0.6717	1.8606	1.8255
0.79865	1.35E-03	600	1154	1.0007	62.96	951.51	0.6719	1.8586	1.8231
0.79499	5.01E-03	600	1154	1.0033	63.12	954.99	0.6728	1.8489	1.8118

the Excel spreadsheet Chapter 9.XLS, with the SS4 equilibrium temperature in column F and X_{CO_2} in column H. If so, the equations of interest are as follows for the first sample with the SS4 equilibrium temperature in cell F2 and X_{CO_2} in cell H2:

$$\begin{aligned}\log(\varphi_{H_2}/\varphi_{H_2O}) = {} & (-0.0000000001849413^*H2^\wedge4 \\ & + 0.0000000002108805^*H2^\wedge3 \\ & - 0.00000000009316924^*H2^\wedge2 \\ & + 0.00000000002142593^*H2 \\ & - 0.000000000002038788)^*F2^\wedge5 \\ & + (0.0000002670383^*H2^\wedge4 \\ & - 0.0000003040473^*H2^\wedge3 \\ & + 0.000000134159^*H2^\wedge2 \\ & - 0.00000003059662^*H2 \\ & + 0.000000002753366)^*F2^\wedge4 \\ & + (-0.0001482871^*H2^\wedge4 \\ & + 0.0001685202^*H2^\wedge3 \\ & - 0.00007423016^*H2^\wedge2 \\ & + 0.00001675254^*H2 \\ & - 0.000001428221)^*F2^\wedge3 \\ & + (0.03961086^*H2^\wedge4 \\ & - 0.04492647^*H2^\wedge3 \\ & + 0.01975374^*H2^\wedge2 \\ & - 0.00440961^*H2 \\ & + 0.0003628211)^*F2^\wedge2 \\ & + (-5.090384^*H2^\wedge4 \\ & + 5.763558^*H2^\wedge3 \\ & - 2.529566^*H^\wedge2 + 0.5591099^*H2 \\ & - 0.04451479)^*F2 \\ & + (251.8967^*H2^\wedge4 \\ & - 284.7872^*H2^\wedge3 \\ & + 124.7784^*H2^\wedge2 - 27.33983^*H2 \\ & + 2.119306)\end{aligned} \tag{2}$$

$$\begin{aligned}\log(\varphi_{H_2}/\varphi_{H_2O}) = {} & (-0.000001467493^*H2^\wedge4 \\ & + 0.000001462963^*H2^\wedge3 \\ & - 0.0000004562762^*H2^\wedge2 \\ & - 0.000000001616483^*H2 \\ & + 0.00000004005742)^*F2^\wedge3 \\ & + (0.00262969^*H2^\wedge4 \\ & - 0.002653236^*H2^\wedge3 \\ & + 0.0008580888^*H2^\wedge2 \\ & - 0.00001520236^*H2 \\ & - 0.00006824068)^*F2^\wedge2 \\ & + (-1.582929^*H2^\wedge4 \\ & + 1.61913^*H2^\wedge3 - 0.545578^*H2^\wedge2 \\ & + 0.02274471^*H2 \\ & + 0.03743424)^*F2 + (320.776^*H2^\wedge4 \\ & - 333.534^*H2^\wedge3 + 117.8856^*H2^\wedge2 \\ & - 8.43481^*H2 - 5.975348)\end{aligned} \tag{3}$$

Equation (2) is valid for T < 475 °C, whereas Eq. (3) holds true for T $\geq$ 475 °C up to 600 °C. The equal sign and the right-hand side of Eqs. (2) and (3) are written in cells I2 and J2 of the Excel spreadsheet Chapter 9.XLS to compute the $\log(\varphi_{H_2}/\varphi_{H_2O})$ for the first sample. Then, the two equations are quickly copied into the adjacent cells of columns I and J dragging the fill handle down, thus computing the possible $\log(\varphi_{H_2}/\varphi_{H_2O})$ for all the available samples. Finally, in column K, the IF function is used to select the correct $\log(\varphi_{H_2}/\varphi_{H_2O})$ for each sample depending on the value of the SS4 equilibrium temperature. For the first sample, the IF function is as follows: $\log(\varphi_{H_2}/\varphi_{H_2O})$ = IF (F2 < 475;I2;J2).

The absolute error on the $\log(\varphi_{H_2}/\varphi_{H_2O})$ calculated using Eqs. (2) and (3) is

0.00045 ± 0.00044 (1σ) log-units on average, with range 0.0000015 to 0.0018 log-units. Strictly speaking, Eq. (3) can be used up to 600 ° C, but it can be probably applied also at temperatures somewhat higher accepting larger uncertainties.

The R_H parameter (column L of the Excel file Chapter 9.XLS) is computed by adding the log-ratio of the fugacity coefficients of H_2 and H_2O (column K) to the log-ratio of the mole fractions of H_2 and H_2O, assuming that the latter ones undergo nil to negligible variations upon cooling from the deep gas equilibration zone to the surface. Note that the R_H values of Solfatara fluids vary from -3.644 to -2.289.

The R_H values can be easily converted into the corresponding oxygen fugacity values, by means of the relation (T in K):

$$\log f_{O_2} = -2 \cdot R_H + 5.224 - 25432/T. \quad (4)$$

Equation (4) is obtained rearranging the log K of the dissociation reaction of water vapour in gaseous H_2 and O_2:

$$2H_2O = 2H_2 + O_2 \quad (5)$$

computed using the SUPCRT92 code (Johnson et al. 1992). However, as underscored by Giggenbach (1987), the commonly used oxygen fugacity has little physical meaning below 600 ° C, because "*actual O_2 contents, in a gas phase consisting essentially of water vapor, drop below one molecule (not mole) of O_2 per kg of vapor at temperatures of between 350° and 450° (*Wones and Gilbert 1969*) or to less than one molecule of oxygen per Gt (1 km^3 of condensed water) at temperatures between 210 and 280 °C*". In contrast, the R_H value is determinable and, therefore, meaningful under these conditions. Indeed, the $\log f_{O_2}$ values of Solfatara fluids range from −37.66 to −17.91 (column M).

According to the theoretical approach of Werner Friedrich Giggenbach (Giggenbach 1987, 1993, 1996, 1997b), two different redox buffers are active in distinct sectors of hydrothermal-magmatic systems. One is the H_2S-SO_2 magmatic gas buffer, which is controlled by the following reaction:

$$H_2S + 3H_2O = SO_2 + 3H_2 + H_2O \quad (6)$$

involving gas species only. Assuming that the fugacities of H_2S and SO_2, the two main S species of magmatic fluids, are equal (thus providing the best buffer capacity), the following relation is obtained from the log K of reaction (6):

$$R_{H,magm} = 1.2638 - \frac{3726}{T} - \frac{1}{3} \cdot \log f_{H_2O}. \quad (7)$$

Equation (7) shows that that the R_H values controlled by the H_2S-SO_2 magmatic redox buffer depend on temperature (in K) and water fugacity (in bar).

The other redox buffer is the (FeO)–($FeO_{1.5}$) hydrothermal rock buffer, which is controlled by the equilibrium between the aqueous fluid and the rocks, according to the following reaction:

$$H_2O + 2(FeO) = H_2 + 2(FeO_{1.5}), \quad (8)$$

in which (FeO) and ($FeO_{1.5}$) represent divalent and trivalent iron incorporated into unspecified oxide or Al-silicate minerals. Fayalite and hematite can be used as proxies of (FeO) and ($FeO_{1.5}$), respectively. If so, reaction (8) can be rewritten as follows:

$$H_2O + Fe_2SiO_4 = H_2 + Fe_2O_3 + SiO_2. \quad (9)$$

Again, the optimum buffer capacity is achieved for $a_{(FeO_{1.5})}/a_{(FeO)} = 1$ or $a_{(Fe_2O_3)}/a_{(Fe_2SiO_4)} = 1$. According to Giggenbach (1987), the log K of reaction (8) is practically temperature independent and, therefore, the (FeO)–($FeO_{1.5}$) hydrothermal redox buffer fixes the R_H at −2.82 ± 0.02 at all temperatures. Actually, the log K values of the fayalite-hematite-quartz (FHQ) reaction calculated by means of the SUPCRT92 code (Table 7) deviate by 0.1−0.4 log-units from the reference R_H value of −2.82 in the temperature range 200–1200 °C. These small differences are probably due to the use of different thermodynamic data for computing the log K of the FHQ reaction.

Table 7 Log K values of the FHQ reaction (9)[1]

T (°C)	100	200	300	400	500	600	700	800	900	1000	1100	1200
Log (H_2/H_2O)	−2.02	−2.39	−2.59	−2.69	−2.73	−2.73	−2.70	−2.65	−2.61	−2.58	−2.55	−2.53

The R_H values controlled by the H_2S-SO_2 magmatic gas buffer and the (FeO)-($FeO_{1.5}$) hydrothermal rock buffer are shown in Fig. 1 in the 200–1200 °C temperatures range. The FHQ buffer is also shown to appreciate the uncertainty on the rock buffer. Different lines are drawn for the magmatic gas buffer. One (labelled 'brine') refers to H_2O fugacities fixed by boiling aqueous solutions of increasingly high salinities with increasing temperatures, as defined by Eq. (134) in chapter "The Expansion (Decompression) of the Solfatara Fumarolic Fluids", which is valid up to 600 °C approximately. The other lines correspond to different H_2O fugacities, as specified. To be noted that the two buffers overlap at temperatures of 650 °C or higher, depending of the f_{H_2O} value of the magmatic gas buffer. Consequently, it is not possible to understand which buffer controls the redox potential of the fluid at these conditions of high temperature and variable f_{H_2O}.

Also shown in Fig. 1 are the R_H and temperature values of melt inclusions in volcanic rocks erupted by the Stromboli, Vesuvius, Vulcano and Etna volcanoes (Scaillet and Pichavant 2005 and references therein). The R_H and temperature values of the melt inclusions from the active Italian volcanoes are consistent with the R_H and temperature values governed by the H_2S-SO_2 magmatic gas buffer, allowing for H_2O fugacities higher than 1 kbar or for SO_2/H_2S fugacity ratios different from unity. Actually, the $\log(X_{SO_2}/X_{H_2S})$ ratio of the 42 melt inclusion data reported by Scaillet and Pichavant (2005) varies between a minimum of -0.78 and a maximum of 3.62, with a mean of 1.11, a median of 1.20 and a standard deviation of 0.82, whereas the SO_2/H_2S mole fraction log-ratio is close to zero only in two cases, namely, 0.00 and -0.05, both from Mt. Etna. Thus, the equality of the fugacities of SO_2 and H_2S is a useful reference condition, although it appears to be rarely present in the melt inclusions from the active Italian volcanoes.

Furthermore, the SO_2/H_2S fugacity ratio of volcanic gases depends on temperature, f_{H_2O}, and redox conditions, as dictated by the log K of reaction (6), with H_2S stabilized by decreasing temperatures and increasing pressures and SO_2 becoming stable at increasing temperatures and decreasing pressures. In addition, the SO_2/H_2S fugacity ratio is also controlled by the extent of magma degassing, owing to the different solubilities of H_2S and SO_2 in silicate melts (Marini et al. 2011 and references therein).

Returning to consider Fig. 1, due to the temperature dependence of the magmatic gas buffer, magmatic gases are expected to become progressively oxidizing upon cooling, if they follow any intermediate pathway between the two limiting lines for f_{H_2O} fixed by a boiling brine and f_{H_2O} of 1 bar. During the whole cooling path of magmatic gases, the sulfur redox pair may remain active or not, depending on the extent of the scrubbing process (Symonds et al. 2001) through either (i) interaction of magmatic gases with external (non-magmatic) waters or (ii) steam condensation at depth, possibly due to the close-to-isenthalpic expansion of the fluids of interest or both (Caliro et al. 2007; Moretti et al. 2013; Aiuppa et al. 2013; Chiodini et al. 2012, 2021).

In the environments present above a shallow degassing magma batch, where dry (superheated) vapors are presumably present, the magmatic gas buffer remains active until the surface. This is the case of the central magmatic core of the so-called vapor-cored magmatic-hydrothermal systems

[1] The log K values of the FHQ reaction (9) were computed at total pressure of 1 bar, but they are valid at any pressure since the log K of this reaction is pressure independent.

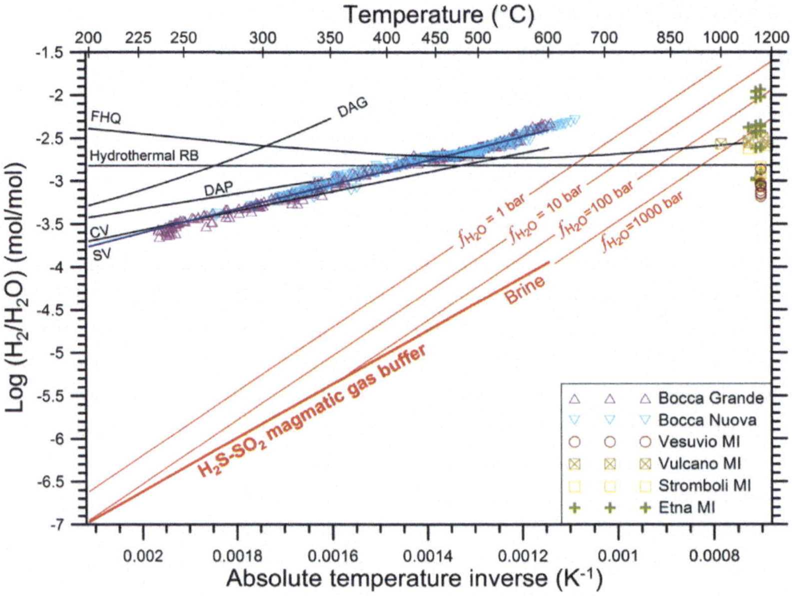

Fig. 1 Correlation plot showing the R_H values of the Solfatara fluids versus the reciprocal of their SS4 temperature (in K) for gas equilibration in a single saturated vapor phase coexisting with a brine containing 21 wt% NaCl (see Sects. 3 and 6.2 in chapter "Calibration and Application of the RWG and SS4 Geoindicators for the Possible Expansion Paths of Solfatara Fluids"). Also shown are the melt inclusion data of active Italian volcanoes (from Scaillet and Pichavant 2005) and different redox buffers active in hydrothermal-magmatic environments Reprinted from Giggenbach (1987), modified,

(Reyes et al. 1993) or "immature" geothermal systems related to active strato-volcanoes (Bogie et al. 2005). Examples are White Island (Giggenbach 1987) and Vulcano Island (Chiodini et al. 1993), as well as several systems of Indonesia and the Philippines (Abiyudo et al. 2016 and references therein), as further discussed in chapter "A Comparison Between the Solfatara and Other Magmatic-hydrothermal Systems".

In contrast, equilibrium with the hydrothermal rock buffer is approached and possibly achieved, where magmatic gases enter a hydrothermal system from below, are completely absorbed into the deep circulating aqueous solutions, and these aqueous solutions are then neutralized through extensive water–rock interaction. This is the case of the numerous hydrothermal-magmatic systems hosting neutral sodium-chloride aqueous solutions (as is the case of the examples mentioned in chapter "A Comparison Between the Solfatara and Other Magmatic-hydrothermal Systems"). Moreover, the hydrothermal rock buffer may be active in the hydrothermal envelope surrounding the central magmatic core in the vapor-cored hydrothermal systems.

Sulfide and sulfate aqueous species and minerals are expected to coexist in the zone where magmatic gases are absorbed in deep waters because sulfur dioxide is involved into a disproportionation reaction with generation of H_2S and sulfate species, comprising SO_4^{2-}, HSO_4^- and H_2SO_4 in variable amounts depending on pH (Murray and Cubicciotti, 1983). Assuming that bisulfate ion is the main sulfate species, the SO_2 disproportionation reaction can be written as:

$$4SO_{2(aq)} + 4H_2O_{(L)} \rightarrow H_2S_{(aq)} + 3HSO^-_{4(aq)} + 3H^+_{(aq)}. \tag{10}$$

Due to reaction (10), redox conditions are governed by coexisting aqueous sulfide and sulfate species instead of dissolved SO_2, whose concentration is decidedly small (see Sect. 3.3).

Before the monumental study of Giggenbach (1987), other hydrothermal rock buffers, also shown in Fig. 1, were previously proposed by D'Amore and Panichi (1980; acronym DAP in Fig. 1):

$$R_{H,DAP} = -885.5/T - 1.552, \quad (11)$$

and D'Amore and Gianelli (1984; acronym DAG in Fig. 1):

$$R_{H,DAG} = 1,037,441/T^2 - 5853/T + 4.452. \quad (12)$$

More recently, Chiodini and Marini (1998) calibrated the redox buffer of the hydrothermal systems associated with the Campanian volcanoes Solfatara, Ischia and Vesuvio (acronym CV in Fig. 1):

$$R_{H,CV} = -1122.5/T - 1.327 \quad (13)$$

using the data acquired in 1983–1996 for the Solfatara system, as already recalled in Sect. 1 in chapter "The geoindicators involving H_2O, CO_2, CO, CH_4 and H_2". Although the equilibrium temperatures of the fumarolic fluids used to calibrate Eq. (13) are lower than 500 °C, this relation is extrapolated up to 600 °C in Fig. 1 to compare it with all the Solfatara data acquired by Chiodini and coworkers (see Chiodini et al. 2021 and references therein). It turns out that the Solfatara fluids of temperature up to 300–350 °C approximately are positioned close to the redox buffer of the hydrothermal systems associated with the Campanian volcanoes, whereas the Solfatara fluids of higher temperature are found more and more above this buffer line.

Since the hydrothermal redox buffer of Campanian volcanoes, that is Eq. (13), was calibrated by Chiodini and Marini (1998) considering the log-ratio of the mole fractions of H_2 and H_2O, but neglecting the log-ratio of the fugacity coefficients of H_2 and H_2O, the hydrothermal redox buffer was recalibrated here, based on all the data available for the Solfatara system, suitably corrected for the deviations from ideality, as expressed by Eqs. (2) and (3), but disregarding the data of Ischia and Vesuvio. In this way, the following relation, which is representative of the hydrothermal redox buffer of Solfatara volcano (acronym SV in Fig. 1), was obtained ($N = 664$; $R^2 = 0.9848$; T in K):

$$R_{H,SV} = -1404.7/T - 0.7904. \quad (14)$$

Equation (14) has slope higher and intercept lower than those of Eq. (13) and fits reasonably well all the Solfatara data, with absolute deviations of 0.029 log-units, on average, a minimum deviation of 0.000048 log-units, and a maximum deviation of 0.116 log-units. Thus, Eq. (14) allows one to estimate the redox conditions in the "hydrothermal" sector of the Solfatara magmatic-hydrothermal system, up to temperatures of 600 °C approximately, although it must be kept in mind that it is constrained by the adoption of the SS4 equilibrium temperature (for gas equilibration in a single saturated vapor phase coexisting with a brine containing 21 wt% NaCl) in its calibration. This discussion will be resumed in Sect. 4.2.

3 The Sulfur-Bearing Gas Species

3.1 Hydrogen Sulfide

According to Giggenbach (1980), hydrothermal gases are characterized by the presence of relatively high concentrations of hydrogen sulfide, which are usually lower than those of carbon dioxide and higher than those of other non-condensable gaseous species, whereas other S-bearing gases are present in undetectable amounts. From this point of view, the Solfatara fumarolic fluids can be considered typical hydrothermal gases having comparatively high H_2S concentrations and undetectable concentrations of other S-bearing gases (see Sects. 3.2, 3.3 and 3.4 for further details). Hydrogen sulfide concentrations of Solfatara fluids vary as follows, based on the data of Chiodini and coworkers (Chiodini et al. 2021 and references therein):

(i) from 960 to 3988 μmol/mol with a mean of 1520 μmol/mol, a median of 1420 μmol/mol, and a standard deviation of 459 μmol/mol at Bocca Grande;

(ii) from 602 to 5014 μmol/mol with a mean of 1150 μmol/mol, a median of 1060 μmol/mol, and a standard deviation of 401 μmol/mol at Bocca Nuova;
(iii) from 467 to 909 μmol/mol with a mean of 599 μmol/mol, a median of 591 μmol/mol, and a standard deviation of 66 μmol/mol at Pisciarelli.

The time changes of H_2S and the relations between H_2S and other gas species are discussed in Sects. 1, 2, and 3 in chapter "Chemical and Isotopic Characteristics of Solfatara-Pisciarelli Fumarolic Fluids", whereas the isotopic composition of H_2S is considered in Sect. 8 in chapter "Chemical and Isotopic Characteristics of Solfatara-Pisciarelli Fumarolic Fluids".

3.2 Sulfur Fugacity

Hydrogen provides a fundamental contribution to the geoindicators discussed in chapters "The Geoindicators Involving H_2O, CO_2, CO, CH_4 and H_2" and "Calibration and Application of the RWG and SS4 Geoindicators for the Possible Expansion Paths of Solfatara Fluids", is implied in the definition of the R_H parameter (see Sect. 2), and is also involved, together with H_2S, in the following reaction:

$$2H_2 + S_2 = 2H_2S. \quad (15)$$

Therefore, the mole fractions of H_2 and H_2S of Solfatara fluids can be used to calculate sulfur fugacity, at any temperature of interest, under the hypothesis of chemical equilibrium among gas species. In fact, rearranging the log K of reaction (15), $\log K_{S_2}$, it is possible to the obtain the relation:

$$\log f_{S_2} = 2 \cdot \log\left(\frac{X_{H_2S}}{X_{H_2}}\right) + 2 \cdot \log\left(\frac{\varphi_{H_2S}}{\varphi_{H_2}}\right) - \log K_{S_2}. \quad (16)$$

The value of the term $2 \cdot \log(\varphi_{H_2S}/\varphi_{H_2})$ varies from −0.014 to −0.219 at X_{CO_2} of 0.05 and from -0.026 to -0.273 at X_{CO_2} of 0.40 and attains the minimum value, corresponding to the maximum deviation from zero, at 400–425 °C. Accepting this uncertainty on computed sulfur fugacities, the term $2 \cdot \log(\varphi_{H_2S}/\varphi_{H_2})$ was disregarded in Eq. (16). On the basis of SUPCRTBL outcomes, the temperature dependence of the $\log K_{S_2}$, at water saturation pressures, is expressed by the following relation (T in K):

$$\log K_{S_2} = -111{,}660/T^2 + 9501.7/T - 5.2385. \quad (17)$$

Based on the analytical mole fractions of H_2 and H_2S of Solfatara fluids, Eqs. (16) and (17) were used to compute sulfur fugacities at both (i) the outlet temperature and (ii) the SS4 temperature, for gas equilibration in a single saturated vapor phase coexisting with a brine containing 21 wt% NaCl (see Sect. 6.2 in chapter "Calibration and Application of the RWG and SS4 Geoindicators for the Possible Expansion Paths of Solfatara Fluids"). The log K values of the reaction:

$$2S_{(s,l)} = S_2 \quad (18)$$

were also calculated by running SUPCRTBL, to obtain the temperature dependence (at water saturation pressures) of sulfur fugacity at saturation with pure elemental sulfur:

$$\log f_{S_2,sat} = -30{,}782{,}980/T^3 + 154{,}736/T^2 - 6838.1/T + 8.5142. \quad (19)$$

It turns out that sulfur fugacity, at both the outlet temperature and the SS4 temperature, is lower than the sulfur fugacity at saturation with elemental sulfur by several orders of magnitude, as shown by the two plots of Fig. 2.

Therefore, the deposition of elemental sulfur (either as liquid above 115.2 °C or as monoclinic solid between 115.2 and 95.2 °C or as orthorhombic solid below 95.2 °C) is not expected to occur neither at the SS4 temperature, nor at the outlet temperature, not at any intermediate temperature along the plumbing system of Solfatara fumarolic fluids. Accepting these findings, it is permissible to use the H_2S

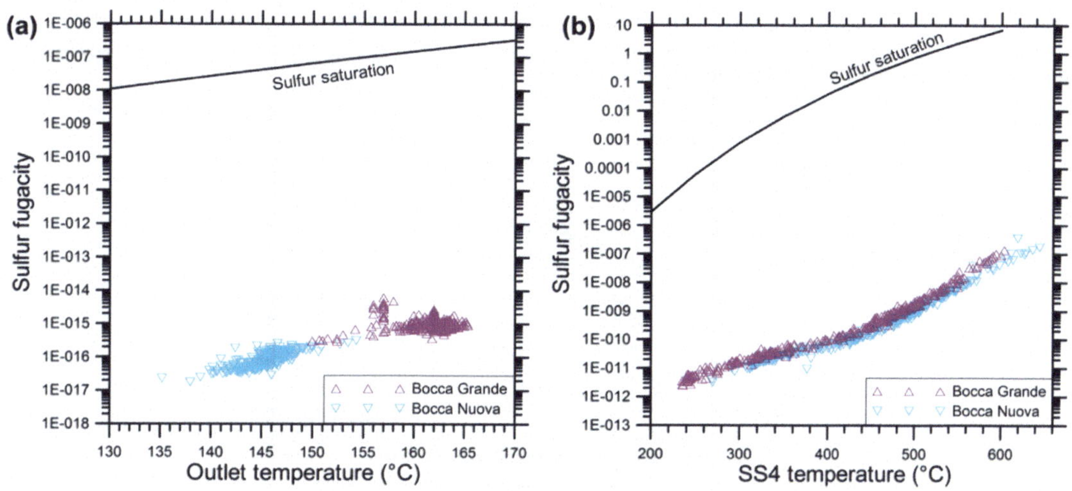

Fig. 2 The sulfur fugacity of Solfatara fluids, calculated from the analytical mole fractions of H_2S and H_2 (symbols, see legend) at both **a** the outlet temperature and **b** the SS4 temperature, for gas equilibration in a single saturated vapor phase coexisting with a brine containing 21 wt% NaCl, is compared with the sulfur fugacity at saturation with pure elemental sulfur (solid line)

concentration of fumarolic fluids and the parameters derived from it, such as the sulfur fugacity calculated from the analytical mole fractions of H_2S and H_2 (see above), to investigate the characteristics of the Solfatara hydrothermal-magmatic system at depth.

The diagram of Fig. 3 shows that the sulfur fugacity of the Solfatara fluids at the SS4 temperature varies from $2.5 \cdot 10^{-12}$ to $1.4 \cdot 10^{-7}$ bar with a mean of $4.3 \cdot 10^{-9}$ bar, a median of $1.8 \cdot 10^{-10}$ bar and a standard deviation of $1.5 \cdot 10^{-8}$ bar for Bocca Grande and from $3.2 \cdot 10^{-12}$ to $3.6 \cdot 10^{-7}$ bar with a mean of $6.2 \cdot 10^{-9}$ bar, a median of $3.4 \cdot 10^{-10}$ bar and a standard deviation of $2.8 \cdot 10^{-8}$ bar for Bocca Nuova.

Moreover, the computed sulfur fugacity is strongly correlated with the SS4 temperature itself, as expected. Also shown in Fig. 3 is the sulfur fugacity fixed by the hydrothermal buffer of D'Amore and Gianelli (1984), which is defined by the relation (T in K):

$$\log f_{S_2,DAG} = -18.67928 + 14.72771 \cdot (1000/T) - 6.139625 \cdot (1000/T)^2. \quad (20)$$

Below 260 °C, the sulfur fugacity of the Solfatara fluids at the SS4 temperature is greater than the sulfur fugacity fixed by the hydrothermal buffer of D'Amore and Gianelli (1984) by more than one order of magnitude, but discrepancies decrease with increasing temperature. Equation (20) was calibrated empirically, based on the chemical composition of the fluids discharged from deep wells of the Larderello, Travale, Bagnore, Piancastagnaio, The Geysers, and Cerro Prieto geothermal fields and related reservoir temperatures up to a maximum value of 350°C.

The sulfur fugacities and temperatures given by Eq. (20) are consistent with the occurrence of pyrite or both pyrite and pyrrhotite in the geothermal reservoir. As underscored by D'Amore and Gianelli (1984), the equilibrium coexistence of pyrite-pyrrhotite does not seem to buffer sulfur fugacity below 250 °C, whereas it closely reproduces the sulfur fugacities encountered at The Geysers and Cerro Prieto above 250 °C, consistent with the occurrence of both pyrite and pyrrhotite in the deepest zones of these two geothermal fields. As shown by the chronogram of Fig. 4, the sulfur fugacity at the

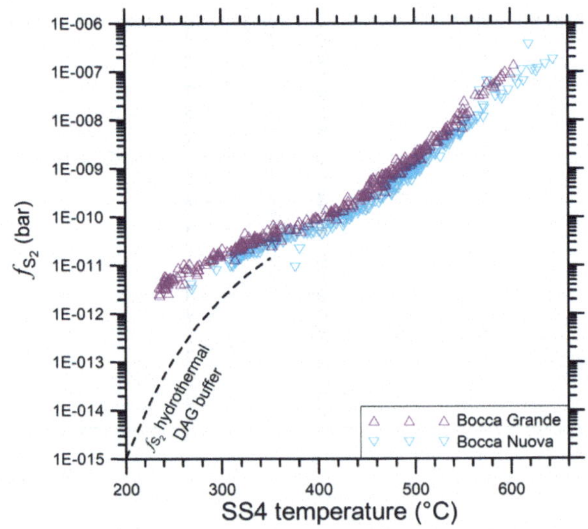

Fig. 3 Diagram of the sulfur fugacity of Solfatara fluids (calculated from the analytical mole fractions of H_2S and H_2) at the SS4 temperature, for gas equilibration in a single saturated vapor phase coexisting with a brine containing 21 wt% NaCl, versus the SS4 temperature (symbols, see legend). Also shown is the sulfur fugacity fixed by the hydrothermal buffer of D'Amore and Gianelli (1984—DAG; dashed line)

SS4 temperature of Bocca Grande and Bocca Nuova fluids increases with time in line with a concurrent increase in the SS4 temperature.

3.3 Sulfur Dioxide Fugacity

Sulfur dioxide is usually the main sulfur-bearing gaseous species in high-temperature magmatic gases, such as those constituting the crater plume

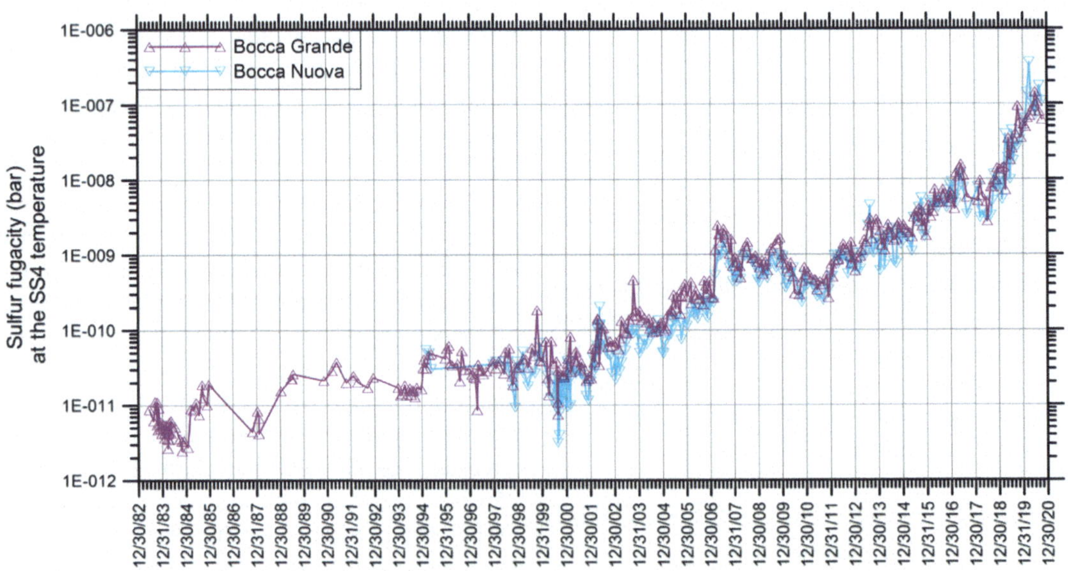

Fig. 4 Chronogram of the sulfur fugacity of Solfatara fluids at the SS4 temperature, calculated from the analytical mole fractions of H_2S and H_2

of active volcanoes and those emitted from the high-temperature fumaroles typically present in the so-called vapor-cored hydrothermal systems (Reyes et al. 1993) or "immature" geothermal systems related to active strato-volcanoes (Bogie et al. 2005), which are also discussed in Sect. 2 and chapter "A Comparison Between the Solfatara and Other Magmatic-hydrothermal Systems". These gaseous mixtures are also characterized by high or relatively high concentrations of HCl, HF, and CO (e.g., Symonds et al. 1994; Fischer and Chiodini 2015 and references therein). As already noted above, SO_2 is present in undetectable concentrations in the hydrothermal gases discharged from the Solfatara fumaroles. Nevertheless, it is worth to calculate the expected SO_2 fugacity in these gas mixtures under the hypothesis of chemical equilibrium at a specified temperature considering the reaction:

$$H_2S + 3H_2O = 3H_2 + SO_2 + H_2O. \quad (21)$$

The decimal logarithm of the thermodynamic equilibrium constant of reaction (21), $\log K_{SO_2}$, depends on temperature according to the relation (T in K):

$$\log K_{SO_2} = 148{,}446/T^2 - 11{,}713/T + 4.3234, \quad (22)$$

based on the results of SUPCRTBL. The following equation is obtained rearranging the $\log K_{SO_2}$ relation:

$$\log f_{SO_2} = \log K_{SO_2} + \log\left(\frac{\varphi_{H_2S}}{\varphi_{H_2O}}\right) - 3\cdot\log\left(\frac{\varphi_{H_2}}{\varphi_{H_2O}}\right) + \log\left(\frac{X_{H_2S}}{X_{H_2O}}\right) - 3\cdot\log\left(\frac{X_{H_2}}{X_{H_2O}}\right). \quad (23)$$

Since the term $\log K_{SO_2} + \log(\varphi_{H_2S}/\varphi_{H_2O}) - 3\cdot\log(\varphi_{H_2}/\varphi_{H_2O})$ of Eq. (23) can be expressed as a function of the temperature and CO_2 mole fraction, it is called $\log K^*_{SO_2}$ and Eq. (23) is rewritten as follows:

$$\log f_{SO_2} = \log K^*_{SO_2} + \log\left(\frac{X_{H_2S}}{X_{H_2O}}\right) - 3\cdot\log\left(\frac{X_{H_2}}{X_{H_2O}}\right). \quad (24)$$

Using the Excel spreadsheet Chapter 9.XLS, with the reciprocal of the absolute SS4 temperature in column G and X_{CO_2} in column H, the equation to compute $\log K^*_{SO_2}$ can be written as follows for the first sample with the reciprocal of the absolute SS4 temperature in cell G2 and X_{CO_2} in cell H2:

$$\begin{aligned}\log K^*_{SO_2} = {} & (174438200000000{}^*H2^\wedge4 - 218759700000000{}^*H2^\wedge3 + 109711400000000{}^*H2^\wedge2 - 34205710000000{}^*H2 + 8533075000000){}^*G2^\wedge4 \\ & + (-1288330000000{}^*H2^\wedge4 + 1611175000000{}^*H2^\wedge3 - 807025200000{}^*H2^\wedge2 + 249352200000{}^*H2 - 63536640000){}^*G2^\wedge3 \\ & + (3499785000{}^*H2^\wedge4 - 4362451000{}^*H2^\wedge3 + 2182018000{}^*H2^\wedge2 - 666859900{}^*H2 + 173521600){}^*G2^\wedge2 \\ & + (-4129481{}^*H2^\wedge4 + 5127563{}^*H2^\wedge3 - 2560137{}^*H2^\wedge2 + 772015{}^*H2 - 215430.2){}^*G2 \\ & + (1776.094{}^*H2^\wedge4 - 2195.816{}^*H2^\wedge3 + 1093.564{}^*H2^\wedge2 - 324.4112{}^*H2 + 89.88207)\end{aligned} \quad (25)$$

The equal sign and the right-hand side of Eq. (25) are written in cell N2 of the Excel spreadsheet to compute the $\log K^*_{SO_2}$ for the first sample. Then, the equation is copied into the adjacent cells of column N dragging the fill handle down, thus computing the $\log K^*_{SO_2}$ for all the considered samples. The mean absolute error on the $\log K^*_{SO_2}$ calculated by means of Eq. (25) is

0.019 ± 0.014 (1σ) log-units, the minimum absolute error is 0.000092 log-units, whereas the maximum absolute error is 0.062 log-units. Strictly speaking, Eq. (25) can be utilized up to 600 °C, but it can be probably extrapolated at somewhat higher temperatures accepting larger errors.

Equations (24) and (25) were used to calculate the SO_2 fugacity of Solfatara fluids at the SS4 temperature (for gas equilibration in a single saturated vapor phase coexisting with a brine containing 21 wt% NaCl; see Sect. 6.2 in chapter "Calibration and Application of the RWG and SS4 Geoindicators for the Possible Expansion Paths of Solfatara Fluids") based on the analytical mole fraction ratios H_2/H_2O and H_2S/H_2O. The computed SO_2 fugacity at the SS4 temperature (column O) is highly correlated with the SS4 temperature itself (Fig. 5) and varies from $7.9 \cdot 10^{-11}$ to $2.0 \cdot 10^{-4}$ bar with a mean of $1.0 \cdot 10^{-5}$ bar, a median of $7.2 \cdot 10^{-7}$ bar, and a standard deviation of $2.5 \cdot 10^{-5}$ bar for Bocca Grande and from $3.3 \cdot 10^{-10}$ to $5.6 \cdot 10^{-4}$ bar with a mean of $2.1 \cdot 10^{-5}$ bar, a median of $3.0 \cdot 10^{-6}$ bar, and a standard deviation of $6.7 \cdot 10^{-5}$ bar for Bocca Nuova.

The low computed SO_2 fugacities of Solfatara fluids at the SS4 temperature might be controlled by the occurrence of the SO_2 disproportionation reaction (10), in a zone of the system upstream of that whose temperature is indicated by the SS4 reaction, with production of HSO_4^- ion, H_2S and H^+, as anticipated in Sect. 2. Accepting that most of the SO_2 presumably present in deep magmatic gases is absorbed by interaction with deep brines, the low computed SO_2 fugacities at the SS4 temperature suggest the presence of such deep brines in a deeper and hotter environment.

To be noted that the computed SO_2 fugacity of Solfatara fluids at the SS4 temperature increases with time, in line with the concurrent increase in the SS4 temperature, as shown by the chronogram of Fig. 6, which is very similar to the chronogram of sulfur fugacity of Fig. 4. To be noted also that further SO_2 can be removed from Solfatara fluids, during their travel towards the surface discharge points, due to scrubbing, as already mentioned in Sects. 3.2 in chapter "The Magmatic–Hydrothermal System Hosted in the Campi Flegrei Caldera with Emphasis on the Solfatara" and 2.

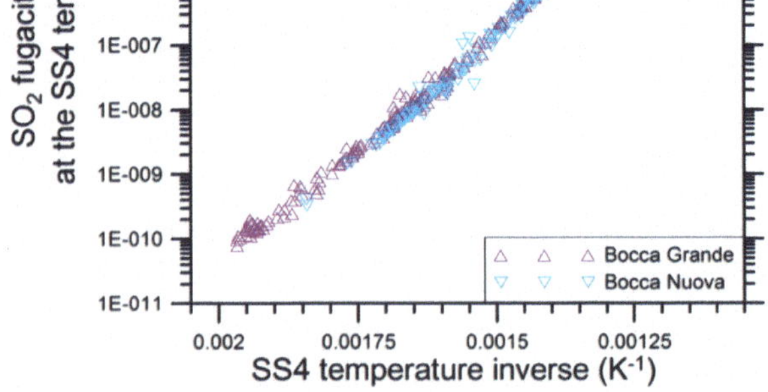

Fig. 5 Diagram of the sulfur dioxide fugacity of Solfatara fluids at the SS4 temperature (calculated from the analytical mole fraction ratios H_2/H_2O and H_2S/H_2O) versus the reciprocal of the SS4 temperature, for gas equilibration in a single saturated vapor phase coexisting with a brine containing 21 wt% NaCl

3.4 Carbonyl Sulfide Concentration

Carbonyl sulfide is a gas species potentially occurring in measurable concentrations in the Solfatara gaseous mixtures, as shown in this section. Unfortunately, COS in hydrothermal gases has received little attention and, so far, it was measured only in the fluids discharged by 12 wells of the Larderello geothermal field (Chiodini et al. 1991), to the best of our knowledge. These analytical COS concentrations vary from 0.005 to 0.1 μmol/mol, as expected for attainment of gas equilibrium in a saturated vapor phase at the temperature and pressure conditions of the Larderello geothermal reservoir.

Since the thermodynamic data of carbonyl sulfide are absent in SUPCRT and SUPCRTBL, the expected concentrations of COS in the Solfatara gas mixtures were calculated referring to the relations derived by Chiodini et al. (1991), who considered the following two reactions:

$$CO_2 + H_2S = COS + H_2O \quad (26)$$

$$CO + H_2S = COS + H_2. \quad (27)$$

The log K of reactions (26) and (27) depend on temperature as defined by the equations (T in K):

$$\log K_{COS-CO_2} = 0.12 - 1575.4/T \quad (28)$$

$$\log K_{COS-CO} = -1.91 + 472.3/T, \quad (29)$$

which were obtained by Chiodini et al. (1991) using the thermodynamic data of Barin and Knacke (1973) and are valid in the temperature range 127 to 327 °C. The log K of reactions (26) and (27), can be rearranged as follows:

$$\log X_{COS} = \log K_{COS-CO_2} - \log(\varphi_{COS}/\varphi_{CO_2}) + \log(\varphi_{H_2S}/\varphi_{H_2O}) + \log(X_{H_2S}/X_{H_2O}) + \log X_{CO_2} \quad (30)$$

$$\log X_{COS} = \log K_{COS-CO} - \log(\varphi_{COS}/\varphi_{CO}) + \log(\varphi_{H_2S}/\varphi_{H_2}) + \log(X_{H_2S}/X_{H_2}) + \log X_{CO}. \quad (31)$$

Since the COS concentration is unknown, it is not possible to compute its fugacity coefficient using the approach of Sects. 5, 7.3, 7.4, 7.5 in

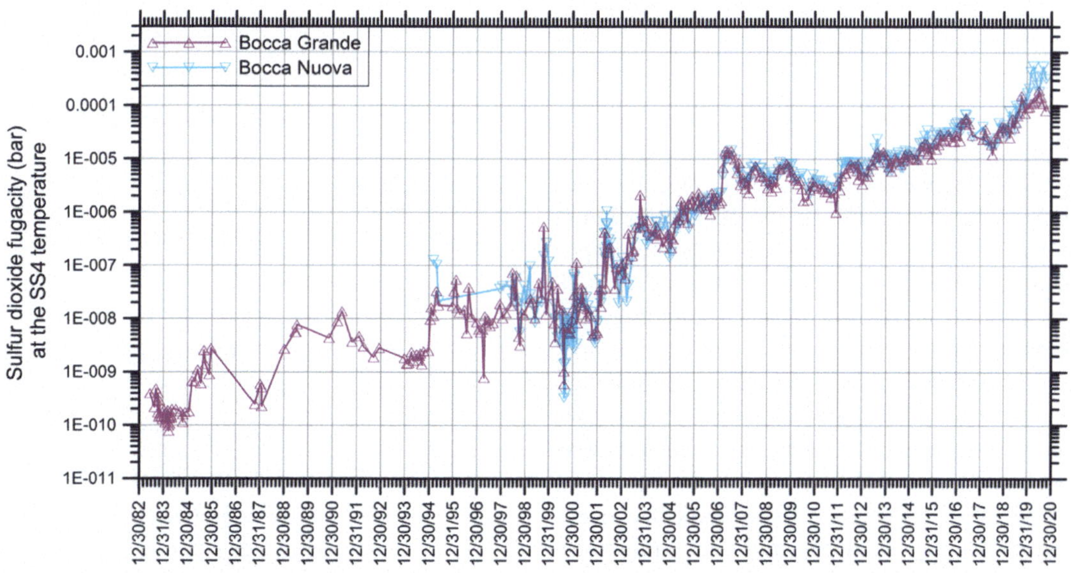

Fig. 6 Chronogram of the SO_2 fugacity of Solfatara fluids at the SS4 temperature, calculated from the analytical mole fractions of H_2S, H_2 and H_2O

chapter "The Expansion (Decompression) of the Solfatara Fumarolic Fluids", and 1. Assuming that the fugacity coefficients of COS, CO, and CO_2 are similar to each other, Eqs. (30) and (31) can be simplified as follows:

$$\log X_{COS} \cong \log K_{COS-CO_2} + \log(\varphi_{H_2S}/\varphi_{H_2O}) + \log(X_{H_2S}/X_{H_2O}) + \log X_{CO_2} \quad (32)$$

$$\log X_{COS} \cong \log K_{COS-CO} + \log(\varphi_{H_2S}/\varphi_{H_2}) + \log(X_{H_2S}/X_{H_2}) + \log X_{CO}. \quad (33)$$

Since the terms $\log K_{COS-CO_2} + \log(\varphi_{H_2S}/\varphi_{H_2O})$ of Eq. (32) and $\log K_{COS-CO} + \log(\varphi_{H_2S}/\varphi_{H_2})$ of Eq. (33) can be expressed as functions of temperature and the CO_2 mole fraction, they are called $\log K^*_{COS-CO_2}$ and $\log K^*_{COS-CO}$, respectively, and Eqs. (32) and (33) are rewritten as follows:

$$\log X_{COS} \cong \log K^*_{COS-CO_2} + \log(X_{H_2S}/X_{H_2O}) + \log X_{CO_2} \quad (34)$$

$$\log X_{COS} \cong \log K^*_{COS-CO} + \log(X_{H_2S}/X_{H_2}) + \log X_{CO}. \quad (35)$$

Thus, provided that $\log K^*_{COS-CO_2}$ and $\log K^*_{COS-CO}$ are calculated, the COS concentration of Solfatara fluids at the SS4 temperature (for gas equilibration in a single saturated vapor phase coexisting with a brine containing 21 wt% NaCl; see Sect. 6.2 in chapter "Calibration and Application of the RWG and SS4 Geoindicators for the Possible Expansion Paths of Solfatara Fluids") can be computed by means of Eqs. (34) taking into account the analytical concentrations of H_2O, CO_2, and H_2S, as well as by Eqs. (35), taking into consideration the analytical concentrations of H_2S and H_2 and the equilibrium concentration of CO at the SS4 temperature instead of the analytical concentration of carbon monoxide due to the fast re-equilibration of this gas species upon cooling (see Sect. 7 in chapter "The Chemical Kinetics of the Reactions Involving H_2O, CO_2, CO, CH_4 and H_2). Owing to the uncertainties related to the calculation of the equilibrium CO concentration and the low enthalpy term of Eq. (29), only Eqs. (34) was used to compute the COS concentration of Solfatara fluids at the SS4 temperature.

Using the Excel spreadsheet Chapter 9.XLS, with the reciprocal of the absolute SS4 temperature in column G and X_{CO_2} in column H, the equation to compute $\log K^*_{COS-CO_2}$ is as follows for the first sample with the reciprocal of the absolute SS4 temperature in cell G2 and X_{CO_2} in cell H2:

$$\begin{aligned}\log K^*_{COS-CO_2} = {} & (-85029570000000^*H2^\wedge 4 + 104237900000000^*H2^\wedge 3 - 52099640000000^*H2^\wedge 2 + 15856810000000^*H2 - 3497329000000)^*G2^\wedge 4 \\ & + (623995600000^*H2^\wedge 4 - 764211000000^*H2^\wedge 3 + 380950200000^*H2^\wedge 2 - 115436200000^*H2 + 25696610000)^*G2^\wedge 3 \\ & + (-1683304000^*H2^\wedge 4 + 2059202000^*H2^\wedge 3 - 1023268000^*H2^\wedge 2 + 308264400^*H2 - 69067950)^*G2^\wedge 2 \\ & + (1970814^*H2^\wedge 4 - 2407723^*H2^\wedge 3 + 1191954^*H2^\wedge 2 - 356184.6^*H2 + 78183.8)^*G2 \\ & + (-840.6462^*H2^\wedge 4 + 1025.338^*H2^\wedge 3 - 505.2078^*H2^\wedge 2 + 149.1864^*H2 - 32.69801)\end{aligned} \quad (36)$$

The equal sign and the right-hand side of Eq. (36) are written in cell P2 of the Excel spreadsheet to compute the $\log K^{*}_{COS-CO_2}$ for the first sample. Then, the relation is copied into the other cells of column P dragging the fill handle down, thus calculating the $\log K^{*}_{COS-CO_2}$ for all the available samples.

The mean absolute error on the $\log K^{*}_{COS-CO_2}$ computed by means of Eq. (36) is 0.0063 ± 0.0051 (1σ) log-units, the minimum absolute error is 0.00028 log-units, whereas the maximum absolute error is 0.022 log-units. Strictly speaking, Eq. (36) can be utilized up to 600 °C, but its use can be probably extended at somewhat higher temperatures accepting larger uncertainties.

The COS concentration of Solfatara fluids at the SS4 temperature turned out to vary from 0.32 to 86.1 μmol/mol with an average of 10.8 μmol/mol, a median of 5.91 μmol/mol and a standard deviation of 14.5 μmol/mol for Bocca Grande and from 0.28 to 106 μmol/mol with an average of 10.3 μmol/mol, a median of 7.05 μmol/mol and a standard deviation of 12.8 μmol/mol for Bocca Nuova (column Q). These values are higher than those measured in the Larderello fluids by Chiodini et al. (1991) and are expected to be measurable, although re-equilibration of COS upon cooling, similar to CO, cannot be excluded.

The diagram of Fig. 7 shows that the computed COS concentration of Solfatara fluids at the SS4 temperature is well correlated with the SS4 temperature itself, whereas the chronogram of Fig. 8 points out that the computed COS concentration at the SS4 temperature increases with time, similar to the calculated sulfur fugacity (Fig. 4) and SO_2 fugacity (Fig. 6).

4 Gas-Solids Equilibria Involving H_2S

As already mentioned in Sect. 2.1, hydrogen sulfide is the only sulfur-bearing gas species for which data are available for the Solfatara fluids. Therefore, it is impossible to investigate the temperatures and pressures at which gaseous H_2S attains or approach equilibrium with other S-bearing gas species by considering homogeneous gas-phase reactions. Instead, it is possible to investigate the temperatures and pressures at which H_2S reaches or comes close to equilibrium with S-bearing solid species, such as pyrite and anhydrite, by taking into account suitable heterogeneous reactions. This approach was already adopted in many studies (e.g., D'Amore and Nuti 1977; D'Amore and Panichi 1980; Giggenbach 1980; Nehring and D'Amore 1984; Arnórsson 1985; Arnórsson and Gunnlaugsson 1985; Stefánsson and Arnórsson 2002).

4.1 Gas-Solids Equilibria Involving Pyrite

The hydrothermal sulfur-bearing mineral most commonly found in drilled geothermal systems is pyrite [Py, FeS_2], as shown by several investigations (e.g., Browne and Ellis 1970; Kristmannsdóttir 1979; Browne 1978; Reyes 1990). For instance, at Broadlands (New Zealand), pyrite is present in the cores from all boreholes in amounts of up to 10% (Browne and Ellis 1970). In the Philippine geothermal systems, pyrite is commonly-occurring and abundant both in the neutral-pH alteration suite and in the acid alteration suite (Reyes 1990). In the Icelandic geothermal fields, pyrite occurs over the whole temperature range, from 100 to 300 °C (Kristmannsdóttir 1979). Similarly, in the wells Mofete 1 and 2, pyrite is stable at almost all depths and temperatures, up to 350 °C (Balducci and Chelini 1992).

Being ubiquitous and abundant, pyrite is the most obvious partner of H_2S. In pyrite, the oxidation state of the Fe atom is +2 and that of each S atom is −1. The adoption of pyrite requires the involvement of a Fe-bearing mineral in the considered heterogeneous reactions to balance the Fe atom of pyrite. Possible choices are pyrrhotite [Po, FeS], fayalite [Fa, Fe_2SiO_4],[2] magnetite [Mt,

[2] Fayalite, in turn, calls for the involvement of a silica mineral, usually quartz [Qz, SiO_2].

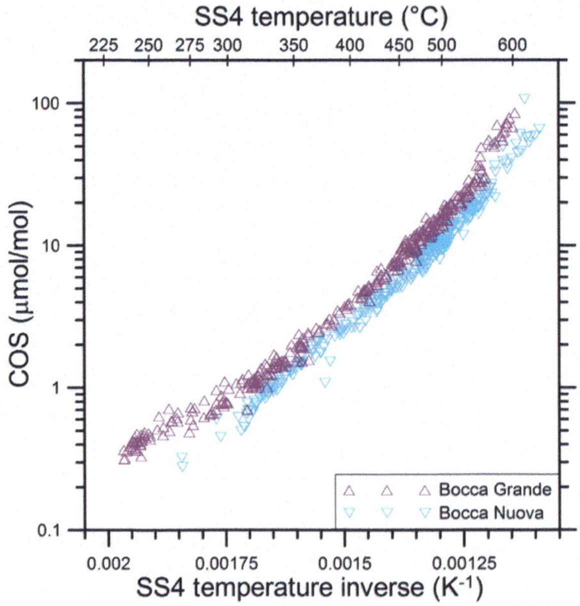

Fig. 7 Diagram of the COS concentration of Solfatara fluids, at the SS4 temperature (calculated from the analytical mole fractions of H_2S, CO_2 and H_2O, versus the reciprocal of the SS4 temperature (for gas equilibration in a single saturated vapor phase coexisting with a brine containing 21 wt% NaCl)

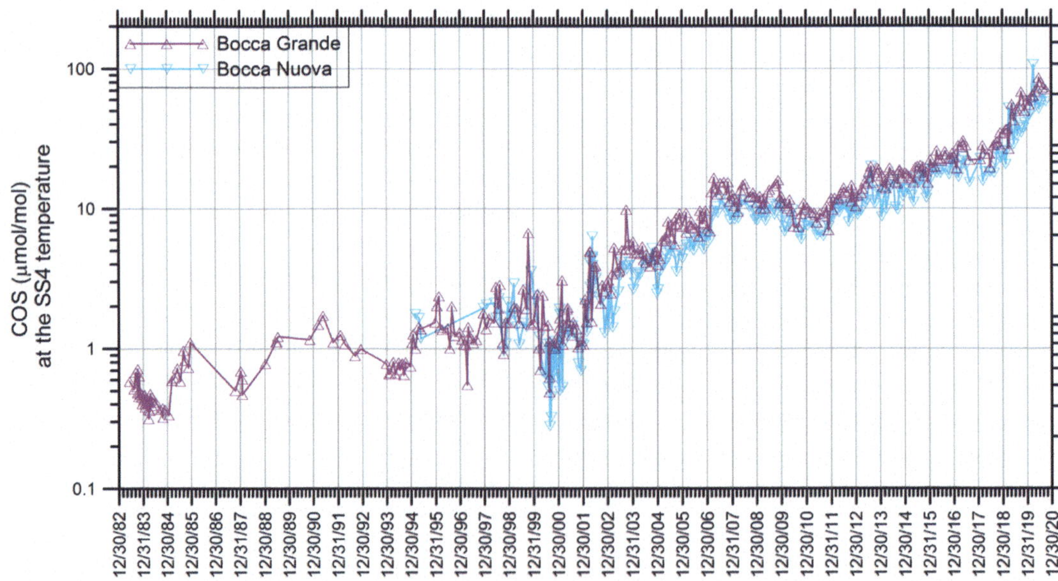

Fig. 8 Chronogram of the COS concentration of Solfatara fluids at the SS4 temperature, calculated from the analytical mole fractions of H_2S, CO_2 and H_2O

Fe_3O_4], and hematite [Hm, Fe_2O_3]. The use of these simple minerals instead of the more complex silicates and Al-silicates occurring as hydrothermal minerals in geothermal systems is an acceptable practice because they are components of the well-known redox buffers (e.g., quartz-fayalite-magnetite, hematite-magnetite, and pyrite-pyrrhotite) which are commonly utilized since 1957, when their usefulness in the experimental and theoretical studies of heterogeneous redox reactions at high pressures and temperatures was underscored by Hans Eugster (1957). Actually, these solid phases, with the possible exception of fayalite, occur as hydrothermal minerals in geothermal systems, including Mofete, as briefly recalled in the following discussion.

At Broadlands, pyrrhotite is much less abundant than pyrite, being present only in half of the boreholes studied by Browne and Ellis (1970), usually in fine-grained sediments containing organic matter, under relatively low P_{H_2S}/P_{H_2} conditions typical of impermeable zones, in poor steam producers, and at temperatures higher than ~150 °C. Pyrrhotite occurs only in the neutral-pH suite of some Philippine geothermal systems but sparsely (Reyes 1990). In the geothermal fields of Iceland, pyrrhotite is less common than pyrite and is absent in the aquifer zones hosting high-salinity waters (Kristmannsdóttir 1979). In wells Mofete 1 and 2, pyrrhotite is present only in the propylitic and thermometamorphic zones, at temperatures of 250–350 °C (Balducci and Chelini 1992). Pyrrhotite is expected to be stable under relatively reducing conditions, since the oxidation state of its Fe atom is + 2 (as in pyrite), while the oxidation state of its S atom is −2 (against −1 in pyrite).

Since the oxidation state of the two Fe atoms in fayalite is +2, fayalite is expected to be stable under conditions somewhat more oxidizing than those of pyrrhotite. Magnetite occurs both in the neutral-pH suite and in the contact metamorphic aureole of some Philippine geothermal systems but sparsely (Reyes 1990). Magnetite is stable under relatively oxidizing conditions, since one of its Fe atoms has oxidation state + 2 and the other two Fe atoms have oxidation state +3. In the Philippine geothermal systems, hematite is commonly-occurring and abundant in the neutral-pH suite only (Reyes 1990). Hematite requires conditions even more oxidizing than those stabilizing magnetite because its two Fe atoms have oxidation state + 3. At Darajat, magnetite is more common than hematite, being present in all wells studied by Herdianita (2012), including DRJ 13 and DRJ 17, below 1250 m. However, hematite is associated with magnetite, at about 1250 m depth in well DRJ S-2, and, hematite becomes dominant, at 900–1000 m depth in wells DRJ S-1 and S-3, suggesting the presence of more oxidizing conditions with increasing depth.

Let us take into consideration the following pyrite-involving reactions, which are listed in order of increasing redox potentials:

$$FeS_2 + H_2 = FeS + H_2S \tag{37}$$

$$FeS_2 + 0.5SiO_2 + H_2 + H_2O = 0.5Fe_2SiO_4 + 2H_2S \tag{38}$$

$$1.5FeS_2 + H_2 + 2H_2O = 0.5Fe_3O_4 + 3H_2S \tag{39}$$

$$2FeS_2 + H_2 + 3H_2O = Fe_2O_3 + 4H_2S. \tag{40}$$

The corresponding log K are pressure-independent and, if all solid phases are assumed to be pure, they are described by the following temperature functions (T in K), which were obtained by fitting the log K values computed by SUPCRT92:

$$\log K_{Py-Po} = \log\left(\frac{f_{H_2S}}{f_{H_2}}\right) = -2683.8 / T + 4.4362 \tag{41}$$

$$\log K_{Py-Fa-Qz} = \log\left(\frac{f_{H_2S}}{f_{H_2}}\right) + \log\left(\frac{f_{H_2S}}{f_{H_2O}}\right) = -4569.6 / T + 4.6247 \tag{42}$$

$$\log K_{Py-Mt} = \log\left(\frac{f_{H_2S}}{f_{H_2}}\right) + 2 \cdot \log\left(\frac{f_{H_2S}}{f_{H_2O}}\right) = -6392.3/T + 5.2635 \tag{43}$$

$$\log \mathrm{K}_{\mathrm{Py-Hm}} = \log\left(\frac{f_{\mathrm{H_2S}}}{f_{\mathrm{H_2}}}\right) + 3 \cdot \log\left(\frac{f_{\mathrm{H_2S}}}{f_{\mathrm{H_2O}}}\right) = -8434.4 / \mathrm{T} + 5.5003 \quad (44)$$

It is convenient to rearrange Eqs. (41) to (44) as was done in Sect. 1 in chapter "Calibration and Application of the RWG and SS4 Geoindicators for the Possible Expansion Paths of Solfatara Fluids", thus obtaining the following relations:

$$\log\left(\frac{\mathrm{X_{H_2S}}}{\mathrm{X_{H_2}}}\right) = \log \mathrm{K}_{\mathrm{Py-Po}} - \log\left(\frac{\varphi_{\mathrm{H_2S}}}{\varphi_{\mathrm{H_2}}}\right) \quad (45)$$

$$\log\left(\frac{\mathrm{X_{H_2S}}}{\mathrm{X_{H_2}}}\right) + \log\left(\frac{\mathrm{X_{H_2S}}}{\mathrm{X_{H_2O}}}\right) = \log \mathrm{K}_{\mathrm{Py-Fa-Qz}} - \log\left(\frac{\varphi_{\mathrm{H_2S}}}{\varphi_{\mathrm{H_2}}}\right) - \log\left(\frac{\varphi_{\mathrm{H_2S}}}{\varphi_{\mathrm{H_2O}}}\right) \quad (46)$$

$$\log\left(\frac{\mathrm{X_{H_2S}}}{\mathrm{X_{H_2}}}\right) + 2 \cdot \log\left(\frac{\mathrm{X_{H_2S}}}{\mathrm{X_{H_2O}}}\right) = \log \mathrm{K}_{\mathrm{Py-Mt}} - \log\left(\frac{\varphi_{\mathrm{H_2S}}}{\varphi_{\mathrm{H_2}}}\right) - 2 \cdot \log\left(\frac{\varphi_{\mathrm{H_2S}}}{\varphi_{\mathrm{H_2O}}}\right) \quad (47)$$

$$\log\left(\frac{\mathrm{X_{H_2S}}}{\mathrm{X_{H_2}}}\right) + 3 \cdot \log\left(\frac{\mathrm{X_{H_2S}}}{\mathrm{X_{H_2O}}}\right) = \log \mathrm{K}_{\mathrm{Py-Hm}} - \log\left(\frac{\varphi_{\mathrm{H_2S}}}{\varphi_{\mathrm{H_2}}}\right) - 3 \cdot \log\left(\frac{\varphi_{\mathrm{H_2S}}}{\varphi_{\mathrm{H_2O}}}\right). \quad (48)$$

In Eqs. (45) to (48), the left-hand side comprises one or two mole fraction ratios, which are obtained from the analysis of the Solfatara fluids, whereas the right-hand side includes a log K, which is a simple temperature function (see above) and one or more ratios of fugacity coefficients, which can be expressed as function of the temperature and CO_2 mole fraction.

Using the Excel spreadsheet Chapter 9.XLS, with the values of $\log(\mathrm{X_{H_2S}}/\mathrm{X_{H_2}})$ in column R and $\mathrm{X_{CO_2}}$ in column H, the equation to compute the pyrite-pyrrhotite equilibrium temperature, $\mathrm{T_{Py-Po}}(^\circ\mathrm{C})$, is as follows for the first sample with the value of $\log(\mathrm{X_{H_2S}}/\mathrm{X_{H_2}})$ in cell R2 and $\mathrm{X_{CO_2}}$ in cell H2:

$$\begin{aligned} T_{Py-Po}(^\circ C) = {} & 1/((-0.00007640952^{*}H2^{\wedge}4 \\ & +0.00007705252^{*}H2^{\wedge}3 \\ & -0.00004567308^{*}H2^{\wedge}2 \\ & +0.000007522511^{*}\mathrm{H2} \\ & -0.000008866179)^{*}\mathrm{R2}^{\wedge}3+ \\ & (-0.0001182057^{*}\mathrm{H2}^{\wedge}4 \\ & +0.00009858571^{*}\mathrm{H2}^{\wedge}3 \\ & -0.000063468^{*}\mathrm{H2}^{\wedge}2 \\ & +0.000006258986^{*}\mathrm{H2} \\ & -0.00001098045)^{*}\mathrm{R2}^{\wedge}2 \\ & +(0.0000897619^{*}\mathrm{H2}^{\wedge}4 \\ & -0.0001135452^{*}\mathrm{H2}^{\wedge}3 \\ & +0.00006698667^{*}\mathrm{H2}^{\wedge}2 \\ & -0.000008401262 * \mathrm{H2} \\ & -0.0003500853)^{*}\mathrm{R2} \\ & +(0.0001271429^{*}\mathrm{H2}^{\wedge}4 \\ & -0.00009230952^{*}\mathrm{H2}^{\wedge}3 \\ & +0.00007365^{*}\mathrm{H2}^{\wedge}2 \\ & +0.000009849524^{*}H2 \\ & +0.001681375)) - 273.15 \end{aligned} \quad (49)$$

The equal sign and the right-hand side of Eq. (49) are written in cell S2 of the Excel spreadsheet to compute the pyrite-pyrrhotite equilibrium temperature for the first sample. Then, the relation is copied into the other cells of column S dragging the fill handle down, thus calculating the pyrite-pyrrhotite equilibrium temperature for all the available samples. The mean absolute error on the pyrite- pyrrhotite equilibrium temperature computed by means of Eq. (49) is 0.92 °C ± 0.48 °C (1σ), the minimum absolute error is 0.072 °C, whereas the maximum absolute error is 2.2 °C.

Utilizing the Excel spreadsheet Chapter 9. XLS, with the values of $\log(\mathrm{X_{H_2S}}/\mathrm{X_{H_2}}) + \log(X_{H_2S}/X_{H_2O})$ in column T and $\mathrm{X_{CO_2}}$ in column H, the equation to compute the pyrite-fayalite-quartz equilibrium temperature, $T_{Py-Fa-Qz}(^\circ C)$, is as follows for the first sample with the value of $\log(X_{H_2S}/X_{H_2}) + \log(X_{H_2S}/X_{H_2O})$ in cell T2 and X_{CO_2} in cell H2:

$$T_{Py-Fa-Qz}(^\circ C) = 1/((0.000109434^*H2^\wedge4 - 0.0001316183^*H2^\wedge3 + 0.00006018245^*H2^\wedge2 - 0.0000165768^*H2 + 0.000002680795)^*T2^\wedge3 + (0.001197873^*H2^\wedge4 - 0.001437238^*H2^\wedge3 + 0.0006435427^*H2^\wedge2 - 0.000170371^*H2 + 0.00002316502)^*T2^\wedge2 + (0.003668457^*H2^\wedge4 - 0.004370024^*H2^\wedge3 + 0.001871565^*H2^\wedge2 - 0.0004419177^*H2 - 0.0001957832)^*T2 + (0.002459505^*H2^\wedge4 - 0.002816838^*H2^\wedge3 + 0.001015807^*H2^\wedge2 - 0.0000808639 * H2 + 0.0008895927)) - 273.15 \quad (50)$$

The equal sign and the right-hand side of Eq. (50) are written in cell U2 of the Excel spreadsheet to compute the pyrite-fayalite-quartz equilibrium temperature for the first sample. Then, the equation is copied into the other cells of column U dragging the fill handle down, thus calculating the pyrite-fayalite-quartz equilibrium temperature for all the other samples. The mean absolute error on the pyrite-fayalite-quartz equilibrium temperature computed by means of Eq. (50) is 1.96 °C ± 1.77 °C (1σ), the minimum absolute error is 0.011 °C, whereas the maximum absolute error is 8.6 °C.

Using the Excel spreadsheet Chapter 9.XLS, with the values of $\log(X_{H_2S}/X_{H_2}) + 2 \cdot \log(X_{H_2S}/X_{H_2O})$ in column V and X_{CO_2} in column H, the equation to compute the pyrite-magnetite equilibrium temperature, $T_{Py-Mt}(^\circ C)$, is as follows for the first sample with the value of $\log(X_{H_2S}/X_{H_2}) + 2 \cdot \log(X_{H_2S}/X_{H_2O})$ in cell V2 and X_{CO_2} in cell H2:

$$T_{Py-Mt}(^\circ C) = 1/((0.0000393821^*H2^\wedge4 - 0.00004879165^*H2^\wedge3 + 0.00002488189^*H2^\wedge2 - 0.000007338693^*H2 + 0.000001348355)^*V2^\wedge4 + (0.001123319^*H2^\wedge4 - 0.001385642^*H2^\wedge3 + 0.000700085^*H2^\wedge2 - 0.0002035986^*H2 + 0.0000368431)^*V2^\wedge3 + (0.01154752^*H2^\wedge4 - 0.01417647^*H2^\wedge3 + 0.007086862^*H2^\wedge2 - 0.002024711^*H2 + 0.0003569913)^*V2^\wedge2 + (0.050069^*H2^\wedge4 - 0.06114115^*H2^\wedge3 + 0.03018728^*H2^\wedge2 - 0.008416803^*H2 + 0.001251926)^*V2 + (0.07563224^*H2^\wedge4 - 0.09175409^*H2^\wedge3 + 0.04461624^*H2^\wedge2 - 0.01196619^*H2 + 0.002576017)) - 273.15 \quad (51)$$

The equal sign and the right-hand side of Eq. (51) are written in cell W2 of the Excel spreadsheet to calculate the pyrite-magnetite equilibrium temperature for the first sample. Then, the expression is copied into the other cells of column W dragging the fill handle down, thus computing the pyrite-magnetite equilibrium temperature for all the considered samples. The mean absolute error on the pyrite-magnetite equilibrium temperature computed by means of Eq. (51) is 1.33 °C ± 1.28 °C (1σ), the minimum absolute error is 0.051 °C, whereas the maximum absolute error is 5.8 °C.

Two equations were developed to calculate the pyrite-hematite equilibrium temperature, $T_{Py-Hm}(^{\circ}C)$, adopting the same Excel spreadsheet Chapter 9.XLS, with the values of $\log(X_{H_2S}/X_{H_2}) + 3 \cdot \log(X_{H_2S}/X_{H_2O})$ in column X and X_{CO_2} in column H. For the first sample, with the value of $\log(X_{H_2S}/X_{H_2}) + 3 \cdot \log(X_{H_2S}/X_{H_2O})$ in cell X2 and X_{CO_2} in cell H2, the two equations are as follows:

$$\begin{aligned} T_{Py-Hm}(^{\circ}C) = 1/((&-0.000004755114^{*}H2^{\wedge}4 \\ &+0.000006730664^{*}H2^{\wedge}3 \\ &-0.00000463062^{*}H2^{\wedge}2 \\ &+0.000002558058^{*}H2 \\ &-0.0000009322103)^{*}X2^{\wedge}3 \\ &+(-0.0001744486^{*}H2^{\wedge}4 \\ &+0.0002476936^{*}H2^{\wedge}3 \\ &-0.0001702425^{*}H2^{\wedge}2 \\ &+0.00009501663^{*}H2 \\ &-0.0000360857)^{*}X2^{\wedge}2 \\ &+(-0.00210881^{*}H2^{\wedge}4 \\ &+0.00300926^{*}H2^{\wedge}3 \\ &-0.002065033^{*}H2^{\wedge}2 \\ &+0.001168183 * H2 \\ &-0.0005866398) * X2 \\ &+(-0.008386829^{*}H2^{\wedge}4 \\ &+0.01206681^{*}H2^{\wedge}3 \\ &-0.00825789^{*}H2^{\wedge}2 \\ &+0.004757456^{*}H2 \\ &-0.00139406)) - 273.15 \end{aligned} \tag{52}$$

$$\begin{aligned} T_{Py-Hm}(^{\circ}C) = 1/((&-0.00183927^{*}H2^{\wedge}4 \\ &+0.002105799^{*}H2^{\wedge}3 \\ &-0.0009196818^{*}H2^{\wedge}2 \\ &+0.0001922329^{*}H2 \\ &-0.00001709551)^{*}X2^{\wedge}3 \\ &+(-0.03982462^{*}H2^{\wedge}4 \\ &+0.04535639^{*}H2^{\wedge}3 \\ &-0.01962118^{*}H2^{\wedge}2 \\ &+0.004011997^{*}H2 \\ &-0.0003350631)^{*}X2^{\wedge}2 \\ &+(-0.2829093^{*}H2^{\wedge}4 \\ &+0.3204814^{*}H2^{\wedge}3 \\ &-0.1373093^{*}H2^{\wedge}2 \\ &+0.02746167^{*}H2 \\ &-0.00228143)^{*}X2 \\ &+(-0.6606229^{*}H2^{\wedge}4 \\ &+0.7444759^{*}H2^{\wedge}3 \\ &-0.3160783^{*}H2^{\wedge}2 \\ &+0.06203628^{*}H2 \\ &-0.004143932)) - 273.15 \end{aligned} \tag{53}$$

Equation (52) is valid for $T_{Py-Hm} < 375$ °C, whereas Eq. (53) holds true for $T_{Py-Hm} \geq 375$ °C. The equal sign and the right-hand side of Eqn. (52), (53) are written in cells Y2 and Z2, respectively, of the Excel spreadsheet to compute the two possible pyrite-hematite equilibrium temperatures for the first sample. Then, the IF function is used to select the correct pyrite-hematite equilibrium temperature for the first sample, writing it as follow in cell AA2: = IF (Y2 < 375;Y2;Z2).

Finally, the two Eqs. (52), (53) are copied into the other cells of columns Y and Z and the IF function is copied into the other cells of column AA, dragging the fill handle down, thus calculating the pyrite-hematite equilibrium temperatures for all the samples of interest. The mean absolute error on the pyrite-hematite equilibrium temperatures computed by means of Eqs. (52), (53) is 0.25 °C ± 0.21 °C (1σ), the minimum absolute error is 0.0026 °C, whereas the maximum absolute error is 0.84 °C.

The pyrite-pyrrhotite geothermometer, is scarcely affected by X_{CO_2}(Fig. 9a), because the maximum value of $\log(\varphi_{H_2S}/\varphi_{H_2})$ is -0.136. In contrast, the pyrite-fayalite-quartz, the pyrite-magnetite, and the pyrite-hematite geothermometers are more and more affected by X_{CO_2}, because the maximum values of the sums of log ratios involving $\log(\varphi_{H_2S}/\varphi_{H_2O})$ are (i) 0.480 for $\log(\varphi_{H_2S}/\varphi_{H_2}) + \log(\varphi_{H_2S}/\varphi_{H_2O})$, linked to the pyrite-fayalite-quartz geothermometer, (ii) 1.03 for $\log(\varphi_{H_2S}/\varphi_{H_2}) + 2 \cdot \log(\varphi_{H_2S}/\varphi_{H_2O})$, connected to the pyrite-magnetite geothermometer, and (iii) 1.59 for $\log(\varphi_{H_2S}/\varphi_{H_2}) + 3 \cdot \log(\varphi_{H_2S}/\varphi_{H_2O})$, related to the pyrite-hematite geothermometer. However, this effect is significant above 300–350 °C only, as shown by Figs. 9b and 10a, b, respectively.

The pyrite-pyrrhotite equilibrium temperature varies from 358 to 493 °C for Bocca Grande and from 337 to 410 °C for Bocca Nuova, as indicated by the analytical values of $\log(X_{H_2S}/X_{H_2})$, which are plotted in the diagram of Fig. 9a. The pyrite-fayalite-quartz equilibrium temperature ranges from 382 to 501 °C for Bocca Grande and from 339 to 509 °C for Bocca Nuova, as pointed out by the analytical values of $\log(X_{H_2S}/X_{H_2}) + \log(X_{H_2S}/X_{H_2O})$, which are reported in the graph of Fig. 9b. The pyrite-magnetite equilibrium temperature oscillates from 348 to 474 °C for Bocca Grande and from 298 to 506 °C for Bocca Nuova, as indicated by the analytical values of $\log(X_{H_2S}/X_{H_2}) + 2 \cdot \log(X_{H_2S}/X_{H_2O})$, which are displayed in the plot of Fig. 10a. The pyrite-hematite equilibrium temperature fluctuates from 372 to 525 °C for Bocca Grande and from 315 to 558 °C for Bocca Nuova, as pointed out by the analytical values of $\log(X_{H_2S}/X_{H_2}) + 3 \cdot \log(X_{H_2S}/X_{H_2O})$, which are shown in the diagram of Fig. 10b.

The four H_2S-pyrite temperatures are at variance with the SS4 temperature for Bocca Grande and Bocca Nuova fluids as shown by the diagrams of Figs. 9 and 10, as well as by the

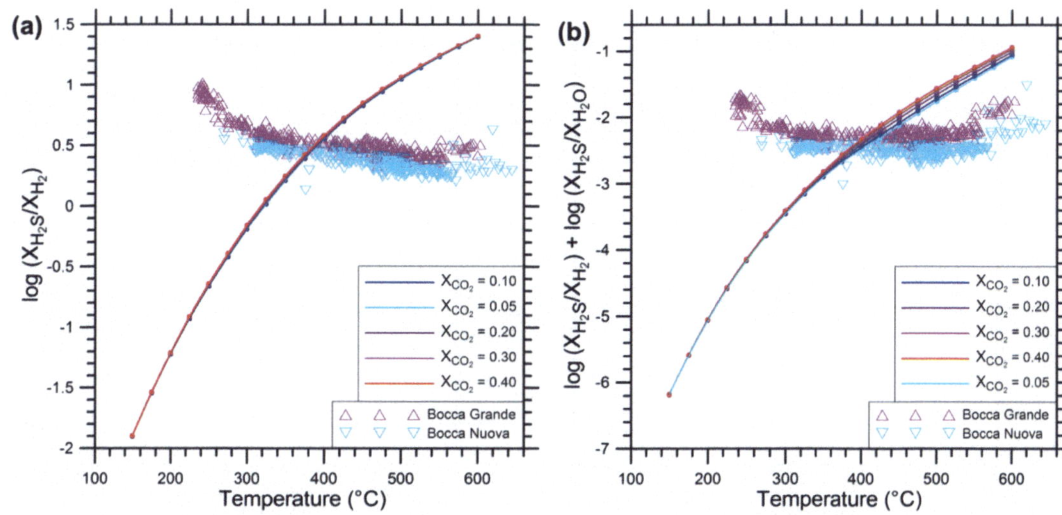

Fig. 9 Temperature dependence of **a** the $\log(X_{H_2S}/X_{H_2})$, related to the pyrite-pyrrhotite geothermometer and **b** $\log(X_{H_2S}/X_{H_2}) + \log(X_{H_2S}/X_{H_2O})$, related to the pyrite-fayalite-quartz geothermometer, at CO_2 mole fraction in the vapor phase of 0.05, 0.10, 0.20, 0.30, and 0.40, for the saturation decompression path of Solfatara fluids involving a brine containing 21 wt% NaCl and the related vapor phase. In both diagrams, the corresponding analytical values of Bocca Grande and Bocca Nuova fumarolic fluids (data from Chiodini et al. 2021 and references therein) are plotted against the SS4 temperature at saturation with a brine containing 21 wt% NaCl

chronograms of Figs. 11 and 12, indicating that the H_2S-pyrite temperatures are generally higher than the SS4 temperatures, before 2003–2004, whereas the H_2S-pyrite temperatures are lower than the SS4 temperatures afterwards.

The time changes of the SS4 temperature, at least since September 1984, imply the provenance of fluids either from progressively deeper zones of the Solfatara magmatic-hydrothermal system, whose temperature and pressure increase with depth, or from the same deep permeable zone that experienced a progressive heating and pressurization with time. Irrespective of the assumed model, H_2S would have equilibrated slower than CH_4, from September 1984 to 2003–2004, whereas H_2S would have equilibrated faster than CH_4, after 2003–2004.

Since there is no plausible reason for these changes in the kinetic behavior of these two gas species, it seems likely that the temperatures indicated by the pyrite-pyrrhotite, pyrite-fayalite-quartz, pyrite-magnetite, and pyrite-hematite geothermometers have no physical significance and that the H_2S concentration of Solfatara fluids is controlled by other reactions. This possibility will be investigated in the following section.

4.2 Gas-Solids Equilibria Involving Anhydrite

Anhydrite is a hydrothermal mineral often present in several geothermal systems (e.g., Browne 1978). In the geothermal wells Mofete 1 and 2, anhydrite was encountered at temperatures of 125 to 250 °C (Balducci and Chelini 1992), whereas it was found almost at all depths in well Mofete 5. Due to the retrograde solubility of anhydrite, its precipitation at Reykjanes was ascribed to progressive heating of seawater, either in the descending leg of the hydrothermal circulation systems or to occasional inflow of seawater into the hydrothermal system, when the cap rock was fractured by earthquakes (Tomasson and Kristmannsdóttir 1972).

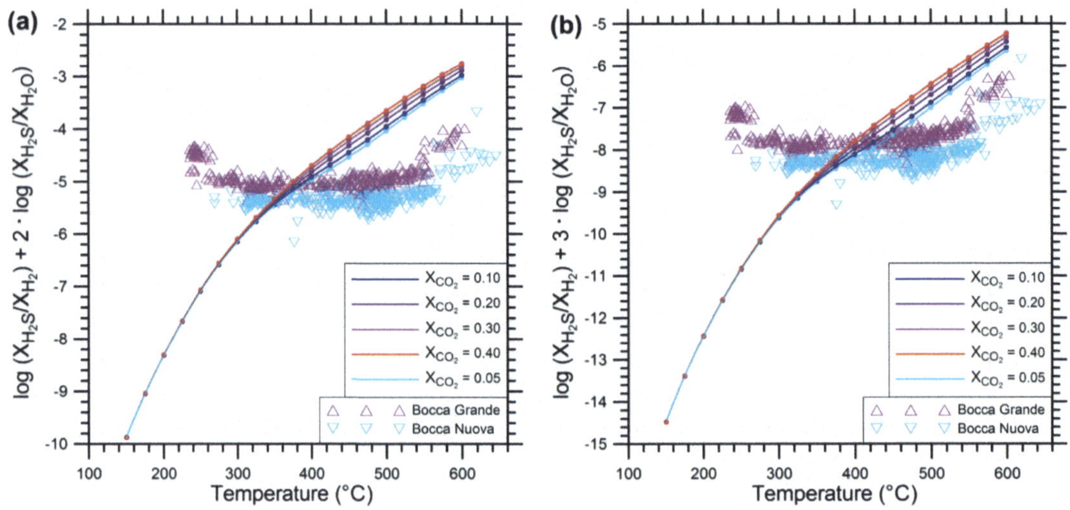

Fig. 10 Temperature dependence of **a** $\log(X_{H_2S}/X_{H_2}) + 2 \cdot \log(X_{H_2S}/X_{H_2O})$, related to the pyrite-magnetite geothermometer and **b** $\log(X_{H_2S}/X_{H_2}) + 3 \cdot \log(X_{H_2S}/X_{H_2O})$, related to the pyrite-hematite geothermometer, at CO_2 mole fraction in the vapor phase of 0.05, 0.10, 0.20, 0.30, and 0.40, for the saturation decompression path of Solfatara fluids involving a brine containing 21 wt% NaCl and the related vapor phase. In both diagrams, the corresponding analytical values of Bocca Grande and Bocca Nuova fumarolic fluids (data from Chiodini et al. 2021 and references therein) are plotted against the SS4 temperature at saturation with a brine containing 21 wt% NaCl

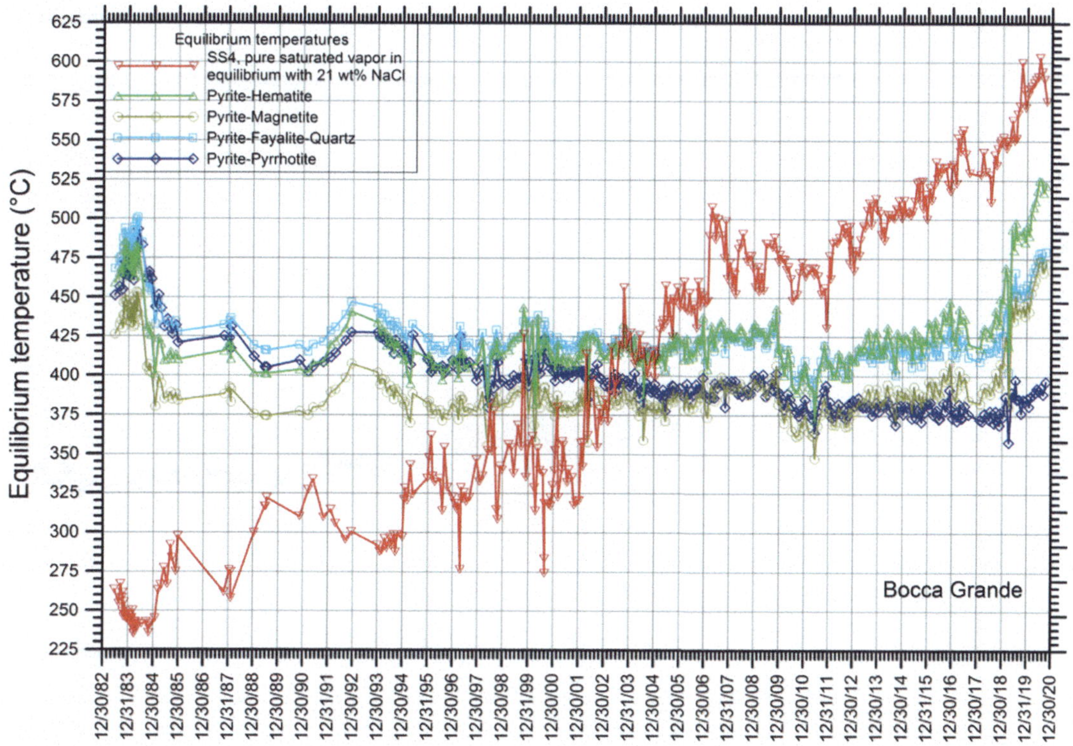

Fig. 11 Chronogram of the pyrite-pyrrhotite, pyrite-fayalite-quartz, pyrite-magnetite, and pyrite-hematite temperatures of Bocca Grande fluids. Also shown is the SS4 temperature, for the saturation decompression path, with a brine containing 21 wt% NaCl and the related vapor phase

In the Philippine geothermal systems, anhydrite is frequently-occurring and abundant both in the neutral-pH suite and in the acid alteration suite (Reyes 1990) and fluids are saturated with anhydrite at temperatures ≥ 180 °C (Reyes 1986). In particular, in the acid alteration suite, anhydrite is produced by both ascending acid fluids with probable magmatic components and descending steam-heated acid waters (Reyes 1990). In the geothermal fields of Bacon-Manito and Tongonan, anhydrite is more abundant than calcite in the upflow zone and, vice versa, calcite becomes more abundant than anhydrite moving away from the upflow zone (Reyes 1990). In some parts of the Bacon-Manito and Palinpinon geothermal fields, the pattern is complicated by the local occurrence of limestone or by the superposition of different hydrothermal parageneses (Reyes 1990).

In the geothermal systems of the Taupo Volcanic Zone, New Zealand, anhydrite is commonly found close to the surface or at depths lower than 500–600 m (Steiner and Rafter 1966; Browne 1978). This shallow anhydrite is interpreted to be produced through the mixing of upflowing hot fluids with descending steam-heated waters which are enriched in sulfate because of H_2S oxidation occurring close to the surface. Deep anhydrite is rare in the geothermal systems of the Taupo Volcanic Zone. Its origin was related to: (1) sulfate in equilibrium with deep H_2S, (2) descending sulfate produced by oxidation of H_2S or pyrite close to the surface, and (3) less likely, disproportionation of magmatic SO_2 (Chambefort 2021).

Anhydrite is present at all depths in the geothermal wells Nisyros 1 and 2 and anhydrite veins are present in the lava block ejecta of hydrothermal eruptions (Marini et al. 1993). In particular, the abundance of anhydrite and pyrite in the marbles positioned below the thick limestone formation encountered in well Nisyros 1

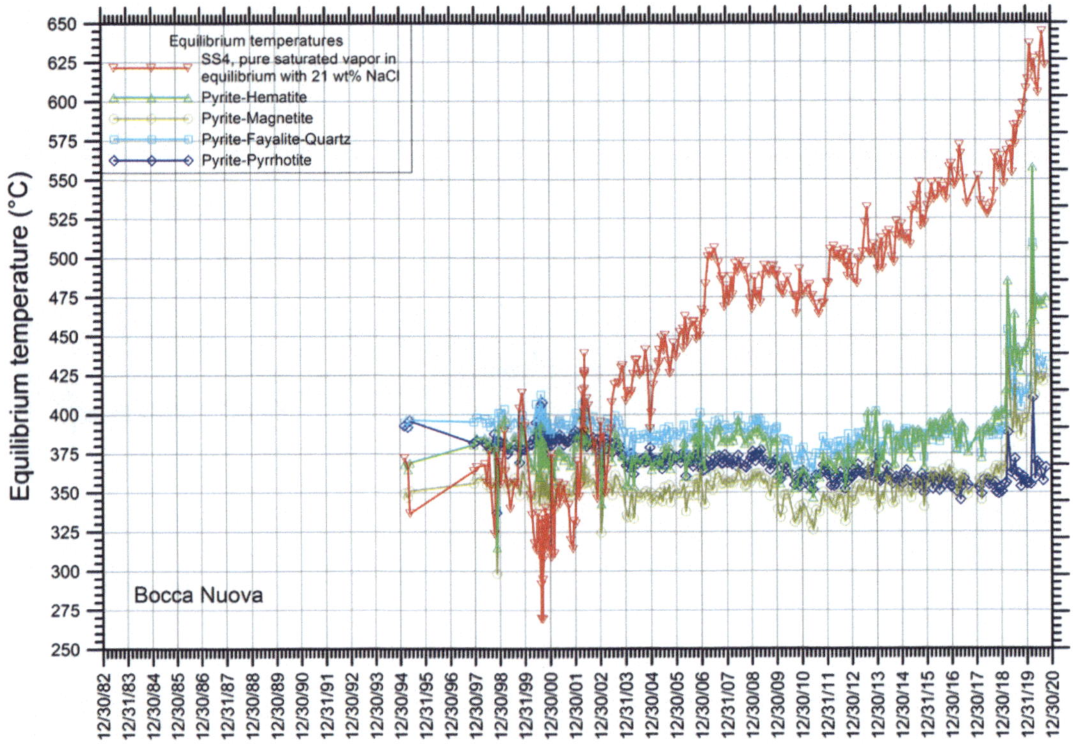

Fig. 12 Chronogram of the pyrite-pyrrhotite, pyrite-fayalite-quartz, pyrite-magnetite, and pyrite-hematite temperatures of Bocca Nuova fluids. Also shown is the SS4 temperature, for the saturation decompression path, with a brine containing 21 wt% NaCl and the related vapor phase

indicates a considerable and gradual increase in temperature, as well as the input of S-bearing fluids (Ambrosio et al. 2010).

In the geothermal wells of Latera, both anhydrite and calcite occur as very abundant authigenic minerals in three distinct associations, namely: (i) a contact-metasomatism paragenesis, together with diopside, grossular-garnet, phlogopite, wollastonite or monticellite; (ii) a high-temperature hydrothermal paragenesis, together with adularia, vesuvianite, melanite-garnet, tourmaline, amphibole, epidote, pyrite, and pyrrhotite; (iii) a low-temperature hydrothermal paragenesis, together with adularia, illite-celadonite, and Fe-sulfides (Cavarretta et al. 1985). The widespread occurrence of anhydrite and calcite is not surprising as the Latera geothermal reservoir is hosted in carbonate rocks (see also Marini and Chiodini 1994). Since carbonate rocks are presumably present at depth within the Campi Flegrei caldera (see Sect. 2.2 in chapter "The Magmatic–Hydrothermal System Hosted in the Campi Flegrei Caldera with Emphasis on the Solfatara"), it is interesting to investigate, as possible gas-geothermometer, the following heterogeneous redox reaction:

$$CaSO_{4(s)} + CO_2 + 4H_2 = CaCO_{3(s)} + H_2S + 3H_2O. \quad (54)$$

If both calcite and anhydrite are assumed to be pure minerals (which is a reasonable hypothesis), the log K of reaction (9.54) can be written as follows (see Sect. 1 in chapter "Calibration and Application of the RWG and SS4 Geoindicators for the Possible Expansion Paths of Solfatara Fluids" for the treatment of the H_2 fugacity term):

$$\log K_{H_2S-CO_2} = \log\left(\frac{X_{H_2S}}{X_{CO_2}}\right) - 4 \cdot \log\left(\frac{X_{H_2}}{X_{H_2O}}\right) + \log\left(\frac{\varphi_{H_2S}}{\varphi_{CO_2}}\right) - 3 \cdot \log\left(\frac{\varphi_{H_2}}{\varphi_{H_2O}}\right) - \log \varphi_{H_2} - \log P_{H_2O} \quad (55)$$

Separating the terms that can be expressed as function of temperature and X_{CO_2} from the analytical terms, Eq. (55) is rearranged as:

$$\log\left(\frac{X_{H_2S}}{X_{CO_2}}\right) - 4 \cdot \log\left(\frac{X_{H_2}}{X_{H_2O}}\right) = \log K(H_2S - CO_2) - \log\left(\frac{\varphi_{H_2S}}{\varphi_{CO_2}}\right) + 3 \cdot \log\left(\frac{\varphi_{H_2}}{\varphi_{H_2O}}\right) + \log \varphi_{H_2} + \log P_{H_2O}. \quad (56)$$

In preliminary calculations, the values of $\log K_{H_2S-CO_2}$ were computed by using SUPCRTBL at the pressures and temperatures fixed by coexistence of the brine containing 21 wt% NaCl and the related vapor phase, that is, up to 1539 bar and 600 °C. These $\log K_{H_2S-CO_2}$ values were then used, together with the fugacity coefficients of pertinent gas species and water partial pressure, to obtain the right-hand term of Eq. (56), which was compared with the left-hand (analytical) term of the same equation. It was thus evident that Solfatara fluids achieve the gas–solid equilibrium condition of reaction (54), if any, at temperatures and pressures higher than 600 °C and 1539 bar, respectively.

Therefore, pressures and temperatures were assumed to be constrained by coexistence of the brine containing 33.5 wt% NaCl and the related vapor phase, thus attaining maximum pressures and temperatures of 3627 bar and 1000 °C, respectively. The values of $\log K_{H_2S-CO_2}$ were calculated by utilizing SUPCRTBL (Table 8), while the fugacity coefficient of H_2S was computed by means of the Peng-Robinson EOS and merged with those of other relevant gas species adopting the same approach of Sect. 1 (Tables 9, 10, 11, 12 and 13).

Using the Excel spreadsheet Chapter 9.XLS, with the values of $\log(X_{H_2S}/X_{CO_2}) \; -4 \cdot \log(X_{H_2}/X_{H_2O})$ in column AB and X_{CO_2} in column H, the equation to compute the H_2S-CO_2 equilibrium temperature, $T_{H_2S-CO_2}$, is as follows for the first sample with the value of $\log(X_{H_2S}/X_{CO_2}) \; -4 \cdot \log(X_{H_2}/X_{H_2O})$ in cell AB2 and X_{CO_2} in cell H2:

$$\begin{aligned} T_{H_2S-CO_2}(^\circ C) =\ & 1/((-0.0000622619^*\mathrm{H2}^\wedge 4 + 0.00006685541^*\mathrm{H2}^\wedge 3 - 0.00002653208^*\mathrm{H2}^\wedge 2 + 0.000003056223^*\mathrm{H2} \\ & + 0.000001021845)^*AB2^\wedge 4 + (0.00280445^*\mathrm{H2}^\wedge 4 - 0.002992959^*\mathrm{H2}^\wedge 3 + 0.001167822^*\mathrm{H2}^\wedge 2 \\ & - 0.0001254123^*\mathrm{H2} - 0.00004531143)^*\mathrm{AB2}^\wedge 3 + (-0.0470429^*\mathrm{H2}^\wedge 4 + 0.04991906^*\mathrm{H2}^\wedge 3 \\ & - 0.01917336^*\mathrm{H2}^\wedge 2 + 0.001920842^*\mathrm{H2} + 0.0007510058)^*\mathrm{AB2}^\wedge 2 + (0.3483213^*\mathrm{H2}^\wedge 4 \\ & - 0.367639^*\mathrm{H2}^\wedge 3 + 0.1391247^*\mathrm{H2}^\wedge 2 - 0.01297486^*\mathrm{H2} - 0.005414418)^*\mathrm{AB2} \\ & + (-0.96068^*\mathrm{H2}^\wedge 4 + 1.008786^*\mathrm{H2}^\wedge 3 - 0.3764276^*\mathrm{H2}^\wedge 2 + 0.03258146^*\mathrm{H2} \\ & + 0.01498928)) - 273.15 \end{aligned} \quad (57)$$

The equal sign and the right-hand side of Eq. (57) are written in cell AC2 of the Excel spreadsheet to compute the H_2S-CO_2 equilibrium temperature

Table 8 Decimal logarithm of the thermodynamic equilibrium constant of the reaction (54), involving calcite, anhydrite, H_2S, CO_2, H_2, and H_2O, at the pressures and temperatures constrained by coexistence of the brine containing 33.5 wt% NaCl and the related vapor phase with X_{CO_2} of 0.05, 0.10, 0.20, 0.3, and 0.40

T (°C)	$X_{CO_2} = 0.05$		$X_{CO_2} = 0.10$		$X_{CO_2} = 0.20$		$X_{CO_2} = 0.30$		$X_{CO_2} = 0.40$	
	P_{tot} (bar)	log K	P_{tot} (bar)	log K	P_{tot} (bar)	log K	P_{tot} (bar)	log K	P_{tot} (bar)	log K
150	3.39	16.567	3.58	16.567	4.03	16.567	4.60	16.567	5.37	16.567
200	11.40	14.854	12.04	14.854	13.54	14.855	15.48	14.855	18.06	14.855
250	30.41	13.452	32.10	13.452	36.12	13.453	41.27	13.453	48.15	13.454
300	67.61	12.28	71.36	12.281	80.28	12.281	91.75	12.282	107.04	12.284
350	129.81	11.285	137.02	11.285	154.15	11.287	176.17	11.289	205.53	11.291
400	226.49	10.428	239.07	10.429	268.95	10.431	307.37	10.434	358.60	10.438
450	360.95	9.682	381.00	9.684	428.63	9.687	489.86	9.691	571.51	9.697
500	528.20	9.026	557.54	9.028	627.23	9.033	716.84	9.038	836.31	9.046
550	716.84	8.443	756.66	8.446	851.25	8.451	972.85	8.459	1134.99	8.468
600	913.73	7.920	964.49	7.923	1085.05	7.930	1240.06	7.939	1446.73	7.951
650	1108.10	7.448	1169.67	7.451	1315.87	7.459	1503.86	7.470	1754.50	7.483
700	1293.67	7.018	1365.54	7.022	1536.23	7.031	1755.70	7.042	2048.31	7.058
750	1468.59	6.625	1550.17	6.629	1743.94	6.639	1993.08	6.651	2325.26	6.668
800	1634.39	6.264	1725.19	6.268	1940.84	6.279	2218.10	6.292	2587.79	6.309
850	1794.72	5.931	1894.43	5.936	2131.23	5.946	2435.69	5.960	2841.64	5.979
900	1954.25	5.623	2062.81	5.628	2320.67	5.639	2652.19	5.654	3094.22	5.673
950	2118.00	5.338	2235.67	5.343	2515.13	5.355	2874.43	5.370	3353.50	5.39
1000	2291.05	5.075	2418.33	5.080	2720.62	5.092	3109.28	5.108	3627.49	5.129

for the first sample. Then, the relation is copied into the other cells of column AC dragging the fill handle down, thus calculating the H_2S-CO_2 equilibrium temperature for all the available samples. The mean absolute error on the H_2S-CO_2 equilibrium temperature computed by means of Eq. (57) is 1.2 °C ± 0.8 °C (1σ), the minimum absolute error is 0.13 °C, whereas the maximum absolute error is 3.6 °C. Equation (57) is valid from 450 to 1000 °C. No relation was derived for temperatures lower than 450 °C because there is no reason to perform these geothermometric calculations at lower temperatures based on preliminary results (see above). Furthermore, the comparison between the theoretical and analytical terms of Eq. (56) indicates that Solfatara fluids attain the equilibrium condition with the gas–solid reaction (54) at temperatures of ca. 600 to over 1000 °C (Fig. 13).

Assuming that CH_4 is generated by the SS4 reaction at the SS4 equilibrium temperature, the analytical molar fractions of hydrogen, $X_{H_2,m}$, carbon dioxide, $X_{CO_2,m}$, and water, $X_{H_2O,m}$, were corrected for the stoichiometry of this reaction, as follows:

$$X_{H_2,c} = X_{H_2,m} + 4 \cdot X_{CH_4,m} \tag{58}$$

$$X_{CO_2,c} = X_{CO_2,m} + X_{CH_4,m} \tag{59}$$

$$X_{H_2O,c} = X_{H_2O,m} - 2 \cdot X_{CH_4,m}. \tag{60}$$

This correction is important for H_2 but negligible for CO_2 and H_2O. The corrected concentrations were used in the calculation of the H_2S-CO_2 equilibrium temperature by means of Eq. (57).

In detail, the H_2S-CO_2 equilibrium temperature varies from 618 to 1039 °C, with a mean of 796 °C, a median of 798 °C, and a standard

Table 9 Compressibility factor, volume, and fugacity of the gas mixture with CO_2 mole fraction of 0.05 and fugacity coefficients of H_2O, CO_2, H_2S, H_2, and CH_4, in the same gas mixture, computed by means of the Peng-Robinson EOS at the P,T conditions fixed by coexistence with a brine containing 33.5 wt% NaCl

P	T	Z_{mixt}	V_{mixt}	f_{mixt}	φ_{H_2O}	φ_{CO_2}	φ_{H_2S}	φ_{H_2}	φ_{CH_4}
bar	°C		cm^3/mol	bar					
3.39	150	0.9803	10,167.72	3.33	0.9795	1.0025	1.0080	1.0199	1.0145
11.4	200	0.9506	3279.01	10.87	0.9497	1.0100	1.0230	1.0533	1.0403
30.4	250	0.8985	1284.94	27.58	0.9006	1.0310	1.0568	1.1218	1.0957
67.6	300	0.8212	578.86	57.15	0.8343	1.0823	1.1275	1.2516	1.2069
130	350	0.7264	289.90	100.98	0.7603	1.1939	1.2671	1.4858	1.4217
226	400	0.6356	157.07	161.57	0.6877	1.4157	1.5295	1.8903	1.8284
361	450	0.6159	102.59	242.32	0.6393	1.6784	1.8204	2.2747	2.2905
528	500	0.6703	81.57	349.75	0.6270	1.8429	1.9732	2.3854	2.5253
717	550	0.7516	71.76	487.84	0.6436	1.9394	2.0354	2.3432	2.5969
914	600	0.8361	66.43	655.83	0.6795	2.0057	2.0611	2.2559	2.5975
1108	650	0.9147	63.36	849.28	0.7272	2.0536	2.0693	2.1613	2.5656
1294	700	0.9843	61.56	1062.16	0.7812	2.0870	2.0672	2.0719	2.5182
1469	750	1.0446	60.51	1289.10	0.8377	2.1092	2.0592	1.9917	2.4650
1634	800	1.0969	59.88	1526.98	0.8943	2.1241	2.0491	1.9220	2.4127
1795	850	1.1425	59.44	1775.59	0.9496	2.1355	2.0403	1.8634	2.3662
1954	900	1.1833	59.06	2037.59	1.0033	2.1472	2.0354	1.8156	2.3286
2118	950	1.2207	58.61	2318.10	1.0555	2.1623	2.0367	1.7781	2.3019
2291	1000	1.2561	58.03	2624.33	1.1068	2.1831	2.0458	1.7505	2.2875

Table 10 Compressibility factor, volume, and fugacity of the gas mixture with CO_2 mole fraction of 0.10 and fugacity coefficients of H_2O, CO_2, H_2S, H_2, and CH_4, in the same gas mixture, computed by means of the Peng-Robinson EOS at the P,T conditions fixed by coexistence with a brine containing 33.5 wt% NaCl

P	T	Z_{mixt}	V_{mixt}	f_{mixt}	φ_{H_2O}	φ_{CO_2}	φ_{H_2S}	φ_{H_2}	φ_{CH_4}
bar	°C		cm^3/mol	bar					
3.58	150	0.9804	9633.71	3.51	0.9784	1.0018	1.0069	1.0197	1.0141
12.0	200	0.9511	3108.05	11.47	0.9472	1.0081	1.0201	1.0527	1.0391
32.1	250	0.9002	1219.60	29.15	0.8956	1.0271	1.0503	1.1200	1.0930
71.4	300	0.8263	551.79	60.54	0.8262	1.0742	1.1138	1.2455	1.1999
137	350	0.7403	279.91	107.46	0.7496	1.1744	1.2359	1.4621	1.3986
239	400	0.6682	156.42	173.41	0.6762	1.3579	1.4456	1.7979	1.7399
381	450	0.6629	104.61	262.89	0.6295	1.5669	1.6669	2.0919	2.1033
558	500	0.7199	83.00	382.70	0.6194	1.7199	1.8058	2.1954	2.3184
757	550	0.8012	72.47	536.92	0.6382	1.8293	1.8857	2.1837	2.4160
964	600	0.8856	66.65	724.51	0.6765	1.9146	1.9367	2.1298	2.4521
1170	650	0.9639	63.25	940.30	0.7267	1.9821	1.9703	2.0640	2.4541

(continued)

Table 10 (continued)

P	T	Z_{mixt}	V_{mixt}	f_{mixt}	φ_{H_2O}	φ_{CO_2}	φ_{H_2S}	φ_{H_2}	φ_{CH_4}
bar	°C		cm^3/mol	bar					
1366	700	1.0331	61.21	1177.33	0.7833	2.0337	1.9914	1.9978	2.4362
1550	750	1.0929	59.97	1429.48	0.8423	2.0722	2.0037	1.9359	2.4075
1725	800	1.1445	59.19	1693.27	0.9014	2.1011	2.0109	1.8808	2.3754
1894	850	1.1896	58.64	1968.58	0.9593	2.1247	2.0168	1.8338	2.3452
2063	900	1.2299	58.15	2258.58	1.0155	2.1468	2.0245	1.7952	2.3210
2236	950	1.2670	57.63	2569.21	1.0703	2.1710	2.0367	1.7653	2.3056
2418	1000	1.3022	57.00	2908.81	1.1243	2.2002	2.0557	1.7440	2.3010

Table 11 Compressibility factor, volume, and fugacity of the gas mixture with CO_2 mole fraction of 0.20 and fugacity coefficients of H_2O, CO_2, H_2S, H_2, and CH_4, in the same gas mixture, computed by means of the Peng-Robinson EOS at the P,T conditions fixed by coexistence with a brine containing 33.5 wt% NaCl

P	T	Z_{mixt}	V_{mixt}	f_{mixt}	φ_{H_2O}	φ_{CO_2}	φ_{H_2S}	φ_{H_2}	φ_{CH_4}
bar	°C		cm^3/mol	bar					
4.03	150	0.9805	8564.27	3.95	0.9761	1.0002	1.0043	1.0194	1.0132
13.5	200	0.9519	2764.86	12.92	0.9415	1.0042	1.0136	1.0519	1.0369
36.1	250	0.9032	1087.71	32.87	0.8847	1.0190	1.0363	1.1171	1.0879
80.3	300	0.8361	496.28	68.56	0.8093	1.0582	1.0857	1.2351	1.1871
154	350	0.7670	257.77	122.82	0.7283	1.1397	1.1773	1.4220	1.3592
269	400	0.7253	150.93	201.51	0.6557	1.2743	1.3180	1.6677	1.6166
429	450	0.7443	104.40	311.76	0.6142	1.4263	1.4639	1.8616	1.8740
627	500	0.8112	83.13	461.95	0.6097	1.5636	1.5822	1.9476	2.0588
851	550	0.8971	72.12	656.76	0.6340	1.6867	1.6794	1.9654	2.1809
1085	600	0.9844	65.86	894.54	0.6779	1.7979	1.7626	1.9522	2.2610
1316	650	1.0647	62.10	1168.15	0.7339	1.8950	1.8328	1.9255	2.3106
1536	700	1.1349	59.77	1468.15	0.7964	1.9762	1.8901	1.8931	2.3373
1744	750	1.1951	58.29	1786.31	0.8615	2.0424	1.9359	1.8596	2.3479
1941	800	1.2466	57.31	2118.14	0.9265	2.0962	1.9731	1.8278	2.3493
2131	850	1.2914	56.58	2463.75	0.9903	2.1420	2.0056	1.7998	2.3474
2321	900	1.3315	55.96	2827.61	1.0525	2.1840	2.0370	1.7771	2.3473
2515	950	1.3687	55.34	3217.93	1.1135	2.2263	2.0706	1.7607	2.3531
2721	1000	1.4044	54.64	3646.06	1.1739	2.2725	2.1096	1.7512	2.3677

Table 12 Compressibility factor, volume, and fugacity of the gas mixture with CO_2 mole fraction of 0.30 and fugacity coefficients of H_2O, CO_2, H_2S, H_2, and CH_4, in the same gas mixture, computed by means of the Peng-Robinson EOS at the P,T conditions fixed by coexistence with a brine containing 33.5 wt% NaCl

P	T	Z_{mixt}	V_{mixt}	f_{mixt}	φ_{H_2O}	φ_{CO_2}	φ_{H_2S}	φ_{H_2}	φ_{CH_4}
bar	°C		cm^3/mol	bar					
4.60	150	0.9804	7492.93	4.51	0.9733	0.9983	1.0013	1.0194	1.0123
15.5	200	0.9522	2420.00	14.76	0.9348	0.9999	1.0060	1.0515	1.0348
41.3	250	0.9055	954.24	37.63	0.8722	1.0104	1.0204	1.1154	1.0832
91.7	300	0.8454	439.09	78.83	0.7906	1.0424	1.0555	1.2272	1.1759
176	350	0.7931	233.23	142.57	0.7066	1.1094	1.1215	1.3903	1.3277
307	400	0.7780	141.66	237.87	0.6373	1.2157	1.2194	1.5789	1.5345
490	450	0.8193	100.55	375.68	0.6029	1.3420	1.3285	1.7201	1.7401
717	500	0.8997	80.68	567.48	0.6058	1.4751	1.4382	1.7952	1.9113
973	550	0.9947	69.98	819.63	0.6375	1.6115	1.5487	1.8295	2.0507
1240	600	1.0887	63.73	1129.90	0.6891	1.7455	1.6568	1.8424	2.1634
1504	650	1.1739	59.91	1488.38	0.7530	1.8697	1.7572	1.8431	2.2506
1756	700	1.2475	57.49	1881.82	0.8236	1.9789	1.8461	1.8362	2.3146
1993	750	1.3100	55.91	2298.69	0.8967	2.0721	1.9227	1.8249	2.3595
2218	800	1.3631	54.83	2732.78	0.9697	2.1510	1.9888	1.8122	2.3913
2436	850	1.4090	54.02	3184.42	1.0415	2.2197	2.0479	1.8006	2.4162
2652	900	1.4501	53.33	3660.12	1.1117	2.2830	2.1040	1.7920	2.4400
2874	950	1.4884	52.66	4171.63	1.1808	2.3455	2.1611	1.7882	2.4678
3109	1000	1.5256	51.94	4735.13	1.2500	2.4118	2.2231	1.7902	2.5037

Table 13 Compressibility factor, volume, and fugacity of the gas mixture with CO_2 mole fraction of 0.40 and fugacity coefficients of H_2O, CO_2, H_2S, H_2, and CH_4, in the same gas mixture, computed by means of the Peng-Robinson EOS at the P,T conditions fixed by coexistence with a brine containing 33.5 wt% NaCl

P	T	Z_{mixt}	V_{mixt}	f_{mixt}	φ_{H_2O}	φ_{CO_2}	φ_{H_2S}	φ_{H_2}	φ_{CH_4}
bar	°C		cm^3/mol	bar					
5.37	150	0.9800	6419.72	5.26	0.9698	0.9962	0.9975	1.0196	1.0115
18.1	200	0.9518	2073.49	17.22	0.9267	0.9949	0.9968	1.0518	1.0328
48.2	250	0.9070	819.29	43.93	0.8573	1.0010	1.0017	1.1153	1.0792
107	300	0.8545	380.39	92.43	0.7693	1.0265	1.0219	1.2223	1.1669
206	350	0.8200	206.71	168.96	0.6838	1.0827	1.0661	1.3661	1.3034
359	400	0.8312	129.72	287.12	0.6201	1.1734	1.1367	1.5159	1.4792
572	450	0.8957	94.22	463.87	0.5951	1.2912	1.2289	1.6272	1.6605
836	500	0.9934	76.35	716.49	0.6079	1.4314	1.3409	1.6993	1.8331
1135	550	1.1019	66.44	1055.09	0.6500	1.5879	1.4690	1.7484	1.9956
1447	600	1.2066	60.54	1477.46	0.7124	1.7504	1.6046	1.7838	2.1433
1754	650	1.3001	56.87	1969.73	0.7878	1.9074	1.7380	1.8086	2.2705

(continued)

Table 13 (continued)

P	T	Z_{mixt}	V_{mixt}	f_{mixt}	φ_{H_2O}	φ_{CO_2}	φ_{H_2S}	φ_{H_2}	φ_{CH_4}
bar	°C		cm^3/mol	bar					
2048	700	1.3800	54.51	2512.48	0.8702	2.0505	1.8619	1.8247	2.3744
2325	750	1.4472	52.94	3088.60	0.9550	2.1765	1.9732	1.8342	2.4567
2588	800	1.5037	51.84	3688.76	1.0397	2.2861	2.0727	1.8398	2.5222
2842	850	1.5524	51.01	4313.53	1.1229	2.3835	2.1635	1.8441	2.5776
3094	900	1.5960	50.31	4972.79	1.2048	2.4740	2.2500	1.8499	2.6297
3354	950	1.6368	49.63	5684.37	1.2859	2.5633	2.3371	1.8594	2.6847
3627	1000	1.6769	48.93	6472.91	1.3677	2.6570	2.4297	1.8743	2.7481

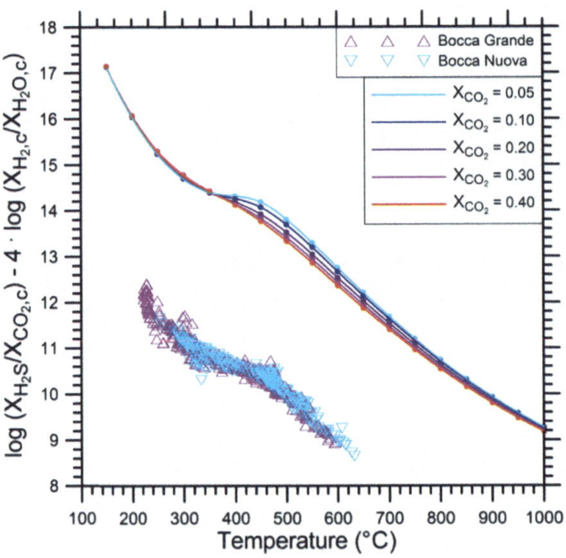

Fig. 13 Temperature dependence of the H_2S-CO_2 gas-geothermometer, at CO_2 mole fraction in the vapor phase of 0.05, 0.10, 0.20, 0.30, and 0.40, for the saturation decompression path of Solfatara fluids involving a brine containing 33.5 wt% NaCl and the related vapor phase. Also shown are the corresponding analytical values of Bocca Grande and Bocca Nuova fumarolic fluids (data from Chiodini et al. 2021 and references therein) against the SS4 temperature at saturation with a brine containing 33.5 wt% NaCl

deviation of 84 °C, for Bocca Grande fluids, and from 693 to 1086 °C, with a mean of 831 °C, a median of 819 °C, and a standard deviation of 75 °C, for Bocca Nuova fluids.[3]

Further indications are given by the chronograms of the H_2S-CO_2 equilibrium temperature of Bocca Grande and Bocca Nuova fluids (Fig. 14), also showing the chronograms of the SS4 equilibrium temperature at saturation with a brine containing 21 wt% NaCl. The H_2S-CO_2 equilibrium temperature was 667 ± 24 °C in June 1983 – July 1984 and increased almost continuously with time, at least since 2000–2001, with average values of 744 ± 14 °C in 2000–2001, 812 ± 15 °C in 2005–2011, 889 ± 18 °C in 2016–2017, and 990 ± 30 °C in 2019–2020, for Bocca Grande, whereas the corresponding values for Bocca Nuova are 745 ± 16 °C in

[3] Using the analytical concentrations instead of the corrected concentrations in calculations, the computed H_2S-CO_2 equilibrium temperature are lower by 34 °C on average for Bocca Grande and by 25 °C on average for Bocca Nuova, with minimum differences of 13 – 11 °C and maximum differences of 73 – 53 °C.

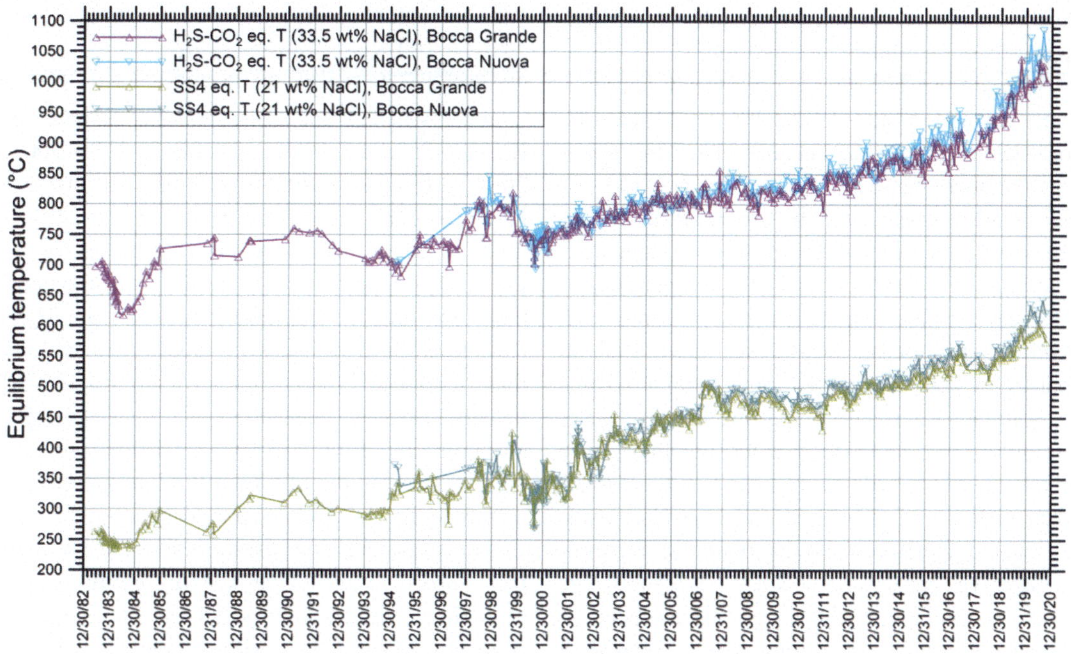

Fig. 14 Chronograms of the H_2S-CO_2 equilibrium temperature of Bocca Grande and Bocca Nuova fluids, at saturation with a brine containing 33.5 wt% NaCl. Also shown are the chronograms of the SS4 equilibrium temperature of Bocca Grande and Bocca Nuova fluids at saturation with a brine containing 21 wt% NaCl

2000–2001, 820 ± 15 °C in 2005–2011, 913 ± 18 °C in 2016–2017, and 1010 ± 37 °C in 2019–2020.

At a given time, there are moderate differences, in the order of some tens of degrees, between the H_2S-CO_2 equilibrium temperature of Bocca Grande and that of Bocca Nuova fluids, with the second usually greater than the first. Furthermore, the chronograms of the H_2S-CO_2 equilibrium temperature parallel those of the SS4 equilibrium temperature, with an almost constant shift of 350–400 °C approximately. This similarity is not surprising because both equilibrium temperatures are largely controlled by the H_2/H_2O ratio, although the SS4 equilibrium temperature involves the analytical H_2/H_2O ratio, whereas the H_2S-CO_2 equilibrium temperature was computed using the H_2/H_2O ratio corrected according to Eqs. (58) and (60).

One might wonder whether the computed H_2S-CO_2 equilibrium temperatures are meaningful or whether they are numbers without physical meaning. Two points have to be underscored, in this regard. First, the SO_2 disproportionation reaction (10), occurring upon SO_2 dissolution in the brines present at depth, is expected to trigger the deposition of sulfate minerals, such as anhydrite or alunite, and to contribute to the acidity of the brines which are also affected by the absorption of magmatic HCl and HF (Giggenbach 1987). Moreover, these initially acidic brines are expected to be rapidly neutralized by interaction with the carbonate rocks presumably present below the Campi Flegrei caldera, with the consequent precipitation of calcite and further anhydrite, similar to what was observed at Latera and Nisyros (see above). In other terms, due to these fast processes, reaction

(54) is expected to quickly reach or at least approach the equilibrium condition.

Second, the H_2S-CO_2 equilibrium temperatures calculated up to October 2020 can be considered reasonable because they are lower than magmatic temperatures, such as (i) 1145 °C, the mean homogenization temperature of the Minopoli 1 melt inclusions in olivine (Cannatelli et al. 2007), and (ii) 1120 °C, the temperature of the trachybasaltic magma presumably present at depth (Caliro et al. 2014), see Sect. 10.2 in chapter "Chemical and Isotopic Characteristics of Solfatara-Pisciarelli Fumarolic Fluids". It is currently impossible to predict whether these threshold values will be exceeded in the future. Even in this case, however, the H_2S-CO_2 equilibrium temperatures would maintain a semi-quantitative meaning, because it cannot be excluded that the H_2S-CO_2 geothermometer is affected by calibration errors, mainly related to the calculation of the average fugacity coefficients of relevant gas species in the gaseous mixture.

Accepting the meaningfulness of the H_2S-CO_2 equilibrium temperatures, the corresponding total fluid pressures, given by Eq. (24) in chapter "Calibration and Application of the RWG and SS4 Geoindicators for the Possible Expansion Paths of Solfatara Fluids", vary from 1117 to 3280 bar, with a mean of 1953 bar, a median of 1906 bar, and a standard deviation of 444 bar, for Bocca Grande fluids, and from 1401 to 3408 bar, with a mean of 2099 bar, a median of 2051 bar, and a standard deviation of 418 bar, for Bocca Nuova fluids.

Similar to the H_2S-CO_2 equilibrium temperature, also the H_2S-CO_2 total fluid pressure experienced a considerable increase with time for both Bocca Grande and Bocca Nuova fluids, as shown by the chronograms of Fig. 15, also reporting the total fluid pressure indicated by the SS4 reaction, at saturation with a brine containing 21 wt% NaCl. The H_2S-CO_2 total fluid pressure increased almost constantly with time, with approximate values of 1308 ± 102 bar in June 1983–July 1984, 1625 ± 60 bar in 2000–2001, 2035 ± 98 bar in 2005–2011, 2479 ± 100 bar in 2016–2017, and 2958 ± 146 bar in 2019–2020, for Bocca Grande, whereas the corresponding values for Bocca Nuova are 1618 ± 74 bar in 2000–2001, 2044 ± 111 bar in 2005–2011, 2591 ± 101 bar in 2016–2017, and 3034 ± 183 bar in 2019–2020. Similar to what was observed above for the H_2S-CO_2 temperature: (i) at a given moment, there are limited differences between the H_2S-CO_2 total fluid pressure of Bocca Grande and that of Bocca Nuova fluids, with the second usually greater than the first and (ii) the chronograms of the H_2S-CO_2 total fluid pressure parallel those of the SS4 total fluid pressure, with a shift of 1.0 - 1.5 kbar approximately.

Let us assume that: (a) the transition between the hydrostatic and the lithostatic regime occurs at a depth of 2.5 km, based on the evidence provided by well San Vito 1 (see Sect. 1.2 in chapter "The Magmatic–Hydrothermal System Hosted in the Campi Flegrei Caldera with Emphasis on the Solfatara"), and (b) reaction (54), involving calcite, anhydrite, H_2S, CO_2, H_2, and H_2O, attains thermo-chemical equilibrium in the deepest part of the carbonate layer present below the Campi Flegrei caldera (see Sect. 2.2 in chapter "The Magmatic–Hydrothermal System Hosted in the Campi Flegrei Caldera with Emphasis on the Solfatara"), let us say at a depth of 6–7 km, to be considered as an educated guess. The external (overburden) pressure at this depth is expected to be of 1330 ± 135 bar, as indicated by the sky blue strip in Fig. 15.

Therefore, in June 1983–July 1984, total fluid pressure was nearly equal to the external pressure, at 6–7 km depth, and a very small flow of magmatic gases could come from the carbonate layer and flush the overlying system. Then, since September 1984, fluid pressure began to exceed the external pressure, although the excess fluid pressure (that is, the difference between the fluid pressure and the overburden pressure) remained relatively low until 2000–2001. In contrast, a continuous increase in the excess fluid pressure occurred during the last twenty years and its

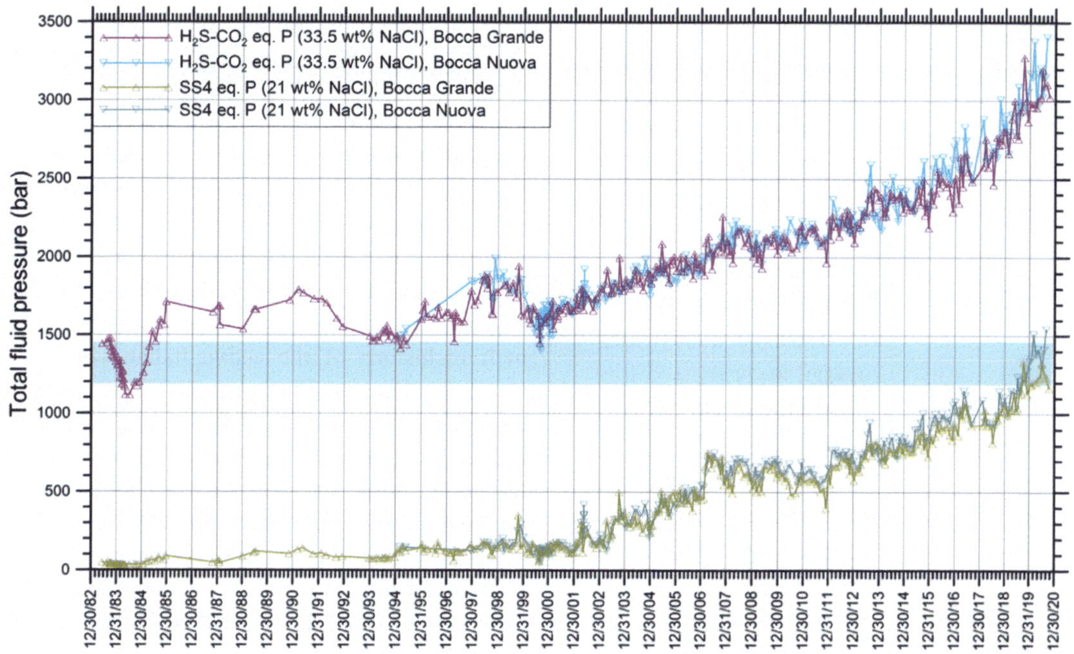

Fig. 15 Chronograms of the H_2S-CO_2 total fluid pressure of Bocca Grande and Bocca Nuova fluids at saturation with a brine containing 33.5 wt% NaCl. Also shown are the chronograms of the SS4 total fluid pressure of Bocca Grande and Bocca Nuova fluids at saturation with a brine containing 21 wt% NaCl. The sky blue strip indicates the external pressure expected at a depth of 6–7 km, where reaction (54), involving calcite, anhydrite, H_2S, CO_2, H_2, and H_2O, presumably attains the condition of thermo-chemical equilibrium

growth was especially significant during the last 3–4 years. This increment in the excess fluid pressure at 6–7 km depth with time might explain the increasing input, into the overlying system, of fluids coming from the carbonate layer, since September 1984, the growth in the degassing process observed at the surface, representing the subject of chapter "Soil CO_2 Diffuse Degassing and Thermal Energy Release as Indicators of Volcanic Unrest in the Solfatara-Pisciarelli Area", and, even more important, could play a fundamental role in governing the resurgence currently affecting the whole Campi Flegrei caldera.

It is now instructive to redraw the first diagram of this section, that is, the correlation plot of R_H values versus the reciprocal of the absolute temperature, considering the equilibrium temperature of Bocca Grande and Bocca Nuova fluids at saturation with a brine containing 33.5 wt% NaCl instead of the SS4 equilibrium temperature at saturation with a brine containing 21 wt% NaCl. Therefore, the R_H values were recomputed using the H_2 and H_2O concentrations corrected according to Eqs. (58) and (60), respectively, and the the ratio between the fugacity coefficients of H_2 and H_2O at the H_2S-CO_2 equilibrium temperature for saturation with a brine containing 33.5 wt% NaCl. The value of $\log(\varphi_{H_2}/\varphi_{H_2O})$ is computed using the Excel spreadsheet Chapter 9.XLS, with the H_2S-CO_2 equilibrium temperature in column AC and X_{CO_2} in column H. For the first sample, with the H_2S-CO_2 equilibrium temperature in cell AC2 and X_{CO_2} in cell H2, the equation of interest is:

$$\begin{aligned}\log(\varphi_{H_2}/\varphi_{H_2O}) = &(0.0000000002217428{}^*H2^\wedge 4\\ &- 0.0000000002409409{}^*H2^\wedge 3\\ &+0.00000000008993057{}^*H2^\wedge 2\\ &- 0.000000000004584406{}^*H2\\ &- 0.0000000000966278){}^*AC2^\wedge 4\\ &+(-0.0000007372114{}^*H2^\wedge 4\\ &+0.0000008066481{}^*H2^\wedge 3\\ &- 0.0000003078955{}^*H2^\wedge 2\\ &+0.00000001902425 * H2\\ &+0.0000000315995){}^*AC2^\wedge 3\\ &+(0.0009159429{}^*H2^\wedge 4\\ &- 0.0010109{}^*H2^\wedge 3\\ &+0.0003967025{}^*H2^\wedge 2\\ &- 0.00003019783{}^*H2\\ &- 0.0000375318){}^*AC2^\wedge 2\\ &+(-0.5049971{}^*H2^\wedge 4\\ &+0.5634355{}^*H2^\wedge 3\\ &- 0.2290065{}^*H2^\wedge 2\\ &+0.02200912{}^*H2\\ &+0.01828961){}^*AC2\\ &+(104.4709{}^*H2^\wedge 4\\ &- 118.3179{}^*H2^\wedge 3\\ &+50.28985{}^*H2^\wedge 2\\ &- 6.420118{}^*H2 - 2.487764)\end{aligned} \quad (61)$$

Equation (61) is valid in the temperatures range 500 to 1000 °C. The equal sign and the right-hand side of Eqns. (61) are written in cell AF2 to compute $\log(\varphi_{H_2}/\varphi_{H_2O})$ for the first sample. Then, the equation is quickly copied into the adjacent cells of column AF dragging the fill handle down, thus computing $\log(\varphi_{H_2}/\varphi_{H_2O})$ for all the available samples. The absolute error on $\log(\varphi_{H_2}/\varphi_{H_2O})$ calculated using Eq. (61) is 0.00070 ± 0.00044 (1σ) log-units on average, with range 0.0000046 to 0.0020 log-units. The R_H parameter (column AG) is calculated by adding the log-ratio of the fugacity coefficients of H_2 and H_2O (column AF) to the log-ratio of the corrected mole fractions of H_2 and H_2O (column AE).

In the correlation plot of R_H values versus the reciprocal of the absolute temperature of Fig. 16, the Bocca Grande and Bocca Nuova samples are situated in the area of the magmatic gas buffer at progressively higher f_{H_2O} values with time. It must be kept in mind that the f_{H_2O} values of the magmatic gas buffer are approximated because this buffer assumes $f_{H_2S}/f_{SO_2} = 1$, whereas the f_{H_2S}/f_{SO_2} ratio of Solfatara fluids is probably different from unity. To be noted also that the R_H values and H_2S-CO_2 equilibrium temperature of Solfatara fluids, especially during the last 3–4 years, are similar to those recorded by some melt inclusions of the active Italian volcanoes (from Scaillet and Pichavant 2005), in spite of the considerable scatter of these data. Both evidences support the magmatic provenance of Solfatara fluids and the growing influence of magmatic fluids from the 1980s to today.

At this point, it is worth to take into account a reaction similar to (54), but involving CH_4 instead of CO_2 and H_2, that is, the heterogeneous redox reaction:

$$CaSO_{4(s)} + CH_4 = CaCO_{3(s)} + H_2S + H_2O. \quad (62)$$

Again, if both calcite and anhydrite are hypothesized to be pure solid phases, the log K of reaction (62) can be written as:

$$\log K_{H_2S-CH_4} = \log\left(\frac{X_{H_2S}}{X_{CH_4}}\right) + \log\left(\frac{\varphi_{H_2S}}{\varphi_{CH_4}}\right) + \log \varphi_{H_2O} + \log P_{H_2O}. \quad (63)$$

Separating the analytical terms from the terms which are function of temperature, X_{CO_2}, and P_{H_2O}, Eq. (63) is reorganized as follows:

$$\log\left(\frac{X_{H_2S}}{X_{CH_4}}\right) = \log K_{H_2S-CH_4} - \log\left(\frac{\varphi_{H_2S}}{\varphi_{CH_4}}\right) - \log \varphi_{H_2O} - \log P_{H_2O}. \quad (64)$$

The values of $\log K_{H_2S-CH_4}$ were calculated by means of SUPCRTBL, at pressures and temperatures constrained by coexistence of the brine containing 33.5 wt% NaCl and the related vapor

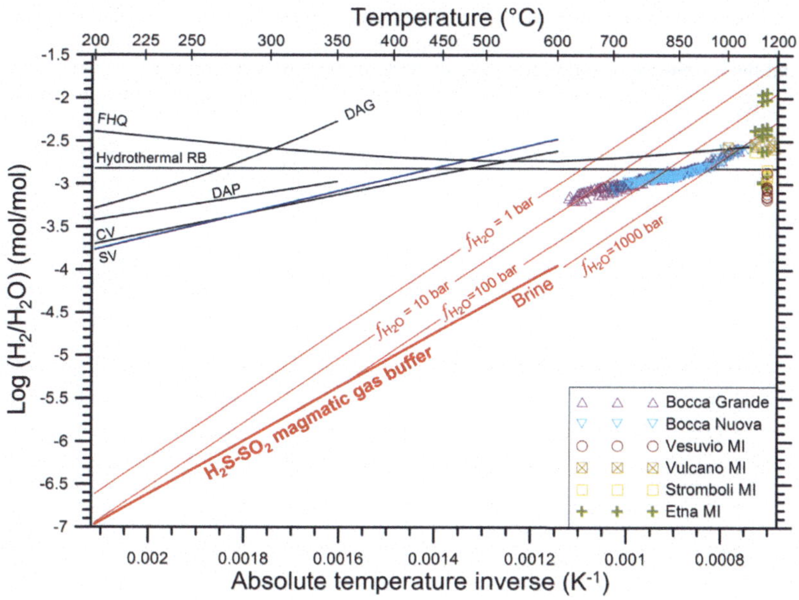

Fig. 16 Correlation plot showing the R_H values of Solfatara fluids at the H_2S-CO_2 equilibrium temperature, for gas equilibration in a single saturated vapor phase coexisting with a brine containing 33.5 wt% NaCl, versus the reciprocal of this temperature (in K). Also shown are the melt inclusion data of active Italian volcanoes (from Scaillet and Pichavant 2005) and different redox buffers. Reprinted from Giggenbach (1987), modified, Copyright © 1987, with permission from Elsevier

phase (Table 14). Then, the $\log K_{H_2S-CH_4}$ values were used, together with the fugacity coefficients of relevant gas species and water partial pressure, to compute the right-hand term of Eq. (64), which shows limited variations in the temperature and pressure range of interest.

In the diagram of Fig. 17, these theoretical values are compared with the analytical X_{H_2S}/X_{CH_4} log-ratios (that is, the left-hand term of the same equation), which are plotted against the H_2S-CO_2 equilibrium temperature at saturation with a brine containing 33.5 wt% NaCl.

Inspection of Fig. 17 shows that the analytical X_{H_2S}/X_{CH_4} log-ratios are lower than the theoretical counterparts by over three log-units, suggesting that the analytical CH_4 concentrations of Solfatara fluids are too high (or, less likely, H_2S concentrations are too low) for the attainment of equilibrium with anhydrite and calcite according to reaction (62). Accepting that analytical CH_4 concentrations are too high, it is possible to use Eq. (64) for estimating the CH_4 concentration in the equilibration zone of reaction (54), which turns out to be 0.0055 ± 0.0015 μmol/mol on average, with a minimum value of 0.0035 μmol/mol and a maximum value of 0.0137 μmol/mol. These very low CH_4 concentrations are expected in magmatic fluids, presumably somewhat modified by absorption in deep brines and interaction with carbonate rocks. In spite of the uncertainties on these computed methane concentrations, it is instructive to inspect also their time changes, which are reported in Fig. 18 and, of course, are similar to those of H_2S (see Figs. 6 and 7 in chapter "Chemical and Isotopic Characteristics of Solfatara-Pisciarelli Fumarolic Fluids"). The reasons for the marked increase in 2019–2020 are unclear.

All in all, it seems likely that H_2S, CO_2, H_2, and possibly CH_4 equilibrate with anhydrite and calcite at temperatures of ca. 600 to over 1000 °C and pressures of ca. 1 to 3 kbar, according to reaction (54), thus largely fixing the R_H parameter of Solfatara fluids. Under these conditions, methane concentration is very low. However,

Table 14 Decimal logarithm of the thermodynamic equilibrium constant of the reaction (58), involving calcite, anhydrite, H_2S, CH_4, and H_2O, at the pressures and temperatures constrained by coexistence of the brine containing 33.5 wt% NaCl and the related vapor phase for X_{CO_2} of 0.05, 0.10, 0.20, 0.3, and 0.40

T (°C)	X_{CO_2} = 0.05		X_{CO_2} = 0.10		X_{CO_2} = 0.20		X_{CO_2} = 0.30		X_{CO_2} = 0.40	
	P_{tot} (bar)	log K	P_{tot} (bar)	log K	P_{tot} (bar)	log K	P_{tot} (bar)	log K	P_{tot} (bar)	log K
150	3.39	5.348	3.58	5.348	4.03	5.348	4.60	5.348	5.37	5.349
200	11.40	5.887	12.04	5.887	13.54	5.888	15.48	5.888	18.06	5.888
250	30.41	6.329	32.10	6.330	36.12	6.330	41.27	6.330	48.15	6.331
300	67.61	6.698	71.36	6.699	80.28	6.700	91.75	6.701	107.04	6.702
350	129.81	7.011	137.02	7.012	154.15	7.013	176.17	7.015	205.53	7.017
400	226.49	7.28	239.07	7.281	268.95	7.283	307.37	7.286	358.60	7.290
450	360.95	7.513	381.00	7.515	428.63	7.518	489.86	7.522	571.51	7.528
500	528.20	7.718	557.54	7.719	627.23	7.724	716.84	7.730	836.31	7.737
550	716.84	7.897	756.66	7.899	851.25	7.905	972.85	7.912	1134.99	7.922
600	913.73	8.054	964.49	8.057	1085.05	8.064	1240.06	8.072	1446.73	8.084
650	1108.10	8.192	1169.67	8.195	1315.87	8.203	1503.86	8.213	1754.50	8.227
700	1293.67	8.313	1365.54	8.317	1536.23	8.326	1755.70	8.337	2048.31	8.353
750	1468.59	8.421	1550.17	8.425	1743.94	8.434	1993.08	8.447	2325.26	8.463
800	1634.39	8.516	1725.19	8.520	1940.84	8.530	2218.10	8.543	2587.79	8.561
850	1794.72	8.600	1894.43	8.605	2131.23	8.615	2435.69	8.629	2841.64	8.648
900	1954.25	8.676	2062.81	8.680	2320.67	8.692	2652.19	8.706	3094.22	8.726
950	2118.00	8.744	2235.67	8.749	2515.13	8.761	2874.43	8.776	3353.50	8.796
1000	2291.05	8.808	2418.33	8.813	2720.62	8.825	3109.28	8.841	3627.49	8.862

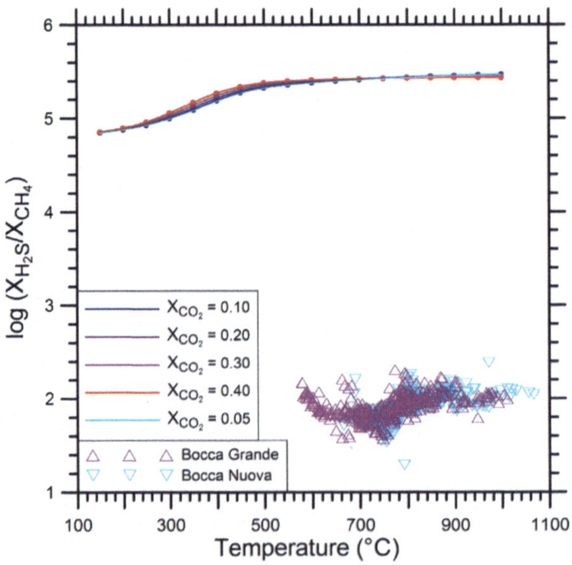

Fig. 17 Temperature dependence of the H_2S-CH_4 gas-geothermometer, at X_{CO_2} in the vapor phase of 0.05, 0.10, 0.20, 0.30, and 0.40, for the saturation decompression path of Solfatara fluids involving a brine containing 33.5 wt% NaCl and the related vapor phase. Also shown are the analytical values X_{H_2S}/X_{CH_4} log-ratios of the Bocca Grande and Bocca Nuova fluids plotted against the H_2S-CO_2 equilibrium temperature at saturation with a brine containing 33.5 wt% NaCl

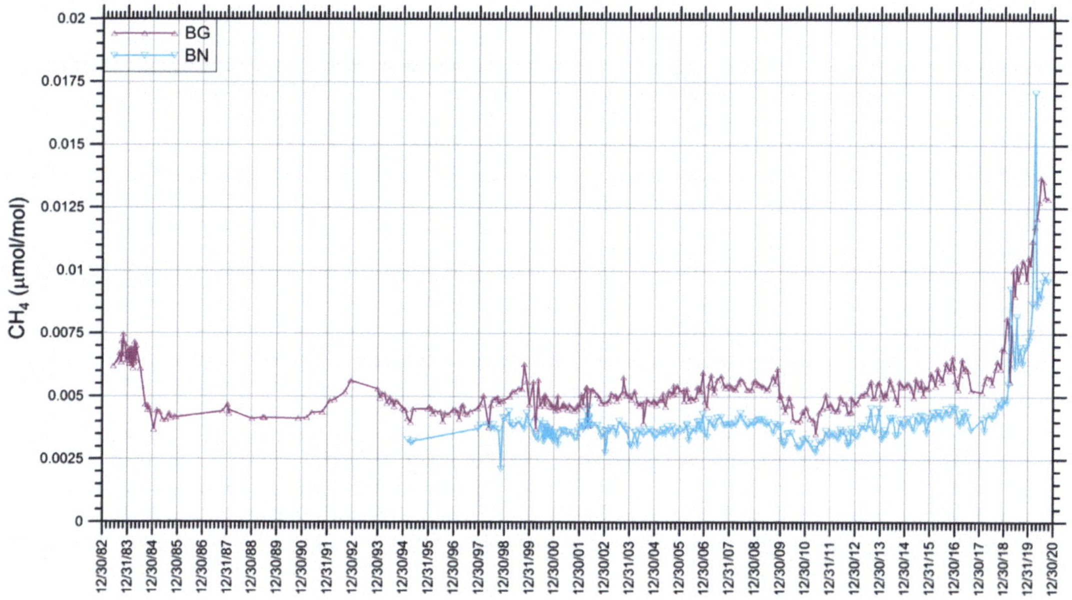

Fig. 18 Chronograms of the estimated methane concentration of Solfatara fluids at the H_2S-CO_2 equilibrium temperature, for gas equilibration in a single saturated vapor phase coexisting with a brine containing 33.5 wt% NaCl

upon cooling-depressurization of the gas mixture to temperatures of ca. 250 to 600 °C and pressures of few tens of bar to ca. 1.5 kbar, CH_4 is produced via the SS4 reaction.

References

Abiyudo R, Hadi J, Cumming W, Marini L (2016) Conceptual model assessment of vapor core geothermal system for exploration. Mt. Bromo case study. In: Proceedings of the 4th Indonesia international geothermal convention & exhibition 2016. Cendrawasih Hall-Jakarta Convention Center, Indonesia, 10-12 Aug 2016

Aiuppa A, Tamburello G, Di Napoli R, Cardellini C, Chiodini G, Giudice G, Grassa F, Pedone M (2013) First observations of the fumarolic gas output from a restless caldera: Implications for the current period of unrest (2005–2013) at Campi Flegrei. Geochem Geophys Geosyst 14(10):4153–4169

Ambrosio M, Doveri M, Fagioli MT., Marini L, Principe C, Raco B (2010) Water–rock interaction in the magmatic-hydrothermal system of Nisyros Island (Greece). Journal of Volcanology and Geothermal Research, 192(1-2):7-68.

Arnórsson S (1985) Gas pressures in geothermal systems. Chem Geol 49(1–3):319–328

Arnórsson S, Gunnlaugsson E (1985) New gas geothermometers for geothermal exploration-Calibration and application. Geochim Cosmochim Acta 49:1307–1325

Balducci S, Chelini W (1992) Hydrothermal equilibria in the active Mofete geothermal system, Phlegraean Fields, Naples, Italy. Acta Vulcanologica, Marinelli 2:17–34

Barin I, Knacke O (1973) Thermochemical properties of inorganic substances. Springer, Berlin

Bogie I, Lawless JV, Rychagov S, Belousov V (2005) Magmatic-related hydrothermal systems: classification of the types of geothermal systems and their ore mineralization. In: Proceedings of geoconference in Russia, Kuril

Browne PRL (1978) Hydrothermal alteration in active geothermal fields. Annu Rev Earth Pl Sc 6:229–248

Browne PRL, Ellis AJ (1970) The Ohaaki-Broadlands hydrothermal area, New Zealand: mineralogy and related geochemistry. Am J Sci 269:97–131

Caliro S, Chiodini G, Moretti R, Avino R, Granieri D, Russo M, Fiebig J (2007) The origin of the fumaroles of La Solfatara (Campi Flegrei, south Italy). Geochim Cosmochim Acta 71(12):3040–3055

Caliro S, Chiodini G, Paonita A (2014) Geochemical evidences of magma dynamics at Campi Flegrei (Italy). Geochim Cosmochim Acta 132:1–15

Cannatelli C, Lima A, Bodnar RJ, De Vivo B, Webster JD, Fedele L (2007) Geochemistry of melt inclusions from the Fondo Riccio and Minopoli 1 eruptions at Campi Flegrei (Italy). Chem Geol 237(3–4):418–432

Cavarretta G, Gianelli G, Scandiffio G, Tecce F (1985) Evolution of the Latera geothermal system II:

metamorphic, hydrothermal mineral assemblages and fluid chemistry. J Volcanol Geoth Res 26(3–4):337–364

Chambefort I (2021) Sulfur in New Zealand geothermal systems: insights from stable isotope and trace element analyses of anhydrite from Rotokawa and Ngatamariki geothermal fields, Taupo Volcanic Zone. New Zealand Journal of Geology and Geophysics, 64(2-3):372-388.

Chiodini G, Caliro S, Avino R, Bini G, Giudicepietro F, De Cesare W, Ricciolino P, Aiuppa A, Cardellini C, Petrillo Z, Selva J, Siniscalchi A, Tripaldi S (2021) Hydrothermal pressure-temperature control on CO_2 emissions and seismicity at Campi Flegrei (Italy). J Volcanol Geoth Res 414:107245

Chiodini G, Cioni R, Raco B, Scandiffio G (1991) Carbonyl sulphide (COS) in geothermal fluids: an example from the Larderello field (Italy). Geothermics 20(5-6):319-327.

Chiodini G, Marini L (1998) Hydrothermal gas equilibria: The H_2O-H_2-CO_2-CO-CH_4 system. Geochim. Cosmochim. Acta 62:2673-2687

Chiodini G, Cioni R, Marini L (1993) Reactions governing the chemistry of crater fumaroles from Vulcano Island, Italy, and implications for volcanic surveillance. Appl Geochem 8:357–371

Chiodini G, Caliro S, De Martino P, Avino R, Gherardi F (2012) Early signals of new volcanic unrest at Campi Flegrei caldera? Insights from geochemical data and physical simulations. Geology 40(10):943–946

D'Amore F, Gianelli G (1984) Mineral assemblages and oxygen and sulfur fugacities in natural water-rock interaction processes. Geochim Cosmochim Acta 48:847–857

D'Amore F, Nuti S (1977) Notes on the chemistry of geothermal gases. Geothermics 6:39–45

D'Amore F, Panichi C (1980) Evaluation of deep temperature of hydrothermal systems by a new gas-geothermometer. Geochim Cosmochim Acta 44:549–556

Elliott JR, Lira CT (2012) Introductory chemical engineering thermodynamics, 2nd edn, 876 p. Prentice Hall, Upper Saddle River, NJ. ISBN 978-0-13-606854-9.

Eugster HP (1957) Heterogeneous reactions involving oxidation and reduction at high pressures and temperatures. J Chem Phys 26(6):1760–1761

Evans KA, Powell R, Holland TJB (2010) Internally consistent data for sulphur-bearing phases and application to the construction of pseudosections for mafic greenschist facies rocks in Na_2O–CaO–K_2O–FeO–MgO–Al_2O_3–SiO_2–CO_2–O–S–H_2O. J Metamorph Geol 28(6):667–687

Fischer TP, Chiodini G (2015) Volcanic, magmatic and hydrothermal gases. In: Sigurdsson H, Houghton B, McNutt S, Rymer H, Stix J (eds) The encyclopedia of volcanoes. Academic Press-Elsevier, Second Edition, pp 779–797

Giggenbach WF (1980) Geothermal gas equilibria. Geochim Cosmochim Acta 44:2021–2032

Giggenbach WF (1987) Redox processes governing the chemistry of fumarolic gas discharges from White Island, New Zeland. Appl Geochem 2:143–161

Giggenbach WF (1993) Redox control of gas compositions in Philippine volcanic-hydrothermal systems. Geothermics 22:575–587

Giggenbach WF (1996) Chemical composition of volcanic gases. In: Scarpa R, Tilling RI (eds) Monitoring and mitigation of volcano hazards. Springer, pp 221–256

Giggenbach WF (1997) The origin and evolution of fluids in magmatic-hydrothermal systems. In: Barnes HL (ed) Geochemistry of hydrothermal ore deposits, 3d edn. Wiley, pp 737–796

Herdianita NR (2012) The evolution of the Darajat geothermal system, West Java–Indonesia. Ph.D. Thesis, University of Auckland, New Zealand, 264 p

Holland TJB, Powell R (1998) An internally consistent thermodynamic data set for phases of petrological interest. J Metamorph Geol 16(3):309–343

Holland TJB, Powell R (2011) An improved and extended internally consistent thermodynamic dataset for phases of petrological interest, involving a new equation of state for solids. Journal of Metamorphic Geology 29 (3): 333-383

Johnson JW, Oelkers EH, Helgeson HC (1992) SUPCRT92: A software package for calculating the standard molal thermodynamic properties of minerals, gases, aqueous species, and reactions from 1 to 5000 bar and 0 to 1000°C. Computers & Geosciences, 18 (7):899-947.

Kristmannsdóttir H (1979) Alteration of basaltic rocks by hydrothermal activity at 100–300°C. In: Mortland MM, Farmer VC (eds) International clay conference 1978. Developments in Sedimentology, Elsevier, 27, pp 359–367

Marini L, Chiodini G (1994) The role of carbon dioxide in the carbonate-evaporite geothermal systems of Tuscany and Latium (Italy). Acta Vulcanol 5:95–104

Marini L, Principe C, Chiodini G, Cioni R, Fytikas M, Marinelli G (1993) Hydrothermal eruptions of Nisyros (Dodecanese, Greece): past events and present hazard. J Volcanol Geoth Res 56:71–94

Marini L, Moretti R, Accornero M (2011) Sulfur isotopes in magmatic-hydrothermal systems, melts, and magmas. In: Behrens H, Webster JD (eds) Sulfur in magmas and melts: its importance for natural and technical processes (Rev Mineral Geochem 73 (1):423–492)

Moretti R, Arienzo I, Civetta L, Orsi G, Papale P (2013) Multiple magma degassing sources at an explosive volcano. Earth Planet Sci Lett 367:95–104

Murray C, Cubicciotti D (1983) Thermodynamics of aqueous sulfur species to 300°C and potential-pH diagrams. J Electrochem Soc 38:863–869

Nehring NL, D'Amore F (1984) Gas chemistry and thermometry of the Cerro Prieto, Mexico, geothermal field. Geothermics 13:75–89

Reyes AG (1986) Acid hydrothermal alteration in the Philippines. In: Fifth international symposium on

water-rock interaction extended abstracts, Reykjavik, Iceland, pp 461–465
Reyes AG (1990) Petrology of Philippine geothermal systems and the application of alteration mineralogy to their assessment. J Volcanol Geoth Res 43:279–309
Reyes AG, Giggenbach WF, Saleras JR, Salonga ND, Vergara MC (1993) Petrology and geochemistry of Alto Peak, a vapor-cored hydrothermal system, Leyte Province Philippines. Geothermics 22(5–6):479–519
Scaillet B, Pichavant M (2005) A model of sulphur solubility for hydrous mafic melts: application to the determination of magmatic fluid compositions of Italian volcanoes. Annals of Geophysics, 48(4-5):671-698.
Steiner A, Rafter TA (1966) Sulfur isotopes in pyrite, pyrrhotite, alunite and anhydrite from steam wells in the Taupo volcanic zone, New Zealand. Economic Geology, 61(6):1115-1129.
Stefánsson A, Arnórsson S (2002) Gas pressures and redox reactions in geothermal fluids in Iceland. Chem Geol 190(1–4):251–271
Symonds RB, Rose WI, Bluth GJ, Gerlach TM (1994) Volcanic-gas studies: methods, results, and applications. In: Carroll MR, Holloway JR (eds) Volatiles in magmas (Rev Mineral 30:1–66)
Symonds RB, Gerlach TM, Reed MH (2001) Magmatic gas scrubbing: implications for volcano monitoring. J Volcanol Geoth Res 108(1–4):303–341
Tomasson J, Kristmannsdóttir H (1972) High temperature alteration minerals and thermal brines, Reykjanes, Iceland. Contributions to Mineralogy and Petrology, 36(2), 123-134.
Wones DR, Gilbert MC (1969) The fayalite-magnetite-quartz assemblage between 600 and 800°C. Am J Sci 267:480–488
Zimmer K, Zhang Y, Lu P, Chen Y, Zhang G, Dalkilic M, Zhu C (2016) SUPCRTBL: a revised and extended thermodynamic dataset and software package of SUPCRT92. Comput Geosci 90:97–111

Soil CO_2 Diffuse Degassing and Thermal Energy Release as Indicators of Volcanic Unrest in the Solfatara-Pisciarelli Area

Abstract

The Solfatara-Pisciarelli Diffuse Degassing Structure (DDS) has been the perfect natural laboratory to test the accumulation chamber method for measuring the diffuse CO_2 flux from soil which was developed by the Pisa group led by Roberto Cioni (Chiodini et al. 1998). The effectiveness of the simple graphic-statistical technique for the processing and interpretation of the obtained data (Sinclair 1974) was later proven by Chiodini et al. (2008), who measured both the CO_2 flux from soil and the carbon isotopes of soil CO_2, again at Solfatara-Pisciarelli. Besides, Chiodini et al. (2008) estimated the upper limit of biogenic degassing from soil, 50 $g \cdot m^{-2} \cdot day^{-1}$, which represents a very important figure for the interpretation of local CO_2 flux data. Based on the H_2O/CO_2 ratio of Solfatara fumarolic fluids and assuming that they are unaffected by steam condensation, it was possible to convert the total CO_2 flow emitted from the Solfatara-Pisciarelli DDS in the associated steam flow and heat flow (Chiodini et al. 2001). Owing to the periodic measurement of CO_2 diffuse degassing from soil since 1998, a twenty years long dataset is available today for the Solfatara-Pisciarelli DDS. Based on these data, Chiodini et al. (2021) estimated that the total CO_2 flow from the entire Solfatara-Pisciarelli DDS increased from about 1000 $t \cdot day^{-1}$ in 2008-2010 to ca. 3000-4000 $t \cdot day^{-1}$ in 2019-2020. The increase in surface gas emission is considered to be controlled by the repeated inputs of magmatic gases in the hydrothermal system, causing the increase of its fluid pressure and temperature and promoting the Campi Flegrei seismicity.

As in most active volcanic-hydrothermal systems, diffuse soil degassing is the main form of volcanic degassing also at Solfatara-Pisciarelli. Moreover, as typical of high-temperature areas, a considerable amount of thermal energy coming from deep sources (e.g., the magma body and the overlying hydrothermal aquifers) is transferred to the atmosphere together with the fluids, which are mainly constituted by H_2O and CO_2.

Changes in natural diffuse degassing may be caused by variations of several parameters, such as pressure, temperature, and permeability, at different depths and in distinct locations, which, in turn, are possibly triggered by variations in volcano activity. This is the main reason why the monitoring of diffuse degassing is a powerful method to detect changes in volcanic activity before, during, and after a volcanic unrest. Nevertheless, long time series of diffuse degassing observations are scarce or incomplete for most active or quiescent volcanic systems. Solfatara volcano represents an exception because measurements of CO_2 diffuse degassing by means of the accumulation chamber method

L. Marini et al., *The Solfatara Magmatic-Hydrothermal System*, Advances in Volcanology,
https://doi.org/10.1007/978-3-030-98471-7_10

(Chiodini et al. 1998) and soil temperature were periodically carried out since 1998 (Chiodini et al. 1998, 2001, 2008, 2016, 2021; Granieri et al. 2003, 2009; Bruno et al. 2007; Bagnato et al. 2014; Cardellini et al. 2017). Thus, a twenty years long dataset is available today. It is one of the longest dataset available for a single volcanic system characterized by a long period of unrest.

Chiodini et al. (2001) reported the results and elaboration of 406 measurements of CO_2 diffuse degassing, performed from December 16 to 18, 1998 over an area of 1.16 km × 1.06 km (Fig. 1). Soil CO_2 fluxes (Φ_{CO_2}) range from 2 to $52,000\ g \cdot m^{-2} \cdot day^{-1}$, with an average of $1270\ g \cdot m^{-2} \cdot day^{-1}$. As shown by Fig. 1, high Φ_{CO_2} values are mainly distributed in two areas, one centered on the Solfatara crater, the other corresponding with the Pisciarelli fumarolic manifestations, situated to the North-East of it. The high Φ_{CO_2} values cover a total surface of about 0.5 km^2, named Solfatara-Pisciarelli Diffuse Degassing Structure (DDS). The two high-Φ_{CO_2} areas are separated by a low-Φ_{CO_2} area, elongated in a NW–SE direction, where the impervious products of the Astroni volcano crop out (see Fig. 5 in chapter "Summary of the Campi Flegrei Geology"). As noted by Chiodini et al. (2001), the highest Φ_{CO_2} values coincide with alignments of faults and fractures, implying a strong relation between the degassing process and the tectonic setting. The graphic-statistical approach of Chiodini et al. (1998) was applied to the dataset of 406 Φ_{CO_2} data. A bimodal frequency distribution was outlined, due to the combination of two different and partly overlapping log-normal populations, one related to the biological background, characterized by relatively low Φ_{CO_2} values (population LF) and the other linked to the hydrothermal source, characterized by comparatively high Φ_{CO_2} values (population HF; see Fig. 2). The main statistical parameters of the two populations, obtained by means of the the partitioning procedure of Sinclair (1974), are reported in Table 1, together with the results of similar surveys performed afterwards.

The total diffuse hydrothermal CO_2 flow of $1524\ t \cdot day^{-1}$ was computed by Chiodini et al. (2001) for the central square of 1 km^2 enclosing 333 measurements and comprising the whole

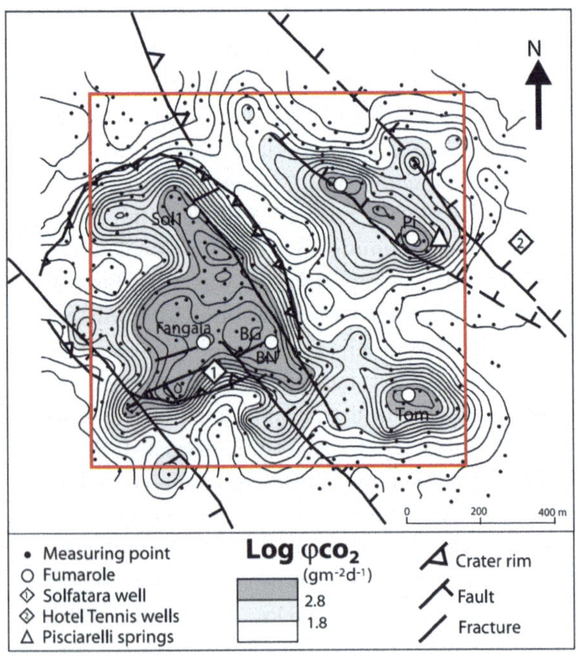

Fig. 1 Contour map of soil CO_2 fluxes (iso-line spacing of 0.2 log-units) obtained by Chiodini et al. (2001) through ordinary Kriging. The Solfatara-Pisciarelli Diffuse Degassing Structure (DDS) is highlighted by light gray ($\log \Phi_{CO_2} > 1.8 \log g \cdot m^{-2} \cdot day^{-1}$) and dark gray ($\log \Phi_{CO_2} > 2.8 \log g \cdot m^{-2} \cdot day^{-1}$) areas. The central squared area of 1 km^2 delimited by the red line includes the 333 data used to compute the total CO_2 flow from the Solfatara DDS by Chiodini et al. (2001). Reprinted from Granieri et al. (2003), Copyright © 2003, with permission from Elsevier

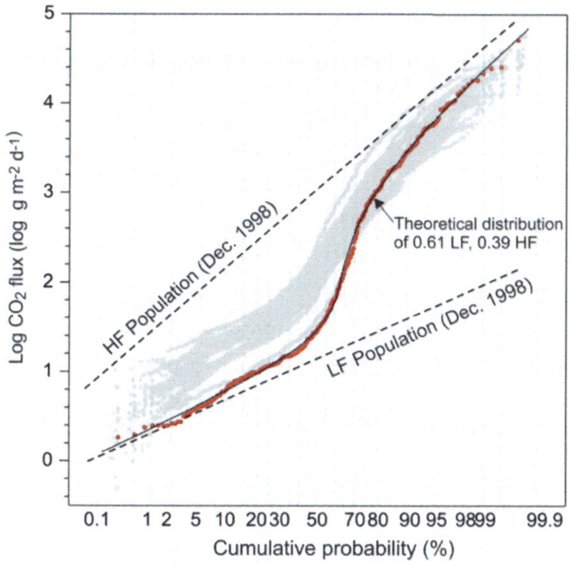

Fig. 2 Log probability plot of the soil CO_2 fluxes measured at Solfatara-Pisciarelli. The grey dots show the data of all the surveys performed from July 2000 to June 2016, whereas the red dots indicate the December 1998 dataset, whose theoretical distribution results from mixing of the hydrothermal high-flux population HF with the background low-flux population LF, in the proportions 61% and 39%, respectively. From Cardellini et al. (2017), which is licensed under a Creative Commons Attribution 4.0 CC BY International License

Table 1 Main statistical parameters for the Φ_{CO_2} measurements performed at Solfatara crater and neighboring areas (from Cardellini et al. 2017, modified)

N° measurements	Survey period	Population	Relative proportion	Mean Φ_{CO_2} ± SD ($g \cdot m^{-2} \cdot day^{-1}$)	Total CO_2 flow ±SD ($t \cdot day^{-1}$)	DDS surface (m^2)
406	Dec. 1998	LF	0.61	19.4 ± 1.26	1329 ± 122	454,585
		HF	0.39	3730 ± 980		
414	Jul. 2000	LF	0.61	23.9 ± 1.50	1513 ± 146	455,188
		HF	0.39	4590 ± 1351		
398	Feb. 2003	LF	0.40	14.4 ± 1.17	784 ± 85	545,655
		HF	0.60	2248 ± 1116		
391	Jul. 2003	LF	0.40	32.7 ± 2.41	745 ± 61	843,293
		HF	0.60	1274 ± 283		
413	Apr. 2004	LF	0.41	46.5 ± 2.38	1351 ± 164	990,716
		HF	0.59	2778 ± 775		
404	Aug. 2004	LF	0.35	46.5 ± 3.29	889 ± 96	961,003
		HF	0.65	1421 ± 323		
423	Mar. 2005	LF	0.43	42.5 ± 2.71	941 ± 89	890,046
		HF	0.57	1397 ± 273		

(continued)

Table 1 (continued)

N° measurements	Survey period	Population	Relative proportion	Mean Φ_{CO_2} ± SD ($g \cdot m^{-2} \cdot day^{-1}$)	Total CO_2 flow ±SD ($t \cdot day^{-1}$)	DDS surface (m^2)
408	Oct. 2005	LF	0.63	88.0 ± 8.19	1054 ± 103	914,215
		HF	0.37	2814 ± 452		
403	May 2006	LF	0.60	65.6 ± 4.51	1155 ± 147	978,181
		HF	0.40	1559 ± 217		
400	Oct. 2006	LF	0.53	74.7 ± 6.34	1242 ± 135	917,084
		HF	0.47	2629 ± 566		
372	Mar. 2007	LF	0.46	46.9 ± 2.58	1202 ± 120	923,259
		HF	0.54	2254 ± 544		
427	Jun. 2008	LF	0.44	57.6 ± 3.40	1099 ± 107	1,058,003
		HF	0.56	1550 ± 277		
473	Mar. 2009	LF	0.50	40.8 ± 2.49	1501 ± 171	823,329
		HF	0.50	3970 ± 1297		
503	Jul. 2009	LF	0.42	42.4 ± 3.01	1201 ± 107	949,466
		HF	0.58	1952 ± 342		
451	Nov. 2009	LF	0.52	56.7 ± 5.33	1271 ± 150	870,470
		HF	0.48	3041 ± 697		
505	May 2010	LF	0.49	41.5 ± 2.71	888 ± 97	788,067
		HF	0.51	1885 ± 442		
424	Mar. 2011	LF	0.57	41.3 ± 2.87	1027 ± 100	677,936
		HF	0.43	2603 ± 512		
470	Jun. 2012	LF	0.48	83.8 ± 6.93	1325 ± 135	1,139,904
		HF	0.52	2831 ± 458		
396	Oct. 2012	LF	0.52	98.7 ± 7.52	1524 ± 164	1,229,796
		HF	0.48	3709 ± 773		
438	Feb. 2013	LF	0.59	85.8 ± 7.40	1523 ± 151	1,029,707
		HF	0.41	3917 ± 683		
463	May 2013	LF	0.47	50.6 ± 2.69	1404 ± 170	1,095,011
		HF	0.53	2266 ± 398		
416	Jul. 2013	LF	0.50	74.6 ± 4.79	1162 ± 125	1,130,882
		HF	0.50	2791 ± 580		
422	Sept. 2013	LF	0.43	56.1 ± 4.13	1116 ± 138	1,090,043
		HF	0.57	2316 ± 455		
468	Jun. 2014	LF	0.40	63.2 ± 3.29	1113 ± 149	1,113,823
		HF	0.60	2359 ± 363		
398	Sept. 2014	LF	0.60	121 ± 12.2	1530 ± 169	1,151,089
		HF	0.40	4450 ± 704		

(continued)

Table 1 (continued)

N° measurements	Survey period	Population	Relative proportion	Mean Φ_{CO_2} ± SD ($g \cdot m^{-2} \cdot day^{-1}$)	Total CO_2 flow ±SD ($t \cdot day^{-1}$)	DDS surface (m^2)
397	Gen. 2015	LF	0.54	117 ± 13.8	2815 ± 318	1,168,794
		HF	0.46	6579 ± 964		
434	Mar. 2015	LF	0.60	76.6 ± 5.45	1922 ± 210	996,530
		HF	0.40	5478 ± 825		
508	Nov. 2015	LF	0.51	65.4 ± 4.33	1343 ± 153	919,573
		HF	0.49	3846 ± 684		
583	Gen. 2016	LF	0.54	70.4 ± 6.54	1750 ± 221	869,627
		HF	0.46	5538 ± 1156		
554	Jun. 2016	LF	0.60	136 ± 10.3	1563 ± 165	1,176,213
		HF	0.40	5235 ± 801		

DDS (Fig. 1). They also underscored that this CO_2 flow value compares with the mean CO_2 flow of 648 $t \cdot day^{-1}$ computed from the crater plume data of 101 non-erupting volcanoes (Williams et al. 1992) and the mean CO_2 flow of 1436 $t \cdot day^{-1}$, calculated from the crater plume data of 20 active volcanoes, excluding the four major CO_2-producers, namely Etna, Oldoinyo Lengai, Nyiragongo, and Nyamuragira (Brantley and Koepenick 1995), thus stressing the significance of diffuse degassing from the Solfatara volcano.

Based on the H_2O/CO_2 ratio of Bocca Grande (BG) and Bocca Nuova (BN) fumarolic fluids, which were assumed to be unaffected by steam condensation, a steam flow of 3353 $t \cdot day^{-1}$ and a heat flow of 1.19×10^{13} $J \cdot day^{-1}$ (138 MW) were estimated for the central sector of the Solfatara-Pisciarelli DDS (Chiodini et al. 2001).

Chiodini et al. (2001) underscored also that there is a strong correlation between the diffuse degassing process at the Solfatara-Pisciarelli DDS and the bradyseism, because the release of hydrothermal gases from the DDS area is the main process through which the magmatic-hydrothermal system dissipate the excess of energy deriving from its pressurization.

Granieri et al. (2003) reported on (i) the results of the continuous Φ_{CO_2} monitoring performed during four years of observation, from 1998 to 2002, by means of the automatic station FLXOV1, which is positioned in a selected site inside the crater of Solfatara, as well as (ii) the outcomes of periodic Φ_{CO_2} measurements carried out at 30 fixed stations (or measuring points) distributed in the crater floor (Fig. 3). The number of fixed stations inside the crater was increased in the period May 2002–June 2004, reaching the total of 71, and the grid thus obtained was used to perform over 200 periodic Φ_{CO_2} surveys (Granieri et al. 2009). Using the mean Φ_{CO_2} values for each station, an estimated mean CO_2 flow of 677 $t \cdot day^{-1}$ was calculated for the inner crater area (Granieri et al. 2009). The continuous monitoring of CO_2 soil diffuse degassing shows the presence of long-term trends and short-term periodicities, which are correlated to ground deformation and variations of environmental parameters, respectively (Granieri et al. 2003). According to Granieri and coworkers, the deep origin of the CO_2 emitted from soil at the FLXOV1 station is suggested by the high Φ_{CO_2} values measured, spanning the range 315–4724 $g \cdot m^{-2} \cdot day^{-1}$, for a total of 15,655 available data.

Assuming that ground deformation is controlled by fluid pressure within the hydrothermal system, as suggested by geochemical and geophysical data (e.g., Chiodini et al. 2003; De Natale et al. 2006; Todesco et al. 2003), the

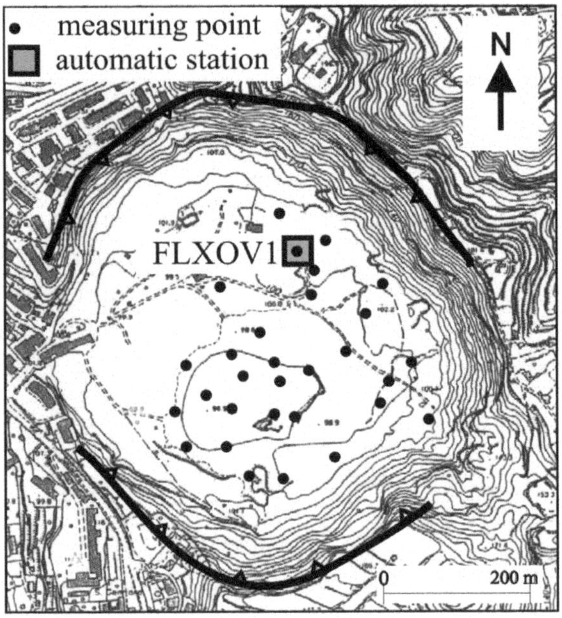

Fig. 3 Location of the automatic station FLXOV1 and the 30 fixed stations inside the Solfatara crater. Reprinted from Granieri et al. (2003), Copyright © 2003, with permission from Elsevier

variations in Φ_{CO_2} can be related to fluid pressure changes at depth and to vertical soil movements (Granieri et al. 2003, 2009). The relations between changes in fluid pressure at depth, variations in fluid degassing activity, and ground deformation were also recognized by Bruno et al. (2007), through the integration of (1) different, high-resolution geophysical data, that is, electromagnetic and electrical data, active seismic data, microseismic noise, and gravity data and (2) hydrogeological data, namely, the water table level and temperature of about 30 boreholes situated at Solfatara and nearby areas.

Chiodini et al. (2008) combined, for the first time at Solfatara, the Φ_{CO_2} data with the carbon isotopes of soil CO_2, in order to identify the possible C sources contributing to diffuse degassing of CO_2 from soil. A total of 373 Φ_{CO_2} measurements and carbon isotopic determinations were performed in March 2007, obtaining the mean Φ_{CO_2} and total CO_2 flow reported in Table 1. Based on the isotopic composition of carbon, three populations were recognized, namely: (i) Hydrothermal, comprising 49% of the data, with mean $\delta^{13}C$ of -2.3 ± 0.9 (1σ) ‰ versus VPDB, and mean Φ_{CO_2} of 2875 $g \cdot m^{-2} \cdot day^{-1}$; (ii) Biogenic, including 20% of the measurements, with mean $\delta^{13}C$ of -19.4 ± 2.1 ‰, and mean Φ_{CO_2} of 26 $g \cdot m^{-2} \cdot day^{-1}$; (iii) Mixed, comprising 31% of the data, with mean $\delta^{13}C$ of -9.8 ± 3.7 ‰, and mean Φ_{CO_2} of 73 $g \cdot m^{-2} \cdot day^{-1}$. In contrast, two populations were recognized by means of the graphic-statistical approach of Chiodini et al. (1998), namely (a) the HF population, comprising 54% of the measurements, with mean Φ_{CO_2} of 2228 $g \cdot m^{-2} \cdot day^{-1}$, and (b) the LF population, including 46% of the data, with mean Φ_{CO_2} of 47 $g \cdot m^{-2} \cdot day^{-1}$. The results of the two partitioning approaches are compared in the log-probability diagram of Fig. 4.

The upper limit of the biogenic Φ_{CO_2} was estimated to be either (a) 140 $g \cdot m^{-2} \cdot day^{-1}$, corresponding to the 95th percentile of the LF population separated through the graphic-statistical approach or (b) 50 $g \cdot m^{-2} \cdot day^{-1}$, corresponding to the 95th percentile of the biogenic (B) population recognized based on isotopic evidence. The probability that the Φ_{CO_2} is higher than each of these two thresholds was estimated for each station and mapped, obtaining the two maps of Fig. 5. In both maps, the DDS is

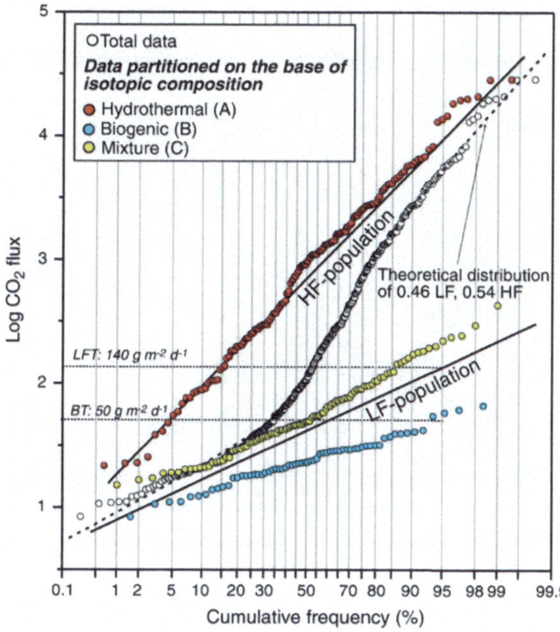

Fig. 4 Log-probability diagram of the CO_2 fluxes from soil. The empty black circles identify the 373 data constraining the cumulative curve. The black solid lines refer to the HF and LF populations recognized by means of the graphic-statistical approach of Chiodini et al. (1998), whereas the black dashed line denotes the theoretical distribution obtained by combining the HF and LF populations in the proportions 0.46 LF plus 0.54 HF. The filled red, blue, and yellow circles display the three populations separated on the basis of the isotopic composition of carbon. The BT and LFT values refer to the upper limit of the biogenic CO_2 flux based on the graphic-statistical approach and the isotopic data of carbon, respectively. Reprinted from Chiodini et al. (2008), Copyright © 2008, with permission from Elsevier

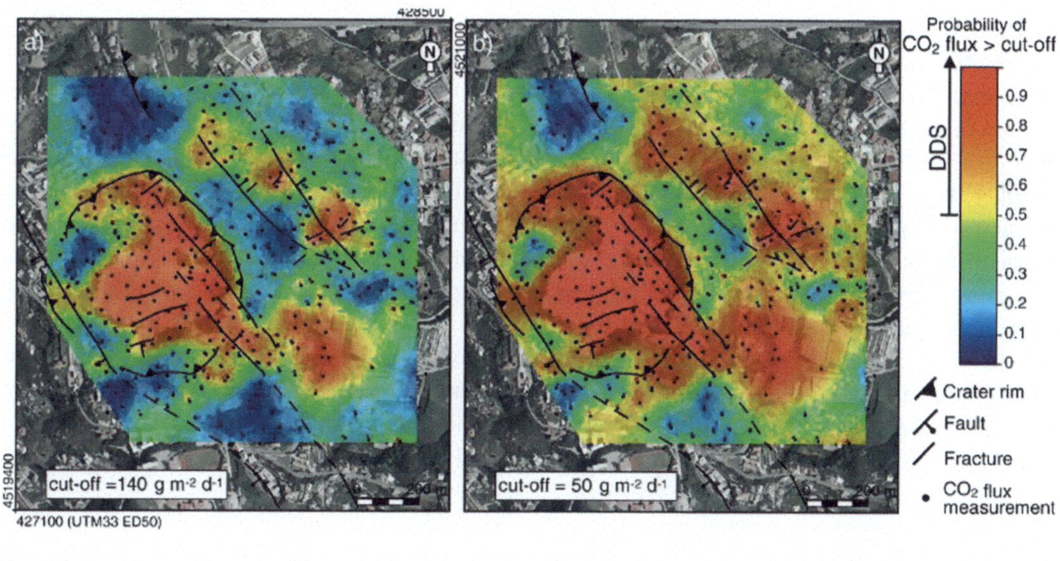

Fig. 5 Probability maps of the Φ_{CO_2} spatial distribution based on two different cut-off values: **a** $140\ g \cdot m^{-2} \cdot day^{-1}$, statistically derived, and **b** $50\ g \cdot m^{-2} \cdot day^{-1}$, isotopically derived. The maps were prepared to select the best estimation of the upper limit of the purely biogenic Φ_{CO_2}. The different colors represent the probability that the Φ_{CO_2} at each location is higher than the considered threshold. Reprinted from Chiodini et al. (2008), Copyright © 2008, with permission from Elsevier

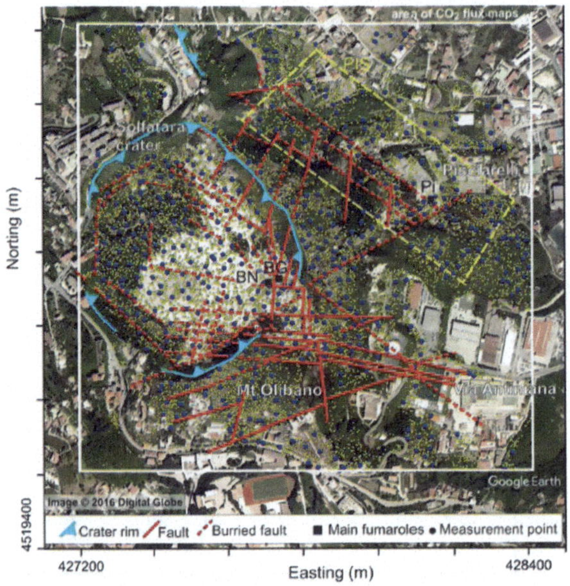

Fig. 6 Location map of all Φ_{CO_2} stations (yellow dots) and the stations of the 2016 survey (blue dots). Also shown are the Bocca Nuova (BN), Bocca Grande (BG), and Pisciarelli (PI) fumaroles as well as the main tectonic structures. The white box encloses the area considered for Φ_{CO_2} mapping, whereas the dashed yellow box (PIS) encompasses the area adopted for the computation of the total CO_2 flow from the Pisciarelli DDS. Modified from Cardellini et al. (2017), which is licensed under a Creative Commons Attribution 4.0 CC BY International License

defined by probabilities $\geq 50\%$, that is, by the yellow- and red-colored areas. According to Chiodini et al. (2008), the best estimation of the upper limit of the biogenic Φ_{CO_2} is the isotopically derived value of 50 $g \cdot m^{-2} \cdot day^{-1}$, representing an important figure for the interpretation of Φ_{CO_2} measurements. Adopting this cut-off, the extension of the Solfatara-Pisciarelli DDS area in March 2007 results to be 0.8 km^2.

Further Φ_{CO_2} measurements inside the Solfatara crater were performed on April 2012 at 144 stations distributed over an area of 46,000 m^2 (Tassi et al. 2013, 2015) and September 2013 at 116 stations over an area of about 90,000 m^2 (Bagnato et al. 2014). Tassi et al. (2013) and Bagnato et al. (2014) estimated a mean Φ_{CO_2} value of 946 and 1977 $g \cdot m^{-2} \cdot day^{-1}$, respectively.

Detailed reviews of the results obtained by periodic Φ_{CO_2} surveys were performed by Chiodini et al. (2010) and Cardellini et al. (2017). In the last work, 30 different surveys were compiled and compared, each comprising 372–583 Φ_{CO_2} measurements, for a total number of 13,158 data, distributed over an area of about 1.2 km × 1.2 km (Fig. 6). The Φ_{CO_2} data of each survey show a bimodal frequency distribution (Fig. 2), suggesting the presence of two different CO_2 sources, in spite of the possible influence of different gas transport mechanisms, varying soil permeability, and other factors as well.

The Φ_{CO_2} data confirm the presence of the two populations identified by Chiodini et al. (2001), one characterized by relatively high Φ_{CO_2} values, named HF population, and the other characterized by comparatively low Φ_{CO_2} values, named LF population (Cardellini et al. 2017). The main statistical parameters of both populations are reported in Table 1, together with the estimation of the total CO_2 flow and the extension of the DDS surface. After 2003, even the mean Φ_{CO_2} of the LF population, representative of the biogenic contribution, experienced a considerable increase, reaching the peak value of 88 $g \cdot m^{-2} \cdot day^{-1}$ in October 2005. Based on the C isotope data of soil CO_2 of Chiodini et al. (2008), the increase in the Φ_{CO_2} value of the LF population after 2003 was related to a new arrival of deep CO_2 (Chiodini et al. 2010; Cardellini et al. 2017).

This interpretation is also supported by the marked increase in the areal extension of the DDS surface, from 0.45×10^6 m^2 in 1998 to over 1×10^6 m^2 for different surveys performed after 2012 (Cardellini et al. 2017). The enlargement of the DDS surface is particularly evident towards the East (Fig. 7) and the possible opening of deep fractures enhancing the ascent of magmatic CO_2-rich fluids, especially along the Pisciarelli faults, was suggested by Chiodini et al. (2010).

The areal extension of the Solfatara-Pisciarelli DDS reached maximum values in January-March 2015. Also the total flow of diffuse soil CO_2 experienced significant changes, varying from $745\ (\pm 61)\ t \cdot day^{-1}$ to $2,815\ (\pm 318)\ t \cdot day^{-1}$, with the peak value again in January 2015 (Table 1). Repeated episodes of magmatic fluids input in the hydrothermal system of the Solfatara are considered to be the common controlling factor of the CO_2 release from soil, seismicity and ground deformation (Cardellini et al. 2017). The Φ_{CO_2} values calculated simulating the injection of magmatic fluids into a virtual hydrothermal system (Chiodini et al. 2016) show the same increasing trend of the Φ_{CO_2} measurements, in spite of some minor discrepancies (Cardellini et al. 2017).

As recently reported by Chiodini et al. (2021), in the period April 2004–October 2020, 149 Φ_{CO_2} surveys were carried out in the so-called Solfatara crater target area, over a grid of 63 points, whose location remained the same in the time interval of interest (Fig. 8a). For each survey, the total CO_2 flow and its uncertainty were calculated by means of a geostatistical approach.

Based on these numerous data and the less frequent surveys over the whole Solfatara-Pisciarelli DDS, Chiodini et al. (2021) estimated that the total CO_2 flow from the whole Solfatara-Pisciarelli DDS increased from about 1000 $t \cdot day^{-1}$ in 2008–2010 to ca. 3000–4000 $t \cdot d^{-1}$ in 2019–2020 (Fig. 8b). Taking into account that the fumarolic vents experienced a similar increase in fluid emission (Tamburello et al. 2019), the total CO_2 flow from the Solfatara-Pisciarelli area, in 2019–2020, was estimated to be close to 5000 $t \cdot day^{-1}$ by Chiodini et al. (2021).

Using a multivariate analysis, Chiodini et al. (2021) suggested also that the increase in surface gas release could be the result of repeated inputs of magmatic gases in the hydrothermal system, causing the increase of its fluid pressure and temperature, and promoting the Campi Flegrei seismicity, as already recognized by previous studies (e.g., Chiodini et al. 2003; Granieri et al. 2003, 2009; Todesco et al. 2003; De Natale et al. 2006).

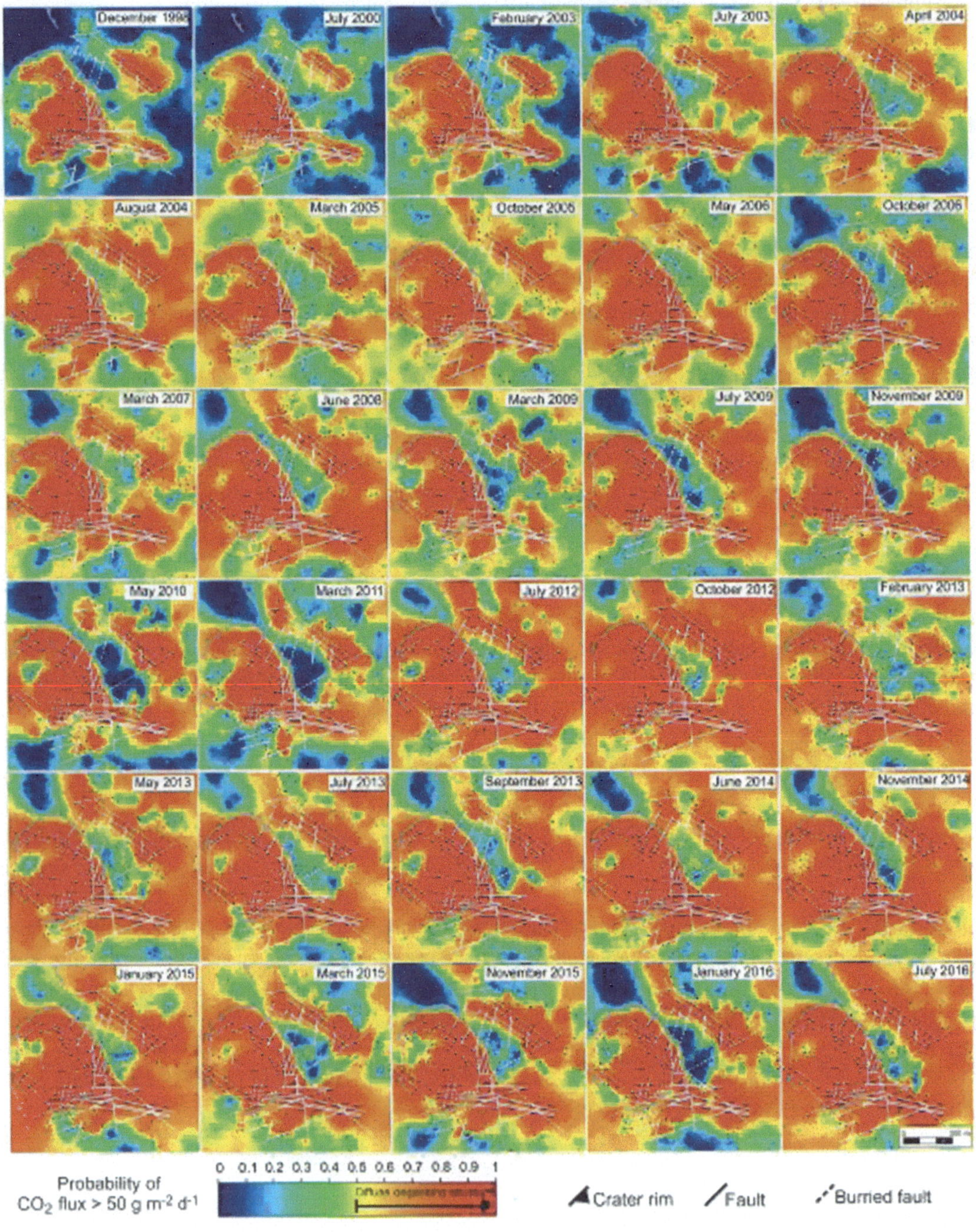

Fig. 7 Maps of the Solfatara-Pisciarelli DDS for each survey performed from 1998 to 2016. The threshold for a pure biogenic Φ_{CO_2} was set at $50\ g \cdot m^2 \cdot day^{-1}$. The yellow and red colors represent the probability that the Φ_{CO_2} is greater than $50\ g \cdot m^2 \cdot day^{-1}$, indicating the contribution of CO_2 from deep sources. From Cardellini et al. (2017), which is licensed under a Creative Commons Attribution 4.0 CC BY International License

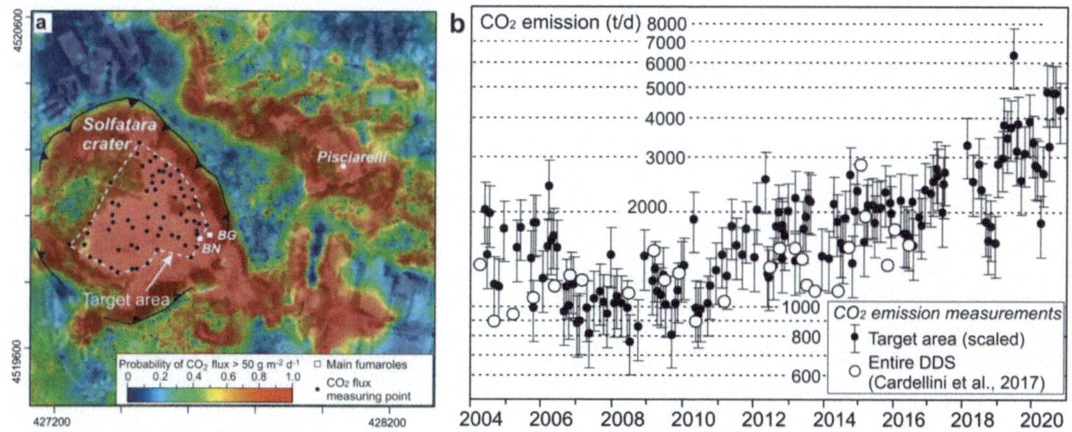

Fig. 8 **a** Map of the Solfatara-Pisciarelli diffuse degassing structure (DDS) showing the location of the 63 measurement points (black dots) of the Solfatara crater target area, which is delimited by the dashed white line. The color map is based on 13,158 CO_2 flux data obtained from 1998 to 2016 (Cardellini et al. 2017) and displays the probability that the simulated CO_2 flux is > 50 g m^{-2} day^{-1}, representing the upper threshold of the pure biogenic CO_2 flux. **b** CO_2 flow emitted from the Solfatara-Pisciarelli DDS in the period 2004–2020. The white circles represent the CO_2 flow emitted from the whole DDS for the thirty surveys reported by Cardellini et al. (2017), whereas the black circles (with error bars) identify the CO_2 flow from the whole DDS computed based on the CO_2 flow from the Solfatara crater target area, adopting a correction factor of 4.7 ± 1.1. Reprinted from Chiodini et al. (2021), Copyright © 2021, with permission from Elsevier

References

Bagnato E, Barra M, Cardellini C, Chiodini G, Parello F, Sprovieri M (2014) First combined flux chamber survey of mercury and CO_2 emissions from soil diffuse degassing at Solfatara of Pozzuoli crater, Campi Flegrei (Italy): Mapping and quantification of gas release. J Volcanol Geotherm Res 289:26–40

Brantley SL, Koepenick KW (1995) Measured carbon dioxide emissions from Oldoinyo Lengai and the skewed distribution of passive volcanic fluxes. Geology 23(10):933–936

Bruno PG, Ricciardi GP, Petrillo Z, Di Fiore V, Troiano A, Chiodini G (2007) Geophysical and hydrogeological experiments from a shallow hydrothermal system at Solfatara Volcano, Campi Flegrei, Italy: response to caldera unrest. J Geophys Res 112:B06201

Cardellini C, Chiodini G, Frondini F, Avino R, Bagnato E, Caliro S, Lelli M, Rosiello A (2017) Monitoring diffuse volcanic degassing during volcanic unrests: the case of Campi Flegrei (Italy). Sci Rep 7(1):6757. https://doi.org/10.1038/s41598-017-06941-2

Chiodini G, Caliro S, Avino R, Bini G, Giudicepietro F, De Cesare W, Ricciolino P, Aiuppa A, Cardellini C, Petrillo Z, Selva J, Siniscalchi A, Tripaldi S (2021) Hydrothermal pressure-temperature control on CO_2 emissions and seismicity at Campi Flegrei (Italy). J Volcanol Geotherm Res 414, 107245

Chiodini G, Cioni R, Guidi M, Raco B, Marini L (1998) Soil CO_2 flux measurements in volcanic and geothermal areas. Appl Geochem 13(5):543–552

Chiodini G, Frondini F, Cardellini C, Granieri D, Marini L, Ventura G (2001) CO_2 degassing and energy release at Solfatara volcano, Campi Flegrei, Italy. J Geophys Res 106:16213–16222

Chiodini G, Todesco M, Caliro S, Del Gaudio C, Macedonio G, Russo M (2003) Magma degassing as a trigger of bradyseismic events: The case of Phlegrean Fields (Italy). Geophys Res Lett 30(8):1434. https://doi.org/10.1029/2002GL016790

Chiodini G, Caliro S, Cardellini C, Avino R, Granieri D, Schmidt A (2008) Carbon isotopic composition of soil CO_2 efflux, a powerful method to discriminate different sources feeding soil CO_2 degassing in volcanic-hydrothermal areas. Earth Planet Sci Lett 274:372–379

Chiodini G, Caliro S, Cardellini C, Granieri D, Avino R, Baldini A, Donnini M, Minopoli C (2010) Long-term variations of the Campi Flegrei, Italy, volcanic system as revealed by the monitoring of hydrothermal activity. J Geophys Res: Solid Earth 115(B3):B03205. https://doi.org/10.1029/2008JB006258

Chiodini G, Paonita A, Aiuppa A, Costa A, Caliro S, De Martino P, Acocella V, Vandemeulebrouck J (2016) Magmas near the critical degassing pressure drive volcanic unrest towards a critical state. Nat Commun 7 (1):1–9

De Natale G, Troise C, Pingue F, Mastrolorenzo G, Pappalardo L, Battaglia M, Boschi E (2006) The Campi Flegrei caldera: unrest mechanisms and hazards. Geol Soc, London, Special Publications 269 (1):25–45

Granieri D, Chiodini G, Marzocchi W, Avino R (2003) Continuous monitoring of CO_2 soil diffuse degassing at Phlegraean Fields (Italy): influence of environmental and volcanic parameters. Earth Planet Sci Lett 212:167–179

Granieri D, Avino R, Chiodini G (2009) Carbon dioxide diffuse emission from the soil: ten years of observations at Vesuvio and Campi Flegrei (Pozzuoli), and linkages with volcanic activity. Bull Volcanol 72:103–118

Sinclair AJ (1974) Selection of threshold values in geochemical data using probability graphs. J. Geochem. Explor. 3:129-149.

Tamburello G, Caliro S, Chiodini G, De Martino P, Avino R, Minopoli C, Carandente A, Rouwet D, Aiuppa A, Costa A, Bitetto M, Giudice G, Francofonte V, Ricci T, Sciarra A, Bagnato E, Capecchiacci F (2019) Escalating CO_2 degassing at the Pisciarelli fumarolic system, and implications for the ongoing Campi Flegrei unrest. J Volcanol Geoth Res 384:151–157

Tassi F, Nisi B, Cardellini C, Capecchiacci F, Donnini M, Vaselli O, Avino R, Chiodini G (2013) Diffuse soil emission of hydrothermal gases (CO_2, CH_4 and C_6H_6) at Solfatara crater (Campi Flegrei, southern Italy). App Geochem 35:142–153

Tassi F, Venturi S, Cabassi J, Capecchiacci F, Nisi B, Vaselli O (2015) Volatile organic compounds (VOCs) in soil gases from Solfatara crater (Campi Flegrei, southern Italy): Geogenic source (s) vs. biogeochemical processes. Appl Geochem 56:37–49

Todesco M, Chiodini G, Macedonio G (2003) Monitoring and modelling hydrothermal fluid emission at La Solfatara (Phlegrean Fields, Italy). An interdisciplinary approach to the study of diffuse degassing. J Volcanol Geotherm Res 125(1–2):57–79

Williams SN, Schaefer SJ, Calvache VML, Lopez D (1992) Global carbon dioxide emission to the atmosphere by volcanoes. Geochim Cosmochim Acta 56 (4):1765–1770

A Comparison Between the Solfatara and Other Magmatic-Hydrothermal Systems

Abstract

It is common practice to distinguish the mature magmatic-hydrothermal systems from the vapor-cored or immature magmatic-hydrothermal systems, based on the chemistry of fumaroles and hot springs. In fact, the first ones are characterized by the absence of highly soluble magmatic gas species in fumarolic fluids and the consequent presence of acid-SO_4 waters in mud pools, whereas the second ones are characterized by the presence of SO_2, HCl and HF in fumarolic fluids and the consequent occurrence of acid SO_4-Cl to acid Cl-SO_4 waters in hot springs and crater lakes. This distinction makes sense, in spite of possible complications due to scrubbing of acidic magmatic gas species, but the conceptual models of mature and immature magmatic-hydrothermal systems can be generalized in that both comprise a deeper magmatic sub-system and a shallower hydrothermal sub-system fighting against each other. Actually, Fournier (1999) proposed a general conceptual model of volcano-hosted magmatic-hydrothermal systems with an impermeable self-sealed zone, at temperature close to 400°C, separating an upper section hosting meteoric hydrothermal fluids at hydrostatic pressure, from a lower section storing magmatic, hypersaline brines and gases at lithostatic pressure. Fournier (1999) also recognized that episodically, the self-sealed zone is breached, in response to different processes, and the hypersaline brines and gases are discharged into the overlying hydrothermal domain, determining an increase in the rate of discharge of hydrothermal-magmatic fluids at the surface. Since the Solfatara magmatic-hydrothermal system is experiencing considerable and relatively fast changes during the on-going unrest period and is one of the most studied case histories worldwide, its evolution is of general interest to understand the competition occurring between the magmatic and the hydrothermal sub-systems in several active volcanoes. ■■■

In volcanology, from its very beginning up to the first half of the twentieth century, types of volcanoes and volcanic activities were defined taking as reference the best-known examples at the time and indicating each new type of volcano or volcanic activity with a name related to the reference case. For a number of reasons, before the big explosive eruptions of the first decades of the twentieth century, the best-known volcanic activity was that of the Italian active volcanoes (e.g. Sigurdsson 1999). For these reasons, different types of volcanic activity took names as Plinian, Strombolian, Vulcanian, and so on. This nomenclature to some extent continues to influence the present classification (Cas and Wright 1987).

L. Marini et al., *The Solfatara Magmatic-Hydrothermal System*, Advances in Volcanology,
https://doi.org/10.1007/978-3-030-98471-7_11

Due to these historical reasons, the term "solfatara" is frequently used to indicate surface manifestations similar to those of the Solfatara, that is, including fumaroles and mud pools and characterized by the deposition of elemental sulfur, sulfate minerals (such as alunite, natroalunite, jarosite, gypsum, and a range of usually ephemeral hydrous sulfates), as well as kaolinite, dickite, pyrophyllite, diaspore, and amorphous silica (Hochstein and Browne 2000). The process ultimately leading to the formation of all these authigenic minerals (except elemental sulfur, see below) is the O_2-driven oxidation of H_2S (and, if present, of SO_2 as well) to SO_3, whose hydration generates sulfuric acid, which promotes the incongruent dissolution of rock-forming minerals. Furthermore, sulfuric acid is often accompanied by hydrochloric and hydrofluoric acids and these mixed acids constitute a very aggressive cocktail. Elemental sulfur is produced by oxidation of H_2S, also driven by atmospheric oxygen. In some places, such as Tatun (Taiwan) and Kawah Ijen (Java, Indonesia), elemental sulfur is mined being deposited in large amounts (Hochstein and Browne 2000).

The occurrence of a "solfatara" represents a useful indication for the reconstruction of the conceptual model of the magmatic-hydrothermal system of interest (e.g., Cumming 2009, 2016), but it is also very important to distinguish if H_2S is the only S-bearing gas species present in fumarolic fluids, as typical of geothermal/hydrothermal steam, or if H_2S is accompanied by SO_2 as well as by HCl and HF, as typical of magmatic fluids (e.g., Giggenbach 1988; Fournier 1999; Fischer and Chiodini 2015).

If H_2S is the only S-bearing gas species, it can be deduced that the highly soluble magmatic gases (i.e., SO_2, HCl, and HF) are totally absorbed by the hydrothermal aquifer situated above the degassing magma batch present at depth. This means that the system of interest is (or is similar to) a mature magmatic-hydrothermal system associated to a stratovolcano, according to the classification of Bogie et al. (2005) (Fig. 1). In this setting, neutral sodium-chloride waters are

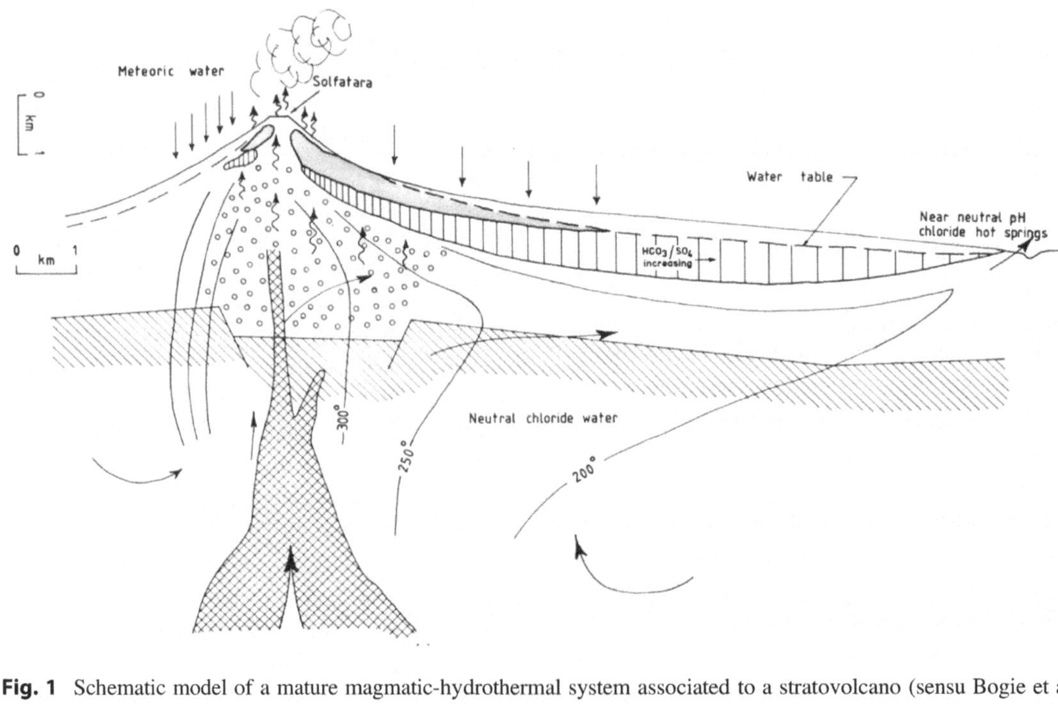

Fig. 1 Schematic model of a mature magmatic-hydrothermal system associated to a stratovolcano (sensu Bogie et al. 2005). Reprinted from Henley and Ellis (1983), Copyright © 1983, with permission from Elsevier

generally stored in the hydrothermal aquifer (or geothermal reservoir), as is the case of the geothermal systems situated along convergent plate boundaries, such as those of Central America (e.g., Miravalles and Berlin), Japan (e.g., Mori-Nigorikawa, Sumikawa, Uenotai-Wasabizawa, Onikobe, Oku-Aizu, Takigami, Oguni, and Fushime), the Philippines (e.g., Bacon-Manito, Tongonan-Mahanagdong, Alto Peak, and Palinpinon), and New Zealand (e.g., Ngawha, Kawerau, Rotorua, Waiotapu, Orakeikorako, Mokai, Ngatamariki, Broadlands-Ohaaki, Wairakei, and Rotokawa), as discussed by Cioni and Marini (2020) and references therein. Most of these neutral sodium-chloride waters are generated though complete neutralization of the initially acidic aqueous solutions produced through absorption of magmatic gases in deep circulating meteoric waters (Giggenbach 1988, 1992), whereas seawater may be involved in the recharge of geothermal reservoirs in islands and coastal areas such as Nisyros Island, Greece (Marini and Fiebig 2005) and the Campi Flegrei (see Sect. 1.3 in chapter "The Magmatic–Hydrothermal System Hosted in the Campi Flegrei Caldera with Emphasis on the Solfatara").

Alternatively, if H_2S is accompanied by SO_2, HCl and HF, it can be inferred that highly soluble magmatic gases reach the surface and, therefore, a continuous column of magmatic gases or a so-called "magmatic vapor core", connects the degassing magma batch present at depth and the "solfatara" occurring at the surface. In this setting, the summit of the volcanic edifice is often characterized by the presence of an active crater lake, hosting acidic waters with high concentrations of SO_4, Cl, and F as well as of SiO_2, Al and Fe in addition to the major cations, such as at the already mentioned Kawah Ijen (Delmelle and Bernard 1994; Delmelle et al. 2000; van Hinsberg et al. 2010a,b), Ruapehu, New Zealand (Christenson 2000), and Poás, Costa Rica (Martínez Cruz 2008; Rouwet et al. 2019). Similar acidic Cl–SO_4 to SO_4–Cl waters, usually more dilute than those of the active crater lake (if any), are discharged by the springs present on the flanks of several volcanoes in Indonesia, being the result of shallow downflows of acidic mixed condensate and meteoric water that are progressively neutralized by rock dissolution, as pointed out by Hochstein and Sudarman (2015). Nevertheless, neutral chloride waters are (or may be) present also in these magmatic-hydrothermal systems, away from the central magmatic vapor core. These systems are called vapor-cored magmatic-hydrothermal systems (Reyes et al. 1993) or immature magmatic-hydrothermal systems associated to stratovolcanoes, according to the classification of Bogie et al. (2005) (Fig. 2). Examples are Mt. Apo, in the southeastern part of the Mindanao Island (Ramos-Candelaria et al. 1995), Alto Peak in Leyte Province (Reyes et al. 1993), and Biliran in the homonymous island at the northern tip of Leyte (Apuada and Sigurjonsson 2008), all three in the Philippines, several systems in Indonesia, such as Tangkuban Perahu, Kawah Ijen, Mt. Bromo, Mt. Galunggung, Arjuno-Welirang and Mt. Slamet (Abiyudo et al. 2016).

The first conceptual "gasometer-type" model of vapor-cored magmatic-hydrothermal systems comprising a bellows-like envelope of boiling brine surrounding the central magmatic gas column was proposed by Giggenbach (1975, 1982, 1983) to explain cyclic changes in the chemical and isotopic characteristics of White Island fumaroles, New Zealand. The same model was found suitable for the magmatic-hydrothermal system of the La Fossa cone, Vulcano Island, Italy by Cioni and D'Amore (1984), who recognized that fumarolic fluids are mixtures of magmatic gases and vapors generated by total evaporation of brines coming from lateral aquifers probably recharged by seawater and seeping into the fumarolic plumbing system. Giggenbach (1987) performed a thermodynamic assessment of ca. 250 chemical analyses of gases discharged from White Island fumaroles, at outlet temperatures of 100–800 °C and found that fumarolic fluids are constituted by two distinct components: (a) an SO_2-rich magmatic component ascending directly and quickly from the underlying degassing magma and (b) a hydrothermal component moving slowly upwards and coming from a bi-phase (brine + vapor) envelope surrounding the central magmatic system (Fig. 3).

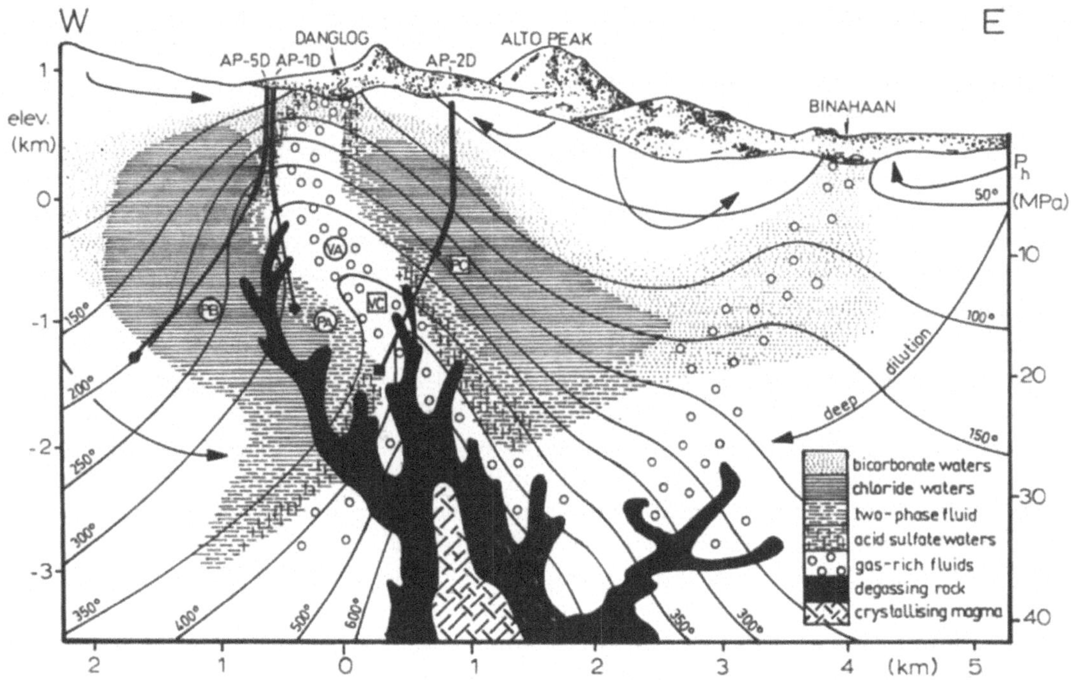

Fig. 2 Schematic model of a vapor-cored magmatic-hydrothermal system associated to a stratovolcano (Reyes et al. 1993) or immature magmatic-hydrothermal system associated to a stratovolcano (sensu Bogie et al. 2005). Reprinted from Reyes et al. (1993), Copyright © 1993, with permission from Elsevier

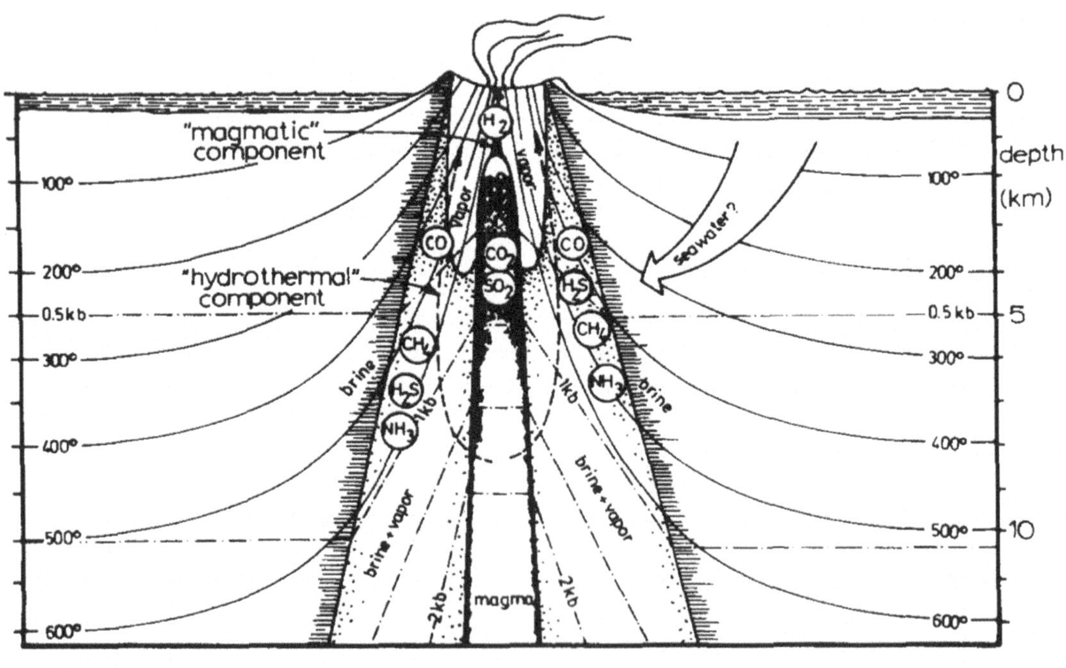

Fig. 3 Conceptual model of the magmatic-hydrothermal system of White Island, New Zealand. Reprinted from Giggenbach (1987), modified, Copyright © 1987, with permission from Elsevier

Giggenbach (1987) recognized the presence of two redox buffers (see Sect. 2 in chapter "The Redox Potential and Sulfur Gas Species"), estimated the temperatures and pressures of equilibration for the sub-systems H_2–H_2O, CO–CO_2, CH_4–CO_2, and NH_3–N_2, and used these outcomes to infer the approximate places of the equilibration zones of these gas sub-systems, which are reported in the conceptual model of Fig. 3. Both ammonia and methane are produced at depth within the hydrothermal sectors of the system, where redox conditions are controlled by the Fe(II)–Fe(III) rock-system buffer and are transported at shallow levels by the slowly uprising hydrothermal vapor plume. Then, NH_3 and CH_4 are entrained in variable amounts into the fast-rising dry (superheated) magmatic gas flow, where redox conditions are fixed by the SO_2–H_2S magmatic gas buffer, but neither NH_3 nor CH_4 are able to re-equilibrate owing to the very short travel times. In contrast, the fast-reacting H_2–H_2O and possibly CO–CO_2 sub-systems are able to respond, readjusting at the pressure and temperature which are present close to the surface discharge.

A few years later, Chiodini et al. (1993) found that the same processes affecting the fumarolic fluids of White Island take place also in the fumarolic fluids discharged from La Fossa crater area on Vulcano Island, thus confirming the groundbreaking findings of Giggenbach (1987). The fumaroles of the Baia di Levante beach, at a distance of ca. 500 m from the northern foot of the La Fossa crater, discharge steam without highly soluble magmatic gases coming from a boiling hydrothermal aquifer which was met by the shallow boreholes drilled in the 50s. The beach fumaroles have experienced relatively important time changes in flow rate and weak variations in gas chemistry, while temperature has remained practically constant at ca. 100 °C (Cioni and D'Amore 1984 and references therein). In contrast, crater fumaroles have experienced conspicuous time changes in flow rate, temperature, and gas chemistry. These different time changes of the two groups of fumaroles are probably due to the role of the hydrothermal aquifer. On the one hand, the hydrothermal aquifer is maintained at high temperature by the input of deep magmatic fluids, possibly the same as those discharged by the crater fumaroles. On the other hand, the hydrothermal aquifer buffers the time changes in the flow, chemistry, and temperature of the deep magmatic fluids entering it from below mainly due to hydrodynamic dispersion.

Based on previous discussion, it is evident that it makes sense to distinguish (1) the mature magmatic-hydrothermal systems, which are characterized by the absence of highly soluble magmatic gases in fumarolic fluids and the consequent presence of acid-SO_4 waters in mud pools, from (2) the immature magmatic-hydrothermal systems, which are characterized by the presence of SO_2, HCl and HF in fumarolic fluids and the consequent occurrence of acid SO_4–Cl to acid Cl–SO_4 waters in related hot springs and crater lakes (if any). Nevertheless, owing to their high solubility in water, SO_2, HCl and HF are easily scrubbed from magmatic gases by steam condensation or interaction with external waters (Symonds et al. 2001; Marini and Gambardella 2005). In other words, SO_2, HCl and HF are carried by dry (superheated) steam but are not transported by saturated steam because of their preferential partition in the liquid phase. Thus, the high solubility of SO_2, HCl and HF may complicate the distinction between mature and immature geothermal systems. Indeed, this was the case of Alto Peak, where fumarole gas chemistry is typical of a mature system, due to the occurrence of steam condensation, but the magmatic vapor chimney was met at depth upon drilling (Reyes et al. 1993). As already recalled in Sect. 3.2 in chapter "The Magmatic–Hydrothermal System Hosted in the Campi Flegrei Caldera with Emphasis on the Solfatara", the same might occur also at Solfatara, where the acidic-magmatic gas species, namely SO_2, HCl and HF, are practically absent possibly because of removal by scrubbing upon minor steam condensation at depth during the close-to-isenthalpic decompression of the uprising gases (Caliro et al. 2007; Moretti et al. 2013; Aiuppa et al. 2013; Chiodini et al. 2012, 2021). Further complications are posed by the possible conversion of H_2S in SO_2 upon depressurization and cooling. In fact, SO_2 is stabilized by decreasing pressure and increasing temperatures, whereas H_2S is favored by increasing pressures and decreasing temperatures (Marini et al. 2011).

In addition, it must be noted that the conceptual models of mature and immature magmatic-hydrothermal systems can be generalized in that both comprise two contenders fighting against each other, namely: (1) a degassing magma batch situated at depth and (2) a hydrothermal aquifer hosting neutral Na-Cl brines positioned above it. The hydrothermal aquifer may prevail absorbing all the gases released from the degassing magma, as is the case of the mature magmatic-hydrothermal systems or, alternatively, the flow of uprising magmatic gases may pierce and push away the hydrothermal aquifer, as is the case of the immature magmatic-hydrothermal systems. The winner of this competition depends on several controlling factors, such as the flow of ascending magmatic gases, the flow of meteoric (or marine) waters recharging the hydrothermal aquifer, the thermodynamic conditions in different parts of the magmatic-hydrothermal system, the permeability of the distinct rock formations, the deposition of authigenic minerals locally reducing rock permeability, and seismic activity locally increasing rock permeability in zones affected by fracturing.

Thus, the classification of a given magmatic-hydrothermal system as mature or immature might be correct, in spite of the controversial aspect concerning the fate of highly soluble magmatic gases (see above), but it refers to a given time. However, magmatic-hydrothermal systems may experience significant changes due to log-term variations in one or more of the controlling factors recalled above or, in a short-term and in a catastrophic way, due to the occurrence of an eruption (either hydrothermal or phreatomagmatic or magmatic) or an external event, such as a landslide removing the cap-rock of the hydrothermal aquifer and killing it.

Actually, as already recalled in chapter "The Magmatic–Hydrothermal System Hosted in the Campi Flegrei Caldera with Emphasis on the Solfatara", https://doi.org/10.1007/978-3-030-98471-7_3 a general conceptual model of volcano-hosted magmatic-hydrothermal systems, without distinguishing the mature and immature types, was proposed by Fournier (1999). In the Fournier's model, a thin, relatively impermeable zone generated through self-sealing (mainly consisting in quartz deposition) at temperature close to 400 °C separates an upper section, in which hydrothermal fluids of meteoric derivation circulate through brittle rocks at hydrostatic pressure, from a lower section, in which hypersaline brines and gases coming from the underlying crystallizing magma are stored at lithostatic pressure within plastic rocks. Fournier (1999) also recognized that episodically, the self-sealed zone is breached in response to different processes and the hypersaline brines and gases are discharged into the brittle domain at hydrostatic pressure, ultimately causing an increase in the rate of discharge of hydrothermal-magmatic fluids at the surface. Indeed, Fournier (1999) provided a very detailed description of the competition between the magmatic and hydrothermal contenders.

Whatever the reason for the changes occurring in magmatic-hydrothermal systems, it is important to realize that sooner or later they will occur, altering the state of the system of interest, which is not immutable over time. If we accept this point of view, the distinction between mature and immature systems, although real in a given moment, fades away in the long term, which is probably much greater than our intrinsic measurement unit, that is, our life or a part of it. Since the Solfatara system is evolving very fast during the current unrest period and is one of the most studied case histories worldwide, it is extremely interesting not only to the geochemists and geoscientists monitoring it for volcanic surveillance, but also to the geochemists and geoscientists working on other volcanoes in order to better understand the competition between the magmatic sub-system and the hydrothermal subsystem.

References

Abiyudo R, Hadi J, Cumming W, Marini L (2016) Conceptual model assessment of vapor core geothermal system for exploration. Mt. Bromo case study. In: Proceedings of the 4th Indonesia international geothermal convention and exhibition 2016, 10–12 August 2016, Cendrawasih Hall, Jakarta Convention Center, Indonesia

Aiuppa A, Tamburello G, Di Napoli R, Cardellini C, Chiodini G, Giudice G, Grassa F, Pedone M (2013) First observations of the fumarolic gas output from a restless caldera: implications for the current period of unrest (2005–2013) at Campi Flegrei. Geochem Geophys Geosyst 14(10):4153–4169

Apuada NA, Sigurjonsson GF (2008) The geothermal potential of Biliran Island, Philippines. In: Proceedings of the 8th Asian geothermal symposium, 73–77

Bogie I, Lawless JV, Rychagov S, Belousov V (2005) Magmatic-related hydrothermal systems: classification of the types of geothermal systems and their ore mineralization. In: Proceedings of geoconference in Russia, Kuril

Caliro S, Chiodini G, Moretti R, Avino R, Granieri D, Russo M, Fiebig J (2007) The origin of the fumaroles of La Solfatara (Campi Flegrei, south Italy). Geochim Cosmochim Acta 71(12):3040–3055

Cas RAF, Wright JV (1987) Volcanic successions. Modern and ancient. Allen and Unwin, London, p 528

Christenson BW (2000) Geochemistry of fluids associated with the 1995–1996 eruption of Mt. Ruapehu, New Zealand: signatures and processes in the magmatic-hydrothermal system. J Volcanol Geotherm Res 97(1–4):1–30

Chiodini G, Cioni R, Marini L (1993) Reactions governing the chemistry of crater fumaroles from Vulcano Island, Italy, and implications for volcanic surveillance. Appl Geochem 8:357–371

Chiodini G, Caliro S, De Martino P, Avino R, Gherardi F (2012) Early signals of new volcanic unrest at Campi Flegrei caldera? Insights from geochemical data and physical simulations. Geology 40(10):943–946

Chiodini G, Caliro S, Avino R, Bini G, Giudicepietro F, De Cesare W, Ricciolino P, Aiuppa A, Cardellini C, Petrillo Z, Selva J, Siniscalchi A, Tripaldi S (2021) Hydrothermal pressure-temperature control on CO_2 emissions and seismicity at Campi Flegrei (Italy). J Volcanol Geotherm Res 414:107245

Cioni R, D'Amore F (1984) A genetic model for the crater fumaroles of Vulcano Island (Sicily, Italy). Geothermics 13(4):375–384

Cioni R, Marini L (2020) A thermodynamic approach to water geothermometry. Springer geochemistry. Springer Nature, Cham, Switzerland, pp 415

Cumming W (2009) Geothermal resource conceptual models using surface exploration data. In: Proceedings of the 34th workshop on geothermal reservoir engineering. Stanford University, Stanford, California, February 9–11, SGP-TR-187, pp 6

Cumming W (2016) Resource conceptual models of volcano-hosted geothermal reservoirs for exploration well targeting and resource capacity assessment: construction, pitfalls and challenges. Geotherm Resour Counc Trans 40:623–637

Delmelle P, Bernard A (1994) Geochemistry, mineralogy, and chemical modeling of the acid crater lake of Kawah Ijen Volcano, Indonesia. Geochimica Et Cosmochimica Acta 58(11):2445–2460

Delmelle P, Bernard A, Kusakabe M, Fischer TP, Takano B (2000) Geochemistry of the magmatic–hydrothermal system of Kawah Ijen volcano, East Java, Indonesia. J Volcanol Geoth Res 97(1–4):31–53

Fischer TP, Chiodini G (2015) Volcanic, magmatic and hydrothermal gases. In: Sigurdsson H, Houghton B, McNutt S, Rymer H, Stix J (eds) The encyclopedia of volcanoes, 2nd edn. Academic Press-Elsevier, pp 779–797

Fournier RO (1999) Hydrothermal processes related to movement of fluid from plastic into brittle rock in the magmatic-epithermal environment. Econ Geol 94 (8):1193–1211

Giggenbach WF (1975) Variations in the carbon, sulfur and chlorine contents of volcanic gas discharges from White Island, New Zealand. Bull Volcanol 39(1):15–27

Giggenbach WF (1982) The chemical and isotopic composition of gas discharges from New Zealand andesitic volcanoes. Bull Volcanol 45(3):253–255

Giggenbach WF (1983) Chemical surveillance of active volcanoes in New Zealand. In: Tazieff H, Sabroux JC (eds) Forecasting volcanic events. Elsevier, Amsterdam, pp 311–322

Giggenbach WF (1987) Redox processes governing the chemistry of fumarolic gas discharges from White Island, New Zeland. Appl Geochem 2:143–161

Giggenbach WF (1988) Geothermal solute equilibria. Derivation of Na-K-Mg-Ca geoindicators. Geochimica et Cosmochimica Acta 52(12):2749–2765

Giggenbach WF (1992) Isotopic shifts in waters from geothermal and volcanic systems along convergent plate boundaries and their origin. Earth Planet Sci Lett 113(4):495–510

Henley RW, Ellis AJ (1983) Geothermal systems ancient and modern: a geochemical review. Earth Sci Rev 19 (1):1–50

Hochstein MP, Browne PR (2000) Surface manifestations of geothermal systems with volcanic heat sources. In: Sigurdsson H, Houghton BF, McNutt SR, Rymer H, Stix J (eds) Encyclopedia of volcanoes. Academic Press, San Diego, pp 835–855

Hochstein MP, Sudarman S (2015) Indonesian volcanic geothermal systems. In: Proceedings world geothermal congress 2015, Melbourne, Australia, 19–25 April 2015, pp 11

Martínez Cruz M (2008) Geochemical evolution of the acid crater lake of Poás volcano (Costa Rica): insights into volcanic-hydrothermal processes. PhD Thesis, Utrecht University, pp 161

Marini L, Fiebig J (2005) Fluid geochemistry of the magmatic-hydrothermal system of Nisyros (Greece). In: Hunziker JC, Marini L (eds) The geology, geochemistry and evolution of Nisyros Volcano (Greece). Implications for the volcanic hazards, Memoires de Géologie (Lausanne), 44, pp 121–163

Marini L, Gambardella B (2005) Geochemical modeling of magmatic gas scrubbing. Ann Geophys 48:739–753

Marini L, Moretti R, Accornero M (2011) Sulfur isotopes in magmatic-hydrothermal systems, melts, and magmas. In: Behrens H, Webster JD (eds) Sulfur in magmas and melts: its importance for natural and technical processes. Rev Mineral Geochem 73 (1):423–492

Moretti R, Arienzo I, Civetta L, Orsi G, Papale P (2013) Multiple magma degassing sources at an explosive volcano. Earth Planet Sci Lett 367:95–104

Ramos-Candelaria M, Sanchez DR, Salonga ND (1995) Magmatic contributions to Philippine hydrothermal systems. In: Proceedings of the world geothermal congress, vol 2. Firenze, Italy, pp 1337–1341

Rouwet D, Amador RAM, Sandri L, Ramírez-Umaña C, González G, Pecoraino G, Capaccioni B (2019) 39 years of geochemical monitoring of Laguna Caliente Crater Lake, Poás: patterns from the past as keys for the future. In: Poás Volcano. Springer, Cham pp 213–233. https://doi.org/10.1007/978-3-319-02156-0_9

Reyes AG, Giggenbach WF, Saleras JR, Salonga ND, Vergara MC (1993) Petrology and geochemistry of Alto Peak, a vapor-cored hydrothermal system, Leyte Province, Philippines. Geothermics 22(5–6):479–519

Sigurdsson H (1999) Melting the earth: the history of ideas on volcanic eruptions. Oxford University Press, New York, p 260

Symonds RB, Gerlach TM, Reed MH (2001) Magmatic gas scrubbing: implications for volcano monitoring. J Volcanol Geoth Res 108(1–4):303–341

Van Hinsberg V, Berlo K, Sumarti S, Van Bergen M, Williams-Jones A (2010b) Extreme alteration by hyperacidic brines at Kawah Ijen volcano, East Java, Indonesia: II: metasomatic imprint and element fluxes. J Volcanol Geotherm Res 196(3–4):169–184

Van Hinsberg V, Berlo K, van Bergen M, Williams-Jones A (2010a) Extreme alteration by hyperacidic brines at Kawah Ijen volcano, East Java, Indonesia: I. Textural and mineralogical imprint. J Volcanol Geotherm Res 198(1–2):253–263

Conclusive Remarks

Abstract

In the previous chapters of this book, new gas geoindicators involving the fugacity coefficients of relevant gas species, applicable up to ca. 1000°C and ca. 3 kbar, have been developed and applied to Solfatara fluids to answer the urgent need to estimate the temperatures and pressures present at depth, in different parts of the magmatic-hydrothermal system of Solfatara volcano, during its on-going unrest. In this chapter, the obtained geothermometric-geobarometric results are summarized and discussed. We used the new geoindicators developed in this work to monitor the temperature and pressure in different portions of the Solfatara magmatic-hydrothermal system, namely: (a) in the shallow steam zone, where CO equilibrates at 215-226°C and 16.1-28.2 bar, (b) at intermediate depths, where CH_4 attains equilibrium at T, P conditions increasing with time, namely, from 235-267°C, 31-54 bar, in June 1983 – July 1984, to 576-604°C, 1167-1326 bar in 2020 for Bocca Grande, and up to 605-645°C, 1293-1543 bar in 2020, for Bocca Nuova, assuming saturation with a brine containing 21 wt% NaCl, and (c) in the carbonate layer at 6–7 km depth, where H_2S and CO_2 equilibrate with anhydrite and calcite, at T, P conditions increasing with time, namely, from 667±24°C, 1307±102 bar in June 1983 – July 1984 to 990±30°C, 2957 ±146 bar in 2019-2020, for Bocca Grande, and up to 1010±37°C, 3033±182 bar in 2019-2020 for Bocca Nuova, assuming saturation with a brine containing 33.5 wt% NaCl. Accepting that the fluid pressure in the carbonate layer at 6–7 km depth is governed by magmatic degassing and that the external (overburden) pressure is 1330±135 bar, the latter value was very similar to the total fluid pressure of 1307±102 bar at 6-7 km depth, in June 1983 – July 1984, and a very small flow of magmatic gases flushed the system at that time. Since September 1984, fluid pressure at 6-7 km depth began to exceed the external pressure, making possible the increasing input of fluids coming from the carbonate layer into the overlying system and the breaching of the self-sealed zone separating the upper hydrothermal part from the underlying magmatic portion of the system. The excess fluid pressure increased during the last twenty years, especially during the last 3-4 years, explaining the growth in the degassing process detected at the surface. Consequently: (1) in June 1983 – July 1984, the residence time of fluids within the shallow steam zone was long enough for the attainment of full equilibrium in the H_2O-CO_2-CO-CH_4-H_2 system, in spite of the relatively low temperatures, whereas (2) since September 1984, the residence time of fluids within the shallow steam zone decreased,

L. Marini et al., *The Solfatara Magmatic-Hydrothermal System*, Advances in Volcanology,
https://doi.org/10.1007/978-3-030-98471-7_12

preventing CH_4 re-equilibration, but still allowing CO re-equilibration at local P-T conditions. ■■■

The high-quality geochemical data acquired by Chiodini and coworkers (Chiodini et al. 2021 and references therein) unequivocally demonstrate that Solfatara volcano is experiencing an unrest episode since at least the bradyseismic crisis of 1982–1984, as pointed out by the important time changes in the chemical composition of fumarolic fluids (see chapter "Chemical and Isotopic Characteristics of Solfatara-Pisciarelli Fumarolic Fluids"), in soil CO_2 flux, and related thermal energy release from the Solfatara-Pisciarelli diffuse degassing structure (see chapter "Soil CO_2 Diffuse Degassing and Thermal Energy Release as Indicators of Volcanic Unrest in the Solfatara-Pisciarelli Area"), as well as in geodetic and geophysical parameters, which are beyond the aims of this book. In contrast, the outlet temperature of the two main fumaroles, Bocca Grande and Bocca Nuova, underwent nil to negligible time variations, being fixed at 161.4 ± 2.5 °C (1σ) and 145.4 ± 2.3 °C, respectively (see chapter "The Expansion (Decompression) of the Solfatara Fumarolic Fluids"). This lack of time changes in the outlet temperature is probably due by the little variations in the temperature and pressure of the shallow steam zone directly connected with the two fumaroles (see below).

Given this situation, there is an urgent need to estimate the high temperatures and pressures presumably present at greater depths (i.e., below this shallow steam zone) along the expansion path of fumarolic fluids. Therefore, the approach to geothermometry and geobarometry of fumarolic gases was thoroughly revised in this book, elaborating new geoindicators controlled by the re-equilibration of the following gas constituents:

(1) carbon monoxide, based on the reverse water–gas shift reaction (indicated by the acronym RWG):

$$CO_2 + H_2 = CO + H_2O \quad (1)$$

(2) methane, on the basis of the Sabatier-Senderens reduction with stoichiometric coefficient of H_2 equal to 4 (indicated by the acronym SS4):

$$CO_2 + 4H_2 = CH_4 + 2H_2O \quad (2)$$

(3) hydrogen sulfide and carbon dioxide, as well as hydrogen and water, as dictated by the heterogeneous reaction involving anhydrite and calcite:

$$CaSO_{4(s)} + CO_2 + 4H_2 = CaCO_{3(s)} + H_2S + 3H_2O \quad (3)$$

As discussed in chapters "Calibration and Application of the RWG and SS4 Indicators for the Possible Expansion Paths of Solfatara Fluids" and "The Redox Potential and Sulfur Gas Species", these new geothermometric-geobarometric techniques were calibrated taking into account the deviations from the ideal gas behavior in a simplified and generalizable form and different expansion (decompression) paths, namely:

(a) the isenthalpic path,
(b) two saturation paths, one involving a liquid phase containing 21 wt% NaCl, the other comprising a liquid phase containing 33.5 wt % NaCl, and
(c) the linear P–T path, in which both total fluid pressure and temperature increase linearly from the critical point of water to 2879 bar, 1120 °C, that is, the total fluid pressure and temperature of the magmatic endmember M1 of Caliro et al. (2014), whereas below the critical point of water, H_2O partial pressure and temperature are assumed to be constrained by the saturation condition for pure water.

The reasons for treating separately CH_4 and CO in reactions (1) and (2) and related implications are discussed in chapter "The Geoindicators Involving H_2O, CO_2, CO, CH_4 and H_2", whereas the possible reactive effects during the cooling of the Solfatara gas mixtures were investigated in chapter "The Chemical Kinetics of the Reactions Involving H_2O, CO_2, CO, CH_4 and H_2", assuming that the kinetics of relevant reactions involving H_2O, CO_2, CO, CH_4 and H_2 is conveniently described by the rate equations proposed by Xu and Froment (1989) and one or more unknown but effective catalysts are present in the natural system of interest.

The application of the new geothermometric-geobarometric techniques derived in this book to the Solfatara fumarolic fluids gives the outcomes summarized here below (see chapters "Calibration and Application of the RWG and SS4 Indicators for the Possible Expansion Paths of Solfatara Fluids" and "The Redox Potential and Sulfur Gas Species" for further details), considering the analytical concentrations of gas species to compute the RWG and SS4 equilibrium temperatures, but using the corrected concentrations of H_2, CO_2, and H_2O, according to Eqs. (58), (59), and (60) in chapter "The Redox Potential and Sulfur Gas Species", to calculate the H_2S-CO_2 equilibrium temperature.

The equilibrium temperatures and total fluid pressures indicated by the RWG reaction are relatively low and do not show important changes with time, with mean values of 215–223 °C and 16.1–27.3 bar for Bocca Grande and 217–226 °C and 16.7–28.2 bar for Bocca Nuova. These T, P values are representative of the shallow permeable zone, in which carbon monoxide attains the last equilibrium condition. The depth of this shallow permeable zone is probably close to 160 to 280 m below the local water table.

In contrast, the equilibrium temperatures and total fluid pressures indicated by the SS4 reaction exhibit significant increments with time, namely: (a) from 235–267 °C, 31–54 bar in June 1983–July 1984 to 576–604 °C, 1167–1326 bar in 2020 for Bocca Grande and up to 605–645 °C, 1293–1543 bar in 2020 for Bocca Nuova, for the saturation path involving 21 wt% NaCl; (b) from 222–252 °C, 17–29 bar in June 1983–July 1984 to 565–595 °C, 1028–1198 bar in 2020 for Bocca Grande and up to 595–632 °C, 1160–1393 bar in 2020 for Bocca Nuova, for the saturation path involving 33.5 wt% NaCl.

Thus, from June 1983 to July 1984, the SS4 reaction equilibrated under the same P–T conditions of the RWG reaction, as already recognized by Cioni et al. (1984), suggesting that the residence time of fluids within the shallow permeable zone was long enough for the attainment of the overall equilibrium condition in the H_2O-CO_2-CO-CH_4-H_2 system, in spite of the comparatively low temperatures. Assuming that the system was close to steady state, the long residence time of fluids, τ, implies that the flow of magmatic gases, Q, entering this shallow permeable zone from below, was probably very small, as dictated by the well-known relation (e.g., Kazemi et al. 2006 and references therein):

$$\tau = \frac{V}{Q} \tag{4}$$

where V is the volume of the shallow permeable zone.

Since September 1984, an important change in gas chemistry was recorded, with a considerable increase in the N_2/CO_2 ratio and a significant decrease in the H_2S/CO_2 ratio. Thanks to the availability of He and Ar data since June 1989 and July 1998, respectively, it was ascertained that the increase in the N_2/CO_2 ratio was accompanied by a decrease in the N_2/He and Ar/CO_2 ratios (Chiodini et al. 2012) and these changes were ascribed to degassing of a trachybasalt magma (Caliro et al. (2014). Thus, a considerable flow of deep magmatic gases began to flush the shallow permeable zone since September 1984, possibly due to breaching of the thin, relatively impermeable self-sealed zone presumably separating the upper brittles rocks from the underlying plastic region, according to the general conceptual model of volcano-hosted magmatic-hydrothermal systems proposed by

Fournier (1999).[1] Consequently, the residence time of fluids within the shallow permeable zone decreased, preventing CH_4 re-equilibration, but still allowing CO re-equilibration at local P–T conditions. In fact, since September 1984, the SS4 reaction began to indicate equilibrium P–T conditions higher than those pointed out by the RWG reaction (assuming that gas equilibration is attained in a single saturated vapor phase) and the difference between these two equilibrium temperatures increased progressively with time. The increasing SS4 equilibrium temperature with time allows for two distinct implications, either (1) fumarolic fluids came from progressively deeper zones of the Solfatara magmatic-hydrothermal system, characterized by gradually higher fluid pressures and temperatures, or (2) fumarolic fluids came from the same deep permeable zone which experienced a progressive increase in fluid pressure and temperature with time.

Further indications are given by the H_2S-CO_2 equilibrium temperature and fluid pressure fixed by calcite-anhydrite coexistence, although these outcomes must be considered with some caution because of the heterogeneous nature of the reaction controlling this geothermometer and the possible involvement of solid phases other than anhydrite as partner of H_2S. Nevertheless, pyrite does not seem to be a plausible partner of H_2S, as discussed in Sect. 4.1 in chapter "The Redox Potential and Sulfur Gas Species". Similar to the SS4 equilibrium temperature and fluid pressure, also the H_2S-CO_2 equilibrium temperature and fluid pressure fixed by calcite-anhydrite coexistence, for the saturation path involving 33.5 wt% NaCl, increase almost continuously with time, from 667 ± 24 °C, 1307 ± 102 bar in June 1983–July 1984 to 990 ± 30 °C, 2957 ± 146 bar in 2019–2020, for Bocca Grande, and up to 1010 ± 37 °C, 3033 ± 182 bar in 2019–2020 for Bocca Nuova. To gain more mileage from these figures, let us assume that: (a) the transition between the hydrostatic and the lithostatic regime occurs at a depth of 2.5 km, and (b) equilibrium among calcite, anhydrite, H_2S, CO_2, H_2, and H_2O is attained in the deepest part of the carbonate layer present below the Campi Flegrei caldera, at a depth of 6–7 km, by analogy with the abundance of authigenic anhydrite in carbonate geothermal reservoirs (e.g., Latera; Cavarretta et al. 1985). The external (overburden) pressure at this depth (to be considered as an educated guess) is expected to be of 1330 ± 135 bar. Therefore, in June 1983–July 1984, the total fluid pressure of 1307 ± 102 bar was very similar to the external pressure at 6–7 km depth, in line with the very small flow of magmatic gases flushing the system (see above). Since September 1984, fluid pressure at 6–7 km depth began to exceed the external pressure, making possible the increasing input, into the overlying system, of fluids coming from the carbonate layer, and the excess fluid pressure (that is, the difference between the fluid pressure and the overburden pressure) increased during the last twenty years, especially during the last 3–4 years. This increment in the excess fluid pressure with time explains the progress in the degassing process detected at the surface (see chapter "Soil CO_2 Diffuse Degassing and Thermal Energy Release as Indicators of Volcanic Unrest in the Solfatara-Pisciarelli Area "), and, even more important, could play a pivotal role in controlling the bradyseism resurgence currently affecting the entire Campi Flegrei caldera. However, the fluid pressure in the carbonate layer at 6–7 km depth, in turn, is controlled by magmatic degassing and the fluid pressure of the underlying magma layer, which has been obviously higher than the fluid pressure in the carbonate layer at 6–7 km depth at least since September 1984.

[1] Actually, abundant hydrothermal quartz was found in well San Vito 1, at depths between ca. 2500 and 2800 m and temperatures of ca. 360 to 385 °C, at the transition between the Ca-Al-silicate (propylitic) zone (above) and the thermometamorphic zone (below), in agreement with the Fournier's conceptual model (see Sect. 1.2 in chapter "The Magmatic–hydrothermal System Hosted in the Campi Flegrei Caldera with Emphasis on the Solfatara"). Moreover, well San Vito 1 has produced gas-rich brines (see Sect. 1.3 in chapter "The Magmatic–hydrothermal System Hosted in the Campi Flegrei Caldera with Emphasis on the Solfatara"), probably coming from the magmatic reservoir marked by thermo-metamorphic alteration, where fluid pressure is controlled by the lithostatic regime.

All in all, the new geothermometric-geobarometric techniques implemented in this book are effective tools to monitor the temperature and fluid pressure in different zones of the Solfatara magmatic-hydrothermal system, namely: (a) in the shallow steam zone, where CO equilibrates, (b) at intermediate depths, where CH_4 attains the equilibrium condition, and (c) in the carbonate layer at 6–7 km depth, where H_2S and CO_2 equilibrates with anhydrite and calcite, thus fixing the R_H parameter, although these outcomes must be considered with some caution for the reasons mentioned above.

Although the prediction of the future evolution of the current unrest episode is a difficult task, two possible limiting scenarios can be envisaged, assuming that no ascent of magma will occur, that is:

(1) further pressurization of the permeable zones (hosting aqueous fluids) interposed between the magma layer and the surface or
(2) dry out of the plumbing system underneath the fumarolic vents, extending until the degassing magma, and consequent depressurization, similar to what might have occurred on January 15, 1935 (see Sect. 3 in chapter "Historical Data").

Magmatic degassing is probably the main controlling factor that will possibly cause the system to evolve towards one or the other of these two limiting scenarios. Monitoring the temperature and fluid pressure in the three different zones of the Solfatara magmatic-hydrothermal system (see above) should provide an effective contribution to understand towards which scenario the system will evolve in the future.

Having shown that it is possible to implement geochemical tools to estimate the T, P conditions present at different depths in the Solfatara magmatic-hydrothermal system, we trust that this type of geochemical research can continue with the improvement of new geochemical modeling tools, based on the solid pillars of thermodynamics and chemical kinetics.

References

Caliro S, Chiodini G, Paonita A (2014) Geochemical evidences of magma dynamics at Campi Flegrei (Italy). Geochim Cosmochim Acta 132:1–15

Cavarretta G, Gianelli G, Scandiffio G, Tecce F (1985) Evolution of the Latera geothermal system II: metamorphic, hydrothermal mineral assemblages and fluid chemistry. J Volcanol Geoth Res 26(3–4):337–364

Chiodini G, Caliro S, De Martino P, Avino R, Gherardi F (2012) Early signals of new volcanic unrest at Campi Flegrei caldera? Insights from geochemical data and physical simulations. Geology 40(10):943–946

Chiodini G, Caliro S, Avino R, Bini G, Giudicepietro F, De Cesare W, Ricciolino P, Aiuppa A, Cardellini C, Petrillo Z, Selva J, Siniscalchi A, Tripaldi S (2021) Hydrothermal pressure-temperature control on CO_2 emissions and seismicity at Campi Flegrei (Italy). J Volcanol Geothermal Res 414:107245

Cioni R, Corazza E, Marini L (1984) The gas/steam ratio as indicator of heat transfer at the Solfatara fumaroles, Phlegraean Fields (Italy). Bulletin of Volcanology 47 (2):295-302.

Fournier RO (1999) Hydrothermal processes related to movement of fluid from plastic into brittle rock in the magmatic-epithermal environment. Econ Geol 94 (8):1193–1211

Kazemi GA, Lehr JH, Perrochet P (2006) Groundwater age. Wiley, New Jersey, p 325

Xu J, Froment GF (1989) Methane steam reforming, methanation and water-gas shift: I Intrinsic Kinetics. Aiche J 35(1):88–96

The manufacturer's authorised representative in the EU is Springer
Nature Customer Service Centre GmbH, Europaplatz 3, 69115 Heidelberg,
Germany. If you have any concerns regarding our products, please
contact ProductSafety@springernature.com

Printed and bound by CPI Group (UK) Ltd, Croydon, CR0 4YY
30/09/2026
02225498-0020